T0291687

CAMBRIDGE LIBRARY COLLECTION

Books of enduring scholarly value

Mathematical Sciences

From its pre-historic roots in simple counting to the algorithms powering modern desktop computers, from the genius of Archimedes to the genius of Einstein, advances in mathematical understanding and numerical techniques have been directly responsible for creating the modern world as we know it. This series will provide a library of the most influential publications and writers on mathematics in its broadest sense. As such, it will show not only the deep roots from which modern science and technology have grown, but also the astonishing breadth of application of mathematical techniques in the humanities and social sciences, and in everyday life.

Mathematical Theory of Electricity and Magnetism

Sir James Jeans (1877–1946) is regarded as one of the founders of British cosmology, and was the first to suggest (in 1928) the steady state theory, which assumes a continuous creation of matter in the universe. He made many major contributions over a wide area of mathematical physics, but was also well known as an accessible writer for the non-specialist. This well-known treatise, first published in 1908, covers the topics in electromagnetic theory required by every non-specialist physicist at that time. It provides the relevant mathematical analysis and was therefore useful to those with only limited mathematical knowledge, as well as to more advanced physicists, engineers and applied mathematicians. It includes a large number of examples, and provides an interesting snapshot of the state of the discipline a century ago.

Cambridge University Press has long been a pioneer in the reissuing of out-of-print titles from its own backlist, producing digital reprints of books that are still sought after by scholars and students but could not be reprinted economically using traditional technology. The Cambridge Library Collection extends this activity to a wider range of books which are still of importance to researchers and professionals, either for the source material they contain, or as landmarks in the history of their academic discipline.

Drawing from the world-renowned collections in the Cambridge University Library, and guided by the advice of experts in each subject area, Cambridge University Press is using state-of-the-art scanning machines in its own Printing House to capture the content of each book selected for inclusion. The files are processed to give a consistently clear, crisp image, and the books finished to the high quality standard for which the Press is recognised around the world. The latest print-on-demand technology ensures that the books will remain available indefinitely, and that orders for single or multiple copies can quickly be supplied.

The Cambridge Library Collection will bring back to life books of enduring scholarly value (including out-of-copyright works originally issued by other publishers) across a wide range of disciplines in the humanities and social sciences and in science and technology.

Mathematical Theory of Electricity and Magnetism

JAMES JEANS

CAMBRIDGE
UNIVERSITY PRESS

CAMBRIDGE UNIVERSITY PRESS

Cambridge, New York, Melbourne, Madrid, Cape Town, Singapore,
São Paolo, Delhi, Dubai, Tokyo

Published in the United States of America by Cambridge University Press, New York

www.cambridge.org
Information on this title: www.cambridge.org/9781108005616

© in this compilation Cambridge University Press 2009

This edition first published 1925
This digitally printed version 2009

ISBN 978-1-108-00561-6

THE MATHEMATICAL THEORY

OF

ELECTRICITY AND MAGNETISM

THE
MATHEMATICAL THEORY
OF
ELECTRICITY AND
MAGNETISM

BY

SIR JAMES JEANS

FIFTH EDITION

CAMBRIDGE
AT THE UNIVERSITY PRESS
1963

CAMBRIDGE UNIVERSITY PRESS
Cambridge, New York, Melbourne, Madrid, Cape Town, Singapore, São Paulo, Delhi

Cambridge University Press
The Edinburgh Building, Cambridge CB2 8RU, UK

Published in the United States of America by Cambridge University Press, New York

www.cambridge.org
Information on this title: www.cambridge.org/9780521091664

First Edition 1908
Second Edition 1911
Third Edition 1915
Fourth Edition 1920
Reprinted 1923
Fifth Edition 1925
Reprinted 1927, 1933, 1941, 1943, 1946, 1948, 1951, 1958, 1960, 1963
Re-issued in this digitally printed version 2008

A catalogue record for this publication is available from the British Library

ISBN 978-0-521-09166-4 paperback

PREFACE

[TO THE FIRST EDITION]

THERE is a certain well-defined range in Electromagnetic Theory, which every student of physics may be expected to have covered, with more or less of thoroughness, before proceeding to the study of special branches of developments of the subject. The present book is intended to give the mathematical theory of this range of electromagnetism, together with the mathematical analysis required in its treatment.

The range is very approximately that of Maxwell's original Treatise, but the present book is in many respects more elementary than that of Maxwell. Maxwell's Treatise was written for the fully-equipped mathematician: the present book is written more especially for the student, and for the physicist of limited mathematical attainments.

The questions of mathematical analysis which are treated in the text have been inserted in the places where they are first needed for the development of the physical theory, in the belief that, in many cases, the mathematical and physical theories illuminate one another by being studied simultaneously. For example, brief sketches of the theories of spherical, zonal and ellipsoidal harmonics are given in the chapter on Special Problems in Electrostatics, interwoven with the study of harmonic potentials and electrical applications: Stokes' Theorem is similarly given in connection with the magnetic vector-potential, and so on. One result of this arrangement is to destroy, at least in appearance, the balance of the amounts of space allotted to the different parts of the subject. For instance, more than half the book appears to be devoted to Electrostatics, but this space will, perhaps, not seem excessive when it is noticed how many of the pages in the Electrostatic part of the book are devoted to non-electrical subjects in applied mathematics (potential-theory, theory of stress, etc.), or in pure mathematics (Green's Theorem, harmonic analysis, complex variable, Fourier's series, conjugate functions, curvilinear coordinates, etc.).

A number of examples, taken mainly from the usual Cambridge examination papers, are inserted. These may provide problems for the mathematical student, but it is hoped that they may also form a sort of compendium of results for the physicist, shewing what types of problem admit of exact mathematical solution.

It is again a pleasure to record my thanks to the officials of the University Press for their unfailing vigilance and help during the printing of the book.

J. H. JEANS.

PRINCETON,
December, 1907.

[TO THE SECOND EDITION]

The second Edition will be found to differ only very slightly from the first in all except the last few chapters. The chapter on Electromagnetic Theory of Light has, however, been largely rewritten and considerably amplified, and two new chapters appear in the present edition, on the Motion of Electrons and on the General Equations of the Electromagnetic Field. These last chapters attempt to give an introduction to the more recent developments of the subject. They do not aim at anything like completeness of treatment, even in the small parts of the subjects with which they deal, but it is hoped they will form a useful introduction to more complete and specialised works and monographs.

J. H. JEANS.

CAMBRIDGE,
August, 1911.

[TO THE THIRD EDITION]

In preparing a third Edition I have made only a few changes in the latter chapters, which were necessary to bring the book up to date.

J. H. JEANS.

LONDON,
November, 1914.

[TO THE FOURTH EDITION]

It will be found that the main changes in the fourth Edition consist in a rearrangement of the later chapters and the addition of a wholly new chapter on the Theory of Relativity. It need hardly be said that no attempt is made to give a full account of the Theory; I have tried to present its broad outlines in the simplest possible way, and in striving after simplicity I have intentionally omitted all elaboration and detail. It is hoped that the new chapter will provide a suitable introduction to the Theory of Relativity for the student who approaches the subject for the first time, equipped with such knowledge of general electrical theory as can be gained from the rest of the book.

J. H. JEANS.

DORKING,
December, 1919.

Preface

[TO THE FIFTH EDITION]

In preparing a Fifth Edition I have introduced the changes that seemed to be called for by the now established position of the new theories of relativity and quanta. I have not attempted any detailed account of the theory of quanta but have added a chapter on "The Electrical Structure of Matter" which will introduce the reader to this theory.

It is a pleasure to record my thanks to friends and correspondents who have helped me by making suggestions and pointing out errors and misprints in earlier editions. My thanks are especially due to Dr A. Russell, F.R.S., Dr Harold Jeffreys, F.R.S., Professor E. P. Adams, Dr R. E. Baynes, Mr L. A. Pass and Dr H. L. Curtis.

J. H. JEANS.

DORKING,
March, 1925.

CONTENTS

INTRODUCTION

ELECTROSTATICS AND CURRENT ELECTRICITY

MAGNETISM

ELECTROMAGNETISM

INTRODUCTION

THE THREE DIVISIONS OF ELECTROMAGNETISM

1. THE fact that a piece of amber, on being rubbed, attracted to itself other small bodies, was known to the Greeks, the discovery of this fact being attributed to Thales of Miletus (640–548 B.C.). A second fact, namely, that a certain mineral ore (lodestone) possessed the property of attracting iron, is mentioned by Lucretius. These two facts have formed the basis from which the modern science of Electromagnetism has grown. It has been found that the two phenomena are not isolated, but are insignificant units in a vast and intricate series of phenomena. To study, and as far as possible interpret, these phenomena is the province of Electromagnetism. And the mathematical development of the subject must aim at bringing as large a number of the phenomena as possible within the power of exact mathematical treatment.

2. The first great branch of the science of Electromagnetism is known as Electrostatics. The second branch is commonly spoken of as Magnetism, but is more accurately described as Magnetostatics. We may say that Electrostatics has been developed from the single property of amber already mentioned, and that Magnetostatics has been developed from the single property of the lodestone. These two branches of Electromagnetism deal solely with states of rest, not with motion or changes of state, and are therefore concerned only with phenomena which can be described as statical. The developments of the two statical branches of Electromagnetism, namely Electrostatics and Magnetostatics, are entirely independent of one another. The science of Electrostatics could have been developed if the properties of the lodestone had never been discovered, and similarly the science of Magnetostatics could have been developed without any knowledge of the properties of amber.

The third branch of Electromagnetism, namely, Electrodynamics, deals with the *motion* of electricity and magnetism, and it is in the development of this branch that we first find that the two groups of phenomena of electricity and magnetism are related to one another. The relation is a reciprocal relation: it is found that magnets in motion produce the same effects as electricity at rest, while electricity in motion produces the same effects as magnets at rest. The third division of Electromagnetism, then, connects the two former divisions of Electrostatics and Magnetostatics, and is in a sense symmetrically placed with regard to them. Perhaps we may compare the whole structure of Electromagnetism to an arch made of three stones. The two side stones can be placed in position independently, neither in any way resting on the other, but the third cannot be placed in position until the two side stones are securely fixed. The third stone rests equally on the two other stones and forms a connection between them.

3. In the present book these three divisions will be developed in the order in which they have been mentioned, namely Electrostatics, Magnetostatics, Electrodynamics. The earlier chapters will give an explanation of the physical ideas adopted by Maxwell in his Treatise on Electricity and Magnetism side by side with a purely mathematical theory. Maxwell's treatment of Electrical Science was differentiated from that of other writers by his insistence on Faraday's conception of electric and magnetic energy as residing in the medium. According to this view, the forces acting on electrified or magnetised bodies did not form the whole system of forces in action, but served only to reveal the presence of a vastly more intricate system of forces, which acted throughout the ether by which the material bodies were supposed to be surrounded. It was only through the presence of matter that the supposed system of forces became perceptible to human observation, so that it was necessary to try to reconstruct the whole system of forces from no data except those given by the resultant effect of the forces on matter, where matter was present. As might be expected, these data proved insufficient to give full and definite knowledge of the system of ethereal forces; it was found that a great number of systems of ethereal forces could be constructed, each of which would produce the same effects on matter as are observed. Of these systems, however, a single one seemed so very much more probable than any of the others, that it was unhesitatingly adopted both by Maxwell and by Faraday.

As soon as the step had been taken of attributing the mechanical forces acting on matter to a system of forces acting throughout the whole ether, a further physical development was made not only possible but also necessary. A stress in the ether might be supposed to represent either an electric or a magnetic force, but could not be both. Faraday supposed a stress in the ether to be identical with electrostatic force. There was no longer any possibility, in this scheme of the universe, of regarding magnetostatic forces as evidence of simple stresses in the ether.

It has, however, been said that magnetostatic forces are found to be produced by the motion of electric charges. Now if electric charges at rest produce simple stresses in the ether, the motion of electric charges must obviously be accompanied by *changes* in the stresses in the ether. It accordingly became possible to identify magnetostatic force with change in the system of stresses in the ether. This interpretation of magnetic force formed an essential part of Maxwell's theory. Comparing the ether to an elastic material medium, we may say that the electric forces were interpreted as the statical pressures and strains which accompanied the compression, dilatation or displacement of the medium, while magnetic forces were interpreted as the pressures and strains in the medium caused by its motion and momentum. Thus electrostatic energy was regarded as the potential energy of the medium, while magnetic energy was regarded as its kinetic energy. Maxwell shewed that the whole series of known electrostatic and magnetostatic phenomena might be consistently interpreted as phenomena produced by the stresses and motion of a medium, this motion being in conformity with the laws of dynamics. This hypothesis is examined in the earlier chapters of the book, although, as will be seen later, recent developments call for at least a drastic modification, and more probably for the complete abandonment of the whole hypothesis.

4. The observational fact that magnetostatic forces were produced by the motion of electric charges inevitably raised the question of the interpretation of general magnetic phenomena in electrical terms. A solution of the problem suggested by Ampère and Weber needs but little modification to represent the answer to which modern investigations have led. Recent experimental researches shew that all matter must be supposed to consist solely of electrically charged particles, and it seems highly probable that all magnetic phenomena can be explained by the motion of these charges. If the motion of the charges is governed by a regularity of a certain kind, the body as a whole will shew magnetic properties. If this regularity does not obtain, the magnetic forces produced by the motions of the individual charges will on the whole neutralise one another, and the body will appear to be non-magnetic. On this view the electricity and magnetism which at first sight appeared to exist independently in the universe, are resolved into electricity alone—electricity and magnetism become electricity at rest and electricity in motion.

This discovery of the ultimate identity of electricity and magnetism is by no means the last word of the science of Electromagnetism. As far back as the time of Maxwell and Faraday, it was recognised that the forces at work in chemical phenomena must be regarded largely, if not entirely, as electrical forces. Later, Maxwell shewed light to be an electromagnetic phenomenon, so that the whole science of Optics became a branch of Electromagnetism.

Gradually the conviction grew that all physical forces, with the possible exception of Gravitation, would prove to be ultimately of Electromagnetic origin, so that by the end of the nineteenth century most scientists believed that the science of Electromagnetism would advance along the road opened out by Maxwell until the whole physical universe had been explained in the terms of electromagnetic theory. Recently this belief has experienced two very severe checks.

If, as Maxwell believed, the ultimate seat of electromagnetic and optical phenomena is the ether, it ought to be possible to find out something about the ether by electromagnetic and optical means. It ought, for instance, at least to be possible to determine the velocity with which we move through the ether. A series of experiments devised to this end have one and all failed to disclose this velocity. To every experimental enquiry, Nature seems to give the answer either that there is no ether or that natural phenomena go on exactly as if there were no ether. If this view is finally established, and at present there seems only a very meagre chance of any alternative, Maxwell's theory of the electromagnetic ether must necessarily fall out of science; it will have served its purpose as a scaffolding which will have enabled the structure of electromagnetic theory to have been built in perfect form, but it will not be part of that structure. Nevertheless the time for finally deciding how much of Maxwell's theory is scaffolding and how much is part of the essential structure has hardly yet come, so that in the present book we shall first develop the theory along the general lines initiated by Maxwell, and then shall devote a chapter to the development of a more modern theory and to a discussion of how far the existence of an ether is essential to electromagnetic theory.

The second check to Maxwell's theory has originated from the study of radiation and the ultimate electrical structure of matter; phenomena of primary importance have been found not to be reconcileable with Maxwell's original theory. In a sense the new facts hardly cut at the roots of the theory; they must rather be thought of as restricting the spread of the branches. There is no question that the electrical phenomena of everyday life, thunderstorms, telephones and dynamos, are all governed by Maxwell's laws; it is only when we pass to the phenomena arising from the most intimate electrical structure of matter that Maxwell's laws appear to be inadequate. Our final chapter will contain an explanation of the failure of Maxwell's Electrodynamics to deal with these problems, and a very brief introduction to the new theory which has taken its place.

CHAPTER I

PHYSICAL PRINCIPLES

THE FUNDAMENTAL CONCEPTIONS OF ELECTROSTATICS

I. *State of Electrification of a Body.*

5. WE proceed to a discussion of the fundamental conceptions which form the basis of Electrostatics. The first of these is that of a state of electrification of a body. When a piece of amber has been rubbed so that it attracts small bodies to itself, we say that it is in a state of electrification— or, more shortly, that it is electrified.

Other bodies besides amber possess the power of attracting small bodies after being rubbed, and are therefore susceptible of electrification. Indeed it is found that all bodies possess this property, although it is less easily recognised in the case of most bodies, than in the case of amber. For instance a brass rod with a glass handle, if rubbed on a piece of silk or cloth, will shew the power to a marked degree. The electrification here resides in the brass; as will be explained immediately, the interposition of glass or some similar substance between the brass and the hand is necessary in order that the brass may retain its power for a sufficient time to enable us to observe it. If we hold the instrument by the brass rod and rub the glass handle we find that the same power is acquired by the glass.

II. *Conductors and Insulators.*

6. Let us now suppose that we hold the electrified brass rod in one hand by its glass handle, and that we touch it with the other hand. We find that after touching it its power of attracting small bodies will have completely disappeared. If we immerse it in a stream of water or pass it through a flame we find the same result. If on the other hand we touch it with a piece of silk or a rod of glass, or stand it in a current of air, we find that its power of attracting small bodies remains unimpaired, at any rate for a time. It appears therefore that the human body, a flame or water have the power of destroying the electrification of the brass rod when placed in contact with it, while silk and glass and air do not possess this property. It is for this reason that in handling the electrified brass rod, the substance in direct contact with the brass has been supposed to be glass and not the hand.

In this way we arrive at the idea of dividing all substances into two classes according as they do or do not remove the electrification when touching the electrified body. The class which remove the electrification are called *conductors*, for as we shall see later, they conduct the electrification away from the electrified body rather than destroy it altogether; the class which allow the electrified body to retain its electrification are called *nonconductors* or *insulators*. The classification of bodies into conductors and insulators appears to have been first discovered by Stephen Gray (1696–1736).

At the same time it must be explained that the difference between insulators and conductors is one of degree only. If our electrified brass rod were left standing for a week in contact only with the air surrounding it and the glass of its handle, we should find it hard to detect traces of electrification after this time—the electrification would have been conducted away by the air and the glass. So also if we had been able to immerse the rod in a flame for a billionth of a second only, we might have found that it retained considerable traces of electrification. It is therefore more logical to speak of good conductors and bad conductors than to speak of conductors and insulators. Nevertheless the difference between a good and a bad conductor is so enormous, that for our present purpose we need hardly take into account the feeble conducting power of a bad conductor, and may without serious inconsistency, speak of a bad conductor as an insulator. There is, of course, nothing to prevent us imagining an ideal substance which has no conducting power at all. It will often simplify the argument to imagine such a substance, although we cannot realise it in nature.

It may be mentioned here that of all substances the metals are by very much the best conductors. Next come solutions of salts and acids, and lastly as very bad conductors (and therefore as good insulators) come oils, waxes, silk, glass and such substances as sealing wax, shellac, indiarubber. Gases under ordinary conditions are good insulators. Indeed it is worth noticing that if this had not been so, we should probably never have become acquainted with electric phenomena at all, for all electricity would be carried away by conduction through the air as soon as it was generated. Flames, however, conduct well, and, for reasons which will be explained later, all gases become good conductors when in the presence of radium or of so-called radio-active substances. Distilled water is an almost perfect insulator, but any other sample of water will contain impurities which generally cause it to conduct tolerably well, and hence a wet body is generally a bad insulator. So also an electrified body suspended in air loses its electrification much more rapidly in damp weather than in dry, owing to conduction by water-particles in the air.

When the body is in contact with insulators only, it is said to be "insulated." The insulation is said to be good when the electrified body retains its electrification for a long interval of time, and is said to be poor when the electrification disappears rapidly. Good insulation will enable a body to retain most of its electrification for some days, while with poor insulation the electrification will last only for a few minutes or seconds.

III.   *Quantity of Electricity.*

7. We pass next to the conception of a definite quantity of electricity, this quantity measuring the degree of electrification of the body with which it is associated. It is found that the quantity of electricity associated with any body remains constant except in so far as it is conducted away by conductors. To illustrate, and to some extent to prove this law, we may use an instrument known as the gold-leaf electroscope. This consists of a glass vessel, through the top of which a metal rod is passed, supporting at its lower end two gold-leaves which under normal conditions hang flat side by side, touching one another throughout their length. When an electrified body touches or is brought near to the brass rod, the two gold-leaves are seen to separate, for reasons which will become clear later (§ 21), so that the instrument can be used to examine whether or not a body is electrified.

Let us fix a metal vessel on the top of the brass rod, the vessel being closed but having a lid through which bodies can be inserted. The lid must be supplied with an insulating handle for its manipulation. Suppose that we have electrified some piece of matter—to make the picture definite, suppose that we have electrified a small brass rod by rubbing it on silk—and let us suspend this body inside the vessel by an insulating thread in such a manner that it does not touch the sides of the vessel. Let us close the lid of the vessel, so that the vessel entirely surrounds the electrified body, and note the amount of separation of the gold-leaves of the electroscope. Let us try the experiment any number of times, placing the electrified body in different positions inside the closed vessel, taking care only that it does not come into contact with the sides of the vessel or with any other conductors. We shall find that in every case the separation of the gold-leaves is exactly the same.

Fig. 1.

In this way then, we get the idea of a definite quantity of electrification associated with the brass rod, this quantity being independent of the position of the rod inside the closed vessel of the electroscope. We find, further, that the divergence of the gold-leaves is not only independent of the position of the rod inside the vessel, but is independent of any changes of state which the rod may have experienced between successive insertions in the vessel, provided only that it has not been touched by conducting bodies. We might for instance heat the rod, or, if it was sufficiently thin, we might bend it into a different shape, and on replacing it inside the vessel we should find that it produced exactly the same deviation of the gold-leaves as before. We may, then, regard the electrical properties of the rod as being due to a *quantity* of electricity associated with the rod, this quantity remaining permanently the same, except in so far as the original charge is lessened by contact with conductors, or increased by a fresh supply.

8. We can regard the electroscope as giving an indication of the magnitude of a quantity of electricity, two charges being equal when they produce the same divergence of the leaves of the electroscope.

In the same way we can regard a spring-balance as giving an indication of the magnitude of a weight, two weights being equal when they produce the same extension of the spring.

The question of the actual quantitative measurement of a quantity of electricity as a multiple of a specified unit has not yet been touched. We can, however, easily devise means for the exact quantitative measurement of electricity in terms of a unit. We can charge a brass rod to any degree we please, and agree that the charge on this rod is to be taken to be the standard unit charge. By rubbing a number of rods until each produces exactly the same divergence of the electroscope as the standard charge, we can prepare a number of unit charges, and we can now say that a charge is equal to n units, if it produces the same deviation of the electroscope as would be produced by n units all inserted in the vessel of the electroscope at once. This method of measuring an electric charge is of course not one that any rational being would apply in practice, but the object of the present explanation is to elucidate the fundamental principles, and not to give an account of practical methods.

9. *Positive and Negative Electricity.* Let us suppose that we insert in the vessel of the electroscope the piece of silk on which one of the brass rods has been supposed to have been rubbed in order to produce its unit charge. We shall find that the silk produces a divergence of the leaves of the electroscope, and further that this divergence is exactly equal to that which is produced by inserting the brass rod alone into the vessel of the electroscope. If, however, we insert the brass rod and the silk together into the electroscope, no deviation of the leaves can be detected.

Again, let us suppose that we charge a brass rod A with a charge which the divergence of the leaves shews to be n units. Let us rub a second brass rod B with a piece of silk C until it has a charge, as indicated by the electroscope, of m units, m being smaller than n. If we insert the two brass rods together, the electroscope will, as already explained, give a divergence corresponding to $n + m$ units. If, however, we insert the rod A and the silk C together, the deviation will be found to correspond to $n - m$ units.

In this way it is found that a charge of electricity must be supposed to have sign as well as magnitude. As a matter of convention, we agree to speak of the m units of charge on the silk as m *positive* units, or more briefly as a charge $+ m$, while we speak of the charge on the brass as m *negative* units, or a charge $- m$.

10. Generation of Electricity. It is found to be a general law that, on rubbing two bodies which are initially uncharged, equal quantities of positive and negative electricity are produced on the two bodies, so that the total charge generated, measured algebraically, is *nil*.

We have seen that the electroscope does not determine the sign of the charge placed inside the closed vessel, but only its magnitude. We can, however, determine both the sign and magnitude by two observations. Let us first insert the charged body alone into the vessel. Then if the divergence of the leaves corresponds to m units, we know that the charge is either $+ m$ or $- m$, and if we now insert the body in company with another charged body, of which the charge is known to be $+ n$, then the charge we are attempting to measure will be $+ m$ or $- m$ according as the divergence of the leaves indicates $n + m$ or $n \sim m$ units. With more elaborate instruments to be described later (electrometers) it is possible to determine both the magnitude and sign of a charge by one observation.

11. If we had rubbed a rod of glass, instead of one of brass, on the silk, we should have found that the silk had a *negative* charge, and the glass of course an equal *positive* charge. It therefore appears that the sign of the charge produced on a body by friction depends not only on the nature of the body itself, but also on the nature of the body with which it has been rubbed.

The following is found to be a general law: If rubbing a substance A on a second substance B charges A positively and B negatively, and if rubbing the substance B on a third substance C charges B positively and C negatively, then rubbing the substance A on the substance C will charge A positively and C negatively.

It is therefore possible to arrange any number of substances in a list such that a substance is charged with positive or negative electricity when rubbed with a second substance, according as the first substance stands above or below the second substance on the list. The following is a list of this kind, which includes some of the most important substances:

Cat's skin, Glass, Ivory, Silk, Rock crystal, The Hand, Wood, Sulphur, Flannel, Cotton, Shellac, Caoutchouc, Resins, Guttapercha, Metals, Guncotton.

A substance is said to be electropositive or electronegative to a second substance according as it stands above or below it on a list of this kind. Thus of any pair of substances one is always electropositive to the other, the other being electronegative to the first. Two substances, although chemically the same, must be regarded as distinct for the purposes of a list such as the above, if their physical conditions are different; for instance, it is found that a hot body must be placed lower on the list than a cold body of the same chemical composition.

IV.　*Attraction and Repulsion of Electric Charges.*

12. A small ball of pith, or some similarly light substance, coated with gold-leaf and suspended by an insulating thread, forms a convenient instrument for investigating the forces, if any, which are brought into play by the presence of electric charges. Let us electrify a pith ball of this kind positively and suspend it from a fixed point. We shall find that when we bring a second small body charged with positive electricity near to this first body the two bodies tend to repel one another, whereas if we bring a negatively charged body near to it, the two bodies tend to attract one another. From this and similar experiments it is found that two small bodies charged with electricity of the same sign repel one another, and that two small bodies charged with electricity of different signs attract one another.

This law can be well illustrated by tying together a few light silk threads by their ends, so that they form a tassel, and allowing the threads to hang vertically. If we now stroke the threads with the hand, or brush them with a brush of any kind, the threads all become positively electrified, and therefore repel one another. They consequently no longer hang vertically but spread themselves out into a cone. A similar phenomenon can often be noticed on brushing the hair in dry weather. The hairs become positively electrified and so tend to stand out from the head.

13. On shaking up a mixture of powdered red lead and yellow sulphur, the particles of red lead will become positively electrified, and those of the sulphur will become negatively electrified, as the result of the friction which has occurred between the two sets of particles in the shaking. If some of this powder is now dusted on to a positively electrified body, the particles of sulphur will be attracted and those of red lead repelled. The red lead will therefore fall off, or be easily removed by a breath of air, while the sulphur particles will be retained. The positively electrified body will therefore assume a yellow colour on being dusted with the powder, and similarly a negatively electrified body would become red. It may sometimes be convenient to use this method of determining whether the electrification of a body is positive or negative.

14. The attraction and repulsion of two charged bodies is in many respects different from the force between one charged and one uncharged body. The latter force, as we have explained, was known to the Greeks: it must be attributed, as we shall see, to what is known as "electric induction," and is invariably *attractive*. The forces between two bodies both of which are charged, forces which may be either attractive or repulsive, seem hardly to have been noticed until the eighteenth century.

The observations of Robert Symmer (1759) on the attractions and repulsions of charged bodies are at least amusing. He was in the habit of wearing two pairs of stockings simultaneously, a worsted pair for comfort and a silk pair for appearance. In pulling off his stockings he noticed that they gave a crackling noise, and sometimes that they even emitted sparks when taken off in the dark. On taking the two stockings off together from the foot and then drawing the one from inside the other, he found that both became inflated so as to reproduce the shape of the foot, and exhibited attractions and repulsions at a distance of as much as a foot and a half.

"When this experiment is performed with two black stockings in one hand, and two white in the other, it exhibits a very curious spectacle; the repulsion of those of the same colour, and the attraction of those of different colours, throws them into an agitation that is not unentertaining, and makes them catch each at that of its opposite colour, and at a greater distance than one would expect. When allowed to come together they all unite in one mass. When separated, they resume their former appearance, and admit of the repetition of the experiment as often as you please, till their electricity, gradually wasting, stands in need of being recruited."

The Law of Force between charged Particles.

15. *The Torsion Balance.* Coulomb (1785) devised an instrument known as the Torsion Balance, which enabled him not only to verify the laws of attraction and repulsion qualitatively, but also to form an estimate of the actual magnitude of these forces.

The apparatus consists essentially of two light balls A, C, fixed at the two ends of a rod which is suspended at its middle point B by a very fine thread of silver, quartz or other material. The upper end of the thread is fastened to a movable head D, so that the thread and the rod can be made to rotate by screwing the head. If the rod is acted on only by its weight, the condition for equilibrium is obviously that there shall be no torsion in the thread. If, however, we fix a third small ball E in the same plane as

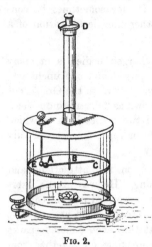

the other two, and if the three balls are electrified, the forces between the fixed ball and the movable ones will exert a couple on the moving rod, and the condition for equilibrium is that this couple shall exactly balance that due to the torsion. Coulomb found that the couple exerted by the torsion of the thread was exactly proportional to the angle through which one end of the thread had been turned relatively to the other, and in this way was enabled to measure his electric forces. In Coulomb's experiments one only of the two movable balls was electrified, the second serving merely as a counterpoise, and the fixed ball was at the same distance from the torsion thread as the two movable balls.

Fig. 2.

Suppose that the head of the thread is turned to such a position that the balls when uncharged rest in equilibrium, just touching one another without pressure. Let the balls receive charges e, e', and let the repulsion between them result in the bar turning through an angle θ. The couple exerted on the bar by the torsion of the thread is proportional to θ, and may therefore be taken to be $\kappa\theta$. If a is the radius of the circle described by the movable ball, we may regard the couple acting on the rod from the electric forces as made up of a force F, equal to the force of repulsion between the two balls, multiplied by $a\cos\frac{1}{2}\theta$, the arm of the moment. The condition for equilibrium is accordingly

$$aF\cos\tfrac{1}{2}\theta = \kappa\theta.$$

Let us now suppose that the torsion head is turned through an angle ϕ in such a direction as to make the two charged balls approach each other; after the turning has ceased, let us suppose that the balls are allowed to come to rest. In the new position of equilibrium, let us suppose that the two charged balls subtend an angle θ' at the centre, instead of the former angle θ. The couple exerted by the torsion thread is now $\kappa(\theta' + \phi)$, so that if F' is the new force of repulsion we must have

$$aF'\cos\tfrac{1}{2}\theta' = \kappa(\theta' + \phi).$$

By observing the value of ϕ required to give definite values to θ' we can calculate values of F' corresponding to any series of values of θ'. From a series of experiments of this kind it is found that so long as the charges on the two balls remain the same, F' is proportional to $\operatorname{cosec}^2\frac{1}{2}\theta'$, from which it is easily seen to follow that the force of repulsion varies inversely as the square of the distance. And when the charges on the two balls are varied it is found that the force varies as the product of the two charges, so long as their distance apart remains the same. As the result of a series of experiments conducted in this way Coulomb was able to enunciate the law:

The force between two small charged bodies is proportional to the product of their charges, and is inversely proportional to the square of their distance apart, the force being one of repulsion or attraction according as the two charges are of the same or of opposite kinds.

16. In mathematical language we may say that there is a force of repulsion of amount

$$\frac{cee'}{r^2} \quad \ldots\ldots\ldots\ldots\ldots\ldots\ldots\ldots\ldots\ldots\ldots(1)$$

where e, e' are the charges, r their distance apart, and c is a positive constant.

If e, e' are of opposite signs the product ee' is negative, and a negative repulsion must be interpreted as an attraction.

Although this law was first published by Coulomb, it subsequently appeared that it had been discovered at an earlier date by Cavendish, whose experiments were much more refined than those of Coulomb. Cavendish was able to satisfy himself that the law was certainly intermediate between the inverse $2 + \frac{1}{50}$ and $2 - \frac{1}{50}$th power of the distance (see below, §§ 46—48). Unfortunately his researches remained unknown until his manuscripts were published in 1879 by Clerk Maxwell.

The experiments of Coulomb and Cavendish, it need hardly be said, were very rough compared with those which are rendered possible by modern refinements of theory and practice, so that these experiments are no longer the justification for using the law expressed by formula (1) as the basis of the Mathematical Theory of Electricity. More delicate experiments with the apparatus used by Cavendish, which will be explained later, have, however, been found to give a complete confirmation of Coulomb's Law, so long as the charged bodies may both be regarded as infinitely small compared with their distance apart. Any deviation from the law of Coulomb must accordingly be attributed to the finite sizes of the bodies which carry the charges. As it is only in the case of infinitely small bodies that the symbol r of formula (1) has had any meaning assigned to it, we may regard the law (1) as absolutely true, at any rate so long as r is large enough to be a measurable quantity.

The Unit of Electricity.

17. The law of Coulomb supplies us with a convenient unit in which to measure electric charges.

The unit of mass, the pound or gramme, is a purely arbitrary unit, and all quantities of mass are measured simply by comparison with this unit. The same is true of the unit of space. If it were possible to keep a charge of electricity unimpaired through all time we might take an arbitrary charge of electricity as standard, and measure all charges by comparison with this one standard charge, in the way suggested in § 8. As it is not possible to do this, we find it convenient to measure electricity with reference to the units of mass, length and time of which we are already in possession, and Coulomb's Law enables us to do this. We define as the unit charge a charge such that when two unit charges are placed one on each of two small particles at a distance of a centimetre apart, the force of repulsion between the particles is one dyne. With this definition it is clear that the quantity c in the formula (1) becomes equal to unity, so long as the c.g.s. system of units is used.

In a similar way, if the mass of a body did not remain constant, we might have to define the unit of mass with reference to those of time and length by saying that a mass is a unit mass provided that two such masses, placed at a unit distance apart, produce in each other by their mutual gravitational attraction an acceleration of a centimetre per second per second. In this case we should have the gravitational acceleration f given by an equation of the form

$$f = \frac{m}{r^2} \quad \dots\dots\dots\dots\dots\dots\dots\dots\dots\dots(2),$$

and this equation would determine the unit of mass.

18. *Physical dimensions.* If the unit of mass were determined by equation (2), m would appear to have the dimensions of an acceleration multiplied by the square of a distance, and therefore dimensions

$$L^3 T^{-2}.$$

As a matter of fact, however, we know that mass is something entirely apart from length and time, except in so far as it is connected with them through the law of gravitation. The complete gravitational acceleration is given by

$$f = \gamma \frac{m}{r^2},$$

where γ is the so-called "gravitation constant."

By our proposed definition of unit mass we should have made the value of γ numerically equal to unity; but its physical dimensions are not those of a mere number, so that we cannot neglect the factor γ when equating physical dimensions on the two sides of the equation.

So also in the formula

$$F = \frac{cee'}{r^2} \quad\text{.....................................(3)}$$

we can and do choose our unit of charge in such a way that the *numerical* value of c is unity, so that the numerical equation becomes

$$F = \frac{ee'}{r^2} \quad\text{.....................................(4),}$$

but we must remember that the factor c still retains its physical dimensions. Electricity is something entirely apart from mass, length and time, and it follows that we ought to treat the dimensions of equation (3), by introducing a new unit of electricity E and saying that c is of the dimensions of a force divided by E^2/r^2 and therefore of dimensions

$$ML^3E^{-2}T^{-2}.$$

If, however, we compare dimensions in equation (4), neglecting to take account of the physical dimensions of the suppressed factor c, it appears as though a charge of electricity can be expressed in terms of the units of mass, length and time, just as it might appear from equation (2) as though a mass could be expressed in terms of the units of length and time. The apparent dimensions of a charge of electricity are now

$$M^{\frac{1}{2}}L^{\frac{3}{2}}T^{-1} \quad\text{.................................(5).}$$

It will be readily understood that these dimensions are merely apparent and not in any way real, when it is stated that other systems of units are also in use, and that the apparent physical dimensions of a charge of electricity are found to be different in the different systems of units. The system which we have just described, in which the unit is defined as the charge which makes c numerically equal to unity in equation (3), is known as the Electrostatic system of units.

There will be different electrostatic systems of units corresponding to different units of length, mass and time. In the c.g.s. system these units are taken to be the centimetre, gramme and second. In passing from one system of units to another the unit of electricity will change as if it were a physical quantity having dimensions $M^{\frac{1}{2}}L^{\frac{3}{2}}T^{-1}$, so long as we hold to the agreement that equation (4) is to be numerically true, *i.e.* so long as the units remain electrostatic. This gives a certain importance to the apparent dimensions of the unit of electricity, as expressed in formula (5).

V. *Electrification by Induction.*

19. Let us suspend a metal rod by insulating supports. Suppose that the rod is originally uncharged, and that we bring a small body charged with electricity near to one end of the rod, without allowing the two bodies to touch. We shall find on sprinkling the rod with electrified powder of the kind previously described (§ 13), that the rod is now electrified, the signs of the charges at the two ends being different. This electrification is known as electrification by induction. We speak of the electricity on the rod as an induced charge, and that on the originally electrified body as the inducing or exciting charge. We find that the induced charge at the end of the rod nearest to the inducing charge is of sign opposite to that of the inducing charge, that at the further end of the rod being of the same sign as the inducing charge. If the inducing charge is removed to a great distance from the rod, we find that the induced charges disappear completely, the rod resuming its original unelectrified state.

If the rod is arranged so that it can be divided into two parts, we can separate the two parts before removing the inducing charge, and in this way can retain the two parts of the induced charge for further examination.

If we insert the two induced charges into the vessel of the electroscope, we find that the total electrification is *nil*: in generating electricity by induction, as in generating it by friction, we can only generate equal quantities of positive and negative electricity; we cannot alter the algebraic total charge. Thus the generation of electricity by induction is in no way a violation of the law that the total charge on a body remains unaltered except in so far as it is removed by conduction.

20. If the inducing charge is placed on a sufficiently light conductor, we notice a violent attraction between it and the rod which carries the induced charge. This, however, as we shall now shew, is only in accordance with Coulomb's Law. Let us, for the sake of argument, suppose that the inducing charge is a positive charge e. Let us divide up that part of the

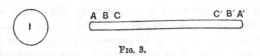

Fig. 3.

rod which is negatively charged into small parts AB, BC, ..., beginning from the end A which is nearest to the inducing charge I, in such a way that each part contains the same small charge $-\epsilon$, of negative electricity. Let us similarly divide up the part of the rod which is positively charged into sections $A'B'$, $B'C'$, ..., beginning from the further end, and such that each of these parts contains a charge $+\epsilon$ of positive electricity. Since the total induced charge is zero, the number of positively charged sections $A'B'$, $B'C'$, ... must be exactly equal to the number of negatively charged sections AB, BC, ... The whole series of sections can therefore be divided into a series of pairs

$$AB \text{ and } A'B'; \quad BC \text{ and } B'C'; \quad \text{etc.}$$

such that the two sections of any pair contain equal and opposite charges. The charge on $A'B'$ being of the same sign as the inducing charge e, repels the body I which carries this charge, while the charge on AB, being of a different sign from the charge on I, attracts I. Since AB is nearer to I than $A'B'$, it follows from Coulomb's Law that the attractive force $e\epsilon/r^2$ between AB and I is numerically greater than the repulsive force $e\epsilon/r^2$ between $A'B'$ and I, so that the resultant action of the pair of sections AB, $A'B'$ upon I is an attraction. Obviously a similar result is true for every other pair of sections, so that we arrive at the result that the whole force between the two bodies is attractive.

This result fully accounts for the fundamental property of a charged body to attract small bodies to which no charge has been given. The proximity of the charged body induces charges of different signs on those parts of the body which are nearer to, and further away from, the inducing charge, and although the total induced charge is zero, yet the attractions will always outweigh the repulsions, so that the resultant force is always one of attraction.

21. The same conceptions explain the divergence of the gold-leaves of the electroscope which occurs when a charged body is brought near to the plate of the electroscope or introduced into a closed vessel standing on this plate. All the conducting parts of the electroscope—gold-leaves, rod, plate and vessel if any—may be regarded as a single conductor, and of this the gold-leaves form the part furthest removed from the charged body. The leaves accordingly become charged by induction with electricity of the same sign as that of the charged body, and as the charges on the two gold-leaves are of similar sign, they repel one another.

22. On separating the two parts of a conductor while an induced charge is on it, and then removing both from the influence of the induced charge, we gain two charges of electricity without any diminution of the inducing charge. We can store or utilise these charges in any way and on replacing the two parts of the conductor in position, we shall again obtain an induced charge. This again may be utilised or stored, and so on indefinitely. There is therefore no limit to the magnitude of the charges which can be obtained from a small initial charge by repeating the process of induction.

This principle underlies the action of the Electrophorus. A cake of resin is electrified by friction, and for convenience is placed with its electrified surface uppermost on a horizontal table. A metal disc is held by an insulating handle parallel to the cake of resin and at a slight distance above it. The operator then touches the upper surface of the disc with his finger. When the process has reached this stage, the metal disc, the body of the operator and the earth itself form one conductor. The negative electricity on the resin induces a positive charge on the nearer parts of this conductor—primarily on the metal disc—and a negative charge on the more remote parts of the conductor—the further region of the earth. When the operator removes his finger, the disc is left insulated and in possession of a positive charge. As already explained, this charge may be used and the process repeated indefinitely.

In all its essentials, the principle utilised in the generation of electricity by the "influence machines" of Voss, Holtz, Wimshurst and others is identical with that of the electrophorus. The machines are arranged so that by the turning of a handle, the various stages of the process are repeated cyclically time after time.

23. *Electric Equilibrium.* Returning to the apparatus illustrated in fig. 3, p. 16, it is found that if we remove the inducing charge without allowing the conducting rod to come into contact with other conductors, the charge on the rod disappears gradually as the inducing charge recedes, positive and negative electricity combining in equal quantities and neutralising one another. This shews that the inducing charge must be supposed to act upon the electricity of the induced charge, rather than upon the matter of the conductor. Upon the same principle, the various parts of the induced charge must be supposed to act directly upon one another. Moreover, in a conductor charged with electricity at rest, there is no reaction between matter and electricity tending to prevent the passage of electricity through the conductor. For if there were, it would be possible for parts of the induced charge to be retained, after the inducing charge had been removed, the parts of the induced charge being retained in position by their reaction with the matter of the conductor. Nothing of this kind is observed to occur. We conclude then that the elements of electrical charge on a conductor are each in equilibrium under the influence solely of the forces exerted by the remaining elements of charge.

24. An exception occurs when the electricity is actually at the surface of the conductor. Here there is an obvious reaction between matter and electricity—the reaction which prevents the electricity from leaving the surface of the conductor. Clearly this reaction will be normal to the surface, so that the forces acting upon the electricity in directions which lie in the tangent plane to the surface must be entirely forces from other charges of electricity, and these must be in equilibrium. To balance the action of the matter on the electricity there must be an equal and opposite reaction of electricity on matter. This, then, will act normally outwards at the surface of the conductor. Experimentally it is best put in evidence by the electrification of soap-bubbles. A soap-bubble when electrified is observed to expand, the normal reaction between electricity and matter at its surface driving the surface outwards until equilibrium is reestablished (see below, § 94).

25. Also when two conductors of different material are placed in contact, electric phenomena are found to occur which have been explained by Helmholtz as the result of the operation of reactions between electricity and matter at the surfaces of the conductors. Thus, although electricity can pass quite freely over the different parts of the same conductor, it is not strictly true to say that electricity can pass freely from one conductor to another of different material with which it is in contact. Compared, however, with the forces with which we shall in general be dealing in electrostatics, it will be legitimate to disregard entirely any forces of the kind just described. We shall therefore neglect the difference between the materials of different conductors, so that any number of conductors placed in contact may be regarded as a single conductor.

THEORIES TO EXPLAIN ELECTRICAL PHENOMENA.

26. *One-fluid Theory.* Franklin, as far back as 1751, tried to include all the electrical phenomena with which he was acquainted in one simple explanation. He suggested that all these phenomena could be explained by supposing the existence of an indestructible " electric fluid," which could be associated with matter in different degrees. Corresponding to the normal state of matter, in which no electrical properties are exhibited, there is a definite normal amount of "electric fluid." When a body was charged with positive electricity, Franklin explained that there was an excess of " electric fluid" above the normal amount, and similarly a charge of negative electricity represented a deficiency of electric fluid. The generation of equal quantities of positive and negative electricity was now explained: for instance, in rubbing two bodies together we simply transfer " electric fluid " from one to the other. To explain the attractions and repulsions of electrified bodies, Franklin supposed that the particles of ordinary matter repelled one another, while attracting the "electric fluid." In the normal state of matter the quantities of "electric fluid " and ordinary matter were just balanced, so that there was neither attraction nor repulsion between bodies in the normal state. According to a later modification of the theory the attractions just out-balanced the repulsions in the normal state, the residual force accounting for gravitation.

27. *Two-fluid Theory.* A further attempt to explain electric phenomena was made by the two-fluid theory. In this there were three things concerned, ordinary matter and two electric fluids—positive and negative. The degree of electrification was supposed to be the measure of the excess of positive electricity over negative, or of negative over positive, according to the sign of the electrification. The two kinds of electricity attracted and repelled, electricities of the same kind repelling, and of opposite kinds attracting, and in this way the observed attractions and repulsions of electrified bodies were explained without having recourse to systems of forces between electricity and ordinary matter. It is, however, obvious that the two-fluid theory was too elaborate for the facts. On this theory ordinary matter devoid of both kinds of electricity would be physically different from matter possessing equal quantities of the two kinds of electricity, although both bodies would equally shew an absence of electrification. There is no evidence that it is possible to establish any physical difference of this kind between totally unelectrified bodies, so that the two-fluid theory must be dismissed as explaining more than there is to be explained.

28. *Modern view of Electricity.* The two theories which have just been mentioned rested on no experimental evidence except such as is required to establish the phenomena with which they are directly concerned. The modern view of electricity, on the other hand, is based on an enormous mass of experimental evidence, to which contributions are made, not only by the phenomena of electrostatics, but also by the phenomena of almost every branch of physics and chemistry. The modern explanation of electricity is found to bear a very close resemblance to the older explanation of the one-fluid theory—so much so that it will be convenient to explain the modern view of electricity simply by making the appropriate modifications of the one-fluid theory.

We suppose the "electric-fluid" of the one-fluid theory replaced by a crowd of small particles—"electrons," it will be convenient to call them—all exactly similar, and each having exactly the same charge of *negative* electricity permanently attached to it. According to the best recent determinations, the amount of this charge is $4\cdot803 \times 10^{-10}$ electrostatic units, while the mass of each electron is $9\cdot12 \times 10^{-28}$ grammes. These determinations, which are due to Millikan and Bucherer, are probably accurate to about one part in a thousand. To a lower degree of accuracy the radius of the electron is probably about 2×10^{-13} cms. We can form some conception of the intense concentration of mass and electrification in the electron by noticing that a gramme of electrons, crammed together in cubical piling, would occupy only 7×10^{-11} cubic centimetres, while two grammes of electrons placed at a distance of a metre apart would repel one another with a force equal to the weight of 3×10^{22} tons. The electric force of repulsion outweighs the gravitational force of attraction in the ratio of $4\cdot2 \times 10^{42}$ to one.

A piece of ordinary matter in its unelectrified state contains a certain number of electrons of this kind, and this number is just such that two pieces of matter each in this state exert no electrical forces on one another—

this condition in fact defines the unelectrified state. A piece of matter appears to be charged with negative or positive electricity according as the number of negatively-charged electrons it possesses is in excess or defect of the number it would possess in its unelectrified state.

From this it follows that we cannot go on dividing a charge of electricity indefinitely—a natural limit is imposed by the charge of one electron, just as in chemistry we suppose a natural limit to be imposed on the divisibility of matter by the mass of an atom. The modern view of electricity may then be justly described as an "atomic" view. And of all the experimental evidence which supports this view none is more striking than the circumstance that these "atoms" continually reappear in experiments of the most varied kinds, and that the atomic charge of electricity appears always to be precisely the same.

It also follows that in charging a body with electricity we either add to or subtract from its mass according as we charge it with negative electricity (*i.e.*, add to it a number of electrons), or charge it with positive electricity (*i.e.*, remove from it a number of electrons). Since the mass of an electron is so minute in comparison with the charge it carries, it will readily be seen that the change in its mass is very much too small to be perceptible by any methods of measurement which are at our disposal. Maxwell mentions, as an example of a body possessing an electric charge large compared with its mass, the case of a gramme of gold, which may be beaten into a gold-leaf one square metre in area, and can, in this state, hold a charge of 60,000 electrostatic units of negative electricity. The mass of the number of negatively electrified electrons necessary to carry this charge will be found, as the result of a brief calculation from the data already given, to be about 10^{-13} grammes. The change of weight by electrification is therefore one which it is far beyond the power of the most sensitive balance to detect.

On this view of electricity, the electrons must repel one another, and must be attracted by matter which is devoid of electrons, or in which there is a deficiency of electrons. The electrons move about freely through conductors, but not through insulators. The reactions which, as we have seen, must be supposed to occur at the surface of charged conductors between "matter" and "electricity," can now be interpreted simply as systems of forces between the electrons and the remainder of the matter. Up to a certain extent these forces will restrain the electrons from leaving the conductor, but if the electric forces acting on the electrons exceed a certain limit, they will overcome the forces acting between the electrons and the remainder of the conductor, and an electric discharge takes place from the surface of the conductor.

Thus an essential feature of the modern view of electricity is that it regards the flow of electricity as a material flow of charged electrons. Good conductors and good insulators are now seen to mean simply substances in which the electrons move with extreme ease and extreme difficulty respectively.

The law that equal quantities of positive and negative electricity are generated simultaneously means that electrons may flow about, but cannot be created or annihilated.

The modern view enables us also to give a simple physical interpretation to the phenomenon of induction. A positive charge placed near a conductor will attract the electrons in the conductor, and these will flow through the conductor towards the charge until electrical equilibrium is established. There will be then an excess of negative electrons in the regions near the positive charge, and this excess will appear as an induced negative charge. The deficiency of electrons in the more remote parts of the conductor will appear as an induced positive charge. If the inducing charge is negative, the flow of electrons will be in the opposite direction, so that the signs of the induced charges will be reversed. In an insulator, no flow of electrons can take place, so that the phenomenon of electrification by induction does not occur.

On this view of electricity, negative electricity is essentially different in its nature from positive electricity: the difference is something more fundamental than a mere difference of sign. Experimental proof of this difference is not wanting, *e.g.*, a sharply pointed conductor can hold a greater charge of positive than of negative electricity before reaching the limit at which a discharge begins to take place from its surface. But until we come to those parts of electric theory in which the flow of electricity has to be definitely regarded as a flow of electrons, this essential difference between positive and negative electricity will not appear, and the difference between the two will be adequately represented by a difference of sign.

In the last chapter of the book, it will be explained how recent experimental work has traced this essential difference between positive and negative electricity down to its source. We shall see that the positive electricity occurs only in the central cores or "nuclei" of the atom of which matter is constituted, while the outer regions of these atoms consist of negatively-charged particles, the "electrons" already described. For this reason the negative electricity can run about from one atom to another, and even from one conductor to another, but the positive electricity necessarily remains permanently associated with the same atoms of matter.

SUMMARY.

29. It will be useful to conclude the chapter by a summary of the results which are arrived at by experiment, independently of all hypotheses as to the nature of electricity.

These have been stated by Maxwell in the form of laws, as follows:

Law I. The total electrification of a body, or system of bodies, remains always the same, except in so far as it receives electrification from or gives electrification to other bodies.

Law II. When one body electrifies another by conduction, the total electrification of the two bodies remains the same ; that is, the one loses as much positive or gains as much negative electrification as the other gains of positive or loses of negative electrification.

Law III. When electrification is produced by friction, or by any other known method, equal quantities of positive and negative electrification are produced.

Definition. The electrostatic unit of electricity is that quantity of positive electricity which, when placed at unit distance from an equal quantity, repels it with unit of force.

Law IV. The repulsion between two small bodies charged respectively with e and e' units of electricity is numerically equal to the product of the charges divided by the square of the distance.

These are the forms in which the laws are given by Maxwell. Law I, it will be seen, includes II and III. As regards the Definition and Law IV, it is necessary to specify the medium in which the small bodies are placed, since, as we shall see later, the force is different when the bodies are in air, or in a vacuum, or surrounded by other non-conducting media. It is usual to assume, for purposes of the Definition and Law IV, that the bodies are in air. For strict scientific exactness, we ought further to specify the density, the temperature, and the exact chemical composition of the air. Also we have seen that when the electricity is not insulated on small bodies, but is free to move on conductors, the forces of Law IV must be regarded as acting on the charges of electricity themselves. When the electricity is not free to move, there is an action and reaction between the electricity and matter, so that the forces which really act on the electricity appear to act on the bodies themselves which carry the charges.

CHAPTER II

THE ELECTROSTATIC FIELD OF FORCE

CONCEPTIONS USED IN THE SURVEY OF A FIELD OF FORCE

I. *The Intensity at a point.*

30. THE space in the neighbourhood of charges of electricity, considered with reference to the electric phenomena occurring in this space, is spoken of as the electric field.

A new charge of electricity, placed at any point O in an electric field, will experience attractions or repulsions from all the charges in the field. The introduction of a new charge will in general disturb the arrangement of the charges on all the conductors in the field by a process of induction. If, however, the new charge is supposed to be infinitesimal, the effects of induction will be negligible, so that the forces acting on the new charge may be supposed to arise from the charges of the original field.

Let us suppose that we introduce an infinitesimal charge ϵ on an infinitely small conductor. Any charge e_1 in the field at a distance r_1 from the point O will repel the charge with a force $\epsilon e_1/r_1^2$. The charge ϵ will experience a similar repulsion from every charge in the field, so that each repulsion will be proportional to ϵ.

The resultant of these forces, obtained by the usual rules for the composition of forces, will be a force proportional to ϵ—say a force $R\epsilon$ in some direction OP. We define the *electric intensity* at O to be a force of which the magnitude is R, and the direction is OP. Thus

The electric intensity at any point is given, in magnitude and direction, by the force per unit charge which would act on a charged particle placed at this point, the charge on the particle being supposed so small that the distribution of electricity on the conductors in the field is not affected by its presence.

The electric intensity at O, defined in this way, depends only on the permanent field of force, and has nothing to do with the charge, or the size, or even the existence of the small conductor which has been used to explain the meaning of the electric intensity. There will be a definite intensity at every point of the electric field, quite independently of the presence of small charged bodies.

A small charged body might, however, conveniently be used for exploring the electric field and determining experimentally the direction of the electric intensity at any point in the field. For if we suppose the body carrying a charge ϵ to be held by an insulating thread, both the body and thread being so light that their weights may be neglected, then clearly all the forces acting on the charged body may be reduced to two:—

(i) A force $R\epsilon$ in the direction of the electric intensity at the point occupied by ϵ,

(ii) the tension of the thread acting along the thread.

For equilibrium these two forces must be equal and opposite. Hence the direction of the intensity at the point occupied by the small charged body is obtained at once by producing the direction of the thread through the charged body. And if we tie the other end of the thread to a delicate spring balance, we can measure the tension of the spring, and since this is numerically equal to $R\epsilon$, we should be able to determine R if ϵ were known. We might in this way determine the magnitude and direction of the electric intensity at any point in the field.

In a similar way, a float at the end of a fishing-line might be used to determine the strength and direction of the current at any point on a small lake. And, just as with the electric intensity, we should only get the true direction of the current by supposing the float to be of infinitesimal size. We could not imagine the direction of the current obtained by anchoring a battleship in the lake, because the presence of the ship would disturb the whole system of currents.

II. *Lines of Force.*

31. Let us start at any point O in the electric field, and move a short distance OP in the direction of the electric intensity at O. Starting from P let us move a short distance PQ in the direction of the intensity at P,

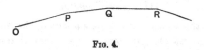

Fig. 4.

and so on. In this way we obtain a broken path $OPQR...$, formed of a number of small rectilinear elements. Let us now pass to the limiting case in which each of the elements OP, PQ, QR, ... is infinitely small. The broken path becomes a continuous curve, and it has the property that at every point on it the electric intensity is in the direction of the tangent to the curve at that point. Such a curve is called a Line of Force. We may therefore define a line of force as follows:—

A line of force is a curve in the electric field, such that the tangent at every point is in the direction of the electric intensity at that point.

If we suppose the motion of a charged particle to be so much retarded by frictional resistance that it cannot acquire any appreciable momentum, then a charged particle set free in the electric field would trace out a line of force. In the same way, we should have lines of current on the surface of a lake, such that the tangent to a line of current at any point coincided with the direction of the current, and a small float set free on the lake would describe a current-line.

32. The resultant of a number of known forces has a definite direction, so that there is a single direction for the electric intensity at every point of the field. It follows that two lines of force can never intersect; for if they did there would be two directions for the electric intensity at the point of intersection (namely, the two tangents to the lines of force at this point) so that the resultant of a number of known forces would be acting in two directions at once. An exception occurs, as we shall see, when the resultant intensity vanishes at any point.

The intensity R may be regarded as compounded of three components X, Y, Z, parallel to three rectangular axes Ox, Oy, Oz.

The magnitude of the electric intensity is then given by

$$R^2 = X^2 + Y^2 + Z^2,$$

and the direction cosines of its direction are

$$\frac{X}{R}, \quad \frac{Y}{R}, \quad \frac{Z}{R}.$$

These, therefore, are also the direction cosines of the tangent at x, y, z to the line of force through the point. The differential equation of the system of lines of force is accordingly

$$\frac{dx}{X} = \frac{dy}{Y} = \frac{dz}{Z}.$$

III. *The Potential.*

33. In moving the small test-charge ϵ about in the field, we may either have to do work against electric forces, or we may find that these forces will do work for us. A small charged particle which has been placed at a point O in the electric field may be regarded as a store of energy, this energy being equal to the work (positive or negative) which has been done in taking the charge to O in opposition to the repulsions and attractions of the field. The energy can be reclaimed by allowing the particle to retrace its path. Assume the charge on the moving particle to be so small that the distribution of electricity on the conductors in the field is not affected by it. Then the work done in bringing the charge ϵ to a point O is proportional to ϵ, and may be taken to be $V\epsilon$. The amount of work done will of course depend on the position from which the charged particle started. It is convenient, in measuring $V\epsilon$, to suppose that the particle started at a point outside the field altogether, *i.e.* from a point so far removed from all the charges of the field that their effect at this point is inappreciable—for brevity, we may say the point at infinity. We now define V to be the potential at the point O. Thus

The potential at any point in the field is the work per unit charge which has to be done on a charged particle to bring it to that point, the charge on the particle being supposed so small that the distribution of electricity on the conductors in the field is not affected by its presence.

In moving the small charge ϵ from x, y, z to $x+dx$, $y+dy$, $z+dz$, we shall have to perform an amount of work

$$- (Xdx + Ydy + Zdz)\,\epsilon,$$

so that in bringing the charge ϵ into position at x, y, z from outside the field altogether, we do an amount of work

$$- \epsilon \int (Xdx + Ydy + Zdz),$$

where the integral is taken along the path followed by ϵ.

Denoting the work done on the charge ϵ in bringing it to any point x, y, z in the electric field by $V\epsilon$, we clearly have

$$V = - \int_{\infty}^{x,\,y,\,z} (Xdx + Ydy + Zdz) \quad \ldots\ldots\ldots\ldots\ldots\ldots (6),$$

giving a mathematical expression for the potential at the point x, y, z.

The same result can be put in a different form. If ds is any element of the path, and if the intensity R at the extremity of this element makes an angle θ with ds, then the component of the force acting on ϵ when moving along ds, resolved in the direction of motion of ϵ, is $R\epsilon \cos \theta$. The work done in moving ϵ along the element ds is accordingly

$$- R\epsilon \cos \theta\, ds,$$

so that the whole work in bringing ϵ from infinity to x, y, z is

$$- \epsilon \int_{\infty}^{x,\,y,\,z} R \cos \theta\, ds,$$

and since this is equal, by definition, to $V\epsilon$, we must have

$$V = - \int_{\infty}^{x,\,y,\,z} R \cos \theta\, ds \quad \ldots\ldots\ldots\ldots\ldots\ldots (7).$$

We see at once that the two expressions (6) and (7) just obtained for V are identical, on noticing that θ is the angle between two lines of which the direction cosines are respectively

$$\frac{X}{R}, \ \frac{Y}{R}, \ \frac{Z}{R} \quad \text{and} \quad \frac{dx}{ds}, \ \frac{dy}{ds}, \ \frac{dz}{ds}.$$

We therefore have 　　　$\cos \theta = \dfrac{X}{R}\dfrac{dx}{ds} + \dfrac{Y}{R}\dfrac{dy}{ds} + \dfrac{Z}{R}\dfrac{dz}{ds},$

so that 　　　　　　$R \cos \theta \, ds = X dx + Y dy + Z dz,$

and the identity of the two expressions becomes obvious.

If the Theorem of the Conservation of Energy is true in the Electrostatic Field, the work done in bringing a small charge e from infinity to any point P must be the same whatever path to P we choose. For if the amounts of work were different on two different paths, let these amounts be $V_P e$ and $V_P' e$, and let the former be the greater. Then by taking the charge from P to infinity by the former path and bringing it back by the latter, we should gain an amount of work $(V_P - V_P') e$, which would be contrary to the Conservation of Energy. Thus V_P and V_P' must be equal, and the potential at P is the same, no matter by what path we reach P. The potential at P will accordingly depend only on the coordinates x, y, z of P.

As soon as we introduce the special law of the inverse square, we shall find that the potential must be a single-valued function of x, y, z, as a consequence of this law (§ 39), and hence shall be able to prove that the Theorem of Conservation of Energy is true in an Electrostatic field. For the moment, however, we assume this.

34. Let us denote by W the work done in moving a charge e from P to Q. In bringing the charge from infinity to P, we do an amount of work

Fig. 5.

which by definition is equal to $V_P e$ where V_P denotes the value of V at the point P. Hence in taking it from infinity to Q, we do a total amount of work $V_P e + W$. This, however, is also equal by definition to $V_Q e$. Hence we have

$$V_P e + W = V_Q e,$$

or 　　　　　　　$W = (V_Q - V_P) e \ \dotfill \ (8).$

35. DEFINITION. *A surface in the electric field such that at every point on it the potential has the same value, is called an Equipotential Surface.*

In discussing the phenomena of the electrostatic field, it is convenient to think of the whole field as mapped out by systems of equipotential surfaces and lines of force, just as in geography we think of the earth's surface as divided up by parallels of latitude and of longitude. A more exact parallel is obtained if we think of the earth's surface as mapped out by "contour-lines" of equal height above sea-level, and by lines of greatest slope. These reproduce all the properties of equipotentials and lines of force, for in point of fact they are actual equipotentials and lines of force for the gravitational field of force.

THEOREM. *Equipotential surfaces cut lines of force at right angles.*

Let P be any point in the electric field, and let Q be an adjacent point on the same equipotential as P. Then, by definition, $V_P = V_Q$, so that by equation (8) $W = 0$, W being the amount of work done in moving a charge ϵ from P to Q. If R is the intensity at Q, and θ the angle which its direction makes with QP, the amount of this work must be $-R\epsilon \cos\theta \times PQ$, so that

$$R\epsilon \cos\theta = 0.$$

Hence $\cos\theta = 0$, so that the line of force cuts the equipotential at right angles. As in a former theorem, an exception has to be made in favour of the case in which $R = 0$.

36. Instead of P, Q being on the same equipotential, let them now be on a line parallel to the axis of x, their coordinates being x, y, z and $x + dx$, y, z respectively. In moving the charge ϵ from P to Q the work done is $-X\epsilon dx$, and by equation (8) it is also $(V_Q - V_P)\epsilon$. Hence

$$-Xdx = V_Q - V_P.$$

Since Q and P are adjacent, we have, from the definition of a differential coefficient,

$$\frac{\partial V}{\partial x} = \frac{V_Q - V_P}{dx} = -X;$$

hence we have the relations

$$X = -\frac{\partial V}{\partial x}, \quad Y = -\frac{\partial V}{\partial y}, \quad Z = -\frac{\partial V}{\partial z} \quad \dots\dots\dots\dots(9),$$

results which are of course obvious on differentiating equation (6) with respect to x, y and z respectively.

Similarly, if we imagine P, Q to be two points on the same line of force we obtain

$$R = -\frac{\partial V}{\partial s},$$

where $\frac{\partial}{\partial s}$ denotes differentiation along a line of force. Since R is necessarily positive, it follows that $\frac{\partial V}{\partial s}$ is negative, *i.e.* V decreases as s increases, or the intensity is in the direction of V decreasing. Thus the lines of force run from higher to lower values of V, and, as we have already seen, cut all equipotentials at right angles.

37. At a point which is occupied by conducting material, the electric charges, as has already been said, must be in equilibrium under the action of the forces from all the other charges in the field. The resultant force from all these charges on any element of charge ϵ is however $R\epsilon$, so that we must have $R = 0$. Hence $X = Y = Z = 0$, so that

$$\frac{\partial V}{\partial x} = \frac{\partial V}{\partial y} = \frac{\partial V}{\partial z} = 0.$$

In other words, V must be constant throughout a conductor for electrostatic equilibrium to be possible. And in particular the surface of a conductor must be an equipotential surface, or part of one. The equipotential of which the surface of a conductor is part has the peculiarity of being three-dimensional instead of two-dimensional, for it occupies the whole interior as well as the surface of the conductor.

In the same way, in considering the analogous arrangement of contour-lines and lines of greatest slope on a map of the earth's surface, we find that the edge of a lake or sea must be a contour-line, but that in strictness this particular contour must be regarded as two-dimensional rather than one-dimensional, since it coincides with the whole surface of the lake or sea.

If V is not constant in any conductor, the intensity is in the direction of V decreasing. Hence positive electricity tends to flow in the direction of V decreasing, and negative electricity in the direction of V increasing. If two conductors in which the potential has different values are joined by a third conductor, the intensity in the third conductor will be in direction from the conductor at higher potential to that at lower potential. Electricity will flow through this conductor, and will continue to flow until the redistribution of potential caused by the transfer of this electricity is such that the potential is the same at all points of the conductors, which may now be regarded as forming one single conductor.

Thus although the potential has been defined only with reference to single points, it is possible to speak of the potential of a whole conductor. In fact, the mathematical expression of the condition that equilibrium shall be possible for a given system of charges is simply that the potential shall be constant throughout each conductor. And when electric contact is established between two conductors, either by joining them by a wire or by other means, the new condition for equilibrium which is made necessary by the new physical condition introduced, is simply that the potentials of the two conductors shall be equal.

The earth is a conductor, and is therefore at the same potential throughout. In all practical applications of electrostatics, it will be legitimate to regard the potential of the earth as zero, a distant point on the earth's surface replacing the imaginary point at infinity, with reference to which potentials have so far been measured. Thus any conductor can be reduced to potential zero by joining it by a metallic wire to the earth.

Mathematical Expressions of the Law of the Inverse Square.

I. *Values of Potential and Intensity.*

38. We now discuss the values of the potential and components of electric intensity when the space between the conductors is air, so that the electric forces are determined by Coulomb's Law.

If we have a single point charge e_1 at a point P, the value of R, the resultant intensity at any point O, is

$$\frac{e_1}{PO^2},$$

and its direction is that of PO. Hence if θ is the angle between OP and

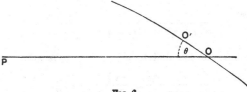

FIG. 6.

OO', the line joining O to an adjacent point O', the work done in moving a charge ϵ from O to O'

$$= \epsilon R \cos \theta . OO'$$
$$= \epsilon R (OP - O'P)$$
$$= - \epsilon R dr,$$

where $OP = r$, $O'P = r + dr$. Hence the work done against the repulsion of the charge e_1 in bringing ϵ from infinity to O' by any path is

$$- \epsilon \int_{r=\infty}^{r=O'P} R dr = - \epsilon \int_{r=\infty}^{r=O'P} \frac{e_1}{r^2} dr = \frac{\epsilon e_1}{r_1},$$

where $r_1 = O'P$.

If there are other charges e_2, e_3, ... the work done against all the repulsions in bringing a charge ϵ to O' will be the sum of terms such as the above, say

$$\epsilon \left(\frac{e_1}{r_1} + \frac{e_2}{r_2} + \frac{e_3}{r_3} + \dots \right),$$

where r_2, r_3, ... are the distances from O' to e_2, e_3, ..., so that by definition

$$V = \frac{e_1}{r_1} + \frac{e_2}{r_2} + \frac{e_3}{r_3} + \dots \quad\dots\dots\dots\dots\dots(10).$$

39. It is now clear that the potential at any point depends only on the coordinates of the point, so that the work done in bringing a small charge from infinity to a point P is always the same, no matter what path we choose, the result assumed in § 33.

It follows that we cannot alter the amount of energy in the field by moving charges about in such a way that the final state of the field is the same as the original state. In other words, the Conservation of Energy is true of the Electrostatic Field.

40. Analytically, let us suppose that the charge e_1 is at x_1, y_1, z_1; e_2 at x_2, y_2, z_2; and so on. The repulsion on a small charge ϵ at x, y, z resulting from the presence of e_1 at x_1, y_1, z_1 is

$$\frac{e_1\epsilon}{(x - x_1)^2 + (y - y_1)^2 + (z - z_1)^2},$$

and the direction-cosines of the direction in which this force acts on the charge ϵ, are

$$\frac{x - x_1}{[(x - x_1)^2 + (y - y_1)^2 + (z - z_1)^2]^{\frac{1}{2}}}, \quad \frac{y - y_1}{[(x - x_1)^2 + (y - y_1)^2 + (z - z_1)^2]^{\frac{1}{2}}}, \text{ etc.}$$

Hence the component parallel to the axis of x is

$$\frac{e_1\epsilon\,(x - x_1)}{[(x - x_1)^2 + (y - y_1)^2 + (z - z_1)^2]^{\frac{3}{2}}}.$$

By adding all such components, we obtain as the component of the electric intensity at x, y, z,

$$X = \Sigma \frac{e_1\,(x - x_1)}{[(x - x_1)^2 + (y - y_1)^2 + (z - z_1)^2]^{\frac{3}{2}}} \quad\dots\dots\dots\dots(11),$$

and there are similar equations for Y and Z.

We have as the value of V at x, y, z, by equation (6),

$$V = -\int_{\infty}^{x, y, z} (X\,dx + Y\,dy + Z\,dz)$$

$$= -\int_{\infty}^{x, y, z} \frac{\Sigma e_1\,\{(x - x_1)\,dx + (y - y_1)\,dy + (z - z_1)\,dz\}}{[(x - x_1)^2 + (y - y_1)^2 + (z - z_1)^2]^{\frac{3}{2}}}$$

$$= \Sigma \frac{e_1}{[(x - x_1)^2 + (y - y_1)^2 + (z - z_1)^2]^{\frac{1}{2}}},$$

giving the same result as equation (10).

41. If the electric distribution is not confined to points, we can imagine it divided into small elements which may be treated as point charges. For instance if the electricity is spread throughout a volume, let the charge on any element of volume $dx'\,dy'\,dz'$ be $\rho\,dx'\,dy'\,dz'$ so that ρ may be spoken of as the "density" of electricity at x', y', z'. Then in formula (11) we can replace e_1 by $\rho\,dx'\,dy'\,dz'$, and x_1, y_1, z_1, by x', y', z'. Instead of summing the charges e_1, ... we of course integrate $\rho\,dx'\,dy'\,dz'$ through all those parts of the space which contain electrical charges. In this way we obtain

$$X = \iiint \frac{\rho\,(x - x')\,dx'\,dy'\,dz'}{[(x - x')^2 + (y - y')^2 + (z - z')^2]^{\frac{3}{2}}}, \text{ etc.,}$$

and

$$V = \iiint \frac{\rho\,dx'\,dy'\,dz'}{[(x - x')^2 + (y - y')^2 + (z - z')^2]^{\frac{1}{2}}}.$$

These equations are one form of mathematical expression of the law of the inverse square of the distance. An attempt to perform the integration, in even a few simple cases, will speedily convince the student that the form is not one which lends itself to rapid progress. A second form of mathematical expression of the law of the inverse square is supplied by a Theorem of Gauss which we shall now prove, and it is this expression of the law which will form the basis of our development of electrostatical theory.

II. *Gauss' Theorem.*

42. THEOREM. *If any closed surface is taken in the electric field, and if N denotes the component of the electric intensity at any point of this surface in the direction of the outward normal, then*

$$\iint N\,dS = 4\pi E,$$

where the integration extends over the whole of the surface, and E is the total charge enclosed by the surface.

Let us suppose the charges in the field, both inside and outside the closed surface, to be e_1 at P_1, e_2 at P_2, and so on. The intensity at any point is the resultant of the intensities due to the charges separately, so that at any point of the surface, we may write

$$N = N_1 + N_2 + \ldots \quad\ldots\ldots\ldots\ldots\ldots\ldots(12),$$

where N_1, N_2, ... are the normal components of intensity due to e_1, e_2, ... separately.

Instead of attempting to calculate $\iint N\,dS$ directly, we shall calculate separately the values of $\iint N_1\,dS$, $\iint N_2\,dS$, ... The value of $\iint N\,dS$ will, by equation (12), be the sum of these integrals.

Let us take any small element dS of the closed surface in the neighbour-hood of a point Q on the surface and join each point of its boundary to the point P_1. Let the small cone so formed cut off an element of area $d\sigma$ from

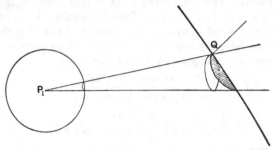

a sphere drawn through Q with P_1 as centre, and an element of area $d\omega$ from a sphere of unit radius drawn about P_1 as centre. Let the normal to the closed surface at Q in the direction away from P_1 make an angle θ with P_1Q.

The intensity at Q due to the charge e_1 at P_1 is e_1/P_1Q^2 in the direction P_1Q, so that the component of the intensity along the normal to the surface in the direction away from P_1 is

$$\frac{e_1}{P_1Q^2}\cos\theta.$$

The contribution to $\iint N_1 dS$ from the element of surface is accordingly

$$\pm\frac{e_1}{P_1Q^2}\cos\theta\,dS,$$

the $+$ or $-$ sign being taken according as the normal at Q in the direction away from P_1 is the outward or inward normal to the surface.

Now $\cos\theta\,dS$ is equal to $d\sigma$, the projection of dS on the sphere through Q having P_1 as centre, for the two normals to dS and $d\sigma$ are inclined at an angle θ. Also $d\sigma = P_1Q^2 d\omega$. For $d\sigma$, $d\omega$ are the areas cut off by the same cone on spheres of radii P_1Q and unity respectively. Hence

$$\frac{e_1}{P_1Q^2}\cos\theta\,dS = \frac{e_1 d\sigma}{P_1Q^2} = e_1 d\omega.$$

If P_1 is inside the closed surface, a line from P_1 to any point on the unit sphere surrounding P_1 may either cut the closed surface only once as at Q (fig. 8)—in which case the normal to the surface at Q in the direction *away* from P_1 is the outward normal to the surface—or it may cut three times, as at Q', Q'', Q'''—in which case two of the normals away from P_1 (those at Q', Q'' in fig. 8) are outward normals to the surface, while the third normal away from P_1 (that at Q'' in the figure) is an inward normal—or it may cut five, seven, or any odd number of times. Thus a cone through a small element of area $d\omega$ on a unit sphere about P_1 may cut the closed surface any odd number of times. However many times it cuts, the first small area cut off will contribute $e_1 d\omega$ to $\iint N_1 dS$, the second and third small areas if they

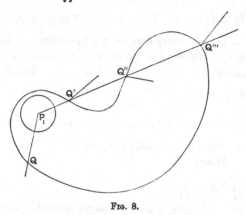

Fig. 8.

occur will contribute $-e_1 d\omega$ and $+e_1 d\omega$ respectively, the fourth and fifth if they occur will contribute $-e_1 d\omega$ and $+e_1 d\omega$ respectively, and so on. The total contribution from the cone surrounding $d\omega$ is, in every case, $+e_1 d\omega$.

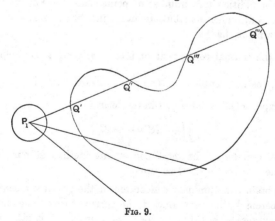

Fig. 9.

Summing over all cones which can be drawn in this way through P_1 we obtain the whole value of $\iint N_1 dS$, which is thus seen to be simply e_1 multiplied by the total surface area of the unit sphere round P_1, and therefore $4\pi e_1$.

On the other hand if P_1 is outside the closed surface, as in fig. 9, the cone through any element of area $d\omega$ on the unit sphere may either not cut the closed surface at all, or may cut twice, or four, six or any even number of times. If the cone through $d\omega$ intersects the surface at all, the first pair of elements of surface which are cut off by the cone contribute $-e_1 d\omega$ and $+e_1 d\omega$ respectively to $\iint N_1 dS$. The second pair, if they occur, make a similar contribution and so on. In every case the total contribution from any small cone through P_1 is *nil*. By summing over all such cones we shall include the contributions from all parts of the closed surface, so that if P_1 is outside the surface $\iint N_1 dS$ is equal to zero.

We have now seen that $\iint N_1 dS$ is equal to $4\pi e_1$ when the charge e_1 is inside the closed surface, and is equal to zero when the charge e_1 is outside the closed surface. Hence

$$\iint N dS = \iint N_1 dS + \iint N_2 dS + \dots$$
$$= 4\pi \times \text{(the sum of all the charges inside the surface)}$$
$$= 4\pi E,$$

which proves the theorem.

Obviously the theorem is true also when there is a continuous distribution of electricity in addition to a number of point charges. For clearly we can divide up the continuous distribution into a number of small elements and treat each as a point charge.

Since N, the normal component of intensity, is equal by § 36 to $-\dfrac{\partial V}{\partial n}$, where $\dfrac{\partial}{\partial n}$ denotes differentiation along the outward normal, it appears that we can also express Gauss' Theorem in the form

$$\iint \frac{\partial V}{\partial n} dS = -4\pi E.$$

Gauss' theorem forms the most convenient method at our disposal, of expressing the law of the inverse square.

We can obtain a preliminary conception of the physical meaning underlying the theorem by noticing that if the surface contains no charge at all, the theorem expresses that the *average* normal intensity is *nil*. If there is a negative charge inside the surface, the theorem shews that the average normal intensity is negative, so that a positively charged particle placed at a point on the imaginary surface will be likely to experience an attraction to the interior of the surface rather than a repulsion away from it, and *vice versâ* if the surface contains a positive charge.

Corollaries to Gauss' Theorem.

43. THEOREM. *If a closed surface be drawn, such that every point on it is occupied by conducting material, the total charge inside it is nil.*

We have seen that at any point occupied by conducting material, the electric intensity must vanish. Hence at every point of the closed surface, $N = 0$, so that $\iint N dS = 0$, and therefore, by Gauss' Theorem, the total charge inside the closed surface must vanish.

The two following special cases of this theorem are of the greatest importance.

44. THEOREM. *There is no charge at any point which is occupied by conducting material, unless this point is on the surface of a conductor.*

For if the point is not on the surface, it will be possible to surround the point by a small sphere, such that every point of this sphere is inside the conductor. By the preceding theorem the charge inside this sphere is *nil*, hence there is no charge at the point in question.

This theorem is often stated by saying :—

The charge of a conductor resides on its surface.

45. THEOREM. *If we have a hollow closed conductor, and place any number of charged bodies inside it, the charge on its inner surface will be equal in magnitude but opposite in sign, to the total charge on the bodies inside.*

For we can draw a closed surface entirely inside the material of the conductor, and by the theorem of § 43, the whole charge inside this surface must be *nil*. This whole charge is, however, the sum of (i) the charge on the inner surface of the conductor, and (ii) the charges on the bodies inside the conductor. Hence these two must be equal and opposite.

This result explains the property of the electroscope which led us to the conception of a definite quantity of electricity. The vessel placed on the plate of the electroscope formed a hollow closed conductor. The charge on the inner surface of this conductor, we now see, must be equal and opposite to the total charge inside, and since the total charge on this conductor is *nil*, the charge on its outer surface must be equal and opposite to that on the inner surface, and therefore exactly equal to the sum of the charges placed inside, independently of the position of these charges.

The Cavendish Proof of the Law of the Inverse Square.

46. We have deduced from the law of the inverse square, that the charge inside a closed conductor is zero. We shall now shew that the converse theorem is also true. Hence, in the known fact, revealed by the observations of Cavendish and Maxwell, that the charge inside a closed conductor is zero, we have experimental proof of the law of the inverse square which admits of much greater accuracy than the experimental proof of Coulomb.

The theorem that if there is no charge inside a spherical conductor the law of force must be that of the inverse square is due to Laplace. We need consider this converse theorem only in its application to a spherical conductor, this being the actual form of conductor used by Cavendish. The apparatus illustrated in fig. 10 is not that used by Cavendish, but is an improved form designed by Maxwell, who repeated Cavendish's experiment in a more delicate form.

Two spherical shells are fixed by a ring of ebonite so as to be concentric

with one another, and insulated from one another. Electrical contact can be established between the two by letting down the small trap-door *B* through which a wire passes, the wire being of such a length as just to establish contact when the trap-door is closed. The experiment is conducted by electrifying the outer shell, opening the trap-door by an insulating thread without discharging the conductor, afterwards discharging the outer conductor and testing whether any charge is to be found on the inner shell by placing it in electrical contact with a delicate electroscope by means of a conducting wire inserted through the trap-door. It is found that there are no traces of a charge on the inner sphere.

Fig. 10.

47. Suppose we start to find the law of electric force such that there shall be no charge on the inner sphere. Let us assume a law of force such that the repulsion between two charges e, e' at distance r apart is $ee'\phi(r)$. The potential, calculated as explained in § 33, is

$$\Sigma e \int_r^\infty \phi(r)\,dr \dots\dots\dots\dots\dots\dots(13),$$

where the summation extends over all the charges in the field.

Let us calculate the potential at a point inside the sphere due to a charge E spread entirely over the surface of the sphere. If the sphere is of radius a, the area of its surface is $4\pi a^2$, so that the amount of charge per unit area is $E/4\pi a^2$, and the expression for the potential becomes

$$V = \iint \frac{E}{4\pi a^2}\left(\int_r^\infty \phi(r)\,dr\right) a^2 \sin\theta\,d\theta\,d\phi\dots\dots\dots\dots(14),$$

the summation of expression (13) being now replaced by an integration which extends over the whole sphere. In this expression r is the distance from the point at which the potential is evaluated, to the element $a^2 \sin \theta \, d\theta \, d\phi$ of spherical surface.

If we agree to evaluate the potential at a point situated on the axis $\theta = 0$ at a distance c from the centre, we may write

$$r^2 = a^2 + c^2 - 2ac \cos \theta.$$

Since c is a constant, we obtain as the relation between dr and $d\theta$, by differentiation of this last equation,

$$r \, dr = ac \sin \theta \, d\theta \dots\dots\dots\dots\dots\dots(15).$$

If we integrate expression (14) with respect to ϕ, the limits being of course $\phi = 0$ and $\phi = 2\pi$, we obtain

$$V = \tfrac{1}{2} E \int_{\theta=0}^{\theta=\pi} \left(\int_r^\infty \phi(r) \, dr \right) \sin \theta \, d\theta,$$

or, on changing the variable from θ to r, by the help of relation (15)

$$V = \tfrac{1}{2} E \int_{r=a-c}^{r=a+c} \left(\int_r^\infty \phi(r) \, dr \right) \frac{r \, dr}{ac}.$$

If we introduce a new function $f(r)$, defined by

$$f(r) = \int \left(\int_r^\infty \phi(r) \, dr \right) r \, dr,$$

we obtain as the value of V,

$$V = \frac{E}{2ac} \{ f(a+c) - f(a-c) \}.$$

If the inner and outer spheres are in electrical contact, their potentials are the same; and if, as experiment shews to be the case, there is no charge on the inner sphere, then the whole potential must be that just found. This expression must, accordingly, have the same value whether c represents the radius of the outer sphere or that of the inner. Since this is true whatever the radius of the inner sphere may be, the expression must be the same for all values of c. We must accordingly have

$$\frac{2ac V}{E} = f(a+c) - f(a-c),$$

where V is the same for all values of c. Differentiating this equation twice with respect to c, we obtain

$$0 = f''(a+c) - f''(a-c).$$

Since by definition, $f(r)$ depends only on the law of force, and not on a or c it follows from the relation

$$f''(a+c) = f''(a-c),$$

that $f''(r)$ must be a constant, say C.

Hence $\qquad\qquad\qquad\qquad f(r) = A + Br + \tfrac{1}{2}Cr^2,$

and by definition $\qquad\qquad f(r) = \int\left(\int_r^\infty \phi(r)\,dr\right)r\,dr,$

so that on equating the two values of $f''(r)$,

$$B + Cr = r\int_r^\infty \phi(r)\,dr.$$

Therefore $\qquad\qquad\qquad \int_r^\infty \phi(r)\,dr = C + \dfrac{B}{r},$

or $\qquad\qquad\qquad\qquad\qquad \phi(r) = \dfrac{B}{r^2},$

so that the law of force is that of the inverse square.

48. Maxwell has examined what charge would be produced on the inner sphere if, instead of the law of force being accurately B/r^2, it were of the form B/r^{2+q}, where q is some small quantity. In this way he found that if q were even so great as $\frac{1}{21600}$, the charge on the inner sphere would have been too great to escape observation. As we have seen, the limit which Cavendish was able to assign to q was $\frac{1}{50}$.

It may be urged that the form B/r^{2+q} is not a sufficiently general law of force to assume. To this Maxwell has replied that it is the most general law under which conductors which are of different sizes but geometrically similar can be electrified similarly, while experiment shews that in point of fact geometrically similar conductors are electrified similarly. We may say then with confidence that the error in the law of the inverse square, if any, is extremely small. It should, however, be clearly understood that experiment has only proved the law B/r^2 for values of r which are great enough to admit of observation. The law of force between two electric charges which are at very small distances from one another still remains entirely unknown to us.

III. *The Equations of Poisson and Laplace.*

49. There is still a third way of expressing the law of the inverse square, and this can be deduced most readily from Gauss' Theorem.

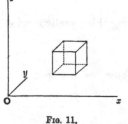

FIG. 11.

Let us examine the small rectangular parallelepiped, of volume $dx\,dy\,dz$, which is bounded by the six plane faces

$$x = \xi \pm \tfrac{1}{2}dx, \qquad y = \eta \pm \tfrac{1}{2}dy, \qquad z = \zeta \pm \tfrac{1}{2}dz.$$

We shall suppose that this element does not contain any point charges of electricity, or part of any charged surface, but for the sake of generality we shall suppose that the whole space is charged with a continuous distribution of electricity, the volume-density of electrification in the neighbourhood of the small element under consideration being ρ. The whole charge contained by the element of volume is accordingly $\rho\,dx\,dy\,dz$, so that Gauss' Theorem assumes the form

$$\iint N dS = 4\pi\rho\,dx\,dy\,dz \quad\ldots\ldots\ldots\ldots\ldots\ldots(16).$$

The surface integral is the sum of six contributions, one from each face of the parallelepiped. The contribution from that face which lies in the plane $x = \xi - \tfrac{1}{2}dx$ is equal to $dy\,dz$, the area of the face, multiplied by the mean value of N over this face. To a sufficient approximation, this may be supposed to be the value of N at the centre of the face, *i.e.* at the point $\xi - \tfrac{1}{2}dx$, η, ζ, and this again may be written

$$\left(\frac{\partial V}{\partial x}\right)_{\xi - \frac{1}{2}dx,\,\eta,\,\zeta},$$

so that the contribution to $\iint N dS$ from this face is

$$dy\,dz\left(\frac{\partial V}{\partial x}\right)_{\xi - \frac{1}{2}dx,\,\eta,\,\zeta}.$$

Similarly the contribution from the opposite face is

$$-dy\,dz\left(\frac{\partial V}{\partial x}\right)_{\xi + \frac{1}{2}dx,\,\eta,\,\zeta},$$

the sign being different because the outward normal is now the positive axis of x, whereas formerly it was the negative axis. The sum of the contributions from the two faces perpendicular to the axis of x is therefore

$$-dy\,dz\left\{\left(\frac{\partial V}{\partial x}\right)_{\xi + \frac{1}{2}dx,\,\eta,\,\zeta} - \left(\frac{\partial V}{\partial x}\right)_{\xi - \frac{1}{2}dx,\,\eta,\,\zeta}\right\} \quad\ldots\ldots\ldots(17)$$

The expression inside curled brackets is the increment in the function $\dfrac{\partial V}{\partial x}$ when x undergoes a small increment dx. This we know is $dx\,\dfrac{\partial}{\partial x}\left(\dfrac{\partial V}{\partial x}\right)$, so that expression (17) can be put in the form

$$-\frac{\partial^2 V}{\partial x^2}\,dx\,dy\,dz.$$

The whole value of $\iint N dS$ is accordingly

$$-\left(\frac{\partial^2 V}{\partial x^2} + \frac{\partial^2 V}{\partial y^2} + \frac{\partial^2 V}{\partial z^2}\right)dx\,dy\,dz,$$

and equation (16) now assumes the form

$$\frac{\partial^2 V}{\partial x^2} + \frac{\partial^2 V}{\partial y^2} + \frac{\partial^2 V}{\partial z^2} = -4\pi\rho \quad\ldots\ldots\ldots\ldots\ldots(18).$$

This is known as Poisson's Equation; clearly if we know the value of the potential at every point, it enables us to find the charges by which this potential is produced.

50. In free space, where there are no electric charges, the equation assumes the form

$$\frac{\partial^2 V}{\partial x^2} + \frac{\partial^2 V}{\partial y^2} + \frac{\partial^2 V}{\partial z^2} = 0 \dots\dots\dots\dots\dots(19),$$

and this is known as Laplace's Equation. We shall denote the operator

$$\frac{\partial^2}{\partial x^2} + \frac{\partial^2}{\partial y^2} + \frac{\partial^2}{\partial z^2}$$

by ∇^2, so that Laplace's equation may be written in the abbreviated form

$$\nabla^2 V = 0 \dots\dots\dots\dots\dots\dots\dots(20).$$

Equations (18) and (20) express the same fact as Gauss' Theorem, but express it in the form of a differential equation. Equation (20) shews that in a region in which no charges exist, the potential satisfies a differential equation which is independent of the charges outside this region by which the potential is produced. It will easily be verified by direct differentiation that the value of V given in equation (10) is a solution of equation (20).

We can obtain an idea of the physical meaning of this differential equation as follows.

Let us take any point O and construct a sphere of radius r about this point. The mean value of V averaged over the surface of the sphere is

$$\overline{V} = \frac{1}{4\pi r^2} \iint V dS$$

$$= \frac{1}{4\pi} \iint V \sin\theta d\theta d\phi,$$

where r, θ, ϕ are polar coordinates, having O as origin. If we change the radius of this sphere from r to $r + dr$, the rate of change of $\overline{V}$ is

$$\frac{\partial \overline{V}}{\partial r} = \frac{1}{4\pi} \iint \frac{\partial V}{\partial r} \sin\theta d\theta d\phi$$

$$= \frac{1}{4\pi r^2} \iint \frac{\partial V}{\partial r} dS$$

$$= 0, \text{ by Gauss' Theorem,}$$

shewing that $\overline{V}$ is independent of the radius r of the sphere. Taking $r = 0$, the value of $\overline{V}$ is seen to be equal to the potential at the origin O.

This gives the following interpretation of the differential equation:

V varies from point to point in such a way that the average value of V taken over any sphere surrounding any point O is equal to the value of V at O.

DEDUCTIONS FROM LAW OF INVERSE SQUARE.

51. THEOREM. *The potential cannot have a maximum or a minimum value at any point in space which is not occupied by an electric charge.*

For if the potential is to be a maximum at any point O, the potential at every point on a sphere of small radius r surrounding O must be less than that at O. Hence the average value of the potential on a small sphere surrounding O must be less than the value at O, a result in opposition to that of the last section.

A similar proof shews that the value of V cannot be a minimum.

52. A second proof of this theorem is obtained at once from Laplace's equation. Regarding V simply as a function of x, y, z, a necessary condition for V to have a maximum value at any point is that $\dfrac{\partial^2 V}{\partial x^2}$, $\dfrac{\partial^2 V}{\partial y^2}$ and $\dfrac{\partial^2 V}{\partial z^2}$ shall each be negative at the point in question, a condition which is inconsistent with Laplace's equation

$$\frac{\partial^2 V}{\partial x^2} + \frac{\partial^2 V}{\partial y^2} + \frac{\partial^2 V}{\partial z^2} = 0.$$

So also for V to be a minimum, the three differential coefficients would have to be all positive, and this again would be inconsistent with Laplace's equation.

53. If V is a maximum at any point O, which as we have just seen must be occupied by an electric charge, then the value of $\dfrac{\partial V}{\partial r}$ must be negative as we cross a sphere of small radius r. Thus $\iint \dfrac{\partial V}{\partial r} dS$ is negative where the integration is taken over a small sphere surrounding O, and by Gauss' Theorem the value of the surface integral is $-4\pi e$, where e is the total charge inside the sphere. Thus e must be positive, and similarly if V is a minimum, e must be negative. Thus:

If V is a maximum at any point, the point must be occupied by a positive charge, and if V is a minimum at any point, the point must be occupied by a negative charge.

54. We have seen (§ 36) that in moving along a line of force we are moving, at every point, from higher to lower potential, so that the potential continually decreases as we move along a line of force. Hence a line of force can end only at a point at which the potential is a minimum, and similarly by tracing a line of force backwards, we see that it can begin only at a point of which the potential is a maximum. Combining this result with that of the previous theorem, it follows that:

Lines of force can begin only on positive charges, and can end only on negative charges.

It is of course possible for a line of force to begin on a positive charge,. and go to infinity, the potential decreasing all the way, in which case the line of force has, strictly speaking, no end at all. So also, a line of force may come from infinity, and end on a negative charge.

Obviously a line of force cannot begin and end on the same conductor, for if it did so, the potential at its two ends would be the same. Hence there can be no lines of force in the interior of a hollow conductor which contains no charges; consequently there can be no charges on its inner surface.

Tubes of Force.

55. Let us select any small area dS in the field, and let us draw the lines of force through every point of the boundary of this small area. If dS is taken sufficiently small, we can suppose the electric intensity to be the same in magnitude and direction at every point of dS, so that the directions of the lines of force at all the points on the boundary will be approximately all parallel. By drawing the lines of force, then, we shall obtain a "tubular" surface—*i.e.*, a surface such that in the neighbourhood of any point the surface may be regarded as cylindrical. The surface obtained in this way is called a "tube of force." A normal cross-section of a "tube of force" is a section which cuts all the lines of force through its boundary at right angles. It therefore forms part of an equipotential surface.

56. THEOREM. *If ω_1, ω_2 be the areas of two normal cross-sections of the same tube of force, and R_1, R_2 the intensities at these sections, then*

$$R_1\omega_1 = R_2\omega_2.$$

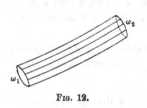

FIG. 12.

Consider the closed surface formed by the two cross-sections of areas ω_1, ω_2, and of the part of the tube of force joining them. There is no charge inside this surface, so that by Gauss' theorem, $\iint N\,dS = 0$.

If the direction of the lines of force is from ω_1 to ω_2, then the outward normal intensity over ω_2 is R_2, so that the contribution from this area to the surface integral is $R_2\omega_2$. So also over ω_1 the outward normal intensity is $-R_1$, so that ω_1 gives a contribution $-R_1\omega_1$. Over the rest of the surface, the outward normal is perpendicular to the electric intensity, so that $N = 0$, and this part of the surface contributes nothing to $\iint N\,dS$. The whole value of this integral, then, is

$$R_2\omega_2 - R_1\omega_1,$$

and since this, as we have seen, must vanish, the theorem is proved.

57. COULOMB'S LAW. *If R is the outward intensity at a point just outside a conductor, then $R = 4\pi\sigma$, where σ is the surface density of electrification on the conductor.*

We have already seen that the whole electrification of a conductor must reside on the surface. Therefore we no longer deal with a volume density of electrification ρ, such that the charge in the element of volume $dx\,dy\,dz$ is $\rho\,dx\,dy\,dz$, but with a surface-density of electrification σ such that the charge on an element dS of the surface of the conductor is $\sigma\,dS$.

The surface of the conductor, as we have seen, is an equipotential, so that by the theorem of p. 29, the intensity is in a direction normal to the surface. Let us draw perpendiculars to the surface at every point on the boundary of a small element of area dS, these perpendiculars each extending a small distance into the conductor in one direction and a small distance away from the conductor in the other direction. We can close the cylindrical surface so formed, by two small plane areas, each equal and parallel to the original element of area dS. Let us now apply Gauss' Theorem to this closed surface. The normal intensity is zero over every part of this surface except over the cap of area dS which is outside the conductor. Over this cap the outward normal intensity is R, so that the value of the surface integral of normal

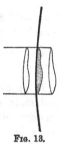

Fig. 13.

intensity taken over the closed surface, consists of the single term $R\,dS$. The total charge inside the surface is $\sigma\,dS$, so that by Gauss' Theorem,

$$R\,dS = 4\pi\sigma\,dS \quad\dots\dots\dots\dots\dots\dots(21),$$

and Coulomb's Law follows on dividing by dS.

58. Let us draw the complete tube of force which is formed by the lines of force starting from points on the boundary of the element dS of the surface of the conductor. Let us suppose that the surface density on this element is positive, so that the area dS forms the normal cross-section at

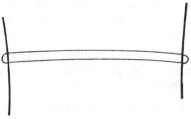

Fig. 14.

the positive end, or beginning, of the tube of force. Let us suppose that at the negative end of the tube of force, the normal cross-section is dS', that the surface density of electrification is σ', σ' being of course negative, and that the intensity in the direction of the lines of force is R'. Then, as in equation (21),

$$R'dS' = -4\pi\sigma'dS',$$

since the outward intensity is now $-R'$.

Since R, R' are the intensities at two points in the same tube of force at which the normal cross-sections are dS, dS', it follows from the theorem of § 56, that

$$RdS = R'dS'$$

and hence, on comparing the values just found for RdS and $R'dS'$, that

$$\sigma dS = -\sigma'dS'.$$

Since σdS and $\sigma'dS'$ are respectively the charges of electricity from which the tube begins and on which it terminates, we see that:

The negative charge of electricity on which a tube of force terminates is numerically equal to the positive charge from which it starts.

If we close the ends of the tube of force by two small caps inside the conductors, as in fig. 14, we have a closed surface such that the normal intensity vanishes at every point. Thus, by Gauss' Theorem, the total charge inside must vanish, giving the result at once.

59. The numerical value of either of the charges at the ends of a tube of force may conveniently be spoken of as the *strength* of the tube. A tube of unit strength is spoken of by many writers as a *unit tube of force*.

The strength of a tube of force is σdS in the notation already used, and this, by Coulomb's Law, is equal to $\dfrac{1}{4\pi} RdS$ where R is the intensity at the end dS of the tube. By the theorem of § 56, RdS is equal to $R_1\omega_1$ where R_1, ω_1 are the intensity and cross-section at any point of the tube. Hence $R_1\omega_1 = 4\pi$ times the strength of the tube. It follows that:

The intensity at any point is equal to 4π times the aggregate strength per unit area of the tubes which cross a plane drawn at right angles to the direction of the intensity.

In terms of unit tubes of force, we may say that the intensity is 4π times the number of unit tubes per unit area which cross a plane drawn at right angles to the intensity.

The conception of tubes of force is due to Faraday: indeed it formed almost his only instrument for picturing to himself the phenomena of the Electric Field. It will be found that a number of theorems connected with the electric field become almost obvious when interpreted with the help of the conception of tubes of force. For instance we proved on p. 37 that when a number of charged bodies are placed inside a hollow conductor, they induce on its inner surface a charge equal and opposite to the sum of all their charges. This may now be regarded as a special case of the obvious theorem that the total charge associated with the beginnings and terminations of any number of tubes of force, none of which pass to infinity, must be *nil*.

EXAMPLES OF FIELDS OF FORCE.

60. It will be of advantage to study a few particular fields of electric force by means of drawing their lines of force and equipotential surfaces.

I. *Two Equal Point Charges.*

61. Let A, B be two equal point charges, say at the points $x = -a, +a$. The equations of the lines of force which are in the plane of x, y are easily found to be

$$\frac{\partial y}{\partial x} = \frac{Y}{X} = \frac{y}{x + a \left(\dfrac{PB^3 - PA^3}{PB^3 + PA^3}\right)} \quad \dots\dots\dots\dots(22),$$

where P is the point x, y.

This equation admits of integration in the form

$$\frac{x+a}{PA} + \frac{x-a}{PB} = \text{cons.} \quad \dots\dots\dots\dots(23).$$

From this equation the lines of force can be drawn, and will be found to lie as in fig. 15.

62. There are, however, only a few cases in which the differential equations of the lines of force can be integrated, and it is frequently simplest to obtain the properties of the lines of force directly from the differential equation. The following treatment illustrates the method of treating lines of force without integrating the differential equation.

From equation (22) we see that obvious lines of force are

(i) $y = 0$, $\dfrac{\partial y}{\partial x} = 0$, giving the axis AB;

(ii) $x = 0$, $PA = PB$, $\dfrac{\partial y}{\partial x} = \infty$, giving the line which bisects AB at right angles.

These lines intersect at C, the middle point of AB. At this point, then, $\dfrac{\partial y}{\partial x}$ has two values, and since $\dfrac{\partial y}{\partial x} = \dfrac{Y}{X}$, it follows that we must have $X = 0$, $Y = 0$. In other words, the point C is a point of equilibrium, as is otherwise obvious.

The same result can be seen in another way. If we start from A and draw a small tube surrounding the line AB, it is clear that the cross-section of the tube, no matter how small it was initially, will have become infinite by the time it reaches the plane which bisects AB at right angles—in fact the cross-section is identical with the infinite plane. Since the product of the cross-section and the normal intensity is constant throughout a tube, it follows that at the point C, the intensity must vanish.

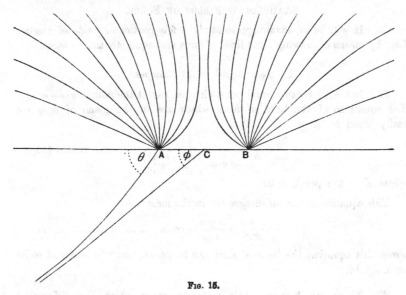

FIG. 15.

At a great distance R from the points A and B, the fraction

$$\frac{PB^3 - PA^3}{PB^3 + PA^3}$$

vanishes to the order of $1/R$, so that

$$\frac{\partial y}{\partial x} = \frac{y}{x},$$

except for terms of the order of $1/R^2$. Thus at infinity the lines of force become asymptotic to straight lines passing through the origin.

Let us suppose that a line of force starts from A making an angle θ with BA produced, and is asymptotic at infinity to a line through C which makes an angle ϕ with BA produced. By rotating this line of force about the axis AB we obtain a surface which may be regarded as the boundary of a bundle of tubes of force. This surface cuts off an area

$$2\pi (1 - \cos \theta) r^2$$

from a small sphere of radius r drawn about A, and at every point of
this sphere the intensity is e/r^2 normal to the sphere. The surface again
cuts off an area

$$2\pi (1 - \cos \phi) R^2$$

from a sphere of very great radius R drawn about C, and at every point
of this sphere the intensity is $2e/R^2$. Hence, applying Gauss' Theorem
to the part of the field enclosed by the two spheres of radii r and R,
and the surface formed by the revolution of the line of force about AB,
we obtain

$$2\pi (1 - \cos \theta)\, r^2 \times \frac{e}{r^2} - 2\pi (1 - \cos \phi)\, R^2 \times \frac{2e}{R^2} = 0,$$

from which follows the relation

$$\sin \tfrac{1}{2} \theta = \sqrt{2} \sin \tfrac{1}{2} \phi.$$

In particular, the line of force which leaves A in a direction perpendicular
to AB is bent through an angle of 30° before it reaches its asymptote at
infinity.

The sections of the equipotentials made by the plane of xy for this case
are shewn in fig. 16 which is drawn on the same scale as fig. 15. The equations of these curves are of course

$$\frac{1}{PA} + \frac{1}{PB} = \text{cons.,}$$

curves of the sixth degree. The equipotential which passes through C is
of interest, as it intersects itself at the point C. This is a necessary conse-

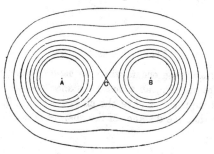

Fig. 16.

quence of the fact that C is a point of equilibrium. Indeed the conditions
for a point of equilibrium, namely

$$\frac{\partial V}{\partial x} = 0, \quad \frac{\partial V}{\partial y} = 0, \quad \frac{\partial V}{\partial z} = 0,$$

may be interpreted as the condition that the equipotential ($V =$ constant)
through the point should have a double tangent plane or a tangent cone at
the point.

II. *Point charges* $+e$, $-e$.

63. Let charges $\pm e$ be at the points $x = \pm a$ (A, B) respectively. The differential equations of the lines of force are found to be

$$\frac{\partial y}{\partial x} = \frac{Y}{X} = \frac{y}{x + a\left(\dfrac{PB^3 + PA^3}{PB^3 - PA^3}\right)},$$

and the integral of this is

$$\frac{x + a}{PA} - \frac{x - a}{PB} = \text{cons.}$$

The lines of force are shewn in fig. 17.

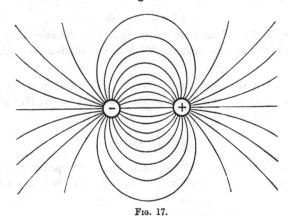

Fig. 17.

III. *Electric Doublet.*

64. An important case occurs when we have two large charges $+e$, $-e$, equal and opposite in sign, at a small distance apart. Taking Cartesian coordinates, let us suppose we have the charge $+e$ at a, 0, 0 and the charge $-e$ at $-a$, 0, 0, so that the distance of the charges is $2a$.

The potential is

$$\frac{e}{\sqrt{(x - a)^2 + y^2 + z^2}} - \frac{e}{\sqrt{(x + a)^2 + y^2 + z^2}},$$

and when a is very small, so that squares and higher powers of a may be neglected, this becomes

$$\frac{2eax}{(x^2 + y^2 + z^2)^{\frac{3}{2}}}.$$

If a is made to vanish, while e becomes infinite, in such a way that $2ea$ retains the finite value μ, the system is described as an electric doublet of strength μ having for its direction the positive axis of x. Its potential is

$$\frac{\mu x}{(x^2 + y^2 + z^2)^{\frac{3}{2}}},$$

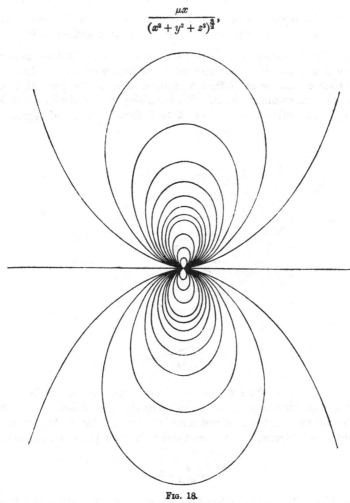

FIG. 18.

or, if we turn to polar coordinates and write $x = r\cos\theta$, is

$$\frac{\mu \cos\theta}{r^2} \quad \dots\dots\dots\dots\dots\dots\dots\dots\dots(24).$$

The lines of force are shewn in fig. 18. Obviously the lines at the centre of this figure become identical with those shewn in fig. 17, if the latter are shrunk indefinitely in size.

IV. *Point charges* $+4e$, $-e$.

65. Fig. 19 represents the distribution of the lines of force when the electric field is produced by two point charges, $+4e$ at A and $-e$ at B.

At infinity the resultant force will be $3e/r^2$, where r is the distance from a point near to A and B. The direction of this force is outwards. Thus no lines of force can arrive at B from infinity, so that all the lines of force which enter B must come from A. The remaining lines of force from A go to infinity. The tubes of force from A to B form a bundle of aggregate

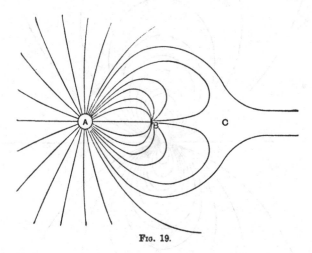

FIG. 19.

strength e, while those from A to infinity have aggregate strength $3e$. The two bundles of tubes of force are separated by the lines of force through C. At C the direction of the resultant force is clearly indeterminate, so that C is a point of equilibrium. As the condition that C is a point of equilibrium we have

$$\frac{4e}{AC^2} - \frac{e}{BC^2} = 0.$$

So that $AB = BC$. At C the two lines of force from A coalesce and then separate out into two distinct lines of force, one from C to B, and the other from C to infinity in the direction opposite to CB.

The equipotentials in this field, the system of curves

$$\frac{4}{PA} - \frac{1}{PB} = \text{cons.},$$

are represented in fig. 20, which is drawn on the same scale as fig. 19.

Since C is a point of equilibrium the equipotential through the point C must of course cut itself at C. . At C the potential

$$\frac{4e}{CA} - \frac{e}{CB} = \frac{e}{AB},$$

since $CA = 2CB$. From the loop of this equipotential which surrounds B, the potential must fall continuously to $- \infty$ as we approach B, since, by the theorem of § 51, there can be no maxima or minima of potential between this loop and the point B. Also no equipotential can intersect itself since there are obviously no points of equilibrium except C. One of the inter-

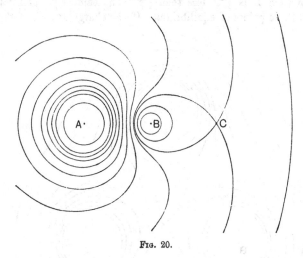

Fig. 20.

mediate equipotentials is of special interest, namely that over which the potential is zero. This is the locus of the point P given by

$$\frac{4}{PA} - \frac{1}{PB} = 0,$$

and is therefore a sphere. This is represented by the outer of the two closed curves which surround B in the figure.

In the same way we see that the other loop of the equipotential through C must be occupied by equipotentials for which the potential rises steadily to the value $+ \infty$ at A. So also outside the equipotential through C, the potential falls steadily to the value zero at infinity. Thus the zero equipotential consists of two spheres—the sphere at infinity and the sphere surrounding B which has already been mentioned.

V.   *Three equal charges at the corners of an equilateral triangle.*

66.   As a further example we may examine the disposition of equipotentials when the field is produced by three point charges at the corners
of an equilateral triangle.   The intersection of these by the plane in which
the charges lie is represented in fig. 21, in which A, B, C are the points at
which the charges are placed, and D is the centre of the triangle ABC.

It will be found that there are three points of equilibrium, one on each
of the lines AD, BD, CD.   Taking $AD = a$, the distance of each point of
equilibrium from D is just less than $\frac{1}{4}a$.   The same equipotential passes
through all three points of equilibrium.   If the charge at each of the points

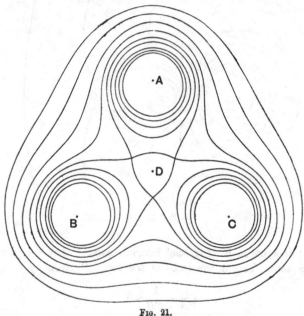

Fɪɢ. 21.

A, B, C is taken to be unity, this equipotential has a potential $\dfrac{3 \cdot 04}{a}$.   The
equipotential has three loops surrounding the points A, B, C.   In each of
these loops the equipotentials are closed curves, which finally reduce to
small circles surrounding the points A, B, C.   Those drawn correspond to
the potentials $\dfrac{3 \cdot 25}{a}$, $\dfrac{3 \cdot 5}{a}$, $\dfrac{3 \cdot 75}{a}$, and $\dfrac{4}{a}$.

Outside the equipotential $\dfrac{3 \cdot 04}{a}$, the equipotentials are closed curves surrounding the former equipotential, and finally reducing to circles at infinity. The curves drawn correspond to potentials $\dfrac{2}{a}$, $\dfrac{2\cdot25}{a}$, $\dfrac{2\cdot5}{a}$, and $\dfrac{2\cdot75}{a}$.

There remains the region between the point D and the equipotential $\dfrac{3\cdot04}{a}$.

At D the potential is $\dfrac{3\cdot00}{a}$, so that the potential falls as we recede from the

equipotential $\dfrac{3\cdot04}{a}$ and reaches its minimum value at D. The potential at D is of course not a minimum for all directions in space: for the potential increases as we move away from D in directions which are in the plane ABC, but obviously decreases as we move away from D in a direction per-

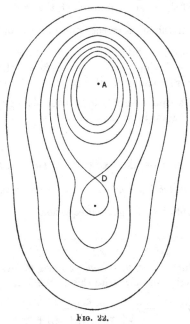

Fig. 22.

pendicular to this plane. Taking D as origin, and the plane ABC as plane of xy, it will be found that near D the potential is

$$V = \frac{3}{a} + \frac{3}{4a^3}(x^2 + y^2 - 2z^2).$$

Thus the equipotential through D is shaped like a right circular cone in the immediate neighbourhood of the point D. From the equation just found, it is obvious that near D the sections of the equipotentials by the plane ABC will be circles surrounding D.

From a study of the section of the equipotentials as shewn in fig. 21, it is easy to construct the complete surfaces. We see that each equipotential for which V has a very high value consists of three small spheres surrounding the points A, B, C. For smaller values of V, which must, however, be greater than $\dfrac{3\cdot04}{a}$, each equipotential still consists of three closed surfaces surrounding A, B, C, but these surfaces are no longer spherical, each one bulging out towards the point D. As V decreases, the surfaces continue to swell out, until, when $V = \dfrac{3\cdot04}{a}$, the surfaces touch one another simultaneously, in a way which will readily be understood on examining the section of this equipotential as shewn in fig. 21. It will be seen that this equipotential is shaped like a flower of three petals from which the centre has been cut away. As V decreases further the surfaces continue to swell, and when $V = \dfrac{3}{a}$, the space at the centre becomes filled up. For still smaller values of V the equipotentials are closed singly-connected surfaces, which finally become spheres at infinity corresponding to the potential $V = 0$.

The sections of the equipotentials by a plane through DA perpendicular to the plane ABC are shewn in fig. 22.

SPECIAL PROPERTIES OF EQUIPOTENTIALS AND LINES OF FORCE.

The Equipotentials and Lines of Force at infinity.

67. In § 40, we obtained the general equation

$$V = \Sigma \frac{e_1}{[(x - x_1)^2 + (y - y_1)^2 + (z - z_1)^2]^{\frac{3}{2}}}.$$

If r denotes the distance of x, y, z from the origin, and r_1 the distance of x_1, y_1, z_1, from the origin, we may write this in the form

$$V = \Sigma \frac{e_1}{[r^2 - 2(xx_1 + yy_1 + zz_1) + r_1^2]^{\frac{3}{2}}}.$$

At a great distance from the origin this may be expanded in descending powers of the distance, in the form

$$V = \Sigma \frac{e_1}{r} \left\{ 1 + \frac{xx_1 + yy_1 + zz_1}{r^2} + \frac{3}{2}\frac{(xx_1 + yy_1 + zz_1)^2}{r^4} - \frac{1}{2}\frac{r_1^2}{r^2} + \dots \right\}.$$

The term of order $\dfrac{1}{r}$ is $\dfrac{\Sigma e_1}{r}$.

The term of order $\dfrac{1}{r^2}$ is $\dfrac{1}{r^3} \Sigma e_1(xx_1 + yy_1 + zz_1)$.

If the origin is taken at the centroid of e_1 at x_1, y_1, z_1, e_2 at x_2, y_2, z_2, etc., we have

$$\Sigma e_1 x_1 = 0, \quad \Sigma e_1 y_1 = 0, \quad \Sigma e_1 z_1 = 0.$$

Thus by taking the origin at this centroid, the term of order $\dfrac{1}{r^2}$ will disappear.

The term of order $\dfrac{1}{r^3}$ is

$$\frac{3}{2r^5} \Sigma e_1 (xx_1 + yy_1 + zz_1)^2 - \frac{1}{2r^3} \Sigma e_1 r_1^2.$$

Let A, B, C, be the moments of inertia about the axes, of e_1 at x_1, y_1, z_1, etc., and let I be the moment of inertia about the line joining the origin to x, y, z; then

$$\Sigma e_1 r_1^2 \qquad\qquad = \tfrac{1}{2}(A + B + C),$$
$$\Sigma e_1 (xx_1 + yy_1 + zz_1)^2 = r^2 (\Sigma e_1 r_1^2 - I),$$

and the terms of order $\dfrac{1}{r^3}$ become

$$\frac{A + B + C - 3I}{2r^3}.$$

Thus taking the centroid of the charges as origin, the potential at a great distance from the origin can be expanded in the form

$$V = \frac{\Sigma e}{r} + \frac{A + B + C - 3I}{2r^3} + \dots.$$

Thus except when the total charge Σe vanishes, the field at infinity is the same as if the total charge Σe were collected at the centroid of the charges. Thus the equipotentials approximate to spheres having this point as centre, and the asymptotes to the lines of force are radii drawn through the centroid. These results are illustrated in the special fields of force considered in §§ 61—66.

The Lines of Force from collinear charges.

68. When the field is produced solely by charges all in the same straight line, the equipotentials are obviously surfaces of revolution about this line, while the lines of force lie entirely in planes through this line. In this important case, the equation of the lines of force admits of direct integration.

Let P_1, P_2, P_3, ... be the positions of the charges e_1, e_2, e_3, ... Let Q, Q' be any two adjacent points on a line of force. Let N be the foot of the perpendicular from Q to the axis $P_1 P_2$, ..., and let a circle be drawn perpendicular to this axis with centre N and radius QN. This circle subtends at P_1 a solid angle

$$2\pi (1 - \cos \theta_1),$$

where θ_1 is the angle QP_1N. Thus the surface integral of normal force arising from e_1, taken over the circle QN, is

$$2\pi e_1 (1 - \cos \theta_1)$$

and the total surface integral of normal force taken over this surface is

$$2\pi \Sigma e_1 (1 - \cos \theta_1).$$

If we draw the similar circle through Q', we obtain a closed surface bounded by these two circles and by the surface formed by the revolution

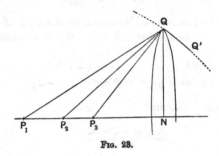

FIG. 23.

of QQ'. This contains no electric charge, so that the surface integral of normal force taken over it must be *nil*. Hence the integral of force over the circle QN must be the same as that over the similar circle drawn through Q'. This gives the equations of the lines of force in the form

(integral of normal force through circle such as QN) = constant,

which as we have seen, becomes

$$\Sigma e_1 \cos \theta_1 = \text{constant.}$$

Analytically, let the point P_1 have coordinates a_1, 0, 0, let P_2 have coordinates a_2, 0, 0, etc. and let Q be the point x, y, z. Then

$$\cos \theta_1 = \frac{x - x_1}{\sqrt{(x - x_1)^2 + y^2 + z^2}},$$

and the equation of the surfaces formed by the revolution of the lines of force is

$$\Sigma \frac{e_1 (x - x_1)}{\sqrt{(x - x_1)^2 + y^2 + z^2}} = \text{constant.}$$

It will easily be verified by differentiation that this is an integral of the differential equation

$$\frac{\partial y}{\partial x} = \frac{Y}{X}.$$

69. We have seen that, in general, the equipotential through any point of equilibrium must intersect itself at the point of equilibrium.

Let x, y, z be a point of equilibrium, and let the potential at this point be denoted by V_0. Let the potential at an adjacent point $x + \xi$, $y + \eta$, $z + \zeta$, be denoted by $V_{\xi, \eta, \zeta}$. By Taylor's Theorem, if $f(x, y, z)$ is any function of x, y, z, we have

$$f(x+\xi, y+\eta, z+\zeta) = f(x, y, z) + \xi \frac{\partial f}{\partial x} + \eta \frac{\partial f}{\partial y} + \zeta \frac{\partial f}{\partial z} + \tfrac{1}{2}\left(\xi^2 \frac{\partial^2 f}{\partial x^2} + 2\xi\eta \frac{\partial^2 f}{\partial x \partial y} + \ldots \right),$$

where the differential coefficients of f are evaluated at x, y, z. Taking $f(x, y, z)$ to be the potential at x, y, z, this of course being a function of the variables x, y, z, the foregoing equation becomes

$$V_{\xi, \eta, \zeta} = V_0 + \xi \frac{\partial V}{\partial x} + \eta \frac{\partial V}{\partial y} + \zeta \frac{\partial V}{\partial z} + \tfrac{1}{2}\left(\xi^2 \frac{\partial^2 V}{\partial x^2} + 2\xi\eta \frac{\partial^2 V}{\partial x \partial y} + \ldots \right) \ldots (25).$$

If x, y, z is a point of equilibrium,

$$\frac{\partial V}{\partial x} = \frac{\partial V}{\partial y} = \frac{\partial V}{\partial z} = 0,$$

so that

$$V_{\xi, \eta, \zeta} = V_0 + \tfrac{1}{2}\left(\xi^2 \frac{\partial^2 V}{\partial x^2} + 2\xi\eta \frac{\partial^2 V}{\partial x \partial y} + \ldots \right).$$

Referred to x, y, z as origin, the coordinates of the point $x + \xi$, $y + \eta$, $z + \zeta$ become ξ, η, ζ, and the equation of the equipotential $V = C$ becomes

$$C - V_0 = \tfrac{1}{2}\left(\xi^2 \frac{\partial^2 V}{\partial x^2} + 2\xi\eta \frac{\partial^2 V}{\partial x \partial y} + \ldots \right).$$

In the neighbourhood of the point of equilibrium, the values of ξ, η, ζ are small, so that in general the terms containing powers of ξ, η, ζ higher than squares may be neglected, and the equation of the equipotential $V = C$ becomes

$$\xi^2 \frac{\partial^2 V}{\partial x^2} + 2\xi\eta \frac{\partial^2 V}{\partial x \partial y} + \ldots = 2(C - V_0).$$

In particular the equipotential $V = V_0$ becomes identical, in the neighbourhood of the point of equilibrium, with the cone

$$\xi^2 \frac{\partial^2 V}{\partial x^2} + 2\xi\eta \frac{\partial^2 V}{\partial x \partial y} + \ldots = 0.$$

Let this cone, referred to its principal axes, become

$$a\xi'^2 + b\eta'^2 + c\zeta'^2 = 0 \ldots\ldots\ldots\ldots\ldots\ldots\ldots(26),$$

then, since the sum of the coefficients of the squares of the variables is an invariant,

$$a + b + c = \frac{\partial^2 V}{\partial x^2} + \frac{\partial^2 V}{\partial y^2} + \frac{\partial^2 V}{\partial z^2} = 0.$$

Now $a + b + c = 0$ is the condition that the cone shall have three perpendicular generators. Hence we see that at the point at which an equipotential cuts itself, we can always find three perpendicular tangents to the equipotential. Moreover we can find these perpendicular tangents in an infinite number of ways.

In the particular case in which the cone is one of revolution (*e.g.*, if the whole field is symmetrical about an axis, as in figures 16 and 20), the equation of the cone must become

$$\xi'^2 + \eta'^2 - 2\zeta'^2 = 0,$$

where the axis of ζ' is the axis of symmetry. The section of the equipotential made by any plane through the axis, say that of $\xi'\zeta'$, must now become

$$\xi'^2 - 2\zeta'^2 = 0$$

in the neighbourhood of the point of equilibrium, and this shews that the tangents to the equipotentials each make a constant angle $\tan^{-1}\sqrt{2}\,(= 54° 44')$ with the axis of symmetry.

In the more general cases in which there is not symmetry about an axis, the two branches of the surface will in general intersect in a line, and the cone reduces to two planes, the equation being

$$a\xi'^2 + b\eta'^2 = 0,$$

where the axis of ζ' is the line of intersection. We now have $a + b = 0$, so that the tangent planes to the equipotential intersect at right angles.

An analogous theorem can be proved when n sheets of an equipotential intersect at a point. The theorem states that the n sheets make equal angles π/n with one another. (Rankin's Theorem, see Maxwell's *Electricity and Magnetism*, § 115, or Thomson and Tait's *Natural Philosophy*, § 780.)

70. A conductor is always an equipotential, and can be constructed so as to cut itself at any angle we please. It will be seen that the foregoing theorems can fail either through the a, b and c of equation (24) all vanishing, or through their all becoming infinite. In the former case the potential near a point at which the conductor cuts itself, is of the form (cf. equation (25)),

$$V_{\xi,\,\eta,\,\zeta} = V_0 + \tfrac{1}{6}\left(\xi^3 \frac{\partial^3 V}{\partial x^3} + 3\xi^2\eta \frac{\partial^3 V}{\partial x^2 \partial y} + \ldots\right),$$

so that the components of intensity are of the forms

$$\frac{\partial V}{\partial \xi} = \tfrac{1}{2}\left(\xi^2 \frac{\partial^3 V}{\partial x^3} + 2\xi\eta \frac{\partial^3 V}{\partial x^2 \partial y} + \ldots\right)$$

$$= \tfrac{1}{2}\frac{\partial}{\partial x}\left(\xi^2 \frac{\partial^2 V}{\partial x^2} + 2\xi\eta \frac{\partial^2 V}{\partial x \partial y} + \ldots\right).$$

The intensity near the point of equilibrium is therefore a small quantity of the second order, and since by Coulomb's Law $R = 4\pi\sigma$, it follows that the surface density is zero along the line of intersection, and is proportional to the square of the distance from the line of intersection at adjacent points.

If, however, a, b and c are all *infinite*, we have the electric intensity also infinite, and therefore the surface density is infinite along the line of intersection.

It is clear that the surface density will vanish when the conducting surface cuts itself in such a way that the angle less than two right angles is external to the conductor; and that the surface density will become infinite when the angle greater than two right angles is external to the conductor. This becomes obvious on examining the arrangement of the lines of force in the neighbourhood of the angle.

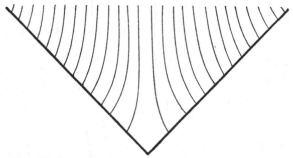

FIG. 24. Angle less than two right angles external to conductor.

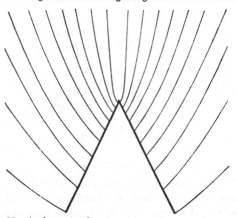

FIG. 25. Angle greater than two right angles external to conductor.

71. The arrangement shewn in fig. 25 is such as will be found at the point of a lightning conductor. The object of the lightning conductor is to ensure that the intensity shall be greater at its point than on any part of the buildings it is designed to protect. The discharge will therefore take place from the point of the lightning conductor sooner than from any part of the building, and by putting the conductor in good electrical communication with the earth, it is possible to ensure that no harm shall be done to the main buildings by the electrical discharge.

An application of the same principle will explain the danger to a human being or animal of standing in the open air in the presence of a thunder cloud, or of standing under an isolated tree. The upward point, whether the head of man or animal, or the summit of the tree, tends to collect the lines of force which pass from the cloud to the ground, so that a discharge of electricity will take place from the head or tree rather than from the ground.

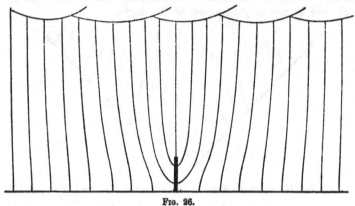

Fig. 26.

72. The property of lines of force of clustering together in this way is utilised also in the manufacture of electrical instruments. A cage of wire is

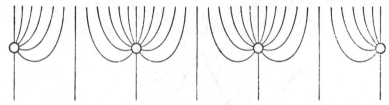

Fig. 27.

placed round the instrument and almost all the lines of force from any charges which there may be outside the instrument will cluster together on the convex surfaces of the wire. Very few lines of force escape through this cage, so that the instrument inside the cage is hardly affected at all by any electric phenomena which may take place outside it. Fig. 27 shews the way in which lines of force are absorbed by a wire grating. It is drawn to represent the lines of force of a uniform field meeting a plane grating placed at right angles to the field of force.

The protection of a wire cage is not adequate for the most sensitive instruments, and it is usual to enclose them entirely in a metal case, except only for one small window through which readings can be taken. When this arrangement is adopted, no lines of force at all can pass from external charges to the instrument inside the metal case except for an infinitesimal number passing through the window. Lines of force which encounter the case terminate on it without in any way affecting the electric field inside, and the instrument is almost perfectly screened from any external electric field. (Cf. § 114 below.)

EXAMPLES.

1. Two particles each of mass m and charged with e units of electricity of the same sign are suspended by strings each of length a from the same point; prove that the inclination θ of each string to the vertical is given by the equation

$$4mga^2 \sin^3 \theta = e^2 \cos \theta.$$

2. Charges $+4e$, $-e$ are placed at the points A, B, and C is the point of equilibrium. Prove that the line of force which passes through C meets AB at an angle of 60° at A and at right angles at C.

3. Find the angle at A (question 2) between AB and the line of force which leaves B at right angles to AB.

4. Two positive charges e_1 and e_2 are placed at the points A and B respectively. Shew that the tangent at infinity to the line of force which starts from e_1 making an angle a with BA produced, makes an angle

$$2 \sin^{-1} \left(\sqrt{\frac{e_1}{e_1 + e_2}} \sin \frac{a}{2} \right)$$

with BA, and passes through the point C in AB such that

$$AC : CB = e_2 : e_1.$$

5. Point charges $+e$, $-e$ are placed at the points A, B. The line of force which leaves A making an angle a with AB meets the plane which bisects AB at right angles, in P. Shew that

$$\sin \frac{a}{2} = \sqrt{2} \sin \frac{PAB}{2}.$$

6. If any closed surface be drawn not enclosing a charged body or any part of one, shew that at every point of a certain closed line on the surface it intersects the equipotential surface through the point at right angles.

7. The potential is given at four points near each other and not all in one plane. Obtain an approximate construction for the direction of the field in their neighbourhood.

8. The potentials at the four corners of a small tetrahedron A, B, C, D are V_1, V_2, V_3, V_4 respectively. G is the centre of gravity of masses M_1 at A, M_2 at B, M_3 at C, M_4 at D. Shew that the potential at G is

$$\frac{M_1 V_1 + M_2 V_2 + M_3 V_3 + M_4 V_4}{M_1 + M_2 + M_3 + M_4}.$$

9. Charges $3e$, $-e$, $-e$ are placed at A, B, C respectively, where B is the middle point of AC. Draw a rough diagram of the lines of force; shew that a line of force which starts from A making an angle a with $AB > \cos^{-1}(-\frac{1}{3})$ will not reach B or C, and shew that the asymptote of the line of force for which $a = \cos^{-1}(-\frac{2}{3})$ is at right angles to AC.

10. If there are three electrified points A, B, C in a straight line, such that $AC = f$, $BC = \dfrac{a^2}{f}$, and the charges are e, $\dfrac{-ea}{f}$ and Va respectively, shew that there is always a spherical equipotential surface, and discuss the position of the points of equilibrium on the line ABC when $V = e\,\dfrac{f+a}{(f-a)^2}$ and when $V = e\,\dfrac{f-a}{(f+a)^2}$.

11. A and C are spherical conductors with charges $e + e'$ and $-e$ respectively. Shew that there is either a point or a line of equilibrium, depending on the relative size and positions of the spheres, and on e'/e. Draw a diagram for each case giving the lines of force and the sections of the equipotentials by a plane through the centres.

12. An electrified body is placed in the vicinity of a conductor in the form of a surface of anticlastic curvature. Shew that at that point of any line of force passing from the body to the conductor, at which the force is a minimum, the principal curvatures of the equipotential surface are equal and opposite.

13. Shew that it is not possible for every family of non-intersecting surfaces in free space to be a family of equipotentials, and that the condition that the family of surfaces

$$f(\lambda, x, y, z) = 0$$

shall be capable of being equipotentials is that

$$\frac{\dfrac{\partial^2 \lambda}{\partial x^2} + \dfrac{\partial^2 \lambda}{\partial y^2} + \dfrac{\partial^2 \lambda}{\partial z^2}}{\left(\dfrac{\partial \lambda}{\partial x}\right)^2 + \left(\dfrac{\partial \lambda}{\partial y}\right)^2 + \left(\dfrac{\partial \lambda}{\partial z}\right)^2}$$

shall be a function of λ only.

14. In the last question, if the condition is satisfied find the potential.

15. Shew that the confocal ellipsoids

$$\frac{x^2}{a^2 + \lambda} + \frac{y^2}{b^2 + \lambda} + \frac{z^2}{c^2 + \lambda} = 1$$

can form a system of equipotentials, and express the potential as a function of λ.

16. If two charged concentric shells be connected by a wire, the inner one is wholly discharged. If the law of force were $\dfrac{1}{r^{2+p}}$, prove that there would be a charge B on the inner shell such that if A were the charge on the outer shell, and f, g the sum and difference of the radii,

$$2gB = -Ap \{(f-g) \log (f+g) - f \log f + g \log g\}$$

approximately.

17. Three infinite parallel wires cut a plane perpendicular to them in the angular points A, B, C of an equilateral triangle, and have charges e, e, $-e'$ per unit length respectively. Prove that the extreme lines of force which pass from A to C make at starting angles $\dfrac{2e-5e'}{6e}\pi$ and $\dfrac{2e+e'}{6e}\pi$ with AC, provided that $e' \not> 2e$.

18. A negative point charge $-e_2$ lies between two positive point charges e_1 and e_3 on the line joining them and at distances a, β from them respectively. Shew that, if the magnitudes of the charges are given by

$$\frac{e_1}{\beta} = \frac{e_3}{a} = \frac{e_2\lambda^3}{a+\beta}, \text{ and if } 1 < \lambda^2 < \left(\frac{a+\beta}{a-\beta}\right)^2,$$

there is a circle at every point of which the force vanishes. Determine the general form of the equipotential surface on which this circle lies.

19. Charges of electricity e_1, $-e_2$, e_3, $(e_3 > e_1)$ are placed in a straight line, the negative charge being midway between the other two. Shew that, if $4e_2$ lie between $(e_3^{\frac{2}{3}} - e_1^{\frac{2}{3}})^3$ and $(e_3^{\frac{2}{3}} + e_1^{\frac{2}{3}})^3$, the number of unit tubes of force that pass from e_1 to e_2 is

$$\tfrac{1}{2}(e_1 + e_2 - e_3) + \frac{3}{4\sqrt{2}}(e_3^{\frac{2}{3}} - e_1^{\frac{2}{3}})(e_1^{\frac{2}{3}} - 2^{\frac{1}{3}}e_2^{\frac{2}{3}} + e_3^{\frac{2}{3}})^{\frac{1}{2}}.$$

CHAPTER III

CONDUCTORS AND CONDENSERS

73. By a conductor, as previously explained, is meant any body or system of bodies, such that electricity can flow freely over the whole. When electricity is at rest on such a conductor, we have seen (§ 44) that the charge will reside entirely on the outer surface, and (§ 37) that the potential will be constant over this surface.

A conductor may be used for the storage of electricity, but it is found that a much more efficient arrangement is obtained by taking two or more conductors—generally thin plates of metal—and arranging them in a certain way. This arrangement for storing electricity is spoken of as a "condenser." In the present Chapter we shall discuss the theory of single conductors and of condensers, working out in full the theory of some of the simpler cases.

CONDUCTORS.

A Spherical Conductor.

74. The simplest example of a conductor is supplied by a sphere, it being supposed that the sphere is so far removed from all other bodies that their influence may be neglected. In this case it is obvious from symmetry that the charge will spread itself uniformly over the surface. Thus if e is the charge, and a the radius, the surface density σ is given by

$$\sigma = \frac{\text{total charge}}{\text{total area of surface}} = \frac{e}{4\pi a^2}.$$

The electric intensity at the surface being, as we have seen, equal to $4\pi\sigma$, is e/a^2.

From symmetry the direction of the intensity at any point outside the sphere must be in a direction passing through the centre. To find the amount of this intensity at a distance r from the centre, let us draw a sphere of radius r, concentric with the conductor. At every point of this sphere the amount of the outward electric intensity is by symmetry the same, say R, and its direction as we have seen is normal to the surface. Applying Gauss' Theorem to this sphere, we find that the surface integral of normal intensity $\iint N\,dS$ becomes simply R multiplied by the area of the surface $4\pi r^2$, so that

$$4\pi r^2 R = 4\pi e,$$

or

$$R = \frac{e}{r^2}.$$

This becomes e/a^2 at the surface, agreeing with the value previously obtained.

Thus the electric force at any point is the same as if the charged sphere were replaced by a point charge e, at the centre of the sphere. And, just as in the case of a single point charge e, the potential at a point outside the sphere, distant r from its centre, is

$$V = \int_{\infty}^{r} \frac{e}{r^2}\, dr = \frac{e}{r},$$

so that at the surface of the sphere the potential is $\dfrac{e}{a}$.

Inside the sphere, as has been proved in § 37, the potential is constant, and therefore equal to e/a, its value at the surface, while the electric intensity vanishes.

As we gradually charge up the conductor, it appears that the potential at the surface is always proportional to the charge of the conductor.

It is customary to speak of the potential at the surface of a conductor as "the potential of the conductor," and the ratio of the charge to this potential is defined to be the "capacity" of the conductor. From a general theorem, which we shall soon arrive at, it will be seen that the ratio of charge to potential remains the same throughout the process of charging any conductor or condenser, so that in every case the capacity depends only on the shape and size of the conductor or condenser in question. For a sphere, as we have seen,

$$\text{capacity} = \frac{\text{charge}}{\text{potential}} = \frac{e}{\dfrac{e}{a}} = a,$$

so that the capacity of a sphere is equal to its radius.

A Cylindrical Conductor.

75. Let us next consider the distribution of electricity on a circular cylinder, the cylinder either extending to infinity, or else having its ends so far away from the parts under consideration that their influence may be neglected.

As in the case of the sphere, the charge distributes itself symmetrically, so that if a is the radius of the cylinder, and if it has a charge e per unit length, we have

$$\sigma = \frac{e}{2\pi a}.$$

To find the intensity at any point outside the conductor, construct a Gauss' surface by first drawing a cylinder of radius r, coaxal with the original cylinder, and then cutting off a unit length by two parallel planes at unit distance apart, perpendicular to the axis. From symmetry the force at every point is perpendicular to the axis of the cylinder, so that the normal intensity vanishes at every point of the plane ends of this Gauss' surface. The surface integral of normal intensity will therefore consist entirely of the contributions from the curved part of the surface, and this curved part consists of a circular band, of unit width and radius r—hence of area $2\pi r$. If R is the outward intensity at every point of this curved surface, Gauss' Theorem supplies the relation

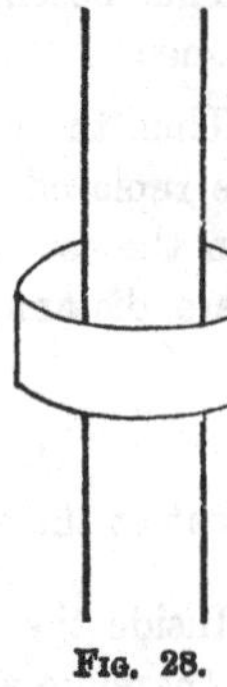

$$2\pi r R = 4\pi e,$$

so that

$$R = \frac{2e}{r}.$$

Fig. 28.

This, we notice, is independent of a, so that the intensity is the same as it would be if a were very small, *i.e.*, as if we had a fine wire electrified with a charge e per unit length.

In the foregoing, we must suppose r to be so small, that at a distance r from the cylinder the influence of the ends is still negligible in comparison with that of the nearer parts of the cylinder, so that the investigation does not hold for large values of r. It follows that we cannot find the potential by integrating the intensity from infinity, as has been done in the cases of the point charge and of the sphere. We have, however, the general differential equation

$$\frac{\partial V}{\partial r} = -R,$$

so that in the present case, so long as r remains sufficiently small

$$\frac{\partial V}{\partial r} = -\frac{2e}{r},$$

giving upon integration

$$V = C - 2e \log r.$$

The constant of integration C cannot be determined without a knowledge of the conditions at the ends of the cylinder. Thus for a long cylinder, the intensity at points near the cylinder is independent of the conditions at the ends, but the potential and capacity depend on these conditions, and are therefore not investigated here.

An Infinite Plane.

76. Suppose we have a plane extending to infinity in all directions, and electrified with a charge σ per unit area. From symmetry it is obvious that the lines of force will be perpendicular to the plane at every point, so that the tubes of force will be of uniform cross-section. Let us take as Gauss' surface the tube of force which has as cross-section any element ω of area of the charged plane, this tube being closed by two cross-sections each of area ω at distance r from the plane. If R is the intensity over either of these cross-sections the contribution of each cross-section to Gauss' integral is $R\omega$, so that Gauss' Theorem gives at once

$$2R\omega = 4\pi\sigma\omega,$$

whence $R = 2\pi\sigma.$

The intensity is therefore the same at all distances from the plane.

The result that at the surface of the plane the intensity is $2\pi\sigma$, may at first seem to be in opposition to Coulomb's Theorem (§ 57) which states that the intensity at the surface of a conductor is $4\pi\sigma$. It will, however, be seen from the proof of this theorem, that it deals only with conductors in which the conducting matter is of finite thickness; if we wish to regard the electrified plane as a conductor of this kind we must regard the total electrification as being divided between the two faces, the surface density being $\frac{1}{2}\sigma$ on each, and Coulomb's Theorem then gives the correct result.

If the plane is not actually infinite, the result obtained for an infinite plane will hold within a region which is sufficiently near to the plane for the edges to have no influence. As in the former case of the cylinder, we can obtain the potential within this region by integration. If r measures the perpendicular distance from the plane

$$-\frac{\partial V}{\partial r} = R = 2\pi\sigma,$$

so that $V = C - 2\pi\sigma r,$

and, as before, the constant of integration cannot be determined without a knowledge of the conditions at the edges.

77. It is instructive to compare the three expressions which have been obtained for the electric intensity at points outside a charged sphere, cylinder and plane respectively. Taking r to be the distance from the centre of the sphere, from the axis of the cylinder, and from the plane, respectively, we have found that

$$\text{outside the sphere,} \quad R \text{ is proportional to } \frac{1}{r^3},$$

$$\text{outside the cylinder,} \quad R \text{ is proportional to } \frac{1}{r},$$

$$\text{outside the plane,} \quad R \text{ is constant.}$$

From the point of view of tubes of force, these results are obvious enough deductions from the theorem that the intensity varies inversely as the cross-section of a tube of force. The lines of force from a sphere meet in a point, the centre of the sphere, so that the tubes of force are cones, with cross-section proportional to the square of the distance from the vertex. The lines of force from a cylinder all meet a line, the axis of the cylinder, at right angles, so that the tubes of force are wedges, with cross-section proportional to the distance from the edge. And the lines of force from a plane all meet the plane at right angles, so that the tubes of force are prisms, of which the cross-section is constant.

78. We may also examine the results from the point of view which regards the electric intensity as the resultant of the attractions or repulsions from different elements of the charged surface.

Let us first consider the charged plane. Let P, P' be two points at distances r, r' from the plane, and let Q be the foot of the perpendicular from either on to the plane. If P is near to Q, it will be seen that almost the whole of the intensity at P is due to the charges in the immediate neighbourhood of Q. The more distant parts contribute forces which make angles with QP nearly equal to a right angle, and after being resolved along QP these forces hardly contribute anything to the resultant intensity at P.

Owing to the greater distance of the point P', the forces from given elements of the plane are smaller at P' than at P, but have to be resolved through a smaller angle. The forces from the regions near Q are greatly diminished from the former cause and are hardly affected by the latter. The forces from remote regions are hardly affected by the former circumstance, but their effect is greatly increased by the latter. Thus on moving

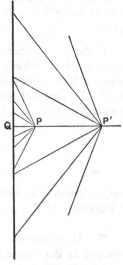

Fig. 29.

from P to P' the forces exerted by regions near Q decrease in efficiency, while those exerted by more remote regions gain. The result that the total resultant intensity is the same at P' as at P, shews that the decrease of the one just balances the gain of the other.

If we replace the infinite plane by a sphere, we find that the force at a near point P is as before contributed almost entirely by the charges in the neighbourhood of Q. On moving from P to P', these forces are diminished just as before, but the number of distant elements of area which now add contributions to the intensity at P' is much less than before. Thus the gain in the contributions from these elements does not suffice to balance the diminution in the contributions from the regions near Q, so that the resultant intensity falls off on withdrawing from P to P'

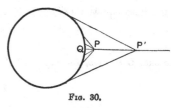

Fig. 30.

The case of a cylinder is of course intermediate between that of a plane and that of a sphere.

CONDENSERS.

Spherical Condenser.

79. Suppose that we enclose the spherical conductor of radius a discussed in § 74, inside a second spherical conductor of internal radius b, the two conductors being placed so as to be concentric and insulated from one another.

It again appears from symmetry that the intensity at every point must be in a direction passing through the common centre of the two spheres, and must be the same in amount at every point of any sphere concentric with the two conducting spheres. Let us imagine a concentric sphere of radius r drawn between the two conductors, and when the charge on the inner sphere is e, let the intensity at every point of the imaginary sphere of radius r be R. Then, as before, Gauss' Theorem, applied to the sphere of radius r, gives the relation

$$4\pi r^2 R = 4\pi e,$$

so that

$$R = \frac{e}{r^2}.$$

This only holds for values of r intermediate between a and b, so that to obtain the potential we cannot integrate from infinity, but must use the differential equation. This is

$$-\frac{\partial V}{\partial r} = R = \frac{e}{r^2},$$

which upon integration gives

$$V = C + \frac{e}{r} \quad\quad\quad\quad\quad\quad\quad\quad\quad\quad\quad\quad\ldots\ldots(27).$$

We can determine the constant of integration as soon as we know the potential of either of the spheres. Suppose for instance that the outer sphere is put to earth so that $V = 0$ over the sphere $r = b$, then we obtain at once from equation (27)

$$0 = C + \frac{e}{b},$$

so that $C = -e/b$, and equation (27) becomes

$$V = \frac{e}{r} - \frac{e}{b}.$$

On taking $r = a$, we find that the potential of the inner sphere is $e\left(\frac{1}{a} - \frac{1}{b}\right)$, and its charge is e, so that the capacity of the condenser is

$$\frac{1}{\dfrac{1}{a} - \dfrac{1}{b}} \quad \text{or} \quad \frac{ab}{b-a}.$$

80. In the more general case in which the outer sphere is not put to earth, let us suppose that V_a, V_b are the potentials of the two spheres of radii a and b, so that, from equation (27)

$$V_a = C + \frac{e}{a},$$

$$V_b = C + \frac{e}{b}.$$

Then we have on subtraction

$$(V_a - V_b) = e\left(\frac{1}{a} - \frac{1}{b}\right),$$

so that the capacity is

$$\frac{e}{V_a - V_b}.$$

The lines of force which start from the inner sphere must all end on the inner surface of the outer sphere, and each line of force has equal and opposite charges at its two ends. Thus if the charge on the inner sphere is e, that on the inner surface of the outer sphere must be $-e$. We can therefore regard the capacity of the condenser as being the charge on either of the two spheres divided by the difference of potential, the fraction being taken always positive. On this view, however, we leave out of account any charge which there may be on the outer surface of the outer sphere: this is not regarded as part of the charge of the condenser.

An examination of the expression for the capacity,

$$\frac{ab}{b-a},$$

will shew that it can be made as large as we please by making $b-a$ sufficiently small. This explains why a condenser is so much more efficient for the storage of electricity than a single conductor.

81. By taking more than two spheres we can form more complicated condensers. Suppose, for instance, we take concentric spheres of radii a, b, c in ascending order of magnitude, and connect both the spheres of radii a and c to earth, that of radius b remaining insulated. Let V be the potential of the middle sphere, and let e_1 and e_2 be the total charges on its inner and outer surfaces. Regarding the inner surface of the middle sphere and the surface of the innermost sphere as forming a single spherical condenser, we have

$$e_1 = \frac{Vab}{b-a},$$

and again regarding the outer surface of the middle sphere and the outermost sphere as forming a second spherical condenser, we have

$$e_2 = \frac{Vbc}{c-b}.$$

Hence the total charge E of the middle sheet is given by

$$E = e_1 + e_2$$
$$= V\left(\frac{ab}{b-a} + \frac{bc}{c-b}\right),$$

so that regarded as a single condenser, the system of three spheres has a capacity

$$\frac{ab}{b-a} + \frac{bc}{c-b},$$

which is equal to the sum of the capacities of the two constituent condensers into which we have resolved the system. This is a special case of a general theorem to be given later (§ 85).

Coaxal Cylinders.

82. A conducting circular cylinder of radius a surrounded by a second coaxal cylinder of internal radius b will form a condenser. If e is the charge on the inner cylinder per unit length, and if V is the potential at any point between the two cylinders at a distance r from their common axis, we have, as in § 75,

$$V = C - 2e \log r,$$

and it is now possible to determine the constant C as soon as the potential of either cylinder is known.

Let V_a, V_b be the potentials of the inner and outer cylinders, so that

$$V_a = C - 2e \log a,$$
$$V_b = C - 2e \log b.$$

By subtraction

$$V_a - V_b = 2e \log \left(\frac{b}{a} \right),$$

so that the capacity is

$$\frac{1}{2 \log \left(\frac{b}{a} \right)},$$

per unit length.

Parallel Plate Condenser.

83. This condenser consists of two parallel plates facing one another, say at distance d apart. Lines of force will pass from the inner face of one to the inner face of the other, and in regions sufficiently far removed from the edges of the plate these lines of force will be perpendicular to the plate throughout their length. If σ is the surface density of electrification of one plate, that of the other will be $-\sigma$. Since the cross-section of a tube remains the same throughout its length, and since the electric intensity varies as the cross-section, it follows that the intensity must be the same throughout the whole length of a tube, and this, by Coulomb's Theorem, will be $4\pi\sigma$, its value at the surface of either plate. Hence the difference of potential between the two plates, obtained by integrating the intensity $4\pi\sigma$ along a line of force, will be

$$4\pi\sigma d.$$

The capacity per unit area is equal to the charge per unit area σ divided by this difference of potential, and is therefore

$$\frac{1}{4\pi d}.$$

The capacity of a condenser formed of two parallel plates, each of area A, is therefore

$$\frac{A}{4\pi d},$$

except for a correction required by the irregularities in the lines of force near the edges of the plates.

Inductive Capacity.

84. It was found by Cavendish, and afterwards independently by Faraday, that the capacity of a condenser depends not only on the shape and size of the conducting plates but also on the nature of the insulating material, or *dielectric* to use Faraday's word, by which they are separated.

It is further found that on replacing air by some other dielectric, the capacity of a condenser is altered in a ratio which is independent of the shape and size of the condenser, and which depends only on the dielectric itself. This constant ratio is called the specific inductive capacity of the dielectric, the inductive capacity of air being taken to be unity.

We shall discuss the theory of dielectrics in a later Chapter. At present it will be enough to know that if C is the capacity of a condenser when its plates are separated by air, then its capacity, when the plates are separated by any dielectric, will be KC, where K is the inductive capacity of the particular dielectric used. The capacities calculated in this Chapter have all been calculated on the supposition that there is air between the plates, so that when the dielectric is different from air each capacity must be multiplied by K.

The following table will give some idea of the values of K actually observed for different dielectrics. For a great many substances the value of K is found to vary widely for different specimens of the material and for different physical conditions.

Sulphur	3·6 to 4·3.	Methyl Alcohol at 13·4° C.	35·4
Mica	5·7 to 7·0.	Water at 17° C.	81
Glass	5 to 10.	Ice at −2° C.	93·9
Paraffin wax	2·0 to 2·3.	Ice at −200° C.	2·43

The values of K for some gases are given on p. 132.

Compound Condensers.

Condensers in Parallel.

85. Let us suppose that we take any number of condensers of capacities $C_1, C_2, \ldots$ and connect all their high potential plates together by a conducting

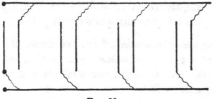

Fig. 31.

wire, and all their low potential plates together in the same way. This is known as connecting the condensers *in parallel*.

The high potential plates have now all the same potential, say V_1, while the low potential plates have all the same potential, say V_0. If $e_1, e_2, \ldots$ are the charges on the separate high potential plates, we have

$$e_1 = C_1 (V_1 - V_0),$$
$$e_2 = C_2 (V_1 - V_0), \text{ etc.},$$

and the total charge E is given by

$$E = e_1 + e_2 + \dots$$
$$= (C_1 + C_2 + \dots)(V_1 - V_0).$$

Thus the system of condensers behaves like a single condenser of capacity

$$C_1 + C_2 + C_3 + \dots.$$

It will be noticed that the compound condenser discussed in § 81 consisted virtually of two simple spherical condensers connected in parallel.

Condensers in Cascade.

86. We might, however, connect the low potential plate of the first to the high potential plate of the second, the low potential plate of the second to the high potential plate of the third, and so on. This is known as arranging the condensers *in cascade*.

FIG. 32.

Suppose that the high potential plate of the first has a charge e. This induces a charge $-e$ on the low potential plate, and since this plate together with the high potential plate of the second condenser now form a single insulated conductor, there must be a charge $+e$ on the high potential plate of the second condenser. This induces a charge $-e$ on the low potential plate of this condenser, and so on indefinitely; each high potential plate will have a charge $+e$, each low potential plate a charge $-e$.

Thus the difference of potential of the two plates of the first condenser will be e/C_1, that of the second condenser will be e/C_2, and so on, so that the total fall of potential from the high potential plate of the first to the low potential plate of the last will be

$$e\left(\frac{1}{C_1} + \frac{1}{C_2} + \dots\right).$$

We see that the arrangement acts like a single condenser of capacity

$$\frac{1}{\dfrac{1}{C_1} + \dfrac{1}{C_2} + \dots}.$$

PRACTICAL CONDENSERS.

Practical Units.

87. As will be explained more fully later, the practical units of electricians are entirely different from the theoretical units in which we have so far supposed measurements to be made. The practical unit of capacity is called the farad, and is equal, very approximately, to 9×10^{11} times the theoretical C.G.S. electrostatic unit, *i.e.*, is equal to the actual capacity of a sphere of radius 9×10^{11} cms. This unit is too large for most purposes, so that it is convenient to introduce a subsidiary unit—the microfarad— equal to a millionth of the farad, and therefore to 9×10^5 C.G.S. electrostatic units. Standard condensers can be obtained of which the capacity is equal to a given fraction, frequently one-third or one-fifth, of the microfarad.

The Leyden Jar.

88. For experimental purposes the commonest form of condenser is the Leyden Jar. This consists essentially of a glass vessel, bottle-shaped, of which the greater part of the surface is coated inside and outside with tinfoil. The two coatings form the two plates of the condenser, contact with the inner coating being established by a brass rod which comes through the neck of the bottle, the lower end having attached to it a chain which rests on the inner coating of tinfoil.

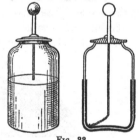

To form a rough numerical estimate of the capacity of a Leyden Jar, let us suppose that the thickness of the glass is $\frac{1}{2}$ cm., that its specific inductive capacity is 7, and that the area covered

FIG. 33.

with tinfoil is 400 sq. cms. Neglecting corrections required by the irregu- larities in the lines of force at the edges and at the sharp angles at the bottom of the jar, and regarding the whole system as a single parallel plate condenser, we obtain as an approximate value for the capacity

$$\frac{KA}{4\pi d} \text{ electrostatic units,}$$

in which we must put $K = 7$, $A = 400$ and $d = \frac{1}{2}$. On substituting these values the capacity is found to be approximately 450 electrostatic units, or about $\frac{1}{2000}$ microfarad.

Parallel Plates.

89. A more convenient condenser for some purposes is a modification of the parallel plate condenser. Let us suppose that we arrange n plates, each of area A, parallel to one another, the distance between any two adjacent plates being d. If alternate plates are joined together so as to be in electrical contact the space between each adjacent pair of plates may be regarded as

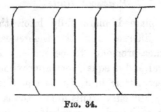

FIG. 34.

forming a single parallel plate condenser of capacity $\dfrac{KA}{4\pi d}$, so that the capacity of the compound condenser is $(n-1)\,KA/4\pi d$. By making n large and d small, we can make this capacity large without causing the apparatus to occupy an unduly large amount of space. For this reason standard condensers are usually made of this pattern.

90. *Guard Ring*. In both the condensers described the capacity can only be calculated approximately. Lord Kelvin has devised a modification of the parallel plate condenser in which the error caused by the irregularities of the lines of force near the edges is dispensed with, so that it is possible accurately to calculate the capacity from measurements of the plates.

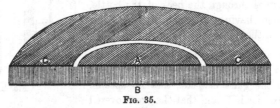

FIG. 35.

The principle consists in making one plate B of the condenser larger than the second plate A, the remainder of the space opposite B being occupied by a "guard ring" C which fits A so closely as almost to touch, and is in the same plane with it. The guard ring C and the plate A, if at the same potential, may without serious error be regarded as forming a single plate of a parallel plate condenser of which the other plate is B. The irregularities in the tubes of force now occur at the outer edge of the guard ring C, while the lines of force from A to B are perfectly straight and uniform. Thus if A is the area of the plate A its capacity may be supposed, with great accuracy, to be

$$\frac{A}{4\pi d},$$

where d is the distance between the plates A and B.

Submarine Cables.

91. Unfortunately for practical electricians, a submarine cable forms a condenser, of which the capacity is frequently very considerable. The effect of this upon the transmission of signals will be discussed later. A cable consists generally of a core of strands of copper wire surrounded by a layer of insulating material, the whole being enclosed in a sheathing of iron wire. This arrangement acts as a condenser of the type of the coaxal cylinders investigated in § 82, the core forming the inner cylinder whilst the iron sheathing and the sea outside form the outer cylinder.

In the capacity formula obtained in § 82, namely

$$\frac{K}{2 \log \left(\frac{b}{a}\right)},$$

let us suppose that $b = 2a$, and that $K = 3\cdot2$, this being about the value for the insulating material generally used. Using the value $\log_e 2 = \cdot69315$, we find a capacity of $2\cdot31$ electrostatic units per unit length. Thus a cable 2000 miles in length has a capacity equal to that of a sphere of radius $2000 \times 2\cdot31$ miles, *i.e.*, of a sphere greater than the earth. In practical units, the capacity of such a cable would be about 827 microfarads.

MECHANICAL FORCE ON A CONDUCTING SURFACE.

92. Let Q be any point on the surface of a conductor, and let the surface-density at the point Q be σ. Let us draw any small area dS

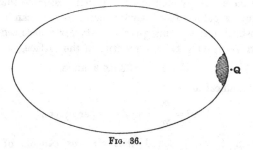

FIG. 36.

enclosing Q. By taking dS sufficiently small, we may regard the area as perfectly plane, and the charge on the area will be σdS. The electricity on the remainder of the conductor will exert forces of attraction or repulsion on the charge σdS, and these forces will shew themselves as a mechanical force acting on the element of area dS of the conductor. We require to find the amount of this mechanical force.

The electric intensity at a point near Q and just outside the conductor is $4\pi\sigma$, by Coulomb's Law, and its direction is normally away from the surface. Of this intensity, part arises from the charge on dS itself, and part from the charges on the remainder of the conductor. As regards the first part, which arises from the charge on dS itself, we may notice that when we are considering a point sufficiently close to the surface, the element dS may be treated as an infinite electrified plane, the electrification being of uniform density σ. The intensity arising from the electrification of dS at such a point is accordingly an intensity $2\pi\sigma$ normally away from the surface. Since the total intensity is $4\pi\sigma$ normally away from the surface, it follows that the intensity arising from the electrification of the parts of the conductor other than dS must also be $2\pi\sigma$ normally away from the surface. It is the forces composing this intensity which produce the mechanical action on dS. The charge on dS being $\sigma\,dS$, the total force will be $2\pi\sigma^2dS$ normally away from the surface. Thus per unit area there is a force $2\pi\sigma^2$ tending to repel the charge normally away from the surface. The charge is prevented from leaving the surface of the conductor by the action between electricity and matter which has already been explained. Action and reaction being equal and opposite, it follows that there is a mechanical force $2\pi\sigma^2$ per unit area acting normally outwards on the material surface of the conductor.

Remembering that $R = 4\pi\sigma$, we find that the mechanical force can also be expressed as $\dfrac{R^2}{8\pi}$ per unit area.

93. Let us try to form some estimate of the magnitude of this mechanical force as compared with other mechanical forces with which we are more familiar. We have already mentioned Maxwell's estimate that a gramme of gold, beaten into a gold-leaf one square metre in area, can hold a charge of 60,000 electrostatic units. This gives 3 units per square centimetre as the charge on each face, giving for the intensity at the surface,

$$R = 4\pi\sigma = 38 \text{ c.g.s. units,}$$

and for the mechanical force

$$2\pi\sigma^2 = \frac{R^2}{8\pi} = 56 \text{ dynes per sq. cm.}$$

Lord Kelvin, however, found that air was capable of sustaining a tension of 9600 grains wt. per sq. foot, or about 700 dynes per sq. cm. This gives $R = 130$, $\sigma = 10$.

Taking $R = 100$ as a *large* value of R, we find $\dfrac{R^2}{8\pi} = 400$ dynes per sq. cm. The pressure of a normal atmosphere is

$$1,013,570 \text{ dynes per sq. cm.,}$$

so that the force on the conducting surface would be only about $\frac{1}{3600}$ of an atmosphere: say ·3 mm. of mercury.

If a gold-leaf is beaten so thin that 1 gm. occupies 1 sq. metre of area, the weight of this is ·0981 dyne per sq. cm. In order that $2\pi\sigma^2$ may be equal to ·0981, we must have $\sigma = ·1249$. Thus a small piece of gold-leaf would be lifted up from a charged surface on which it rested as soon as the surface acquired a charge of about $\frac{1}{8}$ of a unit per sq. cm.

Electrified Soap-Bubble.

94. As has already been said, this mechanical force shews itself well on electrifying a soap-bubble.

Let us first suppose a closed soap-bubble blown, of radius a. If the atmospheric pressure is Π, the pressure inside will be somewhat greater than Π, the resulting outward force being just balanced by the tension of the surface of the bubble. If, however, the bubble is electrified there will be an additional force acting normally outwards on the surface of the bubble, namely the force of amount $2\pi\sigma^2$ per unit area just investigated, and the bubble will expand until equilibrium is reached between this and the other forces acting on the surface.

As the electrification and consequently the radius change, the pressure inside will vary inversely as the volume, and therefore inversely as a^3. Let

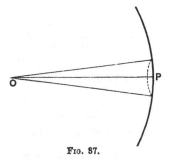

Fig. 37.

us, then, suppose the pressure to be κ/a^3. Consider the equilibrium of the small element of surface cut off by a circular cone through the centre, of small semi-vertical angle θ. This element is a circle of radius $a\theta$, and therefore of area $\pi a^2\theta^2$. The forces acting are:

(i)   The atmospheric pressure $\Pi\pi a^2\theta^2$ normally inwards.

(ii) The internal pressure $\dfrac{\kappa}{a^3}\,\pi a^2\theta^2$ normally outwards.

(iii)   The mechanical force due to electrification, $2\pi\sigma^2 \times \pi a^2\theta^2$ normally outwards.

(iv)   The system of tensions acting in the surface of the bubble across the boundary of the element.

If T is the tension per unit length, the tension across any element of length ds of the small circle will be $T\,ds$ acting at an angle θ with the tangent plane at P, the centre of the circle. This may be resolved into $T\,ds\cos\theta$ in the tangent plane, and $T\,ds\sin\theta$ along PO. Combining the forces all round the small circle of circumference $2\pi a\theta$, we find that the components in the tangent plane destroy one another, while those along PO combine into a resultant $2\pi a\theta \times T\sin\theta$. To a sufficient approximation this may be written as $2\pi a\theta^2 T$.

The equation of equilibrium of the element of area is accordingly

$$\Pi\pi a^2\theta^2 - \frac{\kappa}{a^3}\pi a^2\theta^2 - 2\pi\sigma^2\pi a^2\theta^2 + 2\pi a\theta^2 T = 0,$$

or, simplifying, $\Pi - \dfrac{\kappa}{a^3} - 2\pi\sigma^2 + \dfrac{2T}{a} = 0$ ....(28).

Let a_0 be the radius when the bubble is uncharged, and let the radius be a_1 when the bubble has a charge e, so that

$$\sigma = \frac{e}{4\pi a_1^2}.$$

Then $\Pi - \dfrac{\kappa}{a_0^3} + \dfrac{2T}{a_0} = 0,$

$$\Pi - \frac{\kappa}{a_1^3} - \frac{e^2}{8\pi a_1^4} + \frac{2T}{a_1} = 0.$$

We can without serious error assume T to be the same in the two cases. If we eliminate T from these two equations, we obtain

$$\Pi\,(a_1 - a_0) - \kappa\left(\frac{1}{a_1^2} - \frac{1}{a_0^2}\right) = \frac{e^2}{8\pi a_1^3},$$

giving the charge in terms of the radii in the charged and uncharged states.

95.   We have seen (§ 93) that the maximum pressure on the surface which electrification can produce is only about $\frac{1}{2500}$ atmosphere: thus it is not possible for electrification to change the pressure inside by more than about $\frac{1}{2500}$ atmosphere, so that the increase in the size of the bubble is necessarily very slight.

If, however, the bubble is blown on a tube which is open to the air, equation (28) becomes

$$\pi\sigma^2 = \frac{T}{a}.$$

As a rough approximation, we may still regard the bubble as a uniformly charged sphere, so that if V is its potential,

$$\sigma = V/4\pi a,$$

and the relation is $\qquad V^2 = 16\pi a T,$

giving V in terms of the radius of the bubble, if the tension T is known. In this case the electrification can be made to produce a large change in the radius, by using films for which T is very small.

Energy of Discharge.

96. On discharging a conductor or condenser, a certain amount of energy is set free. This may shew itself in various ways, *e.g.* as a spark or sound (as in lightning and thunder), the heating of a wire, or the piercing of a hole through a solid dielectric. The energy thus liberated has been previously stored up in charging the conductor or condenser.

To calculate the amount of this energy, let us suppose that one plate of a condenser is to earth, and that the other plate has a charge e and is at potential V, so that if C is the capacity of the condenser,

$$e = CV \quad \dots\dots\dots\dots\dots\dots\dots\dots\dots(29).$$

If we bring up an additional charge de from infinity, the work to be done is, in accordance with the definition of potential, Vde. This is equal to dW, where W denotes the total work done in charging the condenser up to this stage, so that

$$dW = Vde$$
$$= \frac{ede}{C} \text{ by equation (29)}.$$

On integration we obtain

$$W = \tfrac{1}{2}\frac{e^2}{C} \quad \dots\dots\dots\dots\dots\dots\dots\dots\dots\dots\dots(30),$$

no constant of integration being added since W must vanish when $e = 0$. This expression gives the work done in charging a condenser, and therefore gives also the energy of discharge, which may be used in creating a spark, in heating a wire, etc.

Clearly an exactly similar investigation will apply to a single conductor, so that expression (30) gives the energy either of a condenser or of a single conductor. Using the relation $e = CV$, the energy may be expressed in any one of the forms

$$W = \tfrac{1}{2}\frac{e^2}{C} = \tfrac{1}{2}eV = \tfrac{1}{2}CV^2 \quad \dots\dots\dots\dots\dots\dots(31).$$

97. As an example of the use of this formula, let us suppose that we have a parallel plate condenser, the area of each plate being A, and the distance of the plates being d, so that $C = A/4\pi d$, by § 83. Let σ be the surface density of the high potential plate, so that $e = \sigma A$. Let the low potential plate be at zero potential, then the potential of the high potential plate is

$$V = \frac{e}{C} = 4\pi d\sigma,$$

and the electrical energy is

$$W = \tfrac{1}{2}eV = 2\pi d\sigma^2 A.$$

Now let us pull the plates apart, so that d is increased to d'. The electrical energy is now $2\pi d'\sigma^2 A$, so that there has been an increase of electrical energy of amount

$$2\pi\sigma^2 A\,(d' - d).$$

It is easy to see that this exactly represents the work done in separating the two plates. The mechanical force on either plate is $2\pi\sigma^2$ per unit area, so that the total mechanical force on a plate is $2\pi\sigma^2 A$. Obviously, then, the above is the work done in separating the plates through a distance $d' - d$.

It appears from this that a parallel plate condenser affords a ready means of obtaining electrical energy at the expense of mechanical. A more valuable property of such a condenser is that it enables us to increase an initial difference of potential. The initial difference of potential

$$4\pi d\sigma$$

is increased, by the separation, to

$$4\pi d'\sigma.$$

By taking d small and d' large, an initial small difference of potential may be multiplied almost indefinitely, and a potential difference which is too small to observe may be increased until it is sufficiently great to affect an instrument. By making use of this principle, Volta first succeeded in detecting the difference of electrostatic potential between the two terminals of an electric battery.

There are practical difficulties which restrict the application of the principle. For if the initial distance d is made too small the condenser may discharge itself by a spark passing directly between the plates, while if d' is made large compared with the size of the plates the formulae we have used are no longer true.

EXAMPLES.

1. The two plates of a parallel plate condenser are each of area A, and the distance between them is d, this distance being small compared with the size of the plates. Find the attraction between them when charged to potential difference V, neglecting the irregularities caused by the edges of the plates. Find also the energy set free when the plates are connected by a wire.

2. A sheet of metal of thickness t is introduced between the two plates of a parallel plate condenser which are at a distance d apart, and is placed so as to be parallel to the plates. Shew that the capacity of the condenser is increased by an amount

$$\frac{t}{4\pi d (d-t)}$$

per unit area. Examine the case in which t is very nearly equal to d.

3. A high-pressure main consists first of a central conductor, which is a copper tube of inner and outer diameters of $\frac{9}{16}$ and $\frac{13}{18}$ inches. The outer conductor is a second copper tube coaxal with the first, from which it is separated by insulating material, and of diameters $1\frac{27}{34}$ and $1\frac{15}{16}$ inches. Outside this is more insulating material, and enclosing the whole is an iron tube of internal diameter $2\frac{1}{16}$ inches. The capacity of the conductor is found to be ·367 microfarad per mile : calculate the inductive capacity of the insulating material.

4. An infinite plane is charged to surface density σ, and P is a point distant half an inch from the plane. Shew that of the total intensity $2\pi\sigma$ at P, half is due to the charges at points which are within one inch of P, and half to the charges beyond.

5. A disc of vulcanite (non-conducting) of radius 5 inches, is charged to a uniform surface density σ by friction. Find the electric intensities at points on the axis of the disc distant respectively 1, 3, 5, 7 inches from the surface.

6. A condenser consists of a sphere of radius a surrounded by a concentric spherical shell of radius b. The inner sphere is put to earth, and the outer shell is insulated. Shew that the capacity of the condenser so formed is $\dfrac{b^2}{b-a}$.

7. Four equal large conducting plates A, B, C, D are fixed parallel to one another. A and D are connected to earth, B has a charge E per unit area, and C a charge E' per unit area. The distance between A and B is a, between B and C is b, and between C and D is c. Find the potentials of B and C.

8. A circular gold-leaf of radius b is laid on the surface of a charged conducting sphere of radius a, a being large compared to b. Prove that the loss of electrical energy in removing the leaf from the conductor—assuming that it carries away its whole charge— is approximately $\frac{1}{4} b^2 E^2 / a^3$, where E is the charge of the conductor, and the capacity of the leaf is comparable to b.

9. Two condensers of capacities C_1 and C_2, and possessing initially charges Q_1 and Q_2, are connected in parallel. Shew that there is a loss of energy of amount

$$\frac{(C_2 Q_1 - C_1 Q_2)^2}{2 C_1 C_2 (C_1 + C_2)}.$$

10. Two Leyden Jars A, B have capacities C_1, C_2 respectively. A is charged and a spark taken : it is then charged as before and a spark passed between the knobs of A and B. A and B are then separated and are each discharged by a spark. Shew that the energies of the four sparks are in the ratio

$$(C_1 + C_2)^2 : (C_1 + C_2) C_2 : C_1^2 : C_1 C_2.$$

11. Assuming an adequate number of condensers of equal capacity C, shew how a compound condenser can be formed of equivalent capacity θC, where θ is any rational number.

12. Three insulated concentric spherical conductors, whose radii in ascending order of magnitude are a, b, c, have charges e_1, e_2, e_3 respectively, find their potentials and shew that if the innermost sphere be connected to earth the potential of the outermost is diminished by

$$\frac{a}{c}\left(\frac{e_1}{a}+\frac{e_2}{b}+\frac{e_3}{c}\right).$$

13. A conducting sphere of radius a is surrounded by two thin concentric spherical conducting shells of radii b and c, the intervening spaces being filled with dielectrics of inductive capacities K and L respectively. If the shell b receives a charge E, the other two being uncharged, determine the loss of energy and the potential at any point when the spheres A and C are connected by a wire.

14. Three thin conducting sheets are in the form of concentric spheres of radii $a+d$, a, $a-c$ respectively. The dielectric between the outer and middle sheet is of inductive capacity K, that between the middle and inner sheet is air. At first the outer sheet is uninsulated, the inner sheet is uncharged and insulated, the middle sheet is charged to potential V and insulated. The inner sheet is now uninsulated without connection with the middle sheet. Prove that the potential of the middle sheet falls to

$$\frac{KVc(a+d)}{Kc(a+d)+d(a-c)}.$$

15. Two insulated conductors A and B are geometrically similar, the ratio of their linear dimensions being as L to L'. The conductors are placed so as to be out of each other's field of induction. The potential of A is V and its charge is E, the potential of B is V' and its charge is E'. The conductors are then connected by a thin wire. Prove that, after electrostatic equilibrium has been restored, the loss of electrostatic energy is

$$\tfrac{1}{2}\frac{(EL'-E'L)(V-V')}{L+L'}.$$

16. If two surfaces be taken in any family of equipotentials in free space, and two metal conductors formed so as to occupy their positions, then the capacity of the condenser thus formed is $\dfrac{C_1 C_2}{C_1-C_2}$, where C_1, C_2 are the capacities of the external and internal conductors when existing alone in an infinite field.

17. A conductor (B) with one internal cavity of radius b is kept at potential U. A conducting sphere (A), of radius a, at great height above B contains in a cavity water which leaks down a very thin wire passing without contact into the cavity of B through a hole in the top of B. At the end of the wire spherical drops are formed, concentric with the cavity ; and, when of radius d, they fall passing without contact through a small hole in the bottom of B, and are received in a cavity of a third conductor (C) of capacity c at a great distance below B. Initially, before leaking commences, the conductors A and C are uncharged. Prove that after the rth drop has fallen the potential of C is

$$\left\{\frac{a^r(b-d)^r}{(ab+bd-ad)^r}-1\right\}\frac{a}{c}U;$$

where the disturbing effect of the wire and hole on the capacities is neglected.

18. An insulated spherical conductor, formed of two hemispherical shells in contact, whose inner and outer radii are b and b', has within it a concentric spherical conductor of radius a, and without it another spherical conductor of which the internal radius is c. These two conductors are earth-connected and the middle one receives a charge. Shew that the two shells will not separate if

$$2ac>bc+b'a.$$

19. Outside a spherical charged conductor there is a concentric insulated but uncharged conducting spherical shell, which consists of two segments. Prove that the two segments will not separate if the distance of the separating plane from the centre is less than

$$\frac{ab}{(a^2+b^2)^{\frac{1}{2}}},$$

where a, b are the internal and external radii of the shell.

20. A soap-bubble of radius a is formed by a film of tension T, the external atmospheric pressure being Π. The bubble is touched by a wire from a large conductor at potential V, and the film is an electrical conductor. Prove that its radius increases to r, given by

$$\Pi(r^3-a^3)+2T(r^2-a^2)=\frac{V^2 r}{8\pi}.$$

21. If the radius and tension of a spherical soap-bubble be a and T respectively, shew that the charge of electricity required to expand the bubble to twice its linear dimensions would be

$$4\sqrt{\pi a^3(6T+7\Pi a)},$$

Π being the atmospheric pressure.

22. A thin spherical conducting envelope, of tension T for all magnitudes of its radius, and with no air inside or outside, is insulated and charged with a quantity Q of electricity. Prove that the total gain in mechanical energy involved in bringing a charge q from an infinite distance and placing it on the envelope, which both initially and finally is in mechanical equilibrium, is

$$\tfrac{3}{2}(2\pi T)^{\frac{1}{3}}\{(Q+q)^{\frac{4}{3}}-Q^{\frac{4}{3}}\}.$$

23. A spherical soap-bubble is blown inside another concentric with it, and the former has a charge E of electricity, the latter being originally uncharged. The latter now has a small charge given to it. Shew that if a and $2a$ were the original radii, the new radii will be approximately $a+x$, $2a+y$, where

$$12y(\Pi a+T)=x\left(24\Pi a+\frac{101}{2}T+\frac{7E^2}{8\pi a^3}\right),$$

where Π is the atmospheric pressure, and T is the surface-tension of each bubble.

24. Shew that the electric capacity of a conductor is less than that of any other conductor which can completely surround it.

25. If the inner sphere of a concentric spherical condenser is moved slightly out of position, so that the two spheres are no longer concentric, shew that the capacity is increased.

CHAPTER IV

SYSTEMS OF CONDUCTORS

98. In the present Chapter we discuss the general theory of an electrostatic field in which there are any number of conductors. The charge on each conductor will of course influence the distribution of charges on the other conductors by induction, and the problem is to investigate the distributions of electricity which are to be expected after allowing for this mutual induction.

We have seen that in an electrostatic field the potential cannot be a maximum or a minimum except at points where electric charges occur. It follows that the highest potential in the field must occur on a conductor, or else at infinity, the latter case occurring only when the potential of every conductor is negative. Excluding this case for the moment, there must be one conductor of which the potential is higher than that anywhere else in the field. Since lines of force run only from higher to lower potential (§ 36), it follows that no lines of force can enter this conductor, there being no higher potential from which they can come, so that lines of force must leave it at every point of its surface. In other words, its electrification must be positive at every point.

So also, except when the potential of every conductor is positive, there must be one conductor of which the potential is lower than that anywhere else in the field, and the electrification at every point of this conductor must be negative.

If the total charge on a conductor is *nil*, the total strength of the tubes of force which enter it must be exactly equal to the total strength of the tubes which leave it. There must therefore be both tubes which enter and tubes which leave its surface, so that its potential must be intermediate between the highest and lowest potentials in the field. For if its potential were the highest in the field, no tubes could enter it, and *vice versâ*. On any such conductor the regions of positive electrification are separated from regions of negative electrification by " lines of no electrification," these lines being loci along which $\sigma = 0$. In general the resultant intensity at any point of a conductor is $4\pi\sigma$. At any point of a line of no electrification, this intensity vanishes, so that every point of a "line of no electrification" is also a point of equilibrium.

At a point of equilibrium we have already seen that the equipotential through the point cuts itself. A line of no electrification, however, lies entirely on a single equipotential, so that this equipotential must cut itself along the line of no electrification. Moreover, by § 69, it must cut itself at right angles, except when it consists of more than two sheets.

99. We can prove the two following propositions:

I. *If the potential of every conductor in the field is given, there is only one distribution of electric charges which will produce this distribution of potential.*

II. *If the total charge of every conductor in the field is given, there is only one way in which these charges can distribute themselves so as to be in equilibrium.*

If proposition I. is not true, let us suppose that there are two different distributions of electricity which will produce the required potentials. Let σ denote the surface density at any point in the first distribution, and σ' in the second. Consider an imaginary distribution of electricity such that the surface density at any point is $\sigma - \sigma'$. The potential of this distribution at any point P is

$$V_P = \iint \frac{\sigma - \sigma'}{r}\, dS,$$

where the integration extends over the surfaces of all the conductors, and r is the distance from P to the element dS. If P is a point on the surface of any conductor,

$$\iint \frac{\sigma}{r}\, dS \quad \text{and} \quad \iint \frac{\sigma'}{r}\, dS$$

are by hypothesis equal, each being equal to the given potential of the conductor on which P lies. Thus

$$V_P = \iint \frac{\sigma}{r}\, dS - \iint \frac{\sigma'}{r}\, dS = 0,$$

so that the supposed distribution of density $\sigma - \sigma'$ is such that the potential vanishes over all the surfaces of the conductors. There can therefore be no lines of force, so that there can be no charges, *i.e.*, $\sigma - \sigma' = 0$ everywhere, so that the two distributions are the same.

And again, if proposition II. is not true, let us suppose that there are two different distributions σ and σ' such that the total charge on each conductor has the assigned value. A distribution $\sigma - \sigma'$ now gives zero as the total charge on each conductor. It follows, as in § 98, that the potential of every conductor must be intermediate between the highest and lowest potentials in the field, a conclusion which is obviously absurd, as it prevents every conductor from having either the highest or the lowest potential. It follows that the potentials of all the conductors must be equal, so that again there can be no lines of force and no charges at any point, *i.e.*, $\sigma = \sigma'$ everywhere.

It is clear from this that the distribution of electricity in the field is fully specified when we know either

(i)   the total charge on each conductor,

or   (ii)   the potential of each conductor.

SUPERPOSITION OF EFFECTS.

100. Suppose we have two equilibrium distributions:

(i)   A distribution of which the surface density is σ at any point, giving total charges E_1, E_2, ... on the different conductors, and potentials V_1, V_2, ...

(ii)   A distribution of surface density σ', giving total charges E_1', E_2', ... and potentials V_1', V_2', ...

Consider a distribution of surface density $\sigma + \sigma'$. Clearly the total charges on the conductors will be $E_1 + E_1'$, $E_2 + E_2'$, ..., and if V_P is the potential at any point P,

$$V_P = \iint \frac{\sigma + \sigma'}{r} \, dS,$$

where the notation is the same as before. If P is on the first conductor, however, we know that

$$\iint \frac{\sigma}{r} \, dS = V_1,$$

$$\iint \frac{\sigma'}{r} \, dS = V_1',$$

so that $V_P = V_1 + V_1'$; and similarly when P is on any other conductor. Thus the imaginary distribution of surface density is an equilibrium distribution, since it makes the surface of each conductor an equipotential, and the potentials are

$$V_1 + V_1', \quad V_2 + V_2', \quad ....$$

The total charges, as we have seen, are $E_1 + E_1'$, $E_2 + E_2'$, ..., and from the proposition previously proved, it follows that the distribution of surface-density $\sigma + \sigma'$ is the only distribution corresponding to these charges.

We have accordingly arrived at the following proposition:

If charges E_1, E_2, ... give rise to potentials V_1, V_2, ..., and if charges

E_1', E_2', ... *give rise to potentials* V_1', V_2', ..., *then charges* $E_1 + E_1'$, $E_2 + E_2'$, ... *will give rise to potentials* $V_1 + V_1'$, $V_2 + V_2'$, ...

In words: if we superpose two systems of charges, the potentials produced can be obtained by adding together the potentials corresponding to the two component systems.

Clearly the proposition can be extended so as to apply to the superposition of any number of systems.

We can obviously deduce the following:

If charges E_1, E_2, ... *give rise to potentials* V_1, V_2, ..., *then charges* KE_1, KE_2, ... *give rise to potentials* KV_1, KV_2, ...

101. Suppose now that we have n conductors fixed in position and uncharged. Let us refer to these conductors as conductor (1), conductor (2), etc. Suppose that the result of placing unit charge on conductor (1) and leaving the others uncharged is to produce potentials

$$p_{11}, \quad p_{12}, \cdots p_{1n},$$

on the n conductors respectively, then the result of placing E_1 on (1) and leaving the others uncharged is to produce potentials

$$p_{11}E_1, \quad p_{12}E_1, \cdots p_{1n}E_1.$$

Similarly, if placing unit charge on (2) and leaving the others uncharged gives potentials

$$p_{21}, \quad p_{22}, \cdots p_{2n},$$

then placing E_2 on (2) and leaving the others uncharged gives potentials

$$p_{21}E_2, \quad p_{22}E_2, \cdots p_{2n}E_2.$$

In the same way we can calculate the result of placing E_3 on (3), E_4 on (4), and so on.

If we now superpose the solutions we have obtained, we find that the effect of simultaneous charges E_1, E_2, ... E_n is to give potentials V_1, V_2, ... V_n, where

$$\left.\begin{aligned} V_1 &= p_{11}E_1 + p_{21}E_2 + p_{31}E_3 + \ldots \\ V_2 &= p_{12}E_1 + p_{22}E_2 + p_{32}E_3 + \ldots \\ \text{etc.} \end{aligned}\right\} \quad \ldots\ldots\ldots\ldots\ldots(32).$$

These equations give the potentials in terms of the charges. The coefficients p_{11}, p_{21}, ... do not depend on either the potentials or charges, being purely geometrical quantities, which depend on the size, shape and position of the different conductors.

Green's Reciprocation Theorem.

102. Let us suppose that charges e_P, e_Q, ... on elements of conducting surfaces at P, Q, ... produce potentials V_P, V_Q, ... at P, Q, ..., and that similarly charges e_P', e_Q', ... produce potentials V_P', V_Q', ... Then Green's Theorem states that

$$\Sigma e_P V_P' = \Sigma e_P' V_P,$$

the summation extending in each case over all the charges in the field.

To prove the theorem, we need only notice that

$$V_P = \Sigma \frac{e_Q}{PQ},$$

the summation extending over all charges except e_P, so that in $\Sigma e_P' V_P$ the coefficient of $\frac{1}{PQ}$ is $e_P' e_Q$ from the term $e_P' V_P$, and $e_P e_Q'$ from the term $e_Q' V_Q$. Thus

$$\Sigma e_P' V_P = \Sigma\Sigma \frac{e_P e_Q' + e_Q e_P'}{PQ}$$

$$= \Sigma e_P V_P', \text{ from symmetry.}$$

103. The following theorem follows at once:

If total charges E_1, E_2 on the separate conductors of a system produce potentials V_1, V_2, ..., and if charges E_1', E_2', ... produce potentials V_1', V_2', ..., then

$$\Sigma E V' = \Sigma E' V \dots\dots\dots\dots\dots\dots(33),$$

the summation extending in each case over all the conductors.

To see the truth of this, we need only divide up the charges E_1, E_2, ... into small charges e_P, e_Q, ... on the different small elements of the surfaces of the conductors, and the proposition becomes identical with that just proved.

104. Let us now consider the special case in which

$$E_1 = 1, \quad E_2 = E_3 = E_4 = \dots = 0,$$

so that $\qquad\qquad V_1 = p_{11}, \quad V_2 = p_{12}, \text{ etc.};$

and $\qquad\qquad E_1' = 0, \quad E_2' = 1, \quad E_3' = E_4' = \dots = 0.$

so that $\qquad\qquad V_1' = p_{21}, \quad V_2' = p_{22}, \text{ etc.}$

Then $\Sigma E V' = p_{21}$ and $\Sigma E' V = p_{12}$, so that the theorem just proved becomes

$$p_{12} = p_{21}.$$

In words: the potential to which (1) is raised by putting unit charge on (2), all the other conductors being uncharged, is equal to the potential to which (2) is raised by putting unit charge on (1), all the other conductors being uncharged.

As a special case, let us reduce conductor (2) to a point P, and suppose that the system contains in addition only one other conductor (1). Then

The potential to which the conductor is raised by placing a unit charge at P, the conductor itself being uncharged, is equal to the potential at P when unit charge is placed on the conductor.

For instance, let the conductor be a sphere, and let the point P be at a distance r from its centre. Unit charge on the sphere produces potential $\dfrac{1}{r}$ at P, so that unit charge at P raises the sphere to potential $\dfrac{1}{r}$.

Coefficients of Potential, Capacity and Induction.

105. The relations $p_{12} = p_{21}$, etc. reduce the number of the coefficients $p_{11}, p_{12}, \dots p_{nn}$, which occur in equations (32), to $\frac{1}{2}n(n+1)$. These coefficients are called the *coefficients of potential* of the n conductors. Knowing the values of these coefficients, equations (31) give the potentials in terms of the charges.

If we know the potentials $V_1, V_2, \dots$, we can obtain the values of the charges by solving equations (32). We obtain a system of equations of the form

$$
\left.
\begin{aligned}
E_1 &= q_{11}V_1 + q_{21}V_2 + \dots \\
E_2 &= q_{12}V_1 + q_{22}V_2 + \dots \\
&\text{etc.}
\end{aligned}
\right\} \quad \dots\dots\dots\dots\dots\dots(34).
$$

The values of the q's obtained by actual solution of the equations (32), are

$$
\frac{q_{11}}{\begin{vmatrix} p_{22}\,p_{32}\,\cdots\,p_{n2} \\ p_{23}\,p_{33}\,\cdots\,p_{n3} \\ \cdots\cdots\cdots \\ p_{2n}\,p_{3n}\,\cdots\,p_{nn} \end{vmatrix}} = \frac{-q_{12}}{\begin{vmatrix} p_{21}\,p_{31}\,\cdots\,p_{n1} \\ p_{23}\,p_{33}\,\cdots\,p_{n3} \\ \cdots\cdots\cdots \\ p_{2n}\,p_{3n}\,\cdots\,p_{nn} \end{vmatrix}} = \frac{1}{\Delta} \quad \dots\dots\dots\dots(35),
$$

where

$$
\Delta = \begin{vmatrix} p_{11}\,p_{21}\,\cdots\,p_{n1} \\ p_{12}\,p_{22}\,\cdots\,p_{n2} \\ \cdots\cdots\cdots\cdots \\ p_{1n}\,p_{2n}\,\cdots\,p_{nn} \end{vmatrix}.
$$

Thus q_{rs} is the co-factor of p_{rs} in Δ, divided by Δ.

The relation $\qquad\qquad q_{rs} = q_{sr}$

follows as an algebraical consequence of the relation $p_{rs} = p_{sr}$, or is at once obvious from the relation

$$\Sigma EV' = \Sigma E'V,$$

and equations (34), on taking the same sets of values as in § 104.

There are n coefficients of the type q_{11}, q_{22}, ... q_{nn}. These are known as *coefficients of capacity*. There are $\frac{1}{2}n(n-1)$ coefficients of the type q_{rs}, and these are known as *coefficients of induction*.

From equations (34), it is clear that q_{11} is the value of E_1 when $V_1 = 1$, $V_2 = V_3 = ... = 0$. This leads to an extended definition of the capacity of a conductor, in which account is taken of the influence of the other conductors in the field. We define the capacity of the conductor 1, when in the presence of conductors 2, 3, 4, ..., to be q_{11}, namely, the charge required to raise conductor 1 to unit potential, all the other conductors being put to earth.

ENERGY OF A SYSTEM OF CHARGED CONDUCTORS.

106. Suppose we require to find the energy of a system of conductors, their charges being E_1, E_2, ... E_n, so that their potentials are V_1, V_2, ... V_n given by equations (32).

Let W denote the energy when the charges are kE_1, kE_2, ... kE_n. Corresponding to these charges, the potentials will be kV_1, kV_2, ... kV_n. If we bring up an additional small charge $dk \cdot E_1$ from infinity to conductor 1, the work to be done will be $dkE_1 \cdot kV_1$; if we bring up dkE_2 to conductor 2 the work will be $dkE_2 kV_2$ and so on. Let us now bring charges dkE_1 to 1, dkE_2 to 2, dkE_3 to 3, ... dkE_n to n. The total work done is

$$k\,dk\,(E_1V_1 + E_2V_2 + ... + E_nV_n) \qquad\qquad (36),$$

and the final charges are

$$(k + dk)\,E_1, \; (k + dk)\,E_2, \; ... \; (k + dk)\,E_n.$$

The energy in this state is the same function of $k + dk$ as W is of k, and may therefore be expressed as

$$W + \frac{\partial W}{\partial k}\,dk.$$

Expression (36), the increase in energy, is therefore equal to $\dfrac{\partial W}{\partial k}\,dk$, whence

$$\frac{\partial W}{\partial k} = k\,(E_1V_1 + E_2V_2 + ... + E_nV_n),$$

so that on integration

$$W = \tfrac{1}{2} k^2\,(E_1V_1 + E_2V_2 + ... + E_nV_n).$$

No constant of integration is added, since W must vanish when $k = 0$. Taking $k = 1$, we obtain the energy corresponding to the final charges E_1, E_2, ... E_n, in the form

$$W = \tfrac{1}{2}\Sigma EV \qquad\qquad\qquad (37).$$

If we substitute for the V's their values in terms of the charges as given by equations (32), we obtain

$$W = \tfrac{1}{2}\,(p_{11}E_1{}^2 + 2p_{12}E_1E_2 + p_{22}E_2{}^2 + \ldots) \quad\ldots\ldots\ldots\ldots(38),$$

and similarly from equations (34),

$$W = \tfrac{1}{2}\,(q_{11}V_1{}^2 + 2q_{12}V_1V_2 + q_{22}V_2{}^2 + \ldots)\quad\ldots\ldots\ldots\ldots(39).$$

107. If W is expressed as a function of the E's, we obtain by differentiation of (38),

$$\frac{\partial W}{\partial E_1} = p_{11}E_1 + p_{12}E_2 + \ldots + p_{1n}E_n$$

$$= V_1 \text{ by equations (32).}$$

This result is clear from other considerations. If we increase the charge on conductor 1 by dE_1, the increase of energy is $\dfrac{\partial W}{\partial E_1}\,dE_1$, and is also $V_1 dE_1$ since this is the work done on bringing up a new charge dE_1 to potential V_1. Thus on dividing by dE_1, we get

$$\frac{\partial W}{\partial E_1} = V_1 \quad\ldots\ldots\ldots\ldots\ldots\ldots\ldots\ldots\ldots\ldots\ldots\ldots\ldots(40).$$

So also

$$\frac{\partial W}{\partial V_1} = E_1 \quad\ldots\ldots\ldots\ldots\ldots\ldots\ldots\ldots\ldots\ldots\ldots\ldots(41)$$

as is at once obvious on differentiation of (39).

108. In changing the charges from E_1, E_2, ... to E_1', E_2', ... let us suppose that the potentials change from V_1, V_2, ... to V_1', V_2', ... The work done, $W' - W$, is given by

$$W' - W = \tfrac{1}{2}\Sigma\,(E'V' - EV).$$

Since, however, by § 103, $\Sigma EV' = \Sigma E'V$, this expression for the work done can either be written in the form

$$\tfrac{1}{2}\Sigma\,\{E'V' - EV - (EV' - E'V)\},$$

which leads at once to

$$W' - W = \tfrac{1}{2}\Sigma\,(E' - E)(V' + V) \quad\ldots\ldots\ldots\ldots(42);$$

or in the form $\quad \tfrac{1}{2}\Sigma\,\{E'V' - EV + (EV' - E'V)\},$

which leads to $\qquad W' - W = \tfrac{1}{2}\Sigma\,(V' - V)(E' + E) \quad\ldots\ldots\ldots\ldots(43).$

109. If the changes in the charges are only small, we may replace E' by $E + dE$, and find that equation (42) reduces to

$$dW = \Sigma V dE,$$

from which equation (40) is obvious, while equation (43) reduces to

$$dW = \Sigma E dV,$$

leading at once to (41).

110. It is worth noticing that the coefficients of potential, capacity and induction can be expressed as differential coefficients of the energy; thus

$$p_{11} = \frac{\partial^2 W}{\partial E_1^2},$$

$$p_{rs} = \frac{\partial^2 W}{\partial E_r \partial E_s},$$

$$q_{rs} = \frac{\partial^2 W}{\partial V_r \partial V_s},$$

and so on.

The last two equations give independent proofs of the relations

$$p_{rs} = p_{sr}, \quad q_{rs} = q_{sr}.$$

PROPERTIES OF THE COEFFICIENTS.

111. A certain number of properties can be deduced at once from the fact that the energy must always be positive. For instance since the value of W given by equation (38) is positive for *all* values of $E_1, E_2, \dots E_n$, it follows at once that

$$p_{11}, p_{22}, p_{33}, \dots \text{ are positive,}$$

that $p_{11}p_{22} - p_{12}^2$ is positive, that

$$\begin{vmatrix} p_{11} & p_{12} & p_{13} \\ p_{12} & p_{22} & p_{23} \\ p_{13} & p_{23} & p_{33} \end{vmatrix} \text{ is positive}$$

and so on. Similarly from equation (39), it follows that

$$q_{11}, q_{22}, q_{33}, \dots \text{ are positive,}$$

and there are other relations similar to those above.

112. More valuable properties can, however, be obtained from a consideration of the distribution of the lines of force in the field.

Let us first consider the field when

$$E_1 = 1, \quad E_2 = E_3 = \dots = 0.$$

The potentials are $\quad V_1 = p_{11}, \quad V_2 = p_{12}, \text{ etc.}$

Since conductors 2, 3, ... are uncharged, their potentials must be intermediate between the highest and lowest potentials in the field. Thus the potential of 1 must be either the highest or the lowest in the field, the other extreme potential being at infinity. It is impossible for the potential of 1 to be the *lowest* in the field; for if it were, lines of force would enter in at every point, and its charge would be negative. Thus the highest potential in the field must be that of conductor 1, and the other potentials must all be intermediate between this potential and the potential at infinity, and must therefore all be positive. Thus $p_{11}, p_{12}, p_{13}, \dots p_{1n}$ *are all positive and the first is the greatest.*

Next let us put
$$V_1 = 1, \quad V_2 = V_3 = \dots = 0,$$
so that the charges are
$$q_{11}, \quad q_{12}, \quad q_{13}, \dots q_{1n}.$$

The highest potential in the field is that of conductor 1. Thus lines of force leave but do not enter conductor 1. The lines may either go to the other conductors or to infinity. No lines can leave the other conductors. Thus the charge on 1 must be positive, and the charges on 2, 3, ... all negative, *i.e.*, q_{11} is positive and $q_{12}, q_{13}, \dots$ are all negative. Moreover the total strength of the tubes arriving at infinity is $q_{11} + q_{12} + q_{13} + \dots + q_{1n}$, so that this must be positive.

113. To sum up, we have seen that

(i)   All the coefficients of potential ($p_{11}, p_{12}, \dots$) are positive,

(ii)   All the coefficients of capacity ($q_{11}, q_{22}, \dots$) are positive,

(iii)   All the coefficients of induction ($q_{12}, q_{13}, \dots$) are negative,

and we have obtained the relations
$$(p_{11} - p_{12}) \text{ is positive,}$$
$$(q_{11} + q_{12} + \dots + q_{1n}) \text{ is positive.}$$

In limiting cases it is of course possible for any of the quantities which have been described as always positive or always negative, to vanish.

Values of the Coefficients in Special Cases.

Electric Screening.

114. The first case in which we shall consider the values of the coefficients is that in which one conductor, say 1, is completely surrounded by a second conductor 2.

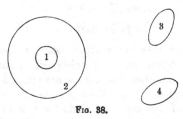

Fig. 38.

If $E_1 = 0$, the conductor 2 becomes a closed conductor with no charge inside, so that the potential in its interior is constant, and therefore $V_1 = V_2$. Putting $E_1 = 0$, the relation $V_1 = V_2$ gives the equation
$$(p_{12} - p_{22}) E_2 + (p_{13} - p_{23}) E_3 + \dots = 0.$$

This being true for all values of E_2, E_3, ... we must have

$$p_{12} = p_{22}, \quad p_{13} = p_{23}, \text{ etc.}$$

Next let us put unit charge on 1, leaving the other conductors uncharged. The energy is $\frac{1}{2}p_{11}$. If we join 1 and 2 by a wire, the conductors 1 and 2 form a single conductor, so that the electricity will all flow to the outer surface. This wire may now be removed, and the energy in the system is $\frac{1}{2}p_{22}$. Energy must, however, have been lost in the flow of electricity, so that p_{22} must be less than p_{11}.

Since we have already seen that $p_{12} = p_{22}$ and $p_{11} - p_{12}$ cannot be negative, it is clear that p_{22} cannot be greater than p_{11}. The foregoing argument, however, goes further and enables us to prove that $p_{11} - p_{22}$ is actually positive.

Let us next suppose that conductor 2 is put to earth, so that $V_2 = 0$. Then if $E_1 = 0$, it follows that $V_1 = 0$. Hence from the equations

$$E_1 = q_{11}V_1 + q_{12}V_2 + \dots + q_{1n}V_n \quad \dots\dots\dots\dots\dots(44)$$

we obtain in this special case that

$$q_{13}V_3 + q_{14}V_4 + \dots + q_{1n}V_n = 0.$$

This is true, whatever the values of V_3, V_4, ..., so that

$$q_{13} = q_{14} = \dots = q_{1n} = 0.$$

Suppose that conductor 1 is raised to unit potential while all the other conductors are put to earth. The aggregate strength of the tubes of force which go to infinity, namely $q_{11} + q_{12} + \dots + q_{1n}$ (§ 112), is in this case zero, so that $q_{12} = -q_{11}$.

The system of equations (44) now reduces, when $V_2 = 0$, to

$$E_1 = q_{11}V_1 \quad \dots\dots\dots\dots\dots\dots\dots\dots(45),$$

$$E_2 = q_{12}V_1 + q_{22}V_3 + q_{24}V_4 + \dots \quad \dots\dots\dots\dots\dots(46),$$

$$\left.\begin{array}{l} E_3 = q_{33}V_3 + q_{34}V_4 + \dots \\ E_4 = q_{34}V_3 + q_{44}V_4 + \dots \end{array}\right\} \quad \dots\dots\dots\dots\dots(47)$$

Equations (47) shew that the relations between charges and potential outside 2 are quite independent of the electrical conditions which obtain inside 2. So also the conditions inside 2 are not affected by those outside 2, as is obvious from equation (45). These results become obvious when we consider that no lines of force can *cross* conductor 2, and that there is no way except by crossing conductor 2 for a line of force to pass from the conductors outside 2 to those inside 2.

An electric system which is completely surrounded by a conductor at potential zero is said to be "electrically screened" from all electric systems outside this conductor; for charges outside this "screen" cannot affect the screened system. The principle of electric screening is utilised in electrostatic instruments, in order that the instrument may not be affected by external electric actions other than those which it is required to observe. As a complete conductor would prevent observation of the working of the instrument, a cage of wire is frequently used as a screen, this being very nearly as efficient as a completely closed conductor (see § 72). In more delicate instruments the screening may be complete except for a small window to admit of observation of the interior.

Spherical Condenser.

115. Let us apply the methods of this Chapter to the spherical condenser described in § 79. Let the inner sphere of radius a be taken to be conductor 1, and the outer sphere of radius b be taken to be conductor 2.

The equations connecting potentials and charges are

$$V_1 = p_{11}E_1 + p_{21}E_2,$$
$$V_2 = p_{12}E_1 + p_{22}E_2.$$

A unit charge placed on 2 raises both 1 and 2 to potential $1/b$, so that on putting $E_1 = 0$, $E_2 = 1$, we must have $V_1 = V_2 = 1/b$. Hence it follows that

$$p_{21} = p_{22} = \frac{1}{b}.$$

If we leave 2 uncharged and place unit charge on 1, the field of force is that investigated in § 79, so that $V_1 = 1/a$, $V_2 = 1/b$. Hence

$$p_{11} = \frac{1}{a}, \qquad p_{12} = \frac{1}{b}.$$

These results exemplify

  (i)　the general relation $p_{12} = p_{21}$,

  (ii)　the relation peculiar to electric screening, $p_{12} = p_{22}$.

The equations now become

$$V_1 = \frac{E_1}{a} + \frac{E_2}{b},$$
$$V_2 = \frac{E_1}{b} + \frac{E_2}{b}.$$

Solving for E_1 and E_2 in terms of V_1 and V_2, we obtain

$$E_1 = \frac{ab}{b-a}V_1 - \frac{ab}{b-a}V_2,$$
$$E_2 = -\frac{ab}{b-a}V_1 + \frac{b^2}{b-a}V_2,$$

so that　　$q_{11} = \dfrac{ab}{b-a}, \qquad q_{12} = q_{21} = -\dfrac{ab}{b-a}, \qquad q_{22} = \dfrac{b^2}{b-a}.$

We notice that $q_{12} = q_{21}$, that the value of each is negative, and that $q_{11} = -q_{12}$, in accordance with § 114. The value of q_{11} is the capacity of sphere 1 when 2 is to earth, and is in agreement with the result of § 79. The capacity of 2 when 1 is to earth, q_{22}, is seen to be $\dfrac{b^2}{b-a}$. This can also be seen by regarding the system as composed of two condensers, the inner sphere and the inner surface of the outer sphere form a single spherical condenser of capacity $\dfrac{ab}{b-a}$, while the outer surface of the outer sphere has capacity b. The total capacity accordingly

$$= \frac{ab}{b-a} + b = \frac{b^2}{b-a}.$$

Two spheres at a great distance apart.

116. Suppose we have two spheres, radii a, b, placed with their centres at a great distance c apart. Let us first place unit charge on the former, the

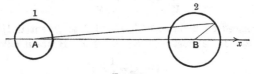

Fig. 39.

charge being placed so that the surface density is constant. This will not produce uniform potential over 2; at a point distant r from the centre of 1 it will produce potential $1/r$. We can, however, adjust this potential to the uniform value $1/c$ by placing on the surface of 2 a distribution of electricity such that it produces a potential $\dfrac{1}{c} - \dfrac{1}{r}$ over this surface.

Take B, the centre of the second sphere, as origin, and AB as axis of x. Then we may write

$$\frac{1}{c} - \frac{1}{r} = \frac{r-c}{cr} = \frac{x}{c^2}, \text{ as far as } \frac{1}{c^2}.$$

Let σ be the surface density required to produce this potential, then clearly σ is an odd function of x, and therefore the total charge, the value of σ integrated over the sphere, vanishes. Thus the potential of 2 can be adjusted to the uniform value $1/c$ without altering the total charge on 2 from zero, neglecting $1/c^3$. The new surface density being of the order of $1/c^2$, the additional potential produced on 1 by it will be at most of order $1/c^3$, so that if we neglect $1/c^3$ we have found an equilibrium arrangement which makes

$$E_1 = 1, \quad E_2 = 0, \quad V_1 = \frac{1}{a}, \quad V_2 = \frac{1}{c}.$$

Substituting these values in the equations

$$V_1 = p_{11}E_1 + p_{12}E_2,$$
$$V_2 = p_{12}E_1 + p_{22}E_2,$$

we find at once that

$$p_{11} = \frac{1}{a} \text{ neglecting } \frac{1}{c^3},$$

$$p_{12} = \frac{1}{r} \qquad \frac{1}{c^3},$$

and similarly we can see that

$$p_{22} = \frac{1}{b} \text{ neglecting } \frac{1}{c^3}.$$

Solving the equations

$$V_1 = \frac{E_1}{a} + \frac{E_2}{c},$$

$$V_2 = \frac{E_1}{c} + \frac{E_2}{b},$$

we find that, neglecting $\frac{1}{c^3}$,

$$q_{11} = \frac{a}{1 - \dfrac{ab}{c^2}},$$

$$q_{12} = q_{21} = -\frac{ab}{c\left(1 - \dfrac{ab}{c^2}\right)} = -\frac{ab}{c} \text{ as far as } \frac{1}{c^3},$$

$$q_{22} = \frac{b}{1 - \dfrac{ab}{c^2}}.$$

We notice that the capacity of either sphere is greater than it would be if the other were removed. This, as we shall see later, is a particular case of a general theorem.

Two conductors in contact.

117. If two conductors are placed in contact, their potentials must be equal. Let the two conductors be conductors 1 and 2, then the equation $V_1 = V_2$ becomes

$$(p_{11} - p_{12}) E_1 + (p_{12} - p_{22}) E_2 + \ldots = 0,$$

or, say,

$$\alpha E_1 + \beta E_2 + \gamma E_3 + \ldots = 0.$$

If we know the total charge E on 1 and 2, we have

$$E_1 + E_2 = E,$$

and on solving these two equations we can obtain E_1 and E_2. We find that

$$\frac{E_1}{E_2} = -\frac{\beta E + \gamma E_3 + \delta E_4 + \ldots}{\alpha E + \gamma E_3 + \delta E_4 + \ldots},$$

giving the ratio in which the charge E will distribute itself between the two conductors 1 and 2. If the conductors 3, 4, ... are either absent or uncharged,

$$\frac{E_1}{E_2} = -\frac{\beta}{\alpha} = \frac{p_{22} - p_{12}}{p_{11} - p_{12}},$$

which is independent of E and always positive. It is to be noticed that E_1 vanishes only if $p_{22} = p_{12}$, *i.e.*, if 2 entirely surrounds 1.

MECHANICAL FORCES ON CONDUCTORS.

118. We have already seen that the mechanical force on a conductor is the resultant of a system of tensions over its surface of amount $2\pi\sigma^2$ per unit area. The results of the present Chapter enable us to find the *resultant* force on any conductor in terms of the electrical coefficients of the system.

Suppose that the positions of the conductors are specified by any co-ordinates $\xi_1, \xi_2, ...$, so that $p_{11}, p_{12}, ..., q_{11}, q_{12}, ...$, and consequently also W, are functions of the ξ's. If ξ_1 is increased to $\xi_1 + d\xi_1$, without the charges on the conductors being altered, the increase in electrical energy is $\dfrac{\partial W}{\partial \xi_1} d\xi_1$, and this increase must represent mechanical work done in moving the conductors. The force tending to increase ξ_1 is accordingly

$$-\frac{\partial W}{\partial \xi_1}$$

Since the charges on the conductors are to be kept constant, it will of course be most convenient to use the form of W given by equation (38), and the force is obtained in the form

$$-\tfrac{1}{2}\left(\frac{\partial p_{11}}{\partial \xi_1} E_1^2 + 2\frac{\partial p_{12}}{\partial \xi_1} E_1 E_2 + ...\right) \quad ....(48).$$

It is however possible, by joining the conductors to the terminals of electric batteries, to keep their potentials constant. In this case, however, we must not use the expression (39) for W, and so obtain for the force

$$-\tfrac{1}{2}\left(\frac{\partial q_{11}}{\partial \xi_1} V_1^2 + 2\frac{\partial q_{12}}{\partial \xi_1} V_1 V_2 + ...\right) \quad ....(49),$$

for the batteries are now capable of supplying energy, and an increase of electrical energy does not necessarily mean an equal expenditure of mechanical energy, for we must not neglect the work done by the batteries. Since the resultant mechanical force on any conductor may be regarded as the resultant of tensions $2\pi\sigma^2$ per unit area acting over its surface, it is clear that this resultant force in any position depends solely on the charges in this position. It is therefore the same whether the charges or potentials are kept constant, and expression (48) will give this force whether the conductors are connected to batteries or not.

119. As an illustration, we may consider the force between the two charged spheres discussed in § 116.

The force tending to increase c, namely $-\dfrac{\partial W}{\partial c}$, is

$$-\tfrac{1}{2}\left(\frac{\partial p_{11}}{\partial c}E_1{}^2 + 2\frac{\partial p_{12}}{\partial c}E_1 E_2 + \frac{\partial p_{22}}{\partial c}E_2{}^2\right),$$

and substituting the values

$$p_{11} = \frac{1}{a} + \text{terms in } \frac{1}{c^3},$$

$$p_{12} = \frac{1}{c} + \quad \text{\textquotedbl} \quad \text{\textquotedbl}$$

$$p_{22} = \frac{1}{b} + \quad \text{\textquotedbl} \quad \text{\textquotedbl}$$

it is found that this force is

$$\frac{E_1 E_2}{c^2} + \text{terms in } \frac{1}{c^4}.$$

Thus, except for terms in c^{-4}, the force is the same as though the charges were collected at the centres of the spheres. Indeed, it is easy to go a stage further and prove that the result is true as far as c^{-4}. We shall, however, reserve a full discussion of the question for a later Chapter.

120. Let us write

$$\tfrac{1}{2}(p_{11}E_1{}^2 + 2p_{12}E_1 E_2 + \ldots) = W_e,$$
$$\tfrac{1}{2}(q_{11}V_1{}^2 + 2q_{12}V_1 V_2 + \ldots) = W_V.$$

Then W_e and W_V are each equal to the electrical energy $\tfrac{1}{2}\Sigma EV$, so that

$$W_e + W_V - \Sigma EV = 0 \quad\ldots\ldots\ldots\ldots\ldots\ldots\ldots(50).$$

In whatever way we change the values of

$$E_1, \quad E_2, \ldots, \quad V_1, \quad V_2, \ldots, \quad \xi_1, \quad \xi_2, \ldots,$$

equation (50) remains true. We may accordingly differentiate it, treating the expression on the left as a function of all the E's, V's and ξ's. Denoting the function on the left-hand of equation (50) by ϕ, the result of differentiation will be

$$\Sigma\frac{\partial\phi}{\partial E_1}\delta E_1 + \Sigma\frac{\partial\phi}{\partial V_1}\delta V_1 + \Sigma\frac{\partial\phi}{\partial \xi_1}\delta\xi_1 = 0.$$

Now $\qquad \dfrac{\partial\phi}{\partial E_1} = \dfrac{\partial W_e}{\partial E_1} - V_1 = 0$, by equation (40),

$$\frac{\partial\phi}{\partial V_1} = \frac{\partial W_V}{\partial V_1} - E_1 = 0, \quad \text{\textquotedbl} \quad \text{\textquotedbl} \quad (41),$$

so that we are left with $\qquad \Sigma\dfrac{\partial\phi}{\partial \xi_1}\delta\xi_1 = 0,$

and since this equation is true for all displacements and therefore for all values of $\delta\xi_1$, $\delta\xi_2$, ..., it follows that each coefficient must vanish separately. Thus $\dfrac{\partial\phi}{\partial\xi_1} = 0$, or

$$\frac{\partial W_e}{\partial\xi_1} + \frac{\partial W_V}{\partial\xi_1} = 0 \quad\ldots\ldots\ldots\ldots\ldots\ldots\ldots\ldots(51).$$

As we have seen, $-\dfrac{\partial W_e}{\partial\xi_1}$ is the mechanical force tending to increase ξ_1, and this has now been shewn to be equal to $\dfrac{\partial W_V}{\partial\xi_1}$, which is expression (49) with the sign reversed. Thus the mechanical force, whether the charges or the potentials are kept constant, is

$$\tfrac{1}{2}\left(\frac{\partial q_{11}}{\partial\xi_1}V_1^2 + 2\frac{\partial q_{12}}{\partial\xi_1}V_1V_2 + \ldots\right) \quad\ldots\ldots\ldots\ldots\ldots(52),$$

a form which is convenient when we know the potentials, but not the charges, of the system.

In making a small displacement of the system such that ξ_1 is changed into $\xi_1 + d\xi_1$, the mechanical work done is $\dfrac{\partial W_e}{\partial\xi_1}d\xi_1$. If the potentials are kept constant the increase in electrical energy is $\dfrac{\partial W_V}{\partial\xi_1}d\xi_1$. The difference of these expressions, namely

$$\left(\frac{\partial W_V}{\partial\xi_1} - \frac{\partial W_e}{\partial\xi_1}\right)d\xi_1,$$

represents energy supplied by the batteries. From equation (51), it appears that this expression is equal to $2\dfrac{\partial W_V}{\partial\xi_1}d\xi_1$, so that the batteries supply energy equal to twice the increase in the electrical energy of the system, and of this energy half goes to an increase of the final electrical energy, while half is expended as mechanical work in the motion of the conductors.

Introduction of a new conductor into the field.

121. When a new conductor is introduced into the field, the coefficients p_{11}, p_{12}, ..., q_{11}, q_{12}, ... are naturally altered.

Let us suppose the new conductor introduced in infinitesimal pieces, which are brought into the field uncharged and placed in position so that they are in every way in their final places except that electric communication is not established between the different pieces. So far no work has been done and the electrical energy of the field remains unaltered.

Now let electric communication be established between the different pieces, so that the whole structure becomes a single conductor. The separate pieces, originally at different potentials, are now brought to the same potential by the flow of electricity over the surface of the conductor. Electricity can only flow from places of higher to places of lower potential, so that electrical energy is lost in this flow. Thus the introduction of the new conductor has diminished the electric energy of the field.

If we now put the new conductor to earth there is in general a further flow of electricity, so that the energy is still further diminished.

Thus the electric energy of any field is diminished by the introduction of a new conductor, whether insulated or not.

Consider the case in which the new conductor remains insulated. Let the energy of the field before the introduction of the new conductor be

$$\tfrac{1}{2}\left(p_{11}E_1^2 + 2p_{12}E_1E_2 + \ldots + p_{nn}E_n^2\right) \ldots\ldots\ldots\ldots\ldots(53).$$

After introduction, the energy may be taken to be

$$\tfrac{1}{2}\left(p_{11}'E_1^2 + 2p_{12}'E_1E_2 + \ldots + p_{nn}'E_n^2\right) \ldots\ldots\ldots\ldots(54),$$

where p_{11}', etc., are the new coefficients of potential. Further coefficients of the type $p_{1,n+1}, p_{2,n+1}, \ldots, p_{n+1,n+1}$ are of course brought into existence, but do not enter into the expression for the energy, since by hypothesis $E_{n+1} = 0$.

Since expression (54) is less than expression (53), it follows that

$$(p_{11} - p_{11}')E_1^2 + 2(p_{12} - p_{12}')E_1E_2 + \ldots$$

is positive for all values of E_1, E_2, ... Hence $p_{11} - p_{11}'$ is positive, and other relations may be obtained, as in § 111.

ELECTROMETERS.

I.   *The Attracted Disc Electrometer.*

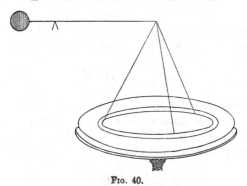

FIG. 40.

122. This instrument is, as regards its essential principle, a balance in which the beam has a weight fixed at one end and a disc suspended from the other. Under normal conditions the fixed weight is sufficiently heavy to outweigh the disc. In using the instrument the disc is made to become one plate of a parallel plate condenser, of which the second plate is adjusted until the electric attraction between the two plates of the condenser is just sufficient to restore the balance.

The inequalities in the distribution of the lines of force which would otherwise occur at the edges of the disc are avoided by the use of a guard-ring (§ 90), so arranged that when the beam of the balance is horizontal the guard-ring and disc are exactly in one plane, and fit as closely as is practicable.

Let us suppose that the disc is of area A and that the disc and guard-ring are raised to potential V. Let the second plate of the condenser be placed parallel to the disc at a distance h from it, and put to earth. Then the intensity between the disc and lower plate is uniform and equal to V/h, so that the surface density on the lower face of the disc is $\sigma = V/4\pi h$. The mechanical force acting on the disc is therefore a force $2\pi\sigma^2 A$ or $V^2 A/8\pi h^2$ acting vertically downwards through the centre of the disc. If this just suffices to keep the beam horizontal, it must be exactly equal to the weight, say W, which would have to be placed on this disc to maintain equilibrium if it were uncharged. This weight is a constant of the instrument, so that the equation

$$\frac{V^2 A}{8\pi h^2} = W$$

enables us to determine V in terms of known quantities by observing h. The instrument is arranged so that the lower plate can be moved parallel to itself by a micrometer screw, the reading of which gives h with great accuracy. We can accordingly determine V in absolute units, from the equation

$$V = h \sqrt{\frac{8\pi W}{A}}.$$

If we wish to determine a difference of potential we can raise the upper plate to one potential V_1, and the lower plate to the second potential V_2, and we then have

$$V_1 - V_2 = h \sqrt{\frac{8\pi W}{A}}.$$

A more accurate method of determining a difference of potential is to keep the disc at a constant potential v, and raise the lower plate successively to potentials V_1 and V_2. If h_1 and h_2 are the values of h which bring the disc to its standard position when the potentials of the lower plate are V_1 and V_2, we have

$$v - V_1 = h_1 \sqrt{\frac{8\pi W}{A}},$$

$$v - V_2 = h_2 \sqrt{\frac{8\pi W}{A}},$$

so that $$V_2 - V_1 = (h_1 - h_2)\sqrt{\frac{8\pi W}{A}}.$$

It is now only necessary to measure $h_1 - h_2$, the distance through which the lower plate is moved forward, and this can be determined with great accuracy, as it depends solely on the *motion* of the micrometer screw.

II. *The Quadrant Electrometer.*

123. *Measurement of Potential Difference.* This instrument is more delicate than the disc electrometer just described, but enables us only to *compare* two potentials, or potential differences; we cannot measure a single potential in terms of known units.

The principal part of the instrument consists of a metal cylinder of height small compared with its radius, divided into four quadrants A, B, C, D by two diameters at right angles. These quadrants are insulated separately, and then opposite quadrants are connected in pairs, two by wires joined to a point E and two by wires joined to some other point F.

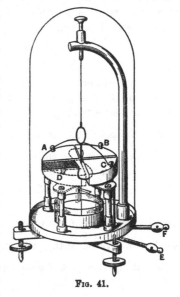

The inside of the cylinder is hollow and inside this a metal disc or " needle " is free to move, being suspended by a delicate fibre, so that it can rotate without touching the quadrants. Before using the instrument the needle is charged to a high potential, say v, either by means of the fibre, if this is a conductor, or by a small conducting

Fig. 41.

wire hanging from the needle which passes through the bottom of the cylinder. The fibre is adjusted so that when the quadrants are at the same potential the needle rests, as shewn in the figure, in a symmetrical position with respect to the quadrants. In this state either surface of the needle and the opposite faces of the quadrants may be regarded as forming a parallel plate condenser.

If, however, the potential of the two quadrants joined to E is different from that of the two quadrants joined to F, there is an electrical force tending to drag the needle under that pair of quadrants of which the potential differs most from v. The needle will accordingly move in this direction until the electric forces are in equilibrium with the torsion of the fibre, and an observation of the angle through which the needle turns will give an indication of the difference of potential between the two pairs of quadrants. This angle is most easily observed by attaching a small mirror to the fibre just above the point at which it emerges from the quadrants.

Let us suppose that when the needle has turned through an angle θ, the total area A of the needle is placed so that an area S is inside the pair of quadrants at potential V_1, and an area $A - S$ inside the pair at potential V_2. Let h be the perpendicular distance from either face of the needle to the faces of the quadrants. Then the system may be regarded as two parallel plate condensers of area S, distance h, and difference of potential $v - V_1$, and two parallel plate condensers for which these quantities have the values $A - S$, h, $v - V_2$. There are two condensers of each kind because there are two faces, upper and lower, to the needle. The electrical energy of this system is accordingly

$$\frac{(v - V_1)^2 S}{4\pi h} + \frac{(v - V_2)^2 (A - S)}{4\pi h}.$$

The energy here appears as a quadratic function of the three potentials concerned: it is expressed in the same form as the W_V of § 120. The mechanical force tending to increase θ, *i.e.*, the moment of the couple tending to turn the needle in the direction of θ increasing, is therefore $\dfrac{\partial W_V}{\partial \theta}$. Now in W_V the only term in the coefficients of the potentials which varies with θ is S, so that on differentiation we obtain

$$\frac{\partial W_V}{\partial \theta} = \frac{(v - V_1)^2 - (v - V_2)^2}{4\pi h} \frac{\partial S}{\partial \theta}.$$

If r is the radius of the needle—measured from its centre, which is under the line of division of the quadrants—we clearly have $\dfrac{\partial S}{\partial \theta} = r^2$, so that we can write the equation just obtained in the form

$$\frac{\partial W_V}{\partial \theta} = \frac{(2v - V_1 - V_2)(V_2 - V_1)}{4\pi h} r^2.$$

In equilibrium this couple is balanced by the torsion couple of the fibre, which tends to decrease θ. This couple may be taken to be $k\theta$, where k is a constant, so that the equation of equilibrium is

$$k\theta = \frac{(2v - V_1 - V_2)(V_2 - V_1) r^2}{4\pi h} \quad\ldots\ldots\ldots\ldots\ldots\ldots(55).$$

For small displacements of the needle, r^2 may be replaced by a^2, the radius of the needle at its centre line. Also v is generally large compared with V_1 and V_2. The last equation accordingly assumes the simpler form

$$k\theta = \frac{va^2}{2\pi h}(V_1 - V_2),$$

shewing that θ is, for small displacements of the needle, approximately proportional to the difference of potential of the two pairs of quadrants. The instrument can be made extraordinarily sensitive owing to the possibility of obtaining quartz-fibres for which the value of k is very small.

If the difference of potential to be measured is large, we may charge the needle simply by joining it to one of the pairs of quadrants, say the pair at potential V_2. We then have $v = V_2$, and equation (55) becomes

$$k\theta = \frac{(V_2 - V_1)^2 a^2}{4\pi h},$$

so that θ is now proportional to the *square* of the potential difference to be measured.

Writing $\dfrac{a^2}{2\pi h k} = C$, so that C is a constant of the instrument, we have, when v is large

$$\theta = Cv (V_1 - V_2) \dots\dots\dots\dots\dots\dots\dots(56),$$

when $v = V_2$,

$$\theta = \tfrac{1}{2}C (V_1 - V_2)^2 \dots\dots\dots\dots\dots\dots(57).$$

124. *Measurement of charge.* Let us speak of the pairs of quadrants at potentials V_1, V_2 as conductors 1, 2 respectively, and let the needle be conductor 3. When the quadrants are to earth and the needle is at potential V_3, the charge E induced on the first pair of quadrants by the charge on the needle will be given by

$$E = q_{13} V_3,$$

where q_{13} is the coefficient of induction. This coefficient is a function of the angle θ which defined the position of the needle. If the instrument is adjusted so that $\theta = 0$ when both pairs of quadrants are to earth, we must use the value of q_{13} corresponding to $\theta = 0$, say $(q_{13})_0$, so that

$$E = (q_{13})_0 V_3 \dots\dots\dots\dots\dots\dots\dots\dots(58).$$

Now suppose that the first pair of quadrants is insulated and receives an additional charge Q, the second pair being still to earth. Let the needle be deflected through an angle θ in consequence. Since the charge on the first pair of quadrants is now $E + Q$, we have

$$E + Q = (q_{11})_\theta V_1 + (q_{13})_\theta V_3.$$

On subtracting equation (58) from this we obtain

$$Q = (q_{11})_\theta V_1 + [(q_{13})_\theta - (q_{13})_0] V_3.$$

If θ is small this may be written

$$Q = q_{11} V_1 + \frac{\partial q_{13}}{\partial \theta} \theta V_3,$$

where q_{11}, $\dfrac{\partial q_{13}}{\partial \theta}$ are supposed calculated for $\theta = 0$. Since $V_2 = 0$, we have from equation (56),

$$\theta = CV_3 V_1,$$

so that

$$Q = \left(\frac{q_{11}}{CV_3} + \frac{\partial q_{13}}{\partial \theta} V_3\right)\theta,$$

shewing that for small values of θ, Q is directly proportional to θ.

Let us suppose that we join the first pair of quadrants (conductor 1) to a condenser of known capacity Γ which is entirely outside the electrometer. Since the needle (3) is entirely screened by the quadrants the value of q_{13} remains unaltered, while q_{11} will become $q_{11} + \Gamma$. If θ' is now the deflection of the needle, we have

$$Q = \left(\frac{q_{11} + \Gamma}{CV_3} + \frac{\partial q_{13}}{\partial \theta} V_3\right)\theta',$$

so that, by combination with the last equation, we have

$$Q\left(\frac{1}{\theta'} - \frac{1}{\theta}\right) = \frac{\Gamma}{CV_3}.$$

If θ'' is the deflection obtained by joining the pairs of quadrants to the terminals of a battery of known potential difference D, we have from equation (56),

$$CV_3 = \frac{\theta''}{D},$$

and on substituting this value for CV_3, our equation becomes

$$Q = \frac{\Gamma D}{\dfrac{\theta''}{\theta'} - \dfrac{\theta''}{\theta}},$$

giving Q in terms of the known quantities Γ, D and the three readings θ, θ' and θ''.

An ordinary quadrant electrometer will measure differences of potential down to about $\frac{1}{10000}$ electrostatic units. Thus in spite of its somewhat high capacity of about 50 electrostatic units, it forms an extremely efficient instrument for the measurement or detection of small electric charges.

An improved form of the instrument has recently been introduced by Dolazalek, in which the electrostatic capacity is very small. This is capable of measuring potential differences down to $\frac{1}{100000}$ electrostatic units, and is correspondingly more sensitive for the measurement of charges.

EXAMPLES.

1. If the algebraic sum of the charges on a system of conductors be positive, then on one at least the surface density is everywhere positive.

2. There are a number of insulated conductors in given fixed positions. The capacities of any two of them in their given positions are C_1 and C_2, and their mutual coefficient of induction is B. Prove that if these conductors be joined by a thin wire, the capacity of the combined conductor is

$$C_1 + C_2 + 2B.$$

3. A system of insulated conductors having been charged in any manner, charges are transferred from one conductor to another till they are all brought to the same potential V. Shew that

$$V = E/(s_1 + 2s_2),$$

where s_1, s_2 are the algebraic sums of the coefficients of capacity and induction respectively, and E is the sum of the charges.

4. Prove that the effect of the operation described in the last question is a decrease of the electrostatic energy equal to what would be the energy of the system if each of the original potentials were diminished by V.

5. Two equal similar condensers, each consisting of two spherical shells, radii a, b, are insulated and placed at a great distance r apart. Charges e, e' are given to the inner shells. If the outer surfaces are now joined by a wire, shew that the loss of energy is approximately

$$\tfrac{1}{4}(e - e')^2 \left(\frac{1}{b} - \frac{1}{r} \right).$$

6. A condenser is formed of two thin concentric spherical shells, radii a, b. A small hole exists in the outer sheet through which an insulated wire passes connecting the inner sheet with a third conductor of capacity c, at a great distance r from the condenser. The outer sheet of the condenser is put to earth, and the charge on the two connected conductors is E. Prove that approximately the force on the third conductor is

$$ac^2E^2 \Big/ \left(\frac{ab}{a-b} + c \right)^2 r^3.$$

7. Two closed equipotentials V_1, V_0 are such that V_1 contains V_0, and V_P is the potential at any point P between them. If now a charge E be put at P, and both equipotentials be replaced by conducting shells and earth-connected, then the charges E_1, E_0 induced on the two surfaces are given by

$$\frac{E_1}{V_0 - V_P} = \frac{E_0}{V_P - V_1} = \frac{E}{V_1 - V_0}.$$

8. A conductor is charged from an electrophorus by repeated contacts with a plate, which after each contact is recharged with a quantity E of electricity from the electrophorus. Prove that if e is the charge of the conductor after the first operation, the ultimate charge is

$$\frac{Ee}{E - e}.$$

9. Four equal uncharged insulated conductors are placed symmetrically at the corners of a regular tetrahedron, and are touched in turn by a moving spherical conductor at the points nearest to the centre of the tetrahedron, receiving charges e_1, e_2, e_3, e_4. Shew that the charges are in geometrical progression.

10. In question 9 replace "tetrahedron" by "square," and prove that

$$(e_1 - e_2)(e_1 e_3 - e_2{}^2) = e_1 (e_2 e_3 - e_1 e_4).$$

11. Shew that if the distance x between two conductors is so great as compared with the linear dimensions of either, that the square of the ratio of these linear dimensions to x may be neglected, then the coefficient of induction between them is $- CC'/x$, where C, C' are the capacities of the conductors when isolated.

12. Two insulated fixed condensers are at given potentials when alone in the electric field and charged with quantities E_1, E_2 of electricity. Their coefficients of potential are p_{11}, p_{12}, p_{22}. But if they are surrounded by a spherical conductor of very large radius R at potential zero with its centre near them, the two conductors require charges E_1', E_2' to produce the given potentials. Prove, neglecting $\dfrac{1}{R^2}$, that

$$\frac{E_1' - E_1}{E_2' - E_2} = \frac{p_{22} - p_{12}}{p_{11} - p_{12}}.$$

13. Shew that the locus of the positions, in which a unit charge will induce a given charge on a given uninsulated conductor, is an equipotential surface of that conductor supposed freely electrified.

14. Prove (i) that if a conductor, insulated in free space and raised to unit potential, produces at any external point P a potential denoted by (P), then a unit charge placed at P in the presence of this conductor uninsulated will induce on it a charge $-(P)$;

(ii) that if the potential at a point Q due to the induced charge be denoted by (PQ), then (PQ) is a symmetrical function of the positions of P and Q.

15. Two small uninsulated spheres are placed near together between two large parallel planes, one of which is charged, and the other connected to earth. Shew by figures the nature of the disturbance so produced in the uniform field, when the line of centres is (i) perpendicular, (ii) parallel to the planes.

16. A hollow conductor A is at zero potential, and contains in its cavity two other insulated conductors, B and C, which are mutually external: B has a positive charge, and C is uncharged. Analyse the different types of lines of force within the cavity which are possible, classifying with respect to the conductor from which the line starts, and the conductor at which it ends, and proving the impossibility of the geometrically possible types which are rejected.

Hence prove that B and C are at positive potentials, the potential of C being less than that of B.

17. A portion P of a conductor, the capacity of which is C, can be separated from the conductor. The capacity of this portion, when at a long distance from other bodies, is c. The conductor is insulated, and the part P when at a considerable distance from the remainder is charged with a quantity e and allowed to move under the mutual attraction up to it; describe and explain the changes which take place in the electrical energy of the system.

18. A conductor having a charge Q_1 is surrounded by a second conductor with charge Q_2. The inner is connected by a wire to a very distant uncharged conductor. It is then disconnected, and the outer conductor connected. Shew that the charges Q_1', Q_2', are now

$$Q_1' = \frac{mQ_1 - nQ_2}{m+n+mn}, \quad Q_2' = \frac{(m+n)Q_2 + mnQ_1'}{m+n},$$

where C, $C(1+m)$ are the coefficients of capacity of the near conductors, and Cn is the capacity of the distant one.

19. If one conductor contains all the others, and there are $n+1$ in all, shew that there are $n+1$ relations between either the coefficients of potential or the coefficients of induction, and if the potential of the largest be V_0, and that of the others V_1, V_2, ... V_n, then the most general expression for the energy is $\frac{1}{2}CV_0^2$ increased by a quadratic function of $V_1 - V_0$, $V_2 - V_0$, ... $V_n - V_0$; where C is a definite constant for all positions of the inner conductors.

20. The inner sphere of a spherical condenser (radii a, b) has a constant charge E, and the outer conductor is at potential zero. Under the internal forces the outer conductor contracts from radius b to radius b_1. Prove that the work done by the electric forces is

$$\tfrac{1}{2}E^2 \frac{b-b_1}{b_1 b}.$$

21. If, in the last question, the inner conductor has a constant potential V, its charge being variable, shew that the work done is

$$\tfrac{1}{2}\frac{V^2 a^2 (b-b_1)}{(b_1-a)(b-a)},$$

and investigate the quantity of energy supplied by the battery.

22. With the usual notation, prove that

$$p_{11} + p_{23} > p_{12} + p_{13}$$
$$p_{11}p_{23} > p_{12}p_{13}.$$

23. Shew that if p_{rr}, p_{rs}, p_{ss} be three coefficients before the introduction of a new conductor, and p_{rr}', p_{rs}', p_{ss}' the same coefficients afterwards, then

$$(p_{rr}p_{ss} - p_{rr}'p_{ss}') \not< (p_{rs} - p_{rs}')^2.$$

24. A system consists of $p+q+2$ conductors, $A_1, A_2, ... A_p, B_1, B_2, ... B_q, C, D$. Prove that when the charges on the A's and on C, and the potentials of the B's and of C are known, there cannot be more than one possible distribution in equilibrium, unless C is electrically screened from D.

25. A, B, C, D are four conductors, of which B surrounds A and D surrounds C.

Given the coefficients of capacity and induction

        (i) of A and B when C and D are removed,

        (ii) of C and D when A and B are removed,

        (iii) of B and D when A and C are removed,

determine those for the complete system of four conductors.

26. Two equal and similar conductors A and B are charged and placed symmetrically with regard to each other; a third moveable conductor C is carried so as to occupy

successively two positions, one practically wholly within A, the other within B, the positions being similar and such that the coefficients of potential of C in either position are p, q, r in ascending order of magnitude. In each position C is in turn connected with the conductor surrounding it, put to earth, and then insulated. Determine the charges on the conductors after any number of cycles of such operations, and shew that they ultimately lead to the ratios

$$1 : -\beta : \beta^2 - 1,$$

where β is the positive root of

$$rx^3 - qx + p - r = 0.$$

27. Two conductors are of capacities C_1 and C_2, when each is alone in the field. They are both in the field at potentials V_1 and V_2 respectively, at a great distance r apart. Prove that the repulsion between the conductors is

$$\frac{C_1 C_2 (r V_1 - C_2 V_2)(r V_2 - C_1 V_1)}{(r^3 - C_1 C_2)^2}.$$

As far as what power of $\dfrac{1}{r}$ is this result accurate?

28. Two equal and similar insulated conductors are placed symmetrically with regard to each other, one of them being uncharged. Another insulated conductor is made to touch them alternately in a symmetrical manner, beginning with the one which has a charge. If e_1, e_2 be their charges when it has touched each once, shew that their charges, when it has touched each r times, are respectively

$$\frac{e_1^2}{2e_1 - e_2}\left\{1 + \left(\frac{e_1 - e_2}{e_1}\right)^{2r-1}\right\} \text{ and } \frac{e_1^2}{2e_1 - e_2}\left\{1 - \left(\frac{e_1 - e_2}{e_1}\right)^{2r}\right\}.$$

29. Three conductors A_1, A_2 and A_3 are such that A_3 is practically inside A_2. A_1 is alternately connected with A_2 and A_3 by means of a fine wire, the first contact being with A_3. A_1 has a charge E initially, A_2 and A_3 being uncharged. Prove that the charge on A_1 after it has been connected n times with A_2 is

$$\frac{E\beta}{a+\beta}\left\{1 + \frac{a(\gamma - \beta)}{\beta(a+\gamma)}\left(\frac{a+\beta}{a+\gamma}\right)^{n-1}\right\},$$

where a, β, γ stand for $p_{11} - p_{12}$, $p_{22} - p_{12}$ and $p_{33} - p_{12}$ respectively.

30. Two spheres, radii a, b, have their centres at a distance c apart. Shew that neglecting $(a/c)^6$ and $(b/c)^6$,

$$p_{11} = \frac{1}{a} - \frac{b^3}{c^4}; \quad p_{12} = \frac{1}{c}; \quad p_{22} = \frac{1}{b} - \frac{a^3}{c^4}.$$

CHAPTER V

DIELECTRICS AND INDUCTIVE CAPACITY

125. MENTION has already been made (§ 84) of the fact, discovered originally by Cavendish, and afterwards rediscovered by Faraday, that the capacity of a conductor depends on the nature of the dielectric substance between its plates.

Let us imagine that we have two parallel plate condensers, similar in all respects except that one has nothing but air between its plates while in the other this space is filled with a dielectric of inductive capacity K. Let us suppose that the two high-potential plates are connected by a wire, and also the two low-potential plates. Let the condensers be charged, the potential of the high-potential plates being V_1, and that of the low-potential plates being V_0.

Then it is found that the charges possessed by the two condensers are not equal. The capacity per unit area of the air-condenser is $1/4\pi d$; that of the other condenser is found to be $K/4\pi d$. Hence the charges per unit area of the two condensers are respectively

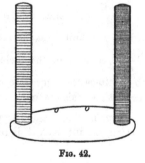

$$\frac{V_1 - V_0}{4\pi d} \text{ and } K\,\frac{V_1 - V_0}{4\pi d}.$$

The work done in taking unit charge from the low-potential plate to the high-potential plate is the same in either condenser, namely $V_1 - V_0$, so that the intensity between the plates in either condenser is the same, namely

$$\frac{V_1 - V_0}{d}.$$

Fig. 42.

In the air-condenser this intensity may be regarded as the resultant of the attraction of the negatively charged plate and the repulsion of the positively charged plate, the law of attraction or repulsion being Coulomb's law $\dfrac{e}{r^2}$.

It is, however, obvious that if we were to calculate the intensity in the second condenser from this law, then the value obtained would be K times that in the first condenser, and would therefore be $K\dfrac{V_1 - V_0}{d}$. In point of fact, the actual value of the intensity is known to be $\dfrac{V_1 - V_0}{d}$.

Thus Faraday's discovery shews that Coulomb's law of force is not of universal validity: the law has only been proved experimentally for air, and it is now found not to be true for dielectrics of which the inductive capacity is different from unity.

This discovery has far-reaching effects on the development of the mathematical theory of electricity. In the present book, Coulomb's law was introduced in § 38, and formed the basis of all subsequent investigations. Thus every theorem which has been proved in the present book from § 38 onwards requires reconsideration.

126. We shall follow Faraday in treating the whole subject from the point of view of lines of force. The conceptions of potential, of intensity, and of lines of force are entirely independent of Coulomb's law, and in the present book have been discussed (§§ 30—37) before the law was introduced. The conception of a tube of force follows at once from that of a line of force, on imagining lines of force drawn through the different points on a small closed curve. Let us extend to dielectrics one form of the definition of the strength of a tube of force which has already been used for a tube in air, and agree that the strength of a tube is to be measured by the charge enclosed by its positive end, whether in air or dielectric.

In the dielectric condenser, the surface density on the positive plate is $K\dfrac{V_1 - V_0}{4\pi d}$, and this, by definition, is also the aggregate strength of the tubes per unit area of cross-section. The intensity in the dielectric is $\dfrac{V_1 - V_0}{d}$, so that in the dielectric the intensity is no longer, as in air, equal to 4π times the aggregate strength of tubes per unit area, but is equal to $4\pi/K$ times this amount.

Thus if P is the aggregate strength of the tubes per unit area of cross-section, the intensity R is related to P by the equation

$$R = \frac{4\pi}{K}P \quad \dotfill (59)$$

in the dielectric, instead of by the equation

$$R = 4\pi P \quad \dotfill (60)$$

which was found to hold in air.

127. Equation (59) has been proved to be the appropriate generalisation of equation (60) only in a very special case. Faraday, however, believed the relation expressed by equation (59) to be universally true, and the results obtained on this supposition are found to be in complete agreement with experiment. Hence equation (59), or some equation of the same significance, is universally taken as the basis of the mathematical theory of dielectrics. We accordingly proceed by assuming the universal truth of equation (59), an assumption for which a justification will be found when we come to study the molecular constitution of dielectrics.

It is convenient to have a single word to express the aggregate strength of tubes per unit area of cross-section, the quantity which has been denoted by P. We shall speak of this quantity as the "polarisation," a term due to Faraday. Maxwell's explanation of the meaning of the term "polarisation" is that "an elementary portion of a body may be said to be polarised when it acquires equal and opposite properties on two opposite sides." Faraday explained the properties of dielectrics by means of his conception that the molecules of the dielectric were in a polarised state, and the quantity P is found to measure the amount of the polarisation at any point in the dielectric. We shall come to this physical interpretation of the quantity P at a later stage: for the present we simply use the term "polarisation" as a name for the mathematical quantity P.

This same quantity is called the "displacement" by Maxwell, and underlying the use of this term also, there is a physical interpretation which we shall come upon later.

128. We now have as the basis of our mathematical theory the following:

DEFINITION. *The strength of a tube of force is defined to be the charge enclosed by the positive end of the tube.*

DEFINITION. *The polarisation at any point is defined to be the aggregate strength of tubes of force per unit area of cross-section.*

EXPERIMENTAL LAW. *The intensity at any point is $4\pi/K$ times the polarisation, where K is the inductive capacity of the dielectric at the point.*

In this last relation, we measure the intensity along a line of force, while the polarisation is measured by considering the flux of tubes of force across a small area perpendicular to the lines of force. Suppose, however, that we take some direction OO' making an angle θ with that of the lines of force. The aggregate strength of the tubes of force which cross an area dS perpendicular to OO' will be $P\cos\theta\,dS$, for these tubes are exactly those which cross an area $dS\cos\theta$ perpendicular to the lines of force. Thus, consistently with the definition of polarisation, we may say that the polarisation in the direction OO' is equal to $P\cos\theta$. Since the polarisation in any direction is equal to P multiplied by the cosine of the angle between this direction and that of the lines of force, it is clear that the polarisation may be regarded as a vector, of which the direction is that of the lines of force, and of which the magnitude is P.

The polarisation having been seen to be a vector, we may speak of its components f, g, h. Clearly f is the number of tubes per unit area which cross a plane perpendicular to the axis of x, and so on.

The result just obtained may be expressed analytically by the equations

$$f = \frac{K}{4\pi} X, \qquad g = \frac{K}{4\pi} Y, \qquad h = \frac{K}{4\pi} Z.$$

129. The polarisation P being measured by the aggregate strength of tubes per unit area of cross-section, it follows that if ω is the cross-section at any point of a tube of strength ϵ, we have $\epsilon = \omega P$. Now we have defined the strength of a tube of force as being equal to the charge at its positive end, so that by definition the strength ϵ of a tube does not vary from point to point of the tube. Thus the product ωP is constant along a tube, or $\omega K R$ is constant along a tube, replacing the result that ωR is constant in air (§ 56).

The value of the product ωP at any point O of a tube, being equal to $\dfrac{\omega K R}{4\pi}$, depends only on the physical conditions prevailing at the point O. It is, however, known to be equal to the charge at the positive end of the tube. Hence it must also, from symmetry, be equal to minus the charge at the negative end of the tube. Thus the charges at the two ends of a tube, whether in the same or in different dielectrics, will be equal and opposite, and the numerical value of either is the strength of the tube.

GAUSS' THEOREM.

130. Let S be any closed surface, and let ϵ be the angle between the direction of the outward normal to any element of surface dS and the direction of the lines of force at the element. The aggregate strength of the tubes of force which cross the element of area dS is $P \cos \epsilon\, dS$, and the integral

$$\iint P \cos \epsilon\, dS,$$

which may be called the surface integral of normal polarisation, will measure the aggregate strength of all the tubes which cross the surface S, the strength of a tube being estimated as positive when it crosses the surface from inside to outside, and as negative when it crosses in the reverse direction.

A tube which enters the surface from outside, and which, after crossing the space enclosed by the surface, leaves it again, will add no contribution to $\iint P \cos \epsilon \, dS$, its strength being counted negatively where it enters the surface, and positively where it emerges. A tube which starts from or ends on a charge e inside the surface S will, however, supply a contribution to $\iint P \cos \epsilon \, dS$ on crossing the surface. If e is positive, the strength of the tube is e; and, as it crosses from inside to outside, it is counted positively, and the contribution to the integral is e. Again, if e is negative, the strength of the tube is $-e$, and this is counted negatively, so that the contribution is again e.

Thus on summing for all tubes,

$$\iint P \cos \epsilon \, dS = E,$$

where E is the total charge inside the surface. The left-hand member is simply the algebraical sum of the strengths of the tubes which begin or end inside the surface; the right-hand member is the algebraical sum of the charges on which these tubes begin or end. Putting

$$P = \frac{K}{4\pi} R$$

the equation becomes $\iint KR \cos \epsilon \, dS = 4\pi E.$

The quantity $R \cos \epsilon$ is, however, the component of intensity along the outward normal, the quantity which has been previously denoted by N, so that we arrive at the equation

$$\iint KN \, dS = 4\pi E \quad \dots\dots\dots\dots\dots\dots(61).$$

When the dielectric was air, Gauss' theorem was obtained in the form

$$\iint N \, dS = 4\pi E.$$

Equation (61) is therefore the generalised form of Gauss' Theorem which must be used when the inductive capacity is different from unity. Since $N = -\dfrac{\partial V}{\partial n}$, the equation may be written in the form

$$\iint K \frac{\partial V}{\partial n} \, dS = -4\pi E.$$

131. The form of this equation shews at once that a great many results which have been shewn to be true for air are true also for dielectrics other than air.

It is obvious, for instance, that V cannot be a maximum or a minimum at a point in a dielectric which is not occupied by an electric charge: as a consequence all lines of force must begin and end on charged bodies, a result which was tacitly assumed in defining the strength of a tube of force.

A number of theorems were obtained in the discussion of the electrostatic field in air, by taking a Gauss' Surface, partly in air and partly in a conductor. Gauss' Theorem was used in the form

$$\iint N\,dS = 4\pi E,$$

but we now see that if the inductive capacity of the conductor were not equal to unity, this equation ought to be replaced by equation (61). It is, however, clear that the difference cannot affect the final result; N is zero inside a conductor, so that it does not matter whether N is multiplied by K or not.

Thus results obtained for systems of conductors in air upon the assumption that Coulomb's law of force holds throughout the field are seen to be true whether the inductive capacity inside the conductors is equal to unity or not.

The Equations of Poisson and Laplace.

132. In § 49, we applied Gauss' theorem to a surface which was formed by a small rectangular parallelepiped, of edges dx, dy, dz, parallel to the axes of coordinates. If we apply the theorem expressed by equation (61) to the same element of volume, we obtain

$$\frac{\partial}{\partial x}\left(K\,\frac{\partial V}{\partial x}\right) + \frac{\partial}{\partial y}\left(K\,\frac{\partial V}{\partial y}\right) + \frac{\partial}{\partial z}\left(K\,\frac{\partial V}{\partial z}\right) = -4\pi\rho \ \ldots\ldots\ldots(62),$$

where ρ is the volume density of electrification. This, then, is the generalised form of Poisson's equation: the generalised form of Laplace's equation is obtained at once on putting $\rho = 0$.

In terms of the components of polarisation, equation (62) may be written

$$\frac{\partial f}{\partial x} + \frac{\partial g}{\partial y} + \frac{\partial h}{\partial z} = \rho \ \ldots\ldots\ldots\ldots\ldots\ldots\ldots(63),$$

while if the dielectric is uncharged,

$$\frac{\partial f}{\partial x} + \frac{\partial g}{\partial y} + \frac{\partial h}{\partial z} = 0 \ \ldots\ldots\ldots\ldots\ldots\ldots\ldots(64).$$

Electric Charges in an infinite homogeneous Dielectric.

133. Consider a charge e placed by itself in an infinite dielectric. If the dielectric is homogeneous, it follows from considerations of symmetry that the lines of force must be radial, as they would be in air. By application of equation (61) to a sphere of radius r, having the point charge as centre, it is found that the intensity at a distance r from the charge is

$$\frac{e}{Kr^2}.$$

The force between two point charges e, e', at distance r apart in a homogeneous unbounded dielectric is therefore

$$\frac{e\,e'}{Kr^2} \quad\dots\dots\dots\dots\dots\dots\dots\dots\dots\dots\dots(65),$$

and the potential of any number of charges, obtained by integration of this expression, is

$$V = \frac{1}{K} \Sigma \frac{e}{r} \quad\dots\dots\dots\dots\dots\dots\dots\dots\dots(66).$$

Coulomb's Equation.

134. The strength of a tube being measured by the charge at its end, it follows that at a point just outside a conductor, P, the aggregate strength of the tubes per unit of cross-section, becomes numerically equal to σ, the surface density. We have also the general relation

$$R = \frac{4\pi}{K}\,P,$$

and on replacing P by σ, we arrive at the generalised form of Coulomb's equation,

$$R = \frac{4\pi\sigma}{K} \quad\dots\dots\dots\dots\dots\dots\dots\dots\dots\dots\dots(67),$$

in which K is the inductive capacity at the point under consideration.

Conditions to be satisfied at the Boundary of a Dielectric.

135. Let us examine the conditions which will obtain at a boundary at which the inductive capacity changes abruptly from K_1 to K_2.

The potential must be continuous in crossing the boundary, for if P, Q, are two infinitely near points on opposite sides of the boundary, the work done in bringing a small charge to P must be the same as that done in bringing it to Q. As a consequence of the potential being continuous, it follows that the tangential components of the intensity must also be continuous. For if P, Q are two very near points on different sides of the boundary, and P', Q' a similar pair of points at a small distance away, we have $V_P = V_Q$, and $V_{P'} = V_{Q'}$, so that

$$\frac{V_P - V_{P'}}{PP'} = \frac{V_Q - V_{Q'}}{QQ'}.$$

The expressions on the two sides of this equation are, however, the two intensities in the direction PP', on the two sides of the boundary, which establishes the result.

Also, if there is no charge on the boundary, the aggregate strength of the tubes which meet the boundary in any small area on this boundary is the same whether estimated in the one dielectric or the other, for the tubes do not alter their strength in crossing the boundary, and none can begin or end in the boundary. Thus the normal component of the polarisation is continuous.

136. If R_1 is the intensity in the first medium of inductive capacity K_1, measured at a point close to the boundary, and if ϵ_1 is the angle which the lines of force make with the normal to the boundary at this point, then the normal polarisation in the first medium is

$$\frac{K_1}{4\pi} R_1 \cos \epsilon_1.$$

Similarly, that in the second medium is

$$\frac{K_2}{4\pi} R_2 \cos \epsilon_2,$$

so that
$$K_1 R_1 \cos \epsilon_1 = K_2 R_2 \cos \epsilon_2 \quad \ldots\ldots\ldots\ldots\ldots\ldots(68).$$

Since, in the notation already used,

$$R_1 \cos \epsilon_1 = N_1 = -\frac{\partial V_1}{\partial n},$$

the equation just obtained may be put in either of the forms

$$K_1 N_1 = K_2 N_2 \quad \ldots\ldots\ldots\ldots\ldots\ldots(69),$$

$$K_1 \frac{\partial V_1}{\partial n} = K_2 \frac{\partial V_2}{\partial n} \ldots\ldots\ldots\ldots\ldots\ldots(70).$$

In these equations, it is a matter of indifference whether the normal is drawn from the first medium to the second or in the reverse direction; it is only necessary that the same normal should be taken on both sides of the equation. Relation (70) is obtained at once on applying the generalised form of Gauss' theorem to a small cylinder having parallel ends at infinitesimal distance apart, one in each medium.

137. To sum up, we have found that in passing from one dielectric to another, the surface of separation being uncharged:

(i) *the tangential components of intensity have the same values on the two sides of the boundary,*

(ii) *the normal components of polarisation have the same values.*

Or, in terms of the potential,

(i) *V is continuous,*

(ii) $K \dfrac{\partial V}{\partial n}$ *is continuous.*

Refraction of the lines of force.

138. From the continuity of the tangential components of intensity, it follows:

(i)   that the directions of R_1 and R_2, the intensities on the two sides of the boundary, must lie in a plane containing the normal, and

(ii)   that $R_1 \sin \epsilon_1 = R_2 \sin \epsilon_2$.

Combining the last relation with equation (68), we obtain

$$K_1 \cot \epsilon_1 = K_2 \cot \epsilon_2 \quad \ldots\ldots\ldots\ldots\ldots\ldots\ldots\ldots(71).$$

From this relation, it appears that if K_1 is greater than K_2, then ϵ_1 is greater than ϵ_2, and *vice versa*. Thus in passing from a smaller value of K to a greater value of K, the lines are bent away from the normal. In illustration of this, fig. 43 shews the arrangement of lines of force when a point charge is placed in front of an infinite slab of dielectric ($K = 7$).

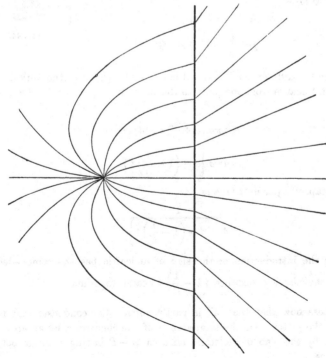

Fig. 43.

A small charged particle placed at any point of this field will experience a force of which the direction is along the tangent to the line of force through the point. The force is produced by the point charge, but its direction will not in general pass through the point charge. Thus we conclude that in a field in which the inductive capacity is not uniform the force between two point charges does not in general act along the line joining them.

139. As an example of the action of a dielectric let us imagine a parallel plate condenser in which a slab of dielectric of thickness t is placed between the plates, its two faces being parallel to the plates and at distances a, b from them, so that $a + b + t = d$, where d is the distance between the plates.

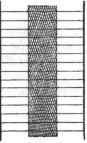

It is obvious from symmetry that the lines of force are straight throughout their path, equation (71) being satisfied by $\epsilon_1 = \epsilon_2 = 0$.

Let σ be the charge per unit area, so that the polarisation is equal to σ everywhere. The intensity, by equation (67), is

$$R = 4\pi\sigma \text{ in air,}$$

and
$$R = \frac{4\pi}{K}\sigma \text{ in dielectric.}$$

FIG. 44.

Hence the difference of potential between the plates, or the work done in taking unit charge from one plate to the other in opposition to the electric intensity,

$$= 4\pi\sigma \cdot a + \frac{4\pi}{K}\sigma \cdot t + 4\pi\sigma \cdot b$$

$$= 4\pi\sigma \left\{ d - \left(1 - \frac{1}{K}\right)t \right\},$$

and the capacity per unit area is

$$\frac{1}{4\pi\left\{ d - \left(1 - \frac{1}{K}\right)t \right\}}.$$

Thus the introduction of the slab of dielectric has the same effect as moving the plates a distance $\left(1 - \frac{1}{K}\right)t$ nearer together.

Suppose now that the slab is partly outside the condenser and partly between the plates. Of the total area A of the condenser, let an area B be occupied by the slab of dielectric, an area $A - B$ having only air between the plates.

The lines of force will be straight, except for those which pass near to the edge of the dielectric slab. Neglecting a small correction required by the curvature of these lines, the capacity C of the condenser is given by

$$C = \frac{B}{4\pi\left\{d - \left(1 - \dfrac{1}{K}\right)t\right\}} + \frac{A - B}{4\pi d}$$

$$= \frac{A}{4\pi d} + \frac{B\left(1 - \dfrac{1}{K}\right)t}{4\pi d\left\{d - \left(1 - \dfrac{1}{K}\right)t\right\}}$$

a quantity which increases as B increases. If V is the potential difference and E the charge, the electrical energy

$$= \tfrac{1}{2}CV^2 = \tfrac{1}{2}\frac{E^2}{C}.$$

If we keep the charge constant, the electrical energy increases as the slab is withdrawn. There must therefore be a mechanical force tending to resist withdrawal: the slab of dielectric will be sucked in between the plates of the condenser. This, as will be seen later, is a particular case of a general theorem that any piece of dielectric is acted on by forces which tend to drag it from the weaker to the stronger parts of an electric field of force.

Charge on the Surface of a Dielectric.

140. Let dS be any small area of a surface which separates two media of inductive capacities K_1, K_2, and let this bounding surface have a charge of electricity, the surface density over dS being σ. If we apply Gauss' Theorem to a small cylinder circumscribing dS we obtain

$$K_1 \frac{\partial V_1}{\partial \nu_1} + K_2 \frac{\partial V_2}{\partial \nu_2} = -4\pi\sigma \quad\ldots\ldots\ldots\ldots(72),$$

where $\dfrac{\partial}{\partial \nu}$ in either medium denotes differentiation with respect to the normal drawn away from dS into the dielectric.

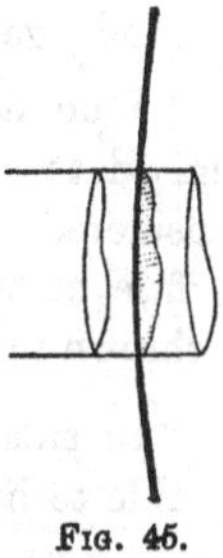

141. As we have seen, the surface of a dielectric may be charged by friction. A more interesting way is by utilising the conducting powers of a flame.

Fig. 46.

Let us place a charge e in front of a slab of dielectric as in fig. 43. A flame issuing from a metal lamp held in the hand may be regarded as a conductor at potential zero. On allowing the flame to play over the surface of the dielectric, this surface is reduced to potential zero, and the distribution of the lines of force is now exactly the same as if the face of the dielectric were replaced by a conducting plane at potential zero. The lines of force from the point charge terminate on this plane, so that there must be a total charge $-e$ spread over it. If the plane were actually a conductor this would be simply an induced charge. If, however, the plane is the boundary of a dielectric, the charge differs from an induced charge on a conductor in that it cannot disappear if the original charge e is removed. For this reason, Faraday described it as a "bound" charge. The charge has of course come to the dielectric through the conducting flame.

MOLECULAR ACTION IN A DIELECTRIC.

142. From the observed influence of the structure of a dielectric upon the electric phenomena occurring in a field in which it was placed, Faraday was led to suppose that the particles of the dielectric themselves took part in this electric action. After describing his researches on the electric action—"induction" to use his own term—in a space occupied by dielectric he says[*]:

"Thus induction appears to be essentially an action of contiguous particles, through the intermediation of which the electric force, originating or appearing at a certain place, is propagated to or sustained at a distance...."

"Induction appears to consist in a certain polarised state of the particles, into which they are thrown by the electrified body sustaining the action, the particles assuming positive and negative points or parts...."

"With respect to the term *polarity*..., I mean at present...a disposition of force by which the same molecule acquires opposite powers on different parts."

And again, later[†],

"I do not consider the powers when developed by the polarisation as limited to two distinct points or spots on the surface of each particle to be considered as the poles of an axis, but as resident on large portions of that surface, as they are upon the surface of a conductor of sensible size when it is thrown into a polar state."

"In such solid bodies as glass, lac, sulphur, etc., the particles appear to be able to become polarised in all directions, for a mass when experimented upon so as to ascertain its inductive capacity in three or more directions, gives no indication of a difference. Now, as the particles are fixed in the mass, and as the direction of the induction through them must change with its charge relative to the mass, the constant effect indicates that they can be polarised electrically in any direction."

---

[*] *Experimental Researches*, 1295, 1298, 1304. (Nov. 1837.)
[†] *Experimental Researches*, 1686, 1688, 1679. (June, 1838.)

"The particles of an insulating dielectric whilst under induction may be compared...to a series of small insulated conductors. If the space round a charged globe were filled with a mixture of an insulating dielectric and small globular conductors, the latter being at a little distance from each other, so as to be insulated, then these would in their condition and action exactly resemble what I consider to be the condition and action of the particles of the insulating dielectric itself. If the globe were charged, these little conductors would all be polar; if the globe were discharged, they would all return to their normal state, to be polarised again upon the recharging of the globe...."

As regards the question of what actually the particles are which undergo this polarisation, Faraday says * :

"An important inquiry regarding the electric polarity of the particles of an insulating dielectric, is, whether it be the molecules of the particular substance acted on, or the component or ultimate particles, which thus act the part of insulated conducting polarising portions."

"The conclusion I have arrived at is, that it is the molecules of the substance which polarise as wholes; and that however complicated the composition of a body may be, all those particles or atoms which are held together by chemical affinity to form one molecule of the resulting body act as one conducting mass or particle when inductive phenomena and polarisation are produced in the substance of which it is a part."

143. A mathematical discussion of the action of a dielectric constructed as imagined by Faraday, has been given by Mossotti, who utilised a mathematical method which had been developed by Poisson for the examination of a similar question in magnetism. For this discussion the molecules are represented provisionally as conductors of electricity.

To obtain a first idea of the effect of an electric field on a dielectric of the kind pictured by Faraday, let us consider a parallel plate condenser,

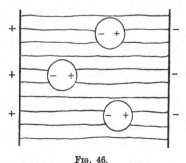

Fig. 46.

* *Experimental Researches*, 1699, 1700.

having a number of insulated uncharged conducting molecules in the space between the plates. Imagine a tube of strength ϵ meeting a molecule. At the point where this occurs, the tube terminates by meeting a conductor, so that there must be a charge $-\epsilon$ on the surface of the molecule. Since the total charge on the molecule is *nil* there must be a corresponding charge on the opposite surface, and this charge may be regarded as a point of restarting of the tube. The tube then may be supposed to be continually stopped and restarted by molecules as it crosses from one plate of the condenser to the other. At each encounter with a molecule there are induced charges $-\epsilon, +\epsilon$ on the surface of the molecule. Any such pair of charges, being at only a small distance apart, may be regarded as forming a small doublet, of the kind of which the field of force was investigated in § 64.

144. We have now replaced the dielectric by a series of conductors, the medium between which may be supposed to be air or ether. In the space between these conductors the law of force will be that of the inverse square. In calculating the intensity at any point from this law we have to reckon the forces from the doublets as well as the forces from the original charges on the condenser-plates. A glance at fig. 46 will shew that the forces from the doublets act in opposition to the original forces. Thus for given charges on the condenser-plates the intensity at any point between the plates is *lessened* by the presence of conducting molecules.

This general result can be seen at once from the theorem of § 121. The introduction of new conductors (the molecules) lessens the energy corresponding to given charges on the plates, *i.e.* increases the capacity of the condenser, and so lessens the intensity between the plates.

145. In calculating that part of the intensity which arises from the doublets, it will be convenient to divide the dielectric into concentric spherical shells having as centre the point at which the intensity is required. The volume of the shell of radii r and $r + dr$ is $4\pi r^2 dr$, so that the number of doublets included in it will contain $r^2 dr$ as a factor. The potential produced by any doublet at a point distant r from it is $\dfrac{\mu \cos \theta}{r^2}$, so that the intensity will contain a factor $\dfrac{1}{r^3}$. Thus the intensity arising from all the doublets in the shell of radii r, $r + dr$ will depend on r through the factor $\dfrac{1}{r^3} \cdot r^2 dr$ or $\dfrac{dr}{r}$

The importance of the different shells is accordingly the same, as regards comparative orders of magnitude, as that of the corresponding contributions to the integral $\int \dfrac{dr}{r}$. The value of this integral is $\log r + $ a constant, and this is infinite when $r = 0$ and when $r = \infty$. Thus the important contributions come from very small and very large values of r. It can however be seen that the contributions from large values of r neutralise one another, for the term $\cos \theta$ in the potentials of the different doublets will be just as often positive as negative.

Hence it is necessary only to consider the contributions from shells for which r is very small, so that the whole field at any point may be regarded as arising entirely from the doublets in the immediate neighbourhood of the point. The force will obviously vary as we move in and out amongst the molecules, depending largely on the nearness and position of the nearest molecules. If, however, we average this force throughout a small volume, we shall obtain an average intensity of the field produced by the doublets, and this will depend only on the strength and number of the doublets in and near to this element of volume. Obviously this average intensity near any point will be exactly proportional to the average strength of the doublets near the point, and this again will be exactly proportional to the strength of the inducing field by which the doublets are produced, so that at any point we may say that the average field of the doublets stands to the total field in a ratio which depends only on the structure of the medium at the point.

146. Now suppose that our measurements are not sufficiently refined to enable us to take account of the rapid changes of intensity of the electric field which must occur within small distances of molecular order of magnitude. Let us suppose, as we legitimately may, that the forces which we measure are forces averaged through a distance which contains a great number of molecules. Then the force which we measure will consist of the sum of the average force produced by the doublets, and of the force produced by the external field. The field which we observe may accordingly be regarded as the superposition of two fields, or what amounts to the same thing, the observed intensity R may be regarded as the resultant of two intensities R_1, R_2, where

R_1 is the average intensity arising from the neighbouring doublets,

R_2 is the intensity due to the charges outside the dielectric, and to the distant doublets in the dielectric.

These forces, as we have seen, must be proportional to one another, so that each must be proportional to the polarisation P. It follows that P is proportional to R, the ratio depending only on the structure of the medium at the point. If we take the relation to be

$$R = \frac{4\pi}{K} P \dots\dots\dots\dots\dots\dots(73),$$

then K is the inductive capacity at the point, and the relation between R and P is exactly the relation upon which our whole theory has been based.

147.   The theory could accordingly be based on Mossotti's theory, instead of on Faraday's assumption, and from the hypothesis of molecular polarisation we should be able to deduce all the results of the theory, by first deducing equation (73) from Mossotti's hypothesis, and then the required results from equation (73) in the way in which they have been deduced in the present chapter.

Thus the influence of the conducting molecules produces physically the same result as if the properties of the medium were altered in the way suggested by Faraday, and mathematically the properties of the medium are in either case represented by the presence of the factor K in equation (73).

Relation between Inductive Capacity and Structure of Medium.

148.   The electrostatic unit of force was defined in such a way that the inductive capacity of air was taken as unity.   It is now obvious that it would have been more scientific to take empty space as standard medium, so that the inductive capacity of every medium would have been greater than unity. Unfortunately, the practice of referring all inductive capacities to air as standard has become too firmly established for this to be possible.   The difference between the two standards is very slight, the inductive capacity of normal air in terms of empty space being $1\cdot00059$.   Thus the inductive capacity of a vacuum may be taken to be $\cdot99941$ referred to air.

So long as the molecules are at distances apart which are great compared with their linear dimensions, we may neglect the interaction of the charges induced on the different molecules, and treat their effects as additive.   It follows that in a gas $K - K_0$, where K_0 is the inductive capacity of free ether, ought to be proportional to the density of the gas.   This law is found to be in exact agreement with experiment*.

149.   It is, however, possible to go further and calculate the actual value of the ratio of $K - K_0$ to the density.   We have seen that this will be a constant for a given substance, so that we shall determine its value in the simplest case: we shall consider a thin slab of the dielectric placed in a parallel plate condenser, as described in § 139.   Let this slab be of thickness ϵ, and let it coincide with the plane of yz.   Let the dielectric contain n molecules per unit volume.

The element $dy\,dz$ will contain $n\epsilon\,dy\,dz$ molecules.   If each of these is a doublet of strength μ, the element $dy\,dz$ will have a field which will be equivalent at all distant points to that of a single doublet of strength $n\mu\epsilon\,dy\,dz$.   This is exactly the field which would be produced if the two faces of the slab were charged with electricity of surface density $\pm\,n\mu$.

---

* Boltzmann, *Wiener Sitzungsber.* 69, p. 812.

We can accordingly at once find the field produced by these doublets—it is the same as that of a parallel plate condenser, in which the plates are at distance ϵ apart and are charged to surface density $\pm n\mu$. There is no intensity except between the plates, and here the intensity of the field is $4\pi n\mu$.

Thus if R is the total intensity outside the slab, that inside will be $R - 4\pi n\mu$. If K is the inductive capacity of the material of the slab, and K_0 that of the free ether outside the slab, we have

$$K_0 R = K (R - 4\pi n\mu),$$

so that
$$\frac{K - K_0}{K} = \frac{4\pi n\mu}{R} \quad\dots\dots\dots\dots\dots\dots(74).$$

It remains to determine the ratio μ/R. The potential of a doublet is $\frac{\mu x}{r^3}$ while that of the field R may be taken to be $- Rx + C$. Thus the total potential of a single doublet and the external field is

$$\frac{\mu x}{r^3} - Rx + C,$$

and this makes the surface $r = a$ an equipotential if $\dfrac{\mu}{a^3} = R$. Thus the surfaces of the molecules will be equipotentials if we imagine the molecules to be spheres of radius a, and the centres of the doublets to coincide with the centres of the spheres, the strength of each doublet being Ra^3.

Putting $\mu = Ra^3$, equation (74) becomes*

$$\frac{K - K_0}{K} = 4\pi n a^3.$$

Now in unit volume of dielectric, the space occupied by the n molecules is $\dfrac{4\pi}{3} na^3$. Calling this quantity θ, we have $\dfrac{K - K_0}{K} = 3\theta$, or, since our calculations only hold on the hypothesis that θ is small,

$$\frac{K}{K_0} = 1 + 3\theta \quad\dots\dots\dots\dots\dots\dots(75).$$

If the lines of force went straight across from one plate of the condenser

---

* Clausius (*Mech. Wärmetheorie*, 2, p. 94) has obtained the relation

$$\frac{K - K_0}{K + 2K_0} = \frac{4\pi}{3} na^3,$$

by considering the field inside a sphere of dielectric. The value of K must of course be independent of the shape of the piece of the dielectric considered. The apparent discrepancy in the two values of K obtained, is removed as soon as we reflect that both proceed on the assumption that $K - K_0$ is small, for the results agree as far as first powers of $K - K_0$. Pagliani (*Accad. dei Lincei*, 2, p. 48) finds that in point of fact the equation

$$\frac{K - K_0}{K} = 4\pi na^3$$

agrees better with experiment than the formula of Clausius.

to the other, the proportion of the length of each which would be inside a conductor would, on the average, be θ. Since there is no fall of a potential inside a conductor, the total fall of potential from one plate to the other would be only $1 - \theta$ times what it would be if the molecules were absent, and the ratio K/K_0 would be $1/(1 - \theta)$ or, if θ is small, $1 + \theta$. Since, however, the lines of force tend to run through conductors wherever possible, there is more shortening of lines of force than is shewn by this simple calculation. Equation (75) shews that when the molecules are spherical the effect is three times that given by this simple calculation. For other shapes of molecules the multiplying factor might of course be different.

Equation (75) gives at once a method of determining θ for substances for which θ is small, namely gases, but, owing to the unwarranted assumption that the molecules are spherical, the results will be true as regards order of magnitude only. If the dielectric is a gas at atmospheric pressure, the value of n is known, being about $2 \cdot 685 \times 10^{19}$, and this enables us to calculate the value of a.

150. The following table gives series of values of $\dfrac{K}{K_0}$ for gases at atmospheric pressure:

Gas		$\dfrac{K}{K_0}$ observed	Autho-rity*	a calculated (Mossotti's Theory)	a calculated (Theory of Gases)†
Helium ...	He	1·000074	4	·598 × 10⁻⁸	1·09 × 10⁻⁸
Hydrogen ...	H₂	1·000264	1	·919 × 10⁻⁸	1·36 × 10⁻⁸
Oxygen	O₂	1·000543	3	1·17 × 10⁻⁸	1·81 × 10⁻⁸
Argon	Ar	1·000566	3	1·18 × 10⁻⁸	1·83 × 10⁻⁸
Air	—	1·000586	2	1·19 × 10⁻⁸	1·87 × 10⁻⁸
Nitrogen ...	N₂	1·000594	3	1·20 × 10⁻⁸	1·89 × 10⁻⁸
Carbon Monoxide	CO	1·000695	2	1·27 × 10⁻⁸	1·89 × 10⁻⁸
Carbon Dioxide	CO₂	1·000985	2	1·42 × 10⁻⁸	2·33 × 10⁻⁸
Nitrous Oxide ...	N₂O	1·00099	2	1·43 × 10⁻⁸	2·33 × 10⁻⁸
Ethylene ...	C₂H₄	1·00146	2	1·63 × 10⁻⁸	2·77 × 10⁻⁸

* Authorities:—1. Boltzmann, 1875.
          2. Klemenčič, 1885.
          3. Calculated from refractive index for Sodium Light.
          4. Hockheim, 1908.
† Jeans, *Introduction to the Kinetic Theory of Gases*, p. 183.

The last two columns give respectively the values of a calculated from equation (75), and the value of a given by the Theory of Gases. The two sets of values do not agree exactly—this could not be expected when we remember the magnitude of the errors introduced in treating the molecules as spherical. But what agreement there is supplies very significant evidence as to the truth of the theory of molecular polarisation.

151. It still remains to explain what physical property of the molecule justifies us in treating its surface as a perfect conductor. It has already been explained that all matter contains a number of negatively charged particles or electrons. These form the outer layers of the atoms and molecules and it is by their motion that the conduction of electricity is effected. In a dielectric there is no conduction, so that each electron must remain permanently associated with the same molecule. There is, however, plenty of evidence that the electrons are not rigidly fixed to the molecules but are free to move within certain limits. The molecule may be regarded as consisting partially or wholly of a cluster of electrons, normally at rest in positions of equilibrium under the various attractions and repulsions present, but capable of vibrating about these positions. Under the influence of an external field of force, the electrons will move slightly from their equilibrium positions—we may imagine that a kind of tidal motion of electrons takes place in the molecule. Obviously, by the time that equilibrium is attained, the outer surface of the molecule must be an equipotential. This, however, is exactly what is required for Mossotti's hypothesis. We may accordingly abandon the conception of conducting spheres, which was only required to make the surface of the molecule an equipotential, and may, without impairing the power of Mossotti's explanation, replace these conducting spheres by shells of electrons. If in some way we can further replace these shells by rings of electrons in rapid orbital motion, the modified hypothesis will be in very close agreement with modern beliefs as to the structure of matter.

On this view, the quantity a tabulated in the sixth column of the table on p. 132, will measure the radius of the outermost shell of electrons. Even outside this outermost shell, however, there will be an appreciable field of force, so that when two molecules of a gas collide there will in general be a considerable distance between their outermost layers of electrons. Thus if the collisions of molecules in a gas are to be regarded as the collisions of elastic spheres, the radius of these spheres must be supposed to be considerably greater than a. Now it is the radius of these imaginary elastic spheres which we calculate in the Kinetic Theory of Gases: there is therefore no difficulty in understanding the differences between the two sets of values for a given in the table of p. 132.

It is known that molecules are not in general spherical in shape, but, as we shall see below, there is no difficulty in extending Mossotti's theory to cover the case of non-spherical molecules.

ANISOTROPIC MEDIA.

152. There are some dielectrics, generally of crystalline structure, in which Faraday's relation between polarisation and intensity is found not to be true. The polarisation in such dielectrics is not, in general, in the same direction as the intensity, and the angle between the polarisation and intensity and also the ratio of these quantities are found to depend on the direction of the field relatively to the axes of the crystal. We shall find that the conception of molecular action accounts for these peculiarities of crystalline dielectrics.

Let us consider an extreme case in which the spherical molecules of fig. 46 are replaced by a number of very elongated or needle-shaped bodies. The lines of force will have their effective lengths shortened by an amount which depends on whether much or little of them falls within the material of the needle-shaped molecules, and, as in § 149, there will be an equation of the form

$$\frac{K}{K_0} = 1 + s\theta,$$

where θ is the aggregate volume of the number of molecules which occur in a unit volume of the gas, and s is a numerical multiplier. But it is at once clear that the value of s will depend not only on the shape but also on the orientation of the molecules. Clearly the value of s will be greatest when the needles are placed so that their greatest length lies in the direction of

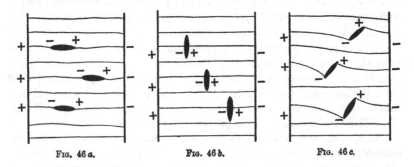

FIG. 46 *a*.   FIG. 46 *b*.   FIG. 46 *c*.

the lines of force, as in fig. 46 *a*, and will be least when the needles lie at right angles to this position, as in fig. 46 *b*. Or to put the matter in another way, a piece of dielectric in which the molecules are needle-shaped and parallel will exhibit different values of K according as the field of force is parallel or at right angles to the lengths of the needles.

This extreme case illustrates the fundamental property of crystalline dielectrics, but it ought to be understood that in actual substances the values of K do not differ so much for different directions as this extreme case might be supposed to suggest. For instance for quartz, one of the substances in which the difference is most marked, Curie finds the extreme values of K to be 4·55 and 4·49.

Before attempting to construct a mathematical theory of the behaviour of a crystalline dielectric we may examine the case of a dielectric having needle-shaped molecules placed parallel to one another, but so as to make any angle θ with the direction of the lines of force, as in fig. 46 c.

It is at once clear that not only are the effective lengths of the lines of force shortened by the presence of the molecules, but also the directions of the lines of force are twisted. It follows that the polarisation, regarded as a vector as in § 128, must in general have a direction different from that of the average intensity R of the field.

To analyse such a case we shall, as in § 146, regard the field near any point as the superposition of two fields:

(i) the field which arises from the doublets on the neighbouring molecules, say a field of components of intensity X_1, Y_1, Z_1;

(ii) the field caused by the doublets arising from the distant molecules and from the charges outside the dielectric, say a field of components of intensity X_2, Y_2, Z_2.

Clearly in the case we are now considering, the intensities R_1, R_2 of these fields will not be in the same direction.

The components of intensity of the whole field are given by

$$X = X_1 + X_2, \text{ etc.}$$

To discuss the first part of the field, let us regard the whole field as the superposition of three fields, having respectively components $(X, 0, 0)$, $(0, Y, 0)$ and $(0, 0, Z)$. If the molecules are spherical, or if, not being spherical, their orientations in space are distributed at random, then clearly the field of components $(X, 0, 0)$ will induce doublets which will produce simply a field of components $(K'X, 0, 0)$ where K' is a constant. But if the molecules are neither spherical in shape nor arranged at random as regards their orientations in space, it will be necessary to assume that the induced doublets give rise to a field of components

$$K'_{11}X, \qquad K'_{12}X, \qquad K'_{13}X.$$

On superposing the doublets induced by the three fields $(X, 0, 0)$, $(0, Y, 0)$ and $(0, 0, Z)$, we obtain

$$\left. \begin{aligned} X_1 &= K'_{11}X + K'_{21}Y + K'_{31}Z \\ Y_1 &= K'_{12}X + K'_{22}Y + K'_{32}Z \\ Z_1 &= K'_{13}X + K'_{23}Y + K'_{33}Z \end{aligned} \right\} \quad \ldots\ldots\ldots\ldots\ldots\ldots (76).$$

Thus we have relations of the form

$$\left. \begin{aligned} 4\pi f &= K_{11}X + K_{21}Y + K_{31}Z \\ 4\pi g &= K_{12}X + K_{22}Y + K_{32}Z \\ 4\pi h &= K_{13}X + K_{23}Y + K_{33}Z \end{aligned} \right\} \quad \ldots\ldots\ldots\ldots\ldots\ldots (77),$$

expressing the relations between polarisation and intensity.

These are the general equations for crystalline media. We shall shortly prove (§ 170) that

$$K_{12} = K_{21}, \quad K_{23} = K_{32}, \quad K_{31} = K_{13} \quad \ldots\ldots\ldots\ldots\ldots (78),$$

so that there are not nine, but only six, independent constants.

NON-SPHERICAL MOLECULES.

152 A. A medium in which the molecules are not spherical but are oriented at random can be discussed in a similar way. The whole field (X, Y, Z) may be regarded as the superposition of three fields $(X, 0, 0)$, $(0, Y, 0)$ and $(0, 0, Z)$. The induced doublets produced by the first field will produce a field of components

$$(K'X, 0, 0),$$

the components along $0y$ and $0z$ necessarily vanishing on account of the random orientation of the molecules. The other fields similarly produce induced fields

$$(0, K'Y, 0) \text{ and } (0, 0, K'Z),$$

whence we readily obtain equations of the form

$$4\pi f = KX, \quad 4\pi g = KY, \quad 4\pi h = KZ.$$

Thus Mossotti's theory can readily be extended to non-spherical molecules, but the difficulty remains that according to modern views, a molecule does not consist of layers of electrons at rest, but of systems of electrons in orbital motion. It will not be possible to make the appropriate modification in the theory until the exact nature of this orbital motion is known.

EXAMPLES.

1. A spherical condenser, radii a, b, has air in the space between the spheres. The inner sphere receives a coat of paint of uniform thickness t and of a material of which the inductive capacity is K. Find the change produced in the capacity of the condenser.

2. A conductor has a charge e, and V_1, V_2 are the potentials of two equipotential surfaces completely surrounding it ($V_1 > V_2$). The space between these two surfaces is now filled with a dielectric of inductive capacity K. Shew that the change in the energy of the system is

$$\tfrac{1}{2}e(V_1 - V_2)(K-1)/K.$$

3. The surfaces of an air-condenser are concentric spheres. If half the space between the spheres be filled with solid dielectric of specific inductive capacity K, the dividing surface between the solid and the air being a plane through the centre of the spheres, shew that the capacity will be the same as though the whole dielectric were of uniform specific inductive capacity $\tfrac{1}{2}(1+K)$.

4. The radii of the inner and outer shells of two equal spherical condensers, remote from each other and immersed in an infinite dielectric of inductive capacity K, are respectively a and b, and the inductive capacities of the dielectric inside the condensers are K_1, K_2. Both surfaces of the first condenser are insulated and charged, the second being uncharged. The inner surface of the second condenser is now connected to earth, and the outer surface is connected to the outer surface of the first condenser by a wire of negligible capacity. Shew that the loss of energy is

$$\frac{Q^2\{2(b-a)K+aK_2\}}{2Kb\{(b-a)K+aK_2\}},$$

where Q is the quantity of electricity which flows along the wire.

5. The outer coating of a long cylindrical condenser is a thin shell of radius a, and the dielectric between the cylinders has inductive capacity K on one side of a plane through the axis, and K' on the other side. Shew that when the inner cylinder is connected to earth, and the outer has a charge q per unit length, the resultant force on the outer cylinder is

$$\frac{4q^2(K-K')}{\pi a(K+K')^2}$$

per unit length.

6. A heterogeneous dielectric is formed of n concentric spherical layers of specific inductive capacities K_1, K_2, ... K_n, starting from the innermost dielectric, which forms a solid sphere; also the outermost dielectric extends to infinity. The radii of the spherical boundary surfaces are a_1, a_2, ... a_{n-1} respectively. Prove that the potential due to a quantity Q of electricity at the centre of the spheres at a point distant r from the centre in the dielectric K_s is

$$\frac{Q}{K_s}\left(\frac{1}{r}-\frac{1}{a_s}\right)+\frac{Q}{K_{s+1}}\left(\frac{1}{a_s}-\frac{1}{a_{s+1}}\right)+\ldots+\frac{Q}{K_n}\frac{1}{a_{n-1}}.$$

7. A condenser is formed by two rectangular parallel conducting plates of breadth b and area A at distance d from each other. Also a parallel slab of a dielectric of thickness t and of the same area is between the plates. This slab is pulled along its length from between the plates, so that only a length x is between the plates. Prove that the electric force sucking the slab back to its original position is

$$\frac{2\pi E^2 d b t' (d - t')}{\{A (d - t') + x b t'\}^2},$$

where $t' = t (K - 1)/K$, K is the specific inductive capacity of the slab, E is the charge, and the disturbances produced by the edges are neglected.

8. Three closed surfaces 1, 2, 3 are equipotentials in an electric field. If the space between 1 and 2 is filled with a dielectric K, and that between 2 and 3 is filled with a dielectric K', shew that the capacity of a condenser having 1 and 3 for faces is C, given by

$$\frac{1}{C} = \frac{1}{AK} + \frac{1}{BK'},$$

where A, B are the capacities of air-condensers having as faces the surfaces 1, 2 and 2, 3 respectively.

9. The surface separating two dielectrics (K_1, K_2) has an actual charge σ per unit area. The electric forces on the two sides of the boundary are F_1, F_2 at angles c_1, c_2 with the common normal. Shew how to determine F_2, and prove that

$$K_2 \cot c_2 = K_1 \cot c_1 \left(1 - \frac{4\pi\sigma}{K_1 F_1 \cos c_1} \right).$$

10. The space between two concentric spheres radii a, b which are kept at potentials A, B, is filled with a heterogeneous dielectric of which the inductive capacity varies as the nth power of the distance from their common centre. Shew that the potential at any point between the surfaces is

$$\frac{A a^{n+1} - B b^{n+1}}{a^{n+1} - b^{n+1}} - \frac{a^{n+1} b^{n+1}}{r^{n+1}} \frac{A - B}{a^{n+1} - b^{n+1}}.$$

11. A condenser is formed of two parallel plates, distant h apart, one of which is at zero potential. The space between the plates is filled with a dielectric whose inductive capacity increases uniformly from one plate to the other. Shew that the capacity per unit area is

$$\frac{K_2 - K_1}{4\pi h \log K_2/K_1},$$

where K_1 and K_2 are the values of the inductive capacity at the surfaces of the plate. The inequalities of distribution at the edges of the plates are neglected.

12. A spherical conductor of radius a is surrounded by a concentric spherical conducting shell whose internal radius is b, and the intervening space is occupied by a dielectric whose specific inductive capacity at a distance r from the centre is $\frac{c + r}{r}$. If the inner sphere is insulated and has a charge E, the shell being connected with the earth, prove that the potential in the dielectric at a distance r from the centre is $\frac{E}{c} \log \frac{b(c + r)}{r(c + b)}$.

13. A spherical conductor of radius a is surrounded by a concentric spherical shell of radius b, and the space between them is filled with a dielectric of which the inductive capacity at distance r from the centre is $\mu e^{-p^2} p^{-3}$ where $p = r/a$. Prove that the capacity of the condenser so formed is

$$2\mu a \left(e^{\frac{b^2}{a^2}} - e\right)^{-1}.$$

14. If the specific inductive capacity varies as $e^{-\frac{r}{a}}$, where r is the distance from a fixed point in the medium, verify that a solution of the differential equation satisfied by the potential is

$$\left(\frac{a}{r}\right)^2 \left[e^{\frac{r}{a}} - 1 - \frac{r}{a} - \frac{r^2}{2a^2}\right] \cos \theta,$$

and hence determine the potential at any point of a sphere, whose inductive capacity is the above function of the distance from the centre, when placed in a uniform field of force.

15. Shew that the capacity of a condenser consisting of the conducting spheres $r = a$, $r = b$, and a heterogeneous dielectric of inductive capacity $K = f(\theta, \phi)$, is

$$\frac{ab}{4\pi (b-a)} \iint f(\theta, \phi) \sin \theta \, d\theta \, d\phi.$$

16. In an imaginary crystalline medium the molecules are discs placed so as to be all parallel to the plane of xy. Shew that the components of intensity and polarisation are connected by equations of the form

$$4\pi f = K_{11} X + K_{21} Y; \qquad 4\pi g = K_{12} X + K_{22} Y; \qquad 4\pi h = K_{33} Z.$$

CHAPTER VI

THE STATE OF THE MEDIUM IN THE ELECTROSTATIC FIELD

153. THE whole electrostatic theory has so far been based simply upon Coulomb's Law of the inverse square of the distance. We have supposed that one charge of electricity exerts certain forces upon a second distant charge, but nothing has been said as to the mechanism by which this action takes place. In handling this question there are two possibilities open. We may either assume "action at a distance" as an ultimate explanation—*i.e.* simply assert that two bodies act on one another across the intervening space, without attempting to go any further towards an explanation of how such action is brought about—or we may tentatively assume that some medium connects the one body with the other, and examine whether it is possible to ascribe properties to this medium, such that the observed action will be transmitted by the medium. Faraday and Maxwell followed the latter course. They refused to admit "action at a distance" as an ultimate explanation of electric phenomena, finding such action unthinkable unless transmitted by an intervening medium.

154. It is worth enquiring whether there is any valid *à priori* argument which compels us to resort to action through a medium. Some writers have attempted to use the phenomenon of Inductive Capacity to prove that the energy of a condenser must reside in the space between the charged plates, rather than on the plates themselves—for, they say, change the medium between the plates, keeping the plates in the same condition, and the energy is changed. A study of Faraday's molecular explanation of the action in a dielectric will shew that this argument proves nothing as to the real question at issue. It goes so far as to prove that when there are molecules placed between electric charges, these molecules themselves acquire charges, and so may be said to be new stores of energy, but it leaves untouched the question of whether the energy resides in the charges on the molecules or in the ether between them.

Again, the phenomenon of induction is sometimes quoted against action at a distance—a small conductor placed at a point P in an electrostatic field shews phenomena which depend on the electric intensity at P. This is taken to shew that the state of the ether at the point P before the introduction of the conductor was in some way different from what it would have been if there had not been electric charges in the neighbourhood. But all that is proved is that the state of the point P *after* the introduction of the conductor will be different from what it would have been if there had not been electric charges in the neighbourhood, and this can be explained equally well either by action at a distance or action through a medium. The new conductor is a collection of positive and negative charges : the phenomena under question are produced by these charges being acted upon by the other charges in the field, but whether this action is action at a distance or action through a medium cannot be told.

Indeed, it will be seen that, viewed in the light of the electron-theory and of Faraday's theory of dielectric polarisation, electrical action stands on just the same level as gravitational action. In each case the system of forces to be explained may be regarded as a system of forces between indestructible centres, whether of electricity or of matter, and the law of force is the law of the inverse square, independently of the state of the space between the centres. Now no scientist would claim that there is any *à priori* proof that gravitation is transmitted through a medium—indeed the trend of opinion at present is quite in the opposite direction—and this fact in itself suffices to shew that there is no *à priori* means of establishing that electrical action is transmitted through a medium.

Failing an *à priori* argument, an attempt may be made to disprove action at a distance, or rather to make it improbable, by an appeal to experience. It may be argued that as all the forces of which we have experience in every-day life are forces between substances in contact, therefore it follows by analogy that forces of gravitation, electricity and magnetism, must ultimately reduce to forces between substances in contact—*i.e.* must be transmitted through a medium. Upon analysis, however, it will be seen that this argument divides all forces into two classes :

(a) Forces of gravitation, electricity and magnetism, which appear to act at a distance.

(β) Forces of pressure and impact between solid bodies, hydrostatic pressure, etc. which appear to act through a medium.

The argument is now seen to be that because class (β) *appear* to act through a medium, therefore class (a) must *in reality* act through a medium. The argument could, with equal logical force, be used in the exactly opposite direction : indeed it has been so used by the followers of Boscovitch. The Newtonian discovery of gravitation, and of apparent action at a distance, so occupied the attention of scientists at the time of Boscovitch that it seemed natural to regard action at a distance as the ultimate basis of force, and to try to interpret action through a medium in terms of action at a distance. The reversion from this view came, as has been said, with Faraday.

Hertz's subsequent discovery of the finite velocity of propagation of electric action, which had previously been predicted by Maxwell's theory, came to the support of Faraday's view. To see exactly what is meant by this finite velocity of propagation, let us imagine that we place two uncharged conductors A, B at a distance r from one another. By charging A, and so performing work at A, we can induce charges on conductor B, and when this has been done, there will be an attraction between conductors A and B. We can suppose that conductor A is held fast, and that conductor B is allowed to move towards A, work being performed by the attraction from conductor A. We are now recovering from B work which was originally performed at A. The experiments of Hertz shew that a finite time is required before any of the work spent at A becomes available at B. A natural explanation is to suppose that work spent on A assumes the form of energy which spreads itself out through the whole of space, and that the finite time observed before energy becomes available at B is the time required for the first part of the advancing energy to travel from A to B. This explanation involves regarding energy as a definite physical entity, capable of being localised in space.   It ought to be noticed that our senses give us no knowledge of energy as a physical entity : we experience force, not energy.   And the fact that energy appears to be propagated through space with finite velocity does not justify us in concluding that it has a real physical existence, for, as we shall see, the potential appears to be propagated in the same way, and the potential can only be regarded as a convenient mathematical fiction.

155.   Although no sufficient reason has been found compelling us to ascribe electric action to the presence of an intervening medium, we are still free to assume, as a hypothesis, that such a medium exists and that electric action is transmitted through this medium.   As various electric and electro-magnetic phenomena are discussed we shall examine what properties would have to be attributed to the medium to account for these properties.   If it is found that contradictory properties would have to be ascribed to the medium, then the hypothesis of action through an intervening medium will have to be abandoned.   If the properties are found to be consistent, then the hypotheses of action at a distance and action through a medium are still both in the field, but the latter becomes more or less probable just in proportion as the properties of the hypothetical medium seem probable or improbable.   We shall return to the general question of the existence of a medium in Chapter XX.

156.   Since electric action takes place even across the most complete vacuum obtainable, we conclude that if this action is transmitted by a medium, this medium must be the ether.   Assuming that the action is transmitted by the ether, we must suppose that at any point in the electro-static field there will be an action and reaction between the two parts of the ether at opposite sides of the point.   The ether, in other words, is in a state of stress at every point in the electrostatic field.   Before discussing the particular system of stresses appropriate to an electrostatic field, we shall investigate the general theory of stresses in a medium at rest.

GENERAL THEORY OF STRESSES IN A MEDIUM AT REST.

157.   Let us take a small area dS in the medium perpendicular to the axis of x.   Let us speak of that part of the medium near to dS for which x is greater than its value over dS as x_+, and that for which x is less than this value as x_-, so that the area dS separates the two regions x_+ and x_-. Those parts of the medium by which these two regions are occupied exert forces upon one another across dS, and this system of forces is spoken of as the stress across dS.   Obviously this stress will consist of an action and reaction, the two being equal and opposite.   Also it is clear that the amount of this stress will be proportional to dS.

Let us assume that the force exerted by x_+ on x_- has components

$$P_{xx}dS, \quad P_{xy}dS, \quad P_{xz}dS,$$

then the force exerted by x_- on x_+ will have components

$$-P_{xx}dS, \quad -P_{xy}dS, \quad -P_{xz}dS.$$

The quantities P_{xx}, P_{xy}, P_{xz} are spoken of as the components of stress perpendicular to Ox. Similarly there will be components of stress P_{yx}, P_{yy}, P_{yz} perpendicular to Oy, and components of stress P_{zx}, P_{zy}, P_{zz} perpendicular to Oz.

Let us next take a small parallelepiped in the medium, bounded by planes

$$x = \xi, \quad x = \xi + dx;$$
$$y = \eta, \quad y = \eta + dy;$$
$$z = \zeta, \quad z = \zeta + dz.$$

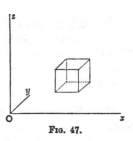

Fig. 47.

The stress acting upon the parallelepiped across the face of area $dy\,dz$ in the plane $x = \xi$ will have components

$$-(P_{xx})_{x=\xi}\,dy\,dz, \quad -(P_{xy})_{x=\xi}\,dy\,dz, \quad -(P_{xz})_{x=\xi}\,dy\,dz,$$

while the stress acting upon the parallelepiped across the opposite face will have components

$$(P_{xx})_{x=\xi+dx}dy\,dz, \quad (P_{xy})_{x=\xi+dx}dy\,dz, \quad (P_{xz})_{x=\xi+dx}dy\,dz.$$

Compounding these two stresses, we find that the resultant of the stresses acting upon the parallelepiped across the pair of faces parallel to the plane of yz, has components

$$\frac{\partial P_{xx}}{\partial x}dx\,dy\,dz, \quad \frac{\partial P_{xy}}{\partial x}dx\,dy\,dz, \quad \frac{\partial P_{xz}}{\partial x}dx\,dy\,dz.$$

Similarly from the other pairs of faces, we get resultant forces of components

$$\frac{\partial P_{yx}}{\partial y}dx\,dy\,dz, \quad \frac{\partial P_{yy}}{\partial y}dx\,dy\,dz, \quad \frac{\partial P_{yz}}{\partial y}dx\,dy\,dz,$$

and

$$\frac{\partial P_{zx}}{\partial z}dx\,dy\,dz, \quad \frac{\partial P_{zy}}{\partial z}dx\,dy\,dz, \quad \frac{\partial P_{zz}}{\partial z}dx\,dy\,dz.$$

For generality, let us suppose that in addition to the action of these stresses the medium is acted upon by forces acting from a distance, of amount Ξ, H, Z per unit volume. The components of the forces acting on the parallelepiped of volume $dx\,dy\,dz$ will be

$$\Xi\,dx\,dy\,dz, \quad H\,dx\,dy\,dz, \quad Z\,dx\,dy\,dz.$$

Compounding all the forces which have been obtained, we obtain as equations of equilibrium

$$\Xi + \frac{\partial P_{xx}}{\partial x} + \frac{\partial P_{yx}}{\partial y} + \frac{\partial P_{zx}}{\partial z} = 0 \quad\quad\quad\quad (79)$$

and two similar equations.

158. These three equations ensure that the medium shall have no motion of translation, but for equilibrium it is also necessary that there should be no rotation. To a first approximation, the stress across any face may be supposed to act at the centre of the face, and the force Ξ, H, Z at the centre of the parallelepiped. Taking moments about a line through the centre parallel to the axis of Ox, we obtain as the equation of equilibrium

$$P_{yz} - P_{zy} = 0 \quad \dots\dots\dots\dots\dots\dots\dots\dots\dots(80).$$

This and the two similar equations obtained by taking moments about lines parallel to Oy, Oz ensure that there shall be no rotation of the medium. Thus the necessary and sufficient condition for the equilibrium of the medium is expressed by three equations of the form of (79), and three equations of the form of (80).

159. Suppose next that we take a small area dS anywhere in the medium. Let the direction cosines of the normal to dS be $\pm l$, $\pm m$, $\pm n$. Let the parts of the medium close to dS and on the two sides of it be spoken of as S_+ and S_-, these being named so that a line drawn from dS with direction cosines $+l$, $+m$, $+n$ will be drawn into S_+, and one with direction cosines $-l$, $-m$, $-n$ will be drawn into S_-. Let the force exerted by S_+ on S_- across the area dS have components

Fig. 48.

$$F dS, \quad G dS, \quad H dS,$$

then the force exerted by S_- on S_+ will have components

$$-F dS, \quad -G dS, \quad -H dS.$$

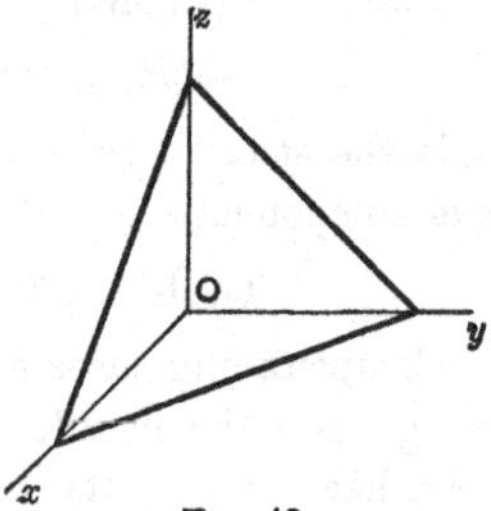

The quantities F, G, H are spoken of as the components of stress across a plane of direction cosines l, m, n.

To find the values of F, G, H, let us draw a small tetrahedron having three faces parallel to the coordinate planes and a fourth having direction cosines l, m, n. If dS is the area of the last face, the areas of the other faces are ldS, mdS, ndS and the volume of the tetrahedron is $\frac{1}{3}\sqrt{2lmn}\,(dS)^{\frac{3}{2}}$. Resolving parallel to Ox, we have, since the medium inside this tetrahedron is in equilibrium,

$$\tfrac{1}{3}\sqrt{2lmn}\,(dS)^{\frac{3}{2}}\,\Xi - ldS P_{xx} - mdS P_{yx} - ndS P_{zx} + F dS = 0,$$

giving, since dS is supposed vanishingly small,

$$F = l P_{xx} + m P_{yx} + n P_{zx} \quad \dots\dots\dots\dots\dots\dots\dots(81)$$

and there are two similar equations to determine G and H.

160. Assuming that equation (80) and the two similar equations are satisfied, the normal component of stress across the plane of which the direction cosines are l, m, n is

$$lF + mG + nH = l^2 P_{xx} + m^2 P_{yy} + n^2 P_{zz} + 2mn\,P_{yz} + 2nl\,P_{zx} + 2lm\,P_{xy}.$$

The quadric

$$x^2 P_{xx} + y^2 P_{yy} + z^2 P_{zz} + 2yz\,P_{yz} + 2zx\,P_{zx} + 2xy\,P_{xy} = 1 \quad\ldots\ldots\ldots(82)$$

is called the stress-quadric. If r is the length of its radius vector drawn in the direction l, m, n, we have

$$r^2\,(l^2 P_{xx} + m^2 P_{yy} + n^2 P_{zz} + 2mn\,P_{yz} + 2nl\,P_{zx} + 2lm\,P_{xy}) = 1.$$

It is now clear that the normal stress across any plane l, m, n is measured by the reciprocal of the square of the radius vector of which the direction cosines are l, m, n. Moreover the direction of the stress across any plane l, m, n is that of the normal to the stress-quadric at the extremity of this radius vector. For r being the length of this radius vector, the coordinates of its extremity will be rl, rm, rn. The direction cosines of the normal at this point are in the ratio

$$rlP_{xx} + rm P_{xy} + rn P_{zx} : rl P_{xy} + rm P_{yy} + rn P_{yz} : rl P_{zx} + rm P_{yz} + rn P_{zz}$$

or $F : G : H$, which proves the result.

The stress-quadric has three principal axes, and the directions of these are spoken of as the axes of the stress. Thus the stress at any point has three axes, and these are always at right angles to one another. If a small area be taken perpendicular to a stress axis at any point, the stress across this area will be normal to the area. If the amounts of these stresses are P_1, P_2, P_3, then the equation of the stress-quadric referred to its principal axes will be

$$P_1\xi^2 + P_2\eta^2 + P_3\zeta^2 = 1.$$

Clearly a positive principal stress is a simple tension, and a negative principal stress is a simple pressure.

As simple illustrations of this theory, it may be noticed that

(i) For a simple hydrostatic pressure P, the stress-quadric becomes an imaginary sphere

$$P\,(\xi^2 + \eta^2 + \zeta^2) = -1.$$

The pressure is the same in all directions, and the pressure across any plane is at right angles to the plane (for the tangent plane to a sphere is at right angles to the radius vector).

(ii) For a simple pull, as in a rope, the stress-quadric degenerates into two parallel planes

$$P\xi^2 = 1.$$

THE STRESSES IN AN ELECTROSTATIC FIELD.

161.   If an infinitesimal charged particle is introduced into the electric field at any point, the phenomena exhibited by it must, on the present view of electric action, depend solely on the state of stress at the point.   The phenomena must therefore be deducible from a knowledge of the stress-quadric at the point.   The only phenomenon observed is a mechanical force tending to drag the particle in a certain direction—namely, in the direction of the line of force through the point.   Thus from inspection of the stress-quadric, it must be possible to single out this one direction.   We conclude that the stress-quadric must be a surface of revolution, having this direction for its axis.   The equation of the stress-quadric at any point, referred to its principal axes, must accordingly be

$$P_1 \xi^2 + P_2 (\eta^2 + \zeta^2) = 1 \quad ....(83),$$

where the axis of ξ coincides with the line of force through the point.   Thus the system of stresses must consist of a tension P_1 along the lines of force, and a tension P_2 perpendicular to the lines of force—and if either of the quantities P_1 or P_2 is found to be negative, the tension must be interpreted as a pressure.

Since the electrical phenomena at any point depend only on the stress-quadric, it follows that R must be deducible from a knowledge of P_1 and P_2. Moreover, the only phenomena known are those which depend on the magnitude of R, so that it is reasonable to suppose that the only quantity which can be deduced from a knowledge of P_1 and P_2 is the quantity R—in other words, that P_1 and P_2 are functions of R only.   We shall for the present assume this as a provisional hypothesis, to be rejected if it is found to be incapable of explaining the facts.

162.   The expression of P_1 as a function of R can be obtained at once by considering the forces acting on a charged conductor.   Any element dS of surface experiences a force $\dfrac{R^2}{8\pi} dS$ urging it normally away from the conductor.   On the present view of the origin of the forces in the electric field, we must interpret this force as the resultant of the ether-stresses on its two sides.   Thus, resolving normally to the conductor, we must have

$$\frac{R^2}{8\pi} dS = (P_1)_R \, dS - (P_1)_0 \, dS,$$

where $(P_1)_R$, $(P_1)_0$ denote the values of P_1 when the intensity is R and 0 respectively.   Inside the conductor there is no intensity, so that the stress-quadrics become spheres, for there is nothing to differentiate one direction from another.   Any value which $(P_1)_0$ may have accordingly arises simply from a hydrostatic pressure or tension throughout the medium, and this cannot influence the forces on conductors. Leaving any such hydrostatic pressure out of account, we may take $(P_1)_0 = 0$, and so obtain $(P_1)_R$ in the form

$$P_1 = \frac{R^2}{8\pi} \quad \dots\dots\dots\dots\dots\dots\dots(84).$$

163. We can most easily arrive at the function of R which must be taken to express the value of P_2, by considering a special case.

Consider a spherical condenser formed of spheres of radii a, b. If this condenser is cut into two equal halves by a plane through its centre, the two halves will repel one another. This action must now be ascribed to the stresses in the medium across the plane of section. Since the lines of force are radial these stresses are perpendicular to the lines of force, and we see at once that the stress perpendicular to the lines of force is a *pressure*. To calculate the function of R which expresses this pressure, we may suppose $b - a$ equal to some very small quantity c, so that R may be regarded as constant along the length of a line of force. The area over which this pressure acts is $\pi(b^2 - a^2)$, and since the pressure per unit area in the medium perpendicular to a line of force is $-P_2$, the total repulsion between the two halves of the condenser will be $-P_2\pi(b^2 - a^2)$.

The whole force on either half of the condenser is however a force $2\pi\sigma^2$ per unit area over each hemisphere, normal to its surface. The resultant of all the forces acting on the inner hemisphere is $\pi a^2 \times 2\pi\sigma^2$, or putting $2\pi a^2\sigma = E$, so that E is the charge on either hemisphere, this force is $E^2/2a^2$. Similarly, the force on the hemisphere of radius b is $E^2/2b^2$. Thus the resultant repulsion on the complete half of the condenser is $\frac{1}{2}E^2\left(\dfrac{1}{a^2} - \dfrac{1}{b^2}\right)$. Since this has been seen to be also equal to $-P_2\pi(b^2 - a^2)$, we have

$$P_2 = -\frac{\frac{1}{2}E^2}{\pi a^2 b^2} = -2\pi\sigma^2 = -\frac{R^2}{8\pi}$$

on taking $a = b$ in the limit.

Thus in order that the observed actions may be accounted for, it is necessary that we have

$$P_1 = \frac{R^2}{8\pi}, \qquad P_2 = -\frac{R^2}{8\pi}.$$

Moreover, if these stresses exist, they will account for all the observed mechanical action on conductors, for the stresses result in a mechanical force $2\pi\sigma^2$ per unit area on the surface of every conductor.

164. It remains to examine whether these stresses are such as can be transmitted by an ether at rest.

As a preliminary we must find the values of the stress-components P_{xx}, P_{xy}, ... referred to fixed axes Ox, Oy, Oz.

The stress-quadric at any point in the ether, referred to its principal axes, is seen on comparison with equation (83) to be

$$\frac{R^2}{8\pi}(\xi^2 - \eta^2 - \zeta^2) = 1 \quad \dots\dots\dots\dots\dots(85).$$

Here the axis of ξ is in the direction of the line of force at the point. Let the direction-cosines of this direction be l_1, m_1, n_1. Then on transforming to axes Ox, Oy, Oz we may replace ξ by $l_1 x + m_1 y + n_1 z$.

Equation (85) may be replaced by

$$\frac{R^2}{8\pi}\{2\xi^2 - (\xi^2 + \eta^2 + \zeta^2)\} = 1,$$

and on transforming axes $\xi^2 + \eta^2 + \zeta^2$ transforms into $x^2 + y^2 + z^2$. Thus the transformed equation of the stress-quadric is

$$\frac{R^2}{8\pi}\{2(l_1 x + m_1 y + n_1 z)^2 - (x^2 + y^2 + z^2)\} = 1.$$

Comparing with equation (82), we obtain

$$P_{xx} = \frac{R^2}{8\pi}(2l_1^2 - 1) \quad \dots\dots\dots\dots\dots(86),$$

$$P_{xy} = \frac{R^2}{8\pi}(2l_1 m_1) \quad \dots\dots\dots\dots\dots(87),$$

and similar values for the remaining components of stress.

Or again, since $\quad X = l_1 R, \quad Y = m_1 R, \quad Z = n_1 R,$

these equations may be expressed in the form

$$P_{xx} = \frac{1}{8\pi}(X^2 - Y^2 - Z^2),$$

$$P_{xy} = \frac{XY}{4\pi}.$$

In this system of stress-components, the relations $P_{xy} = P_{yx}$ are satisfied, as of course they must be since the system of stresses has been derived by assuming the existence of a stress-quadric. Thus the stresses do not set up rotations in the ether (cf. equation (80)).

In order that there may be also no tendency to translation, the stress-components must satisfy equations of the type

$$\frac{\partial P_{xx}}{\partial x} + \frac{\partial P_{xy}}{\partial y} + \frac{\partial P_{xz}}{\partial z} = 0 \quad \dots\dots\dots\dots\dots(88),$$

expressing that no forces beyond these stresses are required to keep the ether at rest (cf. equation (79)).

On substituting the values of the stress-components, we have

$$\frac{\partial P_{xx}}{\partial x} + \frac{\partial P_{xy}}{\partial y} + \frac{\partial P_{xz}}{\partial z}$$

$$= \frac{1}{8\pi} \left\{ \frac{\partial}{\partial x} (X^2 - Y^2 - Z^2) + \frac{\partial}{\partial y} (2XY) + \frac{\partial}{\partial z} (2XZ) \right\}$$

$$= \frac{1}{8\pi} \left\{ 2X \left(\frac{\partial X}{\partial x} + \frac{\partial Y}{\partial y} + \frac{\partial Z}{\partial z} \right) + 2Y \left(\frac{\partial X}{\partial y} - \frac{\partial Y}{\partial x} \right) + 2Z \left(\frac{\partial X}{\partial z} - \frac{\partial Z}{\partial x} \right) \right\}.$$

On putting

$$X = -\frac{\partial V}{\partial x}, \qquad Y = -\frac{\partial V}{\partial y}, \qquad Z = -\frac{\partial V}{\partial z},$$

we find at once that

$$\frac{\partial X}{\partial y} - \frac{\partial Y}{\partial x} = -\frac{\partial^2 V}{\partial x \partial y} + \frac{\partial^2 V}{\partial x \partial y} = 0,$$

$$\frac{\partial X}{\partial z} - \frac{\partial Z}{\partial x} = -\frac{\partial^2 V}{\partial x \partial z} + \frac{\partial^2 V}{\partial x \partial z} = 0,$$

$$\frac{\partial X}{\partial x} + \frac{\partial Y}{\partial y} + \frac{\partial Z}{\partial z} = -\left(\frac{\partial^2 V}{\partial x^2} + \frac{\partial^2 V}{\partial y^2} + \frac{\partial^2 V}{\partial z^2} \right) = 0,$$

shewing that equation (88) is satisfied.

165. Thus, to recapitulate, we have found that a system of stresses consisting of

(i)   a tension $\frac{R^2}{8\pi}$ per unit area in the direction of the lines of force,

(ii)   a pressure $\frac{R^2}{8\pi}$ per unit area perpendicular to the lines of force,

is one which can be transmitted by the medium, in that it does not tend to set up motions in the ether, and is one which will explain the observed forces in the electrostatic field. Moreover it is the only system of stresses capable of doing this, which is such that the stress at a point depends only on the electric intensity at that point.

Examples of Stress.

166. Assuming this system of stresses to exist, it is of value to try to picture the actual stresses in the field in a few simple cases.

Consider first the field surrounding a point charge. The tubes of force are cones. Let us consider the equilibrium of the ether enclosed by a frustum of one of these cones which is bounded by two ends p, q. If ω_p, ω_q are the areas of these ends, we find that there are tensions of amounts $\dfrac{R_p{}^2\omega_p}{8\pi}$, $\dfrac{R_q{}^2\omega_q}{8\pi}$. Since $R_p\omega_p = R_q\omega_q$, the former is the greater.

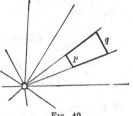

so that the forces on the two ends have as resultant a force tending to move the ether inwards towards the charge. This tendency is of course balanced by the pressures acting on the curved surface, each of which has a component tending to press the ether inside the frustum away from the charge.

Fig. 49.

167. A more complex example is afforded by two equal point charges, of which the lines of force are shewn in fig. 50.

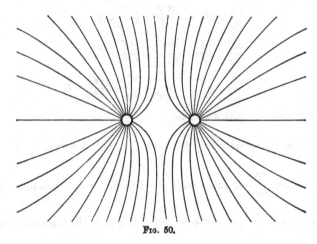

Fig. 50.

The lines of force on either charge fall thickest on the side **furthest** removed from the other charge, so that their resultant action on the charges amounts to a traction on the surface of each tending to drag it away from the other, and this traction appears to us as a repulsion between the bodies.

We can examine the matter in a different way by considering the action and reaction across the two sides of the plane which bisects the line joining the two charges. No lines of force cross this plane, which is accordingly made up entirely of the side walls of tubes of force. Thus there is a pressure $\dfrac{R^2}{8\pi}$ per unit area acting across this plane at every point. The resultant of all these pressures, after transmission by the ether from the plane to the charges immersed in the ether, appears as a force of repulsion exerted by the charges on one another.

ENERGY IN THE MEDIUM.

168. In setting up the system of stresses in a medium originally unstressed, work must be done, analogous to the work done in compressing a gas. This work must represent the energy of the stressed medium, and this in turn must represent the energy of the electrostatic field. Clearly, from the form of the stresses, the energy per unit volume of the medium at any point must be a function of R only. To determine the form of this function, we may examine the simple case of a parallel plate condenser, and we find at once that the function must be $\dfrac{R^2}{8\pi}$.

We have now to examine whether the energy of any electrostatic field can be regarded as made up of a contribution of amount $\dfrac{R^2}{8\pi}$ per unit volume from every part of the field.

In fig. 51, let PQ be a tube of force of strength e, passing from P at potential V_P to Q at potential V_Q. The ether inside this tube of force being supposed to possess energy $\dfrac{R^2}{8\pi}$ per unit volume, the total energy enclosed by the tube will be

$$\int_P^Q \frac{R^2}{8\pi}\, \omega\, ds,$$

where ω is the cross section at any point, and the integration is along the tube. Since $R\omega = 4\pi e$, this expression

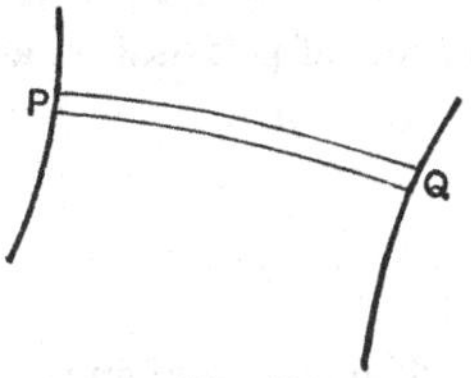

Fig. 51.

$$= \tfrac{1}{2} e \int_P^Q R\, ds$$

$$= -\tfrac{1}{2} e \int_P^Q \frac{\partial V}{\partial s}\, ds$$

$$= \tfrac{1}{2} e\,(V_P - V_Q).$$

This, however, is exactly the contribution made by the charges $\pm e$ at P, Q to the expression $\tfrac{1}{2}\Sigma eV$. Thus on summing over all tubes of force, we find that the total energy of the field $\tfrac{1}{2}\Sigma eV$ may be obtained exactly, by assigning energy to the ether at the rate of $\dfrac{R^2}{8\pi}$ per unit volume.

Energy in a Dielectric.

169. By imagining the parallel plate condenser of § 168 filled with dielectric of inductive capacity K, and calculating the energy when charged, we find that the energy, if spread through the dielectric, must be $\dfrac{KR^2}{8\pi}$ per unit volume.

Let us now examine whether the total energy of any field can be regarded as arising from a contribution of this amount per unit volume. The energy contained in a single tube of force, with the notation already used, will be

$$\int_P^Q \frac{KR^2}{8\pi}\, \omega\, ds,$$

or, since $\dfrac{KR}{4\pi} = P$, where P is the polarisation, this energy

$$= \int_P^Q \tfrac{1}{2} RP\omega\, ds$$
$$= \tfrac{1}{2} e \int_P^Q R\, ds$$
$$= \tfrac{1}{2} e\, (V_P - V_Q),$$

so that the total energy is $\tfrac{1}{2}\Sigma e V$, as before. Thus a distribution of energy of amount $\dfrac{KR^2}{8\pi}$ per unit volume will account for the energy of any field.

Crystalline dielectrics.

170. We have seen (§ 152) that in a crystalline dielectric, the components of polarisation and of electric intensity will be connected by equations of the form

$$\left.\begin{aligned} 4\pi f &= K_{11}X + K_{21}Y + K_{31}Z \\ 4\pi g &= K_{12}X + K_{22}Y + K_{32}Z \\ 4\pi h &= K_{13}X + K_{23}Y + K_{33}Z \end{aligned}\right\} \quad \dots\dots\dots\dots(89).$$

The energy of any distribution of electricity, no matter what the dielectric may be, will be $\tfrac{1}{2}\Sigma E V$. If V_1, V_2 are the potentials at the two ends of a unit tube, the part of this sum which is contributed by the charges at the ends of this tube will be $\tfrac{1}{2}(V_1 - V_2)$. If $\partial/\partial s$ denote differentiation along the tube, this may be written $-\tfrac{1}{2}\int \dfrac{\partial V}{\partial s}\, ds$, or again $-\tfrac{1}{2}\int \dfrac{\partial V}{\partial s} P\omega\, ds$, where P is the polarisation, and ω the cross section of the tube. Thus the energy may be supposed to be distributed at the rate of $-\tfrac{1}{2}\dfrac{\partial V}{\partial s} P$ per unit volume. If ϵ is the angle between the direction of the polarisation and that of the electric intensity, we have $-\dfrac{\partial V}{\partial s} = R\cos\epsilon$, so that the energy per unit volume

$$= \tfrac{1}{2} RP\cos\epsilon = \tfrac{1}{2}(fX + gY + hZ) \quad \dots\dots\dots\dots(90).$$

In a slight increase to the electric charges, the change in the energy of the system is, by § 109, equal to $\Sigma V\delta E$, so that the change in the energy per unit volume of the medium is

$$\delta W = X\delta f + Y\delta g + Z\delta h.$$

Thus $$\frac{\partial W}{\partial f} = X, \quad \frac{\partial W}{\partial g} = Y, \quad \frac{\partial W}{\partial h} = Z \dots\dots\dots\dots(91).$$

From formulae (89) and (90), we must have

$$W = \tfrac{1}{2}(fX + gY + hZ)$$

$$= \frac{1}{8\pi}\{K_{11}X^2 + (K_{12} + K_{21})XY + \ldots\},$$

from which

$$\frac{\partial W}{\partial X} = \frac{1}{4\pi}\{K_{11}X + \tfrac{1}{2}(K_{12} + K_{21})Y + \tfrac{1}{2}(K_{13} + K_{31})Z\}.$$

We must also have

$$\frac{\partial W}{\partial X} = \frac{\partial W}{\partial f}\frac{\partial f}{\partial X} + \frac{\partial W}{\partial g}\frac{\partial g}{\partial X} + \frac{\partial W}{\partial h}\frac{\partial h}{\partial X}$$

$$= \frac{1}{4\pi}\{K_{11}X + K_{21}Y + K_{31}Z\}.$$

Comparing these expressions, we see that we must have

$$K_{12} = K_{21}, \quad K_{13} = K_{31}, \quad K_{23} = K_{32}.$$

The energy per unit volume is now

$$W = \frac{1}{8\pi}(K_{11}X^2 + 2K_{12}XY + \ldots) \quad\ldots\ldots\ldots\ldots\ldots(92).$$

MAXWELL'S DISPLACEMENT THEORY.

171. Maxwell attempted to construct a picture of the phenomena occurring in the electric field by means of his conception of "electric displacement." Electric intensity, according to Maxwell, acting in any medium— whether this medium be a conductor, an insulator, or free ether—produces a motion of electricity through the medium. It is clear that Maxwell's conception of electricity, as here used, must be wider than that which we have up to the present been using, for electricity, as we have so far understood it, is incapable of moving through insulators or free ether. Maxwell's motion of electricity in conductors is that with which we are already familiar. As we have seen, the motion will continue so long as the electric intensity continues to exist. According to Maxwell, there is also a motion in an insulator or in free ether, but with the difference that the electricity cannot travel indefinitely through these media, but is simply displaced a small distance within the medium in the direction of the electric intensity, the extent of the displacement in isotropic media being exactly proportional to the intensity, and in the same direction.

The conception will perhaps be understood more clearly on comparing a conductor to a liquid and an insulator to an elastic solid. A small particle immersed in a liquid will continue to move through the liquid so long as there is a force acting on it, but a particle immersed in an elastic solid will be merely "displaced" by a force acting on it. The amount of this displacement will be proportional to the force acting, and when the force is removed, the particle will return to its original position.

Thus at any point in any medium the displacement has magnitude and direction. The displacement, then, is a vector, and its component in any direction may be measured by the total quantity of electricity per unit area which has crossed a small area perpendicular to this direction, the quantity being measured from a time at which no electric intensity was acting.

172. Suppose, now, that an electric field is gradually brought into existence, the field at any instant being exactly similar to the final field except that the intensity at each point is less than the final intensity in some definite ratio κ. Let the displacement be c times the intensity, so that when the intensity at any point is κR, the displacement is $c\kappa R$. The direction of this displacement is along the lines of force, so that the electricity may be regarded as moving through the tubes of force: the lines of force become identical now with the current-lines of a stream, to which they have already been compared.

Let us consider a small element of volume cut off by two adjacent equipotentials and a tube of force. Let the cross section of the tube of force be ω, and the normal distance between the equipotentials where they meet the tube of force be ds, so that the element under consideration is of volume $\omega\, ds$. On increasing the intensity from κR to $(\kappa + d\kappa)\,R$, there is an increase of displacement from $c\kappa R$ to $c(\kappa + d\kappa)\,R$, and therefore an additional displacement of electricity of amount $c\bar{R}d\kappa$ per unit area.

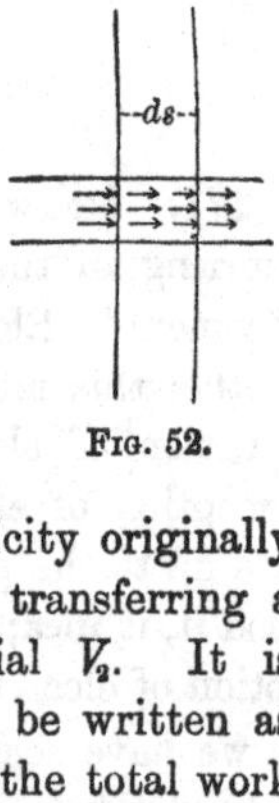

Thus of the electricity originally inside the small element of volume, a quantity $cR\omega\,d\kappa$ flows *out* across one of the bounding equipotentials, whilst an equal quantity flows *in* across the other. Let V_1, V_2 be the potentials of these surfaces, then the whole work done in displacing the electricity originally inside the element of volume $\omega\, ds$, is exactly the work of transferring a quantity $cR d\kappa$ of electricity from potential V_1 to potential V_2. It is therefore $cR\omega\,(V_2 - V_1)\,d\kappa$ and, since $V_2 - V_1 = \kappa R ds$, this may be written as $cR^2\omega\,ds\,\kappa\,d\kappa$. Thus as the intensity is increased from 0 to R, the total work spent in displacing the electricity in the element of volume $\omega\, ds$

$$= \int_0^1 cR^2\,(\omega\,ds)\,\kappa\,d\kappa = \tfrac{1}{2}cR^2\,.\,\omega\,ds.$$

This work, on Maxwell's theory, is simply the energy stored up in the element of volume $\omega\, ds$ of the medium, and is therefore equal to $\dfrac{R^2}{8\pi}\,\omega\,ds$.

Thus c must be taken equal to $\dfrac{1}{4\pi}$, and the displacement at any point is measured by

$$\frac{R}{4\pi}.$$

Fig. 52.

If the element of volume is taken in a dielectric of inductive capacity K, the energy is $\dfrac{KR^2}{8\pi}$, so that $c = \dfrac{K}{4\pi}$, and the displacement is

$$\frac{KR}{4\pi}.$$

173. It is now evident that Maxwell's "displacement" is identical in magnitude and direction with Faraday's "polarisation" introduced in Chap. v.

Denoting either quantity by P, we had the relation

$$\iint P \cos \epsilon \, dS = E \quad\text{.............................(93)},$$

expressing that the normal component of P integrated over any closed surface is equal to the total charge inside. On Maxwell's interpretation of the quantity P, the surface integral $\iint P \cos \epsilon \, dS$ simply measures the total quantity of electricity which has crossed the surface from inside to outside. Thus equation (93) expresses that *the total outward displacement across any closed surface is equal to the total charge inside.*

If we now follow Maxwell in supposing that electricity is of two kinds,

(i) the kind which appears as a charge on an electrified body,

(ii) the kind which Maxwell imagines to occupy the whole of space, and to undergo displacement when electric action takes place,

then it appears that any increase of electricity of kind (i) inside any closed surface is accompanied by an exactly equal decrease of electricity of kind (ii). In other words the sum total of the two kinds of electricity inside any closed surface remains constant.

174. It will be understood that Maxwell's theory of electrical displacement attempts to give a physical picture of the processes of the electric field, but that the truth of the picture is by no means essential to the mathematical theory of electricity. The displacement theory is historically important because it led Maxwell to the hypothesis of displacement currents which form the foundation of his electromagnetic theory of light (Chap. xvii). But we shall see later that the general electromagnetic theory can be developed without the preliminary displacement theory. The displacement theory has served as part of the scaffolding by which the electromagnetic theory was constructed; whether the scaffolding ought now to be discarded remains an open question.

CHAPTER VII

GENERAL ANALYTICAL THEOREMS

Green's Theorem.

175. A THEOREM, first given by Green, and commonly called after him, enables us to express an integral taken over the surfaces of a number of bodies as an integral taken through the space between them. This theorem naturally has many applications to Electrostatic Theory. It supplies a means of handling analytically the problems which Faraday treated geometrically with the help of his conception of tubes of force.

176. THEOREM. *If u, v, w are continuous functions of the Cartesian coordinates x, y, z, then*

$$\Sigma \iint (lu + mv + nw)\, dS = -\iiint \left(\frac{\partial u}{\partial x} + \frac{\partial v}{\partial y} + \frac{\partial w}{\partial z}\right) dx\, dy\, dz \quad \ldots\ldots(94).$$

Here Σ denotes that the surface integrals are summed over any number of closed surfaces, which may include as special cases either

(i) one of finite size which encloses all the others, or

(ii) an imaginary sphere of infinite radius,

and l, m, n are the direction-cosines of the normal drawn in every case from the element dS into the space between the surfaces. The volume integral is taken throughout the space between the surfaces.

Consider first the value of $\iint \frac{\partial u}{\partial x}\, dx\, dy\, dz$. Take any small prism with its axis parallel to that of x, and of cross section $dy\, dz$. Let it meet the surfaces at P, Q, R, S, T, U, ... (fig. 53), cutting off areas dS_P, dS_Q, dS_R, ...

The contribution of this prism to $\iiint \frac{\partial u}{\partial x}\, dx\, dy\, dz$ is $dy\, dz \int \frac{\partial u}{\partial x}\, dx$, where the integral is taken over those parts of the prism which are between the surfaces.

Thus
$$\int \frac{\partial u}{\partial x}\, dx = \int_P^Q \frac{\partial u}{\partial x}\, dx + \int_R^S \frac{\partial u}{\partial x}\, dx + \ldots$$
$$= -u_P + u_Q - u_R + u_S - \ldots,$$

where u_P, u_Q, u_R,... are the values of u at P, Q, R,.... Also, since the projection of each of the areas dS_P, dS_Q,... on the plane of yz is $dy\,dz$, we have

$$dy\,dz = l_P dS_P = - l_Q dS_Q = l_R dS_R = \ldots,$$

where l_P, l_Q, l_R,... are the values of l at P, Q, R,.... The signs in front of l_P, l_Q, l_R,... are alternately positive and negative, because, as we proceed along $PQR...$, the normal drawn *into* the space between the surfaces makes angles which are alternately acute and obtuse with the positive axis of x.

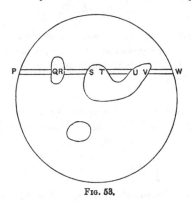

Fig. 53.

Thus

$$dy\,dz \int \frac{\partial u}{\partial x}\,dx = \quad dy\,dz\,(-u_P + u_Q - u_R + \ldots)$$

$$= - l_P u_P dS_P - l_Q u_Q dS_Q - l_R u_R dS_R - \ldots \quad \ldots\ldots\ldots(95),$$

and on adding the similar equations obtained for all the prisms we obtain

$$\iiint \frac{\partial u}{\partial x}\,dx\,dy\,dz = - \Sigma \iint lu\,dS \quad \ldots\ldots\ldots\ldots\ldots\ldots(96),$$

the terms on the right-hand sides of equations of the type (95) combining so as exactly to give the term on the right-hand side of (96).

We can treat the functions v and w similarly, and so obtain altogether

$$\iiint \left(\frac{\partial u}{\partial x} + \frac{\partial v}{\partial y} + \frac{\partial w}{\partial z} \right) dx\,dy\,dz = - \Sigma \iint (lu + mv + nw)\,dS,$$

proving the theorem.

177. If u, v, w are the three components of any vector $\mathbf{F}$, then the expression

$$\frac{\partial u}{\partial x} + \frac{\partial v}{\partial y} + \frac{\partial w}{\partial z}$$

is denoted, for reasons which will become clear later, by div $\mathbf{F}$. If N is the component of the vector in the direction of the normal (l, m, n) to dS, then

$$N = lu + mv + nw.$$

Thus Green's Theorem assumes the form

$$\iiint \operatorname{div} \mathbf{F}\, dx\, dy\, dz = - \Sigma \iint N\, dS \quad \ldots\ldots\ldots\ldots\ldots (97).$$

A vector F which is such that $\operatorname{div} \mathbf{F} = 0$ at every point within a certain region is said to be "solenoidal" within that region. If $\mathbf{F}$ is solenoidal within any region, Green's Theorem shews that

$$\iint N\, dS = 0,$$

where the integral is taken over any closed surface inside the region within which $\mathbf{F}$ is solenoidal. Two instances of a solenoidal vector have so far occurred in this book—the electric intensity in free space, and the polarisation in an uncharged dielectric.

178. *Integration through space external to closed surfaces.* Let the outer surface be a sphere at infinity, say a sphere of radius r, where r is to be made infinite in the limit. The value of

$$\iint (lu + mv + nw)\, dS$$

taken over this sphere will vanish if u, v, and w vanish more rapidly at infinity than $\dfrac{1}{r^2}$. Thus, if this condition is satisfied, we have that

$$\iiint \left(\frac{\partial u}{\partial x} + \frac{\partial v}{\partial y} + \frac{\partial w}{\partial z} \right) dx\, dy\, dz = - \Sigma \iint (lu + mv + nw)\, dS,$$

where the volume integration is taken through all space external to certain closed surfaces, and the surface integration is taken over these surfaces, l, m, n being the direction-cosines of the outward normal.

179. *Integration through the interior of a closed surface.* Let the inner surfaces in fig. 53 all disappear, then we have

$$\iiint \left(\frac{\partial u}{\partial x} + \frac{\partial v}{\partial y} + \frac{\partial w}{\partial z} \right) dx\, dy\, dz = - \iint (lu + mv + nw)\, dS,$$

where the volume integration is throughout the space inside a closed surface, and the surface integration is over this area, l, m, n being the direction-cosines of the inward normal to the surface.

180. *Integration through a region in which u, v, w are discontinuous.* The only case of discontinuity of u, v, w which possesses any physical importance is that in which u, v, w change discontinuously in value in crossing certain surfaces, these being finite in number. To treat this case, we enclose each surface of discontinuity inside a surface drawn so as to fit it closely on both sides. In the space left, after the interiors of such closed surfaces have been excluded, the functions u, v, w are continuous. We may accordingly apply Green's Theorem, and obtain

$$\iiint\left(\frac{\partial u}{\partial x} + \frac{\partial v}{\partial y} + \frac{\partial w}{\partial z}\right) dx\,dy\,dz = -\Sigma \iint (lu + mv + nw)\, dS$$

$$-\Sigma' \iint (lu + mv + nw)\, dS \dots\dots(98),$$

where Σ denotes summation over the closed surfaces by which the original space was limited, and Σ' denotes summation over the new closed surfaces which surround surfaces of discontinuity of u, v, w. Now corresponding to any element of area dS on a surface of discontinuity there will be two elements of area of the enclosing surface. Let the direction-cosines of the two normals to dS be l_1, m_1, n_1 and l_2, m_2, n_2, so that $l_1 = -l_2$, $m_1 = -m_2$, and $n_1 = -n_2$. Let these direction-cosines be those of normals drawn from dS to the two sides of the surface, which we shall denote by 1 and 2, and let the values of u, v, w on the two sides of the surface of discontinuity at the element dS be u_1, v_1, w_1 and u_2, v_2, w_2. Then clearly the two elements of the enclosing surface, which fit against the element dS of the original surface of discontinuity, will contribute to

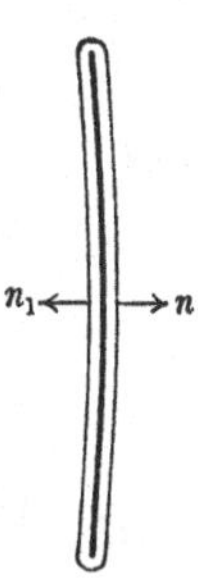

Fig. 54.

$$\Sigma' \iint (lu + mv + nw)\, dS$$

an amount
$$dS\,[(l_1 u_1 + m_1 v_1 + n_1 w_1) + (l_2 u_2 + m_2 v_2 + n_2 w_2)]$$

or
$$\{l_1(u_1 - u_2) + m_1(v_1 - v_2) + n_1(w_1 - w_2)\}\, dS.$$

Thus the whole value of $\Sigma' \iint (lu + mv + nw)\, dS$ may be expressed in the form

$$\Sigma'' \iint \{l_1(u_1 - u_2) + m_1(v_1 - v_2) + n_1(w_1 - w_2)\}\, dS,$$

where the integration is now over the actual surfaces of discontinuity. Thus Green's Theorem becomes

$$\iiint\left(\frac{\partial u}{\partial x} + \frac{\partial v}{\partial y} + \frac{\partial w}{\partial z}\right) dx\,dy\,dz$$

$$= -\Sigma \iint (lu + mv + nw)\, dS$$

$$-\Sigma'' \iint \{l_1(u_1 - u_2) + m_1(v_1 - v_2) + n_1(w_1 - w_2)\}\, dS \dots\dots(99).$$

Special Form of Green's Theorem.

181. An important case of the theorem occurs when u, v, w have the special values

$$u = \Phi \frac{\partial \Psi}{\partial x},$$

$$v = \Phi \frac{\partial \Psi}{\partial y},$$

$$w = \Phi \frac{\partial \Psi}{\partial z},$$

where Φ and Ψ are any functions of x, y and z. The value of $(lu + mv + nw)$ is now

$$\Phi \left(l \frac{\partial \Psi}{\partial x} + m \frac{\partial \Psi}{\partial y} + n \frac{\partial \Psi}{\partial z} \right)$$

or

$$\Phi \frac{\partial \Psi}{\partial n},$$

where $\frac{\partial}{\partial n}$ denotes differentiation along the normal, of which the direction-cosines are l, m, n.

We also have

$$\frac{\partial u}{\partial x} + \frac{\partial v}{\partial y} + \frac{\partial w}{\partial z} = \frac{\partial}{\partial x} \left\{ \Phi \frac{\partial \Psi}{\partial x} \right\} + \frac{\partial}{\partial y} \left\{ \Phi \frac{\partial \Psi}{\partial y} \right\} + \frac{\partial}{\partial z} \left\{ \Phi \frac{\partial \Psi}{\partial z} \right\}$$

$$= \frac{\partial \Phi}{\partial x} \frac{\partial \Psi}{\partial x} + \frac{\partial \Phi}{\partial y} \frac{\partial \Psi}{\partial y} + \frac{\partial \Phi}{\partial z} \frac{\partial \Psi}{\partial z} + \Phi \left(\frac{\partial^2 \Psi}{\partial x^2} + \frac{\partial^2 \Psi}{\partial y^2} + \frac{\partial^2 \Psi}{\partial z^2} \right).$$

Thus the theorem becomes

$$\iiint \left\{ \Phi \nabla^2 \Psi + \frac{\partial \Phi}{\partial x} \frac{\partial \Psi}{\partial x} + \frac{\partial \Phi}{\partial y} \frac{\partial \Psi}{\partial y} + \frac{\partial \Phi}{\partial z} \frac{\partial \Psi}{\partial z} \right\} dx\, dy\, dz = - \Sigma \iint \Phi \frac{\partial \Psi}{\partial n}\, dS \dots (100).$$

This theorem is true for all values of Φ and Ψ, so that we may interchange Φ and Ψ, and the equation remains true. Subtracting the equation so obtained from equation (100), we get

$$\iiint (\Phi \nabla^2 \Psi - \Psi \nabla^2 \Phi)\, dx\, dy\, dz = - \Sigma \iint \left(\Phi \frac{\partial \Psi}{\partial n} - \Psi \frac{\partial \Phi}{\partial n} \right) dS \dots (101).$$

APPLICATIONS OF GREEN'S THEOREM.

182. In equation (101), put $\Phi = 1$ and $\Psi = V$, where V denotes the electrostatic potential. We obtain

$$\iiint \nabla^2 V\, dx\, dy\, dz = - \Sigma \iint \frac{\partial V}{\partial n}\, dS \dots (102).$$

Let us divide the sum on the right into I_1, the integral over a single closed surface enclosing any number of conductors, and I_2, the integrals over the surfaces of the conductors. Thus

$$I_1 = - \iint \frac{\partial V}{\partial n}\, dS,$$

where $\dfrac{\partial}{\partial n}$ denotes differentiation along the normal drawn *into* the surface.

Thus $-\dfrac{\partial V}{\partial n}$ is equal to the component of intensity along this normal, and therefore to $-N$, where N is the component along the *outward* normal. Hence

$$I_1 = - \iint N\, dS.$$

At the surface of a conductor $\dfrac{\partial V}{\partial n} = -4\pi\sigma$, so that

$$I_2 = 4\pi \Sigma \iint \sigma\, dS \text{ over conductors}$$

$$= 4\pi \times \text{total charge on conductors.}$$

If there is any volume electrification, $\nabla^2 V = -4\pi\rho$, so that

$$\iiint \nabla^2 V\, dx\, dy\, dz = -4\pi \iiint \rho\, dx\, dy\, dz,$$

and the integral on the right represents the total volume electrification.

Thus equation (102) becomes

$$\iint N\, dS = 4\pi \times \text{(total charge on conductors + total volume electrification)},$$

so that the theorem reduces to Gauss' Theorem.

183. Next put Φ and Ψ each equal to V. Then equation (100) becomes

$$\iiint V \nabla^2 V\, dx\, dy\, dz + \iiint \left\{ \left(\frac{\partial V}{\partial x}\right)^2 + \left(\frac{\partial V}{\partial y}\right)^2 + \left(\frac{\partial V}{\partial z}\right)^2 \right\}\, dx\, dy\, dz$$
$$+ \Sigma \iint V \frac{\partial V}{\partial n}\, dS = 0.$$

Take the surfaces now to be the surfaces of conductors, and a sphere of radius r at infinity. At infinity V is of order $\dfrac{1}{r}$, so that $\dfrac{\partial V}{\partial n}$ is of order $\dfrac{1}{r^2}$, and hence $V\dfrac{\partial V}{\partial n}$, integrated over the sphere at infinity, vanishes (§ 178).

The equation becomes

$$-4\pi \iiint \rho V\, dx\, dy\, dz + \iiint R^2\, dx\, dy\, dz - 4\pi \iint V\sigma\, dS = 0.$$

The first and last terms together give $-4\pi \times \Sigma eV$, where e is any element of charge, either of volume-electrification or surface-electrification. Thus the whole equation becomes

$$\tfrac{1}{2}\Sigma eV = \iiint \frac{R^2}{8\pi}\, dx\,dy\,dz,$$

shewing that the energy may be regarded as distributed through the space outside the conductors, to the amount $\dfrac{R^2}{8\pi}$ per unit volume—the result already obtained in § 168.

184. In Green's Theorem, take

$$u = \Phi \left(K \frac{\partial \Psi}{\partial x}\right),$$

$$v = \Phi \left(K \frac{\partial \Psi}{\partial y}\right),$$

$$w = \Phi \left(K \frac{\partial \Psi}{\partial z}\right).$$

Here K is ultimately to be taken to be the inductive capacity, which may vary discontinuously on crossing the boundary between two dielectrics. We accordingly suppose u, v, w to be discontinuous, and use Green's Theorem in the form given in § 180. We have then

$$\iiint K \left\{\frac{\partial \Phi}{\partial x}\frac{\partial \Psi}{\partial x} + \frac{\partial \Phi}{\partial y}\frac{\partial \Psi}{\partial y} + \frac{\partial \Phi}{\partial z}\frac{\partial \Psi}{\partial z}\right\} dx\,dy\,dz$$

$$+ \iiint \Phi \left\{\frac{\partial}{\partial x}\left(K \frac{\partial \Psi}{\partial x}\right) + \frac{\partial}{\partial y}\left(K \frac{\partial \Psi}{\partial y}\right) + \frac{\partial}{\partial z}\left(K \frac{\partial \Psi}{\partial z}\right)\right\} dx\,dy\,dz$$

$$= -\Sigma \iint K\Phi \left(l \frac{\partial \Psi}{\partial x} + m \frac{\partial \Psi}{\partial y} + n \frac{\partial \Psi}{\partial z}\right) dS$$

$$- \Sigma'' \iint \left\{l_1 \left(K_1 \Phi_1 \frac{\partial \Psi_1}{\partial x} - K_2 \Phi_2 \frac{\partial \Psi_2}{\partial x}\right) + \dots\right\} dS$$

$$= -\Sigma \iint K\Phi \frac{\partial \Psi}{\partial n}\, dS$$

$$- \Sigma'' \iint \left(K_1 \Phi_1 \frac{\partial \Psi_1}{\partial \nu_1} + K_2 \Phi_2 \frac{\partial \Psi_2}{\partial \nu_2}\right) dS \dots\dots\dots\dots\dots\dots\dots(103),$$

where $\dfrac{\partial}{\partial \nu_1}, \dfrac{\partial}{\partial \nu_2}$ have the meanings assigned to them in § 140.

If we put $\Phi = 1$, $\Psi = V$, in this equation, it reduces, as in § 130, to

$$\iint K \frac{\partial V}{\partial n}\, dS = -4\pi \times \text{total charge inside surface},$$

so that the result is that of the extension of Gauss' Theorem. Again, if we put $\Phi = \Psi = V$, the equation becomes

$$\iiint \frac{KR^2}{8\pi}\, dx\,dy\,dz = \tfrac{1}{2}\Sigma eV,$$

and the result is that of § 169.

Green's Reciprocation Theorem.

185. In equation (101), put $\Phi = V$, $\Psi = V'$, where V is the potential of one distribution of electricity, and V' is that of a second and independent distribution. The equation becomes

$$\iiint (\rho V' - \rho' V)\, dx\, dy\, dz + \Sigma \iint (\sigma V' - \sigma' V)\, dS = 0,$$

which is simply the theorem of § 102, namely

$$\Sigma e V' = \Sigma e' V \dots\dots\dots\dots\dots\dots(104).$$

If we assign the same values to Φ, Ψ in equation (103), we again obtain equation (104), which is now seen to be applicable when dielectrics are present.

UNIQUENESS OF SOLUTION.

186. We can use Green's Theorem to obtain analytical proofs of the theorems already given in § 99.

THEOREM. *If the value of the potential V is known at every point on a number of closed surfaces by which a space is bounded internally and externally, there is only one value for V at every point of this intervening space, which satisfies the condition that $\nabla^2 V$ either vanishes or has an assigned value, at every point of this space.*

For, if possible, let V, V' denote two values of the potential, both of which satisfy the requisite conditions. Then $V' - V = 0$ at every point of the surfaces, and $\nabla^2 (V' - V) = 0$ at every point of the space. Putting Φ and Ψ each equal to $V' - V$ in equation (100), we obtain

$$\iiint \left\{ \left(\frac{\partial (V' - V)}{\partial x} \right)^2 + \left(\frac{\partial (V' - V)}{\partial y} \right)^2 + \left(\frac{\partial (V' - V)}{\partial z} \right)^2 \right\} dx\, dy\, dz = 0,$$

and this integral, being a sum of squares, can only vanish through the vanishing of each term. We must therefore have

$$\frac{\partial}{\partial x} (V' - V) = \frac{\partial}{\partial y} (V' - V) = \frac{\partial}{\partial z} (V' - V) = 0 \dots\dots\dots\dots(105),$$

or $V' - V$ equal to a constant. And since $V' - V$ vanishes at the surfaces, this constant must be zero, so that $V = V'$ everywhere, *i.e.* the two solutions V and V' are identical: there is only one solution.

187. THEOREM. *Given the value of $\dfrac{\partial V}{\partial n}$ at every point of a number of closed surfaces, there is only one possible value for V (except for additive constants), at each point of the intervening space, subject to the condition that $\nabla^2 V = 0$ throughout this space, or has an assigned value at each point.*

The proof is almost identical with that of the last theorem, the only difference being that at every point of the surfaces we have

$$\frac{\partial}{\partial n}(V' - V) = 0,$$

instead of the former condition $V' - V = 0$. We still have

$$\Sigma \iint (V' - V) \frac{\partial}{\partial n}(V' - V)\, dS = 0,$$

so that equation (105) is true, and the result follows as before, except that V and V' may now differ by a constant.

188. Theorems exactly similar to these last two theorems are easily seen to be true when the dielectric is different from air.

For, let V, V' be two solutions, such that

$$\frac{\partial}{\partial x}\left\{K \frac{\partial}{\partial x}(V - V')\right\} + \frac{\partial}{\partial y}\left\{K \frac{\partial}{\partial y}(V - V')\right\} + \frac{\partial}{\partial z}\left\{K \frac{\partial}{\partial z}(V - V')\right\} = 0$$

at all points of the space, and at the surface either $V - V' = 0$, or $\frac{\partial}{\partial n}(V - V') = 0$.

By Green's Theorem

$$\iiint K \left[\left\{\frac{\partial (V - V')}{\partial x}\right\}^2 + \left\{\frac{\partial (V - V')}{\partial y}\right\}^2 + \left\{\frac{\partial (V - V')}{\partial z}\right\}^2\right] dx\,dy\,dz$$

$$= -\iiint (V - V') \left[\frac{\partial}{\partial x}\left\{K \frac{\partial}{\partial x}(V - V')\right\} + \frac{\partial}{\partial y}\left\{K \frac{\partial}{\partial y}(V - V')\right\}\right.$$

$$\left. + \frac{\partial}{\partial z}\left\{K \frac{\partial}{\partial z}(V - V')\right\}\right] dx\,dy\,dz$$

$$+ \Sigma \iint K(V - V') \frac{\partial}{\partial n}(V - V')\, dS$$

$= 0$ by hypothesis.

Equation (105) now follows as before, so that the result is proved.

COMPARISONS OF DIFFERENT FIELDS.

189. Theorem. *If any number of surfaces are fixed in position, and a given charge is placed on each surface, then the energy is a minimum when the charges are placed so that every surface is an equipotential.*

Let V' be the actual potential at any point of the field, and V the potential when the electricity is arranged so that each surface is an equipotential. Calling the corresponding energies W' and W, we have

$$W' - W = \frac{1}{8\pi} \iiint \left\{ \left(\frac{\partial V'}{\partial x}\right)^2 - \left(\frac{\partial V}{\partial x}\right)^2 + \ldots \right\} dx\,dy\,dz$$

$$= \frac{1}{8\pi} \iiint \left\{ \left(\frac{\partial V'}{\partial x} - \frac{\partial V}{\partial x}\right)^2 + \ldots \right\} dx\,dy\,dz$$

$$+ \frac{1}{8\pi} \iiint \left\{ 2\left(\frac{\partial V'}{\partial x} - \frac{\partial V}{\partial x}\right)\frac{\partial V}{\partial x} + \ldots \right\} dx\,dy\,dz.$$

If we put $\Phi = V$, $\Psi = V' - V$, in equation (100), we find that the last integral becomes

$$-\Sigma \frac{1}{4\pi} \iint V \left(\frac{\partial V'}{\partial n} - \frac{\partial V}{\partial n}\right) dS,$$

or, since V is by hypothesis constant over each conductor,

$$\Sigma V \iint (\sigma' - \sigma)\, dS,$$

and this vanishes since each total charge $\iint \sigma' dS$ is the same as the corresponding total charge $\iint \sigma dS$. Thus

$$W' - W = \frac{1}{8\pi} \iiint \left\{ \left(\frac{\partial V'}{\partial x} - \frac{\partial V}{\partial x}\right)^2 + \ldots \right\} dx\,dy\,dz.$$

This integral is essentially positive, so that W' is greater than W, which proves the theorem.

If any distribution is suddenly set free and allowed to flow so that the surface of each conductor becomes an equipotential, the loss of energy $W' - W$ is seen to be equal to the energy of a field of potential $V' - V$ at any point.

190. THEOREM. *The introduction of a new conductor lessens the energy of the field.*

Let accented symbols refer to the field after a new conductor S has been introduced, insulated and uncharged. Then

$$W - W' = \frac{1}{8\pi} \iiint R^2 dx\,dy\,dz \text{ through the field before } S \text{ is introduced}$$

$$- \frac{1}{8\pi} \iiint R'^2 dx\,dy\,dz \text{ through the field after } S \text{ is introduced}$$

$$= \frac{1}{8\pi} \iiint R^2 dx\,dy\,dz \text{ through the space ultimately occupied by } S$$

$$+ \frac{1}{8\pi} \iiint (R^2 - R'^2) \text{ through the field after } S \text{ is introduced.}$$

The last integral

$$= \frac{1}{8\pi} \iiint \left\{ \left(\frac{\partial V}{\partial x}\right)^2 - \left(\frac{\partial V'}{\partial x}\right)^2 + \ldots \right\} dx\,dy\,dz,$$

and this, as in the last theorem, is equal to

$$\frac{1}{8\pi} \iiint \left\{ \left(\frac{\partial V}{\partial x} - \frac{\partial V'}{\partial x}\right)^2 + \ldots \right\} dx\,dy\,dz$$

$$+ \frac{1}{4\pi} \Sigma V' \iint \left(\frac{\partial V}{\partial n} - \frac{\partial V'}{\partial n}\right) dS,$$

where Σ denotes summation over all conductors, including S.

This last sum of surface integrals vanishes, so that

$$W - W' = \frac{1}{8\pi} \iiint R^2 dx\,dy\,dz \text{ through } S$$

$$+ \frac{1}{8\pi} \iiint \left\{ \left(\frac{\partial V}{\partial x} - \frac{\partial V'}{\partial x}\right)^2 + \ldots \right\} dx\,dy\,dz \quad \text{through the field after}$$

$$S \text{ has been introduced.}$$

Thus $W - W'$ is essentially positive, which proves the theorem.

On putting the new conductor to the earth, it follows from the preceding theorem that the energy is still further lessened.

191. THEOREM. *Any increase in the inductive capacity of the dielectric between conductors lessens the energy of the field.*

Let the conductors of the field be supposed fixed in position and insulated, so that their total charge remains unaltered. Let the inductive capacity at any point change from K to $K + \delta K$, and as a consequence let the potential change from V to $V + \delta V$, and the total energy of the field from W to $W + \delta W$.

If $E_1, E_2, \ldots$ denote the total charges of the conductors, $V_1, V_2, \ldots$ their potentials, and ρ the volume density at any point,

$$W = \tfrac{1}{2} \Sigma EV + \tfrac{1}{2} \iiint \rho V dx\,dy\,dz,$$

so that, since the E's and ρ remain unaltered by changes in K, we have

$$\delta W = \tfrac{1}{2} \Sigma E \delta V + \tfrac{1}{2} \iiint \rho \delta V dx\,dy\,dz \quad \ldots\ldots\ldots\ldots(106).$$

We also have

$$W = \frac{1}{8\pi} \iiint K \left\{ \left(\frac{\partial V}{\partial x}\right)^2 + \left(\frac{\partial V}{\partial y}\right)^2 + \left(\frac{\partial V}{\partial z}\right)^2 \right\} dx\,dy\,dz,$$

so that

$$\delta W = \frac{1}{8\pi} \iiint \left\{ \left(\frac{\partial V}{\partial x}\right)^2 + \left(\frac{\partial V}{\partial y}\right)^2 + \left(\frac{\partial V}{\partial z}\right)^2 \right\} \delta K \, dx\,dy\,dz$$

$$+ \frac{1}{4\pi} \iiint K \left\{ \frac{\partial V}{\partial x} \frac{\partial \delta V}{\partial x} + \frac{\partial V}{\partial y} \frac{\partial \delta V}{\partial y} + \frac{\partial V}{\partial z} \frac{\partial \delta V}{\partial z} \right\} dx\,dy\,dz \quad \ldots(107).$$

By Green's Theorem, the last line

$$= -\frac{1}{4\pi} \iiint \delta V \left\{ \frac{\partial}{\partial x} \left(K \frac{\partial V}{\partial x} \right) + \frac{\partial}{\partial y} \left(K \frac{\partial V}{\partial y} \right) + \frac{\partial}{\partial z} \left(K \frac{\partial V}{\partial z} \right) \right\} dx\,dy\,dz$$

$$- \Sigma \frac{1}{4\pi} \iint \delta V \left\{ lK \frac{\partial V}{\partial x} + mK \frac{\partial V}{\partial y} + nK \frac{\partial V}{\partial z} \right\} dS,$$

the summation of surface integrals being over the surfaces of all the conductors,

$$= \iiint \rho \, \delta V \, dx\,dy\,dz + \Sigma \iint \sigma \, \delta V \, dS$$

$$= \iiint \rho \, \delta V \, dx\,dy\,dz + \Sigma E \delta V$$

$$= 2\delta W$$

by equation (106). Thus equation (107) becomes

$$\delta W = \frac{1}{8\pi} \iiint R^2 \, \delta K \, dx\,dy\,dz + 2\delta W,$$

so that

$$\delta W = -\frac{1}{8\pi} \iiint R^2 \, \delta K \, dx\,dy\,dz.$$

Thus δW is necessarily negative if δK is positive, proving the theorem.

It is worth noticing that, on the molecular theory of dielectrics, the increase in the inductive capacity of the dielectric at any point will be most readily accomplished by introducing new molecules. If, as in Chap. v, these molecules are regarded as uncharged conductors, the theorem just proved becomes identical with that of § 190.

EARNSHAW'S THEOREM.

192. THEOREM. *A charged body placed in an electric field of force cannot rest in stable equilibrium under the influence of the electric forces alone.*

Let us suppose the charged body A to be in any position, in the field of force produced by other bodies B, B', ... First suppose all the electricity on A, B, B', ... to be fixed in position on these conductors. Let V denote the potential, at any point of the field, of the electricity on B, B', ... Let x, y, z be the coordinates of any definite point in A, say its centre of gravity, and let $x + a$, $y + b$, $z + c$ be the coordinates of any other point. The potential energy of any element of charge e at $x + a$, $y + b$, $z + c$ is eV, where V is evaluated at $x + a$, $y + b$, $z + c$. Denoting eV by w, we clearly have

$$\frac{\partial^2 w}{\partial x^2} + \frac{\partial^2 w}{\partial y^2} + \frac{\partial^2 w}{\partial z^2} = 0,$$

since V is a solution of Laplace's equation.

Let W be the total energy of the body A in the field of force from $B, B', \ldots$. Then $W = \Sigma w$, and therefore

$$\frac{\partial^2 W}{\partial x^2} + \frac{\partial^2 W}{\partial y^2} + \frac{\partial^2 W}{\partial z^2} = 0,$$

i.e. the sum $W = \Sigma w$ satisfies Laplace's equation, because this equation is satisfied by the terms of the sum separately. It follows from this equation, as in § 52, that W cannot be a true maximum or a true minimum for any values of x, y, z. Thus, whatever the position of the body A, it will always be possible to find a displacement—*i.e.* a change in the values of x, y, z—for which W decreases. If, after this displacement, the electricity on the conductors $A, B, B', \ldots$ is set free, so that each surface becomes an equipotential, it follows from § 189 that the energy of the field is still further lessened. Thus a displacement of the body A has been found which lessens the energy of the field, and therefore the body A cannot rest in stable equilibrium.

One physical application of Earnshaw's Theorem is of extreme importance. The theorem shews that an electron cannot rest in stable equilibrium under the forces of attraction and repulsion from other charges, so long as these forces are supposed to obey the law of the inverse square of the distance. Thus, if a molecule is to be regarded as a cluster of electrons and positive charges, as in § 151, then the law of force must be something different from that of the inverse square.

There seems to be no difficulty about the supposition that at very small distances the law of force is different from the inverse square. On the contrary, there would be a very real difficulty in supposing that the law $1/r^2$ held down to zero values of r. For the force between two charges at zero distance would be infinite ; we should have charges of opposite sign continually rushing together and, when once together, no force would be adequate to separate them. Thus the universe would in time consist only of doublets, each consisting of permanently interlocked positive and negative charges. If the law $1/r^2$ held down to zero values of r, the distance apart of the charges would be zero, so that the strength of each doublet would be *nil*, and there would be no way of detecting its presence. Thus the matter in the universe would tend to shrink into nothing or to diminish indefinitely in size. The observed permanence of matter precludes any such hypothesis.

Earnshaw's Theorem accordingly limits us to two alternatives. Either the molecule does not consist of a cluster of electrons in relative rest, or else the law of the inverse square fails at molecular distances.

Recent experimental investigations decide very definitely against the second alternative and in favour of the first. Recent experiments on the deflection of the positively charged *a*-particles by matter indicate that the law of the inverse square holds down to distances of the order of 10^{-11} cms., a distance which is less than a thousandth part of the radius of the hydrogen atom, and a large mass of other evidence suggests, with a probability approximating to certainty, that the electrons in an atom or molecule must be in rapid orbital motion. Thus the problem of the structure of the molecule is removed from the province of Earnshaw's Theorem.

STRESSES IN THE MEDIUM.

193. Let us take any surface S in the medium, enclosing any number of charges at points and on surfaces $S_1, S_2, \ldots$

Let l, m, n be the direction-cosines of the normal at any point of $S_1, S_2, \ldots$ or S, the normal being supposed drawn, as in Green's Theorem, into the space between the surfaces.

The total mechanical force acting on all the matter inside this surface is compounded of a force eR in the direction of the intensity acting on every point charge or element of volume-charge e, and a force $2\pi\sigma^2$ or $\frac{1}{2}\sigma R$ per unit area on each element of conducting surface. If X, Y, Z are the components parallel to the axes of the total mechanical force,

$$X = \Sigma eX + \Sigma \iint \tfrac{1}{2}\sigma X \, dS$$

$$= \iiint \rho X \, dx \, dy \, dz + \Sigma \iint \tfrac{1}{2}\sigma X \, dS,$$

where the surface integral is taken over all conductors $S_1, S_2, \ldots$ inside the surface S, and the volume integral throughout the space between S and these surfaces. Substituting for ρ and σ,

$$X = \frac{1}{4\pi} \iiint \left(\frac{\partial^2 V}{\partial x^2} + \frac{\partial^2 V}{\partial y^2} + \frac{\partial^2 V}{\partial z^2}\right) \frac{\partial V}{\partial x} \, dx \, dy \, dz$$

$$+ \frac{1}{4\pi} \Sigma \iint \tfrac{1}{2} \left(l\frac{\partial V}{\partial x} + m\frac{\partial V}{\partial y} + n\frac{\partial V}{\partial z}\right) \frac{\partial V}{\partial x} \, dS \quad \ldots\ldots\ldots(108).$$

By Green's Theorem,

$$\iiint \frac{\partial^2 V}{\partial x^2} \frac{\partial V}{\partial x} \, dx\,dy\,dz = \tfrac{1}{2} \iiint \frac{\partial}{\partial x}\left(\frac{\partial V}{\partial x}\right)^2 dx\,dy\,dz$$

$$= -\tfrac{1}{2}\Sigma \iint l\left(\frac{\partial V}{\partial x}\right)^2 dS - \tfrac{1}{2}\iint l\left(\frac{\partial V}{\partial x}\right)^2 dS,$$

$$\iiint \frac{\partial^2 V}{\partial y^2} \frac{\partial V}{\partial x} \, dx\,dy\,dz = -\iiint \frac{\partial V}{\partial y} \frac{\partial}{\partial y}\left(\frac{\partial V}{\partial x}\right) dx\,dy\,dz$$

$$- \Sigma \iint m\frac{\partial V}{\partial x}\frac{\partial V}{\partial y} \, dS - \iint m\frac{\partial V}{\partial x}\frac{\partial V}{\partial y} \, dS.$$

Now

$$\iiint \frac{\partial V}{\partial y}\frac{\partial}{\partial y}\left(\frac{\partial V}{\partial x}\right) dx\,dy\,dz = \iiint \tfrac{1}{2}\frac{\partial}{\partial x}\left(\frac{\partial V}{\partial y}\right)^2 dx\,dy\,dz$$

$$= -\Sigma \iint \tfrac{1}{2} l\left(\frac{\partial V}{\partial y}\right)^2 dS - \iint \tfrac{1}{2} l\left(\frac{\partial V}{\partial y}\right)^2 dS,$$

so that the last equation becomes

$$\iiint \frac{\partial^2 V}{\partial y^2} \frac{\partial V}{\partial x} \, dx \, dy \, dz = \Sigma \iint \left\{ \tfrac{1}{2} l \left(\frac{\partial V}{\partial y} \right)^2 - m \frac{\partial V}{\partial x} \frac{\partial V}{\partial y} \right\} dS$$
$$+ \iint \left\{ \tfrac{1}{2} l \left(\frac{\partial V}{\partial y} \right)^2 - m \frac{\partial V}{\partial x} \frac{\partial V}{\partial y} \right\} dS,$$

and there is a similar value for

$$\iiint \frac{\partial^2 V}{\partial z^2} \frac{\partial V}{\partial x} \, dx \, dy \, dz.$$

Substituting these values, equation (108) becomes

$$\mathsf{X} = \frac{1}{4\pi} \Sigma \iint \left\{ \tfrac{1}{2} l \left[\left(\frac{\partial V}{\partial y} \right)^2 + \left(\frac{\partial V}{\partial z} \right)^2 \right] - \tfrac{1}{2} m \frac{\partial V}{\partial x} \frac{\partial V}{\partial y} - \tfrac{1}{2} n \frac{\partial V}{\partial x} \frac{\partial V}{\partial z} \right\} dS$$
$$- \frac{1}{4\pi} \iint \left\{ \tfrac{1}{2} l \left[\left(\frac{\partial V}{\partial x} \right)^2 - \left(\frac{\partial V}{\partial y} \right)^2 - \left(\frac{\partial V}{\partial z} \right)^2 \right] + m \frac{\partial V}{\partial x} \frac{\partial V}{\partial y} + n \frac{\partial V}{\partial x} \frac{\partial V}{\partial z} \right\} dS.$$

Since we have at every point of the surface of a conductor

$$\frac{\frac{\partial V}{\partial x}}{l} = \frac{\frac{\partial V}{\partial y}}{m} = \frac{\frac{\partial V}{\partial z}}{n} \dots\dots\dots\dots\dots\dots(109),$$

it follows that the integral over each conductor vanishes, leaving only the integral with respect to dS, which gives

$$\mathsf{X} = - \iint (l P_{xx} + m P_{xy} + n P_{xz}) \, dS,$$

where

$$P_{xx} = \frac{1}{8\pi} (X^2 - Y^2 - Z^2),$$

$$P_{xy} = \frac{1}{4\pi} X Y,$$

$$P_{xz} = \frac{1}{4\pi} X Z.$$

If we write also

$$P_{yy} = \frac{1}{8\pi} (Y^2 - Z^2 - X^2),$$

$$P_{zz} = \frac{1}{8\pi} (Z^2 - X^2 - Y^2),$$

$$P_{yz} = \frac{1}{4\pi} Y Z,$$

the resultant force parallel to the axis of Y will be

$$\mathsf{Y} = - \iint (l P_{xy} + m P_{yy} + n P_{yz}) \, dS,$$

and there is a similar value for Z. The action is therefore the same (cf. § 159) as if there was a system of stresses of components

$$P_{xx}, \ P_{yy}, \ P_{zz}, \ P_{yz}, \ P_{zx}, \ P_{xy},$$

given by the above equations: *i.e.* these may be regarded as the stresses of the medium.

194. It remains to investigate the couples on the system inside S. If L, M, N are the moments of the resultant couple about the axes of x, y, z, we have

$$L = \iiint \rho\,(yZ - zY)\,dx\,dy\,dz + \tfrac{1}{2}\Sigma \iint \sigma\,(yZ - zY)\,dS$$

$$= \frac{1}{4\pi} \iiint \left(\frac{\partial^2 V}{\partial x^2} + \frac{\partial^2 V}{\partial y^2} + \frac{\partial^2 V}{\partial z^2}\right)\left(y\,\frac{\partial V}{\partial z} - z\,\frac{\partial V}{\partial y}\right) dx\,dy\,dz$$

$$+ \frac{1}{8\pi} \Sigma \iint \left(l\,\frac{\partial V}{\partial x} + m\,\frac{\partial V}{\partial y} + n\,\frac{\partial V}{\partial z}\right)\left(y\,\frac{\partial V}{\partial z} - z\,\frac{\partial V}{\partial y}\right) dS.$$

Now

$$\iiint \frac{\partial^2 V}{\partial x^2}\left(y\,\frac{\partial V}{\partial z} - z\,\frac{\partial V}{\partial y}\right) dx\,dy\,dz$$

$$= -\iiint \frac{\partial V}{\partial x}\,\frac{\partial}{\partial x}\left(y\,\frac{\partial V}{\partial z} - z\,\frac{\partial V}{\partial y}\right) dx\,dy\,dz$$

$$- \Sigma \iint l\,\frac{\partial V}{\partial x}\left(y\,\frac{\partial V}{\partial z} - z\,\frac{\partial V}{\partial y}\right) dS - \iint l\,\frac{\partial V}{\partial x}\left(y\,\frac{\partial V}{\partial z} - z\,\frac{\partial V}{\partial y}\right) dS,$$

so that

$$L = -\frac{1}{4\pi} \iiint \left\{\frac{\partial V}{\partial x}\,\frac{\partial}{\partial x}\left(y\,\frac{\partial V}{\partial z} - z\,\frac{\partial V}{\partial y}\right) + \frac{\partial V}{\partial y}\,\frac{\partial}{\partial y}\left(y\,\frac{\partial V}{\partial z} - z\,\frac{\partial V}{\partial y}\right)\right.$$

$$\left. + \frac{\partial V}{\partial z}\,\frac{\partial}{\partial z}\left(y\,\frac{\partial V}{\partial z} - z\,\frac{\partial V}{\partial y}\right)\right\} dx\,dy\,dz$$

$$- \frac{1}{8\pi} \Sigma \iint \left(l\,\frac{\partial V}{\partial x} + m\,\frac{\partial V}{\partial y} + n\,\frac{\partial V}{\partial z}\right)\left(y\,\frac{\partial V}{\partial z} - z\,\frac{\partial V}{\partial y}\right) dS$$

$$- \frac{1}{4\pi} \iint \left(l\,\frac{\partial V}{\partial x} + m\,\frac{\partial V}{\partial y} + n\,\frac{\partial V}{\partial z}\right)\left(y\,\frac{\partial V}{\partial z} - z\,\frac{\partial V}{\partial y}\right) dS \quad\ldots\ldots\ldots(110).$$

The first term in this expression

$$= -\frac{1}{4\pi} \iiint \left\{y\left(\frac{\partial V}{\partial x}\,\frac{\partial^2 V}{\partial x\,\partial z} + \frac{\partial V}{\partial y}\,\frac{\partial^2 V}{\partial y\,\partial z} + \frac{\partial V}{\partial z}\,\frac{\partial^2 V}{\partial z^2}\right)\right.$$

$$\left. - z\left(\frac{\partial V}{\partial x}\,\frac{\partial^2 V}{\partial x\,\partial y} + \frac{\partial V}{\partial y}\,\frac{\partial^2 V}{\partial y^2} + \frac{\partial V}{\partial z}\,\frac{\partial^2 V}{\partial y\,\partial z}\right)\right\} dx\,dy\,dz$$

$$= -\frac{1}{8\pi} \iiint \left(y\,\frac{\partial R^2}{\partial z} - z\,\frac{\partial R^2}{\partial y}\right) dx\,dy\,dz$$

$$= \frac{1}{8\pi} \Sigma \iint (ynR^2 - zmR^2)\,dS + \frac{1}{8\pi} \iint (ynR^2 - zmR^2)\,dS \ldots\ldots\ldots(111).$$

The second term in expression (110) for L may, in virtue of the relations (109), be expressed in the form

$$- \frac{1}{8\pi} \Sigma \iint (ynR^2 - zmR^2)\,dS,$$

which is exactly cancelled by the first term in expression (111).

We are accordingly left with

$$\mathsf{L} = \frac{1}{8\pi} \iint \left\{ (ynR^2 - zmR^2) - 2\left(l\frac{\partial V}{\partial x} + m\frac{\partial V}{\partial y} + n\frac{\partial V}{\partial z} \right)\left(y\frac{\partial V}{\partial z} - z\frac{\partial V}{\partial y} \right) \right\} d\mathsf{S}$$

$$= -\iint \{ y\,(lP_{xz} + mP_{yz} + nP_{zz}) - z\,(lP_{xy} + mP_{yy} + nP_{yz}) \}\, d\mathsf{S},$$

verifying that the couples are also accounted for by the supposed system of ether-stresses.

195.　Thus the stresses in the ether are identical with those already found in Chapter VI, and these, as we have seen, may be supposed to consist of a *tension* $\dfrac{R^2}{8\pi}$ per unit area across the lines of force, and a *pressure* $\dfrac{R^2}{8\pi}$ per unit area in directions perpendicular to the lines of force.

MECHANICAL FORCES ON DIELECTRICS IN THE FIELD.

196.　Let us begin by considering a field in which there are no surface charges, and no discontinuities in the structure of the dielectrics. We shall afterwards be able to treat surface-charges and discontinuities as limiting cases.

Let us suppose that the mechanical forces on material bodies are Ξ, H, Z per unit volume at any typical point x, y, z of this field.

Let us displace the material bodies in the field in such a way that the point x, y, z comes to the point $x + \delta x$, $y + \delta y$, $z + \delta z$. The work done in the whole field will be

$$= -\iiint (\Xi\delta x + \mathrm{H}\delta y + \mathrm{Z}\delta z)\, dx\,dy\,dz \quad\ldots\ldots\ldots\ldots(112),$$

and this must shew itself in an equal increase in the electric energy. The electric energy W can be put in either of the forms

$$W = W_1 \equiv \tfrac{1}{2}\iiint \rho V\,dx\,dy\,dz,$$

$$W = W_2 \equiv \frac{1}{8\pi}\iiint K\left\{ \left(\frac{\partial V}{\partial x}\right)^2 + \left(\frac{\partial V}{\partial y}\right)^2 + \left(\frac{\partial V}{\partial z}\right)^2 \right\} dx\,dy\,dz.$$

When the displacement takes place, there will be a slight variation in the distribution of electricity and a slight alteration of the potential. There is also a slight change in the value of K at any point owing to the motion of the dielectrics in the field. Thus we can put

$$\delta W = \delta W_1 = (\delta W_1)_\rho + (\delta W_1)_V,$$

$$\delta W = \delta W_2 = (\delta W_2)_K + (\delta W_2)_V,$$

where $(\delta W_1)_\rho$ denotes the change produced in the function W_1 by the variation of electrical density alone, $(\delta W_1)_V$ that produced by the variation of potential alone, and so on.

We have

$$(\delta W_1)_V = \tfrac{1}{2} \iiint \rho\, \delta V\, dx\, dy\, dz,$$

$$(\delta W_2)_V = \frac{1}{4\pi} \iiint K \left(\frac{\partial V}{\partial x} \frac{\partial \delta V}{\partial x} + \frac{\partial V}{\partial y} \frac{\partial \delta V}{\partial y} + \frac{\partial V}{\partial z} \frac{\partial \delta V}{\partial z} \right) dx\, dy\, dz.$$

By Green's Theorem, the last expression transforms into

$$(\delta W_2)_V = -\frac{1}{4\pi} \iiint \delta V \left\{ \frac{\partial}{\partial x} \left(K \frac{\partial V}{\partial x} \right) + \frac{\partial}{\partial y} \left(K \frac{\partial V}{\partial y} \right) + \frac{\partial}{\partial z} \left(K \frac{\partial V}{\partial z} \right) \right\} dx\, dy\, dz$$

$$= \iiint \rho\, \delta V\, dx\, dy\, dz,$$

so that $\qquad\qquad 2\,(\delta W_1)_V = (\delta W_2)_V.$

We accordingly have

$$\delta W = 2\delta W_1 - \delta W_2 = 2\,(\delta W_1)_\rho - (\delta W_2)_K,$$

the variation produced by alterations in V no longer appearing.

Now $\qquad (\delta W_1)_\rho = \tfrac{1}{2} \iiint \delta \rho\, V\, dx\, dy\, dz,$

$$(\delta W_2)_K = \frac{1}{8\pi} \iiint \delta K \left\{ \left(\frac{\partial V}{\partial x} \right)^2 + \left(\frac{\partial V}{\partial y} \right)^2 + \left(\frac{\partial V}{\partial z} \right)^2 \right\} dx\, dy\, dz,$$

so that $\qquad\qquad \delta W = \iiint \left\{ V \delta \rho - \frac{R^2}{8\pi} \delta K \right\} dx\, dy\, dz \quad\ldots\ldots\ldots\ldots(113).$

The change in ρ is due to two causes. In the first place, the electrification at x, y, z was originally at $x - \delta x, y - \delta y, z - \delta z$, so that $\delta \rho$ has as part of its value

$$-\frac{\partial \rho}{\partial x} \delta x - \frac{\partial \rho}{\partial y} \delta y - \frac{\partial \rho}{\partial z} \delta z \quad\ldots\ldots\ldots\ldots\ldots\ldots(114).$$

Again, the element of volume $dx\, dy\, dz$ becomes changed by displacement into an element

$$\left\{ dx + \frac{\partial}{\partial x} (\delta x)\, dx \right\} \left\{ dy + \frac{\partial}{\partial y} (\delta y)\, dy \right\} \left\{ dz + \frac{\partial}{\partial z} (\delta z)\, dz \right\},$$

or $\qquad\qquad dx\, dy\, dz \left(1 + \frac{\partial \delta x}{\partial x} + \frac{\partial \delta y}{\partial y} + \frac{\partial \delta z}{\partial z} \right) \quad\ldots\ldots\ldots\ldots(115),$

so that, even if there were no motion of translation, an original charge $\rho\, dx\, dy\, dz$ would after displacement occupy the volume given by expression (115), and this would give an increase in ρ of amount

$$-\rho \left(\frac{\partial \delta x}{\partial x} + \frac{\partial \delta y}{\partial y} + \frac{\partial \delta z}{\partial z} \right) \quad\ldots\ldots\ldots\ldots\ldots\ldots(116).$$

Combining the two parts of $\delta\rho$ given by expressions (114) and (115), we find

$$\delta\rho = -\left\{\frac{\partial}{\partial x}(\rho\,\delta x) + \frac{\partial}{\partial y}(\rho\,\delta y) + \frac{\partial}{\partial z}(\rho\,\delta z)\right\}.$$

The change in K is also due to two causes. In the first place the point which in the displaced position is at x, y, z was originally at $x - \delta x$, $y - \delta y$, $z - \delta z$. Hence as part of the value in δK we have

$$-\frac{\partial K}{\partial x}\,\delta x - \frac{\partial K}{\partial y}\,\delta y - \frac{\partial K}{\partial z}\,\delta z.$$

Also, with the displacement, the density of the medium is changed, so that its molecular structure is changed, and there is a corresponding change in K. If we denote the density of the medium by τ, and the increase in τ produced by the displacement by $\delta\tau$, the increase in K due to this cause will be

$$\frac{\partial K}{\partial \tau}\,\delta\tau,$$

and we know, as in equation (116), that

$$\delta\tau = -\tau\left(\frac{\partial\,\delta x}{\partial x} + \frac{\partial\,\delta y}{\partial y} + \frac{\partial\,\delta z}{\partial z}\right).$$

We now have, as the total value of δK,

$$\delta K = -\frac{\partial K}{\partial x}\,\delta x - \frac{\partial K}{\partial y}\,\delta y - \frac{\partial K}{\partial z}\,\delta z$$
$$-\tau\frac{\partial K}{\partial \tau}\left(\frac{\partial\,\delta x}{\partial x} + \frac{\partial\,\delta y}{\partial y} + \frac{\partial\,\delta z}{\partial z}\right),$$

and hence, on substituting in equation (113) for $\delta\rho$ and δK,

$$\delta W = -\iiint V\left\{\frac{\partial(\rho\,\delta x)}{\partial x} + \frac{\partial(\rho\,\delta y)}{\partial y} + \frac{\partial(\rho\,\delta z)}{\partial z}\right\}dx\,dy\,dz$$
$$+\iiint \frac{R^2}{8\pi}\left(\frac{\partial K}{\partial x}\,\delta x + \frac{\partial K}{\partial y}\,\delta y + \frac{\partial K}{\partial z}\,\delta z\right)dx\,dy\,dz$$
$$+\iiint \frac{R^2}{8\pi}\tau\frac{\partial K}{\partial \tau}\left(\frac{\partial\,\delta x}{\partial x} + \frac{\partial\,\delta y}{\partial y} + \frac{\partial\,\delta z}{\partial z}\right)dx\,dy\,dz.$$

Integrating by parts, this becomes

$$\delta W = \iiint\left(\frac{\partial V}{\partial x}\rho\,\delta x + \frac{\partial V}{\partial y}\rho\,\delta y + \frac{\partial V}{\partial z}\rho\,\delta z\right)dx\,dy\,dz$$
$$+\iiint \frac{R^2}{8\pi}\left(\frac{\partial K}{\partial x}\,\delta x + \frac{\partial K}{\partial y}\,\delta y + \frac{\partial K}{\partial z}\,\delta z\right)dx\,dy\,dz$$
$$-\iiint\left\{\frac{\partial}{\partial x}\left(\frac{R^2}{8\pi}\tau\frac{\partial K}{\partial \tau}\right)\delta x + \frac{\partial}{\partial y}\left(\frac{R^2}{8\pi}\tau\frac{\partial K}{\partial \tau}\right)\delta y + \frac{\partial}{\partial z}\left(\frac{R^2}{8\pi}\tau\frac{\partial K}{\partial \tau}\right)\delta z\right\}dx\,dy\,dz,$$

or, rearranging the terms,

$$\delta W = \iiint \left\{ \left[\rho \frac{\partial V}{\partial x} + \frac{R^2}{8\pi} \left(\frac{\partial K}{\partial x} \right) - \frac{\partial}{\partial x} \left(\frac{R^2}{8\pi} \tau \frac{\partial K}{\partial \tau} \right) \right] \delta x + \left[\ldots \right] \delta y + \left[\ldots \right] \delta z \right\} dx\,dy\,dz.$$

Comparing with expression (112), we obtain

$$\Xi = -\rho \frac{\partial V}{\partial x} - \frac{R^2}{8\pi} \frac{\partial K}{\partial x} + \frac{\partial}{\partial x} \left(\frac{R^2}{8\pi} \tau \frac{\partial K}{\partial \tau} \right) \quad \ldots\ldots\ldots\ldots(117),$$

etc., giving the body forces acting on the matter of the dielectric.

197. This may be written in the form

$$\Xi = \rho X - \frac{R^2}{8\pi} \frac{\partial K}{\partial x} + \frac{\partial}{\partial x} \left(\frac{R^2}{8\pi} \tau \frac{\partial K}{\partial \tau} \right).$$

Thus in addition to the force of components $(\rho X, \rho Y, \rho Z)$ acting on the charges of the dielectric, there is an additional force of components

$$-\frac{R^2}{8\pi} \frac{\partial K}{\partial x}, \quad -\frac{R^2}{8\pi} \frac{\partial K}{\partial y}, \quad -\frac{R^2}{8\pi} \frac{\partial K}{\partial z}$$

arising from variations in K, and also a force of components

$$\frac{\partial}{\partial x} \left(\frac{R^2}{8\pi} \tau \frac{\partial K}{\partial \tau} \right), \quad \frac{\partial}{\partial y} \left(\frac{R^2}{8\pi} \tau \frac{\partial K}{\partial \tau} \right), \quad \frac{\partial}{\partial z} \left(\frac{R^2}{8\pi} \tau \frac{\partial K}{\partial \tau} \right),$$

which occurs when either the intensity of the field or the structure of the dielectric varies from point to point.

STRESSES IN DIELECTRIC MEDIA.

198. Replacing ρ by its value, as given by Laplace's equation, we obtain equation (117) in the form

$$
\begin{aligned}
\Xi = \frac{1}{8\pi} \Bigg\{ & 2 \frac{\partial V}{\partial x} \left[\frac{\partial}{\partial x} \left(K \frac{\partial V}{\partial x} \right) + \frac{\partial}{\partial y} \left(K \frac{\partial V}{\partial y} \right) + \frac{\partial}{\partial z} \left(K \frac{\partial V}{\partial z} \right) \right] \\
& - \frac{\partial K}{\partial x} \left[\left(\frac{\partial V}{\partial x} \right)^2 + \left(\frac{\partial V}{\partial y} \right)^2 + \left(\frac{\partial V}{\partial z} \right)^2 \right] + \frac{\partial}{\partial x} \left(R^2 \tau \frac{\partial K}{\partial \tau} \right) \Bigg\} \\
= \frac{1}{8\pi} \Bigg\{ & - \frac{\partial}{\partial x} \left[K \left(\left(\frac{\partial V}{\partial x} \right)^2 + \left(\frac{\partial V}{\partial y} \right)^2 + \left(\frac{\partial V}{\partial z} \right)^2 \right) \right] \\
& + 2 \frac{\partial V}{\partial x} \frac{\partial}{\partial x} \left(K \frac{\partial V}{\partial x} \right) + K \frac{\partial}{\partial x} \left(\frac{\partial V}{\partial x} \right)^2 \\
& + 2 \frac{\partial V}{\partial x} \frac{\partial}{\partial y} \left(K \frac{\partial V}{\partial y} \right) + K \frac{\partial}{\partial x} \left(\frac{\partial V}{\partial y} \right)^2 \\
& + 2 \frac{\partial V}{\partial x} \frac{\partial}{\partial z} \left(K \frac{\partial V}{\partial z} \right) + K \frac{\partial}{\partial x} \left(\frac{\partial V}{\partial z} \right)^2 \\
& + \frac{\partial}{\partial x} \left(R^2 \tau \frac{\partial K}{\partial \tau} \right) \Bigg\}
\end{aligned}
$$

$$= \frac{1}{8\pi} \left\{ -\frac{\partial}{\partial x} \left[K \left(\left(\frac{\partial V}{\partial x} \right)^2 + \left(\frac{\partial V}{\partial y} \right)^2 + \left(\frac{\partial V}{\partial z} \right)^2 \right) \right] \right.$$

$$\left. + 2 \frac{\partial}{\partial x} \left(K \left(\frac{\partial V}{\partial x} \right)^2 \right) + 2 \frac{\partial}{\partial y} \left(K \frac{\partial V}{\partial x} \frac{\partial V}{\partial y} \right) + 2 \frac{\partial}{\partial z} \left(K \frac{\partial V}{\partial x} \frac{\partial V}{\partial z} \right) \right.$$

$$\left. + \frac{\partial}{\partial x} \left(R^2 \, \tau \, \frac{\partial K}{\partial \tau} \right) \right\}.$$

If we put

$$P_{xx} = \frac{K}{8\pi} \left\{ \left(\frac{\partial V}{\partial x} \right)^2 - \left(\frac{\partial V}{\partial y} \right)^2 - \left(\frac{\partial V}{\partial z} \right)^2 \right\} + \frac{R^2}{8\pi} \tau \frac{\partial K}{\partial \tau} \quad \ldots\ldots\ldots(118),$$

$$P_{xy} = \frac{K}{4\pi} \frac{\partial V}{\partial x} \frac{\partial V}{\partial y}, \text{ etc.} \quad \ldots\ldots\ldots\ldots\ldots\ldots\ldots\ldots\ldots(119),$$

this becomes
$$\Xi = \frac{\partial P_{xx}}{\partial x} + \frac{\partial P_{xy}}{\partial y} + \frac{\partial P_{xz}}{\partial z}.$$

Let us suppose that a medium is subjected to a system of internal stresses P_{xx}, P_{xy}, etc.; and let it be found that a system of body forces of components Ξ', H', Z' is just sufficient to keep the medium at rest when under the action of these stresses. Then from equation (79) we must have

$$\Xi' = - \left(\frac{\partial P_{xx}}{\partial x} + \frac{\partial P_{xy}}{\partial y} + \frac{\partial P_{xz}}{\partial z} \right).$$

Thus if P_{xx}, P_{xy}, etc. have the values given by equations (118) and (119), we have

$$\Xi' = -\Xi, \text{ etc.}$$

This shews that the mechanical force Ξ, H, Z reversed would just be in equilibrium with the system of stresses P_{xx}, P_{xy}, etc. given by equations (118) and (119). In other words, the mechanical forces which have been found to act on a dielectric can exactly be accounted for by a system of stresses in the medium, these stresses being given by equations (118) and (119).

199. The system of stresses given by equations (118) and (119) can be regarded as the superposition of two systems:

I.　A system in which

$$P_{xx} = \frac{K}{8\pi} (X^2 - Y^2 - Z^2), \quad P_{xy} = \frac{KXY}{4\pi}, \text{ etc.};$$

II.　A system in which

$$P_{xx} = P_{yy} = P_{zz} = \frac{R^2}{8\pi} \tau \frac{\partial K}{\partial \tau},$$

$$P_{xy} = P_{yz} = P_{zx} = 0.$$

The first system is exactly K times the system which has been found to occur in free ether, while the second system represents a hydrostatic pressure of amount

$$-\frac{R^2}{8\pi}\,\tau\,\frac{\partial K}{\partial\tau}.$$

(In general $\dfrac{\partial K}{\partial\tau}$ will be positive, so that this pressure will be negative, and must be interpreted as a tension.)

Hence, as in § 165, the system of stresses may be supposed to consist of:

(i) a tension $\dfrac{KR^2}{8\pi}$ per unit area in the direction of the lines of force;

(ii) a pressure $\dfrac{KR^2}{8\pi}$ per unit area perpendicular to the lines of force;

(iii) a hydrostatic pressure of amount $-\dfrac{R^2}{8\pi}\tau\dfrac{\partial K}{\partial\tau}$ in all directions.

The system of stresses we have obtained was first given by Helmholtz. The system differs from that given by Maxwell by including the pressure $-\dfrac{R^2}{8\pi}\,\tau\,\dfrac{\partial K}{\partial\tau}$. The neglect of this pressure by Maxwell, and by other writers who have followed him, does not appear to be defensible. Helmholtz has shewn that still further terms are required if the dielectric is such that the value of K changes when the medium is subjected to distortion without change of volume.

200. This system of stresses has not been proved to be the only system of stresses by which the mechanical forces can be replaced, and, as we have seen, it is not certain that the mechanical forces must be regarded as arising from a system of stresses at all, rather than from action at a distance.

It may be noticed, however, that whether or not these stresses actually exist, the resultant force on any piece of dielectric must be exactly the same as it would be if the stresses actually existed. For the resultant force on any piece of dielectric has a component X parallel to the axis of x, given by

$$\mathsf{X} = \iiint \Xi\, dx\, dy\, dz$$

$$= \iiint \left(\frac{\partial P_{xx}}{\partial x} + \frac{\partial P_{xy}}{\partial y} + \frac{\partial P_{xz}}{\partial z}\right) dx\, dy\, dz$$

$$= -\iint (l P_{xx} + m P_{xy} + n P_{xz})\, dS$$

by Green's Theorem, and this shews that the actual force is identical with what it would be if these stresses existed (cf. § 193).

Force on a charged conductor.

201. The mechanical force on the surface of a charged conductor immersed in a dielectric can be obtained at once by regarding it as produced by the stresses in the ether. There will be no stresses in the interior of the conductor, so that the force on its surface may be regarded as due to the tensions of the tubes of force in the dielectric. The tension is accordingly of amount

$$\frac{KR^2}{8\pi} + \frac{R^2}{8\pi}\tau\frac{\partial K}{\partial\tau}$$

per unit area, an expression which can be written in the simpler form

$$\frac{R^2}{8\pi}\frac{\partial}{\partial\tau}(K\tau).$$

Force at boundary of a dielectric.

202. Let us consider the equilibrium of a dielectric at a surface of discontinuity, at which the lines of force undergo refraction on passing from one medium of inductive capacity K_1 to a second of inductive capacity K_2.

Let axes be taken so that the boundary is the plane of xy, while the lines of force at the point under consideration lie in the plane of xz. Let the components of intensity in the first medium be $(X_1, 0, Z_1)$, while the corresponding quantities in the second medium are $(X_2, 0, Z_2)$. The boundary conditions obtained in § 137 require that

$$X_1 = X_2, \quad K_1Z_1 = K_2Z_2 = 4\pi h,$$

where h is the normal component of polarisation.

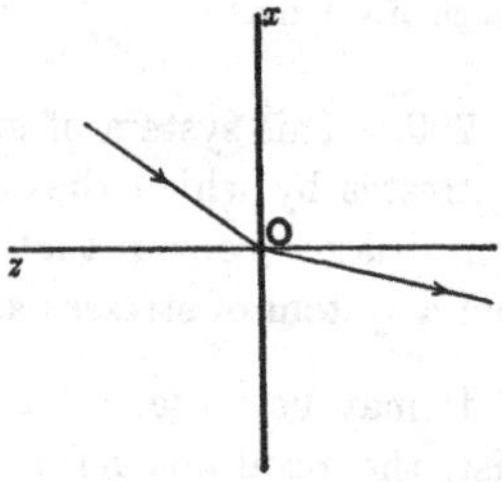

Fig. 55.

In view of a later physical interpretation of the forces, it will be convenient to regard these forces as divided up into the two systems mentioned in § 199, and to consider the contributions from these systems separately.

As regards the contribution from the first system, the force per unit area acting on the dielectric from the first medium has components

$$\frac{K_1}{4\pi}X_1Z_1, \quad 0, \quad \frac{K_1}{8\pi}(Z_1^2 - X_1^2),$$

while that from the second medium has components

$$\frac{K_2}{4\pi}X_2Z_2, \quad 0, \quad \frac{K_2}{8\pi}(Z_2^2 - X_2^2).$$

Since $K_1 X_1 Z_1 = K_2 X_2 Z_2$, it follows that the resultant force on the boundary is parallel to Oz—*i.e.* is normal to the surface. Its amount, measured as a tension dragging the surface in the direction from medium 1 to medium 2

$$= \frac{K_2}{8\pi}(Z_2{}^2 - X_2{}^2) - \frac{K_1}{8\pi}(Z_1{}^2 - X_1{}^2)$$

which after simplification can be shewn to be equal to

$$\left(\frac{X_1{}^2}{8\pi} + \frac{2\pi h^2}{K_1 K_2}\right)(K_1 - K_2).$$

This is always positive if $K_1 > K_2$. Thus this force invariably tends to drag the surface from the medium in which K is greater, to that in which K is less—*i.e.* to increase the region in which K is large at the expense of the region in which K is small. This normal force is exactly similar to the normal force on the surface of a conductor, which tends to increase the volume of the region enclosed by the conducting surface.

On Maxwell's Theory, the forces which have now been considered are the only ones in existence, so that according to this theory the total mechanical force is that just found, and the boundary forces ought always to tend to increase the region in which K is large. This theory, as we have said, is incomplete, so that it is not surprising that the result just stated is not confirmed by experiment.

We now proceed to consider the action of the second system of forces— the system of negative hydrostatic pressures. There are pressures per unit area of amounts

$$-\frac{R_1{}^2}{8\pi}\tau_1 \frac{\partial K_1}{\partial \tau_1}, \qquad -\frac{R_2{}^2}{8\pi}\tau_2 \frac{\partial K_2}{\partial \tau_2}$$

acting respectively on the two sides of the boundary. There is accordingly a resultant tension of amount

$$-\frac{1}{8\pi}\left(R_1{}^2 \tau_1 \frac{\partial K_1}{\partial \tau_1} - R_2{}^2 \tau_2 \frac{\partial K_2}{\partial \tau_2}\right),$$

per unit area, tending to drag the boundary surface from region 1 to region 2.

Thus the total tension per unit area, dragging the surface into region 1, is

$$\left(\frac{X_1{}^2}{8\pi} + \frac{2\pi h^2}{K_1 K_2}\right)(K_1 - K_2) - \frac{1}{8\pi}\left(R_1{}^2 \tau_1 \frac{\partial K_1}{\partial \tau_1} - R_2{}^2 \tau_2 \frac{\partial K_2}{\partial \tau_2}\right)\ \dots\dots(120).$$

In § 139, in considering a parallel plate condenser with a movable dielectric slab, we discovered the existence of a mechanical force tending to drag the dielectric in between the plates. This force is identical with the mechanical force just discussed. But we have now arrived at a mechanical interpretation of this force, for we can regard the pull on the dielectric as the resultant of the pulls of the tubes of force at the different parts of the surface of the dielectric.

Let us attempt to assign physical interpretations to the terms of expression (120) by considering their significance in this particular instance. Consider first a region in the condenser so far removed from the edges of the condenser and of the slab of dielectric, that the field may be treated as absolutely uniform (cf. fig. 44, p. 124). We put $K_2 = 1$, $X_1 = 0$, $R_1 = \dfrac{4\pi h}{K_1}$ in expression (120) and obtain

$$2\pi h^2 \left(\frac{K_1 - 1}{K_1} - \frac{\tau_1}{K_1^2} \frac{\partial K_1}{\partial \tau_1} \right) \dots\dots\dots\dots\dots(121)$$

as the force per unit area on either face of the dielectric, acting normally outwards.

The forces will of course act in such a direction that they tend to decrease the electrostatic energy of the field. Now this energy is made up of contributions $2\pi h^2$ per unit volume from air, and $\dfrac{2\pi h^2}{K_1}$ per unit volume from the dielectric. From the conditions of the problem h must remain unaltered. Thus the total energy can be decreased in either of two ways—by increasing the volume occupied by dielectric and decreasing that occupied by air, or by increasing the value of K in the dielectric. There will therefore be a tendency for the boundary of the dielectric to move in such a direction as to increase the volume occupied by dielectric, and also a tendency for this boundary to move so that K will be increased by the consequent change of density. These two tendencies are represented by the two terms of expression (121).

If $\dfrac{\partial K}{\partial \tau}$ is negative, an expansion of the dielectric will both increase the volume occupied by the dielectric, and will also increase the value of K inside the dielectric. In this case, then, both tendencies act towards an expansion of the dielectric, and we accordingly find that both terms in expression (121) are positive.

If $\dfrac{\partial K}{\partial \tau}$ is positive, the tendency to expansion, represented by the first (positive) term of expression (121) is checked by a tendency to contraction (to increase τ, and therefore K) represented by the second (now negative) term of expression (121). If $\dfrac{\partial K}{\partial \tau}$ is not only positive, but is numerically large, expression (121) may be negative and the dielectric will contract. In this case the decrease in energy resulting on the increase of K produced by contraction will more than outweigh the gain resulting from the diminution of the volume occupied by dielectric.

These considerations enable us to see the physical significance of all the terms in expression (120), except the first term $\frac{X_1^2}{8\pi}(K_1-1)$. To interpret this term we must examine the conditions near the edge of the dielectric slab, for it is only here that X_1 has a value different from zero. We see at once that this term represents a pull at and near the edge of the dielectric, tending to suck the dielectric further between the plates—in fact this force alone gives rise to the tendency to motion of the slab as a whole, which was discovered in § 139.

Returning to the general systems of forces of § 199, we may say that the first system (which as we have seen always tends to drag the surface of the dielectric into the region in which K has the greater value) represents the tendency for the system to decrease its energy by increasing the volume occupied by dielectrics of large inductive capacity, whilst the second system (which tends to compress or expand the dielectric in such a way as to increase its inductive capacity) represents the tendency of the system to decrease its energy by increasing the inductive capacity of its dielectrics. That any increase in the inductive capacity is invariably accompanied by a decrease of energy has already been proved in § 191.

Electrostriction.

203. It will now be clear that the action of the various tractions on the surface of a dielectric must always be accompanied not only by a tendency for the dielectric to move as a whole, but also by a slight change in shape and dimensions of the dielectric as this yields to the forces acting on it. This latter phenomenon is known as electrostriction. It has been observed experimentally by Quincke and others. A convenient way of shewing its existence is to fill the bulb of a thermometer-tube with liquid, and place the whole in an electric field. The pulls on the surface of the glass result in an increase in the volume of the bulb, and the liquid is observed to fall in the tube. From what has already been said it will be clear that a dielectric may either expand or contract under the influence of electric forces.

The stresses in the interior of a dielectric, as given in § 199, may also be accompanied by mechanical deformation. Thus it has been observed by Kerr and others, that a piece of non-crystalline glass acquires crystalline properties when placed in an electric field. Such a piece of glass reflects light like a uniaxal crystal of which the optic axis is in the direction of the lines of force.

GREEN'S EQUIVALENT STRATUM.

204. Let S be any closed surface enclosing a number of electric charges, and let P be any point outside it. The potential at P due to the charges inside S is

$$V_P = \iiint \frac{\rho}{r} \, dx \, dy \, dz,$$

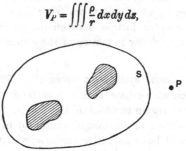

Fig. 56.

where r is the distance from P to the element $dx\,dy\,dz$, and the integration extends throughout S. By Green's Theorem (equation (101))

$$\iiint (U\nabla^2 V - V\nabla^2 U)\, dx\, dy\, dz = \iint \left(U \frac{\partial V}{\partial n} - V \frac{\partial U}{\partial n} \right) dS,$$

where the normal is now drawn outwards from the surface S.

In this equation, put $U = \frac{1}{r}$, then, since $\nabla^2 V = -4\pi\rho$, we have as the value of the first term,

$$\iiint U\nabla^2 V \, dx\, dy\, dz = -4\pi V_P.$$

And since $\nabla^2 U = 0$, the second term vanishes. The equation accordingly becomes

$$-4\pi V_P = \iint \left\{ \frac{1}{r}\left(\frac{\partial V}{\partial n}\right) - V \frac{\partial}{\partial n}\left(\frac{1}{r}\right) \right\} dS \quad \dots\dots\dots(122).$$

205. Suppose, first, that the surface S is an equipotential. Then

$$\iint V \frac{\partial}{\partial n}\left(\frac{1}{r}\right) dS = V \iint \frac{\partial}{\partial n}\left(\frac{1}{r}\right) dS$$

$$= V \iiint \nabla^2\left(\frac{1}{r}\right) dx\, dy\, dz$$

$$= 0,$$

so that equation (122) becomes

$$V_P = \iint \frac{\left(-\dfrac{1}{4\pi}\dfrac{\partial V}{\partial n}\right)}{r} \, dS \quad \dots\dots\dots\dots\dots(123).$$

Thus the potential of any system of charges is the same at every point outside any selected equipotential which surrounds all the charges, as that of a charge of electricity spread over this equipotential, and having surface density $-\dfrac{1}{4\pi}\dfrac{\partial V}{\partial n}$. Obviously, in fact, if the equipotential is replaced by a conductor, this will be the density on its outer surface.

206. If the surface is not an equipotential, the term $\iint V \dfrac{\partial}{\partial n}\left(\dfrac{1}{r}\right) dS$ will not vanish. Since, however, $\mu \dfrac{\partial}{\partial n}\left(\dfrac{1}{r}\right)$ is the potential of a doublet of strength μ and direction that of the outward normal, it follows that $\iint V \dfrac{\partial}{\partial n}\left(\dfrac{1}{r}\right) dS$ is the potential of a system of doublets arranged over the surface S, the direction at every point being that of the outward normal, and the total strength of doublets per unit area at any point being V.

Thus the potential V_P may be regarded as due to the presence on the surface S of

(i)   a surface density of electricity $-\dfrac{1}{4\pi}\dfrac{\partial V}{\partial n}$;

(ii) a distribution of electric doublets, of strength $\dfrac{V}{4\pi}$ per unit area, and direction that of the outward normal.

207. Equation (122) expresses the potential at any point in the space outside S in terms of the values of V and $\dfrac{\partial V}{\partial n}$ over the boundary of this space. We have seen, however, that the value of the potential is uniquely determined by the values *either* of V or of $\dfrac{\partial V}{\partial n}$ over the boundary of the space. In actual electrostatic problems, the boundaries are generally conductors, and therefore equipotentials. In this case equation (123) expresses the values of the potential in terms of $\dfrac{\partial V}{\partial n}$ only, amounting in fact simply to

$$V_P = \iint \frac{\sigma}{r}\, dS.$$

What is generally required is a knowledge of the value of V_P in terms of the values of V over the boundaries, and this the present method is unable to give. For special shapes of boundary, solutions have been obtained by various special methods, and these it is proposed to discuss in the next chapter.

EXAMPLES.

1. If the electricity in the field is confined to a given system of conductors at given potentials, and the inductive capacity of the dielectric is slightly altered according to any law such that at no point is it diminished, and such that the differential coefficients of the increment are also small at all points, prove that the energy of the field is increased.

2. A slab of dielectric of inductive capacity K and of thickness x is placed inside a parallel plate condenser so as to be parallel to the plates. Shew that the surface of the slab experiences a tension

$$2\pi\sigma^2\left\{1 - \frac{1}{K} - x\frac{d}{dx}\left(\frac{1}{K}\right)\right\}.$$

3. For a gas $K = 1 + \theta\rho$, where ρ is the density and θ is small. A conductor is immersed in the gas: shew that if θ^2 is neglected the mechanical force on the conductor is $2\pi\sigma^2$ per unit area. Give a physical interpretation of this result.

CHAPTER VIII

METHODS FOR THE SOLUTION OF SPECIAL PROBLEMS

THE METHOD OF IMAGES.

Charge induced on an infinite uninsulated plane.

208. THE potential at P of charges e at a point A and $-e$ at another point A' is

$$V = \frac{e}{AP} - \frac{e}{A'P} \quad \dots\dots\dots\dots\dots\dots(124),$$

and this vanishes if P is on the plane which bisects AA' at right angles. Call this plane the plane S. Then the above value of V gives $V = 0$ over the plane S, $V = 0$ at infinity, and satisfies Laplace's equation in the region to the right of S, except at the point A, at which it gives a point charge e.

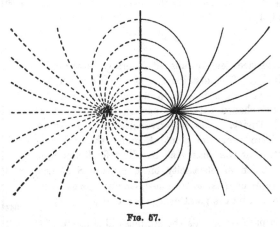

FIG. 57.

These conditions, however, are exactly those which would have to be satisfied by the potential on the right of S if S were a conducting plane at zero potential under the influence of a charge e at A. These conditions amount to a knowledge of the value of the potential at every point on the boundary of a certain region—namely, that to the right of the plane S—and of the charges inside this region. There is, as we know, only one value of the potential inside this region which satisfies these conditions (cf. § 186), so that this value must be that given by equation (124).

To the right of S the potential is the same, whether we have the charge $-e$ at A' or the charge on the conducting plane S. To the left of S in the latter case there is no electric field. Hence the lines of force, when the plane S is a conductor, are entirely to the right of S, and are the same as in the original field in which the two point-charges were present. The lines end on the plane S, terminating of course on the charge induced on S.

We can find the amount of this induced charge at any part of the plane by Coulomb's Law. Taking the plane to be the plane of yz, and the point A to be the point $(a, 0, 0)$ on the axis of x, we have

$$4\pi\sigma = R = -\frac{\partial V}{\partial x}$$

$$= -\frac{\partial}{\partial x}\left\{\frac{e}{\sqrt{(x-a)^2 + y^2 + z^2}} - \frac{e}{\sqrt{(x+a)^2 + y^2 + z^2}}\right\},$$

where the last line has to be calculated at the point on the plane S at which we require the density. We must therefore put $x = 0$ after differentiation, and so obtain for the density at the point $0, y, z$ on the plane S,

$$4\pi\sigma = -\frac{2ae}{(a^2 + y^2 + z^2)^{\frac{3}{2}}},$$

or, if $a^2 + y^2 + z^2 = r^2$, so that r is the distance of the point on the plane S from the point A,

$$\sigma = -\frac{ae}{2\pi r^3}.$$

Thus the surface density falls off inversely as the cube of the distance from the point A. The distribution of electricity on the plane is represented graphically in fig. 58, in which the thickness of the shaded part is proportional to the surface density of electricity. The negative electricity is, so to speak, heaped up near the point A under the influence of the attraction of the charge at A. The field produced by this distribution of electricity on the plane S at any point to the right of S is, as we know, exactly the same as would be produced by the point charge $-e$ at A'.

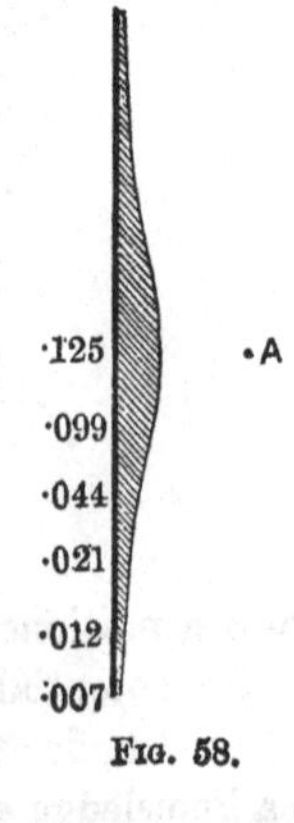

·125   •A

·099

·044

·021

·012

·007

FIG. 58.

209. This problem affords the simplest illustration of a general method for the solution of electrostatic problems, which is known as the "method of images." The principle underlying this method is that of finding a system of electric charges such that a certain surface, ultimately to be made into a conductor, is caused to coincide with the equipotential $V = 0$. We then replace the charges inside this equipotential by the Green's equivalent stratum on its surface (cf. § 204). As this surface is an equipotential, we can imagine it to be replaced by a conductor and the charges on it will be in equilibrium. These charges now become charges induced on a conductor at potential zero by charges outside this conductor.

From the analogy with optical images in a mirror, the system of point charges which have to be combined with the original charges to produce zero potential over a conductor are spoken of as the "electrical images" of the original charges. For instance, in the example already discussed, the field is produced partly by the charge at A, partly by the charge induced on the infinite plane: the method of images enables us to replace the whole charge induced on the plane by a single point charge at A'. So also, if A were a candle placed in front of an infinite plane mirror, the illumination in front of the mirror would be produced partly by the candle at A, partly by the light reflected from the infinite mirror; the method of optical images enables us to replace the whole of this reflected light by the light from a single source at A'.

210.   In an electrostatic field produced by any number of point charges, we can, as we have seen, select *any* equipotential and replace it by a conductor. The charges on either side of this equipotential are then the "images" of those on the other side.

Thus if we can write the equation of any surface in the form

$$\frac{e}{r} + \frac{e'}{r'} + \frac{e''}{r''} + \ldots = 0 \quad\ldots\ldots\ldots\ldots\ldots\ldots(125),$$

where r is the distance from a point outside the surface, and $r', r'', \ldots$ are the distances from points inside the surface, then we may say that charges $e', e'', \ldots$ at these latter points are the images of a charge e at the former point.

The method of images may be applied in a similar way to two-dimensional problems. Suppose that the equation of a cylindrical surface can be expressed in the form

$$c - 2e \log r - 2e' \log r' - 2e'' \log r'' - \ldots = 0,$$

where r is the perpendicular distance from a fixed line on one side of the surface, and $r', r'', \ldots$ are perpendicular distances from fixed lines on the other side. Then line-charges of line-densities $e', e'', \ldots$ at these latter lines may be taken to be the image of a line-charge of line-density e at the former line.

Illustrations of the use of images in three dimensions are given in §§ 211—219. An illustration of the use of a two-dimensional image will be found in § 220.

Charges induced on Intersecting planes.

211. It will be found that charges

$$e \quad \text{at} \quad x, \quad y, \quad 0,$$
$$-e \quad \text{at} \quad -x, \quad y, \quad 0,$$
$$-e \quad \text{at} \quad x, \quad -y, \quad 0,$$
$$e \quad \text{at} \quad -x, \quad -y, \quad 0$$

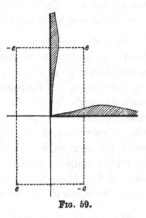

give zero potential over the planes $x = 0$, $y = 0$.
The potential of these charges is therefore the
same, in the quadrant in which x, y are both
positive, as if the boundary of this quadrant
were a conductor put to earth under the influence of a charge e at the point x, y, 0.

It will be found that a conductor consisting
of three planes intersecting at right angles can
be treated in the same way.

Fig. 59.

212. The method of images also supplies a solution when the conductor
consists of two planes intersecting at any angle of the form $\dfrac{\pi}{n}$, where n is

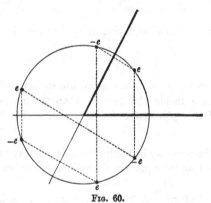

Fig. 60.

any positive integer. If we take polar coordinates, so that the two planes
are $\theta = 0$, $\theta = \dfrac{\pi}{n}$, and suppose the charge to be a charge e at the point r, θ,
we shall find that charges

$$e \quad \text{at} \quad (r, \quad \theta), \quad \left(r, \quad \theta + \frac{2\pi}{n}\right), \quad \left(r, \quad \theta + \frac{4\pi}{n}\right), \quad \dots,$$
$$-e \quad \text{at} \quad (r, -\theta), \quad \left(r, -\left(\theta + \frac{2\pi}{n}\right)\right), \quad \left(r, -\left(\theta + \frac{4\pi}{n}\right)\right), \quad \dots,$$

give zero potential over the planes

$$\theta = 0, \quad \theta = \frac{\pi}{n}.$$

Charge induced on a sphere.

213. The most obvious case, other than the infinite plane, of a surface whose equation can be expressed in the form (125), is a sphere.

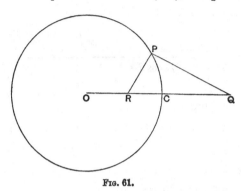

Fig. 61.

If R, Q are any two inverse points in the sphere, and P any point on the surface, we have

$$RP : PQ = OC : OQ,$$

so that

$$\frac{OQ}{PQ} - \frac{OC}{PR} = 0.$$

Thus the image of a charge e at Q is a charge $- e \dfrac{OC}{OQ}$ at R, or the image of any point at a distance f from the centre of a sphere of radius a is a charge $- \dfrac{ea}{f}$ at the inverse point, *i.e.* at a point on the same radius distant $\dfrac{a^2}{f}$ from the centre.

Let us take polar coordinates, having the centre of the sphere for origin and the line OQ as $\theta = 0$. Our result is that at any point S outside the sphere, the potential due to a charge e at Q and the charge induced on the surface of the sphere, supposed put to earth, is

$$V = \frac{e}{QS} - \frac{\dfrac{ea}{f}}{RS}$$

$$= \frac{e}{\sqrt{r^2 + f^2 - 2fr \cos \theta}} - \frac{ea}{f \sqrt{r^2 + \dfrac{a^4}{f^2} - 2 \dfrac{a^2}{f} r \cos \theta}},$$

where r, θ are the coordinates of S.

214. We can now find the surface-density of the induced charge. For at any point on the sphere

$$\sigma = \frac{R}{4\pi} = -\frac{1}{4\pi}\frac{\partial V}{\partial r},$$

in which we have to put $r = a$ after differentiation. Clearly

$$-\frac{\partial V}{\partial r} = \frac{e\,(r - f\cos\theta)}{(r^2 + f^2 - 2fr\cos\theta)^{\frac{3}{2}}} - \frac{ea\left(r - \dfrac{a^2}{f}\cos\theta\right)}{f\left(r^2 + \dfrac{a^4}{f^2} - 2\dfrac{a^2}{f}r\cos\theta\right)^{\frac{3}{2}}}.$$

Putting $r = a$ we obtain

$$\sigma = \frac{e}{4\pi}\left\{ \frac{a - f\cos\theta}{(a^2 + f^2 - 2fa\cos\theta)^{\frac{3}{2}}} - \frac{a^2 f^2 - a^3 f\cos\theta}{(a^2 f^2 + a^4 - 2a^3 f\cos\theta)^{\frac{3}{2}}} \right\}$$

$$= \frac{e}{4\pi}\left\{ \frac{a - f^2/a}{(a^2 + f^2 - 2fa\cos\theta)^{\frac{3}{2}}} \right\}$$

$$= -\frac{e}{4\pi}\frac{(f^2 - a^2)}{a \,.\, SQ^3}.$$

Thus the surface-density varies inversely as SQ^3, so that it is greatest at C and falls off continually as we recede from the radius OC. The total charge on the sphere is $-\dfrac{ea}{f}$, as can be seen at once by considering that the total strength of the tubes of force which end on it is just the same as would

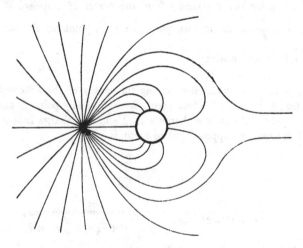

FIG. 62.

be the total strength of the tubes ending on the image at R if the conductor were not present.

Figure 62 shews the lines of force when the strength of the image is a quarter of that of the original charge, so that $f = 4a$. It is obtained from fig. 19 by replacing the spherical equipotential by a conductor, and annihilating the field inside.

Superposition of Fields.

215. We have seen that by adding the potentials of two separate fields at every point, we obtain the potential produced by charges equal to the total charges in the two fields. In this way we can arrive at the field produced by any number of point charges and uninsulated conductors of the kind we have described. The potential of each conductor is zero in the final solution because it is zero for each separate field.

There is also another type of field which may be added to that obtained by the method of images, namely the field produced by raising the conductor or conductors to given potentials, without other charges being present. By superposing a field of this kind we can find the effect of point charges when the conductors are at any potential.

216. For instance, suppose that, as in fig. 62, we have a point charge e and the conductor at potential 0. Let us superpose on to the field of force already found, the field which is obtained by raising the conductor to potential V when the point charge is absent. The charge on the sphere in the second field is aV, so that the total charge is

$$aV - \frac{ea}{f}.$$

By giving different values to V, we can obtain the total field, when the sphere has any given charge or potential.

If the sphere is to be uncharged, we must have $V = \dfrac{e}{f}$, so that a point charge placed at a distance f from the centre of an uncharged sphere raises it to potential $\dfrac{e}{f}$, a result which is also obvious from the theorem of § 104.

Sphere in a uniform field of force.

217. A uniform field of force of which the lines are parallel to the axis of x may be regarded as due to an infinite charge E at $x = R$, and a charge $-E$ at $x = -R$, when in the limit E and R both become infinite. The intensity at any point is

$$-\frac{2E}{R^2}$$

parallel to the axis of x, so that to produce a uniform field in which the intensity is F parallel to the axis of x, we must suppose E and R to become infinite in such a way that

$$\frac{2E}{R^2} = -F.$$

Since, in this case, $F = -\dfrac{\partial V}{\partial x}$, the potential of such a field will clearly be $-Fx + C$.

Suppose that a sphere is placed in a uniform field of force of this kind, its centre being at the origin. We can suppose the charge E at $x = R$ to have an image of strength

$$\frac{Ea}{R} \quad \text{at} \quad x = \frac{a^2}{R},$$

while the other charge has an image

$$\frac{Ea}{R} \quad \text{at} \quad x = -\frac{a^2}{R}.$$

These two images may be regarded as a doublet (cf. § 64) of strength $\dfrac{Ea}{R} \times \dfrac{2a^2}{R}$, and of direction parallel to the negative axis of x. The strength

$$= \frac{2a^3 E}{R^2} = -Fa^3.$$

Thus we may say that the image of a uniform field of force of strength F is a doublet of strength Fa^3 and of direction parallel to that of the intensity of the uniform field.

The potential of this doublet is

$$\frac{Fa^3 \cos \theta}{r^2},$$

and that of the field of original field of force is

$$-Fx + C,$$

or, in polar coordinates, $-Fr \cos \theta + C,$

so that the potential of the whole field

$$= -F \cos\theta \left(r - \frac{a^3}{r^2} \right) + C \quad \dots\dots\dots\dots(126).$$

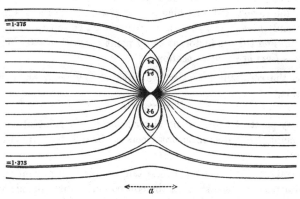

FIG. 63.

As it ought, this gives a constant potential C over the surface of the sphere.

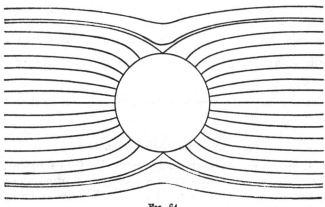

FIG. 64.

The lines of force of the uniform field F disturbed by the presence of a doublet of strength Fa^3 are shewn in fig. 63. On obliterating all the lines of force inside a sphere of radius a, we obtain fig. 64, which accordingly shews the lines of force when a sphere of radius a is placed in a field of intensity F. These figures are taken from Thomson's *Reprint of Papers on Electrostatics and Magnetism* (pp. 488, 489)*.

* I am indebted to Lord Kelvin for permission to use these figures.

218. *Line of no electrification.* The theory of lines of no electrification has already been briefly given in § 98. We have seen that on any conductor on which the total charge is zero, and which is not entirely screened from an electric field, there must be some points at which the surface-density σ is positive, and some points at which it is negative. The regions in which σ is positive and those in which σ is negative must be separated by a line or system of lines on the conductor, at every point of which $\sigma = 0$. These lines are known as *lines of no electrification.*

If R is the resultant intensity, we have at any point on a line of no electrification,

$$R = 4\pi\sigma = 0,$$

so that every point of a line of no electrification is a point of equilibrium. At such a point the equipotential intersects itself, and there are two or more lines of force.

If the conductor possesses a single tangent plane at a point on a line of no electrification, then one sheet of the equipotential through this point will be the conductor itself: by the theorem of § 69, the second sheet must intersect the conductor at right angles.

These results are illustrated in the field of fig. 64. Clearly the line of no electrification on the sphere is the great circle in a plane perpendicular to the direction of the field. The equipotential which intersects itself along the line of no electrification ($V = C$) consists of the sphere itself and the plane containing the line of no electrification. Indeed, from formula (126), it is obvious that the potential is equal to C, either when $\theta = \dfrac{\pi}{2}$, or when $r = a$.

The intersection of the lines of force along the line of no electrification is shewn clearly in fig. 64.

Plane face with hemispherical boss.

219. If we regard the whole equipotential $V = C$ as a conductor, we obtain the distribution of electricity on a plane conductor on which there is a hemispherical boss of radius a. If we take the plane to be $x = 0$, we have, by formula (126),

$$V - C = - F\cos\theta\left(r - \frac{a^3}{r^2}\right) = - Fx\left(1 - \frac{a^3}{r^3}\right).$$

At a point on the plane,

$$\sigma = - \frac{1}{4\pi}\left(\frac{\partial V}{\partial x}\right)_{x=0} = \frac{F}{4\pi}\left\{1 - \frac{a^3}{r^3}\right\},$$

and on the hemisphere

$$\sigma = \frac{1}{4\pi}\left(\frac{\partial V}{\partial r}\right)_{r=a} = \frac{F}{4\pi}\cdot 3\cos\theta.$$

The whole charge on the hemisphere is found on integration to be

$$\int_{\theta=0}^{\theta=\frac{\pi}{2}} \left(\frac{F}{4\pi} 3\cos\theta\right) 2\pi a^2 \sin\theta \, d\theta = \tfrac{3}{4} Fa^2,$$

while, if the hemisphere were not present, the charge on the part of the plane now covered by the base of the hemisphere would be

$$\left(\frac{F}{4\pi}\right)\pi a^2 = \tfrac{1}{4} Fa^2.$$

Thus the presence of the boss results in there being three times as much electricity on this part of the plane as there would otherwise be: this is compensated by the diminution of surface-density on those parts of the plane which immediately surround the boss.

Capacity of a telegraph-wire.

220. An important practical application of the method of images is the determination of the capacity of a long straight wire placed parallel to an infinite plane at potential zero, at a distance h from the plane. This may be supposed to represent a telegraph-wire at height h above the surface of the earth.

Let us suppose that the wire has a charge e per unit length. To find the field of force we imagine an image charged with a charge $-e$ per unit length at a distance h below the earth's surface. The potential at a point at distances r, r' from the wire and image respectively is, by §§ 75 and 100,

$$C - 2e\log r + 2e\log r',$$

and for this to vanish at the earth's surface we must take $C = 0$. Thus the potential is

$$2e\log\frac{r'}{r}.$$

At a small distance a from the line-charge which represents the telegraph-wire, we may put $r' = 2h$, so that the potential is

$$2e\log\frac{2h}{a},$$

from which it appears that a cylinder of small radius a surrounding the wire is an equipotential. We may now suppose the wire to have a finite radius a, and to coincide with this equipotential. Thus the capacity of the wire per unit length is

$$\frac{1}{2\log\dfrac{2h}{a}}.$$

Infinite series of Images.

221. Suppose we have two spheres, centres A, B and radii a, b, of which the centres are at distance c apart, and that we require to find the field when

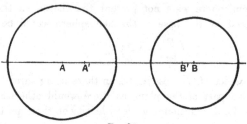

Fig. 65.

both are charged. We can obtain this field by superposing an infinite series of separate fields (cf. § 116).

Suppose first that A is at potential V while B is at potential zero. As a first field we can take that of a charge Va at A. This gives a uniform potential V over A, but does not give zero potential over B. We can reduce the potential over B to zero by superposing a second field arising from the image of the original charge in sphere B, namely a charge $-\dfrac{Vab}{c}$ at B', where $BB' = \dfrac{b^2}{c}$. This new field has, however, disturbed the potential over A. To reduce this to its original value we superpose a new field arising from the image of the charge at B' in A, namely a charge $\dfrac{Vab}{c} \cdot \dfrac{a}{c - \dfrac{b^2}{c}}$ at A',

where $AA' = \dfrac{a^2}{c - \dfrac{b^2}{c}}$. This field in turn disturbs the potential over B, and so we superpose another field, and so on indefinitely. The strengths of the various fields, however, continually diminish, so that although we get an infinite series to express the potential, this series is convergent. As we shall see, this series can be summed as a definite integral, or it may be that a good approximation will be obtained by taking only a finite number of terms.

The total charge on A is clearly the sum of the original charge Va plus the strengths of the images A', A'', ... etc., for this sum measures the aggregate strength of the tubes of force which end on A. Similarly the charge on B is the sum of the strengths of the images at B', B'', ...

To obtain the field corresponding to given potentials of both A and B we superpose on to the field already found, the similar field obtained by raising B to the required potential while that of A remains zero.

If q_{11}, q_{22}, q_{12} are the coefficients of capacity and induction, the total charge on A when B is to earth and $V = 1$ is q_{11}; similarly that on B is q_{12}. In this way we can find the coefficients q_{11}, q_{12} from the series of images already obtained. The result is found to be

$$q_{11} = a + \frac{a^2 b}{c^2 - b^2} + \frac{a^3 b^3}{(c^2 - b^2)^2 - a^2 c^2} + \cdots,$$

$$q_{12} = -\frac{ab}{c} - \frac{a^2 b^3}{c(c^2 - b^2 - a^2)} + \cdots,$$

and from symmetry

$$q_{22} = b + \frac{b^2 a}{c^2 - a^2} + \frac{b^3 a^3}{(c^2 - a^2)^2 - b^2 c^2} + \cdots,$$

As far as $\dfrac{1}{c^3}$, these results clearly agree with those of § 116.

222. The series for q_{11}, q_{12}, q_{22} have been put in a more manageable form by Poisson and Kirchhoff.

Let A_s denote the position of the sth of the series of points A', A'', ..., and B_s the sth of the series B', B'', ...; then A_s is the image of B_s in the sphere of radius a, and similarly B_s is the image of A_{s-1} in the sphere of radius b. Let $a_s = AA_s$, $b_s = BB_s$, and let the charges at A_s, B_s be e_s, e'_s respectively.

Then $\qquad a_s(c - b_s) = a^2$ since A_s is the image of B_s,

$\qquad\qquad b_s(c - a_{s-1}) = b^2 \quad$ „ $\quad B_s \quad$ „ $\quad$ „ $\quad A_{s-1}$.

Further, by comparing the strengths of a charge and its image,

$$e_s = -\frac{a}{c - b_s}e'_s, \qquad e'_s = -\frac{b}{c - a_{s-1}}e_{s-1},$$

so that

$$e_s = \frac{ab}{(c - b_s)(c - a_{s-1})}e_{s-1} \quad\cdots\cdots(127),$$

and similarly

$$e'_s = \frac{ab}{(c - a_{s-1})(c - b_{s-1})}e'_{s-1}.$$

We have therefore

$$\frac{e_s}{e_{s-1}} = \frac{ab}{(c - b_s)(c - a_{s-1})} = \frac{a_s}{a}\frac{b_s}{b} = \frac{a_s c - a^2}{ab}$$

and

$$\frac{e_s}{e_{s+1}} = \frac{(c - b_{s+1})(c - a_s)}{ab} = \frac{c(c - a_s)}{ab} - \frac{b}{a}.$$

By addition we eliminate a_s, and obtain

$$\frac{e_s}{e_{s-1}} + \frac{e_s}{e_{s+1}} = \frac{c^2 - a^2 - b^2}{ab},$$

or, if we put $\dfrac{1}{e_s} = u_s$,

$$u_{s+1} - \frac{c^2 - a^2 - b^2}{ab}u_s + u_{s-1} = 0\cdots\cdots\cdots\cdots(128),$$

and from symmetry it is obvious that the same difference equation must be satisfied by a quantity $u'_s = \dfrac{1}{e'_s}$.

The solution of the difference equation (128) may be taken to be

$$u_s = A\alpha^s + B\beta^s,$$

where α, β are the roots of

$$t^2 - \frac{c^2 - a^2 - b^2}{ab}t + 1 = 0.$$

The product of these roots is unity, so that if a is the root which is less than unity, we can suppose

$$u_s = Aa^s + \frac{B}{a^s},$$

so that

$$e_s = \frac{a^s}{Aa^{2s} + B},$$

and similarly

$$e'_s = \frac{a^s}{A'a^{2s} + B'}.$$

We now have

$$q_{11} = a + e_1 + e_2 + \ldots = a + \sum_1^\infty \frac{a^s}{Aa^{2s} + B},$$

$$q_{12} = e'_1 + e'_2 + \ldots \quad = \sum_1^\infty \frac{a^s}{A'a^{2s} + B'}.$$

To determine A, B, we have

$$e_0 \equiv \frac{1}{A + B} = a,$$

$$e_1 \equiv \frac{a}{Aa^2 + B} = \frac{a^2 b}{c^2 - b^2},$$

so that

$$\frac{A}{-\xi^2} = \frac{B}{1} = \frac{1}{a(1 - \xi^2)},$$

where

$$\xi = \frac{a + ba}{c}.$$

Thus

$$e_s = \frac{aa^s(1 - \xi^2)}{1 - \xi^2 a^{2s}},$$

and

$$q_{11} = a(1 - \xi^2)\left\{\frac{1}{1 - \xi^2} + \frac{a}{1 - \xi^2 a^3} + \frac{a^3}{1 - \xi^2 a^4} + \ldots\right\}.$$

To determine A', B', we have

$$e'_1 = \frac{a}{A'a^2 + B'} = -\frac{ab}{c},$$

$$e'_2 = \frac{a^2}{A'a^4 + B'} = -\frac{a^2 b^2}{c(c^2 - a^2 - b^2)},$$

from which, in the same way,

$$q_{12} = -\frac{ab}{c}(1 - a^2)\left\{\frac{1}{1 - a^2} + \frac{a}{1 - a^4} + \frac{a^3}{1 - a^6} + \ldots\right\}.$$

The value of q_{22} can of course be written down by symmetry from that of q_{11}.

The coefficients each depend on a sum of the type $\sum \dfrac{a^s}{1 - \xi^2 a^{2s}}$. This series has been expressed in terms of definite integrals by Poisson.

From the known formula

$$\int_0^\infty \frac{\sin pt}{e^{2\pi t} - 1} = \frac{1}{4}\left\{\frac{e^p + 1}{e^p - 1}\right\} - \frac{1}{2p}$$

we obtain, on putting $p = \log \xi^2 a^{2s}$,

$$\frac{a^s}{1 - \xi^2 a^{2s}} = \frac{1}{2}a^s - \frac{a^s}{\log \xi^2 a^{2s}} - 2\int_0^\infty \frac{a^s \sin(\log \xi^2 a^{2s})t}{e^{2\pi t} - 1} dt.$$

From this follows

$$\sum \frac{a^s}{1 - \xi^2 a^{2s}} = \frac{1}{2(1 - a)} - \sum \frac{a^s}{2\log \xi + 2s \log a} - 2\int_0^\infty \frac{\sum a^s \sin(2\log \xi + 2s \log a)t}{e^{2\pi t} - 1} dt$$

$$= \frac{1}{2(1 - a)} - \int_0^\infty \frac{\xi^{2t}}{1 - a^{2t+1}} dt - 2\int_0^\infty \frac{\sin(2t \log \xi) - a \sin(2t \log \xi/a)}{(e^{2\pi t} - 1)[1 - 2a \cos(2t \log a) + a^2]} dt.$$

The series has also been expressed in finite terms by E. W. Barnes (*Quart. Journ. Math.* 138 (1903), p. 155) in terms of Double Gamma Functions, but neither of these forms is convenient for numerical computation. A. Russell (*Proc. Phys. Soc.* 23 (1911), p. 352) has shewn how the original series can be rearranged in a rapidly convergent form. If n is an integer, to be chosen subsequently,

$$\sum_{s=0}^{s=\infty} \frac{a^s}{1-\xi^2 a^{2s}} = \sum_{s=0}^{s=n-1} \frac{a^s}{1-\xi^2 a^{2s}} + \sum_{s=n}^{s=\infty} a^s \left(\sum_{p=0}^{p=\infty} \xi^{2p} a^{2ps} \right)$$

$$= \sum_{s=0}^{s=n-1} \frac{a^s}{1-\xi^2 a^{2s}} + a^n \sum_{p=0}^{p=\infty} \frac{(\xi a^n)^{2p}}{1-a^{2p+1}}.$$

The larger n is chosen to be the more rapidly the second series converges, although of course large values for n require the computation of a large number (n) of terms in the original series. As an example, given by Russell, suppose that $a=7r$, $b=r$, $c=10r$; it is sufficient to take $n=1$ and the series are found to be

$$q_{11} = 7r + \tfrac{4}{7}r \{1 + 0.0089509 + 0.0000929 + 0.0000009 + ...\} = 7.5765970r,$$
$$-q_{12} = 0.7r + \tfrac{8}{10}r \{1 + 0.0003580 + 0.0000001 + ...\} = 0.8143266r,$$
$$q_{22} = 1.1601124r.$$

As a second example Russell takes $a=98r$, $b=10.8r$, and $c=a+b+0.2r$, so that the spheres are almost in contact; the values of the coefficients are obtained to seven figures on taking $n=4$ and computing seven terms of the second series.

223. Having calculated the coefficients, we can obtain the relations between the charges and potentials, and can find also the mechanical force between the spheres. If this force is a force of repulsion F, we have

$$F = -\frac{\partial W_E}{\partial c} = -\tfrac{1}{2} \frac{\partial p_{11}}{\partial c} E_1^2 - \frac{\partial p_{12}}{\partial c} E_1 E_2 - \tfrac{1}{2} \frac{\partial p_{22}}{\partial c} E_2^2,$$

or again
$$F = \frac{\partial W_V}{\partial c} = \tfrac{1}{2} \frac{\partial q_{11}}{\partial c} V_1^2 + \frac{\partial q_{12}}{\partial c} V_1 V_2 + \tfrac{1}{2} \frac{\partial q_{22}}{\partial c} V_2^2.$$

The following table, applicable to two spheres of equal radius, taken to be unity, is compiled from materials given by Lord Kelvin[*].

c	$p_{11} (=p_{22})$	p_{12}	$q_{11}(=q_{22})$	q_{12}	$\tfrac{1}{2}\frac{\partial p_{11}}{\partial c}\left(=\frac{\partial p_{22}}{\partial c}\right)$	$-\frac{\partial p_{12}}{\partial c}$	$-\tfrac{1}{2}\frac{\partial q_{11}}{\partial c}$	$\frac{\partial q_{12}}{\partial c}$	Ratio of charges for equilibrium
2·0	·721	·721	∞	− ∞	∞	∞	∞	∞	1
2·1	·915	·509	1·584	− ·884	·154	·453	1·138	2·349	·391
2·2	·939	·475	1·431	− ·724	·0826	·305	·529	1·127	·294
2·5	·968	·406	1·253	− ·525	·0300	·181	·174	·413	·169
3·0	·986	·335	1·146	− ·389	·0101	·115	·066	·186	·089
3·5	·993	·286	1·099	− ·317	·00437	·0825	·0344	·114	·053
4·0	·996	·250	1·072	− ·269	·00216	·0628	·0207	·079	·034
5·0	·998	·200	1·044	− ·209	·00065	·0401	·0096	·048	·016
6·0	·999	·167	1·030	− ·172	·00026	·0278	·0053	·031	·009
∞	1·0	0	1·0	0	0	0	0	0	0

[*] *Papers on Electrostatics and Magnetism*, p. 96, § 142.

Images in dielectrics.

224. The method of images can also be applied to find the field produced by point charges when half of the field is occupied by dielectric, the boundary of the dielectric being an infinite plane.

We begin by considering the field produced by a single charge e at P, it being possible to obtain the most general field by the superposition of simple fields of this kind.

We shall shew that the field in air is the same as that due to a charge e at P and a certain charge e' at P', the image of P, while the field in the dielectric is the same as that due to a certain charge e'' at P, if the whole field were occupied by air.

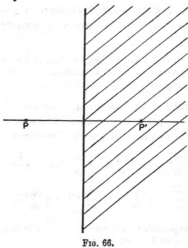

Fig. 66.

Let PP' be taken for axis of x, the origin O being in the boundary of the dielectric, and let $OP = a$. Then we have to shew that the potential V_A in air is

$$V_A = \frac{e}{\sqrt{(x+a)^2 + y^2 + z^2}} + \frac{e'}{\sqrt{(x-a)^2 + y^2 + z^2}},$$

while that in the dielectric is

$$V_D = \frac{e''}{\sqrt{(x+a)^2 + y^2 + z^2}}.$$

These potentials, we notice, satisfy Laplace's equation in each medium, everywhere except at the point P, and they arise from a distribution of charges which consists of a single point charge e at P. The potential in air at the point $0, y, z$ on the boundary is

$$V_A = \frac{e + e'}{\sqrt{a^2 + y^2 + z^2}},$$

while that in the dielectric at the same point is

$$V_D = \frac{e''}{\sqrt{a^2 + y^2 + z^2}}.$$

Thus the condition that the potential shall be continuous at each point of the boundary can be satisfied by taking

$$e'' = e + e' \qquad \dots\dots\dots\dots\dots\dots\dots(129).$$

The remaining condition to be satisfied is that at every point of the boundary, $\frac{\partial V}{\partial x}$ in air shall be equal to $K \frac{\partial V}{\partial x}$ in the dielectric; *i.e.* that

$$K \frac{\partial V_D}{\partial x} = \frac{\partial V_A}{\partial x}, \text{ when } x = 0.$$

Now, when $x = 0$,

$$K \frac{\partial V_D}{\partial x} = -\frac{Ke''a}{(a^2 + y^2 + z^2)^{\frac{3}{2}}},$$

$$\frac{\partial V_A}{\partial x} = -\frac{ea}{(a^2 + y^2 + z^2)^{\frac{3}{2}}} + \frac{e'a}{(a^2 + y^2 + z^2)^{\frac{3}{2}}},$$

so that this last condition is satisfied by taking

$$Ke'' = e - e' \qquad \dots\dots\dots\dots\dots\dots\dots(130).$$

Thus the conditions of the problem are completely satisfied by giving e', e'' values such as will satisfy relations (129) and (130); *i.e.* by taking

$$\left. \begin{array}{l} e'' = \dfrac{2}{1+K}e \\[2mm] e' = -\dfrac{K-1}{1+K}e \end{array} \right\} \qquad \dots\dots\dots\dots\dots\dots(131).$$

225. The pull on the dielectric is that due to the tensions of the lines of force which cross its boundary. In air these lines of force are the same as if we had charges e, e' at P, P' entirely in air, so that the whole tension in the direction $P'P$ of the lines of force in air is

$$-\frac{ee'}{PP'^2},$$

or

$$\frac{e^2 (K-1)}{4a^2 (K+1)}.$$

This system of tensions shews itself as an attraction between the dielectric and the point charge. If the dielectric is free to move and the point charge fixed, the dielectric will be drawn towards the point charge by this force, and conversely if the dielectric is fixed the point charge will be attracted towards the dielectric by this force.

INVERSION.

226. The geometrical method of inversion may sometimes be used to deduce the solution of one problem from that of another problem of which the solution is already known.

Geometrical Theory.

227. Let O be any point which we shall call the centre of inversion, and

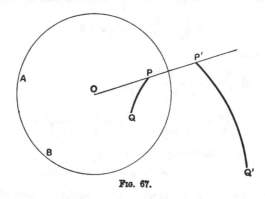

FIG. 67.

let AB be a sphere drawn about O with a radius K which we shall call the radius of inversion.

Corresponding to any point P we can find a second point P', the inverse to P in the sphere. These two points are on the same radius at distances from O such that $OP \cdot OP' = K^2$.

As P describes any surface $PQ \ldots$, P' will describe some other surface $P'Q' \ldots$, each point Q' on the second surface being the inverse of some point Q on the original surface. This second surface is said to be the *inverse* of the original surface, and the process of deducing the second surface from the first is described as *inverting* the first surface.

It is clear that if $P'Q' \ldots$ is the inverse of $PQ \ldots$, then the inverse of $P'Q' \ldots$ will be $PQ \ldots$.

If the polar equation of a surface referred to the centre of inversion as origin be $f(r, \theta, \phi) = 0$, then the equation of its inverse will be $f\left(\dfrac{K^2}{r}, \theta, \phi\right) = 0$. For the polar equation of the inverse surface is by definition $f(r', \theta, \phi) = 0$, where $rr' = K^2$ for all values of θ and ϕ.

Inverse of a sphere. Let chords PP', QQ', ... of a sphere meet in O (fig. 68). Then

$$OP \cdot OP' = OQ \cdot OQ' = \ldots = t^2,$$

where t is the length of the tangent from O to the sphere. Thus, if t is the radius of inversion, the surface $PQ\ldots$ is the inverse of $P'Q'\ldots$, *i.e.* the sphere

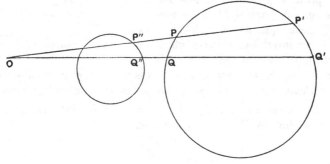

Fig. 68.

is its own inverse. With some other radius of inversion K, let $P''Q''\ldots$ be the inverse of $PQ\ldots$, then

$$OP \cdot OP'' = OQ \cdot OQ'' = \ldots = K^2,$$

so that

$$\frac{OP''}{OP'} = \frac{OQ''}{OQ'} = \ldots = \frac{K^2}{t^2}$$

and the locus of P'', Q'', ... is seen to be a sphere. Thus the inverse of a sphere is always another sphere.

A special investigation is needed when the sphere passes through O. Let OS be the diameter through O, and let S' be the point inverse to S. Then, if P' is the inverse of any point P on the circle,

$$OP \cdot OP' = OS \cdot OS',$$

or

$$\frac{OP}{OS} = \frac{OS'}{OP'},$$

so that POS, $S'OP'$ are similar triangles. Since OPS is a right angle, it follows that $OS'P'$ is a right angle, so that the locus of P' is a plane through S' perpendicular to OS'. Thus the inverse of a sphere which passes through the centre

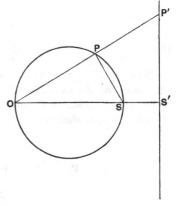

Fig. 69.

of inversion is a plane, and, conversely, the inverse of any plane is a sphere which passes through the centre of inversion.

228. If P, Q are adjacent points on a surface, and P', Q' are the corresponding points on its inverse, then OPQ, $OQ'P'$ are similar triangles, so that PQ, $P'Q'$ make equal angles with OPP'. By making PQ coincide, we find that the tangent plane at P to the surface PQ and the tangent plane at P' to the surface $P'Q'$ make equal angles with OPP'. Hence, if we invert two surfaces which intersect in P, we find that the angle

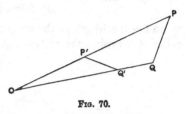

FIG. 70.

between the two inverse surfaces at P' is equal to the angle between the original surfaces at P, *i.e. an angle of intersection is not altered by inversion.*

Also, if a small cone through O cuts off areas dS, dS' from the surface $PQ...$ and its inverse $P'Q'...$, it follows that

$$\frac{dS}{dS'} = \frac{OP^2}{OP'^2}.$$

Electrical Applications.

229. Let PP', QQ' be two pairs of inverse points (fig. 70). Let a charge e at Q produce potential V_P at P, and let a charge e' at Q' produce potential V_P' at P', so that

$$V_P = \frac{e}{PQ}, \qquad V_P' = \frac{e'}{P'Q'},$$

then

$$\frac{V_P'}{V_P} = \frac{e'}{e} \cdot \frac{PQ}{P'Q'} = \frac{e'}{e} \cdot \frac{OP}{OQ'}.$$

Take

$$\frac{e'}{e} = \frac{K}{OQ} = \frac{OQ'}{K},$$

then

$$\frac{V_P'}{V_P} = \frac{OP}{K} = \frac{K}{OP'}.$$

Now let Q be a point of a conducting surface, and replace e by $\sigma\,dS$, the charge on the element of surface dS at Q. Let $\overline{V}_P$ denote the potential of the whole surface at P, and let $\overline{V}_P'$ denote the potential at P' due to a charge e' on each element dS' of the inverse surface, such that

$$\frac{e'}{\sigma\,dS} = \frac{OQ'}{K}.$$

Then, since $V_P' = V_P \dfrac{K}{OP'}$ for each element of charge, we have by addition

$$\overline{V}_P' = \overline{V}_P \frac{K}{OP'}.$$

Thus charges e' on dS', etc. produce a potential

$$\frac{\overline{V}_P K}{OP'} \text{ at } P'.$$

Now suppose that P is a point on the conducting surface Q, so that $\overline{V}_P$ becomes simply the potential of this surface, say V. The charges e' on dS', etc. now produce a potential

$$\frac{VK}{OP'} \text{ at } P',$$

so that if with these charges we combine a charge $- VK$ at O, the potential produced at P' is zero. Thus the given system of charges spread over the surface $P'Q'\ldots$, together with a charge $- VK$ at the origin, make the surface $P'Q'\ldots$ an equipotential of potential zero. In other words, from a knowledge of the distribution which raises $PQ\ldots$ to potential V, we can find the distribution on the inverse surface $P'Q'\ldots$ when it is put to earth under the influence of a charge $- VK$ at the centre of inversion.

If e, e' are the charges on corresponding elements dS, dS' at Q, Q', we have seen that

$$\frac{e'}{e} = \frac{\sigma' dS'}{\sigma dS} = \frac{K}{OQ} = \frac{OQ'}{K} = \sqrt{\frac{OQ'}{OQ}},$$

while

$$\frac{dS'}{dS} = \frac{OQ'^2}{OQ^2}.$$

Hence

$$\frac{\sigma'}{\sigma} = \left(\frac{OQ'}{OQ}\right)^{-\frac{3}{2}} = \frac{K^3}{OQ'^3} \quad\ldots\ldots\ldots\ldots\ldots\ldots(132),$$

giving the ratio of the surface densities on the two conductors.

Conversely, if we know the distribution induced on a conductor $PQ\ldots$ at potential zero by a unit charge at a point O, then by inversion about O we obtain the distribution on the inverse conductor $P'Q'\ldots$ when raised to potential $\frac{1}{K}$. As before, the ratio of the densities is given by equation (132).

Examples of Inversion.

230. Sphere. The simplest electrical problem of which we know the solution is that of a sphere raised to a given potential. Let us examine what this solution becomes on inversion.

If we invert with respect to a point P outside the sphere, we obtain the distribution on another sphere when put to earth under the influence of a point charge P. This distribution has already been obtained in § 214 by the method of images. The result there obtained, that the surface-density varies inversely as the cube of the distance from P, can now be seen at once from equation (132).

So also, if P is inside the sphere, we obtain the distribution on an uninsulated sphere produced by a point charge inside it, a result which can again be obtained by the method of images.

When P is on the sphere, we obtain the distribution on an uninsulated plane, already obtained in § 208.

231.   *Intersecting Planes.* As a more complicated example of inversion, let us invert the results obtained in § 212. We there shewed how to find

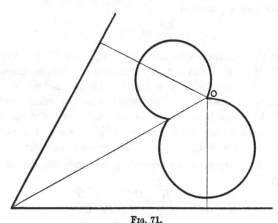

FIG. 71.

the distribution on two planes cutting at an angle $\dfrac{\pi}{n}$, when put to earth under the influence of a point charge anywhere in the acute angle between them. If we invert the solution we obtain the distribution on two spheres, cutting at an angle π/n, raised to a given potential. By a suitable choice of the radius and origin of inversion, we can give any radii we like to the two spheres.

If we take the radius of one to be infinite, we get the distribution on a plane with an excrescence in the form of a piece of a sphere: in the particular case of $n = 2$, this excrescence is hemispherical, and we obtain the distribution of electricity on a plane face with a hemispherical boss. This can, however, be obtained more directly by the method of § 219.

SPHERICAL HARMONICS.

232. The problem of finding the solution of any electrostatic problem is equivalent to that of finding a solution of Laplace's equation

$$\nabla^2 V = 0$$

throughout the space not occupied by conductors, such as shall satisfy certain conditions at the boundaries of this space—*i.e.* at infinity and on the surfaces of conductors. The theory of spherical harmonics attempts to provide a general solution of the equation $\nabla^2 V = 0$.

This is no convenient general solution in finite terms: we therefore examine solutions expressed as an infinite series. If each term of such a series is a solution of the equation, the sum of the series is necessarily a solution.

233. Let us take spherical polar coordinates r, θ, ϕ, and search for solutions of the form

$$V = RS,$$

where R is a function of r only, and S is a function of θ and ϕ only.

Laplace's equation, expressed in spherical polars, can be obtained analytically from the equation

$$\frac{\partial^2 V}{\partial x^2} + \frac{\partial^2 V}{\partial y^2} + \frac{\partial^2 V}{\partial z^2} = 0$$

by changing variables from x, y, z to r, θ, ϕ, but is most easily obtained by applying Gauss' Theorem to the small element of volume bounded by the spheres r and $r + dr$, the cones θ and $\theta + d\theta$, and the diametral planes ϕ and $\phi + d\phi$. The equation is found to be

$$\frac{1}{r^2} \frac{\partial}{\partial r} \left(r^2 \frac{\partial V}{\partial r} \right) + \frac{1}{r^2 \sin\theta} \frac{\partial}{\partial \theta} \left(\sin\theta \frac{\partial V}{\partial \theta} \right) + \frac{1}{r^2 \sin^2\theta} \frac{\partial^2 V}{\partial \phi^2} = 0.$$

Substituting the value $V = RS$, we obtain

$$\frac{S}{r^2} \frac{\partial}{\partial r} \left(r^2 \frac{\partial R}{\partial r} \right) + \frac{R}{r^2 \sin\theta} \frac{\partial}{\partial \theta} \left(\sin\theta \frac{\partial S}{\partial \theta} \right) + \frac{R}{r^2 \sin^2\theta} \frac{\partial^2 S}{\partial \phi^2} = 0,$$

or, simplifying,

$$\frac{1}{R} \frac{\partial}{\partial r} \left(r^2 \frac{\partial R}{\partial r} \right) + \frac{1}{S \sin\theta} \frac{\partial}{\partial \theta} \left(\sin\theta \frac{\partial S}{\partial \theta} \right) + \frac{1}{S \sin^2\theta} \frac{\partial^2 S}{\partial \phi^2} = 0.$$

The first term is a function of r only, while the last two terms are independent of r. Thus the equation can only be satisfied by taking

$$\frac{1}{R} \frac{\partial}{\partial r} \left(r^2 \frac{\partial R}{\partial r} \right) = K \quad\dots\dots\dots\dots\dots\dots(133),$$

$$\frac{1}{S \sin\theta} \frac{\partial}{\partial \theta} \left(\sin\theta \frac{\partial S}{\partial \theta} \right) + \frac{1}{S \sin^2\theta} \frac{\partial^2 S}{\partial \phi^2} = -K \quad\dots\dots\dots\dots(134),$$

where K is a constant. Equation (133), regarded as a differential equation for R, can be solved, the solution being

$$R = Ar^n + \frac{B}{r^{n+1}} \quad\dots\dots\dots\dots\dots\dots(135),$$

where A, B are arbitrary constants, and $n(n + 1) = K$. After simplification equation (134) becomes

$$\frac{\partial}{\sin\theta\, \partial\theta} \left(\sin\theta \frac{\partial S}{\partial \theta} \right) + \frac{1}{\sin^2\theta} \frac{\partial^2 S}{\partial \phi^2} + n(n + 1)S = 0 \quad\dots\dots\dots(136).$$

Any solution of this equation will be denoted by S_n, the solution being a function of n as well as of θ and ϕ. The solution of Laplace's equation we have obtained is now

$$V = RS = \left(Ar^n + \frac{B}{r^{n+1}} \right) S_n,$$

and by the addition of such solutions, the most general solution of Laplace's equation may be reached.

234. DEFINITIONS. *Any solution of Laplace's equation is said to be a spherical harmonic.*

A solution which is homogeneous in x, y, z of dimensions n is said to be a *spherical harmonic of degree n*.

A spherical harmonic of degree n must be of the form r^n multiplied by a function of θ and ϕ, it must therefore be of the form $Ar^n S_n$, where S_n is a solution of equation (136).

Any solution S_n of equation (136) is said to be a *surface-harmonic of degree n*.

235. THEOREM. *If V is any spherical harmonic of degree n, then V/r^{2n+1} is a spherical harmonic of degree $-(n+1)$.*

For V must be of the form $Ar^n S_n$, so that

$$\frac{V}{r^{2n+1}} = \frac{AS_n}{r^{n+1}},$$

which is known to be a solution of Laplace's equation, and is of dimensions $-(n+1)$ in r. Conversely if V is a spherical harmonic of degree $-(n+1)$, then $r^{2n+1}V$ is a spherical harmonic of degree n.

236. THEOREM. *If V is any spherical harmonic of degree n, then*

$$\frac{\partial^{s+t+u}V}{\partial x^s \partial y^t \partial z^u},$$

where $s, t,$ and u are any integers, is a spherical harmonic of degree $n - s - t - u$.

For

$$\frac{\partial^2 V}{\partial x^2} + \frac{\partial^2 V}{\partial y^2} + \frac{\partial^2 V}{\partial z^2} = 0,$$

so that on differentiation s times with respect to x, t times with respect to y, and u times with respect to z,

$$\frac{\partial^{s+t+u+2}V}{\partial x^{s+2}\partial y^t \partial z^u} + \frac{\partial^{s+t+u+2}V}{\partial x^s \partial y^{t+2}\partial z^u} + \frac{\partial^{s+t+u+2}V}{\partial x^s \partial y^t \partial z^{u+2}} = 0,$$

or

$$\nabla^2 \left(\frac{\partial^{s+t+u}V}{\partial x^s \partial y^t \partial z^u} \right) = 0,$$

which proves the theorem.

237. THEOREM. *If S_m, S_n are two surface harmonics of different degrees m, n, then*

$$\iint S_n S_m \, d\omega = 0,$$

where the integration is over the surface of a unit sphere.

In Green's Theorem (§ 181),

$$\iiint (\Phi \nabla^2 \Psi - \Psi \nabla^2 \Phi) \, dx \, dy \, dz = -\iint \left(\Phi \frac{\partial \Psi}{\partial n} - \Psi \frac{\partial \Phi}{\partial n} \right) dS,$$

put $\Phi = r^n S_n$, $\Psi = r^m S_m$, and take the surface to be the unit sphere.

Then $\nabla^2\Phi = 0$, $\nabla^2\Psi = 0$, $\dfrac{\partial\Phi}{\partial n} = -\dfrac{\partial\Phi}{\partial r} = -nr^{n-1}S_n$, and $\dfrac{\partial\Psi}{\partial n} = -mr^{m-1}S_m$.

Thus the volume integral vanishes, and the equation becomes

$$\iint (nr^{m+n-1}S_nS_m - mr^{m+n-1}S_nS_m)\,d\omega = 0,$$

or, since n is, by hypothesis, not equal to m,

$$\iint S_nS_m\,d\omega = 0.$$

Harmonics of Integral Degree.

238. The following table of examples of harmonics of integral degrees $n = 0$, -1, -2, $+1$, is taken from Thomson and Tait's *Natural Philosophy*.

$n = 0.$ 1, $\tan^{-1}\dfrac{y}{x}$, $\log\dfrac{r+z}{r-z}$, $\tan^{-1}\dfrac{y}{x}\log\dfrac{r+z}{r-z}$, $\dfrac{rz(x^2-y^2)}{(x^2+y^2)^2}$, $\dfrac{2rxyz}{(x^2+y^2)^2}$.

Also if V_0 is any one of these harmonics, $\dfrac{\partial V_0}{\partial x}$, $\dfrac{\partial V_0}{\partial y}$, $\dfrac{\partial V_0}{\partial z}$ are harmonics of degree -1, so that $r\dfrac{\partial V_0}{\partial x}$, $r\dfrac{\partial V_0}{\partial y}$, $r\dfrac{\partial V_0}{\partial z}$ are harmonics of degree zero. As examples of harmonics derived in this way may be given

$$\frac{rx}{x^2+y^2},\quad \frac{ry}{x^2+y^2},\quad \frac{zx}{x^2+y^2},\quad \frac{zy}{x^2+y^2},\quad \frac{x}{r+z},\quad \frac{x}{r-z}.$$

By differentiating any harmonic V_0 any number s of times, multiplying by r^{2s-1} and differentiating again $s-1$ times, we obtain more harmonics of degree zero.

$n = -1.$ Any harmonic of degree zero divided by r or differentiated with respect to x, y or z, e.g.

$$\frac{1}{r},\quad \frac{1}{r}\tan^{-1}\frac{y}{x},\quad \frac{1}{r}\log\frac{r+z}{r-z},\quad \frac{x}{x^2+y^2},\quad \frac{x}{r(r+z)}.$$

$n = -2.$ By differentiating harmonics of degree -1 with respect to x, y or z we obtain harmonics of degree -2, e.g.

$$\frac{x}{r^3},\quad \frac{y}{r^3},\quad \frac{z}{r^3},\quad \frac{z}{r^3}\tan^{-1}\frac{y}{x},\quad \frac{z}{r^3}\log\frac{r+z}{r-z}.$$

$n = 1.$ Multiplying harmonics of degree -2 by r^3, we obtain harmonics of degree 1, e.g.

$$x,\ y,\ z,\quad z\tan^{-1}\frac{y}{x},\quad z\log\frac{r+z}{r-z}-2r.$$

Rational Integral Harmonics.

239. An important class of harmonic consists of rational integral algebraic functions of x, y, z. In the most general homogeneous function of x, y, z of degree n there are $\frac{1}{2}(n+1)(n+2)$ coefficients. If we operate with ∇^2 we are left with a homogeneous function of x, y, z of degree $n-2$, and therefore possessing $\frac{1}{2}n(n-1)$ coefficients. For the original function to be a spherical harmonic, these $\frac{1}{2}n(n-1)$ coefficients must all vanish, so that we must have $\frac{1}{2}n(n-1)$ relations between the original $\frac{1}{2}(n+1)(n+2)$ coefficients.

Thus the number of coefficients which may be regarded as independent in the original function, subject to the condition of its being a harmonic, is

$$\tfrac{1}{2}(n+1)(n+2) - \tfrac{1}{2}n(n-1),$$

or $2n+1$. This, then, is the number of independent rational harmonics of degree n.

For instance, when $n=1$ the most general harmonic is

$$Ax + By + Cz,$$

possessing three independent arbitrary constants, and so representing three independent harmonics which may conveniently be taken to be x, y and z.

When $n=2$, the most general harmonic is

$$ax^2 + by^2 + cz^2 + dyz + ezx + fxy,$$

where a, b, c are subject to $a+b+c=0$. The five independent harmonics may conveniently be taken to be

$$yz, \quad zx, \quad xy, \quad x^2 - y^2, \quad x^2 - z^2.$$

When $n=0$, $2n+1=1$. Thus there is only one harmonic of degree zero, and this may be taken to be $V=1$.

Corresponding to a rational integral harmonic V_n of positive degree n, there is the harmonic $\dfrac{V_n}{r^{2n+1}}$ of degree $-(n+1)$. These harmonics of degree $-(n+1)$ are accordingly $2n+1$ in number. Thus the only harmonic of this kind and of degree -1 is $\dfrac{1}{r}$.

Consider now the various expressions of the type

$$\frac{\partial^{s+t+u}}{\partial x^s \partial y^t \partial z^u}\left(\frac{1}{r}\right) \quad\dots\dots\dots\dots\dots\dots\dots(137),$$

where $s+t+u=n$.

These, as we know, are harmonics of degree $-(n+1)$, and from § 235 it is obvious that they must be of the form $\dfrac{V_n}{r^{2n+1}}$, where V_n is a rational integral harmonic of degree n. Since $\dfrac{1}{r}$ is harmonic, $\nabla^2\left(\dfrac{1}{r}\right)=0$, so that

$$\frac{\partial^2}{\partial z^2}\left(\frac{1}{r}\right) = -\left(\frac{\partial^2}{\partial x^2} + \frac{\partial^2}{\partial y^2}\right)\left(\frac{1}{r}\right) \quad\dots\dots\dots\dots\dots(138).$$

The most general harmonic obtained by combining the harmonics of type (137) is

$$\Sigma A_{stu}\frac{\partial^{s+t+u}}{\partial x^s \partial y^t \partial z^u}\left(\frac{1}{r}\right) \quad\dots\dots\dots\dots\dots(139),$$

but by equation (138) this can be reduced at once to the form

$$\frac{\partial}{\partial z}\Sigma A_{pq}\frac{\partial^{p+q}}{\partial x^p \partial y^q}\left(\frac{1}{r}\right) + \Sigma A_{p'q'}\frac{\partial^{p'+q'}}{\partial x^{p'} \partial y^{q'}}\left(\frac{1}{r}\right),$$

where $p + q = n - 1$ and $p' + q' = n$. This again may be replaced by

$$\frac{\partial}{\partial z} \sum_{p=0}^{p=n-1} B_p \frac{\partial^{n-1}}{\partial x^p \partial y^{n-1-p}} \left(\frac{1}{r}\right) + \sum_{p=0}^{p=n} B_p' \frac{\partial^n}{\partial x^p \partial y^{n-p}} \left(\frac{1}{r}\right),$$

so that there are $2n + 1$ arbitrary constants in all, and it is obvious on examination that the harmonics, multiplied by all the coefficients $B_p, \ldots B_p', \ldots$ are independent. Thus, by differentiating $\frac{1}{r}$ n times, we have arrived at $2n + 1$ independent rational integral harmonics, and it is known that this is as many as there are.

Expansion in Rational Integral Harmonics.

240. THEOREM[*]. *The value of any finite single-valued function of position on a spherical surface can be expressed, at every point of the surface at which the function is continuous, as a series of rational integral harmonics, provided the function has only a finite number of lines and points of discontinuity and of maxima and minima on the surface.*

Let F be the arbitrary function of position on the sphere, and let the sphere be supposed of radius a. Let P be any point outside the sphere at a distance f from its centre O, and let Q be any point on the surface of the sphere.

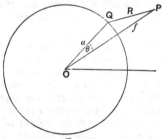

FIG. 72.

Let PQ be equal to R, so that

$$R^2 = f^2 + a^2 - 2af \cos POQ.$$

We have the identity

$$\frac{f^2 - a^2}{4\pi a} \iint \frac{dS}{R^3} = \frac{a}{f} \quad \ldots\ldots\ldots\ldots\ldots\ldots(140),$$

where the integration is taken over the surface of the sphere, a result which it is easy to prove by integration.

A point charge e placed at P induces surface density $-\frac{e}{4\pi a} \frac{f^2 - a^2}{R^3}$ on the surface of the sphere (§ 214), and the total induced charge is $-\frac{ea}{f}$. The identity is therefore obvious from electrostatic principles.

---

[*] The proof of this theorem is stated in the form which seems best suited to the requirements of the student of electricity and makes no pretence at absolute mathematical rigour.

Now introduce a quantity u defined by

$$u = \frac{f^2 - a^2}{4\pi a} \iint \frac{F dS}{R^3} \quad \dots\dots\dots\dots\dots(141),$$

so that u is a function of the position of P. If P is very close to the sphere, $f^2 - a^2$ is small, and the important contributions to the integral arise from those terms for which R is very small: *i.e.* from elements near to P.

If the value of F does not change abruptly near to the point P, or oscillate with infinite frequency, we can suppose that as P approaches the sphere, all elements on the sphere from which the contribution to the integral (141) are of importance, have the same F. This value of F will of course be the value at the point at which P ultimately touches the sphere, say F_P. Thus in the limit we have

$$u = \frac{(f^2 - a^2) F_P}{4\pi a} \iint \frac{dS}{R^3} \quad \dots\dots\dots\dots\dots(142),$$

$$= F_P \frac{a}{f}, \text{ by equation (140)},$$

$$= F_P,$$

when in the limit f becomes equal to a.

If the value of F oscillates with infinite frequency near to the point P, we obviously may not take F outside the sign of integration in passing from equation (141) to equation (142).

If the value of F is discontinuous at the point P' of the sphere with which P ultimately coincides, we again cannot take F outside the sign of integration. Suppose, however, that we take coordinates ρ, ϑ to express the position of a point P'' on the surface of the sphere very near to P', the coordinate ρ being the distance $P'P''$, and ϑ being the angle which $P'P''$ makes with any line through P' in the tangent plane at P'. Then F may be regarded as a function of ρ, ϑ, and the fact that F is discontinuous at P' is expressed by saying that as we approach the limit $\rho = 0$, the limiting value of F (assuming such a limit to exist) is a function of ϑ—*i.e.* depends on the path by which P' is approached. Let $F(\vartheta)$ denote this limit. Then

$$u = \frac{f^2 - a^2}{4\pi a} \int \frac{F(\vartheta) \rho d\rho d\vartheta}{R^3}$$

$$= \frac{f^2 - a^2}{4\pi a} \int F(\vartheta) \left[\int \frac{\rho d\rho}{R^3} \right] d\vartheta$$

$$= \frac{f^2 - a^2}{4\pi a} \int F(\vartheta) \left[\frac{1}{2\pi} \iint \frac{dS}{R^3} \right] d\vartheta$$

$$= \frac{1}{2\pi} \int F(\vartheta) \left[\frac{f^2 - a^2}{4\pi a} \iint \frac{dS}{R^3} \right] d\vartheta$$

$$= \frac{1}{2\pi} \int F(\vartheta) \left(\frac{a}{f} \right) d\vartheta, \text{ by equation (140)}.$$

On passing to the limit and putting $a = f$, we find that

$$u = \frac{1}{2\pi} \int F(\vartheta) d\vartheta \quad \dots\dots\dots\dots\dots(143),$$

i.e. u is the average value of F taken on a small circle of infinitesimal radius surrounding P. In particular, if F changes abruptly on crossing a certain line through P, having a value F_1 on one side, and a value F_2 on the other, then the limiting value of u is

$$u = \tfrac{1}{2}(F_1 + F_2).$$

If we take θ to denote the angle POQ,

$$\frac{1}{R} = (f^2 - 2af \cos\theta + a^2)^{-\frac{1}{2}}$$

$$= \frac{1}{f}\left(1 + \frac{a^2 - 2af\cos\theta}{f^2}\right)^{-\frac{1}{2}}$$

$$= \frac{1}{f}\left[1 - \tfrac{1}{2}\frac{a^2 - 2af\cos\theta}{f^2} + \tfrac{3}{8}\left(\frac{a^2 - 2af\cos\theta}{f^2}\right)^2 - \ldots\right],$$

or, arranging in descending powers of f,

$$\frac{1}{R} = \frac{1}{f}\left[1 + P_1\frac{a}{f} + P_2\frac{a^2}{f^2} + P_3\frac{a^3}{f^3} + \ldots\right] \quad\ldots\ldots\ldots\ldots(144),$$

in which $P_1, P_2, P_3 \ldots$ are functions of θ, being obviously rational integral functions of $\cos\theta$. When $\theta = 0$,

$$\frac{1}{R} = \frac{1}{a - f} = \frac{1}{f}\left(1 + \frac{a}{f} + \frac{a^2}{f^2} + \ldots\right),$$

and when $\theta = \pi$,

$$\frac{1}{R} = \frac{1}{a + f} = \frac{1}{f}\left(1 - \frac{a}{f} + \frac{a^2}{f^2} - \ldots\right),$$

so that when $\theta = 0$,

$$P_1 = P_2 = \ldots = 1,$$

and when $\theta = \pi$,

$$-P_1 = P_2 = -P_3 = \ldots = 1.$$

It is clear, therefore, that the series (144) is convergent for $\theta = 0$ and $\theta = \pi$, and a consideration of the geometrical interpretation of this series will shew that it must be convergent for all intermediate values*.

Differentiating equation (144) with respect to f, we get

$$\frac{a\cos\theta - f}{R^3} = \frac{d\left(\dfrac{1}{R}\right)}{df} = -\frac{1}{f^2} - 2P_1\frac{a}{f^3} - 3P_2\frac{a^2}{f^4} - \ldots \quad\ldots\ldots(145).$$

If we multiply this equation by $2f$, and add corresponding sides to equation (144), we obtain

$$\frac{a^2 - f^2}{R^3} = -\sum_0^\infty (2n + 1)P_n\frac{a^n}{f^{n+1}}.$$

Multiplying this equation by $-\dfrac{F}{4\pi a}$, and integrating over the surface of the sphere, we obtain

$$\frac{f^2 - a^2}{4\pi a}\iint\frac{F\,dS}{R^3} = \sum_0^\infty\frac{2n + 1}{4\pi}\iint FP_n\frac{a^{n-1}}{f^{n+1}}\,dS,$$

---

* Being a power series in $\cos\theta$ it can only have a single radius of convergence, and this cannot be between $\cos\theta = 1$ and $\cos\theta = -1$.

or, by equation (141),

$$u = \frac{1}{4\pi a^2} \overset{\infty}{\underset{0}{\Sigma}} (2n+1) \iint F P_n \left(\frac{a}{f}\right)^{n+1} dS.$$

If the function F is continuous and non-oscillatory at the point P, then on passing to the limit and putting $f = a$, we obtain

$$F = \frac{1}{4\pi a^2} \overset{\infty}{\underset{0}{\Sigma}} (2n+1) \iint F P_n \, dS \quad \dots\dots\dots(146).$$

If F is discontinuous and non-oscillatory, then the value of the series on the right is not F, but is the function defined in equation (143).

Now it is known that $1/r$ is a spherical harmonic, so that we have

$$\nabla^2 \left(\frac{1}{R}\right) = 0,$$

where the differentiation is with respect to the coordinates of Q. Hence $1/R$ must be of the form (cf. § 233)

$$\frac{1}{R} = \Sigma \left(A r^n + \frac{B}{r^{n+1}}\right) S_n \quad \dots\dots\dots(147),$$

where S_n is a surface harmonic of order n. Comparing with equation (144), and remembering that a in this equation is the same as the r of equation (147), we see that P_n, regarded as a function of the position of Q, is a surface harmonic of order n, and we have already seen that it is a series of powers of $\cos \theta$, or of $\frac{x}{r}$, the highest power being the nth, so that $r^n P_n$ is a rational integral harmonic of order n. It follows that

$$\iint F r^n P_n \, dS,$$

being the sum of a number of terms each of the form $r^n P_n$, is also a rational integral harmonic of order n, say V_n. On the surface of the sphere

$$V_n = a^n \iint F P_n \, dS,$$

so that equation (146) becomes

$$F = \frac{1}{4\pi a^2} \overset{\infty}{\underset{0}{\Sigma}} \frac{2n+1}{a^n} V_n \quad \dots\dots\dots(148),$$

which establishes the result in question.

241. THEOREM. *The expansion of an arbitrary function of position on the surface of a sphere as a series of rational integral harmonics is unique.*

For if possible let the same function F be expanded in two ways, say

$$F = \Sigma W_n \quad \dots\dots\dots\dots(149),$$
$$F = \Sigma W_n' \quad \dots\dots\dots\dots(150),$$

where W_n, W_n' are rational integral harmonics of order n. Then the function

$$u = \Sigma (W_n - W_n')$$

is a spherical harmonic, which vanishes at every point of the sphere. Since $\nabla^2 u = 0$ at every point inside the sphere it is impossible for u to have either a maximum or a minimum value inside the sphere (cf. § 52), so that $u = 0$ at every point inside the sphere. Since $W_n - W_n'$ is a harmonic of order n, it must be of the form $r^n S_n$, where S_n is a surface harmonic, so that

$$u = \Sigma r^n S_n = 0.$$

Thus u is a power series in r which vanishes for all values of r from $r = 0$ to $r = a$. Thus $S_n = 0$ for all values of n. Hence $W_n = W_n'$, and the two expansions (149) and (150) are seen to be identical.

242. It is clear that in electrostatics we shall in general only be concerned with functions which are finite and single-valued at every point, and of which the discontinuities are finite in number. Thus the only classes of harmonics which are of importance are rational integral harmonics, and in future we confine our attention to these. We have found that

(i) The rational integral harmonics of degree n are $(2n + 1)$ in number, and may all be derived from the harmonic $\dfrac{1}{r}$ by differentiation.

(ii) Any function of position on a spherical surface, which satisfies the conditions which obtain in a physical problem, can be expanded as a series of rational integral harmonics, and this can be done only in one way.

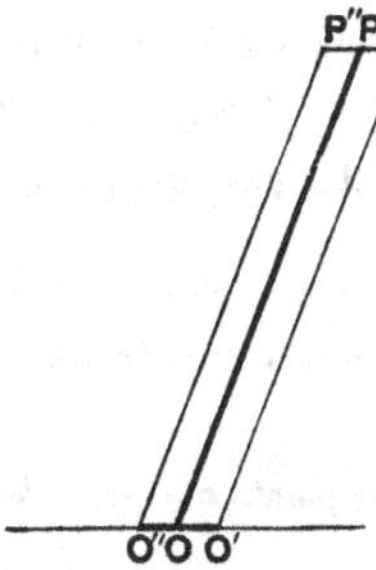

243. Before considering these harmonics in detail, we may try to form some idea of the physical conceptions which lead to them most directly.

The function $\dfrac{1}{r}$ is the potential of a unit charge at the origin. If, as in § 64, we consider two charges $\pm e$ at points O', O'' at equal small distances a, $-a$ from the origin along the axis of x, we obtain as the potential at P,

$$V = \frac{e}{O'P} - \frac{e}{O''P} = \frac{e}{OP''} - \frac{e}{OP'}$$

$$= -e \cdot P'P'' \frac{\partial}{\partial x} \left(\frac{1}{r} \right).$$

If we take $-e \cdot P'P'' = 1$, we have a doublet of strength -1 parallel to the axis of x, and the potential at P is $\dfrac{\partial}{\partial x} \left(\dfrac{1}{r} \right)$. In fact this potential is exactly the same as $-\dfrac{x}{r^3}$ already found in § 64.

Fig. 73.

Thus the three harmonics of order -2 obtained by dividing the rational integral harmonics of order 1 by r^2, namely $\dfrac{\partial}{\partial x}\left(\dfrac{1}{r}\right)$, $\dfrac{\partial}{\partial y}\left(\dfrac{1}{r}\right)$, $\dfrac{\partial}{\partial z}\left(\dfrac{1}{r}\right)$, are simply the potentials of three doublets each of unit strength, parallel to the negative axes of x, y, z respectively.

If in fig. 73 we replace the charge e at O' by a doublet of strength e parallel to the negative axis of x, and the charge $-e$ at O'' by a doublet of strength $-e$ parallel to the negative axis of x, we obtain a potential

$$\frac{\partial^2}{\partial x^2}\left(\frac{1}{r}\right).$$

If instead of the doublets being parallel to the axis of x, we take them parallel to the axis of y, we obtain a potential

$$\frac{\partial^2}{\partial x \partial y}\left(\frac{1}{r}\right).$$

So we can go on indefinitely, for on differentiating the potential of a system with respect to x we get the potential of a system obtained by replacing each unit charge of the original system by a doublet of unit strength parallel to the axis of x. Thus all harmonics of type

$$\frac{\partial^{s+t+u}}{\partial x^s\, \partial y^t\, \partial z^u}\left(\frac{1}{r}\right)$$

(cf. § 236) can be regarded as potentials of systems of doublets at the origin, and, as we have seen (§ 239), it is these potentials which give rise to the rational integral harmonics.

244. For instance in finding a system to give potential $\dfrac{\partial^2}{\partial x^2}\left(\dfrac{1}{r}\right)$, we may replace the charge O in fig. 73 by a charge $\dfrac{1}{2a}$ at distance $2a$ from O and $-\dfrac{1}{2a}$ at O. The charge at O' may be similarly treated, so that the whole system is seen to consist of charges

$$E, \ -2E, \ E,$$

at the points $x = -b, 0, b$ where $b = 2a$, and $E^2 = \dfrac{1}{b^3}$.

A system of this kind placed along each axis gives a charge $-6E$ at the origin and a charge E at each corner of a regular octahedron having the origin as centre. The potential

$$= \frac{\partial^2}{\partial x^2}\left(\frac{1}{r}\right) + \frac{\partial^2}{\partial y^2}\left(\frac{1}{r}\right) + \frac{\partial^2}{\partial z^2}\left(\frac{1}{r}\right)$$
$$= 0,$$

so that such a system sends out no lines of force.

245. The most important class of rational integral harmonics is formed by harmonics which are symmetrical about an axis, say that of x. There is one harmonic of each degree n, namely that derived from the function

$$\frac{\partial^n}{\partial x^n}\left(\frac{1}{r}\right).$$

These harmonics we proceed to investigate.

246. The function

$$\frac{1}{\sqrt{a^2 - 2ar\cos\theta + r^2}} \quad\text{...............................(151)}$$

can, as we have already seen (cf. equation (144)), be expanded in a convergent series in the form

$$\frac{1}{\sqrt{a^2 - 2ar\cos\theta + r^2}} = \frac{1}{a} + P_1\frac{r}{a^2} + P_2\frac{r^2}{a^3} + \ldots + P_n\frac{r^n}{a^{n+1}} + \ldots \quad \ldots(152)$$

if a is greater than r. Here the coefficients $P_1, P_2, \ldots$ are functions of $\cos\theta$, and are known as Legendre's coefficients. When we wish to specify the particular value of $\cos\theta$, we write P_n as $P_n(\cos\theta)$.

Interchanging r and a in equation (152) we find that, if $r > a$,

$$\frac{1}{\sqrt{a^2 - 2ar\cos\theta + r^2}} = \frac{1}{r} + P_1\frac{a}{r^2} + P_2\frac{a^2}{r^3} + \ldots \quad \ldots\ldots\ldots(153).$$

We have already seen that the functions $P_1, P_2, \ldots$ are surface harmonics, each term of the equations (152) and (153) separately satisfying Laplace's equation. The equation satisfied by the general surface harmonic S_n of degree n, namely equation (136), is

$$\frac{\partial}{\sin\theta\,\partial\theta}\left(\sin\theta\,\frac{\partial S_n}{\partial\theta}\right) + \frac{\partial^2 S_n}{\sin^2\theta\,\partial\phi^2} + n(n+1)S_n = 0.$$

In the present case P_n is independent of ϕ, so that the differential equation satisfied by P_n is

$$\frac{\partial}{\sin\theta\,\partial\theta}\left(\sin\theta\,\frac{\partial P_n}{\partial\theta}\right) + n(n+1)P_n = 0,$$

or, if we write μ for $\cos\theta$,

$$\frac{\partial}{\partial\mu}\left\{(1-\mu^2)\,\frac{\partial P_n}{\partial\mu}\right\} + n(n+1)P_n = 0 \quad \ldots\ldots\ldots\ldots(154).$$

This equation is known as Legendre's equation.

247. By actual expansion of expression (151)

$$(a^2 - 2ar\mu + r^2)^{-\frac{1}{2}} = \frac{1}{a}\left\{1 + \frac{1}{2}\left(2\frac{r\mu}{a} - \frac{r^2}{a^2}\right) + \frac{1.3}{2.4}\left(2\frac{r\mu}{a} - \frac{r^2}{a^2}\right)^2 + \ldots\right\},$$

so that on picking out the coefficient of r^n, we obtain

$$P_n = \frac{1.3\ldots2n-1}{n!}\mu^n - \frac{1.3\ldots2n-3}{2.(n-2)!}\mu^{n-2} + \frac{1.3\ldots2n-5}{2.4.(n-4)!}\mu^{n-4} - \ldots$$
$$\ldots\ldots(155).$$

Thus P_n is an even or odd function of μ according as n is even or odd. It will readily be verified that expression (155) is a solution in series of equation (154).

Let us take axes Ox, Oy, Oz, the axis Ox to coincide with the line $\theta = 0$, then $\mu r = r \cos \theta = x$. Then it appears that $P_n r^n$ is a rational integral function of x, y, and z of degree n, and, being a solution of Laplace's equation, it must be a rational integral harmonic of degree n. We have seen that there can only be one harmonic of this type which is also symmetrical about an axis; this, then, must be $P_n r^n$.

248. If we write

$$(a^2 - 2ar\mu + r^2)^{-\frac{1}{2}} = f(a)$$

we have, by Maclaurin's Theorem,

$$f(a) = f(0) + a \left\{ \frac{\partial f(a)}{\partial a} \right\}_{a=0} + \frac{a^2}{2!} \left\{ \frac{\partial^2 f(a)}{\partial a^2} \right\}_{a=0} + \dots \quad \dots\dots(156).$$

If P is the point whose polar coordinates are a, 0 and Q is the point r, θ, then $f(a) = \dfrac{1}{PQ}$. The Cartesian coordinates of P may be taken to be $a, 0, 0$; let those of Q be x, y, z. Then $f(a) = \dfrac{1}{\sqrt{(x-a)^2 + y^2 + z^2}}$, so that as regards differentiation of $f(a)$,

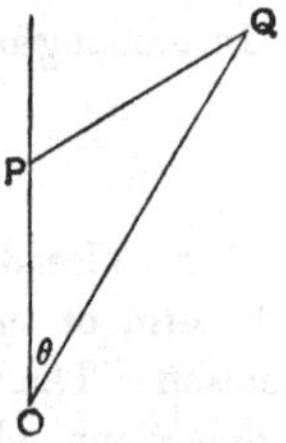

$$\frac{\partial}{\partial a} = -\frac{\partial}{\partial x}.$$

Fig. 74.

Thus

$$\left\{ \frac{\partial^n f(a)}{\partial a^n} \right\}_{a=0} = (-)^n \left\{ \frac{\partial^n f(a)}{\partial x^n} \right\}_{a=0} = (-)^n \frac{\partial^n f(0)}{\partial x^n}$$

$$= (-)^n \frac{\partial^n}{\partial x^n} \frac{1}{\sqrt{x^2 + y^2 + z^2}}$$

$$= (-)^n \frac{\partial^n}{\partial x^n} \left(\frac{1}{r} \right),$$

so that equation (156) becomes

$$f(a) = \frac{1}{r} - a \frac{\partial}{\partial x} \left(\frac{1}{r} \right) + \frac{a^2}{2!} \frac{\partial^2}{\partial x^2} \left(\frac{1}{r} \right) - \dots,$$

and on comparison with expansion (153), we see that

$$P_n = r^{n+1} \frac{(-1)^n}{n!} \frac{\partial^n}{\partial x^n} \left(\frac{1}{r} \right),$$

giving the form for P_n which we have already found to exist in § 245.

249. A more convenient form for P_n can be obtained as follows.

Let

$$1 - hy = (1 - 2h\mu + h^2)^{\frac{1}{2}} \quad \dots\dots\dots\dots\dots(157),$$

so that

$$y = \mu + h \frac{y^2 - 1}{2} \dots\dots\dots\dots\dots(158).$$

From this relation we can expand y by Lagrange's Theorem (cf. Edwards, *Differential Calculus*, § 517) in the form

$$y = \mu + h\frac{\mu^2 - 1}{2} + \dots + \frac{h^n}{n!}\left(\frac{\partial}{\partial\mu}\right)^{n-1}\left(\frac{\mu^2 - 1}{2}\right)^n + \dots$$

Differentiating with respect to μ,

$$\frac{\partial y}{\partial\mu} = 1 + \frac{h}{2}\frac{\partial}{\partial\mu}(\mu^2 - 1) + \dots + \frac{1}{n!}\left(\frac{h}{2}\right)^n\left(\frac{\partial}{\partial\mu}\right)^n(\mu^2 - 1)^n + \dots$$

From equation (157), however, we find

$$\frac{\partial y}{\partial\mu} = (1 - 2h\mu + h^2)^{-\frac{1}{2}} = 1 + hP_1 + \dots + h^n P_n + \dots$$

Equating the coefficients of h^n in the two expansions, we find

$$P_n = \frac{1}{2^n n!}\left(\frac{\partial}{\partial\mu}\right)^n(\mu^2 - 1)^n \quad\dots\dots\dots\dots\dots(159).$$

250. This last formula supplies the easiest way of calculating actual values of P_n. The values of $P_1, P_2, \dots P_7$ are found to be

$$P_1(\mu) = \mu,$$
$$P_2(\mu) = \tfrac{1}{2}(3\mu^2 - 1),$$
$$P_3(\mu) = \tfrac{1}{2}(5\mu^3 - 3\mu),$$
$$P_4(\mu) = \tfrac{1}{8}(35\mu^4 - 30\mu^2 + 3),$$
$$P_5(\mu) = \tfrac{1}{8}(63\mu^5 - 70\mu^3 + 15\mu),$$
$$P_6(\mu) = \tfrac{1}{16}(231\mu^6 - 315\mu^4 + 105\mu^2 - 5),$$
$$P_7(\mu) = \tfrac{1}{16}(429\mu^7 - 693\mu^5 + 315\mu^3 - 35\mu).$$

251. The equation $(\mu^2 - 1)^n = 0$ has $2n$ real roots, of which n may be regarded as coinciding at $\mu = 1$, and n at $\mu = -1$. By a well-known theorem, the first derived equation,

$$\frac{\partial}{\partial\mu}(\mu^2 - 1)^n = 0,$$

will have $2n - 1$ real roots separating those of the original equation. Passing to the nth derived equation, we find that the equation

$$\frac{\partial^n}{\partial\mu^n}(\mu^2 - 1)^n = 0$$

has n real roots, and that these must all lie between $\mu = -1$ and $\mu = +1$. The roots are all separate, for two roots could only be coincident if the original equation $(\mu^2 - 1)^n = 0$ had $n + 1$ coincident roots.

Thus the n roots of the equation $P_n(\mu) = 0$ are all real and separate and lie between $\mu = -1$ and $\mu = +1$.

252. Putting $\mu = 1$, we obtain

$$1 + P_1 h + P_2 h^2 + \ldots = \frac{1}{\sqrt{1 - 2h + h^2}}$$
$$= 1 + h + h^2 + \ldots,$$

so that $P_1 = P_2 = \ldots = 1$. Similarly, when $\mu = -1$, we find (cf. § 240) that

$$-P_1 = +P_2 = -P_3 = \ldots = -1.$$

We can now shew that throughout the range from $\mu = -1$ to $\mu = +1$, the numerical value of P_n is never greater than unity. We have

$$(1 - 2h \cos\theta + h^2)^{-\frac{1}{2}} = (1 - he^{i\theta})^{-\frac{1}{2}} (1 - he^{-i\theta})^{-\frac{1}{2}}$$

$$= \left(1 + \frac{1}{2} he^{i\theta} + \frac{1.3}{2.4} h^2 e^{2i\theta} + \ldots \right)$$

$$\times \left(1 + \frac{1}{2} he^{-i\theta} + \frac{1.3}{2.4} h^2 e^{-2i\theta} + \ldots \right),$$

so that on picking out coefficients of h^n,

$$P_n = \frac{1.3 \ldots 2n-1}{2.4 \ldots 2n} 2 \cos n\theta + \frac{1}{2} \cdot \frac{1.3 \ldots 2n-3}{2.4 \ldots 2n-2} 2 \cos(n-2)\theta + \ldots.$$

Every coefficient is positive, so that P_n is numerically greatest when each cosine is equal to unity, *i.e.* when $\theta = 0$. Thus P_n is never greater than unity.

Fig. 75 shews the graphs of P_1, P_2, P_3, P_4, from $\mu = -1$ to $\mu = +1$, the value of θ being taken as abscissa.

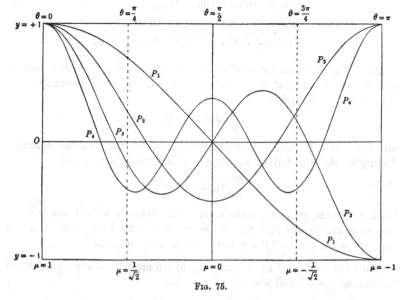

Fig. 75.

Relations between coefficients of different orders.

253. We have

$$(1 - 2h\mu + h^2)^{-\frac{1}{2}} = 1 + \sum_1^\infty h^n P_n \quad\dots\dots\dots\dots(160).$$

Differentiating with regard to h,

$$(\mu - h)(1 - 2h\mu + h^2)^{-\frac{3}{2}} = \sum_1^\infty nh^{n-1}P_n \quad\dots\dots\dots(161),$$

so that

$$(\mu - h)(1 + \sum_1^\infty h^n P_n) = (1 - 2h\mu + h^2)\sum_1^\infty nh^{n-1}P_n.$$

Equating coefficients of h^n, we obtain

$$(n+1)P_{n+1} + nP_{n-1} = (2n+1)\mu P_n \quad\dots\dots\dots(162).$$

This is the difference equation satisfied by three successive coefficients.

Again, if we differentiate equation (160) with respect to μ,

$$h(1 - 2h\mu + h^2)^{-\frac{3}{2}} = \sum h^n \frac{\partial P_n}{\partial \mu},$$

so that, by combining with (161),

$$\sum_1^\infty nh^n P_n = (\mu - h)\sum h^n \frac{\partial P_n}{\partial \mu}.$$

Equating coefficients of h^n,

$$nP_n = \mu \frac{\partial P_n}{\partial \mu} - \frac{\partial P_{n-1}}{\partial \mu} \quad\dots\dots\dots\dots(163).$$

Differentiating (162), we obtain

$$(n+1)\frac{\partial P_{n+1}}{\partial \mu} + n\frac{\partial P_{n-1}}{\partial \mu} = (2n+1)\left(P_n + \mu\frac{\partial P_n}{\partial \mu}\right).$$

Eliminating $\mu\dfrac{\partial P_n}{\partial \mu}$ from this and (163),

$$(2n+1)P_n = \frac{\partial P_{n+1}}{\partial \mu} - \frac{\partial P_{n-1}}{\partial \mu} \quad\dots\dots\dots(164).$$

By integration of this we obtain

$$\int P_n(\mu)\,d\mu = \frac{P_{n+1}(\mu) - P_{n-1}(\mu)}{2n+1} \quad\dots\dots\dots(165),$$

whilst by the addition of successive equations of the type of (164), we obtain

$$\frac{\partial}{\partial \mu}P_n = (2n-1)P_{n-1} + (2n-5)P_{n-3} + \dots \quad\dots\dots\dots(166).$$

254. We have had the general theorem (§ 237)

$$\iint S_n S_m \, d\omega = 0,$$

from which the theorem

$$\iint P_n(\mu) P_m(\mu) \, d\omega = 0$$

follows as a special case. Or since

$$d\omega = \sin\theta \, d\theta \, d\phi = -d\mu \, d\phi,$$

$$\int_{-1}^{+1} P_n(\mu) P_m(\mu) \, d\mu = 0 \ldots\ldots\ldots\ldots\ldots\ldots(167).$$

To find $\int_{-1}^{+1} P_n^2(\mu) \, d\mu$, let us square the equation

$$(1 - 2h\mu + h^2)^{-\frac{1}{2}} = \sum_0^{\infty} h^n P_n,$$

multiply by $d\mu$, and integrate from $\mu = -1$ to $\mu = +1$.

The result is

$$\int_{-1}^{+1} \frac{d\mu}{1 - 2h\mu + h^2} = \int_{-1}^{+1} \left(\sum_0^{\infty} h^n P_n \right)^2 d\mu$$

$$= \int_{-1}^{+1} \sum_0^{\infty} h^{2n} P_n^2 \, d\mu,$$

all products of the form $P_n P_m$ vanishing on integration, by equation (167).

Thus $\int_{-1}^{+1} P_n^2 d\mu$ is the coefficient of h^{2n} in

$$\int_{-1}^{+1} \frac{d\mu}{1 - 2h\mu + h^2},$$

i.e. in

$$-\frac{1}{h} \log \frac{1-h}{1+h},$$

and this coefficient is easily seen to be $\dfrac{2}{2n+1}$.

We accordingly have

$$\int_{-1}^{+1} \{P_n(\mu)\}^2 \, d\mu = \frac{2}{2n+1} \ldots\ldots\ldots\ldots\ldots(168).$$

255. We can obtain this theorem in another way, and in a more general form, by using the expansion of § 240, namely

$$F_P = \frac{1}{4\pi a^2} \sum_0^{\infty} (2s+1) \iint F P_s(\cos\theta) \, dS,$$

where θ is the angle between the point P and the element dS on the sphere. This expansion is true for any function F subject to certain restrictions. Taking F to be a surface harmonic S_n of order n, we obtain

$$(S_n)_P = \frac{1}{4\pi a^2} \sum_{s=0}^{s=\infty} (2s+1) \iint S_n P_s(\cos\theta) \, dS$$

$$= \frac{2n+1}{4\pi a^2} \iint S_n P_n(\cos\theta) \, dS,$$

all other integrals vanishing by the theorem of § 237. Thus

$$\iint S_n P_n (\mu) \, dS = \frac{4\pi a^2}{2n+1} \, (S_n)_{\mu-1}$$

or

$$\iint S_n P_n (\mu) \, d\omega = \frac{4\pi}{2n+1} \, (S_n)_{\mu-1} \quad \dots\dots\dots\dots\dots(169).$$

This is the general theorem, of which equation (168) expresses a particular case. To pass to this particular case, we replace S_n by $P_n (\mu)$ and obtain, instead of equation (169),

$$\iint \{P_n (\mu)\}^2 \sin \theta \, d\theta \, d\phi = \frac{4\pi}{2n+1} \, P_n (1),$$

or, after integrating with respect to ϕ,

$$\int \{P_n (\mu)\}^2 \, d\mu = \frac{2}{2n+1},$$

agreeing with equation (168).

Expansions in Legendre's Coefficients.

256. THEOREM. *The value of any function of θ, which is finite and single-valued from $\theta = 0$ to $\theta = \pi$, and which has only a finite number of discontinuities and of maxima and minima within this range, can be expressed, for every value of θ within this range for which the function is continuous, as a series of Legendre's Coefficients.*

This is simply a particular case of the theorem of § 240. It is therefore unnecessary to give a separate proof of the theorem.

The expansion is easily found. Assume it to be

$$f(\mu) = a_0 + a_1 P_1 + a_2 P_2 + \dots + a_s P_s + \dots \quad \dots\dots\dots\dots(170),$$

then on multiplying by $P_n (\mu) \, d\mu$, and integrating from $\mu = -1$ to $\mu = +1$, we obtain

$$\int_{-1}^{+1} P_n (\mu) f(\mu) \, d\mu = \sum_{s=0}^{s=\infty} a_s \int_{-1}^{+1} P_s (\mu) P_n (\mu) \, d\mu$$

$$= \frac{2a_n}{2n+1},$$

every integral vanishing, except that for which $s = n$. Thus

$$a_n = \frac{2n+1}{2} \int_{-1}^{+1} P_n (\mu) f(\mu) \, d\mu \quad \dots\dots\dots\dots(171),$$

giving the coefficients in the expansion.

If $f(\mu)$ has a discontinuity when $\mu = \mu_0$, the value assumed by the series (168) on putting $\mu = \mu_0$ is, as in § 240, equal to

$$\tfrac{1}{2} \{f_1(\mu_0) + f_2(\mu_0)\} \quad \dots\dots\dots\dots\dots(172),$$

where $f_1(\mu_0)$, $f_2(\mu_0)$ are the values of $f(\mu)$ on the two sides of the discontinuity.

HARMONIC POTENTIALS.

257. We are now in a position to apply the results obtained to problems of electrostatics.

Consider first a sphere having a surface density of electricity S_n. The potential at any internal point P is

$$V_P = \iint \frac{S_n dS}{PQ} = \iint \frac{S_n dS}{\sqrt{a^2 - 2ar \cos \theta + r^2}}$$

$$= \iint \frac{S_n}{a} \left(1 + \frac{r}{a} P_1(\cos \theta) + \frac{r^2}{a^2} P_2(\cos \theta) + \ldots \right) dS$$

$$= \frac{4\pi}{2n + 1} a^2 \frac{r^n}{a^{n+1}} (S_n)_{\cos \theta = 1}, \text{ by the theorems of §§ 237 and 255,}$$

$$= \frac{4\pi}{2n + 1} \frac{r^n S_n}{a^{n-1}} \quad \ldots\ldots\ldots\ldots\ldots\ldots\ldots\ldots\ldots\ldots\ldots\ldots\ldots\ldots\ldots\ldots\ldots\ldots\ldots\ldots\ldots\text{(173),}$$

this expression being evaluated at P.

Similarly the potential at any external point P is

$$V_P = \frac{4\pi a^{n+2} S_n}{(2n + 1) r^{n+1}}.$$

These potentials are obviously solutions of Laplace's equation, and it is easy to verify that they correspond to the given surface density, for

$$\left(\frac{\partial V}{\partial r} \right)_{\text{outside}} - \left(\frac{\partial V}{\partial r} \right)_{\text{inside}} = 4\pi S_n.$$

This gives us the fundamental property of harmonics, on which their application to potential-problems depends: *A distribution of surface density S_n on a sphere gives rise to a potential which at every point is proportional to S_n.*

258. The density of the most general surface distribution can, by the theorem of § 240, be expressed as a sum of surface harmonics, say

$$\sigma = S_0 + S_1 + S_2 + \ldots,$$

in which S_0 is of course simply a constant. The potential, by the results of the last section, is

$$V = 4\pi a \left\{ S_0 + \frac{S_1}{3} \left(\frac{r}{a} \right) + \frac{S_2}{5} \left(\frac{r}{a} \right)^2 + \ldots \right\} \text{ at an internal point } \ldots(174),$$

$$V = 4\pi a \left\{ S_0 \left(\frac{a}{r} \right) + \frac{S_1}{3} \left(\frac{a}{r} \right)^2 + \frac{S_2}{5} \left(\frac{a}{r} \right)^3 + \ldots \right\} \text{ at an external point } \ldots(175).$$

EXAMPLES OF THE USE OF HARMONIC POTENTIALS.

I. *Potential of spherical cap and circular ring.*

259. As a first example, let us find the potential of a spherical cap of angle α—*i.e.* the surface cut from a sphere by a right circular cone of semivertical angle α— electrified to a uniform surface density σ_0.

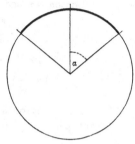

We can regard this as a complete sphere electrified to surface density σ, where

$$\sigma = \sigma_0 \text{ from } \theta = 0 \text{ to } \theta = \alpha,$$
$$\sigma = 0 \text{ from } \theta = \alpha \text{ to } \theta = \pi.$$

The value of σ being symmetrical about the axis $\theta = 0$, let us assume for the value of σ expanded in harmonics

Fig. 76.

$$\sigma = a_0 + a_1 P_1(\cos\theta) + a_2 P_2(\cos\theta) + \ldots,$$

then, by equation (171),

$$a_n = \frac{2n+1}{2} \int_{\theta=\pi}^{\theta=0} \sigma P_n(\cos\theta)\, d(\cos\theta)$$

$$= \frac{2n+1}{2} \sigma_0 \int_{\theta=\alpha}^{\theta=0} P_n(\cos\theta)\, d(\cos\theta)$$

$$= \tfrac{1}{2}\sigma_0 \left\{ P_{n-1}(\cos\alpha) - P_{n+1}(\cos\alpha) \right\}$$

by equation (165), except when $n = 0$. For this case we have

$$a_0 = \tfrac{1}{2}\sigma_0 \int_{\theta=\alpha}^{\theta=0} d(\cos\theta) = \tfrac{1}{2}\sigma_0(1 - \cos\alpha).$$

Thus

$$\sigma = \tfrac{1}{2}\sigma_0 \left[(1 - \cos\alpha) + \sum_{n=1}^{n=\infty} \left\{ P_{n-1}(\cos\alpha) - P_{n+1}(\cos\alpha) \right\} P_n(\cos\theta) \right].$$

It is of interest to notice that when $\theta = \alpha$, the value of σ given by this series is $\sigma = \tfrac{1}{2}\sigma_0$, as it ought to be (cf. expression (172)).

The potential at an external point may now be written down in the form

$$V = 2\pi a\sigma_0 \left[(1 - \cos\alpha)\left(\frac{a}{r}\right) + \sum_{n=1}^{n=\infty} \frac{P_{n-1}(\cos\alpha) - P_{n+1}(\cos\alpha)}{2n+1}\left(\frac{a}{r}\right)^{n+1} P_n(\cos\theta) \right]$$

$$\ldots\ldots(176),$$

and that at an internal point is

$$V = 2\pi a\sigma_0 \left[(1 - \cos\alpha) + \sum_{n=1}^{n=\infty} \frac{P_{n-1}(\cos\alpha) - P_{n+1}(\cos\alpha)}{2n+1}\left(\frac{r}{a}\right)^{n} P_n(\cos\theta) \right]$$

$$\ldots\ldots(177).$$

On differentiating with respect to α, we obtain the potential of a ring of line density $\sigma_0 a\, d\alpha$. At a point at which $r > a$, we differentiate expression (176), and obtain

$$V = 2\pi a \sigma_0 d\alpha \left[\sin\alpha \left(\frac{a}{r}\right) + \sum_{n=1}^{n=\infty} P_n(\cos\alpha)\sin\alpha \left(\frac{a}{r}\right)^{n+1} P_n(\cos\theta) \right],$$

or, putting $a\sigma_0 d\alpha = \tau$ and simplifying,

$$V = 2\pi\tau \sum_{n=0}^{n=\infty} P_n(\cos\alpha)\sin\alpha \left(\frac{a}{r}\right)^{n+1} P_n(\cos\theta)\ldots\ldots\ldots(178).$$

Obviously the potential at a point at which $r < a$ can be obtained on replacing $\left(\frac{a}{r}\right)^{n+1}$ by $\left(\frac{r}{a}\right)^{n}$.

260. These last results can be obtained more directly by considering that at any point on the axis $\theta = 0$ the potential is

$$V = \frac{2\pi a\tau \sin\alpha}{\sqrt{r^2 + a^2 - 2ar\cos\alpha}},$$

or, if $r > a$,

$$V = \frac{2\pi a\tau \sin\alpha}{r} \sum_{n=0}^{n=\infty} P_n(\cos\alpha)\left(\frac{a}{r}\right)^{n},$$

and expression (178) is the only expansion in Lagrange's coefficients which satisfies Laplace's equation and agrees with this expression when $\theta = 0$.

II.   *Uninsulated sphere in field of force.*

261. The method of harmonics enables us to find the field of force produced when a conducting sphere is introduced into any permanent field of force. Let us suppose first that the sphere is uninsulated.

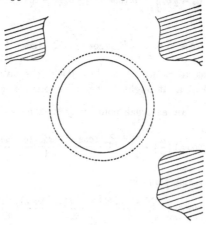

FIG. 77.

Let the sphere be of radius a. Round the centre of the field describe a slightly larger sphere of radius a', so small as not to enclose any of the fixed charges by which the permanent field of force is produced. Between these two spheres the potential of the field will be capable of expression in a series of rational integral harmonics, say

$$V = V_0 + V_1 + V_2 + \dots \quad \dots\dots\dots\dots\dots\dots(179).$$

The problem is to superpose on this a potential, produced by the induced electrification on the sphere, which shall give a total potential equal to zero over the sphere $r = a$. Clearly the only form possible for this new potential is

$$V = - V_0 \left(\frac{a}{r}\right) - V_1 \left(\frac{a}{r}\right)^3 - V_2 \left(\frac{a}{r}\right)^5 - \dots \quad \dots\dots\dots\dots(180).$$

Thus the total potential between the spheres $r = a$ and $r = a'$ is

$$V_0 \left\{1 - \frac{a}{r}\right\} + V_1 \left\{1 - \left(\frac{a}{r}\right)^3\right\} + V_2 \left\{1 - \left(\frac{a}{r}\right)^5\right\} + \dots + V_n \left\{1 - \left(\frac{a}{r}\right)^{2n+1}\right\} + \dots.$$

Putting $V_n = r^n S_n$, the surface density of electrification on the sphere is, by Coulomb's Law,

$$-\frac{1}{4\pi} \Sigma S_n \left\{\frac{\partial}{\partial r}(r^n) - \frac{\partial}{\partial r}\left(\frac{a^{2n+1}}{r^{2n+1}}\right)\right\}_{r=a}$$

$$= -\frac{1}{4\pi} \Sigma a^{n-1}(2n+1) S_n$$

$$= -\frac{1}{4\pi a} \Sigma (2n+1) V_n.$$

This result is indeed obvious from § 258, on considering that the surface electrification must give rise to the potential (180).

If n is different from zero,

$$\iint S_0 S_n dS = 0,$$

where the integration is over any sphere, so that

$$\iint S_n dS = 0 \qquad (n \neq 0),$$

and

$$\iint V_n dS = 0 \qquad (n \neq 0)\dots\dots\dots\dots\dots\dots(181).$$

Thus the total charge on the sphere

$$= \iint \sigma dS$$

$$= -\frac{1}{4\pi a} \Sigma (2n+1) \int V_n dS$$

$$= -\frac{1}{4\pi a} V_0 . 4\pi a^2 = - V_0 a,$$

and V_0 was the potential of the original field at the centre of the sphere.

262. Incidentally we may notice, as a consequence of (181), that the mean value of a potential averaged over the surface of any sphere which does not include any electric charge is equal to the potential at the centre (cf. § 50).

If the sphere is introduced insulated, we superpose on to the field already given, the field of a charge E spread uniformly over the surface of the sphere, and the potential of this field is $\dfrac{E}{r}$. We obtain the particular case of an uncharged sphere by taking $E = V_0 a$, and the potential of this field, namely $V_0\left(\dfrac{a}{r}\right)$, just annihilates the first term in expression (180), to which it has to be added.

It will easily be verified that, on taking the potential of the original field to be $V_1 = Fx$, we arrive at the results already obtained in § 217.

III. *Dielectric sphere in a field of force.*

263. An analogous treatment will give the solution when a homogeneous dielectric sphere is placed in a permanent field of force. The treatment will, perhaps, be sufficiently exemplified by considering the case of the simple field of potential

$$V_1 = Fx = rS_1.$$

Let us assume for the potential V_0 outside the sphere

$$V_0 = rS_1 + \frac{aS_1}{r^2},$$

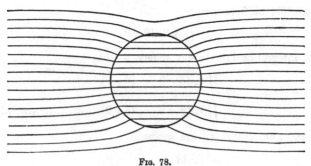

Fig. 78.

and for the potential V_i inside the sphere

$$V_i = \beta rS_1,$$

no term of the form $\dfrac{S_1}{r^2}$ being included in V_i, as it would give infinite potential at the origin. The constants α, β are to be determined from the conditions

$$V_i = V_0 \left.\begin{array}{c} \\ \end{array}\right\}$$
$$K\frac{\partial V_i}{\partial r} = \frac{\partial V_0}{\partial r} \left.\begin{array}{c} \\ \end{array}\right\} \text{at } r = a.$$

These give

$$a + \frac{\alpha}{a^2} = \beta a,$$

$$1 - \frac{2\alpha}{a^3} = K\beta,$$

whence

$$\alpha = -\frac{K-1}{K+2}a^3, \quad \beta = \frac{3}{K+2},$$

so that

$$V_0 = Fx\left\{1 - \frac{K-1}{K+2}\left(\frac{a}{r}\right)^3\right\},$$

$$V_i = \frac{3}{K+2}Fx.$$

Thus the lines of force inside the dielectric are all parallel to those of the original field, but the intensity is diminished in the ratio $\dfrac{3}{K+2}$. The field is shewn in fig. 78.

IV. *Nearly spherical surfaces.*

264. If $r = a$, the surface $r = a + \chi$, where χ is a function of θ and ϕ, will represent a surface which is nearly spherical if χ is small. In this case χ may be regarded as a function of position on the surface of the sphere $r = a$, and expanded in a series of rational integral harmonics in the form

$$\chi = S_0 + S_1 + S_2 + \dots$$

in which S_1, S_2, ... are all small.

The volume enclosed by this surface is

$$\tfrac{1}{3}\iint r^3 \, d\omega$$

$$= \tfrac{1}{3}\iint (a^3 + 3a^2\chi) \, d\omega$$

$$= \frac{4\pi a^3}{3} + a^2 \iint \chi \, d\omega$$

$$= \frac{4\pi a^3}{3} + 4\pi a^2 S_0.$$

If $S_0 = 0$, the volume is that of the original sphere $r = a$.

The following special cases are of importance:

r = a + ϵP₁. To obtain the form of this surface, we pass a distance $\epsilon \cos \theta$ along the radius at each point of the sphere $r = a$. It is easily seen that when ϵ is small the locus of the points so obtained is a sphere of radius a, of which the centre is at a distance ϵ from the origin.

r = a + a₁S₁. The most general form for $a_1 S_1$ is $lx + my + nz$, and this may be expressed as $a\epsilon \cos \theta$, where θ is now measured from the line of which the direction cosines are in the ratio $l : m : n$. Thus the surface is the same as before.

r = a + S₂. Since r is nearly equal to a, this may be written

$$r^2 = a^2 + 2aS_2$$

$$= a^2 + \frac{2}{a} r^2 S_2,$$

or $\qquad x^2 + y^2 + z^2 = a^2 +$ an expression of the second degree.

Thus the surface is an ellipsoid of which the centre is at the origin. It will easily be found that $r = a + \epsilon P_2$ represents a spheroid of semi-axes $a + \epsilon$, $a - \dfrac{\epsilon}{2}$, and therefore of ellipticity $\dfrac{3\epsilon}{2a}$.

265. We can treat these nearly spherical surfaces in the same way in which spherical surfaces have been treated, neglecting the squares of the small harmonics as they occur.

266. As an example, suppose the surface $r = a + S_n$ to be a conductor, raised to unit potential. We assume an external potential

$$V = \frac{A}{r} + BS_n \left(\frac{a}{r}\right)^{n+1},$$

where A and B have to be found from the condition that $V = 1$ when $r = a + S_n$. Neglecting squares of S_n, this gives

$$1 = \frac{A}{a}\left(1 - \frac{S_n}{a}\right) + BS_n,$$

so that $\qquad\qquad A = a, \quad B = \dfrac{1}{a},$

and $\qquad\qquad V = \dfrac{a}{r} + \dfrac{a^n}{r^{n+1}} S_n.$

By applying Gauss' Theorem to a sphere of radius greater than a we readily find that the total charge is a, the coefficient of $\dfrac{1}{r}$. Thus the capacity of the conductor is different from that of the sphere only by terms in $S_n{}^2$, but the surface distribution is different, for

$$4\pi\sigma = -\frac{\partial V}{\partial n} = -\frac{\partial V}{\partial r}, \text{ if we neglect } S_n{}^2,$$

$$= \frac{a}{r^2} + (n+1)\frac{a^n}{r^{n+2}} S_n$$

$$= \frac{a}{a^2}\left(1 - \frac{2S_n}{a}\right) + \left(\frac{n+1}{a^2}\right) S_n$$

$$= \frac{1}{a} + \frac{n-1}{a^2} S_n,$$

the surface density becoming uniform, as it ought, when $n = 1$, *i.e.* when the conductor is still spherical.

267. As a second example, let us examine the field inside a spherical condenser when the two spheres are not quite concentric. Taking the centre of the inner as origin, let the equations of the two spheres be

$$r = a,$$
$$r = b + \epsilon P_1.$$

We have to find a potential which shall have, say, unit value over $r = a$, and shall vanish over $r = b + \epsilon P_1$. Assume

$$V = \frac{A}{r} + \frac{BP_1}{r^2} + C + DP_1 r,$$

when B and D are small, then we must have

$$1 = \frac{A}{a} + \frac{B}{a^2} P_1 + C + DaP_1,$$

$$0 = \frac{A}{b}\left(1 - \frac{\epsilon}{b} P_1\right) + \frac{BP_1}{b^2} + C + DbP_1.$$

These equations must be true all over the spheres, so that the coefficients of P_1 and the terms which do not involve P_1 must vanish separately. Thus

$$\frac{A}{a} + C - 1 = 0; \qquad \frac{B}{a^2} + Da = 0;$$

$$\frac{A}{b} + C = 0, \qquad -\frac{\epsilon A}{b^2} + \frac{B}{b^2} + Db = 0.$$

From the first two equations

$$A = \frac{ab}{b-a},$$

and this being the coefficient of $\dfrac{1}{r}$ in the potential, is the capacity of the condenser. Thus to a first approximation, the capacity of the condenser remains unaltered, but since B and D do not vanish, the surface distribution is altered.

V. *Collection of Electric Charges.*

267 a. If a collection of electric charges are arranged in any way whatever subject only to the condition that none of them lie outside the sphere $r = a$, then the potential at any point outside the sphere must be

$$V = \frac{e}{r} + \frac{S_1}{r^2} + \frac{S_2}{r^3} + \dots,$$

where e is the total charge inside the sphere (cf. § 266) and $S_1, S_2, \dots$ are surface harmonics which depend on the arrangement of the charges inside the sphere.

If the total charge is not zero, the potential can also be treated as in § 67, and on comparing the two expressions obtained for the potential, we can identify the harmonics $S_1, S_2, \dots$. We find that

$$S_1 = \Sigma e_1 \left(x_1 \frac{x}{r} + y_1 \frac{y}{r} + z_1 \frac{z}{r} \right),$$

$$S_2 = \Sigma e_1 \left\{ \tfrac{3}{2} \left(x_1 \frac{x}{r} + y_1 \frac{y}{r} + z_1 \frac{z}{r} \right)^2 - \tfrac{1}{2} (x_1^2 + y_1^2 + z_1^2) \right\},$$

and it will be easily verified by differentiation that the expressions on the right are harmonics.

This example is of some interest in connection with the electron-theory of matter, for a collection of positive and negative charges all collected within a distance a of a centre may give some representation of the structure of a molecule. The total charge on a molecule is zero, so that we must take $e = 0$, and the potential becomes

$$V = \frac{S_1}{r^2} + \frac{S_2}{r^3} + \dots.$$

The most general form for S_1 is (cf. § 239) $\frac{1}{r}(Ax + By + Cz)$, or $\mu \cos \theta$, where θ is the angle between the lines from the origin to the point x, y, z and that to the point A, B, C and μ is $\sqrt{(A^2 + B^2 + C^2)}$.

Thus the term which is important in the potential when r is large is $\frac{\mu \cos \theta}{r^2}$, shewing that at a sufficient distance the molecule has the same field of force as a certain doublet of strength μ. Clearly when μ has any value different from zero, the molecule is "polarised" (cf. § 142) in Faraday's sense. If $\mu = 0$, the potential becomes

$$V = \frac{S_2}{r^3} + \frac{S_3}{r^4} + \dots,$$

shewing that the force now falls off as the inverse fourth power of the distance.

It is worth noticing that the average force at any distance r is always zero, so that to obtain forces which are, on the average, repulsive, we have to assume the presence of terms in the potential which do not satisfy Laplace's equation, and which accordingly are not derivable from forces obeying the simple law e/r^2 (cf. § 192).

FURTHER ANALYTICAL THEORY OF HARMONICS.

General Theory of Zonal Harmonics.

268. The general equation satisfied by a surface harmonic of order n, which is symmetrical about an axis, has already been seen to be

$$\frac{\partial}{\partial \mu}\left\{(1-\mu^2)\frac{\partial S_n}{\partial \mu}\right\} + n(n+1)S_n = 0 \quad \ldots\ldots\ldots\ldots(182).$$

One solution is known to be P_n, so that we can find the other by a known method. Assume $S_n = P_n u$ as a solution, where u is a function of μ. The equation becomes

$$(1-\mu^2)\frac{\partial}{\partial \mu}\left\{\frac{\partial P_n}{\partial \mu}u + P_n\frac{\partial u}{\partial \mu}\right\} - 2\mu\left\{\frac{\partial P_n}{\partial \mu}u + P_n\frac{\partial u}{\partial \mu}\right\} + n(n+1)P_n u = 0 \;\ldots(183),$$

and, since P_n is itself a solution,

$$(1-\mu^2)\frac{\partial}{\partial \mu}\left(\frac{\partial P_n}{\partial \mu}\right) - 2\mu\frac{\partial P_n}{\partial \mu} + n(n+1)P_n = 0.$$

Multiplying this by u and subtracting from (183), we are left with

$$(1-\mu^2)\left\{2\frac{\partial P_n}{\partial \mu}\frac{\partial u}{\partial \mu} + P_n\frac{\partial^2 u}{\partial \mu^2}\right\} - 2\mu P_n\frac{\partial u}{\partial \mu} = 0,$$

or, multiplying by P_n and rearranging,

$$\left((1-\mu^2)\frac{\partial (P_n)^2}{\partial \mu} - 2\mu P_n^2\right)\frac{\partial u}{\partial \mu} + (1-\mu^2)P_n^2\frac{\partial}{\partial \mu}\left(\frac{\partial u}{\partial \mu}\right) = 0,$$

or again

$$\frac{\partial}{\partial \mu}\left\{(1-\mu^2)P_n^2\right\}\frac{\partial u}{\partial \mu} + \left\{(1-\mu^2)P_n^2\right\}\frac{\partial}{\partial \mu}\left(\frac{\partial u}{\partial \mu}\right) = 0.$$

On integration this becomes

$$(1-\mu^2)P_n^2\frac{\partial u}{\partial \mu} = \text{constant}.$$

We may therefore take

$$u = A + B\int\frac{d\mu}{(\mu^2-1)P_n^2},$$

in which the limits may be any we please. If we write

$$Q_n = P_n\int_\mu^\infty\frac{d\mu}{(\mu^2-1)P_n^2} \quad \ldots\ldots\ldots\ldots\ldots(184),$$

the complete solution of equation (182) is

$$S_n = P_n u = AP_n + BQ_n.$$

269. The two solutions P_n and Q_n can be obtained directly by solving the original equation (182) in a series of powers of μ.

Assume a solution

$$S_n = b_0\mu^r + b_1\mu^{r+1} + b_2\mu^{r+2} + \ldots,$$

substitute in equation (182), and equate to zero the coefficients of the different powers of μ. The first coefficient is found to be $b_0 r\,(r-1)$, so that if this is to vanish we must have $r=0$ or $r=1$. The value $r=0$ leads to the solution

$$u_0 = 1 - \frac{n\,(n+1)}{1.2}\,\mu^2 + \frac{(n-2)\,n\,(n+1)\,(n+3)}{1.2.3.4}\,\mu^4 - \ldots$$

while the value $r=1$ leads to the solution

$$u_1 = \mu - \frac{(n-1)\,(n+2)}{1.2.3}\,\mu^3 + \frac{(n-3)\,(n-1)\,(n+2)\,(n+4)}{1.2.3.4.5}\,\mu^5 - \ldots.$$

The complete solution of the equation is therefore

$$\alpha u_0 + \beta u_1.$$

If n is integral one of the two series terminates, while the other does not. If n is even the series u_0 terminates, while if n is odd the terminating series is u_1. But we have already found one terminating series which is a solution of the original equation, namely P_n. Hence in either case the terminating series must be proportional to P_n, and therefore the infinite series must be proportional to Q_n.

270. We can obtain a more useful form for Q_n from expression (184). The roots of $P_n\,(\mu)=0$ are, as we have seen, n in number, all real and separate, and lying between -1 and $+1$. Let us take these roots to be $\alpha_1, \alpha_2, \ldots \alpha_n$. Then

$$\frac{1}{(\mu^2-1)\,\{P_n\,(\mu)\}^2} = \frac{1}{(\mu-1)\,(\mu+1)\,(\mu-\alpha_1)^2\,(\mu-\alpha_2)^2 \ldots (\mu-\alpha_n)^2}$$

$$= \frac{a}{\mu-1} + \frac{b}{\mu+1} + \Sigma \left(\frac{c_s}{\mu-\alpha_s} + \frac{d_s}{(\mu-\alpha_s)^2} \right) \qquad \ldots\ldots\ldots(185),$$

on resolving into partial fractions. Putting $\mu=+1$ and -1, we find at once that $a = \frac{1}{4}$, $b = -\frac{1}{4}$.

In the general fraction

$$\frac{1}{D} \equiv \frac{1}{(x-a_1)\,(x-a_2)\ldots},$$

let us suppose all the factors in the denominator to be distinct, so that we may write

$$\frac{1}{D} = \frac{c_1}{x-a_1} + \frac{c_2}{x-a_2} + \ldots.$$

On putting $x=a_1$, we obtain at once

$$c_1 = \frac{1}{(a_1-a_2)\,(a_1-a_3)\,(a_1-a_4)\ldots},$$

$$c_2 = \frac{1}{(a_2-a_1)\,(a_2-a_3)\,(a_2-a_4)\ldots}, \quad \text{etc.}$$

Now let a_1 and a_2 become very nearly equal, say $a_2 = a_1 + da_1$, then

$$c_1 = - \frac{1}{da_1 (a_1 - a_3)(a_1 - a_4) \ldots},$$

while

$$c_2 = \frac{1}{da_1 (a_2 - a_3)(a_2 - a_4) \ldots}.$$

The fractions

$$\frac{c_1}{x - a_1} + \frac{c_2}{x - a_2}$$

now combine into

$$\frac{(c_1 + c_2) x - (c_1 a_2 - c_2 a_1)}{(x - a_1)^2},$$

and on putting this equal to

$$\frac{c_1'}{x - a_1} + \frac{c_2'}{(x - a_1)^2},$$

it is clear that the value of c_1' must be taken to be $c_1 + c_2$. Now

$$c_1 + c_2 = \frac{1}{da_1} \left\{ \frac{1}{(a_2 - a_3)(a_2 - a_4) \ldots} - \frac{1}{(a_1 - a_3)(a_1 - a_4) \ldots} \right\}$$

$$= \frac{1}{da_1} \left\{ \frac{\partial}{\partial x} \left(\frac{1}{(x - a_3)(x - a_4) \ldots} \right)_{x = a_1} da_1 \right\}$$

$$= \frac{\partial}{\partial x} \left\{ \frac{(x - a_1)^2}{D} \right\}_{x = a_1},$$

and this remains true however many of the roots a_3, a_4 ..., coincide among themselves, so long as they do not coincide with the root a_1. Thus, in expression (185), the value of c_s is

$$c_s = \frac{\partial}{\partial \mu} \left\{ \frac{(\mu - \alpha_s)^2}{(1 - \mu^2) \{P_n (\mu)\}^2} \right\}_{\mu = \alpha_s}.$$

Putting

$$\frac{P_n (\mu)}{\mu - \alpha_s} = R (\mu),$$

we find that

$$c_s = \frac{\partial}{\partial \mu} \left\{ \frac{1}{(1 - \mu^2) \{R (\mu)\}^2} \right\}_{\mu = \alpha_s} = \frac{\partial}{\partial \alpha_s} \left\{ \frac{1}{(1 - \alpha_s^2) \{R (\alpha_s)\}^2} \right\}.$$

Since $(\mu - \alpha_s) R (\mu)$ is a solution of equation (182), we find that

$$\frac{\partial}{\partial \mu} \left[(1 - \mu^2) \left\{ R (\mu) + (\mu - \alpha_s) \frac{\partial R}{\partial \mu} \right\} \right] + n (n + 1) (\mu - \alpha_s) R = 0.$$

On putting $\mu = \alpha_s$, this reduces to

$$\frac{\partial}{\partial \alpha_s} \{(1 - \alpha_s^2) R (\alpha_s)\} + (1 - \alpha_s^2) \frac{\partial R (\alpha_s)}{\partial \alpha_s} = 0,$$

giving, on multiplication by $R (\alpha_s)$,

$$\frac{\partial}{\partial \alpha_s} [(1 - \alpha_s^2) \{R (\alpha_s)\}^2] = 0.$$

Hence $c_s = 0$.

Equation (185) now becomes

$$\frac{1}{(\mu^2-1)\{P_n(\mu)\}^2} = \tfrac{1}{2}\left(\frac{1}{\mu-1} - \frac{1}{\mu+1}\right) + \Sigma\frac{d_s}{(\mu-a_s)^2},$$

so that, on integration,

$$\int_\mu^\infty \frac{d\mu}{(\mu^2-1)\{P_n(\mu)\}^2} = \tfrac{1}{2}\log\frac{\mu+1}{\mu-1} + \Sigma\frac{d_s}{\mu-a_s}.$$

On multiplying by $P_n(\mu)$, we obtain from equation (184),

$$Q_n(\mu) = P_n\int_\mu^\infty \frac{d\mu}{(\mu^2-1)P_n^2} = \tfrac{1}{2}P_n(\mu)\log\frac{\mu+1}{\mu-1} + W_{n-1} \quad\ldots\ldots(186),$$

where W_{n-1} is a rational integral function of μ of degree $n-1$.

It is now clear that $Q_n(\mu)$ is finite and continuous from $\mu = -1$ to $\mu = +1$, but becomes infinite at the actual values $\mu = \pm 1$.

To find the value of W_{n-1} we substitute expression (186) in Legendre's equation, of which it is known to be a solution, and obtain

$$\frac{\partial}{\partial\mu}\left\{(1-\mu^2)\frac{\partial W_{n-1}}{\partial\mu}\right\} + n(n+1)W_{n-1}$$

$$= -\frac{\partial}{\partial\mu}\left\{(1-\mu^2)\frac{\partial}{\partial\mu}\left(\tfrac{1}{2}P_n(\mu)\log\frac{\mu+1}{\mu-1}\right)\right\} - n(n+1)\tfrac{1}{2}P_n(\mu)\log\frac{\mu+1}{\mu-1}$$

$$= -2\frac{\partial P_n}{\partial\mu}$$

$$= -2\{(2n-1)P_{n-1} + (2n-5)P_{n-3} + \ldots\}\ldots\ldots\ldots\ldots\ldots\ldots(187).$$

Since W_{n-1} is a rational integral algebraic function of μ of degree $n-1$, it can be expanded in the form

$$W_{n-1} = a_1 P_{n-1} + a_2 P_{n-2} + \ldots,$$

so that

$$\frac{\partial}{\partial\mu}\left\{(1-\mu^2)\frac{\partial W_{n-1}}{\partial\mu}\right\} + n(n+1)W_{n-1}$$

$$= -\Sigma a_s\left[\frac{\partial}{\partial\mu}\left\{(1-\mu^2)\frac{\partial P_{n-s}}{\partial\mu}\right\} + n(n+1)P_{n-s}\right]$$

$$= -\Sigma a_s\{n(n+1) - (n-s)(n-s+1)\}P_{n-s}.$$

Comparing with (187), we find that $a_s = 0$ when s is odd, and is equal to

$$\frac{2(2n-2s+1)}{s(2n-s+1)}$$

when s is even.

Thus

$$W_{n-1} = -\frac{2n-1}{1.n}P_{n-1} - \frac{2n-5}{3(n-1)}P_{n-3} - \frac{2n-9}{5(n-2)}P_{n-5} - \ldots,$$

and

$$Q_n = \tfrac{1}{2}P_n(\mu)\log\frac{\mu+1}{\mu-1} - \frac{2n-1}{1.n}P_{n-1} - \frac{2n-5}{3(n-1)}P_{n-3} - \ldots.$$

271. When we are dealing with complete spheres it is impossible for the solution Q_n to occur. If the space is limited in such a way that the infinities of the Q_n harmonic are excluded, it may be necessary to take into account both the P_n and Q_n harmonics. An instance of such a case occurs in considering the potential at points outside a conductor of which the shape is that of a complete cone.

Tesseral Harmonics.

272. The equation satisfied by the general surface harmonic S_n is

$$\frac{\partial}{\sin\theta\,\partial\theta}\left(\sin\theta\,\frac{\partial S_n}{\partial\theta}\right) + \frac{1}{\sin^2\theta}\frac{\partial^2 S_n}{\partial\phi^2} + n(n+1)S_n = 0.$$

As a solution, let us examine

$$S_n = \Theta\Phi,$$

where Θ is a function of θ only, and Φ is a function of ϕ only. On substituting this value in the equation, and dividing by $\Theta\Phi/\sin^2\theta$, we obtain

$$\frac{\sin\theta}{\Theta}\frac{\partial}{\partial\theta}\left(\sin\theta\,\frac{\partial\Theta}{\partial\theta}\right) + \frac{1}{\Phi}\frac{\partial^2\Phi}{\partial\phi^2} + n(n+1)\sin^2\theta = 0.$$

We must therefore have

$$\frac{1}{\Phi}\frac{\partial^2\Phi}{\partial\phi^2} = \kappa,$$

$$\frac{\sin\theta}{\Theta}\frac{\partial}{\partial\theta}\left(\sin\theta\,\frac{\partial\Theta}{\partial\theta}\right) + n(n+1)\sin^2\theta = -\kappa.$$

The solution of the former equation is single valued only when κ is of the form $-m^2$, where m is an integer. In this case

$$\Phi = C_m\cos m\phi + D_m\sin m\phi,$$

and Θ is given by

$$\frac{1}{\sin\theta}\frac{\partial}{\partial\theta}\left(\sin\theta\,\frac{\partial\Theta}{\partial\theta}\right) + \left\{n(n+1) - \frac{m^2}{\sin^2\theta}\right\}\Theta = 0,$$

or, in terms of μ,

$$\frac{\partial}{\partial\mu}\left\{(1-\mu^2)\frac{\partial\Theta}{\partial\mu}\right\} + \left\{n(n+1) - \frac{m^2}{1-\mu^2}\right\}\Theta = 0 \quad\ldots\ldots(188),$$

an equation which reduces to Legendre's equation when $m = 0$.

273. To obtain the general solution of equation (188), consider the differential equation

$$(1-\mu^2)\frac{\partial z}{\partial\mu} + 2n\mu z = 0 \ldots\ldots\ldots\ldots\ldots(189),$$

of which the solution is readily seen to be

$$z = C(1-\mu^2)^n \ldots\ldots\ldots\ldots\ldots\ldots(190).$$

If we differentiate equation (189) s times we obtain

$$(1-\mu^2)\frac{\partial^{s+1}z}{\partial\mu^{s+1}} + 2(n-s)\mu\frac{\partial^s z}{\partial\mu^s} + s(2n-s+1)\frac{\partial^{s-1}z}{\partial\mu^{s-1}} = 0 \ldots(191).$$

If in this we put $s = n$, and again differentiate with respect to μ, we obtain

$$\frac{\partial}{\partial \mu}\left\{(1 - \mu^2)\frac{\partial}{\partial \mu}\left(\frac{\partial^n z}{\partial \mu^n}\right)\right\} + n(n+1)\left(\frac{\partial^n z}{\partial \mu^n}\right) = 0 \quad \ldots\ldots\ldots(192),$$

which is Legendre's equation with $\dfrac{\partial^n z}{\partial \mu^n}$ as variable. Thus a solution of this equation is seen to be

$$\frac{\partial^n z}{\partial \mu^n} \quad \text{or} \quad C\left(\frac{\partial}{\partial \mu}\right)^n (1 - \mu^2)^n,$$

giving at once the form for P_n already obtained in § 249. The general solution of equation (192) we know to be

$$\frac{\partial^n z}{\partial \mu^n} = AP_n + BQ_n.$$

If we now differentiate (192) m times, the result is the same as that of differentiating (189) $m + n + 1$ times, and is therefore obtained by putting $s = m + n + 1$ in (191). This gives

$$(1 - \mu^2)\frac{\partial^{m+n+2} z}{\partial \mu^{m+n+2}} - 2(m+1)\mu\frac{\partial^{m+n+1} z}{\partial \mu^{m+n+1}} + (m+n+1)(n-m)\frac{\partial^{m+n} z}{\partial \mu^{m+n}} = 0,$$

or, multiplying by $(1 - \mu^2)^{\frac{m}{2}}$,

$$(1 - \mu^2)^{\frac{m}{2}+1}\frac{\partial^{m+n+2} z}{\partial \mu^{m+n+2}} - 2(m+1)\mu(1 - \mu^2)^{\frac{m}{2}}\frac{\partial^{m+n+1} z}{\partial \mu^{m+n+1}}$$

$$+ (m+n+1)(n-m)(1 - \mu^2)^{\frac{m}{2}}\frac{\partial^{m+n} z}{\partial \mu^{m+n}} = 0 \quad \ldots\ldots(193).$$

Let

$$(1 - \mu^2)^{\frac{m}{2}}\frac{\partial^{m+n} z}{\partial \mu^{m+n}} = v.$$

Then

$$\frac{\partial v}{\partial \mu} = (1 - \mu^2)^{\frac{m}{2}}\frac{\partial^{m+n+1} z}{\partial \mu^{m+n+1}} - m\mu(1 - \mu^2)^{\frac{m}{2}-1}\frac{\partial^{m+n} z}{\partial \mu^{m+n}},$$

and

$$\frac{\partial}{\partial \mu}\left\{(1 - \mu^2)\frac{\partial v}{\partial \mu}\right\} = (1 - \mu^2)^{\frac{m}{2}+1}\frac{\partial^{m+n+2} z}{\partial \mu^{m+n+2}} - (2m+2)\mu(1 - \mu^2)^{\frac{m}{2}}\frac{\partial^{m+n+1} z}{\partial \mu^{m+n+1}}$$

$$- m\left\{(1 - \mu^2)^{\frac{m}{2}} - m\mu^2(1 - \mu^2)^{\frac{m}{2}-1}\right\}\frac{\partial^{m+n} z}{\partial \mu^{m+n}}$$

$$= -v\left\{(m+n+1)(n-m) + m - \frac{m^2\mu^2}{1 - \mu^2}\right\}, \text{ by equation (193)},$$

$$= -v\left\{n(n+1) - \frac{m^2}{1 - \mu^2}\right\}.$$

Thus v satisfies

$$\frac{\partial}{\partial \mu}\left\{(1 - \mu^2)\frac{\partial v}{\partial \mu}\right\} + \left\{n(n+1) - \frac{m^2}{1 - \mu^2}\right\}v = 0,$$

and this is the same as equation (188), which is satisfied by Θ.

274. The solution of equation (188) has now been seen to be

$$\Theta = (1 - \mu^2)^{\frac{m}{2}} \frac{\partial^{m+n} z}{\partial \mu^{m+n}},$$

where

$$\frac{\partial^n z}{\partial \mu^n} = A P_n + B Q_n.$$

Hence

$$\Theta = A (1 - \mu^2)^{\frac{m}{2}} \frac{\partial^m P_n}{\partial \mu^m} + B (1 - \mu^2)^{\frac{m}{2}} \frac{\partial^m Q_n}{\partial \mu^m},$$

The functions

$$(1 - \mu^2)^{\frac{m}{2}} \frac{\partial^m P_n}{\partial \mu^m}, \quad (1 - \mu^2)^{\frac{m}{2}} \frac{\partial^m Q_n}{\partial \mu^m}$$

are known as the associated Legendrian functions of the first and second kinds, and are generally denoted by $P_n^m (\mu)$, $Q_n^m (\mu)$. As regards the former we may replace P_n, from equation (159), by

$$\frac{1}{2^n n!} \frac{\partial^n}{\partial \mu^n} (\mu^2 - 1)^n,$$

and obtain the function in the form

$$P_n^m (\mu) = \frac{1}{2^n n!} (1 - \mu^2)^{\frac{m}{2}} \frac{\partial^{m+n}}{\partial \mu^{m+n}} (\mu^2 - 1)^n \quad \dots\dots\dots(194).$$

It is clear from this form that the function vanishes if $m + n > 2n$, *i.e.* if $m > n$. It is also clear that it is a rational integral function of $\sin \theta$ and $\cos \theta$. From the form of $Q_n (\mu)$, which is not a rational integral function of μ, it is clear that $Q_n^m (\mu)$ cannot be a rational integral function of $\sin \theta$ and $\cos \theta$.

Thus of the solution we have obtained for S_n, only the part

$$P_n^m (\mu) (C_m \cos m\phi + D_m \sin m\phi)$$

gives rise to rational integral harmonics. The terms $P_n^m (\mu) \cos m\phi$ and $P_n^m (\mu) \sin m\phi$ are known as tesseral harmonics.

Clearly there are $(2n + 1)$ tesseral harmonics of degree n, namely

$$P_n (\mu), \quad \cos \phi\, P_n^1 (\mu), \quad \sin \phi\, P_n^1 (\mu), \dots \cos n\phi\, P_n^n (\mu), \quad \sin n\phi\, P_n^n (\mu).$$

These may be regarded as the $(2n + 1)$ independent rational integral harmonics of degree n of which the existence has already been proved in § 239.

Using the formula

$$P_n^m (\mu) = \sin^m \theta \frac{\partial^m P_n (\mu)}{\partial \mu^m}$$

and substituting the value obtained in § 247 for $P_n (\mu)$ (cf. equation (155)), we obtain $P_n^m (\mu)$ in the form

$$P_n^m (\mu) = \frac{(2n)! \sin^m \theta}{2^n n! (n - m)!} \left\{ \cos^{n-m} \theta - \frac{(n - m)(n - m - 1)}{2 (2n - 1)} \cos^{n-m-2} \theta \right.$$
$$\left. + \frac{(n - m)(n - m - 1)(n - m - 2)(n - m - 3)}{2.4 (2n - 1)(2n - 3)} \cos^{n-m-4} \theta - \dots \right\}.$$

The values of the tesseral harmonics of the first four orders are given in the following table.

Order 1. $\cos\theta$, $\sin\theta\cos\phi$, $\sin\theta\sin\phi$.

Order 2. $\frac{1}{2}(3\cos^2\theta-1)$, $3\sin\theta\cos\theta\cos\phi$, $3\sin\theta\cos\theta\sin\phi$,
      $3\sin^2\theta\cos2\phi$, $3\sin^2\theta\sin2\phi$.

Order 3. $\frac{1}{2}(5\cos^3\theta-3\cos\theta)$, $\frac{3}{2}\sin\theta(5\cos^2\theta-1)\cos\phi$,
    $\frac{3}{2}\sin\theta(5\cos^2\theta-1)\sin\phi$, $15\sin^2\theta\cos\theta\cos2\phi$,
  $15\sin^2\theta\cos\theta\sin2\phi$, $15\sin^3\theta\cos3\phi$, $15\sin^3\theta\sin3\phi$.

Order 4. $\frac{1}{8}(35\cos^4\theta-30\cos^2\theta+3)$, $\frac{5}{2}\sin\theta(7\cos^3\theta-3\cos\theta)\cos\phi$,
  $\frac{5}{2}\sin\theta(7\cos^3\theta-3\cos\theta)\sin\phi$, $\frac{15}{2}\sin^2\theta(7\cos^2\theta-1)\cos2\phi$,
   $\frac{15}{2}\sin^2\theta(7\cos^2\theta-1)\sin2\phi$, $105\sin^3\theta\cos\theta\cos3\phi$,
  $105\sin^3\theta\cos\theta\sin3\phi$, $105\sin^4\theta\cos4\phi$, $105\sin^4\theta\sin4\phi$.

275. We have now found that the most general rational integral surface harmonic is of the form

$$S_n = \sum_0^n P_n^m(\mu)(A_m\cos m\phi + B_m\sin m\phi),$$

in which $P_n^m(\mu)$ is to be interpreted to mean $P_n(\mu)$, when $m=0$.

Let us denote any tesseral harmonics of the type

$$P_n^m(\mu)(A\cos m\phi + B\sin m\phi) \text{ by } S_n^m.$$

Then by § 237, $\iint S_n^m S_{n'}^m \, d\omega = 0$

if $n \neq n'$. If $n = n'$, then

$$\iint S_n^m S_n^{m'} = \iint P_n^m(\mu)P_n^{m'}(\mu)(A_m\cos m\phi + B_m\sin m\phi)$$
$$(A_{m'}\cos m'\phi + B_{m'}\sin m'\phi)\,d\omega,$$

and this vanishes except when $m = m'$.

When $n = n'$ and $m = m'$ the value of $\iint S_n^m S_n^{m'}\,d\omega$ clearly depends on that of $\int_{-1}^{+1}\{P_n^m(\mu)\}^2\,d\mu$, and this we now proceed to obtain.

We have

$$\int_{-1}^{+1}\{P_n^m(\mu)\}^2\,d\mu = \int_{-1}^{+1}(1-\mu^2)^m\left(\frac{\partial^m P_n}{\partial\mu^m}\right)^2 d\mu$$

$$= \left[(1-\mu^2)^m\frac{\partial^m P_n}{\partial\mu^m}\frac{\partial^{m-1}P_n}{\partial\mu^{m-1}}\right]_{\mu=-1}^{\mu=+1}$$

$$-\int_{-1}^{+1}\frac{\partial^{m-1}P_n}{\partial\mu^{m-1}}\frac{\partial}{\partial\mu}\left\{(1-\mu^2)^m\frac{\partial^m P_n}{\partial\mu^m}\right\}d\mu \quad \ldots(195).$$

Since $\dfrac{\partial^n z}{\partial \mu^n} = P_n$ is a solution of equation (191), we obtain, on taking $s = m + n$ in this equation, and multiplying throughout by $(1 - \mu^2)^{m-1}$,

$$(1 - \mu)^m \frac{\partial^{m+1} P_n}{\partial \mu^{m+1}} - 2m\mu (1 - \mu^2)^{m-1} \frac{\partial^m P_n}{\partial \mu^m}$$
$$+ (n + m)(n - m + 1)(1 - \mu^2)^{m-1} \frac{\partial^{m-1} P_n}{\partial \mu^{m-1}},$$

which, again, may be written

$$\frac{\partial}{\partial \mu} \left\{ (1 - \mu^2)^m \frac{\partial^m P_n}{\partial \mu^m} \right\} = -(n + m)(n - m + 1)(1 - \mu^2)^{m-1} \frac{\partial^{m-1} P_n}{\partial \mu^{m-1}}.$$

In equation (195) the first term on the right-hand vanishes, so that

$$\int_{-1}^{+1} \{P_n^m (\mu)\}^2 \, d\mu = (n + m)(n - m + 1) \int_{-1}^{+1} (1 - \mu^2)^{m-1} \left(\frac{\partial^{m-1} P_n}{\partial \mu^{m-1}} \right)^2 d\mu$$

$$= (n + m)(n - m + 1) \int_{-1}^{+1} \{P_n^{m-1} (\mu)\}^2 \, d\mu,$$

a reduction formula from which we readily obtain

$$\int_{-1}^{+1} \{P_n^m (\mu)\}^2 \, d\mu = \frac{(n + m)!}{(n - m)!} \int_{-1}^{+1} \{P_n (\mu)\}^2 \, d\mu$$

$$= \frac{2}{2n + 1} \frac{(n + m)!}{(n - m)!}.$$

These results enable us to find any integral of the type $\iint S_n S'_n \, d\omega$.

Biaxal Harmonics.

276. It is often convenient to be able to express zonal harmonics referred to one axis in terms of harmonics referred to other axes—*i.e.* to be able to change the axes of reference of zonal harmonics.

Let P_n be a harmonic having OP as axis. At Q the value of this is $P_n (\cos \gamma)$, where γ is the angle PQ, and our problem is to express this harmonic of order n as a sum of zonal and tesseral harmonics referred to other axes. With reference to these axes, let the coordinates of Q be θ, ϕ, let those of P be Θ, Φ, and let us assume a series of the type

$$P_n (\cos \gamma) = \sum_{s=0}^{s=n} P_n^s (\cos \theta)(A_s \cos s\phi + B_s \sin s\phi).$$

Let us multiply by $P_n^s (\cos \theta) \cos s\phi$ and integrate over the surface of a unit sphere. We obtain

$$\iint P_n (\cos \gamma) \{P_n^s (\cos \theta) \cos s\phi\} \, d\omega = A_s \iint \{P_n^s (\cos \theta)\}^2 \cos^2 s\phi \, d\omega.$$

By equation (169),

$$\iint P_n (\cos \gamma) \{P_n^s (\cos \theta) \cos s\phi\} \, d\omega = \frac{4\pi}{2n+1} \{P_n^s (\cos \theta) \cos s\phi\}_{\gamma=0}$$

$$= \frac{4\pi}{2n+1} P_n^s (\cos \Theta) \cos s\Phi,$$

and

$$\iint \{P_n^s (\cos \theta)\}^2 \cos^2 s\phi \, d\omega = \int_{-1}^{+1} \{P_n^s (\mu)\}^2 \, d\mu \int_0^{2\pi} \cos^2 s\phi \, d\phi$$

$$= \frac{2\pi}{2n+1} \frac{(n+s)!}{(n-s)!}.$$

Thus

$$A_s = 2 \frac{(n-s)!}{(n+s)!} P_n^s (\cos \Theta) \cos s\Phi,$$

and similarly

$$B_s = 2 \frac{(n-s)!}{(n+s)!} P_n^s (\cos \Theta) \sin s\Phi.$$

This analysis needs modification when $s = 0$, but it is readily found that

$$A_0 = P_n (\cos \Theta), \quad B_0 = 0,$$

so that

$$P_n (\cos \gamma) = P_n (\cos \theta) P_n (\cos \Theta) + \sum_{s=1}^{s=\infty} 2 \frac{(n-s)!}{(n+s)!} P_n^s (\cos \theta) P_n^s (\cos \Theta) \cos s (\phi - \Phi)$$

$$\dots\dots\dots(196).$$

GENERAL THEORY OF CURVILINEAR COORDINATES.

277. Let us write

$$\phi (x, y, z) = \lambda,$$

$$\psi (x, y, z) = \mu,$$

$$\chi (x, y, z) = \nu,$$

where ϕ, ψ, χ denote any functions of x, y, z. Then we may suppose a point in space specified by the values of λ, μ, ν at the point, *i.e.* by a knowledge of those members of the three families of surfaces

$$\phi (x, y, z) = \text{cons.}; \quad \psi (x, y, z) = \text{cons.}; \quad \chi (x, y, z) = \text{cons.}$$

which pass through it.

The values of λ, μ, ν are called "curvilinear coordinates" of the point. A great simplification is introduced into the analysis connected with curvilinear coordinates, if the three families of surfaces are chosen in such a way that they cut orthogonally at every point. In what follows we shall suppose this to be the case—the coordinates will be "orthogonal curvilinear coordinates."

The points λ, μ, ν and $\lambda + d\lambda$, μ, ν will be adjacent points, and the distance between them will be equal to $d\lambda$ multiplied by a function of

λ, μ, and ν—let us assume it equal to $\dfrac{d\lambda}{h_1}$. Similarly, let the distance from λ, μ, ν to λ, $\mu + d\mu$, ν be $\dfrac{d\mu}{h_2}$, and let the distance from λ, μ, ν to λ, μ, $\nu + d\nu$ be $\dfrac{d\nu}{h_3}$.

Then the distance ds from λ, μ, ν to $\lambda + d\lambda$, $\mu + d\mu$, $\nu + d\nu$ will be given by

$$(ds)^2 = \frac{(d\lambda)^2}{h_1{}^2} + \frac{(d\mu)^2}{h_2{}^2} + \frac{(d\nu)^2}{h_3{}^2},$$

this being the diagonal of a rectangular parallelepiped of edges

$$\frac{d\lambda}{h_1}, \quad \frac{d\mu}{h_2} \text{ and } \frac{d\nu}{h_3}.$$

Laplace's equation in curvilinear coordinates is obtained most readily by applying Gauss' Theorem to the small rectangular parallelepiped of which the edges are the eight points

$$\lambda \pm \tfrac{1}{2}d\lambda, \quad \mu \pm \tfrac{1}{2}d\mu, \quad \nu \pm \tfrac{1}{2}d\nu.$$

In this way we obtain the relation

$$\iint \frac{\partial V}{\partial n} \, dS = 0 \quad \dotfill \quad (197)$$

in the form

$$\frac{\partial}{\partial \lambda}\left(\frac{h_1}{h_2 h_3}\frac{\partial V}{\partial \lambda}\right) + \frac{\partial}{\partial \mu}\left(\frac{h_2}{h_3 h_1}\frac{\partial V}{\partial \mu}\right) + \frac{\partial}{\partial \nu}\left(\frac{h_3}{h_1 h_2}\frac{\partial V}{\partial \nu}\right) = 0 \quad \dotfill \quad (198),$$

and as we have already seen that equation (197) is exactly equivalent to Laplace's equation $\nabla^2 V = 0$, it appears that equation (198) must represent Laplace's equation transformed into curvilinear coordinates.

In any particular system of curvilinear coordinates the method of procedure is to express h_1, h_2, h_3 in terms of λ, μ and ν, and then try to obtain solutions of equation (198), giving V as a function of λ, μ and ν.

SPHERICAL POLAR COORDINATES.

278. The system of surfaces $r = \text{cons.}$, $\theta = \text{cons.}$, $\phi = \text{cons.}$ in spherical polar coordinates gives a system of orthogonal curvilinear coordinates. In these coordinates equation (198) assumes the form

$$\frac{\partial}{\partial r}\left(r^2 \frac{\partial V}{\partial r}\right) + \frac{1}{\sin \theta}\frac{\partial}{\partial \theta}\left(\sin \theta \frac{\partial V}{\partial \theta}\right) + \frac{1}{\sin^2 \theta}\frac{\partial^2 V}{\partial \phi^2} = 0,$$

already obtained in § 233, which has been found to lead to the theory of spherical harmonics.

CONFOCAL COORDINATES.

279. After spherical polar coordinates, the system of curvilinear coordinates which comes next in order of simplicity and importance is that in which the surfaces are confocal ellipsoids and hyperboloids of one and two sheets. This system will now be examined.

Taking the ellipsoid

$$\frac{x^2}{a^2} + \frac{y^2}{b^2} + \frac{z^2}{c^2} = 1 \qquad \ldots\ldots\ldots\ldots\ldots\ldots(199)$$

as a standard, the conicoid

$$\frac{x^2}{a^2 + \theta} + \frac{y^2}{b^2 + \theta} + \frac{z^2}{c^2 + \theta} = 1 \qquad \ldots\ldots\ldots\ldots\ldots(200)$$

will be confocal with the standard ellipsoid whatever value θ may have, and all confocal conicoids are represented in turn by this equation as θ passes from $-\infty$ to $+\infty$.

If the values of x, y, z are given, equation (200) is a cubic equation in θ. It can be shewn that the three roots in θ are all real, so that three confocals pass through any point in space, and it can further be shewn that at every point these three confocals are orthogonal. It can also be shewn that of these confocals one is an ellipsoid, one a hyperboloid of one sheet, and one a hyperboloid of two sheets.

Let λ, μ, ν be the three values of θ which satisfy equation (200) at any point, and let λ, μ, ν refer respectively to the ellipsoid, hyperboloid of one sheet, and hyperboloid of two sheets. Then λ, μ, ν may be taken to be orthogonal curvilinear coordinates, the families of surfaces $\lambda = $ cons., $\mu = $ cons., $\nu = $ cons. being respectively the system of ellipsoids, hyperboloids of one sheet, and hyperboloids of two sheets, which are confocal with the standard ellipsoid (199).

280. The first problem, as already explained, is to find the quantities which have been denoted in § 277 by h_1, h_2, h_3. As a step towards this, we begin by expressing x, y, z as functions of the curvilinear coordinates λ, μ, ν.

The expression

$$(a^2 + \theta)(b^2 + \theta)(c^2 + \theta)\left[\frac{x^2}{a^2 + \theta} + \frac{y^2}{b^2 + \theta} + \frac{z^2}{c^2 + \theta} - 1\right]$$

is clearly a rational integral function of θ of degree 3, the coefficient of θ^3 being -1. It vanishes when θ is equal to λ, μ or ν, these being the curvilinear coordinates of the point x, y, z. Hence the expression must be equal, identically, to

$$-(\theta - \lambda)(\theta - \mu)(\theta - \nu).$$

Putting $\theta = -a^2$ in the identity obtained in this way, we get the relation

$$x^2(b^2 - a^2)(c^2 - a^2) = (a^2 + \lambda)(a^2 + \mu)(a^2 + \nu),$$

so that x, y, z are given as functions of λ, μ, ν by the relations

$$x^2 = \frac{(a^2+\lambda)(a^2+\mu)(a^2+\nu)}{(b^2-a^2)(c^2-a^2)} \text{ etc.} \quad \ldots\ldots\ldots\ldots(201).$$

281. To examine changes as we move along the normal to the surface $\lambda = \text{cons.}$, we must keep μ and ν constant. Thus we have, on logarithmic differentiation of equation (201),

$$2\frac{dx}{x} = \frac{d\lambda}{a^2+\lambda},$$

and there are of course similar equations giving dy and dz. Thus for the length ds of an element of the normal to $\lambda = \text{constant}$, we have

$$(ds)^2 = (dx)^2 + (dy)^2 + (dz)^2$$

$$= \tfrac{1}{4}\Sigma\left(\frac{x}{a^2+\lambda}\right)^2 (d\lambda)^2$$

$$= \tfrac{1}{4}(d\lambda)^2 \sum_{a,\,b,\,c}\frac{(a^2+\mu)(a^2+\nu)}{(a^2+\lambda)(b^2-a^2)(c^2-a^2)}$$

$$= \tfrac{1}{4}(d\lambda)^2 \frac{(\lambda-\mu)(\lambda-\nu)}{(a^2+\lambda)(b^2+\lambda)(c^2+\lambda)}.$$

The quantity ds is, however, identical with the quantity called $\dfrac{d\lambda}{h_1}$ in § 277, so that we have

$$h_1^2 = \frac{4(a^2+\lambda)(b^2+\lambda)(c^2+\lambda)}{(\lambda-\mu)(\lambda-\nu)} \quad \ldots\ldots\ldots\ldots\ldots(202);$$

and clearly h_2 and h_3 can be obtained by cyclic interchange of the letters λ, μ and ν.

282. If for brevity we write

$$\Delta_\lambda = \sqrt{(a^2+\lambda)(b^2+\lambda)(c^2+\lambda)},$$

we find that

$$\frac{h_1}{h_2 h_3} = \frac{\Delta_\lambda}{2\Delta_\mu\Delta_\nu}(\mu-\nu)\sqrt{-1},$$

so that by substitution in equation (198), Laplace's equation in the present coordinates is seen to be

$$\frac{\partial}{\partial\lambda}\left\{(\mu-\nu)\frac{\Delta_\lambda}{\Delta_\mu\Delta_\nu}\frac{\partial V}{\partial\lambda}\right\} + \frac{\partial}{\partial\mu}\left\{(\nu-\lambda)\frac{\Delta_\mu}{\Delta_\nu\Delta_\lambda}\frac{\partial V}{\partial\mu}\right\} + \frac{\partial}{\partial\nu}\left\{(\lambda-\mu)\frac{\Delta_\nu}{\Delta_\lambda\Delta_\mu}\frac{\partial V}{\partial\nu}\right\} = 0$$

$$\ldots\ldots\ldots\ldots(203).$$

On multiplying throughout by $\Delta_\lambda\Delta_\mu\Delta_\nu$, this equation becomes

$$(\mu-\nu)\Delta_\lambda\frac{\partial}{\partial\lambda}\left(\Delta_\lambda\frac{\partial V}{\partial\lambda}\right) + (\nu-\lambda)\Delta_\mu\frac{\partial}{\partial\mu}\left(\Delta_\mu\frac{\partial V}{\partial\mu}\right) + (\lambda-\mu)\Delta_\nu\frac{\partial}{\partial\nu}\left(\Delta_\nu\frac{\partial V}{\partial\nu}\right) = 0$$

$$\ldots\ldots\ldots\ldots(204).$$

Let us now introduce new variables α, β, γ, given by

$$\alpha = \int^\lambda \frac{d\lambda}{\Delta_\lambda},$$

$$\beta = \int^\mu \frac{d\mu}{\Delta_\mu},$$

$$\gamma = \int^\nu \frac{d\nu}{\Delta_\nu},$$

then we have

$$\frac{\partial}{\partial \alpha} = \Delta_\lambda \frac{\partial}{\partial \lambda};$$

and equation (204) becomes

$$(\mu - \nu)\frac{\partial^2 V}{\partial \alpha^2} + (\nu - \lambda)\frac{\partial^2 V}{\partial \beta^2} + (\lambda - \mu)\frac{\partial^2 V}{\partial \gamma^2} = 0 \quad \ldots\ldots\ldots(205).$$

Distribution of Electricity on a freely-charged Ellipsoid.

283. Before discussing the general solution of Laplace's equation, it will be advantageous to examine a few special problems.

In the first place, it is clear that a particular solution of equation (205) is

$$V = A + B\alpha \ldots\ldots\ldots\ldots\ldots\ldots\ldots\ldots\ldots(206),$$

where A, B are arbitrary constants. The equipotentials are the surfaces $\alpha = $ constant, and are therefore confocal ellipsoids. Thus we can, from this solution, obtain the field when an ellipsoidal conductor is freely electrified.

For instance, if the ellipsoid

$$\frac{x^2}{a^2} + \frac{y^2}{b^2} + \frac{z^2}{c^2} = 1$$

is raised to unit potential, the potential at any external point will be given by equation (206) provided we choose A and B so as to have $V = 1$ when $\lambda = 0$, and $V = 0$ when $\lambda = \infty$. In this way we obtain

$$V = \frac{\displaystyle\int_\lambda^\infty \frac{d\lambda}{\Delta_\lambda}}{\displaystyle\int_0^\infty \frac{d\lambda}{\Delta_\lambda}} \quad \ldots\ldots\ldots\ldots\ldots\ldots\ldots\ldots\ldots(207).$$

The surface density at any point on the ellipsoid is given by

$$4\pi\sigma = -\frac{\partial V}{\partial n} = -\frac{\partial V}{\partial \lambda}\frac{\partial \lambda}{\partial n} = -h_1\frac{\partial V}{\partial \lambda}$$

$$= \frac{h_1\left(\dfrac{1}{\Delta_\lambda}\right)_{\lambda=0}}{\displaystyle\int_0^\infty \frac{d\lambda}{\Delta_\lambda}}$$

$$= \frac{h_1}{abc\displaystyle\int_0^\infty \frac{d\lambda}{\Delta_\lambda}} \quad \ldots\ldots\ldots\ldots\ldots\ldots\ldots\ldots(208).$$

Thus the surface density at different points of the ellipsoid is proportional to h_1.

284. The quantity h_1 admits of a simple geometrical interpretation. Let l, m, n be the direction-cosines of the tangent plane to the ellipsoid at

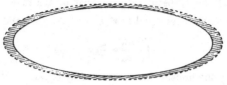

any point λ, μ, ν, and let p be the perpendicular from the origin on to this tangent plane. Then from the geometry of the ellipsoid we have

$$p^2 = (a^2 + \lambda)\, l^2 + (b^2 + \lambda)\, m^2 + (c^2 + \lambda)\, n^2 \quad \ldots\ldots\ldots\ldots(209).$$

Moving along the normal, we shall come to the point $\lambda + d\lambda$, μ, ν. The tangent plane at this point has the same direction-cosines l, m, n as before, but the perpendicular from the origin will be $p + dp$, where $dp = \dfrac{d\lambda}{h_1}$. To obtain dp we differentiate equation (209), allowing λ alone to vary, and so have

$$2p\,dp = d\lambda\,(l^2 + m^2 + n^2) = d\lambda.$$

Comparing this with $dp = \dfrac{d\lambda}{h_1}$, we see that $h_1 = 2p$.

Thus the surface density at any point is proportional to the perpendicular from the centre on to the tangent plane at the point.

In fig. 79, the thickness of the shading at any point is proportional to the perpendicular from the centre on to the tangent plane, so that the shading represents the distribution of electricity on a freely electrified ellipsoid.

It will be easily verified that the outer boundary of this shading must be an ellipsoid, similar to and concentric with the original ellipsoid.

285. Replacing h_1 by $2p$ in equation (208), we find for the total charge E on the ellipsoid,

$$E = \iint \sigma\, dS = \frac{1}{2\pi abc \displaystyle\int_0^\infty \frac{d\lambda}{\Delta_\lambda}} \iint p\, dS.$$

Since $\iint p\, dS$ is three times the volume of the ellipsoid, and therefore equal to $4\pi abc$, this reduces to

$$E = \frac{2}{\displaystyle\int_0^\infty \frac{d\lambda}{\Delta_\lambda}}.$$

Since the ellipsoid is supposed to be raised to unit potential, this quantity E gives the capacity of an ellipsoidal conductor electrified in free space.

The capacity can however be obtained more readily by examining the form of the potential at infinity. At points which are at a distance r from the centre of the ellipsoid so great that a, b, c may be neglected in comparison with r, λ becomes equal to r^2, so that $\Delta_\lambda = r^3$, and

$$\int_\lambda^\infty \frac{d\lambda}{\Delta_\lambda} = \frac{2}{r}.$$

Thus at infinity the limiting form assumed by equation (207) is

$$V = \frac{2}{r \int_0^\infty \frac{d\lambda}{\Delta_\lambda}},$$

and since the value of V at infinity must be $\dfrac{E}{r}$ the value of E follows at once.

A freely-charged spheroid.

286. The integral $\displaystyle\int_0^\infty \frac{d\lambda}{\Delta_\lambda}$ is integrable if any two of the semi-axes become equal to one another.

If $b = c$, the ellipsoid is a prolate spheroid, and its capacity is found to be

$$E = \frac{2}{\displaystyle\int_0^\infty \frac{d\lambda}{(b^2 + \lambda)(a^2 + \lambda)^{\frac{1}{2}}}} = \frac{2ae}{\log\left(\dfrac{1 + e}{1 - e}\right)},$$

where e is the eccentricity.

If $a = b$, the ellipsoid is an oblate spheroid, and its capacity is found to be

$$E = \frac{2}{\displaystyle\int_0^\infty \frac{d\lambda}{(a^2 + \lambda)(c^2 + \lambda)^{\frac{1}{2}}}} = \frac{ae}{\sin^{-1}e}.$$

Elliptic Disc.

287. In the preceding analysis, let a become vanishingly small, then the conductor becomes an elliptic disc of semi-axes b and c.

The perpendicular from the origin on to the tangent-plane is given, as in the ellipsoid, by

$$p^2 = \frac{1}{\dfrac{x^2}{a^4} + \dfrac{y^2}{b^4} + \dfrac{z^2}{c^4}}.$$

and when a is made very small in the limit, this becomes

$$p^2 = \frac{1}{\dfrac{x^2}{a^4}} = \frac{a^2}{1 - \dfrac{y^2}{b^2} - \dfrac{z^2}{c^2}},$$

so that the surface density at any point x, y in the disc is proportional to

$$\left(1 - \frac{y^2}{b^2} - \frac{z^2}{c^2}\right)^{-\frac{1}{2}} \quad\quad\quad\quad\quad\quad(210).$$

Circular Disc.

288. On further simplifying by putting $b = c$, we arrive at the case of a circular disc. The density of electrification is seen at once from expression (210) to be proportional to

$$\left(1 - \frac{r^2}{c^2}\right)^{-\frac{1}{2}},$$

and therefore varies inversely as the shortest chord which can be drawn through the point.

Moreover, when $a = 0$ and $b = c$, we have $\Delta_\lambda = (c^2 + \lambda)\sqrt{\lambda}$, so that

$$\int_\lambda^\infty \frac{d\lambda}{\Delta_\lambda} = \frac{2}{c} \tan^{-1}\left(\frac{c}{\sqrt{\lambda}}\right) \text{ and } \int_0^\infty \frac{d\lambda}{\Delta_\lambda} = \frac{\pi}{c}.$$

Thus the capacity of a circular disc is $\dfrac{2c}{\pi}$, and when the disc is raised to potential unity, the potential at any external point is

$$\frac{2}{\pi} \tan^{-1}\left(\frac{c}{\sqrt{\lambda}}\right),$$

where λ is the positive root of

$$\frac{x^2}{\lambda} + \frac{y^2 + z^2}{c^2 + \lambda} = 1.$$

289. Lord Kelvin* quotes some interesting experiments by Coulomb on the density at different points on a circular plate of radius 5 inches. The results are given in the following table:

Distances from the plate's edge	Observed Densities	Calculated Densities
5 ins.	1	1
4	1·001	1·020
3	1·005	1·090
2	1·17	1·250
1	1·52	1·667
0·5	2·07	2·294
0	2·90	∞

* *Papers on Elect. and Mag.* p. 179.

Much more remarkable is Cavendish's experimental determination of the capacity of a circular disc. Cavendish found this to be $\frac{1}{1\cdot57}$ times that of a sphere of equal radius, while theory shews the true value of the denominator to be $\frac{\pi}{2}$ or 1·5708 !

290. By inverting the distribution of electricity on a circular disc, taking the origin of inversion to be a point in the plane of the disc, Kelvin [*] has obtained the distribution of electricity on a disc influenced by a point charge in its plane, a problem previously solved by another method by Green. The general Green's function for a circular disc has been obtained by Hobson[†].

Spherical Bowl.

291. Lord Kelvin has also, by inversion, obtained the solution for a spherical bowl of any angle freely electrified. Let the bowl be a piece of a sphere of diameter f. Let the distance from the middle point of the bowl to any point of the bowl be r, and let the greatest value of r, *i.e.* the distance from a point on the edge to the middle point of the bowl, be a. Then Kelvin finds for the electric densities inside and outside the bowl:

$$\rho_i = \frac{V}{2\pi^2 f}\left\{\sqrt{\frac{f^2 - a^2}{a^2 - r^2}} - \tan^{-1}\sqrt{\frac{f^2 - a^2}{a^2 - r^2}}\right\},$$

$$\rho_0 = \rho_i + \frac{V}{2\pi f}.$$

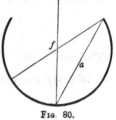

Fig. 80.

Some numerical results calculated from these formulæ are of interest. The six values in the following tables refer to the middle point and the five points dividing the arc from the middle point to the edge into six equal parts.

Plane disc			Curved disc arc 10°			Curved disc arc 20°		
ρ_i	ρ_0	Mean	ρ_i	ρ_0	Mean	ρ_i	ρ_0	Mean
1·00	1·00	1·0000	·91	1·06	1·0000	·86	1·14	1·0000
1·01	1·01	1·0142	·95	1·08	1·0141	·88	1·15	1·0010
1·06	1·06	1·0607	·99	1·13	1·0605	·92	1·20	1·0369
1·15	1·15	1·1547	1·09	1·22	1·1542	1·02	1·29	1·1106
1·34	1·34	1·3416	1·27	1·41	1·3407	1·29	1·56	1·2606
1·81	1·81	1·8091	1·74	1·88	1·8071	1·67	1·94	1·6474

[*] *Papers on Elect. and Mag.* p. 183.
[†] *Trans. Camb. Phil. Soc.* XVIII. p. 277.

Bowl arc 270°   Bowl arc 340°

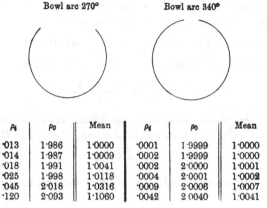

μ	ρ₀	Mean	μ	ρ₀	Mean
·013	1·986	1·0000	·0001	1·9999	1·0000
·014	1·987	1·0009	·0002	1·9999	1·0000
·018	1·991	1·0041	·0002	2·0000	1·0001
·025	1·998	1·0118	·0004	2·0001	1·0002
·045	2·018	1·0316	·0009	2·0006	1·0007
·120	2·093	1·1060	·0042	2·0040	1·0041

Discussing these results, Lord Kelvin says : " It is remarkable how slight an amount of curvature produces a very sensible excess of density on the convex side in the first two cases (10° and 20°), yet how nearly the mean of the densities on the convex and concave sides at any point agrees with that at the corresponding point on a plane disc shewn in the first column. The results for bowls of 270° and 340° illustrate the tendency of the whole charge to the convex surface, as the case of a thin spherical conducting surface with an infinitely small aperture is approached."

ELLIPSOIDAL HARMONICS.

292. We now return to the general equations (205), namely

$$(\mu - \nu)\frac{\partial^2 V}{\partial \alpha^2} + (\nu - \lambda)\frac{\partial^2 V}{\partial \beta^2} + (\lambda - \mu)\frac{\partial^2 V}{\partial \gamma^2} = 0 \quad \ldots \ldots (211),$$

and examine the nature of the general solutions of this equation.

Let us assume a tentative solution

$$V = LMN,$$

in which L is a function of λ only, M a function of μ only, and N a function of ν only. Substituting this solution the equation reduces to

$$(\mu - \nu)\frac{1}{L}\frac{\partial^2 L}{\partial \alpha^2} + (\nu - \lambda)\frac{1}{M}\frac{\partial^2 M}{\partial \beta^2} + (\lambda - \mu)\frac{1}{N}\frac{\partial^2 N}{\partial \gamma^2} = 0.$$

Since α is a function of λ only, $\dfrac{1}{L}\dfrac{\partial^2 L}{\partial \alpha^2}$ is a function of λ only, and the equation may be written in the form

$$(\mu - \nu)f(\lambda) + (\nu - \lambda)F(\mu) + (\lambda - \mu)\Phi(\nu) = 0,$$

where f, F and Φ are functions whose form we have to determine.

This functional equation must hold for all values of λ, μ, ν. Putting $\mu = \nu$ we find that $F(\nu) = \Phi(\nu)$, and since this is true for all values of ν, F and Φ

must be the same function. By a similar procedure, it follows that f must also be the same function, so that the equation can be written

$$(\mu - \nu)f(\lambda) + (\nu - \lambda)f(\mu) + (\lambda - \mu)f(\nu) = 0.$$

To find the form of the function f we put $\lambda = 0$ and obtain

$$\frac{f(\mu) - f(0)}{\mu} = \frac{f(\nu) - f(0)}{\nu}.$$

Thus a function of μ is equal to the same function of ν, so that each must be a constant. Calling this B, and writing A for $f(0)$, we find that

$$f(\lambda) = A + B\lambda.$$

293. Restoring its value to $f(\lambda)$ we see that we must have

$$\frac{\partial^2 L}{\partial \alpha^2} = (A + B\lambda) L \quad ....(212),$$

and similar equations, with the same constants A and B, must be satisfied by M and N.

Equation (212), on substituting for α in terms of λ, becomes

$$\Delta_\lambda \frac{\partial}{\partial \lambda} \left(\Delta_\lambda \frac{\partial L}{\partial \lambda} \right) = (A + B\lambda) L \quad ....(213),$$

a differential equation of the second order in λ, while M and N satisfy equations which are identical except that μ and ν are the variables.

The solution of equation (213) is known as a Lamé's function, or ellipsoidal harmonic. The function is commonly written as $E_n^p(\lambda)$, where p, n are new arbitrary constants, connected with the constants A and B by the relations

$$n(n+1) = B, \quad \text{and} \quad (b^2 + c^2)p = -A.$$

Thus $E_n^p(\lambda)$ is a solution of

$$\frac{\partial^2 L}{\partial \alpha^2} = \{ n(n+1)\lambda - p(b^2 + c^2) \} L,$$

and a solution of equation (211) is

$$V = \Sigma\Sigma E_n^p(\lambda) E_n^p(\mu) E_n^p(\nu) \quad ....(214).$$
$$_{p\ n}$$

294. Equation (213) being of the second order, must have two independent solutions. Denoting one by L, let the other be supposed to be Lu. Then we must have

$$\frac{\partial^2 L}{\partial \alpha^2} = (A + B\lambda) L,$$

$$\frac{\partial^2 (Lu)}{\partial \alpha^2} = (A + B\lambda) Lu;$$

so that on multiplying the former equation by u, and subtracting from the latter,

$$L\frac{\partial^2 u}{\partial \alpha^2} + 2\frac{\partial L}{\partial \alpha}\frac{\partial u}{\partial \alpha} = 0.$$

Thus

$$u = \int \frac{d\alpha}{L^2} = \int \frac{d\lambda}{L^2 \Delta_\lambda},$$

and the complete solution is seen to be

$$CL + DL\int \frac{d\lambda}{L^2 \Delta_\lambda},$$

where C and D are arbitrary constants.

Accordingly, the complete solution of equation (211) can be written as

$$V = \underset{p\ n}{\Sigma\Sigma}\left(C_{np}E_n^p(\lambda) + D_{np}E_n^p(\lambda)\int \frac{d\lambda}{\{E_n^p(\lambda)\}^2\Delta_\lambda}\right)$$

$$\left(C_{np}'E_n^p(\mu) + D_{np}'E_n^p(\mu)\int \frac{d\mu}{\{E_n^p(\mu)\}^2\Delta_\mu}\right)$$

$$\left(C_{np}''E_n^p(\nu) + D_{np}''E_n^p(\nu)\int \frac{d\nu}{\{E_n^p(\nu)\}^2\Delta_\nu}\right).$$

This corresponds exactly to the general solution in rational integral spherical harmonics, namely

$$V = \underset{p\ n}{\Sigma\Sigma}\ (C_{np}r^n + D_{np}r^{-(n+1)})$$

$$(C_{np}'e^{ip\phi} + D_{np}'e^{-ip\phi})$$

$$(C_{np}''P_n^p(\cos\theta) + D_{np}''P_n^p(\cos\theta)).$$

Ellipsoid in uniform field of force.

295. As an illustration of the use of confocal coordinates, let us examine the field produced by placing an uninsulated ellipsoid in a uniform field of force.

The potential of the undisturbed field of force may be taken to be $V = Fx$, or in confocal coordinates (cf. equation (201))

$$V = F\sqrt{\frac{(a^2 + \lambda)(a^2 + \mu)(a^2 + \nu)}{(b^2 - a^2)(c^2 - a^2)}}.$$

This is of the form $V = CLMN$,

where C is the constant $F(b^2 - a^2)^{-\frac{1}{2}}(c^2 - a^2)^{-\frac{1}{2}}$, and L, M, N are functions of λ only, μ only and ν only, respectively, namely $L = \sqrt{a^2 + \lambda}$, etc.

Since $V = LMN$ is a solution of Laplace's equation, there must, as in § 294, be a second solution $V = Lu\,.\,MN$, where

$$u = \int \frac{d\lambda}{L^2\Delta_\lambda} = \int \frac{d\lambda}{(a^2 + \lambda)\Delta_\lambda}.$$

The upper limit of integration is arbitrary: if we take it to be infinite, both u and Lu will vanish at infinity, while M and N are in any case finite at infinity. Thus $Lu \cdot MN$ is a potential which vanishes at infinity and is proportional (since u is a function of λ only) at every point of any one of the surfaces $\lambda = $ cons., to the potential of the original field. Thus the solution

$$V = CLMN + DLu \cdot MN \quad \dots\dots\dots\dots\dots\dots(215)$$

can be made to give zero potential over any one of the surfaces $\lambda = $ cons., by a suitable choice of the constant D.

For instance if the conductor is $\lambda = 0$, we have, on the conductor,

$$u = \int_0^\infty \frac{d\lambda}{(a^2 + \lambda)\Delta_\lambda}.$$

Thus on the conductor we have

$$V = LMN \left(C + D \int_0^\infty \frac{d\lambda}{(a^2 + \lambda)\Delta_\lambda} \right).$$

The condition for this to vanish gives the value of D, and on substituting this value of D, equation (215) becomes

$$V = CLMN \left(1 - \frac{u}{\int_0^\infty \frac{d\lambda}{(a^2 + \lambda)\Delta_\lambda}} \right)$$

$$= Fx \left(1 - \frac{\int_\lambda^\infty \frac{d\lambda}{(a^2 + \lambda)\Delta_\lambda}}{\int_0^\infty \frac{d\lambda}{(a^2 + \lambda)\Delta_\lambda}} \right)$$

$$= Fx \frac{\int_0^\lambda \frac{d\lambda}{(a^2 + \lambda)\Delta_\lambda}}{\int_0^\infty \frac{d\lambda}{(a^2 + \lambda)\Delta_\lambda}} \quad \dots\dots\dots\dots\dots\dots\dots(216).$$

This gives the field when the original field is parallel to the major axis of the ellipsoid. If the original field is in any other direction we can resolve it into three fields parallel to the three axes of the ellipsoid, and the final field is then found by the superposition of three fields of the type of that given by equation (216).

SPHEROIDAL HARMONICS.

296. When any two semi-axes of the standard ellipsoid become equal the method of confocal coordinates breaks down. For the equation

$$\frac{x^2}{a^2 + \theta} + \frac{y^2}{b^2 + \theta} + \frac{z^2}{c^2 + \theta} = 1 \quad \dots\dots\dots\dots\dots(217)$$

reduces to a quadratic, and has therefore only two roots, say λ, μ. The surfaces $\lambda =$ cons. and $\mu =$ cons. are now confocal ellipsoids and hyperboloids of revolution, but obviously a third family of surfaces is required before the position of a point can be fixed. Such a family of surfaces, orthogonal to the two present families, is supplied by the system of diametral planes through the axis of revolution of the standard ellipsoid.

The two cases in which the standard ellipsoid is a prolate spheroid and an oblate spheroid require separate examination.

Prolate Spheroids.

297. Let the standard surface be the prolate spheroid

$$\frac{x^2}{a^2} + \frac{y^2 + z^2}{b^2} = 1,$$

in which $a > b$. If we write

$$y = \varpi \cos \phi, \qquad z = \varpi \sin \phi,$$

then the curvilinear coordinates may be taken to be λ, μ, ϕ, where λ, μ are the roots of

$$\frac{x^2}{a^2 + \theta} + \frac{\varpi^2}{b^2 + \theta} = 1 \quad\quad\quad\quad\dots\dots\dots(218).$$

In this equation, put $a^2 - b^2 = c^2$ and $a^2 + \theta = c^2\theta'^2$, then the equation becomes

$$\frac{x^2}{c^2\theta'^2} + \frac{\varpi^2}{c^2(\theta'^2 - 1)} = 1.$$

If ξ^2, η^2 are the roots of this equation in θ'^2, we readily find that $x^2 = \xi^2\eta^2c^2$, so that we may take

$$x = c\xi\eta \quad\quad\quad\quad\quad\dots\dots\dots\dots\dots\dots(219),$$

$$\varpi = c\sqrt{(1 - \xi^2)(\eta^2 - 1)} \quad\quad\dots\dots\dots\dots\dots(220)$$

in which η is taken to be the greater of the two roots.

The surfaces $\xi =$ cons., $\eta =$ cons. are identical with the surfaces $\theta =$ cons., and are accordingly confocal ellipsoids and hyperboloids. The coordinates ξ, η, ϕ may now be taken to be orthogonal curvilinear coordinates.

It is easily found that

$$h_1 = \frac{1}{c}\sqrt{\frac{1 - \xi^2}{\eta^2 - \xi^2}}, \qquad h_2 = \frac{1}{c}\sqrt{\frac{\eta^2 - 1}{\eta^2 - \xi^2}}, \qquad h_3 = \frac{1}{c\sqrt{(1 - \xi^2)(\eta^2 - 1)}},$$

from which Laplace's equation is obtained in the form

$$\frac{\partial}{\partial \xi}\left\{(1 - \xi^2)\frac{\partial V}{\partial \xi}\right\} + \frac{\partial}{\partial \eta}\left\{(\eta^2 - 1)\frac{\partial V}{\partial \eta}\right\} + \frac{\eta^2 - \xi^2}{(1 - \xi^2)(\eta^2 - 1)}\frac{\partial^2 V}{\partial \phi^2} = 0.$$

298. Let us search for solutions of the form

$$V = \Xi\,H\,\Phi,$$

where Ξ, H, Φ are solutions solely of ξ, η and ϕ respectively. On substituting this tentative solution and simplifying, we obtain

$$\frac{(1-\xi^2)(\eta^2-1)}{\eta^2-\xi^2}\left[\frac{1}{\Xi}\frac{\partial}{\partial\xi}\left\{(1-\xi^2)\frac{\partial\Xi}{\partial\xi}\right\}+\frac{1}{H}\frac{\partial}{\partial\eta}\left\{(\eta^2-1)\frac{\partial H}{\partial\eta}\right\}\right]+\frac{1}{\Phi}\frac{\partial^2\Phi}{\partial\phi^2}=0.$$

As in the theory of spherical harmonics, the only possible solution results from taking

$$\frac{1}{\Phi}\frac{\partial^2\Phi}{\partial\phi^2}=-m^2,$$

where $-m^2$ is a constant, and m must be an integer if the solution is to be single valued. The solution is

$$\Phi = C\cos m\phi + D\sin m\phi \quad\dots\dots\dots\dots\dots(221).$$

We must now have

$$\frac{1}{\Xi}\frac{\partial}{\partial\xi}\left\{(1-\xi^2)\frac{\partial\Xi}{\partial\xi}\right\}+\frac{1}{H}\frac{\partial}{\partial\eta}\left\{(\eta^2-1)\frac{\partial H}{\partial\eta}\right\}=\frac{m^2(\eta^2-\xi^2)}{(1-\xi^2)(\eta^2-1)}$$

$$=\frac{m^2}{1-\xi^2}+\frac{m^2}{\eta^2-1},$$

and this can only be satisfied by taking

$$\frac{\partial}{\partial\xi}\left\{(1-\xi^2)\frac{\partial\Xi}{\partial\xi}\right\}-\frac{m^2\Xi}{1-\xi^2}+s\Xi=0 \quad\dots\dots\dots\dots(222),$$

together with

$$\frac{\partial}{\partial\eta}\left\{(1-\eta^2)\frac{\partial H}{\partial\eta}\right\}-\frac{m^2H}{1-\eta^2}+sH=0 \quad\dots\dots\dots\dots(223).$$

Equations (222) and (223) are identical with the equation already discussed in §§ 273, 274. The solutions are known to be

$$\Xi = AP_n^m(\xi)+BQ_n^m(\xi),$$
$$H = A'P_n^m(\eta)+B'Q_n^m(\eta),$$

where $s=n(n+1)$ and P_n^m, Q_n^m are the associated Legendrian functions already investigated. Combining the values just obtained for Ξ, H with the value for Φ given by equation (221), we obtain the general solution

$$V = \Sigma\Sigma\Xi H\Phi$$
$$= \underset{m\,n}{\Sigma\Sigma}\{AP_n^m(\xi)+BQ_n^m(\xi)\}\{A'P_n^m(\eta)+B'Q_n^m(\eta)\}\{C\cos m\phi+D\sin m\phi\}.$$

At infinity it is easily found that

$$\eta = \infty, \qquad \xi = \frac{x}{\sqrt{x^2+\varpi^2}}=\cos\theta,$$

while at the origin $\qquad \eta = 1, \qquad \xi = 0.$

Thus in the space outside any spheroid, the solution $P_n^m(\xi)\,Q_n^m(\eta)$ is finite everywhere, while, in the space inside, the finite solution is $P_n^m(\xi)\,P_n^m(\eta)$.

Oblate Spheroids.

299. For an oblate spheroid, $a^2 - b^2$ is negative, so that in equation (218) we replace $b^2 - a^2$ by κ^2, so that $\kappa = ic$, and obtain, in place of equations (219) and (220),

$$x = \kappa \xi i \eta,$$

$$\varpi = \kappa \sqrt{(1 - \xi^2)(1 - \eta^2)}.$$

Replacing $i\eta$ by ζ, we may take ξ, ζ and ϕ as real orthogonal curvilinear coordinates, connected with Cartesian coordinates by the relations

$$x = \kappa \xi \zeta,$$

$$\varpi = \kappa \sqrt{(1 - \xi^2)(1 + \zeta^2)}.$$

We proceed to search for solutions of the type

$$V = \Xi Z \Phi,$$

and find that Ξ, Φ must satisfy the same equations as before, while Z must satisfy

$$-\frac{\partial}{\partial \zeta}\left\{(1 + \zeta^2)\frac{\partial Z}{\partial \zeta}\right\} - \frac{m^2}{1 + \zeta^2}Z + n(n + 1)Z = 0.$$

The solution of this is

$$Z = A' P_n^m(i\zeta) + B' Q_n^m(i\zeta),$$

and the most general solution may now be written down as before.

PROBLEMS IN TWO DIMENSIONS.

300. Often when a solution of a three-dimensional problem cannot be obtained, it is found possible to solve a similar but simpler two-dimensional problem, and to infer the main physical features of the three-dimensional problem from those of the two-dimensional problem. We are accordingly led to examine methods for the solution of electrostatic problems in two dimensions.

At the outset we notice that the unit is no longer the point-charge, but the uniform line-charge, a line-charge of line-density σ having a potential (cf. § 75)

$$C - 2\sigma \log r.$$

Method of Images.

301. The method of images is available in two dimensions, but presents no special features. An example of its use has already been given in § 220.

Method of Inversion.

302. In two dimensions the inversion is of course about a *line*. Let this be represented by the point O in fig. 81.

Let PP', QQ' be two pairs of inverse points. Let a line-charge e at Q produce potential V_P at P, and let a line-charge e' at Q produce potential $V_{P'}$ at P', so that

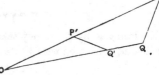

$$V_P = C - 2e \log PQ;$$
$$V_{P'} = C' - 2e' \log P'Q'.$$

FIG. 81.

If we take $e = e'$, we obtain

$$V_P - V_{P'} = C'' - 2e \log \frac{PQ}{P'Q'}$$
$$= C'' - 2e \log \frac{OQ}{OP'} \quad \dots\dots\dots\dots\dots(224).$$

Let P be a point on an equipotential when there are charges e_1 at Q_1, e_2 at Q_2, etc., and let V denote the potential of this equipotential. Let $\overline{V}$ denote the potential at P' under the influence of charges e_1, e_2, ... at the inverse points of Q_1, Q_2, ... Then, by summation of equations such as (224),

$$\overline{V} - V = - \Sigma (2e \log OP') + \Sigma (2e \log OQ) + \text{constants},$$

or

$$\overline{V} = \text{constants} - 2 (\Sigma e) \log OP' \quad \dots\dots\dots\dots(225).$$

The potential at P' of charges e_1, e_2, ... at the inverse points of Q_1, Q_2, ... plus a charge $- \Sigma e$ at O is

$$\overline{V} + C + 2 (\Sigma e) \log OP',$$

and this by equation (225) is a constant. This result gives the method of inversion in two dimensions:

If a surface S is an equipotential under the influence of line-charges e_1, e_2, ... at Q_1, Q_2, ..., then the surface which is the inverse of S about a line O will be an equipotential under the influence of line-charges e_1, e_2, ... on the lines inverse to Q_1, Q_2, ... together with a charge $- \Sigma e$ at the line O.

Two-dimensional Harmonics.

303. A solution of Laplace's equation can be obtained which is the analogue in two dimensions of the three-dimensional solution in spherical harmonics.

In two dimensions we have two coordinates, r, θ, these becoming identical with ordinary two-dimensional polar coordinates. Laplace's equation becomes

$$\frac{1}{r} \frac{\partial}{\partial r} \left(r \frac{\partial V}{\partial r} \right) + \frac{\partial^2 V}{r^2 \partial \theta^2} = 0,$$

and on assuming the form

$$V = R\Theta,$$

in which R is a function of r only, and Θ a function of θ only, we obtain the solution in the form

$$V = \sum_{n=1}^{n=\infty} \left(Ar^n + \frac{B}{r^n} \right) (C \cos n\theta + D \sin n\theta).$$

Thus the "harmonic-functions" in two dimensions are the familiar sine and cosine functions. The functions which correspond to rational integral harmonics are the functions

$$r^n \sin n\theta, \quad r^n \cos n\theta.$$

In x, y coordinates these are obviously rational integral functions of x and y of degree n.

Corresponding to the theorem of § 240, that any function of position on the surface of a sphere can (subject to certain restrictions) be expanded in a series of rational integral harmonics, we have the famous theorem of Fourier, that any function of position on the circumference of a circle can (subject to certain restrictions) be expanded in a series of sines and cosines. In the proof which follows (as also in the proof of § 240), no attempt is made at absolute mathematical rigour: as before, the form of proof given is that which seems best suited to the needs of the student of electrical theory.

Fourier's Theorem.

304. *The value of any function F of position on the circumference of a circle can be expressed, at every point of the circumference at which the function is continuous, as a series of sines and cosines, provided the function is single-valued, and has only a finite number of discontinuities and of maxima and minima on the circumference of the circle.*

Let $P(f, a)$ be any point outside the circle, then if R is the distance from P to the element ds of the circle (a, θ) we have

$$\int \frac{f^2 - a^2}{2\pi a R^2} ds = 1.$$

This result can easily be obtained by integration, or can be seen at once from physical considerations, for the integrand is the charge induced on a conducting cylinder by unit line-charge at P.

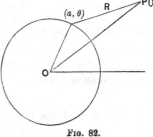

Fig. 82.

Let us now introduce a function u defined by

$$u = \frac{f^2 - a^2}{2\pi a} \int \frac{F}{R^2}\, ds \quad \text{.........................(226).}$$

Then, subject to the conditions stated for F we find, as in § 240, that on the circumference of the circle, the function u becomes identical with F. Also we have

$$\frac{1}{R^2} = \frac{1}{f^2 + a^2 - 2af \cos(\theta - \alpha)}$$

$$= \frac{1}{(f - ae^{i(\theta-\alpha)})(f - ae^{-i(\theta-\alpha)})}$$

$$= \frac{e^{i(\theta-\alpha)}}{f^2 - a^2}\left\{\frac{a}{f - ae^{i(\theta-\alpha)}} - \frac{f}{a - fe^{i(\theta-\alpha)}}\right\}$$

$$= \frac{1}{f^2 - a^2}\left\{1 + \sum_1^\infty \left(\frac{a}{f}\right)^n \left(e^{ni(\theta-\alpha)} + e^{-ni(\theta-\alpha)}\right)\right\}$$

$$= \frac{1}{f^2 - a^2}\left\{1 + 2\sum_1^\infty \left(\frac{a}{f}\right)^n \cos n(\theta - \alpha)\right\}.$$

Hence

$$u = \frac{1}{2\pi a} \int F\left\{1 + 2\sum_1^\infty \left(\frac{a}{f}\right)^n \cos n(\theta - \alpha)\right\} ds$$

$$= \frac{1}{2\pi} \int_{\theta=0}^{\theta=2\pi} F\, d\theta + \frac{1}{\pi}\sum_1^\infty \left(\frac{a}{f}\right)^n \int_{\theta=0}^{\theta=2\pi} F \cos n(\theta - \alpha)\, d\theta,$$

and on passing to the limit and putting $a = f$, this becomes

$$F = \frac{1}{2\pi}\int_{\theta=0}^{\theta=2\pi} F\, d\theta + \frac{1}{\pi}\sum_1^\infty \int_{\theta=0}^{\theta=2\pi} F \cos n(\theta - \alpha)\, d\theta \quad \text{.........(227),}$$

expressing F as a series of sines and cosines of multiples of α.

We can put this result in the form

$$F = \bar{F} + \sum_1^\infty (a_n \cos n\alpha + b_n \sin n\alpha),$$

where

$$a_n = \frac{1}{\pi}\int_0^{2\pi} F \cos n\theta\, d\theta,$$

$$b_n = \frac{1}{\pi}\int_0^{2\pi} F \sin n\theta\, d\theta,$$

and

$$\bar{F} = \frac{1}{2\pi}\int_0^{2\pi} F\, d\theta,$$

so that $\bar{F}$ is the mean value of F.

If F has a discontinuity at any point $\theta = \beta$ of the circle, and if F_1, F_2 are the values of F at the discontinuity, then obviously at the point $\theta = \beta$ on the circle, equation (226) becomes

$$u = \tfrac{1}{2}(F_1 + F_2),$$

so that the value of the series (227) at a discontinuity is the arithmetic mean of the two values of F at the discontinuity (cf. § 256).

305. We could go on to develop the theory of ellipsoidal harmonics etc. in two dimensions, but all such theories are simply particular cases of a very general theory which will now be explained.

CONJUGATE FUNCTIONS.

General Theory.

306. In two-dimensional problems, the equation to be satisfied by the potential is

$$\frac{\partial^2 V}{\partial x^2} + \frac{\partial^2 V}{\partial y^2} = 0 \quad \ldots\ldots\ldots\ldots\ldots\ldots(228);$$

and this has a general solution in finite terms, namely

$$V = f(x + iy) + F(x - iy) \quad \ldots\ldots\ldots\ldots\ldots(229),$$

where f and F are arbitrary functions, in which the coefficients may of course involve the imaginary i.

For V to be wholly real, F must be the function obtained from f on changing i into $- i$. Let $f(x + iy)$ be equal to $u + iv$ where u and v are real, then $F(x + iy)$ must be equal to $u - iv$, so that we must have $V = 2u$. If we introduce a second function U equal to $- 2v$, we have

$$\begin{aligned}
U + iV &= - 2v + 2iu \\
&= 2i(u + iv) \\
&= 2i f(x + iy) \\
&= \phi(x + iy) \quad \ldots\ldots\ldots\ldots\ldots(230),
\end{aligned}$$

where $\phi(x + iy)$ is a completely general function of the single variable $x + iy$.

Thus the most general form of the potential which is wholly real, can be derived from the most general arbitrary function of the single variable $x + iy$, on taking the potential to be the imaginary part of this function.

307. If $\phi(x + iy)$ is a function of $x + iy$, then $i\phi(x + iy)$ will also be a function, and the imaginary part of this function will also give a possible potential. We have, however, from equation (230),

$$\begin{aligned}
i\phi(x + iy) &= i(U + iV) \\
&= - V + iU,
\end{aligned}$$

shewing that U is a possible potential.

Thus when we have a relation of the type expressed by equation (230), either U or V will be a possible potential.

308. Taking V to be the potential, we have by differentiation of equation (230),

$$\frac{\partial U}{\partial x} + i\frac{\partial V}{\partial x} = \phi'(x + iy),$$

$$\frac{\partial U}{\partial y} + i\frac{\partial V}{\partial y} = i\phi'(x + iy),$$

and hence $\qquad i\left(\frac{\partial U}{\partial x} + i\frac{\partial V}{\partial x}\right) = \frac{\partial U}{\partial y} + i\frac{\partial V}{\partial y}.$

Equating real and imaginary parts in the above equation, we obtain

$$\frac{\partial U}{\partial x} = \frac{\partial V}{\partial y},$$

$$\frac{\partial U}{\partial y} = -\frac{\partial V}{\partial x},$$

so that $\qquad \dfrac{\partial U}{\partial x}\dfrac{\partial V}{\partial x} + \dfrac{\partial U}{\partial y}\dfrac{\partial V}{\partial y} = 0 \dots\dots\dots\dots\dots(231).$

This however is the condition that the families of curves $U = \text{cons.}$, $V = \text{cons.}$, should cut orthogonally at every point. Thus the curves $U = \text{cons.}$ are the orthogonal trajectories of the equipotentials—*i.e.* are the lines of force.

Representation of complex quantities.

309. If we write

$$z = x + iy$$

so that z is a complex quantity, we can suppose the position of the point P indicated by the value of the single complex variable z. If z is expressed in Demoivre's form

$$z = re^{i\theta} = r(\cos\theta + i\sin\theta),$$

then we find that $r = \sqrt{x^2 + y^2}$ and $\theta = \tan^{-1}\frac{y}{x}$. The

FIG. 83.

quantity r is known as the *modulus* of z and is denoted by $|z|$, while θ is known as the *argument* of z and is denoted by arg z. The representation of a complex quantity in a plane in this way is known as an Argand diagram.

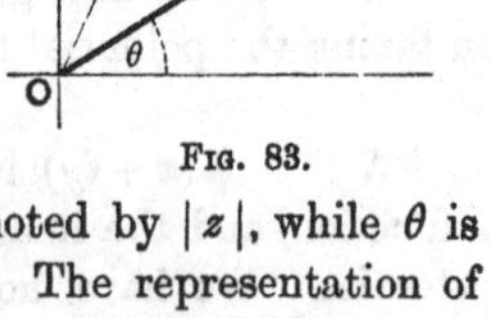

310. *Addition of complex quantities.* Let P be $z = x + iy$, and let P' be $z' = x' + iy'$. The value of $z + z'$ is $(x + x') + i(y + y')$, so that if Q represents the value $z + z'$ it is clear that $OPQP'$ will be a parallelogram. Thus to add together the complex quantities z and z' we complete the parallelogram OPP', and the fourth point of this parallelogram will represent $z + z'$.

The matter may be put more simply by supposing the complex quantity $z = x + iy$ represented by the direction and length of a line, such that its projections on two rectangular axes are x, y. For instance in fig. 83, the value of z will be represented equally by either OP or $P'Q$. We now have the following rule for the addition of complex quantities.

To find $z + z'$, describe a path from the origin representing z in magnitude and direction, and from the extremity of this describe a path representing z'. The line joining the origin to the extremity of this second path will represent $z + z'$

311. Multiplication of complex quantities. If

$$z = x + iy = r \,(\cos \theta + i \sin \theta),$$

and $\quad\quad z' = x' + iy' = r' \,(\cos \theta' + i \sin \theta'),$

then, by multiplication

$$zz' = rr' \,\{\cos (\theta + \theta') + i \sin (\theta + \theta')\},$$

so that $\quad\quad |zz'| = rr' = |z|\,|z'|,$

$$\arg (zz') = \theta + \theta' = \arg z + \arg z',$$

and clearly we can extend this result to any number of factors. Thus we have the important rules:

The modulus of a product is the product of the moduli of the factors.

The argument of a product is the sum of the arguments of the factors.

There is a geometrical interpretation of multiplication.

In fig. 84, let $OA = 1$, $OP = z$, $OP' = z'$ and $OQ = zz'$.

Then the angles QOA, $P'OA$ being equal to $\theta + \theta'$ and θ' respectively, the angle QOP' must be equal to θ, and therefore to POA.

Moreover

$$\frac{OQ}{OP'} = \frac{OP}{OA},$$

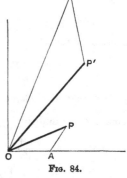

Fig. 84.

each ratio being equal to r, so that the triangles QOP' and POA are similar. Thus to multiply the vector OP' by the vector OP, we simply construct on OP' a triangle similar to AOP.

The same result can be more shortly expressed by saying that to multiply z' ($= OP'$) by z ($= OP$), we multiply the length OP' by $|z|$ and turn it through an angle $\arg z$.

So also to divide by z, we divide the length of the line representing the dividend by $|z|$ and turn through an angle $- \arg z$. In either case an angle is positive when the turning is in the direction which brings us from the axis x to that of y after an angle $\pi/2$.

Conformal Representation.

312. We can now consider more fully the meaning of the relation

$$U + iV = \phi(x + iy).$$

Let us write $z = x + iy$, and $W = U + iV$, z and W being complex imaginaries, which we must now suppose in accordance with equation (230) to be connected by the relation

$$W = \phi(z) \quad \dots\dots\dots\dots\dots\dots\dots(232).$$

We can represent values of z in one Argand diagram, and values of W in another. The plane in which values of z are represented will be called the z-plane, the other will be called the W-plane. Any point P in the z-plane corresponds to a definite value of z and this, by equation (232), may give one or more values of W, according as ϕ is or is not a single-valued function. If Q is a point in the W-plane which represents one of these values of W, the points P and Q are said to correspond.

As P describes any curve S in the z-plane, the point Q in the W-plane which corresponds to P will describe some curve T in the W-plane, and the curve T is said to correspond to the curve S. In particular, corresponding to any infinitesimal linear path PP' in the z-plane, there will correspond a small linear element QQ' in the W-plane. If OP, OP' represent the values z, $z + dz$ respectively, then the element PP' will represent dz. Similarly the element QQ' will represent dW or $\dfrac{dW}{dz} dz$.

Hence we can get the element QQ' from the element PP' on multiplying it by $\dfrac{dW}{dz}$, *i.e.* by $\dfrac{\partial}{\partial z} \phi(z)$, or by $\phi'(x + iy)$. This multiplier depends solely on the position of the point P in the z-plane, and not on the length or direction of the element dz. If we express $\dfrac{dW}{dz}$ or $\phi'(x + iy)$ in the form

$$\frac{dW}{dz} = \phi'(x + iy) = \rho(\cos\chi + i\sin\chi),$$

we find that the element dW can be obtained from the corresponding element dz by multiplying its length by ρ or $\left|\dfrac{dW}{dz}\right|$, and turning it through an angle χ, or $\arg\left(\dfrac{dW}{dz}\right)$. It follows that any element of area in the z-plane is represented in the W-plane by an element of area of which the shape is exactly similar to that of the original element, the linear dimensions are ρ times as great, and the orientation is obtained by turning the original element through an angle χ.

From the circumstance that the shapes of two corresponding elements in the two planes are the same, the process of passing from one plane to the other is known as *conformal representation*.

313. Let us examine the value of the quantity ρ which, as we have seen, measures the linear magnification produced in a small area on passing from the z-plane to the W-plane.

We have
$$\rho (\cos \chi + i \sin \chi) = \frac{dW}{dz} = \phi' (x + iy)$$
$$= \frac{\partial U}{\partial x} + i \frac{\partial V}{\partial x}$$
$$= \frac{\partial V}{\partial y} + i \frac{\partial V}{\partial x},$$

so that
$$\rho = \left| \frac{\partial V}{\partial y} + i \frac{\partial V}{\partial x} \right| = \sqrt{\left(\frac{\partial V}{\partial x}\right)^2 + \left(\frac{\partial V}{\partial y}\right)^2}.$$

The quantity ρ, or $\left| \dfrac{dW}{dz} \right|$, is called the "modulus of transformation." We now see that if V is the potential, this modulus measures the electric intensity R, or $\sqrt{\left(\dfrac{\partial V}{\partial x}\right)^2 + \left(\dfrac{\partial V}{\partial y}\right)^2}$. Since $R = 4\pi\sigma$, this circumstance provides a simple means of finding σ, the surface-density of electricity at any point of a conducting surface.

314. If $\dfrac{\partial}{\partial s}$ denote differentiation along the surface of a conductor, on which the potential V is constant, we have
$$\left| \frac{dW}{dz} \right| = \frac{\partial U}{\partial s},$$

so that
$$\sigma = \frac{1}{4\pi} R = \frac{1}{4\pi} \frac{\partial U}{\partial s}.$$

The total charge on a strip of unit width between any two points P, Q of the conductor is accordingly
$$\int \sigma \, dS = \frac{1}{4\pi} \int_P^Q \frac{\partial U}{\partial s} \, ds = \frac{1}{4\pi} (U_Q - U_P) \quad \ldots\ldots\ldots\ldots(233).$$

315. If, on equating real and imaginary parts of any transformation of the form
$$U + iV = \phi (x + iy) \quad \ldots\ldots\ldots\ldots\ldots(234),$$
it is found that the curve $f(x, y) = 0$ corresponds to the constant value $V = C$, then clearly the general value of V obtained from equation (234) will be a solution of Laplace's equation subject to the condition of having the constant value $V = C$ over the boundary $f(x, y) = 0$. It will therefore be the potential in an electrostatic field in which the curve $f(x, y) = 0$ may be taken to be a conductor raised to potential C.

316. From a given transformation it is obviously always possible to deduce the corresponding electrostatic field, but on being given the conductors and potentials in the field, it is by no means always possible to deduce the required transformation. We shall begin by the examination of a few fields which are given by simple known transformations.

SPECIAL TRANSFORMATIONS.

I. $W = z^n$.

317. Considering the transformation $W = z^n$, we have

$$U + iV = (x + iy)^n = r^n (\cos n\theta + i \sin n\theta),$$

so that $V = r^n \sin n\theta$. Thus any one of the surfaces $r^n \sin n\theta = \text{constant}$ may be supposed to be an equipotential, including as a special case

$$r^n \sin n\theta = 0,$$

in which the equipotential consists of two planes cutting at an angle $\dfrac{\pi}{n}$.

This transformation can be further discussed by assigning particular values to n.

$n = 1$. This gives simply $V = y$, a uniform field of force.

$n = 2$. This gives $V = 2xy$, so that the equipotentials are rectangular hyperbolic cylinders, including as a special case two planes intersecting at right angles (fig. 85).

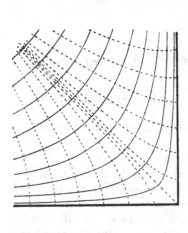

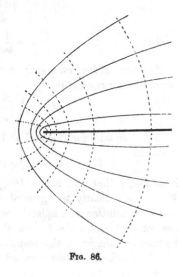

FIG. 85.   FIG. 86.

This transformation gives the field in the immediate neighbourhood of two conducting planes meeting at right angles in any field of force. It also gives the field between two coaxal rectangular hyperbolas.

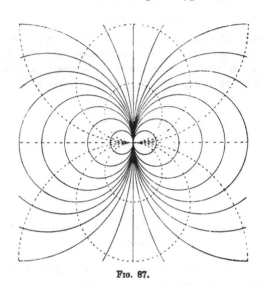

$n = \frac{1}{2}$. This gives $x + iy = (U + iV)^2$, so that

$$x = U^2 - V^2, \quad y = 2UV,$$

and on eliminating U we obtain

$$y^2 = 4V^2 (x + V^2).$$

Thus the equipotentials are confocal and coaxal parabolic cylinders, including as a special case ($V = 0$) a semi-infinite plane bounded by the line of foci.

This transformation clearly gives the field in the immediate neighbourhood of a conducting sharp straight edge in any field of force (fig. 86).

$n = -1$. This gives

$$U + iV = \frac{1}{r}(\cos \theta - i \sin \theta),$$

and the equipotentials are

$$rV = \sin \theta \quad \text{or} \quad x^2 + y^2 - \frac{y}{V} = 0.$$

Thus the equipotentials are a series of circular cylinders, all touching the plane $y = 0$ along the axis $x = 0$, $y = 0$ (fig. 87).

II. $W = \log z$.

318.   The transformation $W = \log z$ gives

$$U + iV = \log r + i\theta,$$

so that the equipotentials are the planes $\theta = $ constant, a system of planes all intersecting in the same line. As a special case, we may take $\theta = 0$ and $\theta = \pi$ to be the conductors, and obtain the field when the two halves of a plane are raised to different potentials. The lines of force, $U = $ constant, are circles (fig. 88).

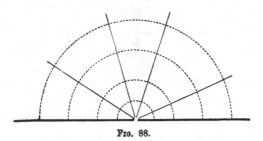

Fɪɢ. 88.

If we take U to be the potential, the equipotentials are concentric circular cylinders, and the field is seen to be simply that due to a uniform line-charge, or uniformly electrified cylinder.

It may be noticed that the transformation

$$W = \log (z - a)$$

gives the transformation appropriate to a line-charge at $z = a$.

Also we notice that

$$W = \log \frac{z - a}{z + a}$$

gives a field equivalent to the superposition of the fields given by

$$W = \log (z - a) \quad \text{and} \quad W = - \log (z + a).$$

This transformation is accordingly that appropriate to two equal and opposite line-charges along the parallel lines $z = a$ and $z = - a$.

This last transformation gives $U = 0$ when $x = 0$, so that it gives the transformation for a line-charge in front of a parallel infinite plane.

GENERAL METHODS.

I. *Unicursal Curves.*

319. Suppose that the coordinates of a point on a conductor can be expressed as real functions of a real parameter, which varies as the point moves over the conductor, in such a way that the whole range of variation of the parameter just corresponds to motion over the whole conductor. In other words, suppose that the coordinates x, y can be expressed in the form

$$x = f(p), \quad y = F(p),$$

and that all real values of p give points on the conductor, while, conversely, all points on the conductor correspond to real values of p.

Then the transformation

$$z = f(W) + iF(W) \quad \dots\dots\dots\dots\dots\dots\dots\dots(235)$$

will give $V = 0$ over the conductor. For on putting $V = 0$ in equation (235) we obtain

$$x + iy = f(U) + iF(U),$$

so that

$$x = f(U), \quad y = F(U),$$

and by hypothesis the elimination of U will lead to the equation of the conductor.

320. For example, consider the parabola (referred to its focus as origin),

$$y^2 = 4a(x + a).$$

We can write the coordinates of any point on this parabola in the form

$$x + a = am^2, \quad y = 2am,$$

and the transformation is seen to be

$$z = aW^2 - a + 2aiW = a(W - i)^2,$$

or

$$W - i = \left(\frac{z}{a}\right)^{\frac{1}{2}}.$$

agreeing with that which has already been seen in § 317 to give a parabola as a possible equipotential.

321. As a second example of this method, let us consider the ellipse

$$\frac{x^2}{a^2} + \frac{y^2}{b^2} = 1.$$

The coordinates of a point on the ellipse may be expressed in the form

$$x = a \cos \phi, \quad y = b \sin \phi,$$

and the transformation is seen to be

$$z = a \cos W + ib \sin W.$$

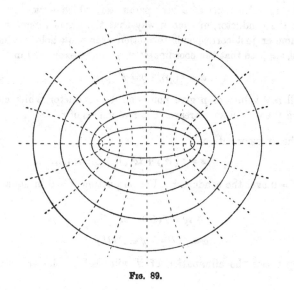

Fig. 89.

We can take $a = c \cosh \alpha$, $b = c \sinh \alpha$, where $c^2 = a^2 - b^2$, and the transformation becomes

$$z = c \cos (W + i\alpha) = c \cos \{U + i(V + \alpha)\}.$$

The same transformation may be expressed in the better known form

$$z = c \cosh W.$$

The equipotentials are the confocal ellipses

$$\frac{x^2}{a^2 + \lambda} + \frac{y^2}{b^2 + \lambda} = 1,$$

while the lines of force are confocal hyperbolic cylinders. On taking V as the potential, we get a field in which the equipotentials are confocal hyperbolic cylinders.

II. *Schwarz's Transformation.*

322. Schwarz has shewn how to obtain a transformation in which one equipotential can be any linear polygon.

At any angle of a polygon it is clear that the property that small elements remain unchanged in shape can no longer hold. The reason is easily seen to be that the modulus of transformation is either infinite or zero (cf. figs. 24 and 25, p. 61). Thus, at the angles of any polygon,

$$\frac{dW}{dz} = 0 \text{ or } \infty.$$

The same result is evident from electrostatic considerations. At an angle of a conductor, the surface-density σ is either infinite or zero (§ 70), while we have the relation (§ 313),

$$\sigma = \frac{R}{4\pi} = \frac{1}{4\pi} \left| \frac{dW}{dz} \right|.$$

Let us suppose that the polygon in the z-plane is to correspond to the line $V = 0$ in the W-plane, and let the angular points correspond to

$$U = u_1, \quad U = u_2, \quad \text{etc.}$$

Then, when

$$W = u_1, \quad W = u_2, \quad \text{etc.},$$

$\frac{dz}{dW}$ must either vanish or become infinite. We must accordingly have

$$\frac{dz}{dW} = F(W - u_1)^{\lambda_1}(W - u_2)^{\lambda_2} \quad \text{.................(236)},$$

where λ_1, λ_2, ... are numbers which may be positive or negative, while F denotes a function, at present unknown, of W.

Suppose that, as we move along the polygon, the values of U at the angular points occur in the order u_1, u_2, ... Then, on passing along the side of the polygon which joins the two angles $U = u_1$, $U = u_2$, we pass along a range for which $V = 0$, and $u_1 < U < u_2$. Thus, along this side of the polygon, $W - u_1$, $W - u_2$, $W - u_3$, etc. are real quantities, positive or negative, which retain the same sign along the whole of this edge. It follows that, as we pass along this edge, the change in the value of $\arg\left(\frac{dz}{dW}\right)$, as given by equation (236), is equal to the change in $\arg F$, the arguments of the factors

$$(W - u_1)^{\lambda_1}(W - u_2)^{\lambda_2} \dots$$

undergoing no change.

Now $\arg\left(\frac{dz}{dW}\right)$ measures the inclination of the axis $V = 0$ to the edge of the polygon at any point, so that if the polygon is to be rectilinear, this must remain constant as we pass along any edge. It follows that there must be no change in $\arg F$ as we pass along any side of the polygon.

This condition can be satisfied by supposing F to be a pure numerical constant. Taking it to be real, we have, from equation (236),

$$\arg\left(\frac{dz}{dW}\right) = \lambda_1 \arg\left(W - u_1\right) + \lambda_2 \arg\left(W - u_2\right) + \ldots \quad \ldots\ldots(237).$$

On passing through the angular point at which $W = u_2$, the quantities $W - u_1$, $W - u_3$, etc. remain of the same sign, while the single quantity $W - u_2$ changes sign. Thus $\arg\left(W - u_2\right)$ increases by π, whence, by equation (237), $\arg\left(\frac{dz}{dW}\right)$ increases by $\lambda_2\pi$.

The axis $V = 0$ does not turn in the W-plane on passing through the value $W = u_2$, while $\arg\left(\frac{dz}{dW}\right)$ measures the inclination of the element of the polygon in the z-plane to the corresponding element of the axis $V = 0$ in the W-plane.

Hence, on passing through the value $W = u_2$, the perimeter of the polygon in the z-plane must turn through an angle equal to the increase in $\arg\left(\frac{dz}{dW}\right)$, namely $\lambda_2\pi$, the direction of turning being from Ox to Oy. Thus $\lambda_1\pi$, $\lambda_2\pi$, ... must be the exterior angles of the polygon, these being positive when the polygon is convex to the axis Ox. Or, if α_1, α_2, ... are the interior angles, reckoned positive when the polygon is concave to the axis of x, we must have

$$\lambda_1 = \frac{\alpha_1}{\pi} - 1, \text{ etc.}$$

Thus the transformation required for a polygon having internal angles α_1, α_2, ... is

$$\frac{dz}{dW} = C\left(W - u_1\right)^{\frac{\alpha_1}{\pi} - 1}\left(W - u_2\right)^{\frac{\alpha_2}{\pi} - 1} \quad \ldots\ldots\ldots\ldots(238),$$

where u_1, u_2, ... are real quantities, which give the values of U at the angular points.

323. As an illustration of the use of Schwarz's transformation, suppose the conducting system to consist of a semi-infinite plane placed parallel to an infinite plane.

In fig. 90, let the conductor be supposed to be a polygon $ABCDE$, which is described by following the dotted line in the direction of the arrows. The points A, B, C, E are all supposed to be at infinity, the points B and C coinciding. Let us take A to be $W = -\infty$, B or C to be $W = 0$, D to be $W = 1$ and E to be $W = +\infty$. The angles of the polygon are zero at (BC) and 2π at D. Thus the transformation is

$$\frac{dz}{dW} = C\frac{W - 1}{W},$$

giving upon integration

$$z = C\{W - \log W + D\} \quad \text{............(239)},$$

where C, D are constants of integration which may be obtained from the

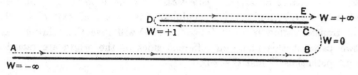

FIG. 90.

condition that the two planes are to be, say, $y = 0$ and $y = h$. From these conditions we obtain $C = \dfrac{h}{\pi}$, $D = i\pi$, so that the transformation is

$$z = \frac{h}{\pi}\{W - \log W + i\pi\} \quad \text{............(240)}.$$

On replacing z, W by $-z$, $-W$, the transformation assumes the simpler form

$$z = \frac{h}{\pi}(W + \log W) \quad \text{............(241)}.$$

III. *Successive Transformations.*

324. If $\zeta = \phi(z)$, $W = f(\zeta)$ are any two transformations, then by elimination of ζ, a relation

$$W = F(z) \quad \text{............(242)}$$

is obtained, which may be regarded as a new transformation.

We may regard the relation $\zeta = \phi(z)$ as expressing a transformation from the z-plane into a ζ-plane, while the second relation $W = f(\zeta)$ expresses a further transformation from the ζ-plane into a W-plane. Thus the final transformation (242) may be regarded as the result of two successive transformations.

Two uses of successive transformations are of particular importance.

325. *Conductor influenced by line-charge.* The transformation

$$W = \log \frac{\zeta - a}{\zeta + a},$$

gives, as we have seen (§ 318) the solution when a line-charge is placed at $\zeta = a$ in front of the plane represented by the real axis of ζ. Let the further transformation $\zeta = f(z)$ transform the real axis of ζ into a surface S, and the point $\zeta = a$ into the point $z = z_0$, so that $a = f(z_0)$. Then the transformation

$$W = \log \frac{f(z) - f(z_0)}{f(z) + f(z_0)}$$

gives the solution when a line-charge is placed at $z = z_0$ in the presence of the surface S. In this transformation it must be remembered that U, and not V, is the potential (cf. § 318).

326.   *Conductors at different potentials.* Let us suppose that the transformation $\zeta = \phi(z)$ transforms a conductor into the real axis of ζ. The further transformation $W = C + D \log \zeta$ (§ 318) will give the solution when the two parts of this plane on different sides of the origin are raised to different potentials C and $C + \pi D$.

Thus the transformation obtained by elimination of ζ, namely

$$W = C + D \log \phi(z),$$

will transform two parts of the same conductor into two parallel planes, and so will give the solution of a problem in which two parts of the same conductor are raised to different potentials.

EXAMPLES OF THE USE OF CONJUGATE FUNCTIONS.

327.   Two examples of practical importance will now be given to illustrate the use of the methods of conjugate functions.

Example I.   Parallel Plate Condenser.

328.   The transformation

$$z = \frac{h}{\pi} (\zeta - \log \zeta + i\pi)$$

has been found to transform the two plates in fig. 90 into the positive and negative parts of the real axis of ζ. The further transformation $W = \log \zeta$ gives the solution when these two parts of the real axis of ζ are at potentials 0 and π respectively (§ 326).

Thus the transformation obtained by the elimination of ζ, namely

$$z = \frac{h}{\pi} (e^W - W + i\pi) \quad \dots\dots\dots\dots\dots\dots(243),$$

will transform the two planes of fig. 90—one infinite and one semi-infinite— into two infinite parallel planes. Thus equation (243) gives the transformation suitable to the case of a semi-infinite plane at distance h from a parallel infinite plane, the potential difference being π.

By the principle of images it is obvious that the distribution on the upper plate is the same as it would be if the lower plate were a semi-infinite plane at distance $2h$ instead of an infinite plane at distance h. The equipotentials and lines of force for either problem are shewn in fig. 91.

Separating real and imaginary parts in equation (243),

$$x = \frac{h}{\pi}(e^U \cos V - U),$$

$$y = \frac{h}{\pi}(e^U \sin V - V + \pi).$$

Thus the equipotential $V = 0$ is the line $y = h$, the equipotential $V = \pi$ is the line $y = 0$.

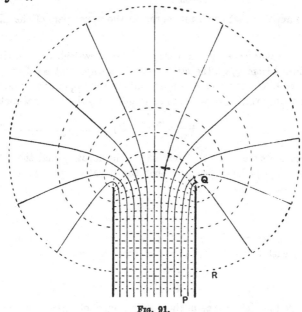

FIG. 91.

On the former equipotential, the relation between x and U is

$$x = \frac{h}{\pi}(e^U - U) \quad\quad\quad\dots\dots\dots\dots\dots\dots(244).$$

When $U = -\infty$, $x = +\infty$; as U increases, x decreases until it reaches a minimum value $x = h/\pi$ when $U = 0$; and as U further increases through positive values x again increases, reaching $x = \infty$ when $U = +\infty$. Thus as U varies while $V = 0$, the path described is the path PQR in fig. 91.

The intensity at any point is

$$R = \left|\frac{dW}{dz}\right| = \frac{\pi}{h\,|e^W - 1|}.$$

At a point on the equipotential $V = 0$, the surface-density is

$$\sigma = \frac{R}{4\pi} = \frac{1}{4h(e^U - 1)}.$$

At P, $U = -\infty$, so that $\sigma = \dfrac{1}{4h}$; as we approach Q, σ increases and finally becomes infinite at Q, while after passing Q and moving along QR, the upper side of the plate, σ decreases, and ultimately vanishes to the order of e^{-U}.

The total charge within any range U_1, U_2 is, by equation (233),

$$\frac{1}{4\pi}(U_2 - U_1).$$

It therefore appears that the total charge on the upper part of the plate QR is infinite.

Let us, however, consider the charges on the two sides of a strip of the plate of width l from Q, *i.e.* the strip between $x = h/\pi$ and $x = l + h/\pi$. The two values of U corresponding to the points in the upper and lower faces at which this strip terminates, are from equation (244) the two real roots of

$$l + \frac{h}{\pi} = \frac{h}{\pi}(e^U - U) \quad \dots\dots\dots\dots\dots\dots(245).$$

Of these roots we know that one, say U_1, is negative and the other (U_2) is positive. If l is large, we find that the negative root U_1 is, to a first approximation, equal to

$$-\frac{\pi}{h}\left(l + \frac{h}{\pi}\right),$$

and this is its actual value when l is very large. Thus the charge on the lower plate within a large distance l of the edge is

$$\frac{h}{4}\left(l + \frac{h}{\pi}\right),$$

and therefore the disturbance in the distribution of electricity as we approach Q results in an increase on the charge of the lower plate equal to what would be the charge on a strip of width h/π in the undisturbed state.

If l is large the positive root of equation (245) is

$$U_2 = \log\left(1 + \frac{l\pi}{h}\right),$$

so that the total charge on a strip of width l of the upper plate approximates, when l is large, to

$$\frac{1}{4\pi}\log\left(1 + \frac{l\pi}{h}\right).$$

Thus although the charge on the upper plate is infinite, it vanishes in comparison with that on the lower plate.

Example II. Bend of a Leyden Jar.

329. The method of conjugate functions enables us to approximate to the correction required in the formula for the capacity of a Leyden Jar, on account of the presence of the sharp bend in the plates.

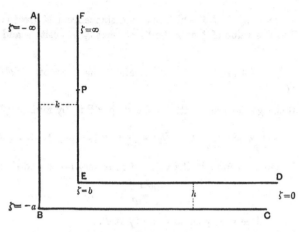

Fig. 92.

As a preliminary, let us find the capacity of a two-dimensional condenser formed of two conductors, each of which consists of an infinite plate, bent into an L-shape, the two L's being fitted into one another as in fig. 92.

Let us assume the five points A, B, (CD), E, F to be $\zeta = -\infty$, $-a$, 0, $+b$, $+\infty$ respectively, and let us for convenience suppose the potential difference which occurs on passing through the value $\zeta = 0$ to be π. Then the transformation is

$$\frac{dz}{d\zeta} = A\,(\zeta + a)^{-\frac{1}{2}}\,\zeta^{-1}\,(\zeta - b)^{\frac{1}{2}},$$

where $W = \log \zeta$ (cf. § 326).

To integrate, we put $u = (\zeta + a)^{-\frac{1}{2}}\,(\zeta - b)^{\frac{1}{2}}$, and obtain

$$z = A \int \frac{d\zeta}{\zeta} \sqrt{\frac{\zeta - b}{\zeta + a}} = A \int u\, d \log \left(\frac{b + au^2}{1 - u^2}\right)$$

$$= -2A \sqrt{\frac{b}{a}} \tan^{-1} \sqrt{\frac{a}{b}}\, u + A \log \frac{1 + u}{1 - u} + C \quad \ldots\ldots\ldots(246),$$

where C is a constant of integration.

To make C vanish, we must have $z = 0$ when $u = 0$, *i.e.* at the point E. We shall accordingly take E as origin, so that $C = 0$.

At B, we now have $\zeta = -a$, $u = \infty$, and therefore

$$z = \pm \pi A \sqrt{\frac{b}{a}} \pm i\pi A.$$

Thus the distances between the pairs of arms are $\pi \sqrt{\dfrac{b}{a}} A$ and πA respectively.

Let P be any point in EF which is at a distance from E great compared with EB. Let the value of ζ at P be ζ_P, so that ζ_P is positive and greater than b.

We have $W = U + iV = \log \zeta$, so that along the conductor FED, $V = 0$ and $U = \log \zeta$.

The total charge per unit width on the strip EP is, by formula (233),

$$\int_E^P \sigma \, dS = \frac{1}{4\pi}(U_P - U_E) = \frac{1}{4\pi}(\log \zeta_P - \log b) \quad \ldots\ldots(247).$$

If P is far removed from E, the value of ζ_P is very great, and since

$$\zeta = \frac{au^2 + b}{1 - u^2} \quad \ldots\ldots\ldots\ldots\ldots\ldots\ldots\ldots(248),$$

the value of u^2 will be nearly equal to unity at P.

From equation (246),

$$z = -2A\sqrt{\frac{b}{a}}\tan^{-1}\sqrt{\frac{a}{b}}\,u + 2A\log(1+u) - A\log(1-u^2),$$

so that $\quad \log(1-u^2) = 2\log(1+u) - 2\sqrt{\dfrac{b}{a}}\tan^{-1}\sqrt{\dfrac{a}{b}}\,u - \dfrac{z}{A} \ \ \ldots..(249),$

in which the terms $\log(1-u^2)$, $-z/A$, are large at P in comparison with the others. Again, from equation (248), we have

$$\log \zeta = \log(au^2 + b) - \log(1 - u^2) \quad \ldots\ldots\ldots\ldots(250),$$

in which $\log \zeta$, $\log(1 - u^2)$ are large at P, in comparison with the term $\log(au^2 + b)$. Combining equations (249) and (250),

$$\log \zeta = \log(au^2 + b) - 2\log(1+u) + 2\sqrt{\frac{b}{a}}\tan^{-1}\sqrt{\frac{a}{b}}\,u + \frac{z}{A}$$
$$\ldots\ldots(251),$$

in which the terms $\log \zeta$ and $\dfrac{z}{A}$ are large at P in comparison with the other terms. At P we may put $u = 1$ in all terms except $\log \zeta$ and z/A, and obtain as an approximation

$$\log \zeta_P = \log(a + b) - 2\log 2 + 2\sqrt{\frac{b}{a}}\tan^{-1}\sqrt{\frac{a}{b}} + \frac{z_P}{A}.$$

The value of z_P is of course $x_P + iy_P$, or EP. Thus, from the equation just obtained, equation (247) may be thrown into the form

$$\int_E^P \sigma \, ds = \frac{1}{4\pi} (\log \zeta_P - \log b)$$

$$= \frac{1}{4\pi} \left\{ \log \left(1 + \frac{a}{b} \right) - 2 \log 2 + 2 \sqrt{\frac{b}{a}} \tan^{-1} \sqrt{\frac{a}{b}} + \frac{EP}{A} \right\} \dots (252).$$

If the lines of force were not disturbed by the bend, we should have

$$\int_E^P \sigma \, ds = \frac{1}{4\pi} \left(\frac{EP}{A} \right).$$

Equation (252) shews that $\int_E^P \sigma \, ds$ is greater than this, by an amount

$$\frac{1}{4\pi} \left\{ \log \left(1 + \frac{a}{b} \right) - 2 \log 2 + 2 \sqrt{\frac{b}{a}} \tan^{-1} \sqrt{\frac{a}{b}} \right\} \dots\dots (253).$$

Let us denote the distances between the plates, namely $\pi A \sqrt{\dfrac{b}{a}}$ and πA,

by h and k respectively, so that $\sqrt{\dfrac{b}{a}} = \dfrac{h}{k}$. Expression (253) now becomes

$$\frac{1}{4\pi} \left\{ \log \frac{h^2 + k^2}{4h^2} + 2 \frac{h}{k} \tan^{-1} \frac{k}{h} \right\},$$

so that the charge on the plate EP is the same as it would be in a parallel plate condenser in which the breadth of the strip was greater than EP by

$$\frac{1}{\pi} \left\{ k \log \frac{h^2 + k^2}{4h^2} + 2h \tan^{-1} \frac{k}{h} \right\}.$$

When $h = k$, this becomes

$$\frac{h}{\pi} \left(\frac{\pi}{2} - \log_e 2 \right) \text{ or } \cdot 279h.$$

MULTIPLE-VALUED POTENTIALS.

330. There are many problems to which mathematical analysis yields more than one solution, although it may be found that only one of these solutions will ultimately satisfy the actual data of the problem. In such a case it will often be of interest to examine what interpretation has to be given to the rejected solutions.

The problem of determining the potential when the boundary conditions are given is not of this class, for it has already been shewn (§§ 186—188) that, subject to specified boundary conditions, the termination of the potential is absolutely unique. But it may happen that, in searching for the required solution, we come upon a multiple-valued solution of Laplace's equation. Only one value can satisfy the boundary conditions, but the interpretation of the other values is of interest, and in this way we arrive at the study of multiple-valued potentials.

Conjugate Functions on a Riemann's Surface.

331. An obvious case of a multiple-valued potential arises from the conjugate function transformation

$$W = \phi(z) \quad \dots\dots\dots\dots\dots\dots\dots(254),$$

when ϕ is not a single-valued function of z. Such cases have already occurred in §§ 317, 320, 323, etc.

The meaning of the multiple-valued potential becomes clear as soon as we construct a Riemann's surface on which $\phi(z)$ can be represented as a single-valued function of position. One point on this Riemann's surface must now correspond to each value of W, and therefore to each point in the W-plane. Thus we see that the transformation (254) transforms the complete W-plane into a complete Riemann's surface. Corresponding to a given value of z there may be many values of the potential, but these values will refer to the different sheets of the Riemann's surface. If any region on this surface is selected, which does not contain any branch points or lines, we can regard this region as a real two-dimensional region, and the corresponding value of the potential, as given by equation (254), will give the solution of an electrostatic problem.

332. To illustrate this by a concrete case, consider the transformation

$$W = z^{\frac{1}{2}} \quad \dots\dots\dots\dots\dots\dots\dots\dots(255),$$

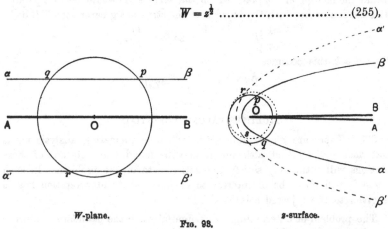

W-plane.   *z*-surface.

Fig. 98.

which has already been considered in § 317. The Riemann's surface appropriate for the representation of the two-valued function $z^{\frac{1}{2}}$ may be supposed to be a surface of two infinite sheets connected along a branch line which extends over the positive half of the real axis of z.

To regard this surface as a deformation of the W-plane, we must suppose that a slit is cut along the line OB (fig. 93) in the W-plane, and that the two edges of the slit are taken and turned so that the angle 2π, which they originally enclosed in the W-plane, is increased to 4π, after which the edges are again joined together.

The upper sheet of the Riemann's surface so formed will now represent the upper half of the W-plane, while the lower sheet will represent the lower half. Two points P_1, P_2, which represent equal and opposite values of W, say $\pm W_0$, will (by equation (255)) be represented by points at which z has the same value; they are accordingly the two points on the upper and lower sheet respectively for which z has the value W_0^2.

A circular path $pqrs$ surrounding O in the W-plane becomes a double circle on the z-surface, one circle being on the upper sheet and one on the lower, and the path being continuous since it crosses from one sheet to the other each time it meets the branch-line.

A line $\alpha\beta$ in the upper half of the W-plane becomes, as we have seen, a parabola $\alpha\beta$ on the upper sheet of the z-surface. Similarly a line $\alpha'\beta'$ in the lower half of the W-plane will become a parabola $\alpha'\beta'$ on the lower sheet of the z-surface. The space outside the parabola $\alpha\beta$ on the upper sheet of the z-surface transforms into a space in the W-plane bounded by the line $\alpha\beta$ and the line at infinity. Consequently the transformation under consideration gives the solution of the electrostatic problem, in which the field is bounded only by a conducting parabola and the region at infinity. The same is not true of the space inside the parabola $\alpha\beta$, for this transforms into a space in the W-plane bounded by both the line $\alpha\beta$ and the axis AOB. It is now clear that the transformation has no application to problems in which the electrostatic field is the space inside a parabola.

In general it will be seen that two points, which are close to one another on one sheet of the z-surface, but are on opposite sides of a branch-line, will transform into two points which are not adjacent to one another in the W-plane, and which therefore correspond to different potentials. Consequently we cannot solve a problem by a transformation which requires a branch-line to be introduced into that part of the Riemann's surface which represents the electrostatic field.

Images on a Riemann's Surface.

333. In the theory of electrical images, a system of imaginary charges is placed in a region which does not form part of the actual electrostatic field. When a two-dimensional problem is solved by a conjugate function transformation, the electrostatic field must, as we have seen, be represented by a region on a single sheet of the corresponding Riemann's surface, and this region must not be broken by branch-lines. The same, however, is not true of the part of the field in which the imaginary images are placed, for this may be represented by a region on one of the other sheets of the Riemann's surface.

To take the simplest possible illustration, suppose that in the ζ-plane we have a line-charge e along the line represented by the point P, in front of

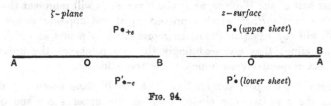

FIG. 94.

the uninsulated conducting plane represented by the real axis AB. The solution, as we know, is obtained by placing a charge $-e$ at the point P', which is the image of P in AOB. The value of the potential (U) is given, as in § 318, by

$$U + iV = A \log \frac{\zeta - \zeta_P}{\zeta - \zeta_{P'}}.$$

Let us now transform this solution by means of the transformation

$$\zeta = z^{\frac{1}{2}} \quad\dots\dots\dots\dots\dots\dots\dots\dots\dots(256).$$

The conducting plane AOB transforms into a semi-infinite plane OB, which may be taken to coincide with the branch-line of the Riemann's surface. The charge e at P becomes a charge at a point P on the upper sheet of the surface, while the image at P' becomes a charge at a point P' on the lower sheet. Thus we can replace the semi-infinite conductor OB in the z-plane by an image at a point P' on the lower sheet of a Riemann's surface, and we obtain the field due to a line-charge and a semi-infinite conductor in an ordinary two-dimensional space.

From the transformation used, the potential is found to be given by

$$U + iV = A \log \frac{\sqrt{z} - \sqrt{a}}{\sqrt{z} - \sqrt{-a}},$$

in which U is the potential, $z = a$ is the point (a, α) on the upper sheet, and $z = -a$ is the image on the lower sheet.

In calculating a potential on a Riemann's surface, we must not assume the potential of a line-charge e at the point (a, α) to be

$$C - 2e \log R \quad\dots\dots\dots\dots\dots\dots\dots\dots(257),$$

where R is the distance from the point (a, α). In fact, this potential would obviously have an infinity both at the point (a, α) on the upper sheet, and also at the point (a, α) on the lower sheet, and O would be the potential of *two* line-charges, one at the point (a, α) on each sheet.

The appropriate potential-function for a single charge can easily be found.

As in the problem just discussed, it is clear that the potential due to the single line-charge at (a, α) on the upper sheet is the value of U given by

$$U + iV = C + A \log (\sqrt{z} - \sqrt{a})$$

$$= C + A \log (r^{\frac{1}{2}} e^{\frac{i\theta}{2}} - a^{\frac{1}{2}} e^{\frac{i\alpha}{2}})$$

$$= C + A \log \left\{ \left(\sqrt{r} \cos \frac{\theta}{2} - \sqrt{a} \cos \frac{\alpha}{2} \right) + i \left(\sqrt{r} \sin \frac{\theta}{2} - \sqrt{a} \sin \frac{\alpha}{2} \right) \right\},$$

so that

$$U = C + \tfrac{1}{2} A \log \left\{ \left(\sqrt{r} \cos \frac{\theta}{2} - \sqrt{a} \cos \frac{\alpha}{2} \right)^2 + \left(\sqrt{r} \sin \frac{\theta}{2} - \sqrt{a} \sin \frac{\alpha}{2} \right)^2 \right\}$$

$$= C + \tfrac{1}{2} A \log \{ r - 2 \sqrt{ar} \cos \tfrac{1}{2} (\theta - \alpha) + a \},$$

and if this is to be the potential due to a line-charge e, it is clear, on examining the value of U near the point (a, α), that the value of A must be $-2e$. Thus the potential function must be

$$C - e \log \{ r - 2 \sqrt{ar} \cos \tfrac{1}{2} (\theta - \alpha) + a \} \dots \dots \dots \dots (258),$$

instead of that given by expression (257), namely,

$$C - e \log \{ r^2 - 2ar \cos (\theta - \alpha) + a^2 \} \dots \dots \dots \dots (259).$$

It will be noticed that both expressions are single-valued for given values of (r, θ), but that for a given value of z, expression (258) has two values, corresponding to two values of θ differing by 2π, while expression (259) has only one value. Or, to state the same thing in other words, the expression (259) is periodic in θ with a period 2π, while expression (258) is periodic with a period 4π.

Potential in a Riemann's Space.

334. Sommerfeld* has extended these ideas so as to provide the solution of problems in three-dimensional space.

His method rests on the determination of a multiple-valued potential function, the function being capable of representation as a single-valued function of position in a "Riemann's space," this space being an imaginary space which bears the same relation to real three-dimensional space as a Riemann's surface bears to a plane.

335. The best introduction to this method will be found in a study of the simplest possible example, and this will be obtained by considering the three-dimensional problem analogous to the two-dimensional problem already discussed in § 333.

* "Ueber verzweigte Potentiale im Raum," *Proc. Lond. Math. Soc.* 28, p. 895, and 30, p. 161.

We suppose that we have a single point-charge in the presence of an uninsulated conducting semi-infinite plane bounded by a straight edge. Let us take cylindrical coordinates r, θ, z, taking the edge of the plane to be $r = 0$, the plane itself to be $\theta = 0$, and the plane through the charge at right angles to the edge of the conductor to be $z = 0$. Let the coordinates of the point-charge be a, α, 0.

The Riemann's space is to be the exact analogue of the Riemann's surface described in § 332. That is to say, it is to be such that one revolution round the line $r = 0$ takes us from one "sheet" to the other of the space, while two revolutions bring us back to the starting-point. Thus, for a function to be a single-valued function of position in this space, it must be a periodic function of θ of period 4π.

Let us denote by $f(r, \theta, z, a, \alpha, 0)$ a function of r, θ, and z which is to satisfy the following conditions :

(i)   it must be a solution of Laplace's equation ;

(ii)   it must be a continuous and single-valued function of position in the Riemann's space ;

(iii) it must have one and only one infinity, this being at the point a, α, 0 on the first "sheet" of the space, and the function approximating near the point to the function $\dfrac{1}{R}$, where R is the distance from this point ;

(iv)   it must vanish when $r = \infty$.

It can be shewn, by a method exactly similar to that used in § 186, that there can be only one function satisfying these conditions. Hence the function $f(r, \theta, z, a, \alpha, 0)$ can be uniquely determined, and when found it will be the potential in the Riemann's space of a point-charge of unit strength at the point a, α, 0.

Consider now the function

$$f(r, \theta, z, a, \alpha, 0) - f(r, \theta, z, a, -\alpha, 0) \ \dots\dots\dots\dots (260),$$

which is of course the potential of equal and opposite point-charges at the point a, α, 0, and at its image in the plane $\theta = 0$, namely, the point a, $-\alpha$, 0.

This function, by conditions (i) and (iv), satisfies Laplace's equation and vanishes at infinity. On the first sheet of the surface, on which α varies from 0 to 2π (or from 4π to 6π, etc.), it has only one infinity, namely, at a, α, 0, at which it assumes the value $\dfrac{1}{R}$.

From the conditions which it satisfies, the function $f(r, \theta, z, a, \alpha, 0)$ must clearly involve θ and α only through $\theta - \alpha$, and must moreover be an even function of $\theta - \alpha$. It follows that, when $\theta = 0$, expression (260) vanishes.

Again, since the function f is periodic in θ with a period 2π, it follows that, when $\theta = -2\pi$, expression (260) may be written in the form

$$f(r, 2\pi, z, a, \alpha, 0) - f(r, -2\pi, z, a, -\alpha, 0),$$

and this clearly vanishes. Thus expression (260) vanishes when $\theta = 0$ and when $\theta = 2\pi$. That is to say, it vanishes on both sides of the semi-infinite conducting plane.

It is now clear that expression (260) satisfies all the conditions which have to be satisfied by the potential. The problem is accordingly reduced to that of the determination of the function $f(r, \theta, z, a, \alpha, 0)$.

336.	Let us write

$$r = e^\rho, \quad a = e^{\rho'},$$

then the distance R from r, θ, z to a, α, 0 is given by

$$R^2 = r^2 - 2ar \cos(\theta - \alpha) + a^2 + z^2$$
$$= 2ar \{\cos i(\rho - \rho') - \cos(\theta - \alpha)\} + z^2.$$

Take new functions R' and $f(u)$ given by

$$R'^2 = 2ar \{\cos i(\rho - \rho') - \cos(\theta - u)\} + z^2,$$
$$f(u) = \frac{ie^{iu}}{e^{iu} - e^{i\alpha}}.$$

The function $f(u)$ has infinities when $u = \alpha$, $\alpha \pm 2\pi$, $\alpha \pm 4\pi$, ..., its residue being unity at each infinity. Also, when $u = \alpha$, the value of R' becomes R. Hence the integral

$$\int \frac{1}{R'} f(u) \, du \quad\dots\dots\dots\dots\dots\dots\dots(261),$$

where the integral is taken round any closed contour in the u-plane which surrounds the value $u = \alpha$, but no other of the infinities of $f(u)$, will have as its value $2i\pi \times \dfrac{1}{R}$. We accordingly have

$$\frac{1}{R} = \frac{1}{2\pi} \int \frac{1}{R'} \frac{e^{iu}}{e^{iu} - e^{i\alpha}} \, du \quad\dots\dots\dots\dots\dots\dots(262).$$

The integral just found gives a form for the potential function in ordinary space which, as we shall now see, can easily be modified so as to give the potential function in the Riemann's space which we are now considering.

We notice first that $\dfrac{1}{R'}$, regarded as a function of r, θ, and z, is a solution of Laplace's equation, whatever value u may have. Hence the integral (261) will be a solution of Laplace's equation for all values of $f(u)$, for each term of the integrand will satisfy the equation separately.

If we take

$$f(u) = \frac{1}{2} \frac{ie^{\frac{iu}{2}}}{e^{\frac{iu}{2}} - e^{\frac{i\alpha}{2}}},$$

we see that the infinities of $f(u)$ occur when $u = \alpha$, $\alpha \pm 4\pi$, $\alpha \pm 8\pi$, etc., and the residue at each is unity. Hence, if we take the integral round one infinity only, say $u = \alpha$, the value of

$$\frac{1}{2i\pi} \int \frac{1}{R'} f(u)\, du \quad \ldots\ldots\ldots\ldots\ldots\ldots(263)$$

will become identical with $\dfrac{1}{R}$ at the point at which $R' = 0$. Moreover, expression (263) is, as we have seen, a solution of Laplace's equation: it is seen on inspection to be a single-valued function of position on the Riemann's surface, and to be periodic in θ with period 4π. Hence it is the potential-function of which we are in search. Thus

$$f(r, \theta, z, a, \alpha, 0) = \frac{1}{4\pi} \int \frac{e^{\frac{iu}{2}}}{e^{\frac{iu}{2}} e^{-\frac{i\alpha}{2}}} \frac{du}{\sqrt{r^2 - 2ar \cos(\theta - u) + a^2 + z^2}}.$$

The details of the integration can be found in Sommerfeld's paper. The value of the integral is found to be

$$\frac{1}{R} \frac{2}{\pi} \tan^{-1} \sqrt{\frac{\sigma + \tau}{\sigma - \tau}},$$

where $\qquad \tau = \cos \tfrac{1}{2}(\phi - \alpha), \quad \sigma = \cos \tfrac{1}{2}(\rho - \rho').$

Other systems of coordinates can be treated in the same way; details will be found in the papers to which reference has already been made.

337. The present chapter has attempted to give an account of the principal methods available for the solution of electrostatic problems. A few examples have been given of each method, but no attempt has been made to enumerate all the problems which can be solved. The reader who wishes to study particular problems more fully may be referred to the following works:

Sir W. THOMSON (Lord KELVIN). *Papers on Electricity and Magnetism.*

In particular a number of examples of images and inversion will be found here, with numerical calculations.

MAXWELL. *Electricity and Magnetism.* Vol. I. (3rd Edn.).

In Chap. IX. the theory of spherical harmonics is developed, and the problem of the distribution of electricity on a nearly spherical conductor free in space, as also that on a nearly spherical conductor enclosed in a nearly spherical and nearly concentric conducting vessel, are solved in detail. The coefficients of capacity and induction of two spherical conductors are investigated by spherical harmonics. Chapter XI. contains examples of the method of images and inversion. Chapter XII. contains a number of examples of conjugate functions, some being of special importance in the theory of electrostatic instruments.

J. J. THOMSON. *Recent Researches in Electricity and Magnetism.*

Chapter III. contains important examples of conjugate function transformations. In particular problems are solved which enable us to estimate the effect on the capacity of a condenser produced by the slit between a guard ring and the moveable plate of the condenser. Transformations are given which solve the problems of (i) a condenser formed by two parallel and equal plates of finite breadth ; (ii) a condenser formed by two parallel and equal strips placed in the same plane; (iii) a pile of plates; (iv) a system of $2n$ plates arranged radially at angles π/n with one another, alternate plates being at the same potential.

KIRCHHOFF. *Gesammelte abhandlungen.*

A formula is given for the capacity of two circular plates of an uniform thickness placed coaxially at any distance apart.

EXAMPLES.

1. An infinite conducting plane at zero potential is under the influence of a charge of electricity at a point O. Shew that the charge on any area of the plane is proportional to the angle it subtends at O.

2. A charged particle is placed in the space between two uninsulated planes which intersect at right angles. Sketch the sections of the equipotentials made by an imaginary plane through the charged particle, at right angles to the planes.

3. In question 2, let the particle have a charge e, and be equidistant from the planes. Shew that the total charge on a strip, of which one edge is the line of intersection of the planes, and of which the width is equal to the distance of the particle from this line of intersection, is $-\frac{1}{4}e$.

4. In question 3, the strip is insulated from the remainder of the planes, these being still to earth, and the particle is removed. Find the potential at the point formerly occupied by the particle, produced by raising the strip to potential V.

5. If two infinite plane uninsulated conductors meet at an angle of 60°, and there is a charge e at a point equidistant from each, and distant r from the line of intersection, find the electrification at any point of the planes. Shew that at a point in a principal plane through the charged point at a distance $r\sqrt{3}$ from the line of intersection, the surface density is

$$-\frac{e}{4\pi r^2}\left(\frac{3}{4}+\frac{1}{7\sqrt{7}}\right).$$

6. Two small pith balls, each of mass m, are connected by a light insulating rod. The rod is supported by parallel threads, and hangs in a horizontal position in front of an infinite vertical plane at potential zero. If the balls when charged with e units of electricity are at a distance a from the plate, equal to half the length of the rod, shew that the inclination θ of the strings to the vertical is given by

$$\tan\theta=\frac{e^2}{4mga^2}\left\{1+\frac{1}{2\sqrt{2}}\right\}.$$

7. What is the least positive charge that must be given to a spherical conductor, insulated and influenced by an external point-charge e at distance r from its centre, in order that the surface density may be everywhere positive?

8. An uninsulated conducting sphere is under the influence of an external electric charge; find the ratio in which the induced charge is divided between the part of its surface in direct view of the external charge and the remaining part.

9. A point-charge e is brought near to a spherical conductor of radius a having a charge E. Shew that the particle will be repelled by the sphere, unless its distance from the nearest point of its surface is less than $\frac{1}{2}a\sqrt{\frac{e}{E}}$, approximately.

10. A hollow conductor has the form of a quarter of a sphere bounded by two perpendicular diametral planes. Find the image of a charge placed at any point inside.

11. A conducting surface consists of two infinite planes which meet at right angles, and a quarter of a sphere of radius a fitted into the right angle. If the conductor is at zero potential, and a point-charge e is symmetrically placed with regard to the planes and the spherical surface at a great distance f from the centre, shew that the charge induced on the spherical portion is approximately $-5ea^3/\pi f^3$.

12. A point-charge is placed in front of an infinite slab of dielectric, bounded by a plane face. The angle between a line of force in the dielectric and the normal to the face of the slab is a; the angle between the same two lines in the immediate neighbourhood of the charge is β. Prove that a, β are connected by the relation

$$\sin\frac{\beta}{2} = \sqrt{\frac{2\kappa}{1+\kappa}} \sin\frac{a}{2}.$$

13. An electrified particle is placed in front of an infinitely thick plate of dielectric. Shew that the particle is urged towards the plate by a force

$$\frac{\kappa-1}{\kappa+1}\frac{e^2}{4d^2},$$

where d is the distance of the point from the plate.

14. Two dielectrics of inductive capacities κ_1 and κ_2 are separated by an infinite plane face. Charges e_1, e_2 are placed at points on a line at right angles to the plane, each at a distance a from the plane. Find the forces on the two charges, and explain why they are unequal.

15. Two conductors of capacities c_1, c_2 in air are on the same normal to the plane boundary between two dielectrics κ_1, κ_2, at great distances a, b from the boundary. They are connected by a thin wire and charged. Prove that the charge is distributed between them approximately in the ratio

$$\kappa_1\left\{\frac{1}{c_2} - \frac{\kappa_1-\kappa_2}{2b(\kappa_1+\kappa_2)} - \frac{2\kappa_2}{(\kappa_1+\kappa_2)(a+b)}\right\} : \kappa_2\left\{\frac{1}{c_1} + \frac{\kappa_1-\kappa_2}{2a(\kappa_1+\kappa_2)} - \frac{2\kappa_1}{(\kappa_1+\kappa_2)(a+b)}\right\}$$

16. A thin plane conducting lamina of any shape and size is under the influence of a fixed electrical distribution on one side of it. If σ_1 be the density of the induced charge at a point P on the side of the lamina facing the fixed distribution, and σ_2 that at the corresponding point on the other side, prove that $\sigma_1-\sigma_2=\sigma_0$, where σ_0 is the density at P of the distribution induced on an infinite plane conductor coinciding with the lamina.

17. An infinite plate with a hemispherical boss of radius a is at zero potential under the influence of a point-charge e on the axis of the boss distant f from the plate. Find the surface density at any point of the plate, and shew that the charge is attracted towards the plate with a force

$$\frac{e^2}{4f^2} + \frac{4e^2a^3f^3}{(f^4-a^4)^2}.$$

18. A conductor is formed by the outer surfaces of two equal spheres, the angle between their radii at a point of intersection being $2\pi/3$. Shew that the capacity of the conductor so formed is

$$\frac{5\sqrt{3}-4}{2\sqrt{3}}a,$$

where a is the radius of either sphere.

19. Within a spherical hollow in a conductor connected to earth, equal point-charges *e* are placed at equal distances *f* from the centre, on the same diameter. Shew that each is acted on by a force equal to

$$e^2\left[\frac{4a^3f^3}{(a^4-f^4)^2}+\frac{1}{4f^2}\right].$$

20. A hollow sphere of sulphur (of inductive capacity 3) whose inner radius is half its outer is introduced into a uniform field of electric force. Prove that the intensity of the field in the hollow will be less than that of the original field in the ratio 27 : 34.

21. A conducting spherical shell of radius *a* is placed, insulated and without charge, in a uniform field of electric force of intensity *F*. Shew that if the sphere be cut into two hemispheres by a plane perpendicular to the field, these hemispheres tend to separate and require forces equal to $\frac{9}{16}a^2F^2$ to keep them together.

22. An uncharged insulated conductor formed of two equal spheres of radius *a* cutting one another at right angles, is placed in a uniform field of force of intensity *F*, with the line joining the centres parallel to the lines of force. Prove that the charges induced on the two spheres are $\frac{11}{8}Fa^2$ and $-\frac{11}{8}Fa^2$.

23. A conducting plane has a hemispherical boss of radius *a*, and at a distance *f* from the centre of the boss and along its axis there is a point-charge *e*. If the plane and the boss be kept at zero potential, prove that the charge induced on the boss is

$$-e\left\{1-\frac{f^2-a^2}{f\sqrt{f^2+a^2}}\right\}.$$

24. A conductor is bounded by the larger portions of two equal spheres of radius *a* cutting at an angle $\frac{1}{3}\pi$, and of a third sphere of radius *c* cutting the two former orthogonally. Shew that the capacity of the conductor is

$$c+a\left(\tfrac{5}{3}-\tfrac{2}{3}\sqrt{3}\right)-ac\left\{2\left(a^2+c^2\right)^{-\frac{1}{2}}-2\left(a^2+3c^2\right)^{-\frac{1}{2}}+\left(a^2+4c^2\right)^{-\frac{1}{2}}\right\}.$$

25. A spherical conductor of internal radius *b*, which is uncharged and insulated, surrounds a spherical conductor of radius *a*, the distance between their centres being *c*, which is small. The charge on the inner conductor is *E*. Find the potential function for points between the conductors, and shew that the surface density at a point *P* on the inner conductor is

$$\frac{E}{4\pi}\left(\frac{1}{a^2}-\frac{3c\cos\theta}{b^3-a^3}\right),$$

where θ is the angle that the radius through *P* makes with the line of centres, and terms in c^2 are neglected.

26. If a particle charged with a quantity *e* of electricity be placed at the middle point of the line joining the centres of two equal spherical conductors kept at zero potential, shew that the charge induced on each sphere is

$$-2em\left(1-m+m^2-3m^3+4m^4\right),$$

neglecting higher powers of *m*, which is the ratio of the radius to the distance between the centres of the spheres.

27. Two insulated conducting spheres of radii *a*, *b*, the distance *c* of whose centres is large compared with *a* and *b*, have charges E_1, E_2 respectively. Shew that the potential energy is approximately

$$\tfrac{1}{2}\left\{\left(\frac{1}{a}-\frac{b^3}{c^4}\right)E_1^2+\frac{2}{c}E_1E_2+\left(\frac{1}{b}-\frac{a^3}{c^4}\right)E_2^2\right\}.$$

28. Shew that the force between two insulated spherical conductors of radius a placed in an electric field of uniform intensity F perpendicular to their line of centres is

$$3F^2 \frac{a^6}{c^4}\left(1 - \frac{2a^3}{c^3} - \frac{8a^6}{c^5} + \ldots\right),$$

c being the distance between their centres.

29. Two uncharged insulated spheres, radii a, b, are placed in a uniform field of force so that their line of centres is parallel to the lines of force, the distance c between their centres being great compared with a and b. Prove that the surface density at the point at which the line of centres cuts the first sphere (a) is approximately

$$\frac{F}{4\pi}\left\{3 + \frac{6b^3}{c^3} + \frac{15ab^3}{c^4} + \frac{28a^2b^3}{c^5} + \frac{57a^3b^3}{c^6} + \ldots\right\}.$$

30. A conducting sphere of radius a is embedded in a dielectric (K) whose outer boundary is a concentric sphere of radius $2a$. Shew that if the system be placed in a uniform field of force F, equal quantities of positive and negative electricity are separated of amount

$$\frac{9Fa^2K}{5K+7}.$$

31. A sphere of glass of radius a is held in air with its centre at a distance c from a point at which there is a positive charge e. Prove that the resultant attraction is

$$\tfrac{1}{2}\beta e^2\frac{a^3}{c^3}\left\{\frac{1+\beta}{c^2-a^2} + \frac{2c^2}{(c^2-a^2)^2} - \frac{c}{a^3}(1-\beta^2)\left(\frac{a}{c}\right)^\beta\int_0^{\frac{a}{c}}\frac{x^{2-\beta}dx}{1-x^2}\right\},$$

where $\beta = (K-1)/(K+1)$.

32. A conducting spherical shell of radius a is placed, insulated and without charge, in a uniform field of force of intensity F. Shew that if the sphere be cut into two hemispheres by a plane perpendicular to the field, a force $\frac{9}{16}a^2F^2$ is required to prevent the hemispheres from separating.

33. A spherical shell, of radii a, b and inductive capacity K, is placed in a uniform field of force F. Shew that the force inside the shell is uniform and equal to

$$\frac{9KF}{9K - 2(K-1)^2(b^3/a^3 - 1)}.$$

34. The surface of a conductor being one of revolution whose equation is

$$\frac{4}{r} + \frac{1}{r'} = \frac{7}{12},$$

where r, r' are the distances of any point from two fixed points at distance 8 apart, find the electric density at either vertex when the conductor has a given charge.

35. The curve

$$\frac{1}{(x^2+y^2)^{\frac{1}{2}}} - \frac{9a}{16}\left\{\frac{a+x}{\{(x+a)^2+y^2\}^{\frac{1}{2}}} + \frac{a-x}{\{(x-a)^2+y^2\}^{\frac{1}{2}}}\right\} = \frac{1}{a},$$

when rotated round the axis of x generates a single closed surface, which is made the bounding surface of a conductor. Shew that its capacity will be a, and that the surface density at the end of the axis will be $e/3\pi a^2$, where e is the total charge.

36. Two equal spheres each of radius a are in contact. Shew that the capacity of the conductor so formed is $2a\log_e 2$.

Examples   291

37. Two spheres of radii a, b are in contact, a being large compared with b. Shew that if the conductor so formed is raised to potential V, the charges on the two spheres are

$$Va\left(1 - \frac{\pi^2 b^2}{6(a+b)^2}\right) \text{ and } Va\left(\frac{\pi^2 b^2}{6(a+b)^2}\right).$$

38. A conducting sphere of radius a is in contact with an infinite conducting plane. Shew that if a unit point-charge be placed beyond the sphere and on the diameter through the point of contact at distance c from that point, the charges induced on the plane and sphere are

$$-\frac{\pi a}{c}\cot\frac{\pi a}{c} \text{ and } \frac{\pi a}{c}\cot\frac{\pi a}{c} - 1.$$

39. Prove that if the centres of two equal uninsulated spherical conductors of radius a be at a distance $2c$ apart, the charge induced on each by a unit charge at a point midway between them is

$$\sum_1^\infty (-1)^n \operatorname{sech} na,$$

where $c = a\cosh a$.

40. Shew that the capacity of a spherical conductor of radius a, with its centre at a distance c from an infinite conducting plane, is

$$a \sinh a \sum_1^\infty \operatorname{cosech} na,$$

where $c = a\cosh a$.

41. An insulated conducting sphere of radius a is placed midway between two parallel infinite uninsulated planes at a great distance $2c$ apart. Neglecting $\left(\frac{a}{c}\right)^3$, shew that the capacity of the sphere is approximately

$$a\left\{1 + \frac{a}{c}\log 2\right\}.$$

42. Two spheres of radii r_1, r_2 touch each other, and their capacities in this position are c_1, c_2. Shew that

$$c_1 = r_2\left\{f^2\sum_1^\infty\frac{1}{n^2} + f^3\sum_1^\infty\frac{1}{n^3} + f^4\sum_1^\infty\frac{1}{n^4} + \dots\right\},$$

where

$$f = \frac{r_1}{r_1 + r_2}.$$

43. A conducting sphere of radius a is placed in air, with its centre at a distance e from the plane face of an infinite dielectric. Shew that its capacity is

$$a\sinh a \sum_1^\infty\left(\frac{K-1}{K+1}\right)^{n-1}\operatorname{cosech} na,$$

where $\cosh a = c/a$.

44. A point-charge e is placed between two parallel uninsulated infinite conducting planes, at distances a and b from them respectively. Shew that the potential at a point between the planes which is at a distance z from the charge and is on the line through the charge perpendicular to the planes is

$$\frac{e}{2(a+b)}\left\{-\frac{\Gamma'\left(\frac{z}{2a+2b}\right)}{\Gamma\left(\frac{z}{2a+2b}\right)} + \frac{\Gamma'\left(\frac{2a-z}{2a+2b}\right)}{\Gamma\left(\frac{2a-z}{2a+2b}\right)} + \frac{\Gamma'\left(\frac{2b+z}{2a+2b}\right)}{\Gamma\left(\frac{2b+z}{2a+2b}\right)} - \frac{\Gamma'\left(\frac{2a+2b-z}{2a+2b}\right)}{\Gamma\left(\frac{2a+2b-z}{2a+2b}\right)}\right\}.$$

45. A spherical conductor of radius a is surrounded by a uniform dielectric K, which is bounded by a sphere of radius b having its centre at a small distance γ from the centre of the conductor. Prove that if the potential of the conductor is V, and there are no other conductors in the field, the surface density at a point where the radius makes an angle θ with the line of centres is

$$\frac{KVb}{4\pi a\{(K-1)a+b\}}\left\{1+\frac{6(K-1)\gamma a^2\cos\theta}{2(K-1)a^3+(K+2)b^3}\right\}.$$

46. A shell of glass of inductive capacity K, which is bounded by concentric spherical surfaces of radii a, b $(a<b)$, surrounds an electrified particle with charge E which is at a point Q at a small distance c from O, the centre of the spheres. Shew that the potential at a point P outside the shell at a distance r from Q is approximately

$$\frac{E}{r}+\frac{2Ec\,(b^3-a^3)(K-1)^2}{2a^3(K-1)^2-b^3(K+2)(2K+1)}\,\frac{\cos\theta}{r^2},$$

where θ is the angle which QP makes with OQ produced.

47. If the centres of the two shells of a spherical condenser be separated by a small distance d, prove that the capacity is approximately

$$\frac{ab}{b-a}\left\{1+\frac{abd^2}{(b-a)(b^3-a^3)}\right\}.$$

48. A condenser is formed of two spherical conducting sheets, one of radius b surrounding the other of radius a. The distance between the centres is c, this being so small that $(c/a)^2$ may be neglected. The surface densities on the inner conductor at the extremities of the axis of symmetry of the instrument are σ_1, σ_2, and the mean surface density over the inner conductor is $\bar{\sigma}$. Prove that

$$\frac{\sigma_2\sim\sigma_1}{\bar{\sigma}}=\frac{6ca^2}{b^3-a^3}.$$

49. The equation of the surface of a conductor is $r=a\,(1+\epsilon P_n)$, where ϵ is very small, and the conductor is placed in a uniform field of force F parallel to the axis of harmonics. Shew that the surface density of the induced charge at any point is greater than it would be if the surface were perfectly spherical, by the amount

$$\frac{3n\epsilon F}{4\pi(2n+1)}\{(n+1)P_{n+1}+(n-2)P_{n-1}\}.$$

50. A conductor at potential V whose surface is of the form $r=a\,(1+\epsilon P_n)$ is surrounded by a dielectric (K) whose boundary is the surface $r=b\,(1+\eta P_n)$, and outside this the dielectric is air. Shew that the potential in the air at a distance r from the origin is

$$\frac{KabV}{(K-1)a+b}\left[\frac{1}{r}+\frac{(2n+1)\,\epsilon a^n b^{2n+1}+(K-1)\,\eta b^n\{nb^{2n+1}+(n+1)a^{2n+1}\}}{(1+n+nK)\,b^{2n+1}+(K-1)(n+1)a^{2n+1}}\,\frac{P_n}{r^{n+1}}\right],$$

where squares and higher powers of ϵ and η are neglected.

51. The surface of a conductor is nearly spherical, its equation being

$$r=a\,(1+\epsilon S_n),$$

where ϵ is small. Shew that if the conductor is uninsulated, the charge induced on it by a unit charge at a distance f from the origin and of angular coordinates θ, ϕ is approximately

$$-\frac{a}{f}\left\{1+\left(\frac{a}{f}\right)^n\epsilon S_n\,(\theta,\phi)\right\}.$$

52. A uniform circular wire of radius a charged with electricity of line density e surrounds an uninsulated concentric spherical conductor of radius c; prove that the electrical density at any point of the surface of the conductor is

$$-\frac{e}{2c}\left(1 - 5\frac{1}{2}\frac{c^2}{a^2}P_2 + 9\frac{1.3}{2.4}\frac{c^4}{a^4}P_4 - 13\frac{1.3.5}{2.4.6}\frac{c^6}{a^6}P_6 + ...\right).$$

53. A dielectric sphere is surrounded by a thin circular wire of larger radius b carrying a charge E. Prove that the potential within the sphere is

$$\frac{E}{b}\left\{1 + \sum_{1}^{\infty}(-1)^n\frac{1+4n}{1+2n(1+K)}\frac{1.3.5...2n-1}{2.4.6...2n}\left(\frac{r}{b}\right)^{2n}P_{2n}\right\}.$$

54. If within a conductor formed by a cone of semi-vertical angle $\cos^{-1}\mu_0$ and two spherical surfaces $r=a$, $r=b$ with centres at the vertex of the cone, a charge q on the axis at distance r' from the vertex gives potential V, and if we write

$$r=ae^{-\lambda}, \quad V=Ue^{\frac{\lambda}{2}}, \quad \lambda_0=\log\frac{a}{b},$$

$$U=\sum_m\sum_n A_{mn}\sin\frac{m\pi\lambda}{\lambda_0}P_n(\mu), \quad V=\sum_n B_nP_n(\mu),$$

the summation with respect to m extending to all positive integers, and that with respect to n to all numbers integral or fractional for which $P_n(\mu_0)=0$, determine A_{mn}. Effecting the summation with respect to m, shew that when $r<r'$,

$$B_n=2q\left(r'^n-\frac{a^{2n+1}}{r'^{n+1}}\right)\left(r^n-\frac{b^{2n+1}}{r^{n+1}}\right)\bigg/(a^{2n+1}-b^{2n+1})\left[(1-\mu^2)\frac{dP_n}{dn}\frac{dP_n}{d\mu}\right]_{\mu=\mu_0},$$

and that when $r>r'$,

$$B_n=2q\left(r^n-\frac{a^{2n+1}}{r^{n+1}}\right)\left(r'^n-\frac{b^{2n+1}}{r'^{n+1}}\right)\bigg/(a^{2n+1}-b^{2n+1})\left[(1-\mu^2)\frac{dP_n}{dn}\frac{dP_n}{d\mu}\right]_{\mu=\mu_0}.$$

55. A spherical shell of radius a with a little hole in it is freely electrified to potential V. Prove that the charge on its inner surface is less than $VS/8\pi a$, where S is the area of the hole.

56. A thin spherical conducting shell from which any portions have been removed is freely electrified. Prove that the difference of densities inside and outside at any point is constant.

57. Electricity is induced on an uninsulated spherical conductor of radius a, by a uniform surface distribution, density σ, over an external concentric non-conducting spherical segment of radius c. Prove that the surface density at the point A of the conductor at the nearer end of the axis of the segment is

$$-\tfrac{1}{2}\sigma\frac{c(c+a)}{a^2}\left(1-\frac{AB}{AD}\right),$$

where B is the point of the segment on its axis, and D is any point on its edge.

58. Two conducting discs of radii a, a' are fixed at right angles to the line which joins their centres, the length of this line being r, large compared with a. If the first have potential V and the second is uninsulated, prove that the charge on the first is

$$\frac{2a\pi r^2 V}{\pi^2 r^2 - 4aa'}.$$

59. A spherical conductor of diameter a is kept at zero potential in the presence of a fine uniform wire, in the form of a circle of radius c in a tangent plane to the sphere with its centre at the point of contact, which has a charge E of electricity; prove that the electrical density induced on the sphere at a point whose direction from the centre of the ring makes an angle ψ with the normal to the plane is

$$-\frac{c^2 E \sec^3 \psi}{4\pi^2 a} \int_0^{2\pi} (a^2 + c^2 \sec^2 \psi - 2ac \tan \psi \cos \theta)^{-\frac{3}{2}} d\theta.$$

60. Prove that the capacity of a hemispherical shell of radius a is

$$a\left(\frac{1}{2} + \frac{1}{\pi}\right).$$

61. Prove that the capacity of an elliptic plate of small eccentricity e and area A is approximately

$$\sqrt{\left(\frac{A}{\pi}\right)} \frac{2}{\pi} \left(1 + \frac{e^4}{64} + \frac{e^6}{64}\right).$$

62. A circular disc of radius a is under the influence of a charge q at a point in its plane at distance b from the centre of the disc. Shew that the density of the induced distribution at a point on the disc is

$$\frac{q}{2\pi^2 R^2} \sqrt{\frac{b^2 - a^2}{a^2 - r^2}},$$

where r, R are the distances of the point from the centre of the disc and the charge.

63. An ellipsoidal conductor differs but little from a sphere. Its volume is equal to that of a sphere of radius r, its axes are $2r(1+a)$, $2r(1+\beta)$, $2r(1+\gamma)$. Shew that neglecting cubes of a, β, γ, its capacity is

$$r\{1 + \tfrac{2}{15}(a^2 + \beta^2 + \gamma^2)\}.$$

64. A prolate conducting spheroid, semi-axes a, b, has a charge E of electricity. Shew that repulsion between the two halves into which it is divided by its diametral plane is

$$\frac{E^2}{4(a^2 - b^2)} \log \frac{a}{b}.$$

Determine the value of the force in the case of a sphere.

65. One face of a condenser is a circular plate of radius a: the other is a segment of a sphere of radius R, R being so large that the plate is almost flat. Shew that the capacity is $\frac{1}{2}KR \log t_1/t_0$ where t_1, t_0 are the thickness of dielectric at the middle and edge of the condenser. Determine also the distribution of the charge.

66. A thin circular disc of radius a is electrified with charge E and surrounded by a spheroidal conductor with charge E_1, placed so that the edge of the disc is the locus of the focus S of the generating ellipse. Shew that the energy of the system is

$$\frac{1}{2} \frac{E^2}{a} B\hat{S}C + \frac{1}{2} \frac{(E+E_1)^2}{a} S\hat{B}C,$$

B being an extremity of the polar axis of the spheroid, and C the centre.

67. If the two surfaces of a condenser are concentric and coaxial oblate spheroids of small ellipticities ϵ and ϵ' and polar axes $2c$ and $2c'$, prove that the capacity is

$$cc'(c'-c)^{-2}\{c'-c+\tfrac{2}{3}(\epsilon c'-\epsilon'c)\},$$

neglecting squares of the ellipticities; and find the distribution of electricity on each surface to the same order of approximation.

68. An accumulator is formed of two confocal prolate spheroids, and the specific inductive capacity of the dielectric is Kl/ϖ, where ϖ is the distance of any point from the axis. Prove that the capacity of the accumulator is

$$\pi Kl/\log\left(\frac{a_1+b_1}{a+b}\right),$$

where a, b and a_1, b_1 are the semi-axes of the generating ellipses.

69. A thin spherical bowl is formed by the portion of the sphere $x^2+y^2+z^2=cz$ bounded by and lying within the cone $\frac{x^2}{a^2}+\frac{y^2}{b^2}=\frac{z^2}{c^2}$, and is put in connection with the earth by a fine wire. O is the origin, and C, diametrically opposite to O, is the vertex of the bowl; Q is any point on the rim, and P is any point on the great circle arc CQ. Shew that the surface density induced at P by a charge E placed at O is

$$-\frac{Ec}{4\pi abI}\frac{CQ}{OP^2(OP^2-OQ^2)^{\frac{1}{2}}},$$

where

$$I=\int_0^{\pi/2}\frac{d\theta}{(a^2\sin^2\theta+b^2\cos^2\theta)^{\frac{1}{2}}}.$$

70. Three long thin wires, equally electrified, are placed parallel to each other so that they are cut by a plane perpendicular to them in the angular points of an equilateral triangle of side $\sqrt{3}c$; shew that the polar equation of an equipotential curve drawn on the plane is

$$r^6+c^6-2r^3c^3\cos3\theta=\text{constant},$$

the pole being at the centre of the triangle and the initial line passing through one of the wires.

71. A flat piece of corrugated metal $(y=a\sin mx)$ is charged with electricity. Find the surface density at any point, and shew that it exceeds the average density approximately in the ratio $my:1$.

72. A long hollow cylindrical conductor is divided into two parts by a plane through the axis, and the parts are separated by a small interval. If the two parts are kept at potentials V_1 and V_2, the potential at any point within the cylinder is

$$\frac{V_1+V_2}{2}+\frac{V_1-V_2}{\pi}\tan^{-1}\frac{2ar\cos\theta}{a^2-r^2},$$

where r is the distance from the axis, and θ is the angle between the plane joining the point to the axis and the plane through the axis normal to the plane of separation.

73. Shew that the capacity per unit length of a telegraph wire of radius a at height h above the surface of the earth is

$$\left[4\tanh^{-1}\sqrt{\frac{h-a}{h+a}}\right]^{-1}.$$

74. An electrified line with charge e per unit length is parallel to a circular cylinder of radius a and inductive capacity K, the distance of the wire from the centre of the cylinder being c. Shew that the force on the wire per unit length is

$$\frac{K-1}{K+1}\frac{4a^2e^2}{c(c^2-a^2)}.$$

75. A cylindrical conductor of infinite length, whose cross-section is the outer boundary of three equal orthogonal circles of radius a, has a charge e per unit length. Prove that the electric density at distance r from the axis is

$$\frac{e}{6\pi a}\frac{(3r^2+a^2)(3r^2-a^2-\sqrt{6}ar)(3r^2-a^2+\sqrt{6}ar)}{r^2(9r^4-3a^2r^2+a^4)}.$$

76. If the cylinder $x^4 + y^4 = a^4$ be freely charged, shew that in free space the resultant force varies as

$$\left(r^4 + 2a^4 \cos 4\theta + \frac{a^8}{r^4} \right)^{-\frac{1}{4}},$$

where $x = r \cos \theta$, $y = r \sin \theta$; and that its direction makes with the axis of x an angle

$$\tfrac{1}{2} \tan^{-1} \left(\frac{r^4 - a^4}{r^4 + a^4} \tan 2\theta \right)^{\frac{1}{2}}.$$

77. If $\phi + i\psi = f(x + iy)$, and the curves for which $\phi =$ constant be closed, shew that the capacity C of a condenser with boundary surfaces $\phi = \phi_1$, $\phi = \phi_0$ is

$$\frac{K[\psi]}{4\pi (\phi_1 - \phi_0)}$$

per unit length, where $[\psi]$ is the increment of ψ on passing once round a ϕ-curve.

78. Using the transformation $x + iy = c \cot \tfrac{1}{2} (U + iV)$, shew that the capacity C per unit length of a condenser formed by two right circular cylinders (radii a, b), one inside the other, with parallel axes at a distance d apart, is given by

$$1/C = 2 \cosh^{-1} \left(\frac{a^2 + b^2 - d^2}{2ab} \right).$$

79. A plane infinite electric grating is made of equal and equidistant parallel thin metal plates, the distance between their successive central lines being π, and the breadth of each plate $2 \sin^{-1} \left(\frac{1}{K} \right)$. Shew that when the grating is electrified to constant potential, the potential and charge functions V, U in the surrounding space are given by the equation

$$\sin (U + iV) = K \sin (x + iy).$$

Deduce that, when the grating is to earth and is placed in a uniform field of force of unit intensity at right angles to its plane, the charge and potential functions of the portion of the field which penetrates through the grating are expressed by

$$U + iV - (x + iy),$$

and expand the potential in the latter problem in a Fourier Series.

80. A cylinder whose cross-section is one branch of a rectangular hyperbola is maintained at zero potential under the influence of a line-charge parallel to its axis and on the concave side. Prove that the image consists of three such line charges, and hence find the density of the induced distribution.

81. A cylindrical space is bounded by two coaxial and confocal parabolic cylinders, whose latera recta are $4a$ and $4b$, and a uniformly electrified line which is parallel to the generators of the cylinder intersects the axes which pass through the foci in points distant c from them $(a > c > b)$. Shew that the potential throughout the space is

$$A \log \frac{\left\{ \cosh \dfrac{\pi r^{\frac{1}{2}} \cos \dfrac{\theta}{2}}{a^{\frac{1}{2}} - b^{\frac{1}{2}}} - \cos \dfrac{\pi \left(r^{\frac{1}{2}} \sin \dfrac{\theta}{2} - c^{\frac{1}{2}} \right)}{a^{\frac{1}{2}} - b^{\frac{1}{2}}} \right\}}{\left\{ \cosh \dfrac{\pi r^{\frac{1}{2}} \cos \dfrac{\theta}{2}}{a^{\frac{1}{2}} - b^{\frac{1}{2}}} + \cos \dfrac{\pi \left(r^{\frac{1}{2}} \sin \dfrac{\theta}{2} + c^{\frac{1}{2}} - a^{\frac{1}{2}} - b^{\frac{1}{2}} \right)}{a^{\frac{1}{2}} - b^{\frac{1}{2}}} \right\}},$$

where r, θ are polar coordinates of a section, the focus being the pole. Determine A in terms of the electrification per unit length of the line.

82. An infinitely long elliptic cylinder of inductive capacity K, given by $\xi = a$ where $x + iy = c \cosh(\xi + i\eta)$, is in a uniform field P parallel to the major axis of any section. Shew that the potential at any point inside the cylinder is

$$- Px \frac{1 + \coth a}{K + \coth a}.$$

83. Two insulated uncharged circular cylinders outside each other, given by $\eta = a$ and $\eta = -\beta$ where $x + iy = c \tan \frac{1}{2}(\xi + i\eta)$, are placed in a uniform field of force of potential Fx. Shew that the potential due to the distribution on the cylinders is

$$- 2Fc \sum_{1}^{\infty} (-)^n \frac{e^{n(\eta-a)} \sinh n\beta + e^{-n(\eta+\beta)} \sinh na}{\sinh n(a+\beta)} \sin n\xi.$$

84. Two circular cylinders outside each other, given by $\eta = a$ and $\eta = -\beta$ where

$$x + iy = c \tan \frac{1}{2}(\xi + i\eta),$$

are put to earth under the influence of a line-charge E on the line $x = 0$, $y = 0$. Shew that the potential of the induced charge outside the cylinders is

$$- 4E \sum \frac{1}{n} \frac{e^{-na} \sinh n(\eta+\beta) + e^{-n\beta} \sinh n(a-\eta)}{\sinh n(a+\beta)} \cos n\xi + \text{constant},$$

the summation being taken for all odd positive integral values of n.

85. The cross-sections of two infinitely long metallic cylinders are the curves

$$(x^2 + y^2 + c^2)^2 - 4c^2 x^2 = a^4 \quad \text{and} \quad (x^2 + y^2 + c^2)^2 - 4c^2 x^2 = b^4,$$

where $b > a > c$. If they are kept at potentials V_1 and V_2 respectively, the intervening space being filled with air, prove that the surface densities per unit length of the electricity on the opposed surfaces are

$$\frac{V_1 - V_2}{4\pi a^2 \log \frac{b}{a}} \sqrt{x^2 + y^2} \quad \text{and} \quad \frac{V_2 - V_1}{4\pi b^2 \log \frac{b}{a}} \sqrt{x^2 + y^2}$$

respectively.

86. What problems are solved by the transformation

$$\begin{cases} \dfrac{d}{dt}(x + iy) = \dfrac{c(t^2 - 1)^{\frac{1}{2}}}{a^2 - t^2}, \\[2mm] \pi(\psi + i\phi) = \log \dfrac{a+t}{a-t} \end{cases}$$

where $a > 1$?

87. What problem in Electrostatics is solved by the transformation

$$x + iy = \operatorname{cn}(\phi + i\psi),$$

where ψ is taken as the potential function, ϕ being the function conjugate to it?

88. One half of a hyperbolic cylinder is given by $\eta = \pm \eta_1$, where $|\eta_1| < \frac{\pi}{2}$, and ξ, η are given in terms of the Cartesian coordinates x, y of a principal section by the transformation

$$x + iy = c \cosh(\xi + i\eta).$$

The half-cylinder is uninsulated and under the influence of a charge of density E per unit length placed along the line of internal foci. Prove that the surface density at any point of the cylinder is

$$- E / \sqrt{2} c\eta_1 \cosh \frac{\pi\xi}{2\eta_1} \sqrt{\cosh 2\xi - \cos 2\eta_1}.$$

89.   Verify that, if r, s be real positive constants, $z = x + iy$, $a = \rho e^{i\beta}$, $\frac{1}{c} = \frac{1}{r} + \frac{1}{s}$, the neld of force outside the conductors $x^2 + y^2 + 2sx = 0$, $x^2 + y^2 - 2rx = 0$ due to a doublet at the point $z = a$, outside both the circles, of strength μ and inclination a to the axis, is given by putting

$$U + iV = \frac{c\mu\pi}{\rho^2} \left\{ e^{i(a-2\beta)} \cot c\pi \left(\frac{1}{z} - \frac{1}{a} \right) - e^{-i(a-2\beta)} \cot c\pi \left(\frac{1}{z} - \frac{1}{a_0} \right) \right\},$$

where $z = a_0$ is the inverse point to $z = a$ with regard to either of the circles.

90.   A very thin indefinitely great conducting plane is bounded by a straight edge of indefinite length, and is connected with the earth. A unit charge is placed at a point P. Prove that the potential at any point Q due to the charge at P and the electricity induced on the conducting plane is

$$\frac{1}{PQ} \frac{1}{\pi} \cos^{-1} \left(-\frac{1}{\sigma} \cos \frac{\phi - \phi'}{2} \right) - \frac{1}{P'Q} \frac{1}{\pi} \cos^{-1} \left(-\frac{1}{\sigma} \cos \frac{\phi + \phi'}{2} \right),$$

where P' is the image of P in the plane, the cylindrical coordinates of Q and P are (r, ϕ, z), (r', ϕ', z'), the straight edge is the axis of z, the angles ϕ, ϕ' lie between 0 and 2π, $\phi = 0$ on the conductor,

$$\sigma = \left\{ \frac{(r+r')^2 + (z-z')^2}{4rr'} \right\}^{\frac{1}{2}},$$

and those values of the inverse functions are taken which lie between $\frac{1}{2}\pi$ and π.

91.   A semi-infinite conducting plane is at zero potential under the influence of an electric charge q at a point Q outside it. Shew that the potential at any point P is given by

$$\frac{q}{\pi\sqrt{2rr_1}} \left[\{\cosh\eta - \cos(\theta - \theta_1)\}^{-\frac{1}{2}} \tan^{-1} \sqrt{\frac{\cosh\frac{1}{2}\eta + \cos\frac{1}{2}(\theta - \theta_1)}{\cosh\frac{1}{2}\eta - \cos\frac{1}{2}(\theta - \theta_1)}} \right.$$
$$\left. - \{\cosh\eta - \cos(\theta + \theta_1)\}^{-\frac{1}{2}} \tan^{-1} \sqrt{\frac{\cosh\frac{1}{2}\eta + \cos\frac{1}{2}(\theta + \theta_1)}{\cosh\frac{1}{2}\eta - \cos\frac{1}{2}(\theta + \theta_1)}} \right],$$

where r, θ, z are the cylindrical coordinates of the point P, $(r_1, \theta_1, 0)$ of the point Q, $\theta = 0$ is the equation of the conducting plane, and

$$2rr_1 \cosh\eta = r^2 + r_1^2 + z^2.$$

Hence obtain the potential at any point due to a spherical bowl at constant potential, and shew that the capacity of the bowl is

$$\frac{a}{\pi} \left\{ 1 + \frac{\pi - a}{\sin a} \right\},$$

where a is the radius of the aperture, and a is the angle subtended by this radius at the centre of the sphere of which the bowl is a part.

92.   A thin circular conducting disc is connected to earth and is under the influence of a charge q of electricity at an external point P. The position of any point Q is denoted by the peri-polar coordinates ρ, θ, ϕ, where ρ is the logarithm of the ratio of the distances from Q to the two points R, S in which a plane QRS through the axis of the disc cuts its rim, θ is the angle RQS, and ϕ is the angle the plane QRS makes with a fixed plane through the axis of the disc, the coordinate θ having values between $-\pi$ and $+\pi$, and changing from $+\pi$ to $-\pi$ in passing through the disc. Prove or verify that the potential of the charge induced on the disc at any point Q (ρ, θ, ϕ) is

$$-\frac{q}{QP} \left[\frac{1}{2} - \frac{1}{\pi} \sin^{-1} \{\cos\frac{1}{2}(\theta - \theta_0) \operatorname{sech}\frac{1}{2}a\} \right] - \frac{q}{QP'} \left[\frac{1}{2} + \frac{1}{\pi} \sin^{-1} \{-\cos\frac{1}{2}(\theta + \theta_0) \operatorname{sech}\frac{1}{2}a\} \right],$$

where ρ_0, θ_0, ϕ_0 are the coordinates of P, θ_0 being positive, the point P' is the optical image of P in the disc, a is given by the equation

$$\cos a = \cosh \rho \cosh \rho_0 - \sinh \rho \sinh \rho_0 \cos (\phi - \phi_0),$$

and the smallest values of the inverse functions are to be taken.

Prove that the total charge on the disc is $-q\theta_0/\pi$.

Explain how to adapt the formula for the potential to the case in which the circular disc is replaced by a spherical bowl with the same rim.

93. Shew that the potential at any point P of a circular bowl, electrified to potential C, is

$$\frac{C}{\pi}\left\{\sin^{-1}\frac{AB}{AP+BP} + \frac{OA}{OP}\sin^{-1}\left(\frac{OP}{OA}\cdot\frac{AB}{AP+BP}\right)\right\},$$

where O is the centre of the bowl, and A, B are the points in which a plane through P and the axis of the bowl cuts the circular rim.

Find the density of electricity at a point on either side of the bowl and shew that the capacity is

$$\frac{a}{\pi}(a+\sin a),$$

where a is the radius of the sphere, and $2a$ is the angle subtended at the centre.

94. Two spheres are charged to potentials V_0 and V_1. The ratio of the distances of any point from the two limiting points of the spheres being denoted by e^η and the angle between them by ξ, prove that the potential at the point ξ, η is

$$V_0\sqrt{\{2(\cosh\eta-\cos\xi)\}}\sum_0^\infty\frac{\sinh(n+\tfrac12)(\beta+\eta)}{\sinh(n+\tfrac12)(\beta+a)}P_n(\cos\xi)e^{-(n+\frac12)a}$$

$$+V_1\sqrt{\{2(\cosh\eta-\cos\xi)\}}\sum_0^\infty\frac{\sinh(n+\tfrac12)(a-\eta)}{\sinh(n+\tfrac12)(\beta+a)}P_n(\cos\xi)e^{-(n+\frac12)\beta},$$

where $\eta=a$, $\eta=-\beta$ are the equations of the spheres. Hence find the charge on either sphere.

CHAPTER IX

STEADY CURRENTS IN LINEAR CONDUCTORS

PHYSICAL PRINCIPLES.

338. IF two conductors charged with electricity to different potentials are connected by a conducting wire, we know that a flow of electricity will take place along the wire. This flow will tend to equalise the potentials of the two conductors, and when these potentials become equal the flow of electricity will cease. If we had some means by which the charges on the conductors could be replenished as quickly as they were carried away by conduction through the wire, then the current would never cease. The conductors would remain permanently at different potentials, and there would be a steady flow of electricity from one to the other. Means are known by which two conductors can be kept permanently at different potentials, so that a steady flow of electricity takes place through any conductor or conductors joining them. We accordingly have to discuss the mathematical theory of such currents of electricity.

We shall begin by the consideration of the flow of electricity in linear conductors, by a linear conductor being meant one which has a definite cross-section at every point. The commonest instance of a linear conductor is a wire.

339. DEFINITION. *The strength of a current at any point in a wire or other linear conductor, is measured by the number of units of electricity which flow across any cross-section of the conductor per unit time.*

If the units of electricity are measured in Electrostatic Units, then the current also will be measured in Electrostatic Units. These, however, as will be explained later, are not the units in which currents are usually measured in practice.

Let P, Q be two cross-sections of a linear conductor in which a steady current is flowing, and let us suppose that no other conductors touch this conductor between P and Q. Then, since the current is, by hypothesis, steady, there must be no accumulation of electricity in the region of the conductor between P and Q. Hence the rate of flow into the section of the conductor across P must be exactly equal to the rate of flow out of this section across Q. Or, the currents at P and Q must be equal. Hence we speak of the current in a conductor, rather than of the current at a point in a conductor. For, as we pass along a conductor, the current cannot change except at points at which the conductor is touched by other conductors.

Ohm's Law.

340. In a linear conductor in which a current is flowing, we have electricity in motion at every point, and hence must have a continuous variation in potential as we pass along the conductor. This is not in opposition to the result previously obtained in Electrostatics, for in the previous analysis it had to be assumed that the electricity was at rest. In the present instance, the electricity is not at rest, being in fact kept in motion by the difference of potential under discussion.

The analogy between potential and height of water will perhaps help. A lake in which the water is at rest is analogous to a conductor in which electricity is in equilibrium. The theorem that the potential is constant over a conductor in which electricity is in equilibrium, is analogous to the hydrostatic theorem that the surface of still water must all be at the same level. A conductor through which a current of electricity is flowing finds its analogue in a stream of running water. Here the level is not the same at all points of the river—it is the difference of level which causes the water to flow. The water will flow more rapidly in a river in which the gradient is large than in one in which it is small. The electrical analogy to this is expressed by Ohm's Law.

OHM'S LAW. *The difference of potential between any two points of a wire or other linear conductor in which a current is flowing, stands to the current flowing through the conductor in a constant ratio, which is called the resistance between the two points.*

It is here assumed that there is no junction with other conductors between these two points, so that the current through the conductor is a definite quantity.

341. Thus if C is the current flowing between two points P, Q at which the potentials are V_P, V_Q, we have

$$V_P - V_Q = CR \quad \dots\dots\dots\dots\dots\dots\dots(264),$$

where R is the resistance between the points P and Q. Very delicate experiments have failed to detect any variation in the ratio

(fall of potential)/(current),

as the current is varied, and this justifies us in speaking of the resistance as a definite quantity associated with the conductor. The resistance depends naturally on the positions of the two points by which the current enters and leaves the conductor, but when once these two points are fixed the resistance is independent of the amount of current. In general, however, the resistance of a conductor varies with the temperature, and for some substances, of which selenium is a notable example, it varies with the amount of light falling on the conductor.

The Voltaic Cell

342. The simplest arrangement by which a steady flow of electricity can be produced is that known as a Voltaic Cell. This is represented diagram-matically in Fig. 95. A voltaic cell consists essentially of two conductors

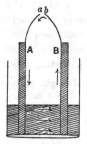

Fig. 95.

A, B of different materials, placed in a liquid which acts chemically on at least one of them. On establishing electrical contact between the two ends of the conductors which are out of the liquid, it is found that a continuous current flows round the circuit which is formed by the two conductors and the liquid, the energy which is required to maintain the current being derived from chemical action in the cell.

To explain the action of the cell, it will be necessary to touch on a subject of which a full account would be out of place in the present book. As an experimental fact it is found that two conductors of dissimilar material, when placed in contact, have different potentials when there is no flow of electricity from one to the other*, although of course the potential over the whole of either conductor must be constant. In the light of this experimental fact, let us consider the conditions prevailing in the voltaic cell before the two ends *a, b* of the conductors are joined.

So long as the two conductors *A, B* and the liquid *C* do not form a closed circuit, there can be no flow of electricity. Thus there is electric equilibrium,

---

* For a long time there has been a divergence of opinion as to whether this difference of potential is not due to the chemical change at the surfaces of the conductors, and therefore dependent on the presence of a layer of air or other third substance between the conductors. It seems now to be almost certain that this is the case, but the question is not one of vital importance as regards the *mathematical* theory of electric currents.

and the three conductors have definite potentials V_A, V_B, V_C. The difference of potential between the two "terminals" a, b is $V_A - V_B$, but the peculiarity of the voltaic cell is that this difference of potential is not equal to the difference of potential between the two conductors when they are placed in contact and are in electrical equilibrium without the presence of the liquid C. Thus on electrically joining the points a, b in the voltaic cell electrical equilibrium is an impossibility, and a current is established in the circuit which will continue until the physical conditions become changed or the supply of chemical energy is exhausted.

Electromotive Force.

343. Let A, B, C be any three conductors arranged so as to form a closed circuit. Let V_{AB} be the contact difference of potential between A and B when there is electric equilibrium, and let V_{BC}, V_{CA} have similar meanings.

If the three substances can be placed in a closed circuit without any current flowing, then we can have equilibrium in which the three conductors will have potentials V_A, V_B, V_C, such that

$$V_A - V_B = V_{AB}; \quad V_B - V_C = V_{BC}; \quad V_C - V_A = V_{CA}.$$

Thus we must have

$$V_{AB} + V_{BC} + V_{CA} = 0,$$

a result known as *Volta's Law.*

If, however, the three conductors form a voltaic cell, the expression on the left-hand of the above equation does not vanish, and its value is called the *electromotive force* of the cell. Denoting the electromotive force by E, we have

$$V_{AB} + V_{BC} + V_{CA} = E \quad\quad\quad\quad\dots\dots\dots\dots\dots(265).$$

We accordingly have the following definition:

DEFINITION. *The Electromotive Force of a cell is the algebraic sum of the discontinuities of potential encountered in passing in order through the series of conductors of which the cell is composed.*

Clearly an electromotive force has direction as well as magnitude. It is usual to speak of the two conductors which pass into the liquid as the high-potential terminal and the low-potential terminal, or sometimes as the positive and negative terminals. Knowing which is the positive or high-potential terminal, we shall of course know the direction of the electromotive force.

344. If the conductors C, A of a voltaic cell ABC are separated, and then joined by a fourth conductor D, such that there is no chemical action between D and the conductors C or A, it will easily be seen that the sum of the discontinuities in the new circuit is the same as in the old.

For by hypothesis CDA can form a closed circuit in which no chemical action can occur, and therefore in which there can be electric equilibrium. Hence we must have

$$V_{CD} + V_{DA} + V_{AC} = 0 \quad \dots\dots\dots\dots\dots\dots(266).$$

Moreover the sum of all the discontinuities in the circuit is

$$V_{AB} + V_{BC} + V_{CD} + V_{DA}$$
$$= V_{AB} + V_{BC} - V_{AC}, \text{ by equation (266)}$$
$$= V_{AB} + V_{BC} + V_{CA}$$
$$= E, \text{ by equation (265)},$$

proving the result. A similar proof shews that we may introduce any series of conductors between the two terminals of a cell, and so long as there is no chemical action in which these new conductors are involved, the sum of all the discontinuities in the circuit will be constant, and equal to the electromotive force of the cell.

Let $ABC\dots MN$ be any series of conductors, including a voltaic cell, and let the material of N be the same as that of A. If N and A are joined we obtain a closed circuit of electromotive force E, such that

$$V_{AB} + V_{BC} + \dots + V_{MN} + V_{NA} = E.$$

Moreover $V_{NA} = 0$, since the material of N and A is the same. Thus the relation may be rewritten as

$$V_{AB} + V_{BC} + \dots + V_{MN} = E \quad \dots\dots\dots\dots\dots(267).$$

In the open series of conductors $ABC\dots MN$, there can be no current, so that each conductor must be at a definite uniform potential. If we denote the potentials by $V_A, V_B, \dots V_M, V_N$, we have

$$V_A - V_B = V_{AB},$$
$$\dots$$
$$V_M - V_N = V_{MN}.$$

Hence equation (267) becomes

$$V_A - V_N = E.$$

We now see that *the electromotive force of a cell is the difference of potential between the ends of the cell when the cell forms an open circuit, and the materials of the two ends are the same.*

A series of cells, joined in series so that the high-potential terminal of one is in electrical contact with the low-potential terminal of the next, and so on, is called a battery of cells, or an " electric battery " arranged in series.

It will be clear from what has just been proved, that the electromotive force of such a battery of cells is equal to the sum of the electromotive forces of the separate cells of the series.

Units.

345. On the electrostatic system, a unit current has been defined to be a current such that an electrostatic unit of electricity crosses any selected cross-section of a conductor in unit time. For practical purposes, a different unit, known as the *ampère*, is in use. The ampère is equal very approximately to 3×10^9 electrostatic units of current (see below, § 587).

To form some idea of the actual magnitude of this unit, it may be stated that the amount of current required to ring an electric bell is about half an ampère. About the same amount is required to light a 200 c.p. 240-volt gas-filled lamp.

As an electromotive force is of the same physical nature as a difference of potential, the electrostatic unit of electromotive force is taken to be the same as that of potential. The practical unit is about $\frac{1}{300}$ of the electrostatic unit, and is known as the *volt* (see below, § 587).

It may be mentioned that the electromotive force of a single voltaic cell is generally intermediate between one and two volts; the electromotive force which produces a perceptible shock in the human body is about 30 volts, while an electromotive force of 500 volts or more is dangerous to life. Both of these latter quantities, however, vary enormously with the condition of the body, and particularly with the state of dryness or moisture of the skin. The electromotive force used to work an electric bell is commonly 6 or 8 volts, while an electric light installation will generally have a voltage of about 240 volts.

The unit of resistance, in all systems of units, is taken to be a resistance such that unit difference of potential between its extremities produces unit current through the conductor. We then have, by Ohm's Law,

$$\text{current} = \frac{\text{difference of potential at extremities}}{\text{resistance}} \quad \dots\dots\dots(268).$$

In the practical system of units, the unit of resistance is called the *ohm*. From what has already been said, it follows that when two points having a potential-difference of one volt are connected by a resistance of one ohm, the current flowing through this resistance will be one ampère. In this case the difference of potential is $\frac{1}{300}$ electrostatic units, and the current is 3×10^9 electrostatic units, so that by relation (268), it follows that one ohm must be equal to $\dfrac{1}{9 \times 10^{11}}$ electrostatic units of resistance (see below, § 587).

Some idea of the amount of this unit may be gathered from the statement that the resistance of a mile of ordinary telegraph wire is about 10 ohms. The resistance of a good telegraph insulator may be billions of ohms.

PHYSICAL THEORIES OF CONDUCTION.

Electron-theory of conduction.

345 a. As has been already explained (§ 28), the modern view of electricity regards a current of electricity as a material flow of electric charges. In all conductors except a small class known as electrolytic conductors (see below, § 345 *b*), these charged bodies are believed to be identical with the electrons.

In a solid some of the electrons are supposed to be permanently bound to particular atoms or molecules, whilst others, spoken of as "free" electrons, move about in the interstices of the solid, continually having their courses changed by collisions with the molecules. Both kinds of electrons will be influenced by the presence of an electric field. It is probable that the restricted motions of the "bound" electrons account for the phenomenon of inductive capacity (§ 151) whilst the unrestricted motion of the free electrons explains the phenomenon of electric conductivity.

Even when no electric forces are applied, the free electrons move about through a solid, but they move at random in all directions, so that as many electrons move from right to left as from left to right and the resultant current is *nil*. If an electric force is applied to the conductor, each electron has superposed on to its random motion a motion impressed on it by the electric force, and the electrons as a whole are driven through the conductor by the continued action of the electric force. If it were not for their collisions with the molecules of the conductor, the electrons would gain indefinitely in momentum under the action of the impressed electric force, but the effect of collisions is continually to check this growth of momentum.

Let us suppose that there are N electrons per unit length of the conductor, and that at any moment these have an average forward velocity u through the material of the conductor. If m is the mass of each electron, the total momentum of the moving electrons will be Nmu. The rate at which this total momentum is checked by collisions will be proportional to N and to u, and may be taken to be $N\gamma u$. The rate at which the momentum is increased by the electric forces acting is NXe, where X is the electric intensity and e is the charge, measured positively, of each electron. Thus we have the equation

$$\frac{d}{dt}(Nmu) = NXe - N\gamma u \quad \dots\dots\dots\dots\dots\dots(a).$$

In unit time the number of electrons which pass any fixed point in the conductor is Nu, so that the total flow of electricity per unit time past any point is Neu. This is by definition equal to the current in the conductor, so that if we call this i, we have

$$Neu = i \quad \dots\dots\dots\dots\dots\dots\dots\dots\dots(b).$$

This enables us to reduce equation (*a*) to the form

$$\frac{di}{dt} = \frac{Ne^2}{m}\left(X - \frac{\gamma}{Ne^2}i\right) \quad \ldots\ldots\ldots\ldots\ldots\ldots(c).$$

The equation shews that if a steady electric force is applied, such that the intensity at any point is X, the current will not increase indefinitely but will remain stationary after it has reached a value i given by

$$i = \frac{Ne^2}{\gamma}X.$$

If V is the potential at any point of a conducting wire, and if s is a coordinate measured along the wire, we have $X = -\frac{\partial V}{\partial s}$, so that

$$-\frac{\partial V}{\partial s} = \frac{\gamma}{Ne^2}i.$$

Integrating between any two points P and Q of the conductor, we have

$$V_P - V_Q = i\int_P^Q \frac{\gamma}{Ne^2}ds.$$

This is the electron-theory interpretation of equation (264), and explains how the truth of Ohm's Law is involved in the modern conception of the nature of an electric current. It will be noticed that on this view of the matter, Ohm's Law is only true for steady currents.

We notice that the resistance of the conductor, on this theory, is γ/Ne^2 per unit length. Thus, generally speaking, bodies in which there are many free electrons ought to be good conductors, and conversely.

The charge on the electron being $4\cdot803 \times 10^{-10}$ electrostatic units, we may notice that a current of one ampère (3×10^9 electrostatic units of current) is one in which $6\cdot3 \times 10^{18}$ electrons pass any given point of the conductor every second. Consider a conductor in which the number of electrons per cubic centimetre is 10^{21} (cf. § 615, below). Then in a wire of 1 square mm. cross-section there are 10^{19} electrons per unit length, so that the average velocity of these when the wire is conveying a current of 1 ampère is of the order of one cm. per sec. This average velocity is superposed on to a random velocity which is known to be of the order of magnitude of 10^7 cms. per sec., so that the additional velocity produced by even a strong current is only very slight in comparison with the normal velocity of agitation of the electrons.

Electrolytic conduction.

345 *b*. Besides the type of electric conduction just explained, there is a second, and entirely different type, known as Electrolytic conduction, the distinguishing characteristic of which is that the passage of a current is accompanied by chemical change in the conductor.

For instance, if a current is passed through a solution of potassium chloride in water, it will be found that some of the salt is divided up by the passage of the current into its chemical constituents, and that the potassium appears solely at the point at which the current leaves the liquid, while the chlorine similarly appears at the point at which the current enters. It thus appears that during the passage of an electric current, there is an actual transport of matter through the liquid, chlorine moving in one direction and potassium in the other. It is moreover found by experiment that the total amount, whether of potassium or chlorine, which is liberated by any current is exactly proportional to the amount of electricity which has flowed through the electrolyte.

These and other facts suggested to Faraday the explanation, now universally accepted, that the carriers of the current are identical with the matter which is transported through the electrolyte. For instance, in the foregoing illustration, each atom of potassium carries a positive charge to the point where the current leaves the liquid, while each atom of chlorine, moving in the direction opposite to that of the current, carries a negative charge. The process is perhaps explained more clearly by regarding the total current as made up of two parts, first a positive current and second a negative current flowing in the reverse direction. Then the atoms of chlorine are the carriers of the negative current, and the atoms of potassium are the carriers of the positive current.

Electrolytes may be solid, liquid, or gaseous, but in most cases of importance they are liquids, being solutions of salts or acids. The two parts into which the molecule of the electrolyte is divided are called the ions (ἰών), that which carries the positive current being called the positive ion, and the other being called the negative ion. The point at which the current enters the electrolyte is called the anode, the point at which it leaves is called the cathode. The two ions are also called the anion or cation according as they give up their charges at the anode or cathode respectively. Thus we have

The anion carries − charge against current, and delivers it at the anode,

The cation carries + charge with current, and delivers it at the cathode.

When potassium chloride is the electrolyte, the potassium atom is the cation, and the chlorine atom is the anion. If experiments are performed with different chlorides (say of potassium, sodium, and lithium), it will be found that the amount of chlorine liberated by a given current is in every case the same, while the amounts of potassium, sodium, or lithium, being exactly those required to combine with this fixed amount of chlorine, are necessarily proportional to their atomic weights. This suggests that each atom of chlorine, no matter what the electrolyte may be in which it occurs, always carries the same negative charge, say $-e$, while each atom of potassium, sodium, or lithium carries the same positive charge, say + *E*. Moreover *E* and *e* must be equal, or else each undissociated molecule of the electrolyte would have to be supposed to carry a charge *E* − *e*, whereas its charge is known to be *nil*.

It is found to be a general rule that every anion which is chemically monovalent carries the same charge − *e*, while every monovalent cation carries a charge + *e*. Moreover divalent ions carry charges ± 2*e*, trivalent ions carry charges ± 3*e*, and so on.

As regards the actual charges carried, it is found that one ampère of current flowing for one second through a salt of silver liberates 0·001118 grammes of silver. Silver is monovalent and its atomic weight is 107·88 (referred to O = 16), so that the amount of any other monovalent element of atomic weight *m* deposited by the same current will be 0·00001036 × *m* grammes. It follows that the passage of one electrostatic unit of electricity will result in the liberation of $\dfrac{0\cdot00001036 \times m}{3 \times 10^9}$, or 3·45 × 10⁻¹⁵ × *m* grammes of the substance.

We can calculate from these data how many ions are deposited by one unit of current, and hence the amount of charge carried by each ion. It is found that, to within the limits of experimental error, the negative charge carried by each monovalent anion is exactly equal to the charge carried by the electron. It follows that each monovalent anion has associated with it one electron in excess of the number required to give it zero charge, while each monovalent cation has a deficiency of one electron; divalent ions have an excess or deficiency of two electrons, and so on.

345 c. Ohm's Law appears, in general, to be strictly true for the resistance of electrolytes. In the light of the explanation of Ohm's Law given in § 345 *a*, this will be seen to suggest that the ions are free to move as soon as an electric intensity, no matter how small, begins to act on them. They must therefore be already in a state of dissociation; no part of the electric intensity is required to effect the separation of the molecule into ions.

Other facts confirm this conclusion, such as for instance the fact that various physical properties—electric conductivity, colour, optical rotatory power, etc.—are additive in the sense that the amount possessed by the whole electrolyte is the sum of the amounts known to be possessed by the separate ions.

We may therefore suppose that as soon as an electric force begins to act, all the positive ions begin to move in the direction of the electric force, while all the negative ions begin to move in the opposite direction. Let us suppose the average velocities of the positive and negative ions to be *u*, *v* respectively, and let us suppose that there are *N* of each per unit length of the electrolyte measured along the path of the current. Then across any cross-section of the electrolyte there pass in unit time *Nu* positive ions each carrying a charge *se* in the direction in which the current is measured, and Nv negative ions each carrying a charge $-se$ in the reverse direction, s being the valency of each ion. It follows that the total current is given by

$$i = Nse\,(u + v) \quad \dots\dots\dots\dots\dots\dots\dots\dots\dots\dots(d).$$

Each unit of time Nu positive ions cross a cross-section close to the anode, having started from positions between this cross-section and the anode. Thus each unit of time Nu molecules are separated in the neighbourhood of the anode, and similarly Nv molecules are separated in the neighbourhood of the cathode. The concentration of the salt is accordingly weakened both at the anode and at the cathode, and the ratio of the amounts of these weakenings is that of $u : v$. This provides a method of determining the ratio of $u : v$.

Also equation (d) provides a method of determining $u + v$, for i can be readily measured, and Nse is the total charge which must be passed through the electrolyte to liberate the ions in unit length, and this can be easily determined.

Knowing $u + v$ and the ratio $u : v$, it is possible to determine u and v. The following table gives results of the experiments of Kohlrausch on three chlorides of alkali metals, for different concentrations, the current in each case being such as to give a potential fall of 1 volt per centimetre.

Concentration	Potassium chloride		Sodium chloride		Lithium chloride	
	u	v	u	v	u	v
0	660	690	450	690	360	690
·0001	654	681	448	681	356	681
·001	643	670	440	670	343	670
·01	619	644	415	644	318	644
·03	597	621	390	623	298	619
·1	564	589	360	592	259	594

[*The unit in every case is a velocity of* 10^{-6} *cms. per second.*]

We notice that when the solution is weak, the velocity of the chlorine ion is the same, no matter which electrolyte it has originated in. This gives, perhaps, the best evidence possible that the conductivity of the electrolyte is the sum of the conductivities of the chlorine and of the metal separately.

By arranging for the ions to produce discoloration of the electrolyte as they move through it, Lodge, Whetham and others have been able to observe the velocity of motion of the ions directly, and in all cases the observed velocities have agreed, within the limits of experimental error, with the theoretically determined values.

Conduction through gases.

345 d. In a gas in its normal state, an electric current cannot be carried in either of the ways which are possible in a solid or a liquid, and it is consequently found that a gas under ordinary conditions conducts electricity only in a very feeble degree. If however Röntgen rays are passed through the gas, or ultra-violet light of very short wave-length, or a stream of the rays from radium or one of the radio-active metals, then it is found that the gas acquires considerable conducting powers, for a time at least. For this kind of conduction it is found that Ohm's Law is not obeyed, the relation between the current and the potential-gradient being an extremely complex one.

The complicated phenomena of conduction through gases can all be explained on the hypothesis that the gas is conducting only when "ionised," and the function of the Röntgen rays, ultra-violet light, etc. is supposed to be that of dividing up some of the molecules into their component ions. The subject of conduction through gases is too extensive to be treated here. In what follows it is assumed that the conductors under discussion are not gases, so that Ohm's Law will be assumed to be obeyed throughout.

KIRCHHOFF'S LAWS.

346. Problems occur in which the flow of electricity is not through a single continuous series of conductors: there may be junctions of three or more conductors at which the current of electricity is free to distribute itself between different paths, and it may be important to determine how the electricity will pass through a network of conductors containing junctions.

The first principle to be used is that, since the currents are supposed steady, there can be no accumulation of electricity at any point, so that the sum of all the currents which enter any junction must be equal to the sum of all the currents which leave it. Or, if we introduce the convention that currents flowing into a junction are to be counted as positive, while those leaving it are to be reckoned negative, then we may state the principle in the form:

The algebraic sum of the currents at any junction must be zero.

From this law it follows that any network of currents, no matter how complicated, can be regarded as made up of a number of closed currents, each of uniform strength throughout its length. In some conductors, two or more of these currents may of course be superposed.

Let the various junctions be denoted by A, B, C, ..., and let their potentials be V_A, V_B, V_C, ... Let R_{AB} be the resistance of any single conductor connecting two junctions A and B, and let C_{AB} be the current flowing through it from A to B. Let us select any path through the network of conductors, such as to start from a junction and bring us back to the starting point, say $ABC...NA$. Then on applying Ohm's Law to the separate conductors of which this path is formed, we obtain (§ 341)

$$V_A - V_B = C_{AB} R_{AB},$$
$$V_B - V_C = C_{BC} R_{BC},$$
$$\dotsc\dotsc\dotsc\dotsc\dotsc$$
$$V_N - V_A = C_{NA} R_{NA}.$$

By addition we obtain $\qquad \Sigma CR = 0$ ....(269),

where the summation is taken over all the conductors which form the closed circuit.

In this investigation it has been assumed that there are no discontinuities of potential, and therefore no batteries, in the selected circuit. If discontinuities occur, a slight modification will have to be made. We shall treat points at which discontinuities occur as junctions, and if A is a junction of this kind, the potentials at A on the two sides of the surface of separation between the two conductors will be denoted by V_A and V_A'. Then, by Ohm's Law, we obtain for the falls of potential in the different conductors of the circuit,

$$V_A' - V_B = C_{AB} R_{AB},$$
$$V_B' - V_C = C_{BC} R_{BC}, \text{ etc.,}$$

and by addition of these equations

$$\Sigma (V_A' - V_A) = \Sigma CR.$$

The left-hand member is simply the sum of all the discontinuities of potential met in passing round the circuit, each being measured with its proper sign. It is therefore equal to the sum of the electromotive forces of all the batteries in the circuit, these also being measured with their proper signs.

Thus we may write $\qquad \Sigma CR = \Sigma E$ ....(270),

where the summation in each term is taken round any closed circuit of conductors, and this equation, together with

$$\Sigma C = 0 \qquad \text{..............................(271),}$$

in which the summation now refers to all the currents entering or leaving a single junction, suffices to determine the current in each conductor of the network.

Equation (271) expresses what is known as Kirchhoff's First Law, while equation (270) expresses the Second Law.

Conductors in Series.

347. When all the conductors form a single closed circuit, the current through each conductor is the same, say C, so that equation (270) becomes

$$C\Sigma R = \Sigma E.$$

The sum ΣR is spoken of as the "resistance of the circuit," so that the current in the circuit is equal to the total electromotive force divided by the total resistance. Conductors arranged in such a way that the whole current passes through each of them in succession are said to be arranged "in series."

Conductors in Parallel.

348. It is possible to connect any two points A, B by a number of conductors in such a way that the current divides itself between all these

Fig. 96.

conductors on its journey from A to B, no part of it passing through more than one conductor. Conductors placed in this way are said to be arranged "in parallel."

Let us suppose that the two points A, B are connected by a number of conductors arranged in parallel. Let R_1, R_2, ... be the resistances of the conductors, and C_1, C_2, ... the currents flowing through them. Then if V_A, V_B are the potentials at A and B, we have, by Ohm's Law,

$$V_A - V_B = C_1 R_1 = C_2 R_2 = \ldots.$$

The total current which enters at A is $C_1 + C_2 + \ldots$, say C. Thus we have

$$V_A - V_B = \frac{C_1}{\dfrac{1}{R_1}} = \frac{C_2}{\dfrac{1}{R_2}} = \ldots = \frac{C}{\dfrac{1}{R_1} + \dfrac{1}{R_2} + \ldots}.$$

The arrangement of conductors in parallel is therefore seen to offer the same resistance to the current as a single conductor of resistance

$$\frac{1}{\dfrac{1}{R_1} + \dfrac{1}{R_2} + \ldots}.$$

The reciprocal of the resistance of a conductor is called the "conductivity" of the conductor. The conductivity of the system of conductors arranged in parallel is $\dfrac{1}{R_1} + \dfrac{1}{R_2} + \ldots$, and is therefore equal to the sum of the conductivities of the separate conductors. Also we have seen that the current divides itself between the different conductors in the ratio of their conductivities

MEASUREMENTS.

The Measurement of Current.

349. The instrument used for measuring the current passing in a circuit at any given instant is called a galvanometer. The theory of this instrument will be given in a later chapter (Chap. XIII).

For measuring the total quantity of electricity passing within a given time an instrument called a voltameter is sometimes used. The current, in passing through the voltameter, encounters a number of discontinuities of potential in crossing which electrical energy becomes transformed into chemical energy. Thus a voltameter is practically a voltaic cell run backwards. On measuring the amount of chemical energy which has been stored in the voltameter, we obtain a measure of the total quantity of electricity which has passed through the instrument.

The Measurement of Resistance.

350. *The Resistance Box.* A resistance box is a piece of apparatus which consists essentially of a collection of coils of wire of known resistances, arranged so that any combination of these coils can be arranged in series. The most usual arrangement is one in which the two extremities of each coil are brought to the upper surface of the box, and are there connected to a thick band of copper which runs over the surface of the box. This

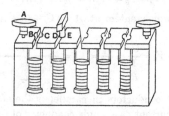

Fig. 97.

band of copper is continuous, except between the two terminals of each coil, and in these places the copper is cut away in such a way that a copper plug can be made to fit exactly into the gap, and so put the two sides of the gap in electrical contact through the plug. The arrangement is shewn diagrammatically in fig. 97. When the plug is inserted in any gap DE, the plug and the coil beneath the gap DE form two conductors in parallel connecting the points D and E. Denoting the resistances of the coil and plug by R_c, R_p, the resistance between D and E will be

$$\frac{1}{\frac{1}{R_c} + \frac{1}{R_p}},$$

and since R_p is very small, this may be neglected. When the plug is removed, the resistance from D to E may be taken to be the resistance of the coil. Thus the resistance of the whole box will be the sum of the resistances of all the coils of which the plugs have been removed.

351. *The Wheatstone Bridge.* This is an arrangement by which it is possible to compare the resistances of conductors, and so determine an unknown resistance in terms of known resistances.

The "bridge" is represented diagrammatically in fig. 98. The current enters it at A and leaves it at D, these points being connected by the lines

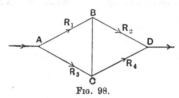

Fig. 98.

ABD, ACD arranged in parallel. The line AD is composed of two conductors AB, BD of resistances R_1, R_2, and the line ACD is similarly composed of two conductors AC, CD of resistances R_3, R_4.

If current is allowed to flow through this arrangement of conductors, it will not in general happen that the points B and C will be at the same potential, so that if B and C are connected by a new conductor, there will usually be a current flowing through BC. The method of using the Wheatstone bridge consists in varying the resistances of one or more of the conductors R_1, R_2, R_3, R_4 until no current flows through the conductor BC.

When the bridge is adjusted in this way, the points B, C must be at the same potential, say v. Let V_A, V_D denote the potentials at A and D, and let the current through ABD be C. Then, by Ohm's Law,

$$V_A - v = CR_1, \qquad v - V_D = CR_2,$$

so that

$$\frac{R_1}{R_2} = \frac{V_A - v}{v - V_D}.$$

From a similar consideration of the flow in ACD, we obtain

$$\frac{R_3}{R_4} = \frac{V_A - v}{v - V_D},$$

so that we must have

$$\frac{R_1}{R_2} = \frac{R_3}{R_4} \quad \dots \dots \dots \dots \dots \dots \dots \dots \dots (272),$$

as the condition to be satisfied between the resistances when there is no current in BC.

Clearly by adjusting the bridge in this way we can determine an unknown resistance R_1 in terms of known resistances R_2, R_3, R_4. In the simplest form of Wheatstone's bridge, the line ACD is a single uniform wire, and the position of the point C can be varied by moving a "sliding contact" along the wire. The ratio of the resistances $R_3 : R_4$ is in this case simply the ratio of the two lengths AC, CD of the wire, so that the ratio $R_1 : R_2$ can be found by sliding the contact C along the wire ACD until there is observed to be no current in BC, and then reading the lengths AC and CD.

EXAMPLES OF CURRENTS IN A NETWORK.

I. *Wheatstone's Bridge not in adjustment.*

352. The condition that there shall be no current in the "bridge" BC in fig. 98 has been seen to be that given by equation (272).

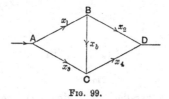

FIG. 99.

Suppose that this condition is not satisfied, and let us examine the flow of currents which then takes place in the network of conductors. Let the conductors AB, BD, AC, CD as before be of resistances R_1, R_2, R_3, R_4, and let the currents flowing through them be denoted by x_1, x_2, x_3, x_4. Let the bridge BC be of resistance R_b, and let the current flowing through it from B to C be x_b.

From Kirchhoff's Laws, we obtain the following equations:

(Law I, point B) $\qquad\qquad x_1 - x_2 - x_b = 0$ ....(273),

(Law I, point C) $\qquad\qquad x_3 - x_4 + x_b = 0$ ....(274),

(Law II, circuit ABC) $\quad x_1 R_1 + x_b R_b - x_3 R_3 = 0$ ....(275),

(Law II, circuit BCD) $\quad x_b R_b + x_4 R_4 - x_2 R_2 = 0$ ....(276).

These four equations enable us to determine the ratios of the five currents x_1, x_2, x_3, x_4, x_b. We may begin by eliminating x_2 and x_4 from equations (273), (274) and (276), and obtain

$$x_b (R_b + R_2 + R_4) + x_3 R_4 - x_1 R_2 = 0,$$

and from this and equation (275),

$$\frac{x_b}{R_2 R_3 - R_1 R_4} = \frac{x_3}{R_1 (R_b + R_2 + R_4) + R_b R_2} = \frac{x_1}{R_3 (R_b + R_2 + R_4) + R_b R_4}$$
$$............(277).$$

The ratios of the other currents can be written down from symmetry.

If the total current entering at A is denoted by X, we have $X = x_1 + x_2$. Thus if each of the fractions of equations (277) is denoted by θ,

$$X = \theta \{(R_1 + R_3)(R_2 + R_4) + R_b(R_1 + R_2 + R_3 + R_4)\} \quad\ldots\ldots(278),$$

and this gives θ, and hence the actual values of the currents, in terms of the total current entering at A.

The fall of potential from A to D is given by

$$V_A - V_D = R_1 x_1 + R_2 x_2,$$

and from equations (277) this is found to reduce to

$$V_A - V_D = \lambda\theta,$$

where

$$\lambda = R_1 R_3 (R_2 + R_4) + R_2 R_4 (R_3 + R_1) + R_b (R_1 R_3 + R_2 R_4 + R_1 R_4 + R_2 R_3),$$

so that λ is the sum of the products of the five resistances taken three at a time, omitting the two products of the three resistances which meet at the points B and C.

There is now a current X flowing through the network, and having a fall of potential $V_A - V_D$. Hence the equivalent resistance of the network

$$= \frac{V_A - V_D}{X}$$

$$= \frac{\lambda\theta}{X}$$

$$= \frac{\lambda}{(R_1 + R_3)(R_2 + R_4) + R_b(R_1 + R_2 + R_3 + R_4)},$$

by equation (278).

II. *Telegraph wire with faults.*

353. As a more complex example of the flow of electricity in a system of linear conductors, we may examine the case of a telegraph wire, in which there are a number of connexions through which the current can leak to earth. Such leaks are technically known as "faults."

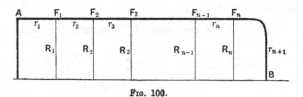

Fɪɢ. 100.

Let AB be the wire, and let $F_1, F_2, \ldots F_{n-1}, F_n$ be the points on it at which faults occur, the resistances through these faults being $R_1, R_2, \ldots$

R_{n-1}, R_n, and the resistances of the sections AF_1, F_1F_2, ... $F_{n-1}F_n$ and F_nB being r_1, r_2, ... r_n, r_{n+1}. Let the end B be supposed put to earth, and let the current be supposed to be generated by a battery of which one terminal is connected to A while the other end is to earth.

The equivalent resistance of the whole network of conductors from A to earth can be found in a very simple way. Current arriving at F_n from the section $F_{n-1}F_n$ passes to earth through two conductors arranged in parallel, of which the resistances are R_n and r_{n+1}. Hence the resistance from F_n to earth is

$$\frac{1}{\dfrac{1}{R_n} + \dfrac{1}{r_{n+1}}},$$

and the resistance from F_{n-1} to earth, through F_n, is

$$r_n + \frac{1}{\dfrac{1}{R_n} + \dfrac{1}{r_{n+1}}} \quad \dots\dots\dots\dots\dots\dots\dots\dots(279).$$

Current reaching F_{n-1} can, however, pass to earth by two paths, either through the fault at F_{n-1}, or past F_n. These paths may be regarded as arranged in parallel, their resistances being R_{n-1} and expression (279) respectively. Thus the equivalent resistance from F_{n-1} is

$$\frac{1}{\dfrac{1}{R_{n-1}} + \dfrac{1}{r_n + \dfrac{1}{\dfrac{1}{R_n} + \dfrac{1}{r_{n+1}}}}},$$

or, written as a continued fraction,

$$\frac{1}{R_{n-1}^{-1}} + \frac{1}{r_n} + \frac{1}{R_n^{-1}} + \frac{1}{r_{n+1}}.$$

We can continue in this way, until finally we find as the whole resistance from A to earth,

$$r_1 + \frac{1}{R_1^{-1}} + \frac{1}{r_2} + \frac{1}{R_2^{-1}} + \dots + \frac{1}{r_n} + \frac{1}{R_n^{-1}} + \frac{1}{r_{n+1}}.$$

If the currents or potentials are required, it will be found best to attack the problem in a different manner.

Let V_A, V_1, V_2, ... be the potentials at the points A, F_1, F_2, ..., then, by Ohm's Law,

$$\text{the current from } F_{s-1} \text{ to } F_s = \frac{V_{s-1} - V_s}{r_s},$$

$$\text{,, \quad ,, \quad ,, } F_s \text{ to } F_{s+1} = \frac{V_s - V_{s+1}}{r_{s+1}},$$

$$\text{,, \quad ,, \quad ,, } F_s \text{ through the fault} = \frac{V_s}{R_s}.$$

Hence, by Kirchhoff's first law,

$$\frac{V_{s-1} - V_s}{r_s} - \frac{V_s - V_{s+1}}{r_{s+1}} - \frac{V_s}{R_s} = 0,$$

or $$V_{s+1} \, r_{s+1}^{-1} - V_s (R_s^{-1} + r_s^{-1} + r_{s+1}^{-1}) + V_{s-1} \, r_s^{-1} = 0,$$

and from this and the system of similar equations, the potentials may be found.

If all the R's are the same, and also all the r's are the same, the equation reduces to a difference equation with constant coefficients. These conditions might arise approximately if the line were supported by a series of similar imperfect insulators at equal distances apart. The difference equation is in this case seen to be

$$V_{s+1} - V_s \left(2 + \frac{r}{R} \right) + V_{s-1} = 0,$$

and if we put $$1 + \frac{r}{2R} = \cosh \alpha,$$

the solution is known to be

$$V_s = A \cosh s\alpha + B \sinh s\alpha \quad \dots\dots\dots\dots\dots(280),$$

in which A and B are constants which must be determined from the conditions at the ends of the line. For instance to express that the end B is to earth, we have $V_{n+1} = 0$, and therefore

$$A = - B \tanh (n + 1) \alpha.$$

III.   *Submarine cable imperfectly insulated.*

354. If we pass to the limiting case of an infinite number of faults, we have the analysis appropriate to a line from which there is leakage at every point. The conditions now contemplated may be supposed to be realised in a submarine cable in which, owing to the imperfection of the insulating sheath, the current leaks through to the sea at every point.

The problem in this form can also be attacked by the methods of the infinitesimal calculus. Let V be the potential at a distance x along the cable, V now being regarded as a continuous function of x. Let the resistance of the cable be supposed to be R per unit length, then the resistance from x to $x + dx$ will be $R\,dx$. The resistance of the insulation from x to $x + dx$, being inversely proportional to dx, may be supposed to be $\dfrac{S}{dx}$.

Let C be the current in the cable at the point x, so that the leak from the cable between the points x and $x + dx$ is $-\dfrac{dC}{dx}\,dx$. This leak is a current which flows through a resistance $\dfrac{S}{dx}$ with a fall of potential V. Hence by Ohm's Law,

$$V = -\frac{dC}{dx}\,dx\left(\frac{S}{dx}\right),$$

or

$$\frac{dC}{dx} = -\frac{V}{S} \quad\dots\dots\dots\dots\dots\dots\dots\dots(281).$$

Also, the fall of potential along the cable from x to $x + dx$ is $-\dfrac{dV}{dx}\,dx$, the current is C, and the resistance is $R\,dx$. Hence by Ohm's Law,

$$-\frac{dV}{dx} = RC \quad\dots\dots\dots\dots\dots\dots\dots\dots(282).$$

Eliminating C from equations (281) and (282), we find as the differential equation satisfied by V,

$$\frac{d}{dx}\left(\frac{1}{R}\frac{dV}{dx}\right) = \frac{V}{S}.$$

If R and S have the same values at all points of the cable, the solution of this equation is

$$V = A\cosh\sqrt{\frac{R}{S}}\,x + B\sinh\sqrt{\frac{R}{S}}\,x,$$

which is easily seen to be the limiting form assumed by equation (280).

Generation of Heat in Conductors.

The Joule Effect.

355. Let P, Q be any two points in a linear conductor, let V_P, V_Q be the potentials at these points, R the resistance between them, and x the current flowing from P to Q. Then, by Ohm's Law,

$$V_P - V_Q = Rx \quad\dots\dots\dots\dots\dots\dots\dots\dots(283).$$

In moving a single unit of electricity from Q to P an amount of work is done against the electric field equal to $V_P - V_Q$. Hence when a unit of electricity passes from P to Q, there is work done on it by the electric field of amount $V_P - V_Q$. The energy represented by the work shews itself in a heating of the conductor.

The electron theory gives a simple explanation of the mechanism of this transformation of energy. The electric forces do work on the electrons in driving them through the field. The total kinetic energy of the electrons can, as we have seen (§ 345 a), be regarded as made up of two parts, the energy of random motion and the energy of forward motion. The work done by the electric field goes directly towards increasing this second part of the kinetic energy of the electrons. But after a number of collisions the direction of the velocity of forward motion is completely changed, and the energy of this motion has become indistinguishable from the energy of the random motion of the electrons. Thus the collisions are continually transforming forward motion into random motion, or what is the same thing, into heat.

We are supposing that x units of electricity pass per unit time from P to Q. Hence the work done by the electric field per unit time within the region PQ is $x(V_P - V_Q)$, and this again, by equation (283), is equal to Rx^2.

Thus in unit time, the heat generated in the section PQ of the conductor represents Rx^2 units of mechanical energy. Each unit of energy is equal to $\dfrac{1}{J}$ units of heat, where J is the "mechanical equivalent of heat." Thus the number of heat-units developed in unit time in the conductor PQ will be

$$\frac{Rx^2}{J} \quad\text{.................................(284).}$$

It is important to notice that in this formula x and R are measured in electrostatic units. If the values of the resistance and current are given in practical units, we must transform to electrostatic units before using formula (284).

Let the resistance of a conductor be R' ohms, and let the current flowing through it be x' ampères. Then, in electrostatic units, the values of the resistance R and the current x are given by

$$R = \frac{R'}{9 \times 10^{11}} \text{ and } x = 3 \times 10^9 x'.$$

Thus the number of heat-units produced per unit time is

$$\frac{Rx^2}{J} = \frac{(3 \times 10^9)^2}{9 \times 10^{11} \cdot J} R' x'^2,$$

and on substituting for J its value $4 \cdot 2 \times 10^7$ in c.g.s.-centigrade units, this becomes

$$0 \cdot 24 \, R' i'^2.$$

Generation of Heat a minimum.

356. In general the solution of any physical problem is arrived at by the solution of a system of equations, the number of these equations being equal to the number of unknown quantities in the problem. The condition that any function in which these unknown quantities enter as variables shall be a maximum or a minimum, is also arrived at by the solution of an equal number of equations. If it is possible to discover a function of the unknown quantities such that the two systems of equations become identical,—*i.e.* if the equations which express that the function is a maximum or a minimum are the same as those which contain the solution of the physical problem— then we may say that the solution of the problem is contained in the single statement that the function in question is a maximum or a minimum.

Examples of functions which serve this purpose are not hard to find. In § 189, we proved that when an electrostatic system is in equilibrium, its potential energy is a minimum. Thus the solution of any electrostatic problem is contained in the single statement that the function which expresses the potential energy is a minimum. Again, the solution of any dynamical problem is contained in the statement that the "action" is a minimum, while in thermodynamics the equilibrium state of any system can be expressed by the condition that the "entropy" shall be a maximum. It will now be shewn that the function which expresses the total rate of generation of heat plays a similar rôle in the theory of steady electric currents.

357. THEOREM. *When a steady current flows through a network of conductors in which no discontinuities of potential occur (and which, therefore, contains no batteries), the currents are distributed in such a way that the rate of generation of heat in the network is a minimum, subject only to the conditions imposed by Kirchhoff's first law; and conversely.*

To prove this, let us select any closed circuit $PQR \ldots P$ in the network, and let the currents and resistances in the sections PQ, QR, ... be x_1, x_2, ... and R_1, R_2, ... Let the currents and resistances in those sections of the network which are not included in this closed circuit be denoted by x_a, x_b, ... and R_a, R_b, ... Then the total rate of production of heat is

$$\Sigma R_a x_a{}^2 + \Sigma R_1 x_1{}^2 \ldots\ldots\ldots\ldots\ldots\ldots(285).$$

A different arrangement of currents, and one moreover which does not violate Kirchhoff's first law, can be obtained in imagination by supposing all the currents in the circuit $PQR \ldots P$ increased by the same amount ϵ. The total rate of production of heat is now

$$\Sigma R_a x_a{}^2 + \Sigma R_1 (x_1 + \epsilon)^2,$$

and this exceeds the actual rate of production of heat, as given by expression (285), by

$$\Sigma R_1 (2x_1\epsilon + \epsilon^2) \ldots\ldots\ldots\ldots\ldots\ldots(286).$$

Now if the original distribution of currents is that which actually occurs in nature, then

$$\Sigma R_1 x_1 = 0,$$

by Kirchhoff's second law. Thus the rate of production of heat, under the new imaginary distribution of currents, exceeds that in the actual distribution by $\epsilon^2 \Sigma R_1$, an essentially positive quantity.

The most general alteration which can be supposed made to the original system of currents, consistently with Kirchhoff's first law remaining satisfied, will consist in superposing upon this system a number of currents flowing in closed circuits in the network. One such current is typified by the current ϵ, already discussed. If we have any number of such currents, the resulting increase in the rate of heat-production

$$= \Sigma R_1 (x_1 + \epsilon + \epsilon' + \epsilon'' + \ldots)^2 - \Sigma R_1 x_1{}^2,$$

where ϵ, ϵ', ϵ'', ... are the additional currents flowing through the resistance R_1. As before this expression

$$= 2\Sigma R_1 x_1 (\epsilon + \epsilon' + \epsilon'' + \dots) + \Sigma R_1 (\epsilon + \epsilon' + \epsilon'' + \dots)^2$$
$$= \Sigma R_1 (\epsilon + \epsilon' + \epsilon'' + \dots)^2,$$

by Kirchhoff's second law. This is an essentially positive quantity, so that any alteration in the distribution of the currents increases the rate of heat-production. In other words, the original distribution was that in which the rate was a minimum.

To prove the converse it is sufficient to notice that if the rate of heat-production is given to be a minimum, then expression (286) must vanish as far as the first power of ϵ, so that we have

$$\Sigma R_1 x_1 = 0,$$

and of course similar equations for all other possible closed circuits. These, however, are known to be the equations which determine the actual distribution.

358. THEOREM. *When a system of steady currents flows through a network of conductors of resistances R_1, R_2, ..., containing batteries of electromotive forces E_1, E_2, ..., the currents x_1, x_2, ... are distributed in such a way that the function*

$$\Sigma R x^2 - 2\Sigma E x \quad \dots\dots\dots\dots\dots\dots\dots\dots(287)$$

is a minimum, subject to the conditions imposed by Kirchhoff's first law; and conversely.

As before, we can imagine the most general variation possible to consist of the superposition of small currents ϵ, ϵ', ϵ'', ... flowing in closed circuits. The increase in the function (287) produced by this variation is

$$\Sigma R \left[(x + \epsilon + \epsilon' + \dots)^2 - x^2\right] - 2\Sigma E \left[(x + \epsilon + \epsilon' + \dots) - x\right]$$
$$= 2\epsilon . (\Sigma R x - \Sigma E) + 2\epsilon' (\dots) + \dots$$
$$+ \Sigma R (\epsilon + \epsilon' + \dots)^2 \dots\dots\dots\dots\dots\dots(288).$$

If the system of currents x, x', ... is the natural system, then the first line of this expression vanishes by Kirchhoff's second law (cf. equations (270)), and the increase in heat-production is the essentially positive quantity

$$\Sigma R (\epsilon + \epsilon' + \dots)^2,$$

shewing that the original value of function (287) must have been a minimum.

Conversely, if the original value of function (287) was given to be a minimum, then expression (288) must vanish as far as first powers of ϵ, ϵ', ..., so that we must have

$$\Sigma R x = E, \text{ etc.,}$$

shewing that the currents x, x', ... must be the natural system of currents.

359. THEOREM. *If two points A, B are connected by a network of conductors, a decrease in the resistance of any one of these conductors will decrease (or, in special cases, leave unaltered) the equivalent resistance from A to B.*

Let x be the current flowing from A to B, R the equivalent resistance of the network, and $V_A - V_B$ the fall of potential. The generation of heat per unit time represents the energy set free by x units moving through a potential-difference $V_A - V_B$. Thus the rate of generation of heat is

$$x(V_A - V_B),$$

or, since $V_A - V_B = Rx$, the rate of generation of heat will be Rx^2.

Let the resistance of any single conductor in the network be supposed decreased from R_1 to R_1', and let x_1 be the current originally flowing through the network. If we imagine the currents to remain unaltered in spite of the change in the resistance of this conductor, then there will be a decrease in the rate of heat-production equal to $(R_1 - R_1') x_1^2$. The currents now flowing are not the natural currents, but if we allow the current entering the network to distribute itself in the natural way, there is, by § 357, a further decrease in the rate of heat-production. Thus a decrease in the resistance of the single conductor has resulted in a decrease in the natural rate of heat-production.

If R, R' are the equivalent resistances before and after the change, the two rates of heat-production are Rx^2 and $R'x^2$. We have proved that $R'x^2 < Rx^2$, so that $R' < R$, proving the theorem.

GENERAL THEORY OF A NETWORK.

360. In addition to depending on the resistances of the conductors, the flow of currents through a network depends on the order in which the conductors are connected together, but not on the geometrical shapes, positions or distances of the conductors. Thus we can obtain the most general case of flow through any network by considering a number of points 1, 2, ... n, connected in pairs by conductors of general resistances which may be denoted by $R_{12}, R_{23}, \ldots$. If, in any special problem, any two points P, Q are not joined by a conductor, we must simply suppose R_{PQ} to be infinite. Discontinuities of potential must not be excluded, so we shall suppose that in passing through the conductor PQ, we pass over discontinuities of algebraic sum E_{PQ}. This is the same as supposing that there are batteries in the arm PQ of total electromotive force E_{PQ}. We shall suppose that the current flowing in PQ from P to Q is x_{PQ}, and shall denote the potentials at the points 1, 2, ... by $V_1, V_2, \ldots$.

The total fall of potential from P to Q is $V_P - V_Q$, but of this an amount

$-E_{PQ}$ is contributed by discontinuities, so that the aggregate fall from P to Q which arises from the steady potential gradient in conductors will be

$$V_P - V_Q + E_{PQ}.$$

Hence, by Ohm's Law,

$$V_P - V_Q + E_{PQ} = R_{PQ} x_{PQ}.$$

If we introduce a symbol K_{PQ} to denote the conductivity $\dfrac{1}{R_{PQ}}$, we have the current given by

$$x_{PQ} = K_{PQ}(V_P - V_Q + E_{PQ}) \quad\dots\dots\dots\dots\dots\dots(289).$$

Suppose that currents X_1, X_2, ... enter the system from outside at the points 1, 2, ..., then we must have

$$X_1 = x_{12} + x_{13} + x_{14} + \dots,$$

since there is to be no accumulation of electricity at the point 1, and so on for the points 2, 3, ... Substituting from equations (289) into the right hand of this equation,

$$\begin{aligned}
X_1 &= K_{12}(V_1 - V_2 + E_{12}) + K_{13}(V_1 - V_3 + E_{13}) + \dots \\
&= V_1(K_{12} + K_{13} + \dots) \\
&\quad - (K_{12}V_2 + K_{13}V_3 + \dots) + K_{12}E_{12} + K_{13}E_{13} + \dots \quad\dots\dots\dots\dots(290).
\end{aligned}$$

The symbol K_{PP} has so far had no meaning assigned to it. Let us use it to denote $-(K_{P1} + K_{P2} + K_{P3} + \dots)$; then equation (290) may be written in the more concise form

$$X_1 = -(K_{11}V_1 + K_{12}V_2 + \dots) + K_{12}E_{12} + K_{13}E_{13} + \dots \quad\dots\dots(291).$$

There are n equations of this type, but it is easily seen that they are not all independent. For if we add corresponding members we obtain

$$X_1 + X_2 + \dots + X_n = -\sum_1^n V_1(K_{11} + K_{12} + \dots + K_{1n}) + \Sigma\Sigma(K_{PQ}E_{PQ} + K_{QP}E_{QP}).$$

The first term on the right vanishes on account of the meaning which has been assigned to K_{11}, etc.; while the second term vanishes because $E_{PQ} = -E_{QP}$, while $K_{PQ} = K_{QP}$. Thus the equation reduces to

$$X_1 + X_2 + \dots + X_n = 0,$$

which simply expresses that the total flow into the network is equal to the total flow out of it, a condition which must be satisfied by X_1, X_2, ... X_n at the outset. Thus we arrive at the conclusion that the equations of system (291) are not independent.

This is as it should be, for if the equations were independent, we should have n equations from which it would be possible to determine the values of V_1, V_2, ... in terms of X_1, X_2, ...; whereas clearly from a knowledge of the currents entering the network, we must be able to determine *differences* of potential only, and not absolute values.

To the right-hand side of equation (291), let us add the expression

$$(K_{11} + K_{12} + \ldots + K_{1n})\, V_n,$$

of which the value is zero by the definition of K_{11}. The equation becomes

$$K_{11}(V_1 - V_n) + K_{12}(V_2 - V_n) + \ldots + K_{1,n-1}(V_{n-1} - V_n)$$
$$= -X_1 + K_{12}E_{12} + K_{13}E_{13} + \ldots + K_{1n}E_{1n}.$$

There are n equations of this type in all. Of these the first $(n-1)$ may be regarded as a system of equations determining

$$V_1 - V_n, \ V_2 - V_n, \ \ldots, \ V_{n-1} - V_n.$$

That these equations are independent will be seen *à posteriori* from the fact that they enable us to determine the values of the $n-1$ independent quantities

$$V_1 - V_n, \ V_2 - V_n, \ \ldots, \ V_{n-1} - V_n.$$

Solving these equations, we have

$$V_1 - V_n$$

$$= \frac{\begin{vmatrix} -X_1 + K_{12}E_{12} + \ldots + K_{1n}E_{1n}, & K_{12}, & K_{13}, & \ldots, & K_{1,n-1} \\ -X_2 + K_{21}E_{21} + \ldots + K_{2n}E_{2n}, & K_{22}, & K_{23}, & \ldots, & K_{2,n-1} \\ \cdots\cdots\cdots\cdots\cdots\cdots\cdots\cdots\cdots\cdots\cdots\cdots\cdots\cdots\cdots\cdots \\ -X_{n-1} + K_{n-1,1}E_{n-1,1} + \ldots + K_{n-1,n}E_{n-1,n}, & K_{n-1,2}, & K_{n-1,3}, & \ldots, & K_{n-1,n-1} \end{vmatrix}}{\begin{vmatrix} K_{11}, & K_{12}, & K_{13}, & \ldots, & K_{1,n-1} \\ K_{21}, & K_{22}, & K_{23}, & \ldots, & K_{2,n-1} \\ \cdots\cdots\cdots\cdots\cdots\cdots\cdots\cdots\cdots\cdots\cdots\cdots \\ K_{n-1,1}, & K_{n-1,2}, & K_{n-1,3}, & \ldots, & K_{n-1,n-1} \end{vmatrix}}$$

The current flowing in conductor $1n$ follows at once from equation (289), and the currents in the other conductors can be written down from symmetry.

If we denote the determinant in the denominator of the foregoing equation by Δ, and the minor of the term K_{PQ} by Δ_{PQ}, we find that the value of $V_1 - V_n$ can be expressed in the form

$$V_1 - V_n = \quad (-X_1 + K_{12}E_{12} + \ldots + K_{1n}E_{1n})\frac{\Delta_{11}}{\Delta}$$

$$+ (-X_2 + K_{21}E_{21} + \ldots + K_{2n}E_{2n})\frac{\Delta_{21}}{\Delta} + \ldots \quad \ldots\ldots(292).$$

361. Suppose first that the whole system of currents in the network is produced by a current X entering at P and leaving at Q, there being no batteries in the network. Then all the E's vanish, and all the X's vanish except X_P and X_Q, these being given by

$$X_P = -X_Q = X.$$

Equation (292) now becomes

$$V_1 - V_n = - X_P \frac{\Delta_{P1}}{\Delta} - X_Q \frac{\Delta_{Q1}}{\Delta}$$

$$= \frac{X}{\Delta} (\Delta_{Q1} - \Delta_{P1}),$$

so that

$$V_1 - V_2 = (V_1 - V_n) - (V_2 - V_n)$$

$$= \frac{X}{\Delta} (\Delta_{Q1} - \Delta_{Q2} - \Delta_{P1} + \Delta_{P2})\ldots\ldots\ldots\ldots\ldots(293).$$

Replacing 1, 2 by P, Q and P, Q by 1, 2, we find that if a current X enters the network at 1 and leaves it at 2, the fall of potential from P to Q is

$$V_P - V_Q = \frac{X}{\Delta} (\Delta_{2P} - \Delta_{2Q} - \Delta_{1P} + \Delta_{1Q}) \ldots\ldots\ldots\ldots(294),$$

and since $\Delta_{rs} = \Delta_{sr}$, it is clear that the right-hand members of equations (293) and (294) are identical.

From this we have the theorem:

The potential-fall from A to B when unit current traverses the network from C to D is the same as the potential-fall from C to D when unit current traverses the network from A to B.

362. Let it now be supposed that the whole flow of current in the network is produced by a battery of electromotive force E placed in the conductor PQ. We now take all the X's equal to zero in equation (292) and all the E's equal to zero except E_{PQ} which we put equal to E, and E_{QP} which we put equal to $-E$. We then have

$$V_1 - V_n = K_{PQ} E_{PQ} \frac{\Delta_{P1}}{\Delta} + K_{QP} E_{QP} \frac{\Delta_{Q1}}{\Delta}$$

$$= \frac{K_{PQ} E}{\Delta} (\Delta_{P1} - \Delta_{Q1}).$$

Hence

$$V_1 - V_2 = \frac{K_{PQ} E}{\Delta} (\Delta_{P1} - \Delta_{P2} - \Delta_{Q1} + \Delta_{Q2}) \ldots\ldots\ldots\ldots(295),$$

and, by equation (289), the current flowing in the arm 12 is

$$x_{12} = \frac{K_{12} K_{PQ} E}{\Delta} (\Delta_{P1} - \Delta_{P2} - \Delta_{Q1} + \Delta_{Q2})\ldots\ldots\ldots\ldots(296).$$

This expression remains unaltered if we replace 1, 2 by P, Q and P, Q by 1, 2. From this we deduce the theorem:

The current which flows from A to B when an electromotive force E is introduced into the arm CD of the network, is equal to the current which flows from C to D when the same electromotive force is introduced into the arm AB.

Conjugate Conductors.

363. The same expression occurs as a factor in the right-hand members of each of the equations (293), (294), (295), and (296), namely,

$$\Delta_{P1} + \Delta_{Q2} - \Delta_{Q1} - \Delta_{P2} \quad\dots\dots\dots\dots\dots\dots(297).$$

If this expression vanishes, the two conductors 12 and PQ are said to be "conjugate."

By examining the form assumed by equations (293) to (296), when expression (297) vanishes, we obtain the following theorems.

THEOREM I. *If the conductors AB and CD are conjugate, a current entering at A and leaving at B will produce no current in CD. Similarly, a current entering at C and leaving at D will produce no current in AB.*

THEOREM II. *If the conductors AB and CD are conjugate, a battery introduced into the arm AB produces no current in CD. Similarly, a battery introduced into the arm CD produces no current in AB.*

As an illustration of two conductors which are conjugate, it may be noticed that when the Wheatstone's Bridge (§ 352) is in adjustment, the conductors AD and BC are conjugate.

Equations expressed in Symmetrical Form.

364. The determinant Δ is not in form a symmetric function of the n points 1, 2, ..., n, so that equations and conditions which must necessarily involve these n points symmetrically have not yet been expressed in symmetrical form.

We have, for instance,

$$\Delta_{13} = \begin{vmatrix} K_{21}, & K_{22}, & K_{24}, & K_{25}, & \dots, & K_{2,n-1} \\ K_{31}, & K_{32}, & K_{34}, & K_{35}, & \dots, & K_{3,n-1} \\ \dots\dots\dots\dots\dots\dots\dots\dots\dots\dots\dots\dots\dots \\ K_{n-1,1}, & K_{n-1,2}, & K_{n-1,4}, & K_{n-1,5}, & \dots, & K_{n-1,n-1} \end{vmatrix},$$

in which the points which enter unsymmetrically are not only 1 and 3, but also n. Similarly, we have

$$\Delta_{14} = - \begin{vmatrix} K_{21}, & K_{22}, & K_{23}, & K_{25}, & \dots, & K_{2,n-1} \\ K_{31}, & K_{32}, & K_{33}, & K_{35}, & \dots, & K_{3,n-1} \\ \dots\dots\dots\dots\dots\dots\dots\dots\dots\dots\dots\dots\dots \\ K_{n-1,1}, & K_{n-1,2}, & K_{n-1,3}, & K_{n-1,5}, & \dots, & K_{n-1,n-1} \end{vmatrix},$$

so that, on subtraction,

$$\Delta_{13} - \Delta_{14} = \begin{vmatrix} K_{21}, & K_{22}, & K_{23}+K_{24}, & K_{25}, & \dots, & K_{2,n-1} \\ K_{31}, & K_{32}, & K_{33}+K_{34}, & K_{35}, & \dots, & K_{3,n-1} \\ \dots\dots\dots\dots\dots\dots\dots\dots\dots\dots\dots\dots\dots \\ K_{n-1,1}, & K_{n-1,2}, & K_{n-1,3}+K_{n-1,4}, & K_{n-1,5}, & \dots, & K_{n-1,n-1} \end{vmatrix}.$$

From the relation

$$K_{P1} + K_{P2} + \ldots + K_{P,n-1} + K_{P,n} = 0 \ldots\ldots\ldots\ldots\ldots(298),$$

it follows that the sum of all the terms in the first row of the above determinant is equal to $-K_{2,n}$, the sum of all the terms in the second row is equal to $-K_{3,n}$, and so on. Thus the equation may be replaced by

$$\Delta_{13} - \Delta_{14} = (-1)^n \begin{vmatrix} K_{21}, & K_{22}, & K_{25}, & \ldots, & K_{2,n-1}, & K_{2,n} \\ K_{31}, & K_{32}, & K_{35}, & \ldots, & K_{3,n-1}, & K_{3,n} \\ \hdotsfor{6} \\ K_{n-1,1}, & K_{n-1,2}, & K_{n-1,5}, & \ldots, & K_{n-1,n-1}, & K_{n-1,n} \end{vmatrix},$$

and similarly,

$$\Delta_{23} - \Delta_{24} = (-1)^{n-1} \begin{vmatrix} K_{11}, & K_{12}, & K_{15}, & \ldots, & K_{1,n} \\ K_{31}, & K_{32}, & K_{35}, & \ldots, & K_{3,n} \\ \hdotsfor{5} \\ K_{n-1,1}, & K_{n-1,2}, & K_{n-1,5}, & \ldots, & K_{n-1,n} \end{vmatrix}.$$

These two determinants differ only in their first row, so that on subtraction,

$$(\Delta_{13} - \Delta_{14}) - (\Delta_{23} - \Delta_{24})$$

$$= (-1)^n \begin{vmatrix} K_{11} + K_{21}, & K_{12} + K_{22}, & K_{15} + K_{25}, & \ldots, & K_{1,n} + K_{2,n} \\ K_{31}, & K_{32}, & K_{35}, & \ldots, & K_{3,n} \\ \hdotsfor{5} \\ K_{n-1,1}, & K_{n-1,2}, & K_{n-1,5}, & \ldots, & K_{n-1,n} \end{vmatrix}$$

$$= \begin{vmatrix} K_{31}, & K_{32}, & K_{35}, & \ldots, & K_{3,n} \\ \hdotsfor{5} \\ K_{n-1,1}, & K_{n-1,2}, & K_{n-1,5}, & \ldots, & K_{n-1,n} \\ K_{n,1}, & K_{n,2}, & K_{n,5}, & \ldots, & K_{n,n} \end{vmatrix} \ldots\ldots\ldots\ldots\ldots(299),$$

the last transformation being effected by the use of relation (298).

The relation which has now been obtained is in a symmetrical shape. If D is a symmetrical determinant given by

$$D \equiv \begin{vmatrix} K_{11}, & K_{12}, & K_{13}, & \ldots, & K_{1,n} \\ K_{21}, & K_{22}, & K_{23}, & \ldots, & K_{2,n} \\ \hdotsfor{5} \\ K_{n,1}, & K_{n,2}, & K_{n,3}, & \ldots, & K_{n,n} \end{vmatrix},$$

then the determinant on the right-hand of equation (299) is obtained from D by striking out the lines and columns which contain the terms K_{13} and K_{24}. Thus equation (299) may be written in the form

$$\Delta_{13} + \Delta_{24} - \Delta_{23} - \Delta_{14} = \frac{\partial^2 D}{\partial K_{13} \partial K_{24}}.$$

Again the determinant Δ given by

$$\Delta = \begin{vmatrix} K_{11}, & K_{12}, & K_{13}, & ..., & K_{1,n-1} \\ K_{21}, & K_{22}, & K_{23}, & ..., & K_{2,n-1} \\ \hdotsfor{5} \\ K_{n-1,1}, & K_{n-1,2}, & K_{n-1,3}, & ..., & K_{n-1,n-1} \end{vmatrix} \quad ....(300)$$

may be written in the form

$$\Delta = \frac{\partial D}{\partial K_{n,n}}.$$

This is not of symmetrical form, for the point n enters unsymmetrically. We can, however, easily shew that the value of Δ is symmetrical, although its form is unsymmetrical.

By application of relation (298), we can transform equation (300) into

$$\Delta = \begin{vmatrix} -K_{n,1}, & -K_{n,2}, & -K_{n,3}, & ..., & -K_{n,n-1} \\ K_{21}, & K_{22}, & K_{23}, & ..., & K_{2,n-1} \\ \hdotsfor{5} \\ K_{n-1,1}, & K_{n-1,2}, & K_{n-1,3}, & ..., & K_{n-1,n-1} \end{vmatrix}$$

$$= (-1)^{n-1} \begin{vmatrix} K_{21}, & K_{22}, & K_{23}, & ..., & K_{2,n-1} \\ \hdotsfor{5} \\ K_{n-1,1}, & K_{n-1,2}, & K_{n-1,3}, & ..., & K_{n-1,n-1} \\ K_{n,1}, & K_{n,2}, & K_{n,3}, & ..., & K_{n,n-1} \end{vmatrix}$$

$$= \begin{vmatrix} K_{22}, & K_{23}, & ..., & K_{2,n-1}, & K_{2,n} \\ \hdotsfor{5} \\ K_{n-1,2}, & K_{n-1,3}, & ..., & K_{n-1,n-1}, & K_{n-1,n} \\ K_{n,2}, & K_{n,3}, & ..., & K_{n,n-1}, & K_{n,n} \end{vmatrix}$$

$$= \frac{\partial D}{\partial K_{11}}.$$

Thus Δ is the differential coefficient of D with respect to either K_{11} or $K_{n,n}$, or of course with respect to any other one of the terms in the leading diagonal of D. Thus, if K denote *any* term in the leading diagonal of D, we have

$$\Delta = \frac{\partial D}{\partial K},$$

and this virtually expresses Δ in a symmetrical form.

We can now express in symmetrical form the relations which have been obtained in §§ 360 to 362, as follows:

I.   (§ 362.) *The conductors 1, 2 and P, Q will be conjugate if*

$$\frac{\partial^2 D}{\partial K_{1,P} \partial K_{2,Q}} = 0.$$

II. (Equation 293.) *If the conductors* 1, 2 *and* P, Q *are not conjugate, a current* X *entering at* P *and leaving at* Q *produces in* 1, 2 *a fall of potential given by*

$$V_1 - V_2 = - X \frac{\dfrac{\partial^2 D}{\partial K_{1,P} \partial K_{2,Q}}}{\dfrac{\partial D}{\partial K}}.$$

III. (Equation 295.) *If the conductors* 1, 2 *and* P, Q *are not conjugate, a battery of electromotive force* E *placed in the arm* PQ *produces in* 1, 2 *a fall of potential given by*

$$V_1 - V_2 = K_{PQ} E \frac{\dfrac{\partial^2 D}{\partial K_{1,P} \partial K_{2,Q}}}{\dfrac{\partial D}{\partial K}},$$

and a current from 1 *to* 2 *given by*

$$x_{12} = E \frac{K_{12} K_{PQ} \dfrac{\partial^2 D}{\partial K_{1,P} \partial K_{2,Q}}}{\dfrac{\partial D}{\partial K}}.$$

All these results and formulæ obtain illustration in the results already obtained for the Wheatstone's Bridge in §§ 351 and 352.

SLOWLY-VARYING CURRENTS.

365. All the analysis of the present chapter has proceeded upon the assumption that the currents are absolutely steady, shewing no variation with the time. Changes in the strength of electric currents are in general accompanied by a series of phenomena, which may be spoken of as "induction phenomena," of which the discussion is beyond the scope of the present chapter. If, however, the rate of change of the strength of the currents is very small, the importance of the induction phenomena also becomes very small, so that if the variation of the currents is slow, the analysis of the present chapter will give a close approximation to the truth. This method of dealing with slowly-varying currents will be illustrated by two examples.

I. *Discharge of a Condenser through a high Resistance.*

366. Let the two plates A, B of a condenser of capacity C be connected by a conductor of high resistance R, and let the condenser be discharged by leakage through this conductor. At any instant let the potentials of the two plates be V_A, V_B, so that the charges on these plates will be $\pm C (V_A - V_B)$. Let i be the current in the conductor, measured in the direction from A to B.

Then, by Ohm's Law,

$$V_A - V_B = Ri,$$

whence we find that the charges on plates A and B are respectively $+CRi$ and $-CRi$. Since i units leave plate A per unit time, we must have

$$\frac{d}{dt}(CRi) = -i,$$

a differential equation of which the solution is

$$i = i_0 e^{-\frac{t}{CR}},$$

where i_0 is the current at time $t=0$. The condition that the strength of the current shall only vary slowly is now seen *à posteriori* to be that CR shall be large.

At time t the charge on the plate A is CRi or

$$CRi_0 e^{-\frac{t}{CR}}.$$

This may be written as

$$Q_0 e^{-\frac{t}{CR}},$$

where Q_0 is the charge at time $t=0$. Thus both the charge and the current are seen to fall off exponentially with the time, both having the same modulus of decay CR.

Later (§ 516) we shall examine the same problem but without the limitation that the current only varies slowly.

II. *Transmission of Signals along a Cable.*

367. It has already been mentioned that a cable acts as an electrostatic condenser of considerable capacity. This fact retards the transmission of signals, and in a cable of high-capacity, the rate of transmission may be so slow that the analysis of the present chapter can be used without serious error.

Let x be a coordinate which measures distances along the cable, let V, i be the potential at x and the current in the direction of x-increasing, and let K and R be the capacity and resistance of the cable per unit length, these latter quantities being supposed independent of x.

The section of the cable between points A and B at distances x and $x + dx$ is a condenser of capacity $K dx$, and is at the same time a conductor of resistance $R\,dx$. The potential of the condenser is V, so that its charge is $VK\,dx$. The fall of potential in the conductor is

$$V_A - V_B = -\frac{\partial V}{\partial x}\,dx,$$

so that by Ohm's Law,

$$-\frac{\partial V}{\partial x}\,dx = iR\,dx \quad \dots\dots\dots\dots\dots\dots\dots(301).$$

The current enters the section AB at a rate i units per unit time, and leaves at a rate of $i + \frac{\partial i}{\partial x}\,dx$ units per unit time. Hence the charge in this section decreases at a rate $\frac{\partial i}{\partial x}\,dx$ per unit time, so that we must have

$$\frac{\partial}{\partial t}\,(VK\,dx) = -\frac{\partial i}{\partial x}\,dx \quad \dots\dots\dots\dots\dots\dots(302).$$

Eliminating i from equations (301) and (302), we obtain

$$\frac{\partial^2 V}{\partial x^2} = KR\,\frac{\partial V}{\partial t} \quad \dots\dots\dots\dots\dots\dots\dots\dots(303).$$

368. This equation, being a partial differential equation of the second order, must have two arbitrary functions in its complete solution. We shall shew, however, that there is a particular solution in which V is a function of the single variable $x/\sqrt{t}$, and this solution will be found to give us all the information we require.

Let us introduce the new variable u, given by $u = x/\sqrt{t}$, and let us assume provisionally that there is a solution V of equation (303) which is a function of u only. For this solution we must have

$$\frac{\partial^2 V}{\partial x^2} = \frac{1}{t}\frac{d^2 V}{du^2},$$

$$\frac{\partial V}{\partial t} = \frac{dV}{du}\frac{\partial u}{\partial t} = -\tfrac{1}{2}\frac{x}{\sqrt{t^3}}\frac{dV}{du},$$

so that equation (303) becomes

$$\frac{d^2 V}{du^2} = tKR\left(-\tfrac{1}{2}\frac{x}{\sqrt{t^3}}\frac{dV}{du}\right)$$

$$= -\tfrac{1}{2}KRu\,\frac{dV}{du} \quad \dots\dots\dots\dots\dots\dots\dots(304).$$

The fact that this equation involves V and u only, shews that there is an integral of the original equation for which V is a function of u only. This integral is easily obtained, for equation (304) can be put in the form

$$\frac{d}{du}\left(\log\frac{dV}{du}\right) = -\tfrac{1}{2}KRu,$$

whence

$$\frac{dV}{du} = Ce^{-\frac{1}{4}KRu^2},$$

in which C is a constant of integration.

Integrating this, we find that the solution for V is

$$V = C \int^u e^{-\frac{1}{4} K R u^2} du,$$

in which the lower limit to the integral is a second constant of integration. Introducing a new variable y such that $y^2 = \frac{1}{4} K R u^2$, and changing the constants of integration, we may write the solution in the form

$$V = V_0 + C' \int_\infty^{y = \frac{1}{2} x \sqrt{KR/t}} e^{-y^2} dy \quad \dots\dots\dots\dots(305).$$

369. We must remember that this is not the general solution of equation (303), but is simply one particular solution. Thus the solution cannot be adjusted to satisfy any initial and boundary conditions we please, but will represent only the solution corresponding to one definite set of initial and boundary conditions. We now proceed to examine what these conditions are.

At time $t = 0$, the value of $x/\sqrt{t}$ is infinite except at the point $x = 0$. Thus except at this point, we have $V = V_0$ when $t = 0$. At this point the value of $x/\sqrt{t}$ is indeterminate at the actual instant $t = 0$, but immediately after this instant assumes the value zero, which it retains through all time. Thus at $x = 0$, the potential has the constant value

$$V = V_0 + C' \int_\infty^0 e^{-y^2} dy,$$

or, say, $V = V_1$, where $C' = \dfrac{2 (V_0 - V_1)}{\sqrt{\pi}}$.

At $x = \infty$, the value of V is $V = V_0$ through all time.

Thus equation (305) expresses the solution for a line of infinite length which is initially at potential $V = V_0$, and of which the end $x = \infty$ remains at this potential all the time, while the end $x = 0$ is raised to potential V_1 by being suddenly connected to a battery-terminal at the instant $t = 0$.

The current at any instant is given by

$$i = -\frac{1}{R} \frac{\partial V}{\partial x}, \text{ from equation (301)},$$

$$= -\frac{C'}{R} \frac{1}{2} \sqrt{\frac{KR}{t}} e^{-\frac{KRx^2}{4t}}, \text{ from equation (305)},$$

$$= (V_0 - V_1) \sqrt{\frac{K}{R\pi t}} e^{-\frac{KRx^2}{4t}} \quad \dots\dots\dots\dots\dots\dots(306).$$

We see that the current vanishes only when $t = 0$ and when $t = \infty$. Thus even within an infinitesimal time of making contact, there will, according to equation (306), be a current at all points along the wire. It must, however, be remembered that equation (306) is only an approximation, holding solely for slowly-varying currents, so that we must not apply the solution at the instant $t = 0$ at which the currents, as given by equation (306), vary with infinite rapidity. For larger values of t, however, we may suppose the current given by equation (306).

The maximum current at any point is found, on differentiating equation (306), to occur at the instant given by

$$t = \tfrac{1}{2} K R x^2 \quad \dots\dots\dots\dots\dots\dots\dots\dots(307),$$

so that the further along the wire we go, the longer it takes for the current to attain its maximum value. The maximum value of this current, when it occurs, is

$$(V - V_0) \sqrt{\frac{2}{R^2 \pi x^2}}\, e^{-\frac{1}{2}} \quad \dots\dots\dots\dots\dots\dots(308),$$

and so is proportional to $\dfrac{1}{x}$. Thus the further we go from the end $x = 0$, the smaller the maximum current will be.

We notice that K occurs in expression (307) but not in (308). Thus the electrostatic capacity of a cable will not interfere with the strength of signals sent along a cable, but will interfere with the rapidity of their transmission.

Equation (307) expresses what is commonly called the "KR law"—the retarding effect is proportional to the product of K and R. The theory just developed is commonly spoken of as the Electrostatic Theory of propagation of signals. It was first given by Lord Kelvin in 1855 in a paper* which is notable as having established the theoretical feasibility of an Atlantic cable.

We shall discuss in a later chapter the more general problem of the transmission of signals along a wire of any kind. It will then be possible to estimate the degree of error involved in the simple assumptions of the Electrostatic Theory.

EXAMPLES.

1. A length $4a$ of uniform wire is bent into the form of a square, and the opposite angular points are joined with straight pieces of the same wire, which are in contact at their intersection. A given current enters at the intersection of the diagonals and leaves at an angular point : find the current strength in the various parts of the network, and shew that its whole resistance is equal to that of a length

$$\frac{a\sqrt{2}}{2\sqrt{2}+1}$$

of the wire.

2. A network is formed of uniform wire in the shape of a rectangle of sides $2a$, $3a$, with parallel wires arranged so as to divide the internal space into six squares of sides a, the contact at points of intersection being perfect. Shew that if a current enter the framework by one corner and leave it by the opposite, the resistance is equivalent to that of a length $121a/69$ of the wire.

* "On the Theory of the Electric Telegraph," *Proc. Roy. Soc.* 1855.

3. A fault of given earth-resistance develops in a telegraph line. Prove that the current at the receiving end, generated by an assigned battery at the signalling end, is least when the fault is at the middle of the line.

4. The resistances of three wires BC, CA, AB, of the same uniform section and material, are a, b, c respectively. Another wire from A of constant resistance d can make a sliding contact with BC. If a current enter at A and leave at the point of contact with BC, shew that the maximum resistance of the network is

$$\frac{(a+b+c)\,d}{a+b+c+4d},$$

and determine the least resistance.

5. A certain kind of cell has a resistance of 10 ohms and an electromotive force of ·85 of a volt. Shew that the greatest current which can be produced in a wire whose resistance is 22·5 ohms, by a battery of five such cells arranged in a single series, of which any element is either one cell or a set of cells in parallel, is exactly ·06 of an ampère.

6. Six points A, A', B, B', C, C' are connected to one another by copper wire whose lengths in yards are as follows: $AA'=16$, $BC=B'C'=1$, $BC'=B'C'=2$, $AB=A'B'=6$, $AC'=A'C'=8$. Also B and B' are joined by wires, each a yard in length, to the terminals of a battery whose internal resistance is equal to that of r yards of the wire, and all the wires are of the same thickness. Shew that the current in the wire AA' is equal to that which the battery would maintain in a simple circuit consisting of $31r+104$ yards of the wire.

7. Two places A, B are connected by a telegraph line of which the end at A is connected to one terminal of a battery, and the end at B to one terminal of a receiver, the other terminals of the battery and receiver being connected to earth. At a point C of the line a fault is developed, of which the resistance is r. If the resistances of AC, CB be p, q respectively, shew that the current in the receiver is diminished in the ratio

$$r\,(p+q) : qr+rp+pq,$$

the resistances of the battery, receiver and earth circuit being neglected.

8. Two cells of electromotive forces e_1, e_2 and resistances r_1, r_2 are connected in parallel to the ends of a wire of resistance R. Shew that the current in the wire is

$$\frac{e_1 r_2 + e_2 r_1}{r_1 R + r_2 R + r_1 r_2},$$

and find the rates at which the cells are working.

9. A network of conductors is in the form of a tetrahedron $PQRS$; there is a battery of electromotive force E in PQ, and the resistance of PQ, including the battery, is R. If the resistances in QR, RP are each equal to r, and the resistances in PS, RS are each equal to $\frac{1}{2}r$, and that in $QS=\frac{2}{3}r$, find the current in each branch.

10. A, B, C, D are the four junction points of a Wheatstone's Bridge, and the resistances c, β, b, γ in AB, BD, AC, CD respectively are such that the battery sends no current through the galvanometer in BC. If now a new battery of electromotive force E be introduced into the galvanometer circuit, and so raise the total resistance in that circuit to a, find the current that will flow through the galvanometer.

11. A cable AB, 50 miles in length, is known to have one fault, and it is necessary to localise it. If the end A is attached to a battery, and has its potential maintained at 200 volts, while the other end B is insulated, it is found that the potential of B when steady is 40 volts. Similarly when A is insulated the potential to which B must be raised to give A a steady potential of 40 volts is 300 volts. Shew that the distance of the fault from A is 19·05 miles.

12. A wire is interpolated in a circuit of given resistance and electromotive force. Find the resistance of the interpolated wire in order that the rate of generation of heat may be a maximum.

13. The resistances of the opposite sides of a Wheatstone's Bridge are a, a' and b, b' respectively. Shew that when the two diagonals which contain the battery and galvanometer are interchanged,

$$\frac{E}{C} - \frac{E}{C'} = \frac{(a - a')(b - b')(G - R)}{aa' - bb'},$$

where C and C' are the currents through the galvanometer in the two cases, G and R are the resistances of the galvanometer and battery conductors, and E the electromotive force of the battery.

14. A current C is introduced into a network of linear conductors at A, and taken out at B, the heat generated being H_1. If the network be closed by joining A, B by a resistance r in which an electromotive force E is inserted, the heat generated is H_2. Prove that

$$\frac{H_1}{C^2 r} + \frac{r H_2}{E^2} = 1.$$

15. A number N of incandescent lamps, each of resistance r, are fed by a machine of resistance R (including the leads). If the light emitted by any lamp is proportional to the square of the heat produced, prove that the most economical way of arranging the lamps is to place them in parallel arc, each arc containing n lamps, where n is the integer nearest to $\sqrt{NR/r}$.

16. A battery of electromotive force E and of resistance B is connected with the two terminals of two wires arranged in parallel. The first wire includes a voltameter which contains discontinuities of potential such that a unit current passing through it for a unit time does p units of work. The resistance of the first wire, including the voltameter, is R: that of the second is r. Shew that if E is greater than $p(B + r)/r$, the current through the battery is

$$\frac{E(R + r) - pr}{Rr + B(R + r)}.$$

17. A system of 30 conductors of equal resistance are connected in the same way as the edges of a dodecahedron. Shew that the resistance of the network between a pair of opposite corners is $\frac{7}{6}$ of the resistance of a single conductor.

18. In a network PA, PB, PC, PD, AB, BC, CD, DA, the resistances are a, β, γ, δ, $\gamma + \delta$, $\delta + a$, $a + \beta$, $\beta + \gamma$ respectively. Shew that, if AD contains a battery of electromotive force E, the current in BC is

$$\frac{P(a\beta + \gamma\delta) \cdot E}{2P^2 Q + (\beta\delta - a\gamma)^2},$$

where
$$P = a + \beta + \gamma + \delta, \quad Q = \beta\gamma + \gamma a + a\beta + a\delta + \beta\delta + \gamma\delta.$$

19. A wire forms a regular hexagon and the angular points are joined to the centre by wires each of which has a resistance $\dfrac{1}{n}$ of the resistance of a side of the hexagon. Shew that the resistance to a current entering at one angular point of the hexagon and leaving it by the opposite point is

$$\frac{2(n + 3)}{(n + 1)(n + 4)}$$

times the resistance of a side of the hexagon.

20. Two long equal parallel wires AB, $A'B'$, of length l, have their ends B, B' joined by a wire of negligible resistance, while A, A' are joined to the poles of a cell whose resistance is equal to that of a length r of the wire. A similar cell is placed as a bridge across the wires at a distance x from A, A'. Shew that the effect of the second cell is to increase the current in BB' in the ratio

$$2\,(2l+r)\,(x+r)/\{r\,(4l+r)+2x\,(2l-r)-4x^2\}.$$

21. There are n points $1, 2, \dots n$, joined in pairs by linear conductors. On introducing a current C at electrode 1 and taking it out at 2, the potentials of these are V_1, V_2, ... V_n. If x_{12} is the actual current in the direction 12, and x_{12}' any other that merely satisfies the conditions of introduction at 1 and abstraction at 2, shew that

$$\Sigma\,(r_{12}x_{12}x_{12}')=(\,V_1-\,V_2)\,C=\Sigma\,(r_{12}x_{12}^2),$$

and interpret the result physically.

If x typify the actual current when the current enters at 1 and leaves at 2, and y typify the actual current when the current enters at 3 and leaves at 4, shew that

$$\Sigma\,(r_{12}x_{12}y_{12})=(X_3-X_4)\,C=(\,Y_1-Y_2)\,C,$$

where the X's are potentials corresponding to currents x, and the Y's are potentials corresponding to currents y.

22. A, B, C are three stations on the same telegraph wire. An operator at A knows that there is a fault between A and B, and observes that the current at A when he uses a given battery is i, i' or i'', according as B is insulated and C to earth, B to earth, or B and C both insulated. Shew that the distance of the fault from A is

$$\{ka-k'b+(b-a)^{\frac{1}{2}}(ka-k'b)^{\frac{1}{2}}\}/(k-k'),$$

where $\qquad\qquad AB=a,\quad BC=b-a,\quad k=\dfrac{i''}{i-i''},\quad k'=\dfrac{i''}{i'-i''}.$

23. Six conductors join four points A, B, C, D in pairs, and have resistances a, a, b, β, c, γ, where a, a refer to BC, AD respectively, and so on. If this network be used as a resistance coil, with A, B as electrodes, shew that the resistance cannot lie outside the limits

$$\left[\frac{1}{c}+\frac{1}{a+b}+\frac{1}{a+\beta}\right]^{-1} \text{ and } \left[\frac{1}{c}+\left\{\left(\frac{1}{a}+\frac{1}{b}\right)^{-1}+\left(\frac{1}{a}+\frac{1}{\beta}\right)^{-1}\right\}^{-1}\right]^{-1}.$$

24. Two equal straight pieces of wire A_0A_n, B_0B_n are each divided into n equal parts at the points $A_1 \dots A_{n-1}$ and $B_1 \dots B_{n-1}$ respectively, the resistance of each part and that of A_nB_n being R. The corresponding points of each wire from 1 to n inclusive are joined by cross wires, and a battery is placed in A_0B_0. Shew that, if the current through each cross wire is the same, the resistance of the cross wire A_sB_s is

$$\{(n-s)^2+(n-s)+1\}\,R.$$

25. If n points are joined two and two by wires of equal resistance r, and two of them are connected to the electrodes of a battery of electromotive force E and resistance R, shew that the current in the wire joining the two points is

$$\frac{2E}{2r+nR}.$$

26. Six points A, B, C, D, P, Q are joined by nine conductors AB, AP, BC, BQ, PQ, QC, PD, DC, AD. An electromotive force is inserted in the conductor AD, and a galvanometer in PQ. Denoting the resistance of any conductor XY by r_{XY}, shew that if no current passes through the galvanometer,

$$(r_{BQ}+r_{BQ}+r_{QQ})\,(r_{AB}r_{DP}-r_{AP}r_{DC})+r_{BC}\,(r_{BQ}r_{DP}-r_{AP}r_{CQ})=0.$$

27. A network is made by joining the five points 1, 2, 3, 4, 5 by conductors in every possible way. Shew that the condition that conductors 23 and 14 are conjugate is

$$(K_{15} + K_{25} + K_{35} + K_{45})(K_{12}K_{34} - K_{13}K_{24})$$
$$= K_{52}(K_{54}K_{13} - K_{34}K_{15}) + K_{53}(K_{24}K_{51} - K_{54}K_{12}),$$

where K_{rs} is conductivity of conductor rs.

28. Two endless wires are each divided into mn equal parts by the successive terminals of mn connecting wires, the resistance of each part being R. There is an identically similar battery in every mth connecting wire, the total resistance of each being the same, and the resistance of each of the other $mn - n$ connecting wires is h. Prove that the current through a connecting wire which is the rth from the nearest battery is

$$\tfrac{1}{2}C(1 - \tan a)(\tan^r a + \tan^{m-r} a)/(\tan a - \tan^m a),$$

where C is the current through each battery, and $\sin 2a = h/(h + R)$.

29. A long line of telegraph wire $AA_1A_2 \ldots A_nA_{n+1}$ is supported by n equidistant insulators at $A_1, A_2, \ldots A_n$. The end A is connected to one pole of a battery of electromotive force E and resistance B, and the other pole of this battery is put to earth, as also the other end A_{n+1} of the wire. The resistance of each portion $AA_1, A_1A_2, \ldots A_nA_{n+1}$ is the same, R. In wet weather there is a leakage to earth at each insulator, whose resistance may be taken equal to r. Shew that the current strength in A_pA_{p+1} is

$$\frac{E \cosh(2n - 2p + 1)a}{B \cosh(2n + 1)a + \sqrt{Rr} \sinh(2n + 2)a},$$

where

$$2 \sinh a = \sqrt{R/r}.$$

30. A regular polygon $A_1A_2 \ldots A_n$ is formed of n pieces of uniform wire, each of resistance σ, and the centre O is joined to each angular point by a straight piece of the same wire. Shew that, if the point O is maintained at zero potential, and the point A_1 at potential V, the current that flows in the conductor A_rA_{r+1} is

$$\frac{2V \sinh a \sinh(n - 2r + 1)a}{\sigma \cosh na},$$

where a is given by the equation

$$\cosh 2a = 1 + \sin\frac{\pi}{n}.$$

31. A resistance network is constructed of $2n$ rectangular meshes forming a truncated cylinder of $2n$ faces, with two ends each in the form of a regular polygon of $2n$ sides. Each of these sides is of resistance r, and the other edges of resistance R. If the electrodes be two opposite corners, then the resistance is

$$\tfrac{1}{4}nr + \tfrac{1}{2}R\frac{\tanh\theta}{\tanh n\theta},$$

where

$$\sinh^2\theta = \frac{r}{2R}.$$

32. A network is formed by a system of conductors joining every pair of a set of n points, the resistances of the conductors being all equal, and there is an electromotive force in the conductor joining the points A_1, A_2. Shew that there is no current in any conductor except those which pass through A_1 or A_2, and find the current in these conductors.

33. Each member of the series of n points A_1, A_2, ... A_n is united to its successor by a wire of resistance ρ, and similarly for the series of n points B_1, B_2, ... B_n. Each pair of points corresponding in the two series, such as A_r and B_r, is united by a wire of resistance R. A steady current i enters the network at A_1 and leaves it at B_n. Shew that the current at A_1 divides itself between $A_1 A_2$ and $A_1 B_1$ in the ratio

$$\sinh a + \sinh (n-1) a + \sinh (n-2) a : \sinh a + \sinh (n-1) a - \sinh (n-2) a,$$

where

$$\cosh a = \frac{R+\rho}{R}.$$

34. An underground cable of length a is badly insulated so that it has faults throughout its length indefinitely near to one another and uniformly distributed. The conductivity of the faults is $1/\rho'$ per unit length of cable, and the resistance of the cable is ρ per unit length. One pole of a battery is connected to one end of a cable and the other pole is earthed. Prove that the current at the farther end is the same as if the cable were free from faults and of total resistance

$$\sqrt{\rho\rho'} \, \sinh\left(a \sqrt{\frac{\rho}{\rho'}}\right)$$

35. Two parallel conducting wires at unit distance are connected by $n+1$ cross pieces of the same wire, so as to form n squares. A current enters by an outer corner of the first square, and leaves by the diagonally opposite corner of the last. Shew that, if the resistance is that of a length $\frac{1}{2} n + a_n$ of the wire,

$$a_{n+1} = \frac{a_n + \frac{1}{2}}{a_n + 2}.$$

36. A, B are the ends of a long telegraph wire with a number of faults, and C is an intermediate point on the wire. The resistance to a current sent from A is R when C is earth connected, but if C is not earth connected the resistance is S or T according as the end B is to earth or insulated. If R', S', T' denote the resistances under similar circumstances when a current is sent from B towards A, shew that

$$T'(R-S) = R'(R-T).$$

37. The inner plates of two condensers of capacities C, C' are joined by wires of resistances R, R' to a point P, and their outer plates by wires of negligible resistance to a point Q. If the inner plates be also connected through a galvanometer, shew that the needle will suffer no sudden deflection on joining P, Q to the poles of a battery, if $CR = C'R'$.

38. An infinite cable of capacity and resistance K and R per unit length is at zero potential. At the instant $t=0$ one end is suddenly connected to a battery for an infinitesimal interval and then insulated. Shew that, except for very small values of t, the potential at any instant at a distance x from this end of the cable will be proportional to

$$\frac{1}{\sqrt{t}} e^{-\frac{KRx^2}{4t}}.$$

CHAPTER X

STEADY CURRENTS IN CONTINUOUS MEDIA

Components of Current.

370. IN the present chapter we shall consider steady currents of electricity flowing through continuous two- and three-dimensional conductors instead of through systems of linear conductors.

We can find the direction of flow at any point P in a conductor by imagining that we take a small plane of area dS and turn it about at the point P until we find the position in which the amount of electricity crossing it per unit time is a maximum. The normal to the plane when in this position will give the direction of the current at P, and if the total amount of electricity crossing this plane per unit time when in this position is CdS, then C may be defined to be the strength of the current at P.

If l, m, n are the direction-cosines of the direction of the current at P, then the current C may be treated as the superposition of three currents lC, mC, nC parallel to the axes. To prove this we need only notice that the flow across an area dS of which the normal makes an angle θ with the direction of the current, and has direction-cosines l', m', n', must be $CdS \cos \theta$, or

$$CdS \, (ll' + mm' + nn').$$

The first term of this expression may be regarded as the contribution from a current lC parallel to the axis Ox, and so on. The quantities lC, mC, nC are called the components of the current at the point P.

Lines and Tubes of Flow.

371. DEFINITION. *A line of flow is a line drawn in a conductor such that at every point its tangent is in the direction of the current at the point.*

DEFINITION. *A tube of flow is a tubular region of infinitesimal cross-section, bounded by lines of flow.*

It is clear that at every point on the surface of a tube of flow, the current is tangential to the surface. Thus no current crosses the boundary of a tube of flow, from which it follows that the aggregate current flowing across all cross-sections of a tube of flow will be the same.

The amount of this current will be called the strength of the tube.

Thus if C is the current at any point of a tube of flow, and if ω is the cross-section of the tube at that point, then $C\omega$ is constant throughout the length of the tube, and is equal to the strength of the tube.

There is an obvious analogy between tubes of flow in current electricity and tubes of force in statical electricity, the current C corresponding to the polarisation P. In current electricity, $C\omega$ is constant and equal to the strength of the tube of flow, while in statical electricity $P\omega$ is constant and equal to the strength of the tube of force (§ 129).

Specific Resistance.

372. The specific resistance of a substance is defined to be the resistance of a cube of unit edge of the substance, the current entering by a perfectly conducting electrode which extends over the whole of one face, and leaving by a similar electrode on the opposite face.

The specific resistances of some substances of which conductors and insulators are frequently made are given in the following table. The units are the centimetre and the ohm.

Silver at 18° C.	$1\cdot66\times10^{-6}$.	Graphite	$0\cdot003$.
Copper at 18° C.	$1\cdot78\times10^{-6}$.	Guttapercha	2×10^{9}.
Iron (pure) at 50° C. ...	$11\cdot5\times10^{-6}$.	Glass (soda-lime)	5×10^{11}.
Steel at 18° C.	$19\cdot9\times10^{-6}$.	„ (pyrex)	10^{14}.
Mercury at 0° C.	$94\cdot07\times10^{-6}$.	Paraffin wax	3×10^{13}.

If τ is the specific resistance of any substance, the resistance of a wire of length l and cross-section s will clearly be $\dfrac{l\tau}{s}$.

Ohm's Law.

373. In a conductor in which a current is flowing, different points will, in general, be at different potentials. Thus there will be a system of equipotentials and of lines of force inside a conductor similar to those in an electrostatic field. It is found, as an experimental fact, that in a homogeneous conductor, the lines of flow coincide with the lines of force—or, in other words, the electricity at every point moves in the direction of the forces acting on it.

In considering the motion of material particles in general it is not usually true that the motion of the particles is in the direction of the forces acting upon them. The velocity of a particle at the end of any small interval of time is compounded of the velocity at the beginning of the interval together with the velocity generated during the interval. The latter velocity is in the direction of the forces acting on the particle, but is generally insignificant in comparison with the original velocity of the particle. In the particular case in which the original velocity of the particle was very small, the direction of motion at the end of a small interval will be that of the force acting on the particle. If the particle moves in a resisting medium, it may be that the velocity of the particle is kept permanently very small by the resistance of the medium : in this case the direction of motion of the particle at every instant, relatively to the medium, may be that of the forces acting on it.

On the modern view of electricity, a current of electricity is composed of electrons which are driven through a conductor by the electric forces acting on them, and in their motion experience frequent collisions with the molecules of the conductor. The effect of these collisions is continually to check the forward velocity of the electrons, so that this forward velocity is kept small just as if they were moving through a resisting medium of the ordinary kind, and so it comes about that the direction of flow of current is in the direction of the electric intensity (cf. § 345 a).

374. Let us select any tube of force of small cross-section inside a conductor, and let P, Q be any two points on this tube of force, at which the potentials are V_P and V_Q, the former being the greater. Let these points be so near together that throughout the range PQ the cross-section of the tube of force may be supposed to have a constant value ω, while the specific resistance of the material of the conductor may be supposed to have a constant value τ.

From what has been said in § 373, it follows that the tube of force under consideration is also a tube of flow. If C denotes the current, then the current flowing through this tube of flow in the direction from P to Q will be $C\omega$. This current may, within the range PQ, be regarded as flowing through a conductor of cross-section ω and of specific resistance τ. The resistance of this conductor from P to Q is accordingly $\dfrac{PQ \cdot \tau}{\omega}$, while the fall of potential is $V_P - V_Q$. Thus by Ohm's Law

$$V_P - V_Q = \frac{PQ \cdot \tau}{\omega} \times C\omega,$$

so that $$\frac{V_P - V_Q}{PQ} = C\tau.$$

If $\dfrac{\partial}{\partial s}$ denotes differentiation along the tube of force, the fraction on the left of the foregoing equation reduces, when P and Q are made to coincide, to $-\dfrac{\partial V}{\partial s}$, so that the equation assumes the form

$$-\frac{\partial V}{\partial s} = C\tau \quad \dotfill \quad (309).$$

Let l, m, n be the direction-cosines of the line of flow at P, and let u, v, w be the components of the current at P, so that $u = lC$, etc. Then

$$\frac{\partial V}{\partial x} = l\frac{\partial V}{\partial s} = -lC\tau = -u\tau, \text{ etc.,}$$

and we see that equation (309) is equivalent to the three equations

$$\left. \begin{array}{l} u = -\dfrac{1}{\tau}\dfrac{\partial V}{\partial x} \\[2mm] v = -\dfrac{1}{\tau}\dfrac{\partial V}{\partial y} \\[2mm] w = -\dfrac{1}{\tau}\dfrac{\partial V}{\partial z} \end{array} \right\} \quad \dots\dots\dots\dots\dots\dots(310).$$

These equations express Ohm's Law in a form appropriate to flow through a solid conductor.

Equation of Continuity.

375. Since we are supposing the currents to be steady, the amount of current which flows into any closed region must be exactly equal to the amount which flows out. This can be expressed by saying that the integral algebraic flow into any closed region must be *nil*.

Let any closed surface S be taken entirely inside a conductor. Let l, m, n be the direction-cosines of the inward normal to any element dS of this surface, and let u, v, w be the components of current at this point. Then the normal component of flow across the element dS is $lu + mv + nw$, and the condition that the integral algebraic flow across the surface S shall be *nil* is expressed by the equation

$$\iint (lu + mv + nw)\, dS = 0.$$

By Green's Theorem (§ 176), this equation may be transformed into

$$\iiint \left(\frac{\partial u}{\partial x} + \frac{\partial v}{\partial y} + \frac{\partial w}{\partial z} \right) dx\, dy\, dz = 0,$$

and since this integral has to vanish, whatever the region through which it is taken, each integrand must vanish separately. Hence at every point inside the conductor, we must have

$$\frac{\partial u}{\partial x} + \frac{\partial v}{\partial y} + \frac{\partial w}{\partial z} = 0 \quad \dots\dots\dots\dots\dots\dots(311).$$

This is the so-called "equation of continuity," expressing that no electricity is created or destroyed or allowed to accumulate during the passage of a steady current through a conductor.

The same equation can be obtained at once on considering the current-flow across the different faces of a small rectangular parallelepiped of edges dx, dy, dz (cf. § 49).

Equation (311) of course expresses that the vector **C** of which the components are u, v, w, must be solenoidal. The equation of continuity can accordingly be expressed in the form

$$\operatorname{div} \mathbf{C} = 0.$$

Equation satisfied by the Potential.

376. On substituting in equation (311) the values for u, v, w given by equations (310), we obtain

$$\frac{\partial}{\partial x}\left(\frac{1}{\tau}\frac{\partial V}{\partial x}\right) + \frac{\partial}{\partial y}\left(\frac{1}{\tau}\frac{\partial V}{\partial y}\right) + \frac{\partial}{\partial z}\left(\frac{1}{\tau}\frac{\partial V}{\partial z}\right) = 0 \quad \ldots\ldots\ldots\ldots(312).$$

The potential must accordingly be a solution of this differential equation. The equation is the same as would be satisfied by the potential in an uncharged dielectric in an electrostatic field, provided the inductive capacity at every point is proportional to $\frac{1}{\tau}$. If the specific resistance of the conductor is the same throughout, the differential equation to be satisfied by the potential reduces to

$$\nabla^2 V = 0.$$

377. We may for convenience suppose that the current enters and leaves by perfectly conducting electrodes, and that the conductor through which the current flows is bounded, except at the electrodes, by perfect insulators. Then, over the surface of contact between the conductor and the electrodes, the potential will be constant. Over the remaining boundaries of the conductor, the condition to be satisfied is that there shall be no flow of current, and this is expressed mathematically by the condition that $\frac{\partial V}{\partial n}$ shall vanish.

Thus the problem of determining the current-flow in a conductor amounts mathematically to determining a function V such that equation (312) is satisfied throughout the volume of the conductor, while either $\frac{\partial V}{\partial n} = 0$, or else V has a specified value, at each point on the boundary. By the method used in § 188, it is easily shewn that the solution of this problem is unique.

It is only in a very few simple cases that an exact solution of the problem can be obtained. There are, however, various artifices by which approximations can be reached, and various ways of regarding the problem from which it may be possible to form some ideas of the physical processes which determine the nature of the flow in a conductor. Some of these will be discussed later (§§ 386—394). At present we consider general characteristics of the flow of currents through conductors.

CONDITIONS TO BE SATISFIED AT THE BOUNDARY OF TWO CONDUCTING MEDIA.

378. The conditions to be satisfied at a boundary at which the current flows from one conductor to another are as follows:

(i) Since there must be no accumulation of electricity at the boundary, the normal flow across the boundary must be the same whether calculated in the first medium or the second. In other words

$$\frac{1}{\tau} \frac{\partial V}{\partial n} \text{ must be continuous,}$$

where $\frac{\partial}{\partial n}$ denotes differentiation along the normal to the boundary.

(ii) The tangential force must be continuous, or else the potential would not be continuous. Thus

$$\frac{\partial V}{\partial s} \text{ must be continuous,}$$

where $\frac{\partial}{\partial s}$ denotes differentiation along any line in the boundary.

These boundary conditions are just the same as would be satisfied in an electrostatical problem at the boundary between two dielectrics of inductive capacities equal to the two values of $\frac{1}{\tau}$. Thus the equipotentials in this electrostatic problem coincide with the equipotentials in the actual current problem, and the lines of force in the electrostatic problem correspond with the lines of flow in the current problem.

Clearly these results could be deduced at once from the differential equation (312) on passing to the limit and making τ become discontinuous on crossing a boundary.

Refraction of Lines of Flow.

379. Let any line of flow cross the boundary between two different conducting media of specific resistances τ_1, τ_2, making angles ϵ_1, ϵ_2 with the normal at the point at which it meets the boundary in the two media respectively. The lines of flow satisfy the same conditions as would be satisfied by electrostatic lines of force crossing the boundary between two dielectrics of inductive capacities $\frac{1}{\tau_1}, \frac{1}{\tau_2}$, so that we must have (cf. equation (71))

$$\frac{1}{\tau_1} \cot \epsilon_1 = \frac{1}{\tau_2} \cot \epsilon_2.$$

Hence $\tau_1 \tan \epsilon_1 = \tau_2 \tan \epsilon_2,$

expressing the law of refraction of lines of flow.

380. As an example of refraction of lines of current flow, we may consider the case of a steady uniform current in a conductor being disturbed by the presence of a sphere of different metal inside the conductor. The lines shewn in fig. 78 will represent the lines of flow if the specific resistance of the sphere is less than that of the main conductor. The lines of flow tend to crowd into the sphere, this being the better conductor—in the language of popular science, the current tends to take the path of least resistance.

Charge on a Surface of Discontinuity.

381. If u is the normal component of current flowing across the boundary between two different conductors, we have by Ohm's Law,

$$u = -\frac{1}{\tau_1}\frac{\partial V_1}{\partial n} = -\frac{1}{\tau_2}\frac{\partial V_2}{\partial n},$$

where $\dfrac{\partial}{\partial n}$ denotes differentiation along the normal which is drawn in the direction in which u is measured (say from (1) to (2)), and V_1, V_2 are the potentials in the two conductors.

If there is no charge on the boundary between the two conductors we must, from equation (70), have the relation

$$K_1\frac{\partial V_1}{\partial n} = K_2\frac{\partial V_2}{\partial n},$$

where K_1, K_2 are the inductive capacities of the two conductors. This condition will, however, in general be inconsistent with the condition which, as we have just seen, is made necessary by the continuity of u. Thus there will in general be a surface charge on the boundary between two conductors of different materials.

The amount of this charge is given at once by equation (72), p. 125. If σ denotes the surface density at any point, we have

$$4\pi\sigma = K_1\frac{\partial V_1}{\partial n} - K_2\frac{\partial V_2}{\partial n}$$

$$= -(K_1\tau_1 - K_2\tau_2)\,u \quad\ldots\ldots\ldots\ldots\ldots\ldots(313).$$

This surface charge is very small compared with the charges which occur in statical electricity. For instance, if we have current of 100 ampères per sq. cm. passing from one metallic conductor to another, we take in formula (313),

$$u = 100 \text{ ampères} = 3 \times 10^{11} \text{ electrostatic units,}$$

$$\tau = 10^{-6} \text{ ohms} = \frac{10^{-6}}{9 \times 10^{11}} \quad \text{''} \quad\quad \text{''}$$

$$K = 1,$$

the last two being true as regards order of magnitude only. The value of $4\pi\sigma$ is of the order of magnitude of $K\tau u$, or $\frac{1}{3} \times 10^{-6}$ in electrostatic units. As has been said, the value of $4\pi\sigma$ at the surface of a conductor charged as highly as possible in air is of the order of 100.

382. As an example of the distribution of a surface charge, we may notice that the surface-density of the charge on the surface of the sphere considered in § 380 will be proportional to either value of $\dfrac{\partial V}{\partial n}$, and therefore to $\cos \theta$, where θ is the angle between the radius through the point and the direction of flow of the undisturbed current.

GENERATION OF HEAT.

383. Consider any small element of a tube of flow, length ds, cross-section ω. The current per unit area is, by equations (310), $-\dfrac{1}{\tau}\dfrac{\partial V}{\partial s}$, so that the current flowing through the tube is $-\dfrac{1}{\tau}\dfrac{\partial V}{\partial s}\omega$. The resistance of the element of the tube under consideration is $\dfrac{\tau ds}{\omega}$. Hence, as in § 355, the amount of heat generated per unit time in this element is

$$\left(\frac{1}{\tau}\frac{\partial V}{\partial s}\omega\right)^2\frac{\tau ds}{\omega} \text{ or } \frac{1}{\tau}\left(\frac{\partial V}{\partial s}\right)^2\omega ds.$$

Thus the heat generated per unit time per unit volume is $\dfrac{1}{\tau}\left(\dfrac{\partial V}{\partial s}\right)^2$, and the total generation of heat per unit time will be

$$\iiint \frac{1}{\tau}\left(\frac{\partial V}{\partial s}\right)^2 dx\,dy\,dz,$$

or
$$\iiint \frac{1}{\tau}\left\{\left(\frac{\partial V}{\partial x}\right)^2+\left(\frac{\partial V}{\partial y}\right)^2+\left(\frac{\partial V}{\partial z}\right)^2\right\} dx\,dy\,dz \quad\ldots\ldots\ldots\ldots(314).$$

Thus the heat generated per unit time is 8π times the energy of the whole field in the analogous electrostatic problem (§ 169).

Rate of generation of heat a minimum.

384. It can be shewn that for a given current flowing through a conductor, the rate of heat generation is a minimum when the current distributes itself as directed by Ohm's Law. To do this we have to compare the rate of heat generation just obtained with the rate of heat generation when the current distributes itself in some other way.

Let us suppose that the components of current at any point have no longer the values

$$-\frac{1}{\tau}\frac{\partial V}{\partial x}, \quad -\frac{1}{\tau}\frac{\partial V}{\partial y}, \quad -\frac{1}{\tau}\frac{\partial V}{\partial z}$$

assigned to them by Ohm's Law, but that they have different values

$$-\frac{1}{\tau}\frac{\partial V}{\partial x}+u, \quad -\frac{1}{\tau}\frac{\partial V}{\partial y}+v, \quad -\frac{1}{\tau}\frac{\partial V}{\partial z}+w.$$

In order that there may be no accumulation at any point under this new distribution, the components of current must satisfy the equation of continuity, so that we must have

$$\frac{\partial u}{\partial x} + \frac{\partial v}{\partial y} + \frac{\partial w}{\partial z} = 0 \quad \dots\dots\dots\dots\dots\dots\dots(315).$$

By the same reasoning as in § 383, we find for the rate at which heat is generated under the new system of currents,

$$\iiint \tau \left\{ \left(-\frac{1}{\tau}\frac{\partial V}{\partial x} + u \right)^2 + \left(-\frac{1}{\tau}\frac{\partial V}{\partial y} + v \right)^2 + \left(-\frac{1}{\tau}\frac{\partial V}{\partial z} + w \right)^2 \right\} dx\,dy\,dz,$$

which, on expanding, is equal to

$$\iiint \frac{1}{\tau} \left\{ \left(\frac{\partial V}{\partial x}\right)^2 + \left(\frac{\partial V}{\partial y}\right)^2 + \left(\frac{\partial V}{\partial z}\right)^2 \right\} dx\,dy\,dz$$

$$-2\iiint \left(u\frac{\partial V}{\partial x} + v\frac{\partial V}{\partial y} + w\frac{\partial V}{\partial z} \right) dx\,dy\,dz$$

$$+ \iiint \tau\,(u^2 + v^2 + w^2)\,dx\,dy\,dz \quad \dots\dots\dots\dots\dots\dots\dots\dots(316).$$

On transforming by Green's Theorem, the second term

$$= 2\iiint V \left(\frac{\partial u}{\partial x} + \frac{\partial v}{\partial y} + \frac{\partial w}{\partial z} \right) dx\,dy\,dz - 2\iint V\,(lu + mv + nw)\,dS.$$

The volume integral vanishes by equation (315), the integrand of the surface integral vanishes over each electrode from the condition that the total flow of current across the electrode is to remain unaltered, and at every point of the insulating boundary from the condition that there is to be no flow across this boundary. Thus the new rate of generation of heat is represented by the first and third terms of expression (316). The first term represents the old rate of generation of heat, the third term is an essentially positive quantity. Thus the rate of heat generation is increased by any deviation from the natural distribution of currents, proving the result.

385. An immediate result of this is that any increase or decrease in the specific resistance of any part of a conductor is accompanied by an increase or decrease of the resistance of the conductor as a whole. For on decreasing the value of τ at any point and keeping the distribution of currents unaltered, the rate of heat production will obviously decrease. On allowing the currents to assume their natural distribution, the rate of heat production will further decrease. Thus the rate of heat production with a natural distribution of currents is lessened by any decrease of specific resistance. But if I is the total current transmitted by the conductor, and R the resistance of the conductor, this rate of heat production is RI^2. Thus R decreases when τ is decreased at any point, and obviously the converse must be true (cf. § 359).

THE SOLUTION OF SPECIAL PROBLEMS.

Current-flow in an Infinite Conductor.

386. A good approximation to the conditions of electric flow can occasionally be obtained by neglecting the restrictive influence of the boundaries of a conductor, and regarding the problem as one of flow between two electrodes in an infinite conductor. For simplicity, we shall consider only the case in which the conductor is homogeneous.

The conditions to be satisfied by the potential V are as follows. We must have $V = V_1$ over one electrode, and $V = V_2$ over the second electrode, while $\dfrac{\partial V}{\partial r}$ must vanish at infinity to a higher order than $\dfrac{1}{r^2}$ and throughout the conductor we must have $\nabla^2 V = 0$ (§ 376). We can easily see (cf. §§ 186, 187) that these conditions determine V uniquely.

Consider now an analogous electrostatic problem. Let the conducting medium be replaced by air, while the electrodes remain conductors. Let the electrodes receive equal and opposite charges of electricity until their difference of potential is $V_1 - V_2$. At this stage let ψ denote the electrostatic potential at any point in the field. Let ψ_1, ψ_2 be the values of ψ over the two electrodes, so that $\psi_1 - \psi_2 = V_1 - V_2$. Then there will be a constant C (namely $V_1 - \psi_1$), such that $\psi + C$ assumes the values V_1, V_2 respectively over the two electrodes. Moreover $\nabla^2 \psi = 0$ throughout the field, so that $\nabla^2 (\psi + C) = 0$ throughout the field, and $\psi = 0$ at infinity except for terms in $\dfrac{1}{r^2}$ (cf. § 67), so that $\dfrac{\partial}{\partial r} (\psi + C)$ vanishes at infinity to a higher order than $\dfrac{1}{r^3}$.

Hence $\psi + C$ satisfies the conditions which, as we have seen, must be satisfied by the potential V in the current problem, and these are known to suffice to determine V uniquely. It follows that the value of V must be $\psi + C$.

Thus the lines of flow in the current problem are identical with the lines of force when the two electrodes are charged to different potentials in air.

The normal current-flow at any point on the surface of an electrode is

$$-\frac{1}{\tau} \frac{\partial V}{\partial n},$$

so that the total flow of current outwards from this electrode

$$= -\frac{1}{\tau} \iint \frac{\partial V}{\partial n} \, dS = -\frac{1}{\tau} \iint \frac{\partial \psi}{\partial n} \, dS.$$

If E is the charge on this electrode in the analogous electrostatic problem we have, by Gauss' Theorem,

$$-\iint \frac{\partial \psi}{\partial n}\, dS = 4\pi E,$$

so that the total flow of current is seen to be $\dfrac{4\pi E}{\tau}$.

If p_{11}, p_{12}, p_{22} are the coefficients of potential in the electrostatic problem

$$\psi_1 = p_{11} E - p_{12} E,$$
$$\psi_2 = p_{12} E - p_{22} E,$$

so that

$$V_1 - V_2 = \psi_1 - \psi_2 = (p_{11} - 2p_{12} + p_{22})\, E.$$

If I is the total current, and R the equivalent resistance between the electrodes, we have just seen that

$$I = \frac{4\pi E}{\tau},$$

so that

$$R = \frac{V_1 - V_2}{I} = \frac{\tau}{4\pi} (p_{11} - 2p_{12} + p_{22}) \quad \ldots\ldots\ldots\ldots\ldots(317).$$

If we regard the two electrodes in air as forming a condenser, and denote its capacity by C, we have

$$V_1 - V_2 = \psi_1 - \psi_2 = \frac{E}{C},$$

so that

$$R = \frac{V_1 - V_2}{I} = \frac{\tau}{4\pi C} \quad \ldots\ldots\ldots\ldots\ldots\ldots\ldots\ldots\ldots(318).$$

387. As instances of the applications of formulae (317) and (318) to special problems, we have the following:

I. The resistance per unit length between two concentric cylinders of radii a, b (as, for instance, the resistance between the core of a submarine cable and the sea), is, by formula (318),

$$\frac{\tau}{2\pi} \log \frac{b}{a}.$$

II. The resistance per unit length between two straight parallel cylindrical wires of radii a, b, placed with their centres at a great distance r apart, in an infinite conducting medium, is, by formula (317),

$$-\frac{\tau}{2\pi} (\log a - 2 \log r + \log b)$$

$$= \frac{\tau}{2\pi} \log \frac{r^2}{ab}.$$

III. The resistance between two spherical electrodes, radii a, b, at a great distance r apart, in an infinite conducting medium, is, by formula (317),

$$\frac{\tau}{4\pi} \left\{ \frac{1}{a} + \frac{1}{b} - \frac{2}{r} \right\}.$$

388. If two electrodes of any shape are placed in an infinite medium at a distance r apart, which is great compared with their linear distances, we may take p_{12} in formula (317) equal, to a first approximation, to $\frac{1}{r}$. This is small compared with p_{11} and p_{22}, so that, to a first approximation, we may replace formula (317) by

$$R = \frac{\tau}{4\pi} (p_{11} + p_{22}).$$

It accordingly appears that the resistance of the infinite medium may be regarded as the sum of two resistances—a resistance $\frac{\tau p_{11}}{4\pi}$ at the crossing of the current from the first electrode to the medium, and a resistance $\frac{\tau p_{22}}{4\pi}$ at the return of the current from the medium to the second electrode. Thus we may legitimately speak of the resistance of a single junction between an electrode and the conducting medium surrounding it.

For instance, suppose a circular plate of radius a is buried deep in the earth, and acts as electrode to distribute a current through the earth. The value of p_{11} for a disc of radius a is $\frac{\pi}{2a}$, so that the resistance of the junction is $\frac{\tau}{8a}$. So also if a disc of radius a is placed on the earth's surface, the resistance at the junction is $\frac{\tau}{4a}$, and clearly this also is the resistance if the electrode is a semicircle of radius a buried vertically in the earth with its diameter in the surface.

Flow in a Plane Sheet of Metal.

389. When the flow takes place in a sheet of metal of uniform thickness and structure, so that the current at every point may be regarded as flowing in a plane parallel to the surface of the sheet, the whole problem becomes two-dimensional. If x, y are rectangular coordinates, the problem reduces to that of finding a solution of

$$\frac{\partial^2 V}{\partial x^2} + \frac{\partial^2 V}{\partial y^2} = 0$$

which shall be such that either V has a given value, or else $\frac{\partial V}{\partial n} = 0$, at every point of the boundary. The methods already given in Chap. VIII for obtaining two-dimensional solutions of Laplace's equation are therefore available for the present problem. The method of greatest value is that of Conjugate Functions.

If the conducting medium extends to infinity, or is bounded entirely by the two electrodes, the transformations will be identical with those already discussed for two conductors at different potentials (§ 386). If the medium has also boundaries at which $\dfrac{\partial V}{\partial n} = 0$, the procedure must be slightly different.

We must try to transform the two electrodes into lines $V = $ constant, and the other boundaries into lines $U = $ constant, so that the whole of the medium becomes transformed into the interior of a rectangle in the U, V plane.

Let $$U + iV = f(x + iy)$$

be a transformation which gives the required value for V over both electrodes, and gives $\dfrac{\partial V}{\partial n} = 0$ over the boundary of a conductor. Then V will be the potential at any point, the lines $V = $ constant will be the equipotentials, and the lines $U = $ constant, being the orthogonal trajectories of the equipotentials, will be the lines of flow.

At any point the direction of the current is normal to the equipotential through the point, and the amount of the current is given by

$$C = \frac{1}{\tau} \frac{\partial V}{\partial n}.$$

But $\dfrac{\partial V}{\partial n}$ is equal to $\dfrac{\partial U}{\partial s}$, where $\dfrac{\partial}{\partial s}$ denotes differentiation in the equipotential. Thus the current flowing across any piece PQ of an equipotential

$$= \int_P^Q C\,ds$$
$$= \int_P^Q \frac{1}{\tau} \frac{\partial U}{\partial s}\,ds = \frac{1}{\tau}(U_Q - U_P).$$

If P, Q are any two points in the conductor, a path from P to Q can be regarded as made up of a piece of an equipotential PN, and a piece of a line of flow NQ. The flow across NQ is zero, that across PN is

$$\frac{1}{\tau}(U_N - U_P).$$

This is accordingly the total flow across PQ, and since $U_N = U_Q$, it may be written as

$$\frac{1}{\tau}(U_Q - U_P).$$

390. As an illustration, let us suppose that the conducting plate is a polygon, two or more edges being the electrodes. We can transform this into the real axis in the ζ-plane by a transformation of the type

$$\frac{\partial z}{\partial \zeta} = (\zeta - a_0)^{\frac{a_0}{\pi} - 1}(\zeta - a_1)^{\frac{a_1}{\pi} - 1}\dots \qquad \dots\dots\dots\dots(319),$$

and this real axis has to be transformed into a rectangle formed (say) by the lines $V = V_0$, $V = V_1$, $U = 0$, $U = C$ in the W-plane. The transformation for this will be

$$\frac{\partial W}{\partial \zeta} = [(\zeta - a_0)(\zeta - a_p)(\zeta - a_q)(\zeta - a_r)]^{-\frac{1}{2}} \quad \ldots\ldots\ldots\ldots(320),$$

where a_0, a_p and a_q, a_r are the points on the real axis of ζ which determine the ends of the electrodes. By elimination of ζ from the integrals of equations (319) and (320) we obtain the transformation required.

391. The following example of this method is taken from a paper by H. F. Moulton (*Proc Lond. Math. Soc.* III. p. 104).

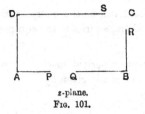

z-plane.
FIG. 101.

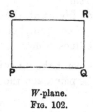

W-plane.
FIG. 102.

In fig. 101, let $ABCD$ be a rectangular plate, the piece PQ of one or more sides being one electrode, and the piece RS of one or more other sides being the other electrode. Let the rectangle $PQRS$ in fig. 102 be its transformation in the W-plane. In the intermediate ζ-plane, let the points A, B, C, D transform to $\zeta = a, b, c, d$ respectively, and let the points P, Q, R, S transform to $\zeta = p, q, r, s$ respectively. Then the transformations are

$$\frac{dz}{d\zeta} = [(\zeta - a)(\zeta - b)(\zeta - c)(\zeta - d)]^{-\frac{1}{2}},$$

$$\frac{dW}{d\zeta} = [(\zeta - p)(\zeta - q)(\zeta - r)(\zeta - s)]^{-\frac{1}{2}}.$$

If we write

$$\frac{(b-c)(a-d)}{(a-c)(b-d)} = \kappa, \quad \frac{(q-r)(p-s)}{(p-r)(q-s)} = \lambda,$$

$$2m = \sqrt{(a-c)(b-d)}, \quad 2m' = \sqrt{(p-r)(q-s)},$$

the integrals are

$$\zeta = \frac{a(b-d) - b(a-d)\,\mathrm{sn}^2\,mz\,(\mathrm{mod}\,\kappa)}{b-d-(a-d)\,\mathrm{sn}^2\,mz\,(\mathrm{mod}\,\kappa)} \quad \ldots\ldots\ldots\ldots(321),$$

$$\zeta = \frac{p(q-s) - q(p-s)\,\mathrm{sn}^2\,m'W\,(\mathrm{mod}\,\lambda)}{q-s-(p-s)\,\mathrm{sn}^2\,m'W\,(\mathrm{mod}\,\lambda)} \quad \ldots\ldots\ldots\ldots(322).$$

The sides AB, AD of the first rectangle are the periods $\dfrac{K}{m}$, $\dfrac{iK'}{m}$ of sn mz (mod κ); the sides PQ, PS of the second rectangle are the periods in W, say $\dfrac{L}{m'}$, $\dfrac{iL'}{m'}$, of sn $m'W$ (mod λ).

In the W-plane, the potential difference of the two electrodes is PS, or $\dfrac{L'}{m'}$, while the current is $\dfrac{1}{\tau} PQ$, or $\dfrac{L'}{m'\tau}$. The equivalent resistance of the plate is accordingly $\tau L'/L$, so that the quantity we are trying to determine is L'/L.

Let the coordinates of P, Q, R, S in the z-plane be z_1, z_2, z_3, z_4. In the ζ-plane the coordinates of these points are p, q, r, s. Hence from equations (321), we have

$$p = \frac{a(b-d) - b(a-d) \operatorname{sn}^2 mz_1 (\operatorname{mod} \kappa)}{(b-d) - (a-d) \operatorname{sn}^2 mz_1 (\operatorname{mod} \kappa)},$$

and similar equations for q, r, s. The ratio L'/L of which we are in search is now given by

$$\frac{L'}{L} = \frac{(q-r)(p-s)}{(p-r)(q-s)} = \frac{(\operatorname{sn}^2 mz_2 - \operatorname{sn}^2 mz_3)(\operatorname{sn}^2 mz_1 - \operatorname{sn}^2 mz_4)}{(\operatorname{sn}^2 mz_1 - \operatorname{sn}^2 mz_3)(\operatorname{sn}^2 mz_2 - \operatorname{sn}^2 mz_4)},$$

the whole being to modulus κ. The values of sn mz can be obtained from Legendre's Tables.

Moulton has calculated the resistance of a square sheet with electrodes, each of length equal to one-fifth of a side, in the following four cases:

(1) Electrodes at middle of two opposite sides, Resistance $= 1·745R$,

(2) Electrodes at ends of two opposite sides and facing one another, Resistance $= 2·408R$,

(3) Electrodes at ends of two opposite sides and not facing one another, Resistance $= 2·589R$,

(4) Electrodes bent equally round two opposite corners of square, Resistance $= 3·027R$,

where R is the resistance of the square when the whole of two opposite sides form the electrodes. A comparison of the results in cases (2) and (3) shews how large a part of the resistance is due to the crowding in of the lines of force near the electrode, and how small a part arises from the uncrowded part of the path.

Limits to the Resistance of a Conductor.

392. The result obtained in § 386 enables us to assign an upper and a lower limit to the resistance of a conductor, when this resistance cannot be calculated accurately. For if any parts of the conductor are made into perfect conductors, the resistance of the whole will be lessened, and it may be possible to change parts of the conductor into perfect conductors in such a way that the resistance of the new conductor can be calculated. This resistance will then be a lower limit to the resistance of the original conductor.

As an illustration, we may examine the case of a straight wire of variable cross-section S. Let us imagine that at small distances along its length we take cross-sections of infinitely small thickness, and make these into perfect conductors. The resistance between two such sections at distance ds apart, will be $\dfrac{\tau ds}{S}$, where S is the cross-section of either. Thus a lower limit to the resistance is supplied by the formula

$$\tau \int \frac{ds}{S}.$$

393. Again, if we replace parts of the conductor by insulators, so causing the current to flow in given channels, the resistance of the whole is increased, and in this way we may be able to assign an upper limit to the resistance of a conductor.

394. As an instance of a conductor to the resistance of which both upper and lower limits can be assigned, let us consider the case of a cylindrical conductor AB terminating in an infinite conductor C of the same material. This example is of practical importance in connection with mercury resistance standards. The appropriate analysis was first given by Lord Rayleigh, discussing a parallel problem in the theory of sound.

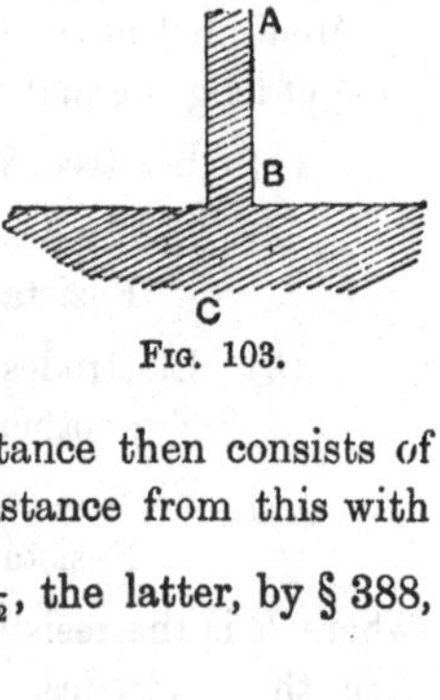

Fig. 103.

Let l be the length and a the radius of the tube. To obtain a lower limit to the resistance, we imagine a perfectly conducting plane inserted at B. The resistance then consists of the resistance to this new electrode at B, plus the resistance from this with the infinite conductor C. The former resistance is $\dfrac{l\tau}{\pi a^2}$, the latter, by § 388, is $\dfrac{\tau}{4a}$, so that a lower limit to the whole resistance is

$$\frac{l\tau}{\pi a^2} + \frac{\tau}{4a},$$

which is the resistance of a length $l + \dfrac{\pi a}{4}$ of the tube.

To obtain an upper limit to the resistance, we imagine non-conducting tubes placed inside the main tube AB, so that the current is constrained to flow in a uniform stream parallel to the axis of the main tube until the end B is reached. After this the current flows through the semi-infinite conductor C as directed by Ohm's Law.

The resistance of the tube AB is, as before, $\dfrac{l\tau}{\pi a^2}$. To obtain the resistance of the conductor C, we must examine the corresponding electrostatic problem. If I is the total current, the flow of current per unit area over the circular mouth at B is $I/\pi a^2$. In order that the potentials in the electrostatic problem may be the same, we must have a uniform surface density of electricity

$$\sigma = \left(\frac{\tau}{4\pi}\right)\left(\frac{I}{\pi a^2}\right) \text{ or } \frac{\tau I}{4\pi^2 a^2},$$

on the surface of the disc.

The heat generated is I^2R, where R is the resistance of the conductor C. It is also

$$\frac{1}{\tau} \iiint \left\{\left(\frac{\partial V}{\partial x}\right)^2 + \left(\frac{\partial V}{\partial y}\right)^2 + \left(\frac{\partial V}{\partial z}\right)^2\right\} dx\,dy\,dz \quad \ldots\ldots\ldots\ldots(323),$$

taken through the conductor C. Now if W is the electrostatic energy of a disc of radius a, having a uniform surface density $\sigma = \dfrac{\tau I}{4\pi^2 a^2}$ on each side, we have

$$W = \frac{1}{8\pi} \iiint \left\{\left(\frac{\partial V}{\partial x}\right)^2 + \left(\frac{\partial V}{\partial y}\right)^2 + \left(\frac{\partial V}{\partial z}\right)^2\right\} dx\,dy\,dz,$$

where the integral is taken through all space, or again,

$$W = \frac{1}{4\pi} \iiint \left\{\left(\frac{\partial V}{\partial x}\right)^2 + \left(\frac{\partial V}{\partial y}\right)^2 + \left(\frac{\partial V}{\partial z}\right)^2\right\} dx\,dy\,dz,$$

where the integral is taken through the semi-infinite space on one side of the disc, *i.e.* through the space C, if the disc is made to coincide with the mouth B. On substituting for the volume integral in expression (323), we find that

$$I^2R = \frac{4\pi W}{\tau} \quad \ldots\ldots\ldots\ldots\ldots\ldots\ldots\ldots\ldots(324).$$

Following Maxwell, we shall find it convenient to calculate W directly from the potential. If a disc of radius r has a uniform surface density σ on each side, the potential at a point P on its edge will be

$$V_P = 2\sigma \iint \frac{dx\,dy}{r},$$

where the integral is taken over one side of the disc, and r is the distance from P to the element $dx\,dy$. Taking polar coordinates, with P as origin, the equation of the circle will be $r = 2b\cos\theta$; we may replace $dx\,dy$ by $r\,dr\,d\theta$, and obtain

$$V_P = 2\sigma \int_{r=0}^{r=2b\cos\theta} \int_{\theta=-\frac{\pi}{2}}^{\theta=\frac{\pi}{2}} dr\,d\theta = 8b\sigma.$$

On increasing the radius of the disc to $b + db$, we bring up a charge $4\pi b\sigma db$ from infinity to potential $8b\sigma$, so that the work done is

$$dW = 32\pi b^3\sigma^2 db,$$

and integrating from $b = 0$ to $b = a$, we find for the potential energy of the complete disc of radius a,

$$W = \tfrac{3}{3}^2\pi a^3\sigma^2.$$

Thus, from equation (324),

$$R = \frac{4\pi W}{I^2\tau} = \frac{128\pi^3 a^3\sigma^2}{3I^2\tau},$$

or, since

$$\sigma = \frac{\tau I}{4\pi^2 a^2},$$

$$R = \frac{8\tau}{3\pi^2 a}.$$

Thus an upper limit to the whole resistance is

$$\frac{l\tau}{\pi a^2} + \frac{8\tau}{3\pi^2 a},$$

which is the resistance of a length $l + \dfrac{8}{3\pi} a$ of the tube.

Thus we may say that the resistance of the whole is that of a length $l + \alpha a$ of the tube, where α is intermediate between $\dfrac{\pi}{4}$ and $\dfrac{8}{3\pi}$, *i.e.* between ·785 and ·849. Lord Rayleigh[*], by more elaborate analysis, has shewn that the upper limit for α must be less than ·8242, and believes that the true value of α must be pretty close to ·82.

The passage of Electricity through Dielectrics.

395. Since even the best insulators are not wholly devoid of conducting power, it is of importance to consider the flow of electricity in dielectrics.

Using the previous notation, we shall denote the potential at any point in the dielectric by V, the specific resistance by τ, and the inductive capacity by K. We shall consider steady flow first.

If the flow is to be steady, the equation of continuity, namely

$$\frac{\partial}{\partial x}\left(\frac{1}{\tau}\frac{\partial V}{\partial x}\right) + \frac{\partial}{\partial y}\left(\frac{1}{\tau}\frac{\partial V}{\partial y}\right) + \frac{\partial}{\partial z}\left(\frac{1}{\tau}\frac{\partial V}{\partial z}\right) = 0 \ldots\ldots\ldots\ldots(325),$$

must be satisfied. Also if there is a volume density of electrification ρ, the potential must satisfy equation (62), namely

$$\frac{\partial}{\partial x}\left(K\frac{\partial V}{\partial x}\right) + \frac{\partial}{\partial y}\left(K\frac{\partial V}{\partial y}\right) + \frac{\partial}{\partial z}\left(K\frac{\partial V}{\partial z}\right) = -4\pi\rho \ldots\ldots(326).$$

[*] *Theory of Sound*, Vol. ii. Appendix A.

From a comparison of equations (325) and (326), it is clear that steady flow will not generally be consistent with having $\rho = 0$. Hence if currents are started flowing through an uncharged dielectric, the dielectric will acquire volume charges before the currents become steady. When the currents have become steady, the value of V will be determined by equation (325) and the boundary conditions, and the value of ρ is then given by equation (326).

From equations (325) and (326), we obtain

$$\rho = -\frac{1}{4\pi\tau}\left\{\frac{\partial V}{\partial x}\frac{\partial}{\partial x}(K\tau) + \frac{\partial V}{\partial y}\frac{\partial}{\partial y}(K\tau) + \frac{\partial V}{\partial z}\frac{\partial}{\partial z}(K\tau)\right\}\ \dots\dots(327).$$

The condition that ρ shall vanish, whatever the value of V, is that $K\tau$ shall be constant throughout the dielectric: if this condition is satisfied the value of ρ necessarily vanishes at every point for all systems of steady currents. The most important case of this condition being satisfied occurs when the dielectric is homogeneous throughout. If $K\tau$ is not constant throughout the dielectric, equation (327) shews that we can have $\rho = 0$ at every point provided the surfaces $V = \text{cons.}$ and $K\tau = \text{cons.}$ cut one another at right angles at every point, *i.e.* provided $K\tau$ is constant along every line of flow.

We have already had an illustration (§ 381) of the accumulation of charge which occurs when the value of $K\tau$ varies in passing along a line of flow.

Time of Relaxation in a Homogeneous Dielectric.

396. Let a homogeneous dielectric be charged so that the volume density at any point is ρ.

If any closed surface is taken inside the dielectric, the total charge inside this surface must be

$$\iiint \rho\, dx\, dy\, dz,$$

while the rate at which electricity flows into the surface will, as in § 375, be

$$\iint (lu + mv + nw)\, dS,$$

where u, v, w are the components of current and l, m, n are the direction cosines of the normal drawn into the surface. Since this rate of flow into the surface must be equal to the rate at which the charge inside the surface increases, we must have

$$\iint (lu + mv + nw)\, dS = \frac{d}{dt}\iiint \rho\, dx\, dy\, dz$$

$$= \iiint \frac{d\rho}{dt}\, dx\, dy\, dz.$$

The integral on the left may, by Green's Theorem, be transformed into

$$- \iiint \left(\frac{\partial u}{\partial x} + \frac{\partial v}{\partial y} + \frac{\partial w}{\partial z} \right) dx\,dy\,dz,$$

and this again is equal, by equations (310), to

$$\iiint \frac{1}{\tau} \left(\frac{\partial^2 V}{\partial x^2} + \frac{\partial^2 V}{\partial y^2} + \frac{\partial^2 V}{\partial z^2} \right) dx\,dy\,dz.$$

Thus we have

$$\iiint \left\{ \frac{1}{\tau} \left(\frac{\partial^2 V}{\partial x^2} + \frac{\partial^2 V}{\partial y^2} + \frac{\partial^2 V}{\partial z^2} \right) - \frac{d\rho}{dt} \right\} dx\,dy\,dz = 0,$$

and since this is true whatever surface is taken, each integrand must vanish separately, and we must have, at every point of the dielectric,

$$\frac{\partial^2 V}{\partial x^2} + \frac{\partial^2 V}{\partial y^2} + \frac{\partial^2 V}{\partial z^2} = \tau \frac{d\rho}{dt}.$$

We have also, as in equation (326),

$$\frac{\partial^2 V}{\partial x^2} + \frac{\partial^2 V}{\partial y^2} + \frac{\partial^2 V}{\partial z^2} = - \frac{4\pi\rho}{K},$$

so that

$$\frac{d\rho}{dt} = - \frac{4\pi}{K\tau} \rho.$$

The integral of this equation is

$$\rho = \rho_0 e^{-\frac{4\pi}{K\tau} t},$$

where ρ_0 is the value of ρ at time $t = 0$.

Thus the charge at every point in the dielectric falls off exponentially with the time, the modulus of decay being $\frac{4\pi}{K\tau}$. The time $\frac{K\tau}{4\pi}$, in which all the charges in the dielectric are reduced to $1/e$ times their original value, is called the "time of relaxation," being analogous to the corresponding quantity in the Dynamical Theory of Gases[*].

The relaxation-time admits of experimental determination, and as τ is easily determined, this gives us a means of determining K experimentally for conductors. In the case of good conductors, the relaxation-time is too small to be observed with any accuracy, but the method has been employed by Cohn and Arons[†] to determine the inductive capacity of water. The value obtained, $K = 73\cdot6$, is in good agreement with the values obtained in other ways (cf. § 84).

---

[*] Cf. Maxwell, *Collected Works*, II. p. 681, or Jeans, *Dynamical Theory of Gases*, p. 294.
[†] *Wied. Ann.* XXVIII. p. 454.

Discharge of a Condenser.

397. Let us suppose that a condenser is charged up to a certain potential, and that a certain amount of leakage takes place through the dielectric between the two plates. Then, as we have just seen, the dielectric will, except in very special cases, become charged with electricity.

Now suppose that the two plates are connected by a wire, so that, in ordinary language, the condenser is discharged. Conduction through the wire is a very much quicker process than conduction through the dielectric, so that we may suppose that the plates of the condenser are reduced to the same potential before the charges imprisoned in the dielectric have begun to move. For simplicity, let us suppose that the plates of the condenser are both reduced to potential zero. Then the surface of the dielectric may, with fair accuracy, be regarded as an equipotential surface, the potential being zero all over it. It follows that there can be no lines of force outside this equipotential: all lines of force which originate on the charges imprisoned in the dielectric, and which do not terminate on similar charges, must terminate on the surface of the dielectric. Thus we shall have a system of charges on the surface of the dielectric, these charges being equal in magnitude but opposite in sign to those of the Green's "equivalent stratum" corresponding to the system of charges imprisoned in the dielectric. This system of charges on the surface of the dielectric is of the kind which Faraday would call a "bound" charge (cf. § 141).

Suppose the plates of the condenser to be again insulated. The system of charges inside the dielectric and at its surface is not an equilibrium distribution, so that currents will be set up in the dielectric, and a general rearrangement of electricity will take place. The potentials throughout the dielectric will change, and in particular the potentials of the condenser-plates at the surface of the dielectric will change. In other words, the charge on these plates is no longer a "bound" charge, but becomes, at least partially, a "free" charge. On joining the two plates by a wire, a new discharge will take place.

This is Maxwell's explanation of the phenomenon of "residual discharge." It is found that, some time after a condenser has been discharged and insulated, a second and smaller discharge can be obtained on joining the plates, after this a third, and so on, almost indefinitely. It should be noticed that, on the explanation which has been given, no residual discharge ought to take place if the dielectric is perfectly homogeneous. It thus becomes possible to test the theory by experiments on homogeneous dielectrics.

Rowland and Nichols[*] tested calcspar, which is a perfectly homogeneous crystal, and found no trace of residual discharge. Hertz[†] found traces of a

---

[*] *Phil. Mag.* [5] vol. II. p. 414 (1881).   [†] *Wied. Ann.* xx. (1883), p. 279.

residual discharge in a homogeneous fluid, benzene, but found that these disappeared as impurities were removed from the fluid; Arons* obtained the same result with paraffin.   Finally Muraoka† experimented with various oils, paraffin, resin, turpentine and xylol.   Residual discharges were not found in the oils singly, but appeared as soon as two or more were mixed together.

These facts are in agreement with Maxwell's theory of residual discharge and afford strong confirmation of the theory.   On the other hand there are a large number of experimental facts which are difficult to explain in terms of Maxwell's theory alone, and which seem to suggest that the theory is incomplete.

EXAMPLES.

1.   The ends of a rectangular conducting lamina of breadth c, length a, and uniform thickness τ, are maintained at different potentials.   If $f(x, y)$ be the specific resistance ρ at a point whose distances from an end and a side are x, y, prove that the resistance of the lamina cannot be less than $\displaystyle\int_0^a \frac{dx}{\tau \displaystyle\int_0^c \frac{dy}{\rho}}$, or greater than $\displaystyle\frac{1}{\tau \displaystyle\int_0^c \frac{dy}{\displaystyle\int_0^a \rho\, dx}}$.

2.   Two large vessels filled with mercury are connected by a capillary tube of uniform bore.   Find superior and inferior limits to the conductivity.

3.   A cylindrical cable consists of a conducting core of copper surrounded by a thin insulating sheath of material of given specific resistance.   Shew that if the sectional areas of the core and sheath are given, the resistance to lateral leakage is greatest when the surfaces of the two materials are coaxal right circular cylinders.

4.   Prove that the product of the resistance to leakage per unit length between two practically infinitely long parallel wires insulated by a uniform dielectric and at different potentials, and the capacity per unit length, is $K\rho/4\pi$, where K is the inductive capacity and ρ the specific resistance of the dielectric.   Prove also that the time that elapses before the potential difference sinks to a given fraction of its original value is independent of the sectional dimensions and relative positions of the wires.

5.   If the right sections of the wires in the last question are semicircles described on opposite sides of a square as diameters, and outside the square, while the cylindrical space whose section is the semicircles similarly described on the other two sides of the square is filled up with a dielectric of infinite specific resistance, and all the neighbouring space is filled up with a dielectric of resistance ρ, prove that the leakage per unit length in unit time is $2V/\rho$, where V is the potential difference.

6.   If $\phi + i\psi = f(x + iy)$, and the curves for which $\phi = \mathrm{cons.}$ be closed curves, shew that the insulation resistance between lengths l of the surfaces $\phi = \phi_0$, $\phi = \phi_1$, is

$$\frac{\rho(\phi_1 \sim \phi_0)}{l[\psi]},$$

where $[\psi]$ is the increment of ψ on passing once round a ϕ-curve, and ρ is the specific resistance of the dielectric.

* *Wied. Ann.* xxxv. (1888), p. 291.   † *Wied. Ann.* xl. (1890), p. 328.

7. Current enters and leaves a uniform circular disc through two circular wires of small radius e whose central lines pass through the edge of the disc at the extremities of a chord of length d. Shew that the total resistance of the sheet is

$$(2\sigma/\pi) \log{(d/e)}.$$

8. Using the transformation

$$\log{(x+iy)} = \xi + i\eta,$$

prove that the resistance of an infinite strip of uniform breadth π between two electrodes distant $2a$ apart, situated on the middle line of the strip and having equal radii δ, is

$$\frac{\sigma}{\pi} \log{\left(\frac{2}{\delta} \tanh a \right)}.$$

9. Shew that the transformation

$$x' + iy' = \cosh{\pi\,(x+iy)/a}$$

enables us to obtain the potential due to any distribution of electrodes upon a thin conductor in the form of the semi-infinite strip bounded by $y=0$, $y=a$, and $x=0$.

If the margin be uninsulated, find the potential and flow due to a source at the point $x=c$, $y=\dfrac{a}{2}$. Shew that if the flows across the three edges are equal, then $\pi c = a \cosh^{-1}{2}$.

10. Equal and opposite electrodes are placed at the extremities of the base of an isosceles triangular lamina, the length of one of the equal sides being a, and the vertical angle $\dfrac{2\pi}{3}$. Shew that the lines of flow and equal potential are given by

$$\sinh^{\frac{2}{3}}{\frac{w}{2}} + 1 = \sqrt{3}\,\frac{1 + \mathrm{cn}\,u}{1 - \mathrm{cn}\,u},$$

where

$$3^{\frac{3}{4}}\Gamma\left(\frac{1}{2}\right) ua = \Gamma\left(\frac{1}{3}\right) \Gamma\left(\frac{1}{6}\right) \left(ze^{-\frac{\pi i}{6}} - a \right),$$

and the modulus of $\mathrm{cn}\,u$ is $\sin 75°$, the origin being at the vertex.

11. A circular sheet of copper, of specific resistance σ_1 per unit area, is inserted in a very large sheet of tinfoil (σ_0), and currents flow in the composite sheet, entering and leaving at electrodes. Prove that the current-function in the tinfoil corresponding to an electrode at which a current e enters the tinfoil is the coefficient of i in the imaginary part of

$$-\frac{\sigma_0 e}{2\pi}\left[\log{(z-c)} + \frac{\sigma_0 - \sigma_1}{\sigma_0 + \sigma_1} \log{\frac{cz}{cz - a^2}} \right],$$

where a is the radius of the copper sheet, z is a complex variable with its origin at the centre of the sheet, and c is the distance of the electrode from the origin, the real axis passing through the electrode.

Generalise the expression for any position of the electrode in the copper or in the tinfoil, and investigate the corresponding expressions determining the lines of flow in the copper.

12. A uniform conducting sheet has the form of the catenary of revolution

$$y^2 + z^2 = c^2 \cosh^2{\frac{x}{c}}.$$

Prove that the potential at any point due to an electrode at x_0, y_0, z_0, introducing a current C, is

$$\text{constant} - \frac{C\sigma}{4\pi} \log{\left(\cosh{\frac{x-x_0}{c}} - \frac{yy_0 + zz_0}{\sqrt{(y^2+z^2)(y_0{}^2 + z_0{}^2)}} \right)}.$$

CHAPTER XI

PERMANENT MAGNETISM

PHYSICAL PHENOMENA.

398. IT is found that certain bodies, known as magnets, will attract or repel one another, while a magnet will also exert forces on pieces of iron or steel which are not themselves magnets, these forces being invariably attractive. The most familiar fact of magnetism, namely the tendency of a magnetic needle to point north and south, is simply a particular instance of the first of the sets of phenomena just mentioned, it being found that the earth itself may be regarded as a vast aggregation of magnets.

The simplest piece of apparatus used for the experimental study of magnetism is that known as a bar-magnet. This consists of a bar of steel which shews the property of attracting to itself small pieces of steel or iron. Usually it is found that the magnetic properties of a bar-magnet reside largely or entirely at its two ends. For instance, if the whole bar is dipped into a collection of iron filings, it is found that the filings are attracted in great numbers to its two ends, while there is hardly any attraction to the middle parts, so that on lifting the bar out from the collection of filings, we shall find that filings continue to cluster round the ends of the bar, while the middle regions will be comparatively free.

Poles of a Magnet.

399. The two ends of a magnet—or, more strictly, the two regions in which the magnetic properties are concentrated—are spoken of as the "poles" of the magnet. If the magnet is freely suspended, it will turn so that the line joining the two poles points approximately north and south. The pole which places itself so as to point towards the north is called the "north-seeking pole," while the other pole, pointing to the south, is called the "south-seeking pole."

By experimenting with two or more magnets, it is found to be a general law that similar poles repel one another, while dissimilar poles attract one another.

The earth may roughly be regarded as a single magnet of which the two magnetic poles are at points near to the geographical north and south poles. Since the northern magnetic pole of the earth attracts the north-seeking pole of a suspended bar-magnet, it is clear that this northern magnetic pole must be a south-seeking pole; and similarly the southern pole of the earth must be a north-seeking pole. Lord Kelvin speaks of a south-seeking pole as a "true north" pole—*i.e.* a pole of which the magnetism is of the kind found in the northerly regions of the earth. But for purposes of mathematical theory it will be most convenient to distinguish the two kinds of pole by the entirely neutral terms, positive and negative. And, as a matter of convention, we agree to call the north-seeking pole positive. Thus we have the following pairs of terms:

$$North\text{-}seeking = True\ South = Positive,$$
$$South\text{-}seeking = True\ North = Negative.$$

Law of Force between Magnetic Poles.

400. By experiments with his torsion-balance, Coulomb established that the force between two magnetic poles varies inversely as the square of the distance between them. It was found also to be proportional to the product of two quantities spoken of as the "strengths" of the poles. Thus if F is the repulsion between two poles of strengths m, m' at a distance r apart, we have

$$F = \frac{cmm'}{r^2}. \quad \dots \dots \dots (328).$$

It is found that c depends on the medium in which the poles are placed, but is otherwise constant. Clearly if we agree that the strength of positive poles is to be reckoned as positive, while that of negative poles is reckoned negative, then c will be a positive quantity

The Unit Magnetic Pole.

401. Just as Coulomb's electrostatic law of force supplied a convenient way of measuring the strength of an electric charge, so the law expressed by equation (328) provides a convenient way of measuring the strength of a magnetic pole, and so gives a system of magnetic units. A system of units, analogous to the electrostatic system (§§ 17, 18) is obtained by defining the unit pole to be such as to make $c = 1$ in equation (328). This system is called the Magnetic (or, more generally, Electromagnetic) system of units. We define a unit pole, in this system, to be a pole of strength such that when placed at unit distance from a pole of equal strength the repulsion between the two poles is one of unit force.

Thus the force F between two poles of strengths m, m', measured in the Electromagnetic system of units, is given by

$$F = \frac{mm'}{r^2} \quad \dots\dots\dots\dots\dots\dots\dots\dots(329).$$

The physical dimensions of the magnetic unit can be discussed in just the same way in which the physical dimensions of the electrostatic unit have already been discussed in § 18.

Moment of a Line-Magnet.

402. It is found that every positive pole has associated with it a negative pole of exactly equal strength, and that these two poles are always in the same piece of matter.

Thus not only are positive and negative magnetism necessarily brought into existence together and in equal quantities, as is the case with positive and negative electricity, but, further, it is impossible to separate the positive and negative magnetism after they have brought into existence, and in this respect magnetism is unlike electricity.

It follows that it is impossible to have a body " charged with magnetism " in the way in which we can have a body charged with electricity. A magnetised body may possess any number of poles, and at each pole there is, in a sense, a charge of magnetism; but the total charge of magnetism in the body will always be zero.

Hence it follows that the simplest and most fundamental piece of matter we can imagine which is of interest for the theory of magnetism, is not a small body carrying a charge of magnetism, but a small body carrying (so to speak) two equal and opposite charges at a certain distance apart.

This leads us to introduce the conception of a line-magnet. A line-magnet is an ideal bar-magnet of which the width is infinitesimal, the length finite, and the poles at the two extreme ends. Thus geometrically the ideal line-magnet is a line, while its poles are points.

The strengths of the two poles of a line-magnet are necessarily equal and opposite. The product of the numerical strength of either pole and the distance between the poles is called the " moment " of the line-magnet.

Magnetic Particle.

403. If we imagine the distance between the two poles of a line-magnet to shrink until it is infinitesimal, the magnet becomes what is spoken of as a magnetic particle. If $\pm m$ are the strengths of its poles and ds is the distance between the two poles, the moment of the magnetic particle is $m\,ds$.

It is easily shewn that, as regards all phenomena occurring at a finite distance, two co-axial magnetic particles have the same effect if their moments are equal; their length and the strengths of their poles separately are of no importance. To see this we need only consider the case of two magnetic particles, each having poles $\pm m$, and length ds, and therefore moment mds. Clearly these will produce the same effect at finite distances whether they are placed end to end or side by side. In the latter case, we have a magnet of length ds, poles $\pm 2m$, while in the former case the two contiguous poles, being of opposite sign, neutralise one another, and the arrangement is in effect a magnet of length $2ds$ and poles $\pm m$. Thus in each case the *moment* is the same, namely $2mds$, while the strengths of the poles and their distances apart are different.

If we place a large number n of similar magnetic particles end to end, all the poles will neutralise one another except those at the extreme ends, so that the arrangement produces the same effect as a line-magnet of length nds. By taking $n = \dfrac{l}{ds}$, where l is a finite length, we see that the effect of a line-magnet of length l can be produced exactly by n magnetic particles of length ds.

The two arrangements will be indistinguishable by their magnetic effects at all external points. There is, however, a way by which it would be easy to distinguish them. If the arrangement were simply two poles $\pm m$, at the ends of a wire of length l, then on cutting the wire into two pieces, we should have one pole remaining in each piece. If, however, the arrangement were

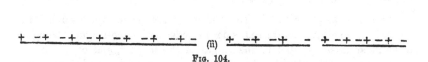

Fig. 104.

that of a series of magnetic particles, we should be able to divide the series *between* two particles, and should in this way obtain two complete magnets. The pair of poles on the two sides of the point of division which have so far been neutralising one another now figure as independent poles.

As a matter of experiment, it is not only found to be possible to produce two complete magnets by cutting a single magnet between its poles, but it is found that two new magnets are produced, no matter at what point the cutting takes place. The inference is not only that a natural magnet must be supposed to consist of magnetic particles, but also that these particles are so small that when the magnet is cut in two, there is no possibility of cutting a magnetic particle in two, so that one pole is left on each side of the division. In other words, we must suppose the magnetic particles either to be identical with the molecules of which the matter is composed or else to be even smaller than these molecules. At the same time, it will not be necessary to limit the magnetic particle of mathematical analysis by assigning this definite meaning to it: any collection of molecules, so small that the whole space occupied by it may be regarded as infinitesimal, will be spoken of as a magnetic particle.

404. *Axis of a magnetic particle.* The axis of a magnetic particle is defined to be the direction of a line drawn from the negative to the positive pole of the particle.

It will be clear, from what has already been said, that the effect of a magnetic particle at all external points is known when we know its position, axis and moment.

Intensity of Magnetisation.

405. In considering a bar-magnet, which must be supposed to have breadth as well as length, we have to consider the magnetic particles as being stacked side by side as well as placed end to end. For clearness, let us suppose that the magnet is a rectangular parallelepiped, its length being parallel to the axis of x, while its height and breadth are parallel to the two other axes. The poles of this bar-magnet may be supposed to consist of a uniform distribution of infinitesimal magnetic poles over each of the two faces parallel to the plane of yz, let us say a distribution of poles of aggregate strength I per unit area at the positive pole, and $-I$ per unit area at the negative pole, so that if A is the area of each of these faces, the poles of the magnet are of strengths $\pm IA$.

As a first step, we may regard the magnet as made up of an infinite number of line-magnets placed side by side, each line-magnet being a rectangular prism parallel to the length of the magnet, and of very small cross-section. Thus a prism of cross-section $dydz$ may be regarded as a line-magnet having poles $\pm I dydz$. This again may be regarded as made up of a number of magnetic particles. As a type, let us consider a particle of length dx, so that the volume of the magnet occupied by this particle is $dxdydz$. The poles of this particle are of strength $\pm I dydz$, so that the moment of the particle is

$$I dxdydz.$$

If we take any small cluster of these particles, occupying a small volume dv, the sum of their moments is clearly $I dv$, and these produce the same magnetic effects at external points as a single particle of moment

$$I dv.$$

The quantity I is called the "intensity of magnetisation" of the magnet. The magnetisation has direction as well as magnitude. In the present instance the direction is that of the axis of x.

406. In general, we define the intensity and direction of magnetisation as follows :

The intensity of magnetisation at any point of a magnetised body is defined to be the ratio of the magnetic moment of any small particle at this point to the volume of the particle.

The direction of magnetisation at any point of a magnetised body is defined to be the direction of the magnetic axis of a small particle of magnetic matter at the point.

Instead of specifying the magnetisation of a body in terms of its poles, it is both more convenient from the mathematical point of view, and more in accordance with truth from the physical point of view, to specify the intensity at every point in magnitude and direction. Thus the bar-magnet which has been under consideration would be specified by the statement that its intensity of magnetisation at every point is I parallel to the axis of x. A body such that the intensity is the same at every point, both in magnitude and direction, is said to be uniformly magnetised.

THE MAGNETIC FIELD OF FORCE.

407. The field of force produced by a collection of magnets is in many respects similar to an electrostatic field of force, so that the various conceptions which were found of use in electrostatic theory will again be employed.

The first of these conceptions was that of electric intensity at a point. In electrostatic theory, the intensity at any point was defined to be the force per unit charge which would act on a small charged particle placed at the point. It was necessary to suppose the charge to be of infinitesimal amount, in order that the charges on the conductors in the field might not be disturbed by induction.

There is, as we shall see later, a phenomenon of magnetic induction, which is in many respects similar to that of electrostatic induction, so that in defining magnetic intensity we have again to introduce a condition to exclude effects of induction.

Also, to avoid confusion between the magnetic intensity and the intensity of magnetisation defined in § 406, it will be convenient to speak of magnetic force at a point, rather than of magnetic intensity. We accordingly have the following definition, analogous to that given in § 30.

The magnetic force at any point is given, in magnitude and direction, by the force per unit strength of pole, which would act on a magnetic pole situated at this point, the strength of the pole being supposed so small that the magnetism of the field is not affected by its presence.

408. The other quantities and conceptions follow in order, as in Chapter II. Thus we have the following definitions:

A line of force is a curve in the magnetic field, such that the tangent at every point is in the direction of the magnetic force at that point (cf. § 31).

The potential at any point in the field is the work per unit strength of pole which has to be done on a magnetic pole to bring it to that point from infinity, the strength of the pole being supposed so small that the magnetism of the field is not affected by its presence (cf. § 33).

Let Ω denote the magnetic potential and α, β, γ the components of magnetic force at any point x, y, z, then we have from this definition (cf. equation (6)),

$$\Omega = -\int_{\infty}^{x, y, z} (\alpha dx + \beta dy + \gamma dz) \quad\dots\dots\dots\dots\dots(330),$$

and the relations (cf. equations (9)),

$$\alpha = -\frac{\partial \Omega}{\partial x}, \quad \beta = -\frac{\partial \Omega}{\partial y}, \quad \gamma = -\frac{\partial \Omega}{\partial z} \quad\dots\dots\dots\dots(331).$$

A surface in the magnetic field such that at every point on it the potential has the same value, is called an Equipotential Surface (cf. § 35).

From this definition, as in § 35, follows the theorem:

Equipotential Surfaces cut lines of force at right angles.

The law of force being the same as in electrostatics, we have as the value of the potential (cf. equation (10)),

$$\Omega = \Sigma \frac{m}{r} \quad\dots\dots\dots\dots\dots\dots\dots\dots\dots(332),$$

where m is the strength of any typical pole, and r is the distance from it to the point at which the potential is being evaluated.

As in § 42, we have Gauss' Theorem:

$$\iint \frac{\partial \Omega}{\partial n} dS = -4\pi \Sigma m \quad\dots\dots\dots\dots\dots\dots(333),$$

where the integration is over any closed surface, and Σm is the sum of the strengths of all the poles inside this surface. If the surface is drawn so as not to cut through any magnetised matter, Σm will be the aggregate strength of the poles of complete magnetic particles, and therefore equal to zero. Thus for a surface drawn in this way

$$\iint \frac{\partial \Omega}{\partial n} dS = 0 \quad\dots\dots\dots\dots\dots\dots\dots\dots(334).$$

If the position of the surface S is determined by geometrical conditions—if, for instance, it is the boundary of a small rectangular element $dx\,dy\,dz$—then we cannot suppose it to contain only complete magnetic particles, and equation (334) will not in general be true.

If there is no magnetic matter present in a certain region, equation (334) is true for any surface in this region, and on applying it to the surface of the small rectangular element $dx\,dy\,dz$, we obtain, as in § 50,

$$\frac{\partial^2\Omega}{\partial x^2} + \frac{\partial^2\Omega}{\partial y^2} + \frac{\partial^2\Omega}{\partial z^2} = 0 \quad\ldots\ldots\ldots\ldots\ldots\ldots(335),$$

the differential equation satisfied by the magnetic potential at every point of a region in which there is no magnetic matter present.

Tubes of Force.

409. A tubular surface bounded by lines of force is, as in electrostatics, called a tube of force. Let ω_1, ω_2 be the areas of any two normal cross-sections of a thin tube of force, and let H_1, H_2 be the values of the intensities at these points. By applying Gauss' Theorem to the closed surface formed by the two cross-sections and the portion of the tube which lies between them, we obtain, as in § 56,

$$H_1\omega_1 - H_2\omega_2 = 0,$$

provided there is no magnetic matter inside this closed surface.

Thus in free space the product $H\omega$ remains constant. The value of this product is called the strength of the tube.

In electrostatics, it was found convenient to define a unit tube to be one which ended on a unit charge, so that the product of intensity and cross-section was not equal to unity but to 4π.

Potential of a Magnetic Particle.

410. Let a magnetic particle consist of a pole of strength $-m_1$ at O, and a pole of strength $+m_1$ at P, the distance OP being infinitesimal.

The potential at any point Q will be

$$\Omega_Q = \frac{m_1}{PQ} - \frac{m_1}{OQ} \quad\ldots\ldots\ldots\ldots(336).$$

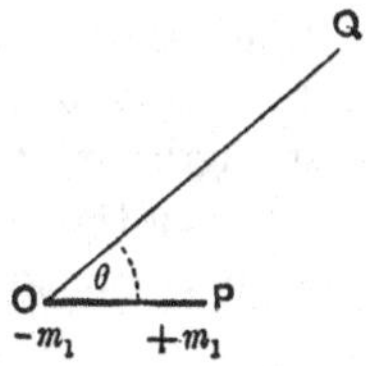

If we put $OQ = r$, and denote the angle POQ by θ, this becomes

$$\Omega_Q = \frac{m_1(OQ - PQ)}{PQ.OQ} = \frac{m_1\,OP\cos\theta}{PQ.OQ} = \frac{\mu\cos\theta}{r^2} \quad\ldots\ldots\ldots(337),$$

where $\mu = m_1 . OP$, the moment of the particle.

Fig. 105.

The analysis here given and the result reached are exactly similar to those already given for an electric doublet in § 64. The same result can also be put in a different form.

Let us put $OP = ds$, and let $\dfrac{\partial}{\partial s}$ denote differentiation in the direction of OP, the axis of the particle. Then equation (336) admits of expression in the form

$$\Omega_Q = m_1 ds \frac{\partial}{\partial s} \left(\frac{1}{r}\right) = \mu \frac{\partial}{\partial s} \left(\frac{1}{r}\right) \quad \text{...................(338).}$$

Let l, m, n be the direction-cosines of the axis of the particle, then formula (338) can also be written

$$\Omega_Q = \mu \left\{ l \frac{\partial}{\partial x}\left(\frac{1}{r}\right) + m \frac{\partial}{\partial y}\left(\frac{1}{r}\right) + n \frac{\partial}{\partial z}\left(\frac{1}{r}\right) \right\} \quad \text{............(339),}$$

where, in differentiation, x, y, z are supposed to be the coordinates of the particle, and not of the point Q.

411. Resolution of a magnetic particle. Equation (339) shews that the potential of the single particle we have been considering is the same as the potential of three separate particles, of strengths μl, μm and μn, and axes in the directions Ox, Oy, Oz respectively. Thus a magnetic particle may be resolved into components, and this resolution follows the usual vector law.

The same result can be seen geometrically.

Let us start from O and move a distance $l\,ds$ parallel to the axis of x, then a distance $m\,ds$ parallel to the axis of y, and then a distance $n\,ds$ parallel to the axis of z. This series of movements brings us from O to P, a distance ds in the direction l, m, n. Let the path be $OqrP$ in fig. 106. The magnetic particle under consideration has poles $-m_1$ at O and $+m_1$ at P. Without altering the field, we can superpose two equal and opposite poles $\pm m_1$ at q, and also two equal and opposite poles $\pm m_1$ at r.

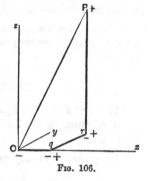

Fig. 106.

The six poles now in the field can be taken in three pairs so as to constitute three doublets of strengths $m_1 . Oq$, $m_1 . qr$ and $m_1\, rP$ respectively along Oq, qr and rP. These, however, are doublets of strengths μl, μm and μn parallel to the coordinate axes.

Potential of a Magnetised Body.

412. Let I be the intensity of magnetisation at any point of a magnetised body, and let l, m, n be the direction-cosines of the direction of magnetisation at this point.

The matter occupying any element of volume $dx\,dy\,dz$ at this point will be a magnetic particle of which the moment is $I\,dx\,dy\,dz$ and the axis is in direction l, m, n. By formula (339), the potential of this particle at any external point is

$$I\left\{l\frac{\partial}{\partial x}\left(\frac{1}{r}\right)+m\frac{\partial}{\partial y}\left(\frac{1}{r}\right)+n\frac{\partial}{\partial z}\left(\frac{1}{r}\right)\right\}dx\,dy\,dz,$$

so that, by integration, we obtain as the potential of the whole body at any external point Q,

$$\Omega_Q=\iiint I\left\{l\frac{\partial}{\partial x}\left(\frac{1}{r}\right)+m\frac{\partial}{\partial y}\left(\frac{1}{r}\right)+n\frac{\partial}{\partial z}\left(\frac{1}{r}\right)\right\}dx\,dy\,dz\ldots\ldots(340)$$

in which r is the distance from Q to the element $dx\,dy\,dz$, and the integration extends over the whole of the magnetised body.

If we introduce quantities A, B, C defined by

$$\left.\begin{array}{l}A=Il\\B=Im\\C=In\end{array}\right\}\qquad\ldots\ldots\ldots\ldots\ldots\ldots\ldots\ldots\ldots(341),$$

then equation (340) can be put in the form

$$\Omega_Q=\iiint\left\{A\frac{\partial}{\partial x}\left(\frac{1}{r}\right)+B\frac{\partial}{\partial y}\left(\frac{1}{r}\right)+C\frac{\partial}{\partial z}\left(\frac{1}{r}\right)\right\}dx\,dy\,dz\ \ldots\ldots(342).$$

The quantities A, B, C are called the components of magnetisation at the point x, y, z. Equation (342) shews that the potential of the original magnet, of magnetisation I, is the same as the potential of three superposed magnets, of intensities A, B, C parallel to the three axes. This is also obvious from the fact that the particle of strength $I\,dx\,dy\,dz$, which occupies the element of volume $dx\,dy\,dz$, may be resolved into three particles parallel to the axes, of which the strengths will be $A\,dx\,dy\,dz$, $B\,dx\,dy\,dz$ and $C\,dx\,dy\,dz$, if A, B, C are given by equations (341).

Potential of a uniformly Magnetised Body.

413. If the magnetisation of any body is uniform, the values of A, B, C are the same at all points of the body.

Let the coordinates of the point Q in equation (342) be x', y', z', so that

$$\frac{1}{r}=[(x-x')^2+(y-y')^2+(z-z')^2]^{-\frac{1}{2}}.$$

Then, clearly,　　　　　$\dfrac{\partial}{\partial x}\left(\dfrac{1}{r}\right)=-\dfrac{\partial}{\partial x'}\left(\dfrac{1}{r}\right)$, etc.

Replacing differentiation with respect to x, y, z by differentiation with respect to x', y', z' in this way, we find that equation (342) assumes the form

$$\Omega_Q = -\left(A\frac{\partial}{\partial x'} + B\frac{\partial}{\partial y'} + C\frac{\partial}{\partial z'}\right)\iiint \frac{1}{r}\,dx\,dy\,dz \quad \ldots\ldots\ldots(343),$$

the quantities A, B, C and the operators $\frac{\partial}{\partial x'}$, $\frac{\partial}{\partial y'}$, $\frac{\partial}{\partial z'}$ being taken outside the sign of integration, since they are not affected by changes in x, y, z.

If V denote the potential of a uniform distribution of electricity of volume density unity throughout the region occupied by the magnet, we have

$$V_Q = \iiint \frac{1}{r}\,dx\,dy\,dz \quad \ldots\ldots\ldots\ldots\ldots\ldots(344),$$

so that equation (343) becomes

$$\Omega_Q = -A\frac{\partial V_Q}{\partial x'} - B\frac{\partial V_Q}{\partial y'} - C\frac{\partial V_Q}{\partial z'} \quad \ldots\ldots\ldots\ldots\ldots(345),$$

or
$$\Omega_Q = AX + BY + CZ,$$

where X, Y, Z are the components of electric intensity at Q produced by this distribution.

Or again if $\frac{\partial}{\partial s'}$ denotes differentiation with respect to the coordinates of Q in a direction parallel to that of the magnetisation of the body, namely that of direction-cosines l, m, n, equation (345) becomes

$$\Omega_Q = -I\frac{\partial V_Q}{\partial s'} \quad \ldots\ldots\ldots\ldots\ldots\ldots(346).$$

414. Yet another expression for the potential of a uniformly magnetised body is obtained on transforming equation (342) by Green's Theorem. If l', m', n' are the direction-cosines of the outward-drawn normal to the magnet at any element dS of its surface, the equation obtained after transformation is

$$\Omega_Q = \iint (Al' + Bm' + Cn')\frac{1}{r}\,dS.$$

By equations (341),
$$Al' + Bm' + Cn' = I(ll' + mm' + nn')$$
$$= I\cos\theta,$$

where θ is the angle between the direction of magnetisation and the outward normal to the element dS of surface. The equation now becomes

$$\Omega_Q = \iint \frac{I\cos\theta}{r}\,dS \ldots\ldots\ldots\ldots\ldots\ldots(347),$$

shewing that the potential at any external point is the same as that of a surface distribution of magnetic poles of density $I\cos\theta$ per unit area, spread over the surface of the magnet.

This distribution is of course simply the "Green's Equivalent Stratum" (§ 204) which is necessary to produce the observed external field.

The bar-magnet already considered in § 405, provides an obvious illustration of these results.

415. *Uniformly magnetised sphere.* A second and interesting example of a uniformly magnetised body is a sphere, magnetised with uniform intensity I. This acquires its interest from the fact that the earth may, to a very rough approximation, be regarded as a uniformly magnetised sphere.

If we follow the method of § 313, we obtain for the value of V_Q, defined by equation (344),

$$V_Q = \tfrac{4}{3}\,\pi a^3 \left(\frac{1}{r}\right),$$

where a is the radius of the sphere. If we suppose the magnetisation to be in the direction of the axis of x, we have

$$\Omega_Q = -I\,\frac{\partial V_Q}{\partial x} = \tfrac{4}{3}\pi a^3 I \left(\frac{x}{r^3}\right)$$

$$= \tfrac{4}{3}\pi a^3 I\,\frac{\cos\theta}{r^2}.$$

Thus the potential at any external point is the same as that of a magnetic particle of moment $\tfrac{4}{3}\pi a^3 I$ at the centre of the sphere.

To treat the problem by the method of § 414, we have to calculate the potential of a surface density $I\cos\theta$ spread over the surface of the sphere. Regarding $\cos\theta$ as the first zonal harmonic $P_1(\cos\theta)$, the result follows at once from § 257.

Poisson's imaginary Magnetic Matter.

416. If the magnetisation of the body is not uniform, the value of Ω_Q given in equation (342) cannot be transformed into a surface integral, so that the potential of the magnet cannot be represented as being due to a surface charge of magnetic matter. If we apply Green's Theorem to the integral which occurs in equation (342), we obtain

$$\Omega_Q = \iiint \left\{ A\,\frac{\partial}{\partial x}\left(\frac{1}{r}\right) + B\,\frac{\partial}{\partial y}\left(\frac{1}{r}\right) + C\,\frac{\partial}{\partial z}\left(\frac{1}{r}\right) \right\} dx\,dy\,dz$$

$$= -\iiint \frac{1}{r}\left(\frac{\partial A}{\partial x} + \frac{\partial B}{\partial y} + \frac{\partial C}{\partial z}\right) dx\,dy\,dz + \iint \frac{1}{r}\,(lA + mB + nC)\,dS,$$

where l, m, n are the direction-cosines of the outward-drawn normal to the element dS of surface.

Thus $$\Omega_Q = \iiint \frac{\rho}{r}\,dx\,dy\,dz + \iint \frac{\sigma}{r}\,dS \quad\text{...............}(348),$$

where ρ, σ are given by

$$\rho = -\left(\frac{\partial A}{\partial x} + \frac{\partial B}{\partial y} + \frac{\partial C}{\partial z}\right) \quad\text{.....................}(349),$$

$$\sigma = lA + mB + nC \quad\text{.......................}(350).$$

Thus the potential of the magnet at any external point Q is the same as if there were a distribution of magnetic charges throughout the interior, of volume-density ρ given by equation (349), together with a distribution over the surface, of surface-density σ given by equation (350).

Potential of a Magnetic Shell.

417. A magnetised body which is so thin that its thickness at every point may be treated as infinitesimal, is called a "magnetic shell." Throughout the small thickness of a shell we shall suppose the magnetisation to remain constant in magnitude and direction, so that to specify the magnetisation of a shell we require to know the thickness of the shell and the intensity and direction of the magnetisation at every point.

Shells in which the magnetisation is in the direction of the normal to the surface of the shell are spoken of as "normally-magnetised shells." These form the only class of magnetic shells of any importance, so that we shall deal only with normally-magnetised shells, and it will be unnecessary to repeat in every case the statement that normal magnetisation is intended.

If I is the intensity of magnetisation at any point inside a shell of this kind, and if τ is its thickness at this point, the product $I\tau$ is spoken of as the "strength" of the shell at this point. Any element dS of the shell will behave as a magnetic particle of moment $I\tau dS$, so that the strength of a shell is the magnetic moment per unit area, just as the intensity of magnetisation of a body is the magnetic moment per unit volume.

Any element dS of a shell of strength ϕ behaves like a magnetic particle of strength ϕdS of which the axis is normal to dS.

The magnetisation of a magnetic shell may often be conveniently pictured as being due to the presence of layers of positive and negative poles on its two faces. Clearly if ϕ is the strength and τ the thickness of a shell at any point, the surface-density of these poles must be taken to be $\pm\dfrac{\phi}{\tau}$.

418. To obtain the potential of a shell at an external point, we regard any element dS of the shell as a magnetic particle of moment ϕdS and axis in the direction of the normal to the shell at this point, it being agreed that this normal must be drawn in the direction of magnetisation of the shell.

The potential of the element dS of the shell at a point Q distant r from dS is then

$$\phi\, dS\, \frac{\partial}{\partial n}\left(\frac{1}{r}\right),$$

so that the potential of the whole shell at Q is given by

$$\Omega_Q = \iint \phi\, \frac{\partial}{\partial n}\left(\frac{1}{r}\right) dS$$

$$= \iint \phi\, \frac{\cos\theta}{r^2}\, dS,$$

where θ is the angle between the normal at dS and the line joining dS to P.

Clearly $dS\cos\theta$ is the projection of the element dS on a plane perpendicular to the line joining dS to Q, so that $\dfrac{dS\cos\theta}{r^2}$ is the solid angle subtended by dS at Q. Denoting this by $d\omega$, we have the potential in the form

$$\Omega_Q = \iint \phi\, d\omega \quad\dotfill\quad(351).$$

419. *Uniform shell.* If the shell is of uniform strength, ϕ may be taken outside the sign of integration in equation (351), so that we obtain

$$\Omega_Q = \phi \iint d\omega = \phi\Omega \quad\dotfill\quad(352),$$

where Ω is the total solid angle subtended by the shell at Q.

POTENTIAL ENERGY OF A MAGNET IN A FIELD OF FORCE.

420. The potential energy of a magnet in an external field of force is equal to the work done in bringing up the magnet from infinity, the field of force being supposed to remain unaltered during the process.

Consider first the potential energy of a single particle, consisting of a pole of strength $-m_1$ at O and a pole of strength $+m_1$ at P. Let the potential of the field of force at O be Ω_O and at P be Ω_P. Then the amounts of work done on the two poles in bringing up this particle from infinity are respectively $-m_1\Omega_O$ and $m_1\Omega_P$, so that the potential energy of the particle when in the position OP

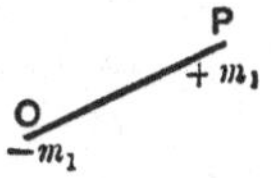

Fig. 107.

$$= m_1(\Omega_P - \Omega_O)$$

$$= m_1 . OP\, \frac{\partial\Omega}{\partial s}, \text{ in the notation already used,}$$

$$= \mu\, \frac{\partial\Omega}{\partial s} = \mu\left(l\, \frac{\partial\Omega}{\partial x} + m\, \frac{\partial\Omega}{\partial y} + n\, \frac{\partial\Omega}{\partial z}\right) \quad\dotfill\quad(353).$$

The potential energy of any magnetised body can be found by integration of expression (353), the body being regarded as an aggregation of magnetic particles.

421. Equation (353) assumes a special form if the magnetic field is due solely to the presence of a second magnetic particle. Let this be of moment μ', its axis having direction cosines l', m', n', and its centre having coordinates x', y', z'. Then we have as the value of Ω, from § 410,

$$\Omega = \mu' \frac{\partial}{\partial s'}\left(\frac{1}{r}\right) = \mu' \left(l' \frac{\partial}{\partial x'} + m' \frac{\partial}{\partial y'} + n' \frac{\partial}{\partial z'}\right)\left(\frac{1}{r}\right).$$

Substituting these values for Ω in the formulae just obtained, we have as the mutual potential energy of the two magnets,

$$W = \mu\mu' \frac{\partial^2}{\partial s \partial s'}\left(\frac{1}{r}\right)$$

$$= \mu\mu' \left(l \frac{\partial}{\partial x} + m \frac{\partial}{\partial y} + n \frac{\partial}{\partial z}\right)\left(l' \frac{\partial}{\partial x'} + m' \frac{\partial}{\partial y'} + n' \frac{\partial}{\partial z'}\right)\left(\frac{1}{r}\right).$$

This is symmetrical with respect to the two magnets, as of course it ought to be—it is immaterial whether we bring the first magnet into the field of the second, or the second into the field of the first.

If we now put

$$\frac{1}{r} = \frac{1}{\{(x-x')^2 + (y-y')^2 + (z-z')^2\}^{\frac{1}{2}}},$$

we obtain on differentiation,

$$\frac{\partial}{\partial x'}\left(\frac{1}{r}\right) = \frac{x-x'}{\{(x-x')^2 + (y-y')^2 + (z-z')^2\}^{\frac{3}{2}}} = \frac{x-x'}{r^3},$$

so that

$$\frac{\partial^2}{\partial x \partial x'}\left(\frac{1}{r}\right) = \frac{1}{r^3} - \frac{3(x-x')^2}{r^5},$$

$$\frac{\partial^2}{\partial y \partial x'}\left(\frac{1}{r}\right) = -\frac{3(x-x')(y-y')}{r^5}, \text{ etc.}$$

Hence we obtain as the value of W,

$$W = \frac{\mu\mu'}{r^3}(ll' + mm' + nn')$$

$$- \frac{3\mu\mu'}{r^5}\{l(x-x') + m(y-y') + n(z-z')\}\{l'(x-x') + m'(y-y') + n'(z-z')\}.$$

Let us now denote the angle between the axes of the two magnets by ϵ, and the angles between the line joining the two magnets and the axes of the first and second magnets respectively by θ and θ'. Then

$$\cos \epsilon = ll' + mm' + nn',$$

$$\cos \theta = \frac{1}{r}\{l(x-x') + m(y-y') + n(z-z')\},$$

$$\cos \theta' = \frac{1}{r}\{l'(x-x') + m'(y-y') + n'(z-z')\}$$

so that W can be expressed in the form

$$W = \frac{\mu\mu'}{r^3} (\cos \epsilon - 3 \cos \theta \cos \theta') \dots\dots\dots\dots(354).$$

If we take the line drawn from the first magnet to the second as pole in spherical polar coordinates, and denote the azimuths of the axes of the two magnets by ψ, ψ', then the polar coordinates of the directions of the axes of the two magnets will be θ, ψ and θ', ψ' respectively, and we shall have

$$\cos \epsilon = \cos \theta \cos \theta' + \sin \theta \sin \theta' \cos (\psi - \psi').$$

On substituting this value for $\cos \epsilon$ in equation (354), we obtain

$$W = \frac{\mu\mu'}{r^3} \{\sin \theta \sin \theta' \cos (\psi - \psi') - 2 \cos \theta \cos \theta'\} \dots\dots(355).$$

422. Knowing the mutual potential energy W, we can derive a knowledge of all the mechanical forces by differentiation. For instance the repulsion between the two magnets, *i.e.* the force tending to increase r, is $-\dfrac{\partial W}{\partial r}$, or

$$\frac{3\mu\mu'}{r^4} \{\sin \theta \sin \theta' \cos (\psi - \psi') - 2 \cos \theta \cos \theta'\}.$$

Thus, whatever the position of the magnets, the force between them varies as the inverse fourth power of the distance.

If the magnets are parallel to one another, $\theta = \theta'$ and $\psi = \psi'$, so that the repulsion

$$= \frac{3\mu\mu'}{r^4} (\sin^2 \theta - 2 \cos^2 \theta).$$

Thus when $\theta = 0$, *i.e.* when the magnets lie along the line joining them, the force is an attractive force $\dfrac{6\mu\mu'}{r^4}$. When $\theta = \dfrac{\pi}{2}$, so that the magnets are at right angles to the line joining them, the force is a repulsive force $\dfrac{3\mu\mu'}{r^4}$. In passing from the one position to the other the force changes from one of attraction to one of repulsion when $\sin^2 \theta - 2 \cos^2 \theta = 0$, *i.e.* when $\theta = \tan^{-1} \sqrt{2}$.

The couples can be found in the same way. If χ is any angle, the couple tending to increase the angle χ is $-\dfrac{\partial W}{\partial \chi}$, or

$$-\frac{\mu\mu'}{r^3} \frac{\partial}{\partial \chi} \{\sin \theta \sin \theta' \cos (\psi - \psi') - 2 \cos \theta \cos \theta'\},$$

so that all the couples vary inversely as the *cube* of the distance.

For instance, taking χ to be the same as ψ, we find that the couple tending to rotate the first magnet about the line joining it to the second, in the direction of ψ increasing

$$= -\frac{\partial W}{\partial \psi} = \frac{\mu\mu'}{r^3} \sin\theta \sin\theta' \sin(\psi - \psi'),$$

so that this couple vanishes if either of the magnets is along the line joining them, or if they are in the same plane, results which are obvious enough geometrically.

Potential Energy of a Shell in a Field of Force.

423. Consider a shell of which the strength at any point is ϕ, placed in a field of potential Ω. The element dS of the shell is a magnetic particle of strength ϕdS, so that its potential energy in the field of force will, by formula (353), be

$$\phi dS \frac{\partial\Omega}{\partial n},$$

where $\frac{\partial}{\partial n}$ denotes differentiation along the normal to the shell. Thus the potential energy of the whole shell will be

$$W = \iint \phi \frac{\partial\Omega}{\partial n} dS \quad\dots\dots\dots\dots\dots\dots(356).$$

If the shell is of uniform strength, this may be replaced by

$$W = \phi \iint \frac{\partial\Omega}{\partial n} dS \quad\dots\dots\dots\dots\dots\dots(357).$$

Since the normal component of force at a point just outside the shell and on its positive face is $-\frac{\partial\Omega}{\partial n}$, it is clear that $\iint \frac{\partial\Omega}{\partial n} dS$ is equal to minus the surface integral of normal force taken over the positive face of the shell, and this again is equal to minus the number of unit tubes of force which emerge from the shell on its positive face. Denoting this number of unit tubes by n, equation (357) may be expressed in the form

$$W = -\phi n \quad\dots\dots\dots\dots\dots\dots\dots(358).$$

Here it must be noticed that we are concerned only with the original field before the shell is supposed placed in position. Or, in other terms, the number n is the number of tubes which would cross the space occupied by the shell, if the shell were annihilated. Since the tubes are counted on the positive face of the shell, we see that n may be regarded as the number of unit tubes of the external field which cross the shell in the direction of its magnetisation.

424. Consider a field consisting only of two shells, each of unit strength. Let n_1 be the number of tubes from shell 1 which cross the area occupied by 2, and let n_2 be the number of tubes from shell 2 which cross the area occupied by 1. The potential energy of the field may be regarded as being either the energy of shell 1 in the field set up by 2, or as the energy of shell 2 in the field set up by 1. Regarded in the first manner, the energy of the field is found to be $-n_2$; regarded in the second manner, the energy is found to be $-n_1$. Hence we see that $n_1 = n_2$. This result, which is of great importance, will be obtained again later (§ 446) by a purely geometrical method.

Potential Energy of any Magnetised Body in a Magnetic Field of Force.

425. Let I be the intensity of magnetisation and l, m, n the direction-cosines of the direction of magnetisation at any point x, y, z of a magnetised body, and let Ω be the potential, at this point, of an external field of magnetic force. The element $dx\,dy\,dz$ of the magnetised body is a magnetic particle of strength $I\,dx\,dy\,dz$, of which the axis is in the direction l, m, n. Thus its potential energy in the field of force is, by formula (353),

$$I\,dx\,dy\,dz\left(l\frac{\partial\Omega}{\partial x} + m\frac{\partial\Omega}{\partial y} + n\frac{\partial\Omega}{\partial z}\right),$$

and by integration the potential of the whole magnet is

$$\iiint I\left(l\frac{\partial\Omega}{\partial x} + m\frac{\partial\Omega}{\partial y} + n\frac{\partial\Omega}{\partial z}\right)dx\,dy\,dz,$$

or

$$\iiint\left(A\frac{\partial\Omega}{\partial x} + B\frac{\partial\Omega}{\partial y} + C\frac{\partial\Omega}{\partial z}\right)dx\,dy\,dz.$$

Force inside a Magnetised Body.

426 So far the magnetic force has been defined and discussed only in regions not occupied by magnetised matter: it is now necessary to consider the more difficult question of the measurement of force at points inside a magnetised body.

At the outset we are confronted with a difficulty of the same kind as that encountered in discussing the measurement of electric force inside a dielectric, on the molecular hypothesis explained in § 143. We found that the molecules of a dielectric could be regarded as each possessing two equal and opposite charges of electricity on two opposite faces. If we replace "electricity" by "magnetism" the state is very similar to what we believe to be the state of the ultimate magnetic particles. In the electric problem a difficulty arose from the fact that the electric force inside matter varied rapidly as we passed from one molecule to another, because the intensity of the field set up by the charges on the molecules nearest to any point was comparable with the whole field. A similar difficulty arises in the magnetic problem, but will be handled in a way slightly different from that previously adopted. There are two reasons for this difference of treatment—in the first place, we are not willing to identify the ultimate magnetic particles with the molecules of the matter, and in the second place, we are not willing to assume that the magnetism of an ultimate particle may be localised in the form of charges on the two opposite faces. We shall follow a method which rests on no assumptions as to the connection between molecular structure and magnetic properties, beyond the well-established fact that on cutting a magnet new magnetic poles appear on the surfaces created by cutting.

427. One way of measuring the force at a point Q inside a magnet will be to imagine a cavity scooped out of the magnetic matter so as to enclose the point Q, and then to imagine the force measured on a pole of unit strength placed at Q. This method of measurement will only determine a definite force at Q if it can be shewn that the force is independent of the position, shape and size of the cavity, and this, as will be obvious from what follows, is not generally the case.

428. Let us suppose that, in order to form a cavity in which to place the imaginary unit pole, we remove a small cylinder of magnetic matter, the axis of this cylinder being in the direction of magnetisation at the point. Let this cylinder be of length l and cross-section S, and let the intensity of magnetisation at the point be I. Let the size of the cylinder be supposed to be very great in comparison with the scale of molecular structure, although very small in comparison with the scale of variation in the magnetisation of the body.

In steel or iron there are roughly 10^{23} molecules to the cubic centimetre, so that a length of 1 millimetre may be regarded as large when measured by the molecular scale, although in most magnets the magnetisation may be treated as constant within a length of a millimetre.

At a point near the centre of this cavity we are at a distance from the nearest magnetic particles, which is, by hypothesis, great compared with molecular dimensions. Hence, by § 416, we may regard the potential at points near the centre of the cavity as being that due to the following distributions of imaginary magnetic matter.—

I. A distribution of surface-density $lA + mB + nC$, spread over the surface of every magnet.

II. A distribution of volume-density

$$-\left(\frac{\partial A}{\partial x} + \frac{\partial B}{\partial y} + \frac{\partial C}{\partial z}\right),$$

spread throughout the whole space which is occupied by magnetic matter after the cavity has been scooped out.

III. A distribution of surface-density $lA + mB + nC$, spread over the walls of the cavity.

From the way in which the cavity has been chosen, it follows that $lA + mB + nC$ vanishes over the side-walls, and is equal to $\pm I$ on the two ends.

The force acting on an imaginary unit pole placed at or near the centre of the cavity may be regarded as the force arising from these three distributions.

429. The force from distribution III can be made to vanish by taking the length of the cavity to be very great in comparison with the linear dimensions of its ends. For the ends of the cavity may then be treated as points, and the force exerted by either end upon a unit pole placed at the centre of the cavity will be

$$\frac{SI}{(\tfrac{1}{2}l)^2},$$

and this will vanish if S is small compared with l^2. The resultant force will therefore arise solely from distributions I and II.

The force arising from distribution II may be regarded as the force arising from a distribution of volume-density

$$-\left(\frac{\partial A}{\partial x} + \frac{\partial B}{\partial y} + \frac{\partial C}{\partial z}\right)$$

spread throughout the whole of the magnetised matter, regardless of the existence of the cavity, together with a distribution of volume-density

$$+\left(\frac{\partial A}{\partial x} + \frac{\partial B}{\partial y} + \frac{\partial C}{\partial z}\right)$$

spread through the space occupied by the cavity. The force from this latter distribution vanishes in the limit when the size of the cavity is infinitesimal, so that the force from distribution II may be regarded as that from a volume-density

$$-\left(\frac{\partial A}{\partial x} + \frac{\partial B}{\partial y} + \frac{\partial C}{\partial z}\right)$$

spread through all the original magnetised matter.

We have now arrived at a force which is independent of the shape, size and position of the cavity, provided only that these satisfy the conditions which have already been laid down. This force we define to be the *magnetic force*, at the point under discussion, inside the magnetised body.

430. In the notation of § 416, the force which has just been defined is due to a distribution of surface-density σ, and a distribution of volume-density

ρ throughout the whole magnetised matter. The potential of these distributions is

$$\iint \frac{\sigma}{r} \, dS + \iiint \frac{\rho}{r} \, dx \, dy \, dz,$$

or Ω_Q if we regard this as *defined* by equation (348). Thus, with this meaning assigned to Ω_Q, the components of force at a point Q inside a magnetic body will be

$$-\frac{\partial \Omega_Q}{\partial x}, \quad -\frac{\partial \Omega_Q}{\partial y}, \quad -\frac{\partial \Omega_Q}{\partial z}.$$

At the same time it must be remembered that Ω_Q has not been shewn to be the true value of the potential except when the point Q is outside the magnetic matter. The true potential inside magnetised matter will vary rapidly as we pass from one magnetic particle to another.

431. Let us next suppose that the length l of the cylindrical cavity is very *small* compared with the linear dimensions of an end. The force, as before, is that due to the distributions I, II and III of § 428. The force from distribution III, however, will no longer vanish, for this distribution consists of distributions $\pm I$ over the ends of the cavity, and the force from these is not now negligible. From

Fig. 108.

analogy with the distribution of electricity on a parallel plate condenser, it is clear that the force arising from distribution III is a force $4\pi I$ in the direction of magnetisation. The forces from distributions I and II are easily seen to be the same as in the former case. Thus the force on a unit pole placed at a point Q inside a cavity of the kind we are now considering is the resultant of

(i) the magnetic force at Q, as defined in § 429,

(ii) a force $4\pi I$ in the direction of the intensity of magnetisation at Q.

The resultant of these forces is called the *magnetic induction* at Q.

432. The magnetic force will be denoted by H, and its components by α, β, γ.

The induction will be denoted by B, and its components by a, b, c.

We have seen that the force B is the resultant of a force H and a force $4\pi I$. The components of this latter force are $4\pi A$, $4\pi B$, $4\pi C$. Hence we have the equations

$$\left. \begin{aligned} a &= \alpha + 4\pi A \\ b &= \beta + 4\pi B \\ c &= \gamma + 4\pi C \end{aligned} \right\} \quad \dots\dots\dots\dots\dots\dots(359).$$

433. Let us next consider the force on a unit pole inside a cylindrical cavity when the cavity is disc-shaped, as in § 431, but its axis is not in the direction of magnetisation. The force can, as in § 428, be regarded as arising from three distributions.

Distributions I and II are the same as before, but distribution III will now consist of charges both on the end and on the side-walls of the cylinder. By making the length of the cylinder small in comparison with the linear dimensions of its cross-section, the force from the distri-

Fig. 109.

bution in the side-walls can be made to vanish. And if θ is the angle between the axis of the cavity and the direction of magnetisation, the distribution on the ends is one of density $\pm I \cos \theta$. Thus the force arising from distribution III is a force $4\pi I \cos \theta$ in the direction of the axis of the cavity.

Thus the force on a pole placed inside this cavity may be regarded as compounded of the force H (arising from distributions I and II), and a force $4\pi I \cos \theta$ in the direction of magnetisation, arising from distribution III.

Let ϵ be the angle between the direction of the force H and the axis of the cavity, then the component force in the direction of the axis of the cavity

$$= H \cos \epsilon + 4\pi I \cos \theta.$$

If l, m, n are the direction-cosines of this last direction,

$$H \cos \epsilon = l\alpha + m\beta + n\gamma,$$

$$4\pi I \cos \theta = 4\pi (lA + mB + nC),$$

so that, by equations (395),

$$H \cos \epsilon + 4\pi I \cos \theta = la + mb + nc.$$

Thus the component of the force in the direction of the axis of the cavity is the same as the component, in the same direction, of the magnetic induction, namely $la + mb + nc$.

434. We are now in a position to understand the importance of the vector which has been called the induction. This arises entirely from the property of the induction which is expressed in the following theorem:

THEOREM. *The surface-integral of the normal component of induction, taken over any surface whatever, vanishes,*

or in other words (cf. § 177),

The induction is a solenoidal vector throughout the whole of the magnetic field.

To prove this let us take any closed surface S in the field, this surface cutting any number of magnetised bodies. Along those parts of the surface which are inside magnetic bodies, let us remove a layer of matter, so that the surface no longer actually passes through any magnetic matter.

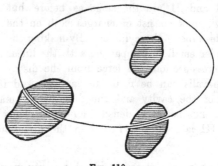

Fig. 110.

Then by Gauss' Theorem (§ 409),

$$\iint N dS = 0 \quad \dots\dots\dots\dots\dots\dots(360),$$

where N is the component of force in the direction of the outward normal to S, acting on a unit pole placed at any point of the surface S. This force, however, is exactly identical with that considered in § 433, and its normal component has been seen to be identical with the normal component of the induction. Thus N, in equation (360), will be the normal component of induction, so that this equation proves the theorem.

Analytically, the theorem may be stated in the form

$$\iint (la + mb + nc)\, dS = 0 \quad \dots\dots\dots\dots\dots(361),$$

and this, by Green's Theorem (§ 179), is identical with

$$\frac{\partial a}{\partial x} + \frac{\partial b}{\partial y} + \frac{\partial c}{\partial z} = 0 \quad \dots\dots\dots\dots\dots(362).$$

435. DEFINITION. *By a line of induction is meant a curve in the magnetic field such that the tangent at every point is in the direction of the magnetic induction at that point.*

DEFINITION. *A tube of induction is a tubular surface of small cross-section, which is bounded entirely by lines of induction.*

By a proof exactly similar to that of § 409, it can be shewn that the product of the induction and cross-section of a tube retains a constant value along the tube. This constant value is called the strength of the tube.

In free space the lines and tubes of induction become identical with the lines and tubes of force, and the foregoing definition of the strength of a tube of induction is such as to make the strengths of the tubes also become identical.

436. At any point of a surface let B be the induction, and let ϵ be the angle between the direction of the induction and the normal to the surface. The aggregate cross-section of all the tubes which pass through an element dS of this surface is $dS \cos \epsilon$, so that the aggregate strength of all these tubes is $B \cos \epsilon dS$. Since $B \cos \epsilon = N$, where N is the normal induction, this may be written in the form $N\,dS$. Thus the aggregate strength of the tubes of induction which cross any area is equal to

$$\iint N\,dS.$$

This, we may say, is the number of unit-tubes of induction which cross this area.

The theorem that $\qquad \iint N\,dS = 0,$

where the integration extends over a closed surface, may now be stated in the form that the number of tubes which enter any closed surface is equal to the number which leave it. This is true no matter where the surface is situated, so that we see that tubes of induction can have no beginning or ending.

437. Let us take any closed circuit s in space, and let n be the number of tubes of induction which pass through this circuit in a specified direction.

Then n will also be the number of tubes which cut any area whatever which is bounded by the circuit s. If S is any such area, this number is known to be $\iint N\,dS$, where the integration is taken over the area S, so that

$$N = \iint N\,dS.$$

The number n, however, depends only on the position of the curve s by which the area S is bounded, so that it must be possible to express n in a form which depends only on the position of the curve s, and not on the area S. In other words, it must be possible to replace $\iint N\,dS$ by an expression which depends only on the boundary of the area s. This we are enabled to do by a theorem due to Stokes.

STOKES' THEOREM.

438. THEOREM. *If X, Y, Z are continuous functions of position in space, then*

$$\int \left(X \frac{dx}{ds} + Y \frac{dy}{ds} + Z \frac{dz}{ds} \right) ds$$

$$= \iint \left\{ l \left(\frac{\partial Z}{\partial y} - \frac{\partial Y}{\partial z} \right) + m \left(\frac{\partial X}{\partial z} - \frac{\partial Z}{\partial x} \right) + n \left(\frac{\partial Y}{\partial x} - \frac{\partial X}{\partial y} \right) \right\} dS \quad \dots(363),$$

where the line integral is taken round any closed curve in space, and the surface integral is taken over any area (or shell) bounded by the contour.

Here l, m, n are the direction-cosines of the normal to the surface. A rule is needed to fix the direction in which the normal is to be drawn. The following is perhaps the simplest. Imagine the shell turned about in space so that the tangent plane at any point P is parallel to the plane of xy, and so that the direction in which the line integral is taken round the contour is the same as that of turning from the axis of x to the axis of y. Then the normal at P must be supposed drawn in the direction of the positive axis of z.

439. To prove the theorem, let us select any two points A, B on the contour, and let us introduce a quantity I defined by

$$I = \int_A^B \left(X \frac{dx}{ds} + Y \frac{dy}{ds} + Z \frac{dz}{ds} \right) ds,$$

the path from A to B being the same as that followed in the integral of equation (363). Let us also introduce a quantity J equal to the same

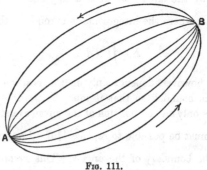

FIG. 111.

integral taken from A to B, but along the opposite edge of the shell. Then the whole integral on the left of equation (363) is equal to $I - J$.

It will be possible to connect A and B by a series of non-intersecting lines drawn in the shell in such a way as to divide the whole shell into narrow strips. Let us denote these lines by the letters a, b, ... n, the lines being taken in order across the shell, starting with the line nearest to that along which we integrate in calculating I. Let us denote the value of

$$\int_A^B \left(X \frac{dx}{ds} + Y \frac{dy}{ds} + Z \frac{dz}{ds} \right) ds$$

taken along the line a by I_a.

Then the left-hand member of equation (363)

$$= I - J$$
$$= (I - I_a) + (I_a - I_b) + (I_b - I_c) + \ldots + (I_n - J).$$

Let us consider the value of any term of this series, say $I_a - I_b$.

Let us take each point on the line a and cause it to undergo a slight displacement, so that the coordinates of any point x, y, z are changed to $x + \delta x$, $y + \delta y$, $z + \delta z$. If δx, δy, δz are continuous functions of x, y, z the result will be to displace the line a into some adjacent position, and by a suitable choice of the values of δx, δy, δz this displaced position of line a can be made to coincide with line b. If this is done, it is clear that the value of I_a, after replacing x, y, z by $x + \delta x$, $y + \delta y$, $z + \delta z$, will be I_b. Hence if we denote this new value of I_a by $I_a + \delta I$, we shall have

$$I_a + \delta I = I_b,$$

so that
$$I_a - I_b = - \delta I$$
$$= - \delta \int_A^B \left(X \frac{dx}{ds} + Y \frac{dy}{ds} + Z \frac{dz}{ds} \right) ds,$$

and the value of this quantity can be obtained by the ordinary rules of the calculus of variations.

We have

$$\delta \int_A^B X \frac{dx}{ds} ds = \int_A^B \delta X \frac{dx}{ds} ds + \int_A^B X \frac{d}{ds} (\delta x) ds$$
$$= \int_A^B \left(\frac{\partial X}{\partial x} \delta x + \frac{\partial X}{\partial y} \delta y + \frac{\partial X}{\partial z} \delta z \right) \frac{dx}{ds} ds + \left[X \delta x \right]_A^B - \int_A^B \frac{dX}{ds} \delta x \, ds,$$

and since δx vanishes both at A and B, the term $\left[X \delta x \right]_A^B$ may be omitted, and the whole expression put equal to

$$\int_A^B \left\{ \left(\frac{\partial X}{\partial x} \delta x + \frac{\partial X}{\partial y} \delta y + \frac{\partial X}{\partial z} \delta z \right) \frac{dx}{ds} - \left(\frac{\partial X}{\partial x} \frac{dx}{ds} + \frac{\partial X}{\partial y} \frac{dy}{ds} + \frac{\partial X}{\partial z} \frac{dz}{ds} \right) \delta x \right\} ds,$$

or again, on simplifying, to

$$\int_A^B \left\{ \frac{\partial X}{\partial y} \left(\delta y \frac{dx}{ds} - \delta x \frac{dy}{ds} \right) - \frac{\partial X}{\partial z} \left(\delta x \frac{dz}{ds} - \delta z \frac{dx}{ds} \right) \right\} ds.$$

This may be written in the form

$$\int_A^B \left\{ \frac{\partial X}{\partial y} \left(\delta y\, dx - \delta x\, dy \right) - \frac{\partial X}{\partial z} \left(\delta x\, dz - \delta z\, dx \right) \right\} \dots\dots\dots(364).$$

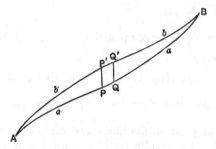

Fig. 112.

Now in fig. 112, let P, Q, P' be the points x, y, z; $x + dx, y + dy, z + dz$; and $x + \delta x$, $y + \delta y$, $z + \delta z$. Let dS denote the area of the parallelogram $PQQ'P'$, and let l, m, n be the direction-cosines of the normal to its plane. Then the projection of the parallelogram on the plane of xy will be of area $n\,dS$, while the coordinates of three of its angular points will be x, y; $x + dx$, $y + dy$; and $x + \delta x$, $y + \delta y$. Using the usual formula for the area, we obtain

$$n\,dS = (\delta y\, dx - \delta x\, dy),$$

and using this relation in expression (364), we obtain

$$\delta \int_A^B X \frac{dx}{ds}\, ds = \int \left(\frac{\partial X}{\partial y} n\,dS - \frac{\partial X}{\partial z} m\,dS \right)\dots\dots\dots\dots(365),$$

the integral denoting summation over all those elements of area of the shell which lie between lines a and b. By summation of three equations of the type of (365), we obtain

$$I_a - I_b = -\delta \int_A^B X \frac{dx}{ds}\, ds - \delta \int_A^B Y \frac{dy}{ds}\, ds - \delta \int_A^B Z \frac{dz}{ds}\, ds$$

$$= \int \left\{ \left(\frac{\partial Z}{\partial y} - \frac{\partial Y}{\partial z} \right) l\,dS + \left(\frac{\partial X}{\partial z} - \frac{\partial Z}{\partial x} \right) m\,dS + \left(\frac{\partial Y}{\partial x} - \frac{\partial X}{\partial y} \right) n\,dS \right\},$$

where the integration has the same meaning as before. If we add a system of equations of this type, one for each strip, the left-hand, as already seen, becomes $I - J$, which is equal to the left-hand member of equation (363), while the right-hand member of the new equation is also the right-hand member of equation (363). This proves the theorem.

440. Stokes' Theorem can be readily expressed in a vector notation. If X, Y, Z are the components of any vector $\mathbf{F}$, it is usual to denote by curl $\mathbf{F}$ the vector of which the components are

$$\frac{\partial Z}{\partial y} - \frac{\partial Y}{\partial z}, \quad \frac{\partial X}{\partial z} - \frac{\partial Z}{\partial x}, \quad \frac{\partial Y}{\partial x} - \frac{\partial X}{\partial y}.$$

Hence Stokes' Theorem assumes the form

$$\int (\text{component of } \mathbf{F} \text{ along } ds)\, ds$$

$$= \int (\text{components of curl } \mathbf{F} \text{ along normal to } dS)\, dS.$$

The theorem enables us to transform any line integral taken round a closed circuit into a surface integral taken over any area by which the circuit can be filled up. The converse operation of changing a surface integral into a line integral may or may not be possible.

441. Theorem. *It will be possible to transform the surface integral*

$$\iint (lu + mv + nw)\, dS \dots\dots\dots\dots\dots\dots(366)$$

into a line integral taken round the contour of the area S if, and only if,

$$\frac{\partial u}{\partial x} + \frac{\partial v}{\partial y} + \frac{\partial w}{\partial z} = 0 \ \dots\dots\dots\dots\dots\dots(367)$$

at every point of the area S.

It is easy to see that this condition is a necessary one. Let S' denote any area having the same boundary as S, and being adjacent to it, but not coinciding with it. Then if I is the line integral into which the surface integral can be transformed, we must have

$$I = \iint (lu + mv + nw)\, dS \dots\dots\dots\dots\dots(368),$$

and also

$$I = \iint (l'u + m'v + n'w)\, dS' \dots\dots\dots\dots(369).$$

On equating these two values for I we obtain an equation which may be expressed in the form

$$\iint (lu + mv + nw)\, d\mathbf{S} = 0 \dots\dots\dots\dots\dots(370),$$

where the integration is over a closed surface bounded by S and S', and l, m, n are the direction-cosines of the outward normal to the surface at any point. From equation (370), the necessity of condition (367) follows at once.

Condition (367) is most easily proved to be sufficient by exhibiting an actual solution of the problem when this condition is satisfied. We have to shew that, subject to condition (367) being satisfied, there are functions X, Y, Z such that

$$\left.\begin{aligned}
\frac{\partial Z}{\partial y} - \frac{\partial Y}{\partial z} &= u \\[4pt]
\frac{\partial X}{\partial z} - \frac{\partial Z}{\partial x} &= v \\[4pt]
\frac{\partial Y}{\partial x} - \frac{\partial X}{\partial y} &= w
\end{aligned}\right\} \quad \dots\dots\dots\dots\dots\dots(371),$$

for if this is so, the required line integral is $\int (lX + mY + nZ)\, dS$.

By inspection a solution of equations (371) is seen to be

$$X = \int v\, dz, \quad Y = -\int u\, dz, \quad Z = 0 \ \dots\dots\dots\dots(372),$$

for it is obvious that the first two equations are satisfied, and on substituting in the third, we obtain

$$\frac{\partial Y}{\partial x} - \frac{\partial X}{\partial y} = \int \left(-\frac{\partial u}{\partial x} - \frac{\partial v}{\partial y} \right) dz = \int \frac{\partial w}{\partial z}\, dz = w,$$

shewing that the proposed solution satisfies all the conditions.

442. The absence of symmetry from solution (372) suggests that this solution is not the most general solution. The most general solution can, however, be easily found. If we assume it to be

$$X = \int v\, dz + X', \quad Y = -\int u\, dz + Y', \quad Z = Z' \ \dots\dots\dots(373),$$

then we find, on substitution in equations (371), that we must have

$$\frac{\partial Z'}{\partial y} = \frac{\partial Y'}{\partial z}, \quad \frac{\partial X'}{\partial z} = \frac{\partial Z'}{\partial x}, \quad \frac{\partial Y'}{\partial x} = \frac{\partial X'}{\partial y} \dots\dots\dots\dots(374),$$

and if we introduce a new variable χ defined by $\chi = \int X'\, dx$, we find at once that

$$X' = \frac{\partial \chi}{\partial x}, \quad Y' = \frac{\partial \chi}{\partial y}, \quad Z' = \frac{\partial \chi}{\partial z},$$

so that the most general solution of equations (371) is

$$X = \int v\, dz + \frac{\partial \chi}{\partial x}, \quad Y = -\int u\, dz + \frac{\partial \chi}{\partial y}, \quad Z = \frac{\partial \chi}{\partial z} \dots\dots\dots(375).$$

Substituting these values, the line integral is found to be

$$\int \left[\left(\int v\, dz \right) \frac{dx}{ds} - \left(\int u\, dz \right) \frac{dy}{ds} \right] ds + \int \frac{d\chi}{ds}\, ds,$$

and the condition that this shall be equal to the surface integral is that

$$\int \frac{d\chi}{ds}\, ds = 0,$$

or that χ shall be single-valued.

Thus if χ is any single-valued function, equations (375) represent a solution, and the most general solution, of equations (371).

VECTOR-POTENTIAL.

443. The discussion as to the transformation from surface to line integrals arose in connection with the integral $\iint N\,dS$ or $\iint (la + mb + nc)\,dS$, in which a, b, c are the components of magnetic induction. Since the condition

$$\frac{\partial a}{\partial x} + \frac{\partial b}{\partial y} + \frac{\partial c}{\partial z} = 0$$

is satisfied throughout all space, it must always be possible (cf. § 441) to transform the surface integral into a line integral by a relation of the form

$$\iint (la + mb + nc)\,dS = \int \left(F\frac{dx}{ds} + G\frac{dy}{ds} + H\frac{dz}{ds} \right) ds.$$

The vector of which the components are F, G, H is known as the *magnetic vector-potential.*

From what has been said in § 442, it is clear that the vector-potential is not fully determined when the magnetic field is given. On the other hand, if the vector-potential is given the magnetic field is fully determined, being given by the equations

$$
\left.
\begin{aligned}
a &= \frac{\partial H}{\partial y} - \frac{\partial G}{\partial z} \\[4pt]
b &= \frac{\partial F}{\partial z} - \frac{\partial H}{\partial x} \\[4pt]
c &= \frac{\partial G}{\partial x} - \frac{\partial F}{\partial y}
\end{aligned}
\right\}
\quad \dots\dots\dots\dots\dots\dots(376).
$$

We shall calculate some possible values of the components of vector-potential in a few simple cases. It must be remembered that the values obtained, although solutions of equations (376), will not be the most general solutions.

Magnetic Particle.

444. Let us first suppose that the field is produced by a single magnetic particle at the point x', y', z' in free space, parallel to the axis of z. Then, by equation (338), $\Omega = \mu \dfrac{\partial}{\partial z'}\left(\dfrac{1}{r}\right)$, so that at any point x, y, z,

$$a = \alpha = -\frac{\partial \Omega}{\partial x} = -\mu \frac{\partial^2}{\partial x\,\partial z'}\left(\frac{1}{r}\right) = \mu \frac{\partial^2}{\partial x\,\partial z}\left(\frac{1}{r}\right),$$

and similarly

$$b = \mu \frac{\partial^2}{\partial y\,\partial z}\left(\frac{1}{r}\right), \qquad c = \mu \frac{\partial^2}{\partial z^2}\left(\frac{1}{r}\right).$$

The equations to be solved (equations (376)) are

$$\frac{\partial H}{\partial y} - \frac{\partial G}{\partial z} = \mu \frac{\partial^2}{\partial x \partial z}\left(\frac{1}{r}\right),$$

$$\frac{\partial F}{\partial z} - \frac{\partial H}{\partial x} = \mu \frac{\partial^2}{\partial y \partial z}\left(\frac{1}{r}\right),$$

$$\frac{\partial G}{\partial x} - \frac{\partial F}{\partial y} = \mu \frac{\partial^2}{\partial z^2}\left(\frac{1}{r}\right),$$

and the simplest solution, similar to that given by equations (372), is

$$F = \mu \frac{\partial}{\partial y}\left(\frac{1}{r}\right), \qquad G = -\mu \frac{\partial}{\partial x}\left(\frac{1}{r}\right), \qquad H = 0.$$

The components of vector-potential for a magnet parallel to the axes of x or y can be written down from symmetry. In terms of the coordinates x', y', z' of the magnetic particle, this solution may be expressed as

$$F = -\mu \frac{\partial}{\partial y'}\left(\frac{1}{r}\right), \qquad G = \mu \frac{\partial}{\partial x'}\left(\frac{1}{r}\right), \qquad H = 0.$$

445. Let us superpose the fields of a magnetic particle of strength $l\mu$ parallel to the axis of x, one of strength $m\mu$ parallel to the axis of y, and one of strength $n\mu$ parallel to the axis of z. Then we obtain the vector-potential at x, y, z due to a magnetic particle of strength μ and axis (l, m, n) at x', y', z' in the forms

$$\left. \begin{array}{l} F = -\mu \left(m \dfrac{\partial}{\partial z} - n \dfrac{\partial}{\partial y} \right)\dfrac{1}{r} = \mu \left(m \dfrac{\partial}{\partial z'} - n \dfrac{\partial}{\partial y'} \right)\dfrac{1}{r} \\[2mm] G = -\mu \left(n \dfrac{\partial}{\partial x} - l \dfrac{\partial}{\partial z} \right)\dfrac{1}{r} = \mu \left(n \dfrac{\partial}{\partial x'} - l \dfrac{\partial}{\partial z'} \right)\dfrac{1}{r} \\[2mm] H = -\mu \left(l \dfrac{\partial}{\partial y} - m \dfrac{\partial}{\partial x} \right)\dfrac{1}{r} = \mu \left(l \dfrac{\partial}{\partial y'} - m \dfrac{\partial}{\partial x'} \right)\dfrac{1}{r} \end{array} \right\} \quad \ldots\ldots(377).$$

The number of lines of induction which cross the circuit from a magnetic particle is (§ 437)

$$\int \left(F \frac{dx}{ds} + G \frac{dy}{ds} + H \frac{dz}{ds} \right) ds,$$

which may be written in the form

$$- \int \begin{vmatrix} \dfrac{dx}{ds}, & \dfrac{dy}{ds}, & \dfrac{dz}{ds} \\[2mm] l, & m, & n \\[2mm] \dfrac{\partial}{\partial x}\left(\dfrac{1}{r}\right), & \dfrac{\partial}{\partial y}\left(\dfrac{1}{r}\right), & \dfrac{\partial}{\partial z}\left(\dfrac{1}{r}\right) \end{vmatrix} \mu\, ds,$$

the integral being taken round the circuit in the direction determined by the rule given in § 438 (p. 388).

Uniform Magnetic Shell.

446. Next let us suppose that the lines of force proceed from a uniform magnetic shell, supposed for simplicity to be of unit strength. Let l', m', n' be the direction-cosines of the normal to any element dS' of this shell. Then the element dS' will be a magnetic particle of moment dS' and of direction-cosines l', m', n'. The element accordingly contributes to F a term which, by equations (377), is seen to be

$$\left(m'\frac{\partial}{\partial z'} - n'\frac{\partial}{\partial y'}\right)\left(\frac{1}{r}\right)dS',$$

where x', y', z' are the coordinates of the element dS'. Thus the whole value of F is

$$F = \iint\left(m'\frac{\partial}{\partial z'} - n'\frac{\partial}{\partial y'}\right)\left(\frac{1}{r}\right)dS'.$$

This surface integral satisfies the condition of § 441, so that it must be possible to transform it into a line integral of the form

$$F = \int\left(f\frac{\partial x'}{\partial s'} + g\frac{\partial y'}{\partial s'} + h\frac{\partial z'}{\partial s'}\right)dS'.$$

The equations giving f, g, h are

$$\frac{\partial h}{\partial y'} - \frac{\partial g}{\partial z'} = 0,$$

$$\frac{\partial f}{\partial z'} - \frac{\partial h}{\partial x'} = \frac{\partial}{\partial z'}\left(\frac{1}{r}\right),$$

$$\frac{\partial g}{\partial x'} - \frac{\partial f}{\partial y'} = -\frac{\partial}{\partial y'}\left(\frac{1}{r}\right).$$

Clearly a solution is

$$f = \frac{1}{r}, \quad g = 0, \quad h = 0,$$

so that on substitution the value of F is

$$F = \int\frac{1}{r}\frac{dx'}{ds'}ds'.$$

Similarly

$$G = \int\frac{1}{r}\frac{dy'}{ds'}ds',$$

$$H = \int\frac{1}{r}\frac{dz'}{ds'}ds'.$$

Thus the number of tubes of induction crossing the circuit s from a magnetic shell of unit strength bounded by the circuit s', is given by

$$n = \int\left(F\frac{dx}{ds} + G\frac{dy}{ds} + H\frac{dz}{ds}\right)ds,$$

or

$$n = \iint\left(\frac{dx}{ds}\frac{dx'}{ds'} + \frac{dy}{ds}\frac{dy'}{ds'} + \frac{dz}{ds}\frac{dz'}{ds'}\right)\frac{1}{r}ds\,ds'.$$

If ϵ is the angle between the two elements ds, ds', the direction of these elements being taken to be that in which the integration takes place, then

$$\frac{dx}{ds}\frac{dx'}{ds'} + \frac{dy}{ds}\frac{dy'}{ds'} + \frac{dz}{ds}\frac{dz'}{ds'} = \cos \epsilon,$$

so that

$$n = \iint \frac{\cos \epsilon}{r}\,ds\,ds'.$$

From the rule as to directions given on p. 388, it will be clear that if the integration is taken in the *same* direction round both circuits, then the direction in which the n lines cross the circuit will be that of the direction of magnetisation of the shell.

Clearly n is symmetrical as regards the two circuits s and s', so that we have the important result:

The number of tubes of induction crossing the circuit s from a shell of unit strength bounded by the circuit s' is equal to the number of tubes of induction crossing the circuit s' from a shell of unit strength bounded by the circuit s.

Here we have arrived at a purely geometrical proof of the theorem already obtained from dynamical principles in § 424.

ENERGY OF A MAGNETIC FIELD.

447. Let $a, b, c, \ldots n$ be a system of magnetised bodies, the magnetisation of each being permanent, and let us suppose that the total magnetic field arises solely from these bodies. Let us suppose that the potential Ω at any point is regarded as the sum of the potentials due to the separate magnets. Denoting these by Ω_a, Ω_b, $\ldots \Omega_n$, we shall have

$$\Omega = \Omega_a + \Omega_b + \ldots + \Omega_n.$$

Let us denote the potential energy of magnet a, when placed in the field of force of potential Ω, by $\Omega(a)$; if placed in the field of force arising from magnet b alone, by $\Omega_b(a)$, etc.

Let us imagine that we construct the magnetic field by bringing up the magnets $a, b, c, \ldots n$ in this order, from infinity to their final positions.

We do no work in bringing magnet a into position, for there are no forces against which work can be done. After the operation of placing a in position, the potential of the field is Ω_a. The operation of bringing magnet a from infinity has of course been simply that of moving a field of force of potential Ω_a from infinity, where this same field of force had previously existed.

On bringing up magnet b, the work done is that of placing magnet b in a field of force of potential Ω_a. The work done is accordingly $\Omega_a(b)$.

The work done in bringing up magnet c is that of placing magnet c in a field of force of potential $\Omega_a + \Omega_b$. It is therefore $\Omega_a(c) + \Omega_b(c)$.

Continuing this process we find that the total work done, W, is given by

$$
\begin{aligned}
W = \quad & \Omega_a(b) \\
& + \Omega_a(c) + \Omega_b(c) \\
& + \Omega_a(d) + \Omega_b(d) + \Omega_c(d) + \text{etc.}
\end{aligned}
$$

If, however, the magnets had been brought up in the reverse order, we should have had

$$
\begin{aligned}
W = \Omega_b(a) + & \Omega_c(a) + \Omega_d(a) + \dots + \Omega_n(a) \\
& + \Omega_c(b) + \Omega_d(b) + \dots + \Omega_n(b) \\
& \qquad\quad + \Omega_d(c) + \dots + \Omega_n(c) \\
& \qquad\qquad\qquad + \text{etc.}
\end{aligned}
$$

so that by addition of these two values for W, we have

$$
\begin{aligned}
2W = \qquad\quad & \Omega_b(a) + \Omega_c(a) + \Omega_d(a) + \dots + \Omega_n(a) \\
+ \Omega_a(b) \qquad\quad & \qquad + \Omega_c(b) + \Omega_d(b) + \dots + \Omega_n(b) \\
+ \Omega_a(c) + \Omega_b(c) \qquad & \qquad\qquad + \Omega_d(c) + \dots + \Omega_n(c) \\
+ \Omega_a(d) + \Omega_b(d) + \Omega_c(d) \qquad & \qquad\qquad\qquad + \dots + \Omega_n(d) \\
+ \text{etc.}
\end{aligned}
$$

The first line is equal to $\Omega(a)$ except for the absence of the term $\Omega_a(a)$, and so on for the other lines. Thus we have

$$
\begin{aligned}
2W = \quad & \Omega(a) - \Omega_a(a) \\
& + \Omega(b) - \Omega_b(b) + \text{etc.} \\
= \quad & \Sigma\Omega(a) - \Sigma\Omega_a(a) \qquad\dots\dots\dots\dots\dots\dots(378).
\end{aligned}
$$

The quantity $\Omega_a(a)$, the potential energy of the magnet a in its own field of force, is purely a constant of the magnet a, being entirely independent of the properties or positions of the other magnets b, c, d, …. Thus in equation (378), we may regard the term $\Sigma\Omega_a(a)$ as a constant, and may replace the equation by

$$
W = \tfrac{1}{2}\Sigma\Omega(a) + \text{constant} \dots\dots\dots\dots\dots\dots(379).
$$

448. If we take the magnets a, b, c, … n to be the ultimate magnetic particles, the values of $\Omega_a(a)$, $\Omega_b(b)$, … etc. all vanish, and their sum also vanishes. Thus equation (379) assumes the form

$$
W = \tfrac{1}{2}\Sigma\Omega(a) \dots\dots\dots\dots\dots\dots\dots\dots(380),
$$

where the standard configuration from which W is measured is one in which the ultimate particles are scattered at infinity. The value of $\Omega(a)$ for a single particle is (cf. § 420)

$$
\mu \left(l\frac{\partial\Omega}{\partial x} + m\frac{\partial\Omega}{\partial y} + n\frac{\partial\Omega}{\partial z} \right).
$$

On replacing μ by $I\,dx\,dy\,dz$, we find for the energy of a system of magnetised bodies

$$W = \tfrac{1}{2} \iiint I \left(l\frac{\partial\Omega}{\partial x} + m\frac{\partial\Omega}{\partial y} + n\frac{\partial\Omega}{\partial z} \right) dx\,dy\,dz$$

$$= \tfrac{1}{2} \iiint \left(A\frac{\partial\Omega}{\partial x} + B\frac{\partial\Omega}{\partial y} + C\frac{\partial\Omega}{\partial z} \right) dx\,dy\,dz \quad \ldots\ldots\ldots(381),$$

the integration being taken throughout all magnetised matter.

449. An alternative proof can be given of equations (380) and (381), following the method of § 106, in which we obtained the energy of a system of electric charges.

Out of the magnetic materials scattered at infinity, it will be possible to construct n systems, each exactly similar as regards arrangement in space to the final system, but of only one-nth the strength of the final system. If n is made very great, it is easily seen that the work done in constructing a single system vanishes to the order of $\dfrac{1}{n^2}$, so that, in the limit when n is very great, the work done in constructing the series of n systems is infinitesimal. Thus the energy of the final system may be regarded as the work done in superposing this series of n systems.

Let us suppose so many of the component systems to have been superposed, that the system in position is κ times its final strength, where κ is a positive quantity less than unity. The potential of the field at any point will be $\kappa\Omega$. On bringing up a new system let us suppose that κ is increased to $\kappa + d\kappa$, so that the strength of the new system is $d\kappa$ times that of the final system. In bringing up the new system, we place a magnet of $d\kappa$ times the strength of a in a field of force of potential $\kappa\Omega$, and so on with the other magnets. Thus the work done is

$$d\kappa \,.\, \kappa\Omega\,(a) + d\kappa \,.\, \kappa\Omega\,(b) + \ldots,$$

and on integration of the work performed, we obtain

$$W = \int_0^1 \kappa\,d\kappa \,\{\Omega\,(a) + \Omega\,(b) + \ldots\}$$

$$= \tfrac{1}{2}\Sigma\Omega\,(a),$$

agreeing with equation (380), and leading as before to equation (381).

450. If the magnetic matter consists solely of normally magnetised shells, we may replace equation (381) by

$$W = \tfrac{1}{2}\Sigma \iiint I \frac{\partial\Omega}{\partial n} \,ds\,dS,$$

where ds denotes thickness and dS an element of area of a shell. Replacing $I\,ds$ by ϕ, so that ϕ is the strength of a shell, we have

$$W = \tfrac{1}{2}\Sigma \iint \phi \frac{\partial\Omega}{\partial n} \,dS.$$

For uniform shells, ϕ may be taken outside the sign of integration, and the equation becomes

$$W = \tfrac{1}{2}\Sigma\phi\iint\frac{\partial\Omega}{\partial n}\,dS = -\tfrac{1}{2}\Sigma\phi n$$

(cf. § 423), where n is the number of lines of induction which cross the shell.

This calculation measures the energy from a standard configuration in which the magnetic materials are all scattered at infinity. To calculate the energy measured from a standard configuration in which the shells have already been constructed and are scattered at infinity as complete shells, we use equation (378), namely

$$W = \tfrac{1}{2}\Sigma\left\{\Omega\left(a\right) - \Omega_a\left(a\right)\right\},$$

from which we obtain $W = \tfrac{1}{2}\Sigma\iint\phi\dfrac{\partial\Omega'}{\partial n}\,dS,$

where $\dfrac{\partial\Omega'}{\partial n}$ denotes the values $\dfrac{\partial\Omega}{\partial n}$ at the surface of any shell if the shell itself is supposed annihilated.

If all the shells are uniform, this may again be written

$$W = -\tfrac{1}{2}\Sigma\phi n' \dotfill (382),$$

where n' is the number of tubes of force from the remaining shells, which cross the shell of strength ϕ. An example of this has already occurred in § 424.

ENERGY IN THE MEDIUM.

451. We have seen that the energy of a magnetic field is given by (cf. equation (381))

$$W = \tfrac{1}{2}\iiint\left(A\frac{\partial\Omega}{\partial x} + B\frac{\partial\Omega}{\partial y} + C\frac{\partial\Omega}{\partial z}\right)dx\,dy\,dz \dotfill (383),$$

the integration being taken over all magnetic matter. As a preliminary to transforming this into an integral taken through all space, we shall prove that

$$\iiint\left(a\alpha + b\beta + c\gamma\right)dx\,dy\,dz = 0 \dotfill (384),$$

the integration being through all space.

The integral on the left can be written as

$$-\iiint\left(a\frac{\partial\Omega}{\partial x} + b\frac{\partial\Omega}{\partial y} + c\frac{\partial\Omega}{\partial z}\right)dx\,dy\,dz,$$

and this, by Green's Theorem, may be transformed into

$$\iiint\Omega\left(\frac{\partial a}{\partial x} + \frac{\partial b}{\partial y} + \frac{\partial c}{\partial z}\right)dx\,dy\,dz - \iint\Omega\left(la + mb + nc\right)dS,$$

the latter integral being taken over a sphere at infinity. Now at infinity Ω is of the order of $\frac{1}{r^2}$ (cf. § 67), while $la + mb + nc$ vanishes, and dS is of the order of r^2, so that the surface integral vanishes on passing to the limit $r = \infty$. Also the volume integral vanishes since

$$\frac{\partial a}{\partial x} + \frac{\partial b}{\partial y} + \frac{\partial c}{\partial z} = 0,$$

and hence the theorem is proved.

Replacing a, b, c by their values, as given by equations (359), we find that equation (384) becomes

$$\iiint (\alpha^2 + \beta^2 + \gamma^2)\, dx\, dy\, dz + 4\pi \iiint (A\alpha + B\beta + C\gamma)\, dx\, dy\, dz = 0 \dots(385).$$

Both integrals are taken through all space, but since $A = B = C = 0$ except in magnetic matter, we can regard the latter integral as being taken only over the space occupied by magnetic matter. This integral is therefore equal, by equation (383), to $-2W$, so that equation (385) becomes

$$W = \frac{1}{8\pi} \iiint (\alpha^2 + \beta^2 + \gamma^2)\, dx\, dy\, dz \dots(386),$$

the integral being taken through all space.

This expression is exactly analogous to that which has been obtained for the energy of an electrostatic system, namely,

$$W = \frac{1}{8\pi} \iiint (X^2 + Y^2 + Z^2)\, dx\, dy\, dz.$$

And, as in the case of an electrostatic system, equation (386) may be interpreted as meaning that the energy may be regarded as spread through the medium at a rate $\frac{1}{8\pi}(\alpha^2 + \beta^2 + \gamma^2)$ per unit volume.

TERRESTRIAL MAGNETISM.

452. The magnetism of the earth is very irregularly distributed and is constantly changing. The simplest and roughest approximation of all to the state of the earth's magnetism is obtained by regarding it as a bar magnet, possessing two poles near to its surface, the position of these in 1906 being as follows:

North Pole 70° 30′ N., 97° 40′ W.

South Pole* 73° 39′ S., 146° 15′ E.

Another approximation, which is better in many ways although still very rough, is obtained by regarding the earth as a uniformly magnetised sphere.

* Sir E. Shackleton gives the position of the South Pole in 1909 as 72° 25′ S., 155° 16′ E.

With the help of a compass-needle, it will be possible to find the direction of the lines of force of the earth's field at any point. It will also be possible to measure the intensity of this field, by comparing it with known magnetic fields, or by measuring the force with which it acts on a magnet of known strength.

453. At any point on the earth, let us suppose that the angle between the line of magnetic force and the horizontal is θ, this being reckoned positive if the line of force points down into the earth, and let the horizontal projection of the line of force make an angle δ with the geographical meridian through the point, this being reckoned positive if this line points west of north. The angle θ is called the *dip* at the point, the angle δ is called the *declination*.

Let H be the horizontal component of force, then the total force may be regarded as made up of three components:

$X = H \cos \delta$, towards the north,

$Y = H \sin \delta$, towards the west,

$Z = H \tan \theta$, vertically downwards.

If Ω is the potential due to the earth's field at a point of latitude l, longitude λ, and at distance r from the centre, we have (cf. equations (331))

$$X = -\frac{1}{r}\frac{\partial \Omega}{\partial l} \qquad Y = -\frac{1}{r \cos l}\frac{\partial \Omega}{\partial \lambda}, \qquad Z = \frac{\partial \Omega}{\partial r} \quad \dots\dots\dots(387).$$

Analysis of Potential of Earth's field.

454. Since Ω is the potential of a magnetic system, the value of Ω in regions in which there is no magnetisation must (by § 408) be a solution of Laplace's equation, and must therefore (by § 233) be capable of expansion in the form

$$\Omega = \left(\frac{S_1}{r^2} + \frac{S_2}{r^3} + \dots\right) + (S_0' + S_1' r + S_2' r^2 + \dots) \quad \dots\dots\dots(388),$$

in which $S_1, S_2, \dots S_0', S_1', S_2', \dots$ are surface harmonics, of degrees indicated by the subscripts.

At the earth's surface, the first term is the part of the potential which arises from magnetism inside the earth, while the second term arises from magnetism outside.

The surface harmonic S_n can, as in § 275, be expanded in the form

$$S_n = \sum_{m=0}^{m=n} P_n^m (\sin l)\,(A_{n,m} \cos m\lambda + B_{n,m} \sin m\lambda),$$

so that Ω can be put in the form

$$\Omega = \sum_{n=0}^{n=\infty} \sum_{m=0}^{m=n} \left\{ \frac{P_n^m (\sin l)}{r^{n+1}} (A_{n,m} \cos m\lambda + B_{n,m} \sin m\lambda) \right.$$
$$\left. + r_n P_n^m (\sin l)\,(A'_{n,m} \cos m\lambda + B'_{n,m} \sin m\lambda) \right\}.$$

Hence from equations (387) we obtain the values of X, Y, Z at any point in terms of the longitude and latitude of the point and the constants such as $A_{n,m}$, $B_{n,m}$, $A'_{n,m}$, $B'_{n,m}$.

By observing the values of X, Y, Z at a great number of points, we obtain a system of equations between the constants $A_{n,m}$, etc., and on solving these we obtain the actual values of the constants, and therefore a knowledge of the potential as expressed by equation (388).

If the magnetic field arose entirely from magnetism inside the earth, we should of course expect to find $S_1' = S_2' = \ldots = 0$, while if the magnetic field arose from magnetism entirely outside the earth, we should find $S_1 = S_2 = \ldots = 0$.

455. The results actually obtained are of extreme interest. The magnetic field of the earth, as we have said, is constantly changing. In addition to a slow, irregular, and so-called "secular" change, it is found that there are periodic changes of which the periods are, in general, recognisable as the periods of astronomical phenomena. For instance there is a daily period, a yearly period, a period equal to the lunar month, a period of about $26\frac{1}{3}$ days (the period of rotation of the inner core of the sun*), a period of about 11 years (the period of sun-spot variations), a period of 19 years (the period of the motion of the lunar nodes), and so on. Thus the potential can be divided up into a number of periodic parts and a residual constant, or slowly and irregularly changing, part. All the periodic parts are extremely small in comparison with the latter. It is found, on analysing the potentials of these different parts of the field, that the constant field arises from magnetisation inside the earth, while the daily variation arises mainly from magnetisation outside the earth. The former result might have been anticipated, but the latter could not have been predicted with any confidence. For the variation might have represented nothing more than a change in the permanent magnetism of the earth due to the cooling and heating of the earth's mass, or to the tides in the solid matter of the earth produced by the sun's attraction.

This daily variation is not such as could be explained by the magnetism of the sun itself; Chree† has found that it cannot be explained by the cooling and heating either of the earth's mass, or of the atmosphere as suggested by Faraday. Balfour Stewart‡ put forward the hypothesis that the daily variation was due mainly to electric currents circulating in the upper atmosphere as a result of the electromotive forces induced by the convective

---

* The outer surface of the sun is not rigid, and rotates at different rates in different latitudes. Thus it is impossible to discover the actual rate of rotation of the inner core except by such indirect methods as that of observing periods of magnetic variation.

† *Roy. Soc. Phil. Trans.*, 202, p. 335.

‡ Art. "Terrestrial Magnetism," in the 9th Edn. of the *Encyc. Brit.* (1882).

motion of the atmosphere across the earth's magnetic field. This hypothesis was examined and developed by Schuster*, who examined the daily variations by the method of harmonic analysis, already explained. Schuster found the origin of the magnetic field to be mainly external; he suggested also that the convection currents indicated by the diurnal barometric changes were ultimately responsible for the phenomenon, and further found that a small part of the field must be attributed to origins inside the earth: these it was suggested might be a system of currents induced in the earth by the atmospheric currents above.

Chapman† has recently reexamined the question, and obtains results in substantial agreement with Schuster's theory. He finds that the contribution from inside the earth is about 28 per cent. of the total diurnal variation. It is supposed that the conducting layer in the upper atmosphere in which the induced currents flow is that of which we already have evidence in the phenomenon of the bending of electromagnetic waves round the earth; this layer is also the seat of the aurora borealis. Chapman finds that the internal magnetic field of induced currents would be explained by assuming that, beneath an upper non-conductive layer of 150 or 200 miles depth, the earth has a specific resistance of about 4×10^{-11} c.g.s. units.

Besides the variation just considered, there is found to be a lunar diurnal variation, of period equal to the apparent period of motion of the moon. This appears to be the result of a semi-diurnal tidal oscillation of the atmosphere the mechanism being otherwise similar to that already explained.

456. The non-periodic part of the earth's field is found to arise entirely from magnetism inside the earth, having a potential of the form

$$\Omega = \frac{S_1}{r^2} + \frac{S_2}{r^3} + \ldots = \sum_{n=1}^{n=\infty} \sum_{m=0}^{m=n} \left\{ \frac{P_n^m (\sin l)}{r^{n+1}} (A_{n,m} \cos m\lambda + B_{n,m} \sin m\lambda) \right\}.$$

This method of analysing the earth's field is due to Gauss, who calculated the coefficients, with such accuracy as was then possible, for the year 1830. The most complete analysis of the field which now exists has been calculated by Neumayer for the year 1885, using observations of the field at 1800 points on the earth's surface.

The first few coefficients obtained by Neumayer are as follows:

$$A_{1,0} = \cdot 3157 \quad \begin{cases} A_{1,1} = \quad \cdot 0248, \\ B_{1,1} = - \cdot 0603, \end{cases}$$

$$A_{2,0} = \cdot 0079 \quad \begin{cases} A_{2,1} = - \cdot 0498, \quad A_{2,2} = - \cdot 0057, \\ B_{2,1} = \quad \cdot 0130, \quad B_{2,2} = - \cdot 0126, \end{cases}$$

$$A_{3,0} = - \cdot 0244 \quad \begin{cases} A_{3,1} = \cdot 0396, \quad A_{3,2} = - \cdot 0279, \quad A_{3,3} = - \cdot 0033, \\ B_{3,1} = \cdot 0074, \quad B_{3,2} = - \cdot 0004, \quad B_{3,3} = - \cdot 0055, \end{cases}$$

$$A_{4,0} = - \cdot 0344 \quad \begin{cases} A_{4,1} = - \cdot 0306, \quad A_{4,2} = - \cdot 0198, \quad A_{4,3} = \cdot 0068, \quad A_{4,4} = - \cdot 0008, \\ B_{4,1} = - \cdot 0119, \quad B_{4,2} = \quad \cdot 0071, \quad B_{4,3} = \cdot 0051, \quad B_{4,4} = \quad \cdot 0010. \end{cases}$$

* *Phil. Trans.* A, 180 (1889), p. 467, and A, 208 (1907), p. 163.
† *Phil. Trans.* A, 213 (1913), p. 279, and A, 218 (1919), p. 1.

457. The simplest approximation is of course obtained by ignoring all harmonics beyond the first. This gives as the magnetic potential

$$\Omega = \frac{1}{r^2}\left\{A_{1,0}P_1(\sin l) + P_1^1(\sin l)(A_{1,1}\cos \lambda + B_{1,1}\sin \lambda)\right\}$$

$$= \frac{1}{r^2}\left\{\cdot 3157\sin l + \cos l(\cdot 0248\cos \lambda - \cdot 0603\sin \lambda)\right\}.$$

The expression in brackets is necessarily a biaxial harmonic of order unity (cf. § 276); it is easily found to be equal to $\cdot 3224\cos \gamma$, where γ is the angular distance of the point (l, λ) from the point

lat. 78° 20′ N., long. 67° 17′ W. ....(389).

The potential is now $\qquad \Omega = \cdot 3224\dfrac{\cos \gamma}{r^2},$

which is the potential of a uniformly magnetised sphere, having as direction of magnetisation the radius through the point (§ 415). Or again, it is the potential of a single magnetic particle at the centre of the earth, pointing in this same direction. It is naturally impossible to distinguish between these two possibilities by a survey of the field outside the earth. Green's theorem has already shewn that we cannot locate the sources of a field inside a closed surface by a study of the field outside the surface.

EXAMPLES.

1. Two small magnets float horizontally on the surface of water, one along the direction of the straight line joining their centres, and the other at right angles to it. Prove that the action of each magnet on the other reduces to a single force at right angles to the straight line joining the centres, and meeting that line at one-third of its length from the longitudinal magnet.

2. A small magnet ACB, free to turn about its centre C, is acted on by a small fixed magnet PQ. Prove that in equilibrium the axis ACB lies in the plane PQC, and that $\tan \theta = -\frac{1}{2}\tan \theta'$, where θ, θ' are the angles which the two magnets make with the line joining them.

3. Three small magnets having their centres at the angular points of an equilateral triangle ABC, and being free to move about their centres, can rest in equilibrium with the magnet at A parallel to BC, and those at B and C respectively at right angles to AB and AC. Prove that the magnetic moments are in the ratios

$$\sqrt{3} : 4 : 4.$$

4. The axis of a small magnet makes an angle ϕ with the normal to a plane. Prove that the line from the magnet to the point in the plane where the number of lines of force crossing it per unit area is a maximum makes an angle θ with the axis of the magnet, such that

$$2\tan \theta = 3\tan 2(\phi - \theta).$$

5. Two small magnets lie in the same plane, and make angles θ, θ' with the line joining their centres. Shew that the line of action of the resultant force between them divides the line of centres in the ratio

$$\tan \theta' + 2 \tan \theta : \tan \theta + 2 \tan \theta'$$

6. Two small magnets have their centres at distance r apart, make angles θ, θ' with the line joining them, and an angle ϵ with each other. Shew that the force on the first magnet in its own direction is

$$\frac{3mm'}{r^4} (5 \cos^2 \theta \cos \theta' - \cos \theta' - 2 \cos \epsilon \cos \theta).$$

Shew that the couple about the line joining them which the magnets exert on one another is

$$\frac{mm'}{r^4} d \sin \epsilon,$$

where d is the shortest distance between their axes produced.

7. Two magnetic needles of moments M, M' are soldered together so that their directions include an angle a. Shew that when they are suspended so as to swing freely in a uniform horizontal magnetic field, their directions will make angles θ, θ' with the lines of force, given by

$$\frac{\sin \theta}{M'} = \frac{\sin \theta'}{M} = \frac{\sin a}{(M^2 + M'^2 + 2MM' \cos a)^{\frac{1}{2}}}.$$

8. Prove that if there are two magnetic molecules, of moments M and M', with their centres fixed at A and B, where $AB = r$, and one of the molecules swings freely, while the other is acted on by a given couple, so that when the system is in equilibrium this molecule makes an angle θ with AB, then the moment of the couple is

$$\tfrac{3}{2} MM' \sin 2\theta / r^3 (3 \cos^2 \theta + 1)^{\frac{1}{2}},$$

where there is no external field.

9. Two small equal magnets have their centres fixed, and can turn about them in a magnetic field of uniform intensity H, whose direction is perpendicular to the line r joining the centres. Shew that the position in which the magnets both point in the direction of the lines of force of the uniform field is stable only if

$$H > 3M/r^3.$$

10. Two magnetic particles of equal moment are fixed with their axes parallel to the axis of z, and in the same direction, and with their centres at the points $\pm a$, 0, 0. Shew that if another magnetic molecule is free to turn about its centre, which is fixed at the point $(0, y, z)$, its axis will rest in the plane $x = 0$, and will make with the axis of z the angle

$$\tan^{-1} \frac{3yz}{2z^2 - a^2 - y^2}.$$

Examine which of the two positions of equilibrium is stable.

11. Prove that there are four positions in which a given bar magnet may be placed so as to destroy the earth's control of a compass-needle, so that the needle can point indifferently in all directions. If the bar is short compared with its distance from the needle, shew that one pair of these positions are about $1\frac{1}{4}$ times more distant than the other pair.

12. Three small magnets, each of magnetic moment μ, are fixed at the angular points of an equilateral triangle ABC, so that their north poles lie in the directions AC, AB, BC respectively. Another small magnet, moment μ', is placed at the centre of the triangle, and is free to move about its centre. Prove that the period of a small oscillation is the same as that of a pendulum of length $Ib^3g/\sqrt{351}\mu\mu'$, where b is the length of a side of the triangle, and I the moment of inertia of the movable magnet about its centre.

13. Three magnetic particles of equal moments are placed at the corners of an equilateral triangle, and can turn about those points so as to point in any direction in the plane of the triangle. Prove that there are four and only four positions of equilibrium such that the angles, measured in the same sense of rotation, between the axes of the magnets and the bisectors of the corresponding angles of the triangle are equal. Also prove that the two symmetrical positions are unstable.

14. Four small equal magnets are placed at the corners of a square, and oscillate under the actions they exert on each other. Prove that the times of vibration of the principal oscillations are

$$2\pi\left\{\frac{Mk^2d^3}{m^2 3\left(2+1/2\sqrt{2}\right)}\right\}^{\frac{1}{2}},$$

$$2\pi\left\{\frac{Mk^2d^3}{m^2\left(3-1/2\sqrt{2}\right)}\right\}^{\frac{1}{2}},$$

$$2\pi\left\{\frac{Mk^2d^3 2\sqrt{2}}{3m^2}\right\}^{\frac{1}{2}},$$

where m is the magnetic moment, and Mk^2 the moment of inertia, of a magnet, and d is a side of the square.

15. A system of magnets lies entirely in one plane and it is found that when the axis of a small needle travels round a contour in the plane that contains no magnetic poles, the needle turns completely round. Prove that the contour contains at least one equilibrium point.

16. Prove that the potential of a body uniformly magnetised with intensity I is, at any external point, the same as that due to a complex magnetic shell coinciding with the surface of the body and of strength Ix, where x is a coordinate measured parallel to the direction of magnetisation.

17. A sphere of hard steel is magnetised uniformly in a constant direction and a magnetic particle is held at an external point with the axis of the particle parallel to the direction of magnetisation of the sphere. Find the couples acting on the sphere and on the particle.

18. A spherical magnetic shell of radius a is normally magnetised so that its strength at any point is S_i, where S_i is a spherical surface harmonic of positive order i. Shew that the potential at a distance r from the centre is

$$-4\pi\,\frac{i+1}{2i+1}\,S_i\left(\frac{r}{a}\right)^i \quad \text{when } r<a,$$

$$4\pi\,\frac{i}{2i+1}\,S_i\left(\frac{a}{r}\right)^{i+1} \quad \text{when } r>a.$$

19. If a small spherical cavity be made within a magnetised body, prove that the components of magnetic force within the cavity are

$$a+\tfrac{4}{3}A,\quad \beta+\tfrac{4}{3}B,\quad \gamma+\tfrac{4}{3}C.$$

20. If the earth were a uniformly magnetised sphere, shew that the tangent of the dip at any point would be equal to twice the tangent at the magnetic latitude.

21. Prove that if the horizontal component, in the direction of the meridian, of the earth's magnetic force were known all over its surface, all the other elements of its magnetic force might be theoretically deduced.

22. From the principle that the line integral of the magnetic force round any circuit ordinarily vanishes, shew that the two horizontal components of the magnetic force at any station may be deduced approximately from the known values for three other stations which lie around it. Shew that these six known elements are not independent, but must satisfy one equation of condition.

23. If the earth were a sphere, and its magnetism due to two small straight bar magnets of the same strength situated at the poles, with their axes in the same direction along the earth's axis, prove that the dip δ in latitude λ would be given by

$$8 \cot\left(\delta + \frac{\lambda}{2}\right) = \cot\frac{\lambda}{2} - 6\tan\frac{\lambda}{2} - 3\tan^3\frac{\lambda}{2}.$$

24. Assuming that the earth is a sphere of radius a, and that the magnetic potential Ω is represented by

$$\Omega = S_1\left(\frac{r}{a}\right) + S_2\left(\frac{r}{a}\right)^2 + S_1{}'\left(\frac{a}{r}\right)^2 + S_2{}'\left(\frac{a}{r}\right)^3,$$

shew that Ω is completely determined by observations of horizontal intensity, declination and dip at four stations, and of dip at four more.

25. Assuming that in the expansion of the earth's magnetic potential the fifth and higher harmonics may be neglected, shew that observations of the resultant magnetic force at eight points are sufficient to determine the potential everywhere.

26. Assuming that the earth's magnetism is entirely due to internal causes, and that in latitude λ the northerly component of the horizontal force is $A\cos\lambda + B\cos^3\lambda$, prove that in this latitude the vertical component reckoned downwards is

$$2\left(A + \tfrac{2}{3}B\right)\sin\lambda - \tfrac{4}{3}B\sin^3\lambda.$$

CHAPTER XII

INDUCED MAGNETISM

Physical Phenomena.

458. Reference has already been made to the well-known fact that a magnet will attract small pieces of iron or steel which are not themselves magnets. Here we have a phenomenon which at first sight does not seem to be explained by the law of the attractions and repulsions of magnetic poles. It is found, however, that the phenomenon is due to a magnetic "induction" of a kind almost exactly similar to the electrostatic induction already discussed. It can be shewn that a piece of iron or steel, placed in the presence of a magnet, will itself become magnetised. Temporarily, this piece of iron or steel will be possessed of magnetic poles of its own, and the system of attractions and repulsions between these and the poles of the original permanent magnet will account for the forces which are observed to act on the metal.

It has, however, been seen that pairs of corresponding positive and negative poles cannot be separated by more than molecular distances, so that we are led to suppose that each particle of the body in which magnetism is induced must become magnetised, the adjacent poles neutralising one another as in a permanent magnet.

Taking this view, it will be seen that the attraction of a magnet for an unmagnetised body is analogous to the attraction of an electrified body for a piece of dielectric (§ 197), rather than to its attraction for an uncharged conductor. The attraction of a charged body for a fragment of a dielectric has been seen to depend upon a molecular phenomenon taking place in the dielectric. Each molecule becomes itself electrified on its opposite faces, with charges of opposite sign, these charges being equal and opposite so that the total charge on any molecule is *nil*. In the same way, when magnetism is induced in any substance, each molecule of the substance must be supposed to become a magnetic particle, the total charge of magnetism on each particle being *nil*. It follows that the attraction of a magnet for a non-magnetic body is merely the aggregate of the attractive forces acting on the different individual particles of the body.

459. Confirmation of this view is found in the fact that the intensity of the attraction exerted by a magnet on a non-magnetised body depends on the material of the latter.   The significance of this fact will, perhaps, best be realised by comparing it with the corresponding fact of electrostatics.   When an uncharged conductor is attracted by a charged body, the phenomena in the former body which lead to this attraction are mass-phenomena: currents of electricity flow through the mass of the body until its surface becomes an equipotential.   Thus the attraction depends solely upon the shape of the body and not upon its structure.   On the other hand, the phenomena which lead to the attraction of a fragment of dielectric are, as we have seen, molecular phenomena.   They are conditioned by the shape and arrangement of the molecules, with the result that the total force depends on the nature of the dielectric material.

All magnetic phenomena occurring in material bodies must be molecular, as a consequence of the fact that corresponding positive and negative poles cannot be separated by more than molecular distances.   Hence we should naturally expect to find, as we do find, that all magnetic phenomena in material bodies, and in particular the attraction of unmagnetised matter by a magnet, would depend on the nature of the matter.   There would be a real difficulty if the attraction were found to depend only on the shape of the bodies.

460.   The amount of the action due to magnetic induction varies enormously more with the nature of the matter than is the case with the corresponding electric action.   Among common substances the phenomenon of magnetic induction is not at all well-marked except in iron and steel.   These substances shew the phenomenon to a degree which appears very surprising when compared with the corresponding electrostatic phenomenon.   After these substances, the next best for shewing the phenomena of induction are nickel and cobalt, although these are very inferior to iron and steel.   It is worth noticing that the atomic weights of iron, nickel and cobalt are very close together*, and that the three elements hold corresponding positions in the table of elements arranged according to the periodic law.

It has recently been found that certain rare metals shew magnetic induction to an extent comparable with iron, and that alloys can be formed to shew great powers of induction although the elements of which these alloys are formed are almost entirely non-magnetic †.

It appears probable that all substances possess some power of magnetic induction, although this is generally extremely feeble in comparison with that of the substances already mentioned.   In some substances, the effect is of the opposite sign from that in iron, so that a fragment of such matter is repelled from a magnetic pole.   Substances in which the effect is of the

---

* Iron = 55·8, nickel = 58·7, cobalt = 58·9.

† For an account of the composition and properties of Heusler's alloys, see a paper by J. C. McLennan, *Phys. Review*, Vol. 24, p. 449.

same kind as in iron are called *paramagnetic*, while substances in which the effect is of the opposite kind are called *diamagnetic*.

The phenomenon of magnetic induction is much more marked in paramagnetic, than in diamagnetic, substances. The most diamagnetic substance known is bismuth, and its coefficient of susceptibility (§ 461, below) is only about $\frac{1}{10^9}$ of that of the most paramagnetic samples of iron.

Coefficients of Susceptibility and Permeability.

461. When a body which possesses no permanent magnetism of its own is placed in a magnetic field, each element of its volume will, for the time it remains under the influence of the magnetic field, be a magnetic particle. If the body is non-crystalline the direction of the induced magnetisation at any point will be that of the magnetic force at the point. Thus if H denote the magnetic force at any point, we can suppose that the induced magnetism, of an intensity I, has its direction the same as that of H.

Thus if α, β, γ are the components of magnetic force, and A, B, C the components of induced magnetisation, we shall have equations of the form

$$\left. \begin{aligned} A &= \kappa\alpha \\ B &= \kappa\beta \\ C &= \kappa\gamma \end{aligned} \right\} \quad \dots\dots\dots\dots\dots\dots(390),$$

the quantity κ being the same in each equation because the directions of I and H are the same.

The quantity κ is called the *magnetic susceptibility*.

If the body has no permanent magnetisation, the whole components of magnetisation are the quantities A, B, C given by equations (390), and the components of induction are given (cf. equations (359)) by

$$a = \alpha + 4\pi A = \alpha(1 + 4\pi\kappa),$$
$$b = \beta + 4\pi B = \beta(1 + 4\pi\kappa),$$
$$c = \gamma + 4\pi C = \gamma(1 + 4\pi\kappa).$$

If we put $$\mu = 1 + 4\pi\kappa \quad \dots\dots\dots\dots\dots\dots(391),$$

we have

$$\left. \begin{aligned} a &= \mu\alpha \\ b &= \mu\beta \\ c &= \mu\gamma \end{aligned} \right\} \quad \dots\dots\dots\dots\dots\dots(392),$$

and μ is called the *magnetic permeability*.

462. The quantities κ and μ are by no means constant for a given substance. Their value depends largely upon the physical conditions, particularly the temperature, of the substance, upon the strength of the magnetic field in which the substance is placed, and upon the previous magnetic experiences of the substance in question.

We pass to the consideration of the way in which the magnetic coefficients vary with some of these circumstances. As κ and μ are connected by a simple relation (equation (391)), it will be sufficient to discuss the variations of one of these quantities only, and the quantity μ will be the most convenient for this purpose. Moreover, as the phenomenon of induced magnetisation is almost insignificant in all substances except iron and steel, it will be sufficient to consider the magnetic phenomena of these substances only.

463. *Dependence of μ on H.* The way in which the value of μ depends on H is, in its main features, the same for all kinds of iron. For small forces, μ is a constant, for larger forces μ increases, finally it reaches a maximum, and after this decreases in such a way that ultimately μH approximates to a constant value, known as the "saturation" value. This is represented graphically in a typical case in fig. 113, which represents the results obtained by Ewing from experiments on a piece of iron wire.

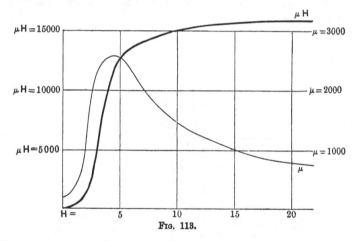

Fig. 113.

The abscissae represent values of H, the ordinate of the thick curve the value of μH, and the ordinate of the thin curve the value of μ. The corresponding numerical values are as follows:

H	μH	μ	H	μH	μ
0·32	40	120	5·17	12680	2450
0·84	170	200	6·20	13640	2200
1·37	420	310	7·94	14510	1830
2·14	1170	550	9·79	14980	1530
2·67	3710	1390	11·57	15230	1320
3·24	7300	2250	15·06	15570	1030
3·89	9970	2560	19·76	15780	800
4·50	11640	2590	21·70	15870	730

464. *Retentiveness and Hysteresis.* It is found that after the magnetising force is removed from a sample of iron, the iron still retains some of its magnetism. Here we have a phenomenon similar to the electrostatic phenomenon of residual charge already described in § 397.

Fig. 114 is taken from a paper by Prof. Ewing (*Phil. Trans. Roy. Soc.* 1885). The abscissae represent values of H, and ordinates values of B, the induction. The magnetic field was increased from $H = 0$ to $H = 22$, and as H increased the value of B increased in the manner shewn by the curve OP of the graph. On again diminishing H from $H = 22$ to $H = 0$, the graph for B was found to be that given by the curve PE. Thus during this operation there was always more magnetisation than at the corresponding stage of the original operation, and finally when the inducing field was entirely removed, there was magnetisation left, of intensity represented by OE. The field was then further decreased from $H = 0$ to $H = -20$, and then increased again from $H = -20$ to $H = 22$. The changes in B are shewn in the graph.

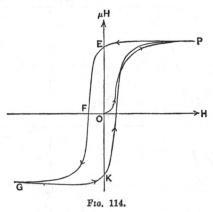

Fig. 114.

465. *Dependence of μ on temperature.* As has already been said, the value of μ depends to a large extent on the temperature of the metal. In general, the value of μ continually increases as the temperature is raised, this increase being slow at first but afterwards more rapid, until a temperature known as the "temperature of recalescence" is reached. This temperature has values ranging from 600° to 700° for steel and from 700° to 800° for iron. This temperature takes its name from the circumstance that a piece of metal cooling through this temperature will sink to a dull glow before reaching it, and will then become brighter again on passing through it.

After passing the temperature of recalescence, the value of μ falls with extreme rapidity, and at a temperature only a few degrees above this temperature, iron appears to be almost completely non-magnetic.

For paramagnetic substances, it appears to be a general law that the susceptibility κ varies inversely as the absolute temperature (Curie's Law).

MATHEMATICAL THEORY.

466. If Ω is the magnetic potential, supposed to be defined at points inside magnetic matter by equation (348), we have, as in equations (341) (cf. § 430), $a = -\dfrac{\partial \Omega}{\partial x}$ etc., so that

$$a = -\mu \frac{\partial \Omega}{\partial x}, \quad b = -\mu \frac{\partial \Omega}{\partial y}, \quad c = -\mu \frac{\partial \Omega}{\partial z}.$$

The quantities a, b, c, as we have seen (§ 434), satisfy

$$\frac{\partial a}{\partial x} + \frac{\partial b}{\partial y} + \frac{\partial c}{\partial z} = 0 \quad \ldots\ldots\ldots\ldots\ldots\ldots\ldots(393)$$

at every point, and

$$\iint (la + mb + nc)\, dS = 0 \ldots\ldots\ldots\ldots\ldots\ldots(394),$$

where the integration is taken over any closed surface. In terms of the potential, equation (393) becomes

$$\frac{\partial}{\partial x}\left(\mu \frac{\partial \Omega}{\partial x}\right) + \frac{\partial}{\partial y}\left(\mu \frac{\partial \Omega}{\partial y}\right) + \frac{\partial}{\partial z}\left(\mu \frac{\partial \Omega}{\partial z}\right) = 0 \quad \ldots\ldots\ldots(395),$$

while equation (394) becomes

$$\iint \mu \frac{\partial \Omega}{\partial n}\, dS = 0 \quad \ldots\ldots\ldots\ldots\ldots\ldots(396).$$

If μ is constant throughout any volume, equation (395) becomes

$$\nabla^2 \Omega = 0.$$

Thus inside a mass of homogeneous non-magnetised matter, the magnetic potential satisfies Laplace's Equation.

467. At a surface at which the value of μ changes abruptly we may take a closed surface formed of two areas fitting closely about an element dS of the boundary, these two areas being on opposite sides of the boundary. On applying equation (396), we obtain

$$\mu_1 \frac{\partial \Omega}{\partial \nu_1} + \mu_2 \frac{\partial \Omega}{\partial \nu_2} = 0 \ldots\ldots\ldots\ldots\ldots\ldots(397),$$

where μ_1, μ_2 are the permeabilities on the two sides, and $\dfrac{\partial}{\partial \nu_1}$, $\dfrac{\partial}{\partial \nu_2}$ denote differentiations with respect to normals to the surface drawn into the two media respectively.

Equations (397) and (395) (or (396)), combined with the condition that Ω must be continuous, suffice to determine Ω uniquely. The equations satisfied by Ω, the magnetic potential, are exactly the same as those which would be satisfied by V, the electrostatic potential, if μ were the Inductive Capacity of a dielectric. Thus the law of refraction of lines of magnetic induction is exactly identical with the law of refraction of lines of electric force investigated in § 138, and figures (43) and (78) may equally well be taken to represent lines of magnetic induction passing from one medium to a second medium of different permeability.

468. At any external point Q, the magnetic potential of the magnetisation induced in a body in which μ and κ have constant values is, by equation (342),

$$\Omega_Q = \iiint \left\{ A \frac{\partial}{\partial x} \left(\frac{1}{r} \right) + B \frac{\partial}{\partial y} \left(\frac{1}{r} \right) + C \frac{\partial}{\partial z} \left(\frac{1}{r} \right) \right\} dx\,dy\,dz$$

$$= -\kappa \iiint \left\{ \frac{\partial \Omega}{\partial x} \frac{\partial}{\partial x} \left(\frac{1}{r} \right) + \frac{\partial \Omega}{\partial y} \frac{\partial}{\partial y} \left(\frac{1}{r} \right) + \frac{\partial \Omega}{\partial z} \frac{\partial}{\partial z} \left(\frac{1}{r} \right) \right\} dx\,dy\,dz \quad ...(398).$$

Transforming by Green's Theorem,

$$\Omega_Q = \kappa \iiint \frac{1}{r} \nabla^2 \Omega - \kappa \iint \left(l \frac{\partial \Omega}{\partial x} + m \frac{\partial \Omega}{\partial y} + n \frac{\partial \Omega}{\partial z} \right) \frac{1}{r} dS$$

$$= -\kappa \iint \left(\frac{\partial \Omega}{\partial n} \right) \frac{1}{r} dS \quad ....(399),$$

shewing that the potential is the same if there were a layer of magnetic matter of surface density $-\kappa \dfrac{\partial \Omega}{\partial n}$ spread over the surface of the body. This is Poisson's expression for the potential due to induced magnetism.

We can also transform equation (398) into

$$\Omega_Q = \kappa \iiint \Omega \nabla^2 \left(\frac{1}{r} \right) - \kappa \iint \Omega \left\{ l \frac{\partial}{\partial x} \left(\frac{1}{r} \right) + m \frac{\partial}{\partial y} \left(\frac{1}{r} \right) + n \frac{\partial}{\partial z} \left(\frac{1}{r} \right) \right\} dS$$

$$= -\kappa \iint \Omega \frac{\partial}{\partial n} \left(\frac{1}{r} \right) dS \quad ....(400),$$

shewing that the potential at any external point Q of the induced magnetism is the same as if there were a magnetic shell of strength $-\kappa\Omega$ coinciding with the surface of the body.

Body in which permanent and induced magnetism coexist.

469. If a permanent magnet has a permeability different from unity, we shall have a magnetisation arising partly from permanent and partly from induced magnetism. If κ is the susceptibility and I the intensity of the permanent magnetisation at any point, the components of the total magnetisation at any point will be

$$A = Il + \kappa a, \text{ etc. } ....(401),$$

and the components of induction are

$$a = \alpha + 4\pi A = 4\pi Il + \mu\alpha, \text{ etc. } \quad\dots\dots\dots\dots\dots(402).$$

For such a substance, it is clear that equations (395) and (396) will not in general be satisfied.

ENERGY OF A MAGNETIC FIELD.

470. To obtain the energy of a magnetic field in which both permanent and induced magnetism may be present, we return to the general equation obtained in § 451,

$$\iiint (a\alpha + b\beta + c\gamma)\, dx\, dy\, dz = 0 \quad\dots\dots\dots\dots\dots(403).$$

On substituting for a, b, c from equations (402), this becomes

$$4\pi \iiint I (l\alpha + m\beta + n\gamma)\, dx\, dy\, dz + \iiint \mu (\alpha^2 + \beta^2 + \gamma^2)\, dx\, dy\, dz = 0 \ \dots(404).$$

Whether or not induced magnetism is present, it is proved, in § 448, that the energy of the field is

$$W = \tfrac{1}{2} \iiint I \left(l\frac{\partial\Omega}{\partial x} + m\frac{\partial\Omega}{\partial y} + n\frac{\partial\Omega}{\partial z} \right) dx\, dy\, dz,$$

where the integral is taken through all space. This is equal to $-\dfrac{1}{8\pi}$ times the first term in equation (404). Thus

$$W = \frac{1}{8\pi} \iiint \mu (\alpha^2 + \beta^2 + \gamma^2)\, dx\, dy\, dz \quad\dots\dots\dots\dots(405).$$

This could have been foreseen from analogy with the formula

$$W = \frac{1}{8\pi} \iiint K (X^2 + Y^2 + Z^2)\, dx\, dy\, dz,$$

which gives the energy of an electrostatic field.

From formula (405) we see that the energy of a magnetic field may be supposed spread throughout the medium, at a rate $\dfrac{\mu H^2}{8\pi}$ per unit volume.

MECHANICAL FORCES IN THE FIELD.

471. The mechanical forces acting on a piece of matter in a magnetic field can be regarded as the superposition of two systems—first, the forces acting on the matter in virtue of its permanent magnetism (if any), and, secondly, the forces acting on the matter in virtue of its induced magnetism (if any).

The problem of finding expressions for the mechanical forces in a magnetic field is mathematically identical with that of finding the forces in an electrostatic field. This is the problem of which the solution has already been given in § 196. The result of the analysis there given may at once be applied to the magnetic problem.

In equation (117), p. 175, we found the value of Ξ, the x-component of the mechanical force per unit volume, in the form

$$\Xi = -\rho\,\frac{\partial V}{\partial x} - \frac{R^2}{8\pi}\frac{\partial K}{\partial x} + \frac{\partial}{\partial x}\left(\frac{R^2}{8\pi}\tau\frac{\partial K}{\partial \tau}\right).$$

To translate this result to the magnetic problem, we must regard ρ as specifying the density of magnetic poles, R must be replaced by H, the magnetic intensity, and K by μ, the magnetic permeability. Also the electrostatic potential V must be replaced by the magnetic potential Ω. We then have, as the value of Ξ in a magnetic field,

$$\Xi = -\rho\,\frac{\partial \Omega}{\partial x} - \frac{H^2}{8\pi}\frac{\partial \mu}{\partial x} + \frac{\partial}{\partial x}\left(\frac{H^2}{8\pi}\tau\frac{\partial \mu}{\partial \tau}\right) \quad\ldots\ldots\ldots\ldots(406).$$

Clearly the first term in the value of Ξ is that arising from the permanent magnetism of the body, while the second and third terms arise from the induced magnetism. The first term can be transformed in the manner already explained in the last chapter. It is with the remaining terms that we are at present concerned. These will represent the forces when no permanent magnetism is present. Denoting the components of this force by Ξ', H', Z', we have

$$\Xi' = -\frac{H^2}{8\pi}\frac{\partial \mu}{\partial x} + \frac{\partial}{\partial x}\left(\frac{H^2}{8\pi}\tau\frac{\partial \mu}{\partial \tau}\right) \quad\ldots\ldots\ldots\ldots(407).$$

472. This general formula assumes a special form in a case which is of great importance, namely when the magnetic medium is a fluid.

All liquid magnetic media in which the susceptibility is at all marked consist of solutions of salts of iron, and the magnetic properties of the liquid arise from the presence of the salts in solution. According to Quincke, the solution having the greatest susceptibility is a solution of chloride of iron in methyl alcohol, and for this the value of $\mu - 1$ is about $\frac{1}{1000}$[*]. In such a liquid, the field arising from the induced magnetism will be small compared with that arising from the original field, so that the magnetisation of any single particle of the salt in the solution may be regarded as produced entirely by the original field. Hence we have conditions similar to those which obtain electrostatically in a gas. The induced field may be regarded simply as the aggregate of the fields arising from the different particles of the magnetic medium, and is therefore jointly proportional to the density of these particles and to the strength of the inducing field. The latter fact shews that, for a given density of the medium, μ ought to be independent of H, a result to which we shall return later. The former fact shews that, as

---

[*] Cf. G. T. Walker, "Aberration" (Cambridge Univ. Press, 1900), p. 76.

the density τ changes, $\mu - 1$ ought to be proportional to τ—a result analogous to the result that $K - 1$ is proportional to the density in a gas. It has been found experimentally by Quincke[*] that $\mu - 1$ is approximately proportional to τ.

In gases we have conditions precisely similar to those which obtain when a gas is placed in an electrostatic field. Hence $\mu - 1$ must, for a gas, be proportional to τ, for exactly the same reason for which $K - 1$ is proportional to τ. This result also has been verified by Quincke[†].

Thus we may say that for fluid media, whether liquid or gaseous, $\mu - 1$ is, in general, proportional to τ, where τ is the density of the magnetic liquid, in the case of a liquid in solution, or of the gas itself, in the case of a gas.

473. If we assume the relation

$$\mu - 1 = c\tau \quad \dots\dots\dots\dots\dots\dots(408),$$

where c is a constant, we find that expression (407) may be put in the simpler form

$$\Xi' = \frac{\mu - 1}{8\pi} \frac{\partial}{\partial x} (H^2),$$

shewing that the whole mechanical force is the same as would be set up by a hydrostatic pressure at every point of the medium of amount $\frac{\mu - 1}{8\pi} H^2$.

If H varies from point to point of the field, the effect of this pressure will clearly be to urge the medium to congregate in the more intense parts of the field. This has been observed by Matteucci[‡] for a medium consisting of drops of chloride of iron dissolved in alcohol placed in a medium of olive oil. The drops of solution were observed to move towards the strongest parts of the field.

Magnetostriction.

474. If a liquid is placed in a magnetic field, it yields under the influence of the mechanical forces acting upon it, so that we have a phenomenon of magnetostriction, analogous to the phenomenon of electrostriction already explained (§ 203). Clearly the liquid will expand until the pressure is decreased by an amount $\frac{\mu - 1}{8\pi} H^2$ at each point, the new pressure and the mechanical forces resulting from the magnetic field now producing equilibrium in the fluid. By measuring the expansion of a liquid placed in a magnetic field Quincke has been able to verify the agreement between theory and experiment.

[*] *Wied. Ann.* 24, p. 347.				[†] *Wied. Ann.* 34, p. 401.
[‡] *Comptes Rendus*, 36, p. 917.

MOLECULAR THEORIES.

Poisson's Molecular Theory of Induced Magnetism.

475. In Chapter V it was found possible to account for all the electro-static properties of a dielectric by supposing it to consist of a number of perfectly conducting molecules. Poisson attempted to apply a similar explanation to the phenomenon of magnetic induction.

Poisson's theory can, however, be disproved at once, by a consideration of the numerical values obtained for the permeability μ. This quantity is analogous to the quantity K of Chapter V, so that its value may be estimated in terms of the molecular structure of the magnetic matter. The fact with respect to which Poisson's theory breaks down is the existence of substances (namely, different kinds of soft iron) for which the value of μ is very large. To understand the significance of the existence of such substances, let us consider the field produced when a uniform infinite slab of such a substance is placed in a uniform field of magnetic force, so that the face of the slab is at right angles to the lines of force. If the value of μ is very large, the fall of potential in crossing the slab is very small. Throughout the supposed perfectly-conducting magnetic molecules the potential would, on Poisson's theory, be constant, so that the fall of potential could occur only in the interstices between the molecules. In these interstices (cf. fig. 46), the fall of potential per unit length would be comparable with that outside the slab. Hence a very large value of μ could be accounted for only by supposing the molecules to be packed together so closely as to leave hardly any interstices. Samples of iron can be obtained for which μ is as large as 4000; it is known, from other evidence, that the molecules of iron are not so close together that such a value of μ could be accounted for in the manner proposed by Poisson.

It is worth noticing, too, that Poisson's theory does not seem able, without modification, to give any reasonable account of the phenomena of saturation, hysteresis, etc.

Weber's Molecular Theory of Induced Magnetism.

476. A theory put forward by Weber shews much more ability than the theory of Poisson to explain the facts of induced magnetism.

Weber supposes that, even in a substance which shews no magnetisation, every molecule is a permanent magnet, but that the effects of these different magnets counteract one another, owing to their axes being scattered at random in all directions. When the matter is placed in a magnetic field each molecule tends, under the influence of the field, to set itself so that its axis is along the lines of force, just as a compass-needle tends to set itself along the lines of force of the earth's magnetic field. The axes of the molecules no longer point in all directions indifferently, so that the magnetic fields of the different molecules no longer destroy one another, and the body as a whole shews magnetisation. This, on Weber's theory, is the magnetisation induced by the external field of force.

Weber supposes that each molecule, in its normal state, is in a position of equilibrium under the influence of the forces from all the neighbouring molecules, and that when it is moved out of this position by the action of an external magnetic field, the forces from the other molecules tend to restore it to its old position. It is, therefore, clear that so long as the external field is small, the angle through which each axis is turned by the action of the field will be exactly proportional to the intensity of the field, so that the magnetisation induced in the body will be just proportional to the strength of the inducing field. In other words, for small values of H, μ must be independent of H.

There is, however, a natural limit imposed upon the intensity of the induced magnetisation. Under the influence of a very intense field all the molecules will set themselves so that their axes are along the lines of force. The magnetisation induced in the body is now of a quite definite intensity, and no increase of the inducing field can increase the intensity of the induced magnetisation beyond this limit. Thus Weber's theory accounts quite satisfactorily for the phenomenon of saturation, a phenomenon which Poisson's theory was unable to explain.

477. In connection with this aspect of Weber's theory, some experiments of Beetz are of great importance. A narrow line was scratched in a coat of varnish covering a silver wire. The wire was placed in a solution of a salt of iron, arranged so that iron could be deposited electrolytically on the wire at the points at which the varnish had been scratched away. The effect was of course to deposit a long thin filament of iron along the scratch. If, however, the experiment was performed in a magnetic field whose lines of force were in the direction of the scratch, it was found not only that the filament of iron deposited on the wire was magnetised, but that its magnetisation was very intense. Moreover, on causing a powerful magnetising force to act in the same direction as the original field, it was found that the increase in the intensity of the induced magnetisation was very small, shewing that the magnetisation had previously been nearly at the point of saturation.

Now if, as Weber supposed, the molecules of iron were already magnets before being deposited on the silver wire, then any magnetic force sufficient to arrange them in order on the wire ought to have produced a filament in a state of magnetic saturation, while if, as Poisson supposed, the magnetism in the molecules was merely induced by the external magnetic field, then the magnetisation of the filament ought to have been proportional to the original field, and ought to have disappeared when the field was destroyed. Thus, as between these two hypotheses, the experiments decide conclusively for the former.

478. Weber's theory is illustrated by the following analysis.

Consider a molecule which, in the normal state of the matter, has its axis in the direction OP, and let the field of force from the neighbouring molecules be a field of intensity D, the direction of the lines of force being of course parallel to OP. Now let an external field of intensity H be applied, its direction being a direction OA making an angle α with OP. The total field acting on the molecule is now compounded of D along OP and H along OA.

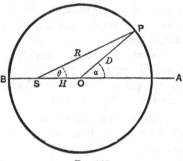

Fig. 115.

In fig. 115, let SO, OP represent H and D in magnitude and direction, then SP will represent the resultant field, so that the new direction of the axis of the molecule will be SP. Suppose that there are n molecules per unit volume, each of moment m. Originally, when the axes of the molecules were scattered indifferently in all directions, the number for which the angle α had a value between α and $\alpha + d\alpha$ was $\frac{1}{2}n \sin \alpha d\alpha$. These molecules now have their axes pointing in the direction SP, and therefore making an angle PSA ($= \theta$, say) with the direction of the external magnetic field. The aggregate moment of all these molecules resolved in the direction of OA is accordingly

$$\tfrac{1}{2} mn \sin \alpha \cos \theta d\alpha,$$

and on integration the aggregate moment of all the molecules per unit volume, which is the same as the intensity of the induced magnetisation I, is given by

$$I = \int_{a=0}^{a=\pi} \tfrac{1}{2} mn \sin \alpha \cos \theta d\alpha \quad \dots\dots\dots\dots\dots\dots\dots(409).$$

If R is the value of SP, measured on the same scale on which SO and OP represent H and D respectively, then

$$R^2 = H^2 + D^2 - 2HD \cos \alpha.$$

so that, on changing the variable from α to R, we must have the relation, obtained by differentiation of the above equation,

$$RdR = HD \sin \alpha d\alpha.$$

We also have $\qquad\cos\theta = \dfrac{R^2 + H^2 - D^2}{2RH}$,

so that equation (409) becomes

$$I = \tfrac{1}{4}mn \int \frac{R^2 + H^2 - D^2}{2H^2 D}\, dR.$$

In fig. 115 the limits of integration for R are $R = D + H$ and $R = D - H$. If, however, $H > D$, then the point S falls outside the circle APB and the limits for R are $R = D + H$ and $R = H - D$.

On integrating, we find as the values of I,

when $X < D$, $\qquad\qquad I = \tfrac{2}{3}mn\, \dfrac{X}{D}$,

„ $X = D$, $\qquad\qquad I = \tfrac{2}{3}mn$,

„ $X > D$, $\qquad\qquad I = mn\left(1 - \dfrac{D^2}{3X^2}\right)$,

„ $X = \infty$, $\qquad\qquad I = mn$.

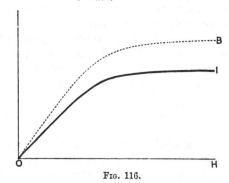

Fig. 116.

In fig. 116, the abscissae represent values of H, the ordinates of the thick curve the values of I, and the ordinates of the dotted curve the values of B or μH, drawn on one-tenth of the vertical scale of the graph for I.

Maxwell's Molecular Theory of Induced Magnetism.

479. It will be seen that Weber's theory fails to account for the *increase* in the value of μ before I reaches its maximum, and also that it gives no account of the phenomenon of retentiveness. Maxwell has shewn how the theory may be modified so as to take account of these two phenomena. He supposes that, so long as the forces acting on the molecules are small, the molecules experience small deflexions as imagined by Weber, but that as soon as these deflexions exceed a certain amount, the molecules are wrenched away entirely from their original positions of equilibrium, and take up positions relative to some new position of equilibrium. It might be, for instance, that originally the molecule had two possible positions of equilibrium, OP and OQ in fig. 117. Suppose the molecule to be in position OP and to be acted upon by a gradually increasing force in some direction OA. At first the molecule will turn from the position OP towards OA. But it may be that, as soon as the molecule passes some position OR, it suddenly swings round and takes up a position in which it

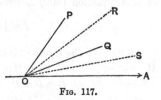

Fig. 117.

must be regarded as being deflected from the position of equilibrium OQ and not from OP. Let its new position be OS, then the deflexion produced is the angle SOP instead of the angle ROP which would be given by Weber's theory.

In Maxwell's original discussion, no distinction was made between the position OR, at which the magnet broke away from its old position of equilibrium, and OS, the new position of equilibrium. Maxwell accordingly had to assume that in some unknown way, the force of restitution broke down as soon as the magnet reached the position OR.

The improvement of distinguishing between the position OR, the limit of stability under the old position of equilibrium, and OS, the new position of equilibrium, was introduced by Ewing. In Ewing's form of the theory, no forces are needed beyond those provided by the mutual action of the magnets upon one another.

On either form of the theory, it is clear that the ratio of I to H will remain approximately constant until the molecules begin to break away from their original positions of equilibrium. As soon as this happens, the induced magnetism will increase more rapidly than the inducing force—*i.e.* μ will increase with H, in agreement with observation.

If the magnetising force is now removed, the molecule in the position OS will not return to its original position OP, but to the position OQ. It will therefore still have a deflexion QOP, called by Maxwell its "permanent set," and this will account for the "retentiveness" of the substance.

No molecular theory of this kind can, however, be regarded as at all complete. We shall return to the discussion of molecular theories of magnetism in Chapter XVI.

EXAMPLES.

1. A small magnet is placed at the centre of a spherical shell of radii a and b. Determine the magnetic force at any point outside the shell.

2. A system of permanent magnets is such that the distribution in all planes parallel to a certain plane is the same. Prove that if a right circular solid cylinder be placed in the field with its axis perpendicular to these planes, the strength of the field at any point inside the cylinder is thereby altered in a constant ratio.

3. A magnetic particle of moment m lies at a distance a in front of an infinite block of soft iron bounded by a plane face, to which the axis of the particle is perpendicular. Find the force acting on the magnet, and shew that the potential energy of the system is

$$- m^2 \left(\mu - 1\right)/8a^3 \left(\mu + 1\right).$$

4. The whole of the space on the negative side of the yz plane is filled with soft iron, and a magnetic particle of moment m at the point $(a, 0, 0)$ points in the direction $(\cos a, 0, \sin a)$. Prove that the magnetic potential at the point x, y, z inside the iron is

$$\frac{2m}{1 + \mu} \frac{z \sin a - (a - x) \cos a}{\{(a - x)^2 + y^2 + z^2\}^{\frac{3}{2}}}.$$

5. A small magnet of moment M is held in the presence of a very large fixed mass of soft iron of permeability μ with a very large plane face : the magnet is at a distance a from the plane face and makes an angle θ with the shortest distance from it to the plane. Shew that a certain force, and a couple

$$(\mu - 1) M^2 \sin \theta \cos \theta / 8 (\mu + 1) a^3,$$

are required to keep the magnet in position.

6. A small sphere of radius b is placed near a circuit which, when carrying unit current, would produce a field of strength H at the point where the centre of the sphere is placed. Shew that if κ is the coefficient of magnetic induction for the sphere, the presence of the sphere increases the self-induction of the wire by, approximately,

$$\frac{8\pi b^3 \kappa \left(3 + 2\pi \kappa\right) H^2}{(3 + 4\pi \kappa)^2}.$$

7. If the magnetic field within a body of permeability μ be uniform, shew that any spherical portion can be removed and the cavity filled up with a concentric spherical nucleus of permeability μ_1 and a concentric shell of permeability μ_2 without affecting the external field, provided μ lies between μ_1 and μ_2, and the ratio of the volume of the nucleus to that of the shell is properly chosen. Prove also that the field inside the nucleus is uniform, and that its intensity is greater or less than that outside according as μ is greater or less than μ_1.

8 A sphere of radius a has at any point (x, y, z) components of permanent magnetisation $(Px, Qy, 0)$, the origin of coordinates being at its centre. It is surrounded by a spherical shell of uniform permeability μ, the bounding radii being a and b. Determine the vector potential at an outside point.

9. A sphere of soft iron of radius a is placed in a field of uniform magnetic force parallel to the axis of z. Shew that the lines of force external to the sphere lie on surfaces of revolution, the equation of which is of the form

$$\left\{ 1 + \frac{2 (\mu - 1)}{\mu + 2} \left(\frac{a}{r}\right)^3 \right\} (x^2 + y^2) = \text{cons.,}$$

r being the distance from the centre of the sphere.

10. A sphere of soft iron of permeability μ is introduced into a field of force in which the potential is a homogeneous polynomial of degree n in x, y, z. Shew that the potential inside the sphere is reduced to its original value multiplied by

$$\frac{2n + 1}{n\mu + n + 1}.$$

11. If a shell of radii a, b is introduced in place of the sphere in the last question, shew that the force inside the cavity is altered in the ratio

$$(2n + 1)^2 \mu : (n\mu + n + 1) (n\mu + n + \mu) - n (n + 1) (\mu - 1)^2 \left(\frac{a}{b}\right)^{2n + 1}.$$

12. An infinitely long hollow iron cylinder of permeability μ, the cross-section being concentric circles of radii a, b, is placed in a uniform field of magnetic force the direction of which is perpendicular to the generators of the cylinder. Shew that the number of lines of induction through the space occupied by the cylinder is changed by inserting the cylinder in the field, in the ratio

$$b^2(\mu+1)^2 - a^2(\mu-1)^2 : 2\mu\{b^2(\mu+1) - a^2(\mu-1)\}.$$

13. A cylinder of iron of permeability μ has for cross-section the curve

$$r = a(1 + \epsilon \cos 2\theta),$$

where ϵ^2 may be neglected. Find the distribution of potential when the cylinder is placed in a field of force of which the potential before the introduction of the cylinder was

$$\Omega = Axy$$

14. An infinite elliptic cylinder of soft iron is placed in a uniform field of potential $-(Xx + Yy)$, the equation of the cylinder being $\dfrac{x^2}{a^2} + \dfrac{y^2}{b^2} = 1$. Shew that the potential of the induced magnetism at any internal point is

$$-(\mu-1)\left(\frac{b}{\mu b + a}Xx + \frac{a}{\mu a + b}Yy\right).$$

15. A solid elliptic cylinder whose equation is $\xi = a$ given by

$$x + iy = c \cosh(\xi + i\eta)$$

is placed in a field of magnetic force whose potential is $A(x^2 - y^2)$. Shew that in the space external to the cylinder the potential of the induced magnetism is

$$-\tfrac{1}{4}Ac^2 \operatorname{cosech} 2(a+\beta)\sin 4a\, e^{2(a-\beta-\xi)}\cos 2\eta,$$

where $\coth 2\beta$ is the permeability.

16. A solid ellipsoid of soft iron, semi-axes a, b, c and permeability μ, is placed in a uniform field of force X parallel to the axis of x, which is the major axis. Verify that the internal and external potentials of the induced magnetisation are

$$\Omega_1 = PA_1 x, \quad \Omega_0 = PA_0 x,$$

where $\quad A_1 = \displaystyle\int_0^\infty \frac{d\psi}{(a^2+\psi)^{\frac{3}{2}}(b^2+\psi)^{\frac{1}{2}}(c^2+\psi)^{\frac{1}{2}}}, \quad A_0 = \displaystyle\int_\lambda^\infty \frac{d\psi}{(a^2+\psi)^{\frac{3}{2}}(b^2+\psi)^{\frac{1}{2}}(c^2+\psi)^{\frac{1}{2}}},$

$$P = (\mu-1)X/\{(\mu-1)A_1 + 2(abc)^{-1}\},$$

and λ is the parameter of the confocal through the point considered.

17. A unit magnetic pole is placed on the axis of z at a distance f from the centre of a sphere of soft iron of radius a. Shew that the potential of the induced magnetism at any external point is

$$-\frac{1}{\pi}\frac{\mu-1}{\mu+1}\frac{a^3}{f^2}\int_0^\pi\int_0^1 \frac{\dfrac{1}{t^{\mu+1}}\,dt\,d\theta}{\left(z + i\varpi\cos\theta - \dfrac{a^2 t}{f}\right)^2},$$

where z, ϖ are the cylindrical coordinates of the point. Find also the potential at an internal point.

18. A magnetic pole of strength m is placed in front of an iron plate of permeability μ and thickness c. If this pole be the origin of rectangular coordinates x, y, and if x be perpendicular and y parallel to the plate, shew that the potential behind the plate is given by

$$\Omega = m(1-\rho^2)\int_0^\infty \frac{e^{-xt}J_0(yt)\,dt}{1-\rho^2 e^{-2ct}}, \quad \text{where} \quad \rho = \frac{\mu-1}{\mu+1}.$$

CHAPTER XIII

THE MAGNETIC FIELD PRODUCED BY ELECTRIC CURRENTS

EXPERIMENTAL BASIS.

480. So far the subjects of electricity and magnetism have been developed as entirely separate groups of physical phenomena. Although the mathematical treatment in the two cases has been on parallel lines, we have not had occasion to deal with any physical links connecting the two series of phenomena.

The first definite link of the kind was discovered by Oersted in 1820. Oersted's discovery was the fact that a current of electricity produced a magnetic field in its neighbourhood.

The nature of this field can be investigated in a simple manner. We first double back on itself a wire in which a current is flowing (fig. 118, 1). It is found that no magnetic field is produced.

Next we open the end into a small plane loop $PQRS$ (fig. 118, 2). It is found that at distances from the loop which are great compared with its linear dimensions, such a loop exercises the same magnetic forces as a magnetic particle of which the axis is perpendicular to the plane $PQRS$,

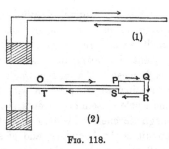

Fig. 118.

and the moment is jointly proportional to the strength of the current and to the area $PQRS$. The single current flowing in the circuit $OPQRST$ is obviously equivalent to two currents of equal strength, the one flowing in the circuit $OPST$ obtained by joining the points P and S, and the other flowing in the closed circuit $PQRSP$. The former current is shewn, by the preliminary experiment, to have no magnetic effects, so that the whole magnetic field may be ascribed to the small closed circuit $PQRS$.

481. Instead of regarding this field as due to a particle of moment jointly proportional to the area *PQRS* and to the current-strength, we may regard it as due to a small magnetic shell, coinciding with the area *PQRS*, and of strength simply proportional to the current flowing in *PQRS*.

482. Next, let us consider the current flowing in a closed circuit of any shape we please, and not necessarily in one plane. Let us cover in the closed circuit by an area of any kind having the circuit for its boundary, and let us cut up this area into infinitely small meshes by two systems of lines. A current of strength *i* flowing round the boundary circuit, is exactly equivalent to a current of strength *i* flowing round each mesh in the same direction as the current in the boundary. For, if we imagine this latter system of currents in existence, any line

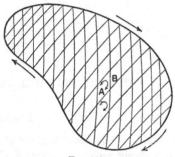

Fig. 119.

such as *AB* in the interior will have two currents flowing through it, one from each of the two meshes which it separates, and these currents will be equal but in opposite directions. Thus all the currents in the lines which have been introduced in the interior of the circuit annihilate one another as regards total effect, while the currents in those parts of the meshes which coincide with the original circuit just combine to reproduce the original current flowing in this circuit.

Thus the original circuit is equivalent, as regards magnetic effect, to a system of currents, one in each mesh. By taking the meshes sufficiently small, we may regard each mesh as plane, so that the magnetic effect of a current circulating in it is known: the magnetic effect of the current in a single mesh is that of a magnetic shell of strength proportional to the current and coinciding in position with the mesh. Thus, by addition, we find that the whole system of currents produces the same magnetic effects as a single magnetic shell coinciding with the surface of which the original current-circuit is the boundary, and of strength proportional to the current. This shell, then, produces the same magnetic effect as the original single current. The magnetic shell is spoken of as the "equivalent magnetic shell."

Thus we have obtained the following result:

"*A current flowing in any closed circuit produces the same magnetic field as a certain magnetic shell, known as the 'equivalent magnetic shell.' This shell may be taken to be any shell having the circuit for its boundary, its strength being uniform and proportional to that of the current.*"

Law of Signs. If an observer is imagined to stand on that side of the "equivalent magnetic shell" which contains the negative poles, the current flows round him in the same direction as that in which the sun moves round an observer standing on the earth's surface in the northern hemisphere.

We can also state the law by saying that to drive an ordinary right-handed screw (*e.g.* a cork-screw) in the direction of magnetisation of the shell, the screw would have to be turned in the direction of the current.

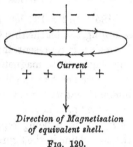

Current

Direction of Magnetisation of equivalent shell.

Fig. 120.

The law of signs expresses a fact of nature, not a mathematical convention. At the same time, it must be noticed that the law does not express that nature shews any preference in this respect for right-handed over left-handed screws. Two conventions have already been made in deciding which are to be called the positive directions of current and of magnetisation, and if either of these conventions had been different, the word "right-handed" in the law of signs would have had to be replaced by "left-handed."

483. Since, by § 346, any system of currents can be regarded as the superposition of a number of simple closed currents, it follows that the magnetic field produced by any system of currents can always be regarded as that produced by a number of magnetic shells, each of uniform strength.

Electromagnetic Unit of Current.

484. If i is the strength of the current flowing in a circuit, and ϕ the strength of the equivalent magnetic shell, then

$$\phi = ki,$$

where k is a constant, which is positive if the law of signs just stated has been obeyed in determining the signs of ϕ and i.

In the system of units known as Electromagnetic, we take $k = 1$, and define a unit current as one such that the equivalent magnetic shell is of unit strength. The strength of a current, in these units, is therefore measured by its magnetic effects. Obviously the strength measured in this way will be entirely different from the strength measured by the number of electrostatic units of electricity which pass a given point. This latter method of measurement is the electrostatic method. A full discussion of systems of units will be given later (§ 585); at present it may be stated that a current which is of unit strength when measured electromagnetically in C.G.S. units is of strength 3×10^{10} (very approximately) when measured electrostatically. The practical unit of current, the ampère, is, as already stated, equal to 3×10^9 electrostatic units of current, so that the electromagnetic unit of current is equal to 10 ampères.

A unit charge of electricity in electromagnetic units will be the amount of electricity that passes a fixed point per unit time in a circuit in which an electromagnetic unit of current is flowing. It is therefore equal to 3×10^{10} electrostatic units.

WORK DONE IN THREADING A CIRCUIT.

485. In fig. 121 let the thick line represent a circuit in which a current is flowing, and let the thin line through the point P represent the outline of any equivalent magnetic shell, P being any point in the shell. Let us imagine that we thread the circuit by any closed path beginning and ending at P, this path being represented by the dotted line in the figure. At every point of this path except P, we have a full knowledge of the magnetic forces.

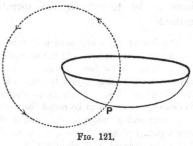

FIG. 121.

It will be convenient to regard the shell as having a definite, although infinitesimal, thickness at P. Let P_+, P_- denote the points in which the path intersects the positive and negative faces of the shell. Then we may say that the forces are known at all points of the path, except over the small range P_+P_-.

The original current can, however, be represented by any number of equivalent magnetic shells, for any shell is capable of representing the current, provided only it has as boundary the circuit in which the current is flowing.

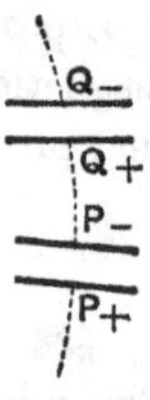

FIG. 122.

Let any other equivalent shell cut the path in the points Q_+Q_-. From our knowledge of the forces exerted by this shell, we can determine the forces exerted by the current at all points of the path except those within the range of Q_+Q_-. In particular we can determine the forces over the range P_+P_-, and it is at once obvious that on passing to the limit and making the range P_+P_- infinitesimal, the forces at the points P_+, P_-, and at all points on the infinitesimal range P_+P_- must be equal. Obviously the forces are also finite.

The work done on a unit pole in taking it round the complete circuit from P_- back to P_-, is accordingly the same as that done in taking it from P_- round the path to P_+. This can be calculated by supposing the forces to be exerted by the first equivalent shell, for the path is entirely outside this shell. If the potential due to the shell is Ω_{P_+} at P_+ and is Ω_{P_-} at P_-, the work done is $\Omega_{P_+} - \Omega_{P_-}$.

Now Ω, the potential of the shell at any point, is, as we know (§ 419), equal to $i\omega$, where ω is the solid angle subtended by the shell and i is the current, measured in electromagnetic units. The change in the solid angle as we pass from P_- to P_+ is, as a matter of geometry, equal to 4π. Thus

$$\Omega_{P_+} - \Omega_{P_-} = 4\pi i \quad \dots\dots\dots\dots\dots\dots(410).$$

The work done in taking a unit pole round the path described is accordingly $4\pi i$.

Magnetic Potential of a Field due to Currents.

486. Let us fix upon a definite equivalent shell to represent a current of strength i. Let us bring a unit pole from infinity to any point A, by a path which cuts the equivalent shell in points $P, Q, \dots Z$. For simplicity, let us at first suppose that at each of these points the path passes from the positive to the negative side of the shell, and let the points on the two sides of the shell be denoted, as before, by P_+, P_-; Q_+, Q_-; and so on.

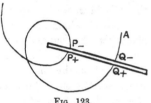

Fig. 123.

Then, if Ω denotes the magnetic potential due to the equivalent shell, the work done in bringing the unit pole from infinity to P_+ will be Ω_{P_+}. In the limit P_+ and P_- are coincident, so that the work in taking the unit pole on from P_+ to P_- is infinitesimal. In taking it from P_- to Q_+ work is done of amount $\Omega_{Q_+} - \Omega_{P_-}$, from Q_+ to Q_-, the work is infinitesimal, and so on, until ultimately we arrive at A. Thus the total work done in bringing the unit pole to A is

$$\Omega_{P_+} + (\Omega_{Q_+} - \Omega_{P_-}) + (\Omega_{R_+} - \Omega_{Q_-}) + \dots + (\Omega_A - \Omega_{Z_-}),$$

or, rearranging, is

$$\Omega_A + (\Omega_{P_+} - \Omega_{P_-}) + (\Omega_{Q_+} - \Omega_{Q_-}) + \dots.$$

Now each of the terms $\Omega_{P+} - \Omega_{P_-}$, $\Omega_{Q_+} - \Omega_{Q_-}$, etc. is equal by equation (410) to $4\pi i$, so that if n is the number of these terms, the whole expression is equal to

$$\Omega_A + 4\pi n i.$$

Replacing Ω_A by $i\omega$, where ω is the solid angle subtended by the shell at A, we find for the potential at A due to the electric current

$$(\omega + 4\pi n)i \quad \dots\dots\dots\dots\dots\dots\dots\dots(411).$$

If the path cuts the equivalent shell n times in the direction from $+$ to $-$, and m times in the opposite direction, the quantity n must be replaced by $n - m$.

Expression (411) shews that the potential at a point is not a single-valued function of the coordinates of the point. The forces, which are obtained by differentiation of this potential, are, however, single-valued.

Current in infinite straight wire.

487.   As an illustration of the results obtained, let us consider the magnetic field produced by a current flowing in a straight wire which is of such great length that it may be regarded as infinite, the return current being entirely at infinity.

Let us take the line itself for axis of z. Any semi-infinite plane terminated by this line may be regarded as an equivalent magnetic shell. Let us fix on any plane and take it as the plane of xz.

Consider any point P such that OP, the shortest distance from P to the axis of z, makes an angle θ with Ox. The cone through P which is subtended by the semi-infinite plane Ox, is bounded by two planes—one a plane through P and the axis of z; the other a plane through P parallel to the plane zOx. These contain an angle $\pi - \theta$, so that the solid angle subtended by the plane zOx at P is $2(\pi - \theta)$. Giving this value to ω in formula (411), we obtain as the magnetic potential at P

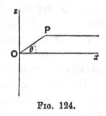

Fig. 124.

$$\Omega = \{2(\pi - \theta) + 4n\pi\}\, i.$$

Since $\dfrac{\partial \Omega}{\partial r} = 0$ it is clear that there is no radial magnetic force, and the force at any point in the direction of θ increasing

$$= -\frac{\partial \Omega}{r\,\partial \theta} = \frac{2i}{r}.$$

This result is otherwise obvious. If the work done in taking a unit pole round a circle of circumference $2\pi r$ is to be $4\pi i$, the tangential force at every point must be $\dfrac{2i}{r}$.

488.   This result admits of a simple experimental confirmation.

Let PQR be a disc suspended in such a way that the only motion of which it is capable is one of pure rotation about a long straight wire in which a current is flowing. On this disc let us suppose that an imaginary unit pole is placed at a distance r from the wire. There will be a couple tending to turn the disc, the moment of this couple being $\dfrac{2i}{r} \times r$ or $2i$. Similarly if we place a unit negative pole on the disc there is a couple $-2i$.

On placing a magnetised body on the disc, there will be a system of couples consisting of one of moment $2i$ for every positive pole and one of moment $-2i$ for every negative pole. Since the total charge

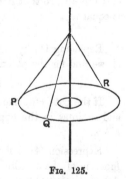

Fig. 125.

in any magnet is *nil*, it appears that the resultant couple must vanish, so that the disc will shew no tendency to rotate. This can easily be verified.

Circular Current.

489. Let us find the potential due to a current of strength i flowing in a circle of radius a. The equivalent magnetic shell may be supposed to be a hemisphere of radius a bounded by this circle.

The potential at any point on the axis of the circle can readily be found. For at a point on the axis distant r from the centre of the circle, the solid angle ω subtended by the circle is given by

$$\omega = 2\pi (1 - \cos \alpha) = 2\pi \left(1 - \frac{r}{\sqrt{a^2 + r^2}}\right),$$

so that the potential at this point is

$$\Omega = 2\pi i \left(1 - \frac{r}{\sqrt{a^2 + r^2}}\right).$$

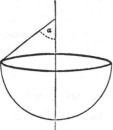

This expression can be expanded in powers of r by the binomial theorem. We obtain the following expansions:

Fig. 126.

if $r < a$.

$$\Omega = 2\pi i \left\{1 - \frac{r}{a} + \frac{1}{2}\frac{r^3}{a^3} - \ldots + (-1)^{n+1}\frac{1.3 \ldots 2n-1}{2.4 \ldots 2n}\left(\frac{r}{a}\right)^{2n+1} + \ldots\right\} \ldots (412),$$

if $r > a$,

$$\Omega = 2\pi i \left\{\frac{1}{2}\frac{a^2}{r^2} - \ldots + (-1)^{n+1}\frac{1.3 \ldots 2n-1}{2.4 \ldots 2n}\left(\frac{a}{r}\right)^{2n} + \ldots\right\} \ldots \ldots \ldots (413).$$

From this it is possible to deduce the potential at any point in space. Let us take spherical polar coordinates, taking the centre of the circle as origin, and the axis of the circle as the initial line $\theta = 0$. Inside the sphere $r = a$, the potential is a solution of $\nabla^2\Omega = 0$ which is symmetrical about the axis $\theta = 0$, and remains finite at the origin. It is therefore capable of expansion in the form

$$\Omega = \sum_0^\infty A_n r^n P_n (\cos \theta).$$

Along the axis we have $\theta = 0$, so that this assumed value of Ω becomes

$$\Omega = \sum_0^\infty A_n r^n,$$

and the coefficients may be determined by comparison with equation (412).

Thus we obtain for the potentials,

$$\Omega = 2\pi i \left\{ 1 - \frac{r}{a} P_1(\cos\theta) + \frac{1}{2}\frac{r^3}{a^3} P_3(\cos\theta) - \ldots \right.$$

$$\left. + (-1)^{n+1}\frac{1.3\ldots 2n-1}{2.4\ldots 2n}\left(\frac{r}{a}\right)^{2n+1} P_{2n+1}(\cos\theta) + \ldots \right\} \ldots(414),$$

when $r < a$, and

$$\Omega = 2\pi i \left\{ \frac{1}{2}\frac{a^2}{r^2} P_1(\cos\theta) - \frac{3}{8}\frac{a^4}{r^4} P_3(\cos\theta) - \ldots \right.$$

$$\left. + (-1)^{n+1}\frac{1.3\ldots 2n-1}{2.4\ldots 2n}\left(\frac{a}{r}\right)^{2n} P_{2n}(\cos\theta) + \ldots \right\} \ldots(415),$$

when $r > a$.

At points so near to the origin that $\frac{r^3}{a^3}$ may be neglected, the potential is

$$\Omega = 2\pi i \left(1 - \frac{r}{a}\cos\theta \right) = 2\pi i \left(1 - \frac{z}{a} \right),$$

where $z = r\cos\theta$, and the magnetic force is a uniform force $-\dfrac{\partial\Omega}{\partial z} = \dfrac{2\pi i}{a}$ parallel to the axis.

Solenoids.

490 A cylinder, wound uniformly with wire through which a current can be sent, is called a "solenoid."

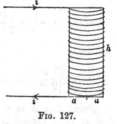

FIG. 127.

Consider first a circular cylinder of radius a and height h, having a wire coiled round it at the uniform rate of n turns per unit length, the wire carrying a current i. Let z be a coordinate measuring the distance of any cross-section from the base of the solenoid. Then the small layer between z and $z + dz$, being of thickness dz, will contain ndz turns of wire. The currents flowing in all these turns may be re-garded as a single current $nidz$ flowing in a circle, this circle being of radius a and at distance z from the base of the solenoid. The magnetic potential of this current may be written down from the formula of the last section, and the potential of the whole solenoid follows by integration.

491. *Endless Solenoid.* In the limiting case in which the solenoid is of infinite length (or in which the ends are so far away that the solenoid may be treated as though it were of infinite length), the field can be determined in a simpler manner.

Consider first the field outside the solenoid. In taking a unit pole round any path outside the solenoid which completely surrounds the solenoid, the work done is, by § 485, $4\pi i$. The current flowing per unit length of the solenoid is ni. In general we are concerned with cases in which this is finite n being very large and i being very small. The quantity $4\pi i$ may accordingly be neglected, and we can suppose that the work done in taking unit pole round the solenoid is zero.

It follows that the force outside the solenoid can have no component at right angles to planes through the axis, and clearly, by a similar argument, the same must be true inside the solenoid. Hence the lines of induction must lie entirely in the planes through the axis of the solenoid. From symmetry, there is no reason why the lines of induction at any point should converge towards, rather than diverge from, the axis, or *vice versa*. Hence the lines of induction will be parallel to the axis, and the force at every point will be entirely parallel to the axis.

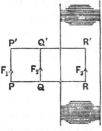

Fig. 128.

Let the lines PQR, $P'Q'R'$ in fig. 128 be radii meeting the axis, the lines PP', QQ', RR' being parallel to the axis and each of length ϵ. Let the magnetic forces along these lines be F_1, F_2 and F_3 respectively.

In taking unit pole round the closed path $PP'Q'QP$ the work done is

$$F_1\epsilon - F_2\epsilon,$$

and since this must vanish, we must have $F_1 = F_2$. Hence the force at all points outside the solenoid must be the same; it must be the same as the force at infinity and must consequently vanish. Thus there is no force at all outside the solenoid.

In taking unit pole round the closed path $PP'R'RP$, the work done is $F_2\epsilon$, and this must be equal to $4\pi ni\epsilon$, so that we must have $F_2\epsilon = 4\pi ni$. Thus the force at any point inside the solenoid is a force $4\pi ni$ parallel to the axis.

Thus the field of force arising from an infinite solenoid consists of a uniform field of strength $4\pi ni$ inside the solenoid, there being no field at all outside. The construction of a solenoid accordingly supplies a simple way of obtaining a uniform magnetic field of any required strength.

GALVANOMETERS.

492. A galvanometer is an instrument for measuring the strength of an electric current, the method of measurement usually being to observe the strength of the magnetic field produced by the current by noting its action on a small movable magnet.

There are naturally various classes and types of galvanometers designed to fulfil various special purposes.

The Tangent Galvanometer.

493. In the tangent galvanometer the current flows in a vertical circular coil, at the centre of which a small magnetic needle is pivoted so as to be free to turn in a horizontal plane.

Before use, the instrument is placed so that the plane of the coil contains the lines of magnetic force of the earth's field. The needle accordingly rests in the plane of the coil. When the current is allowed to flow in the coil a new field is originated, the lines of force being at right angles to the plane of the coil and the needle will now place itself so as to be in equilibrium under the field produced by the superposition of the two fields—the earth's field and the field produced by the current.

As the needle can only move in a horizontal plane, we need consider only the horizontal components of the two fields. Let H, as usual, denote the horizontal component of the earth's field. Let i be the current flowing in the coil, measured in electromagnetic units, let a be the radius and let n be the number of turns of wire. Near the centre of the coil the field produced by the current is, by § 489, a uniform field at right angles to the plane of the coil, of intensity $\dfrac{2\pi i n}{a}$. The total horizontal field is therefore compounded of a field of strength H in the plane of the coil, and a field of strength $\dfrac{2\pi i n}{a}$ at right angles to it.

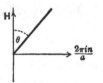

Fig. 129.

The resultant will make an angle θ with the plane of the coil, where

$$\tan \theta = \frac{\left(\dfrac{2\pi i n}{a}\right)}{H} \quad \ldots\ldots\ldots\ldots\ldots\ldots\ldots(416),$$

and the needle will set itself along the lines of force of the field. Thus the needle will, when in equilibrium, make an angle θ with the plane of the coil, where θ is given by equation (416). If we observe θ we can determine i from equation (416). We have

$$i = \frac{H}{G} \tan \theta \quad \ldots\ldots\ldots\ldots\ldots\ldots\ldots(417),$$

where G is a constant, known as the galvanometer constant, its value being $\dfrac{2\pi n}{a}$.

The instrument is called the tangent galvanometer from the circumstance that the current is proportional to the *tangent* of the angle θ.

The tangent galvanometer has the advantage that all currents, no matter how small or how great, can be measured without altering the adjustment of the instrument. A disadvantage is that the readings are not very sensitive when the currents to be measured are large—only a very small change in the reading is produced by a considerable change in the current. Let the current be increased by an amount di, and let the corresponding change in θ be $d\theta$, then from equation (417),

$$d\theta = d \tan^{-1}\left(\frac{iG}{H}\right) = \frac{GH}{H^2 + i^2 G^2}\, di,$$

so that if i is large, $\frac{d\theta}{di}$ is small. Thus, although the instrument may be used for the measurement of large currents, the measurements cannot be effected with much accuracy.

A second defect of the instrument is caused by the circumstance that the field produced by the current is not absolutely uniform near the centre of the coil. If a is the radius of the coil, and b the distance of either pole of the magnet from its centre, the poles will be in a part of the field in which the intensity differs from that at the centre of the coil by terms of the order of $\frac{b^3}{a^3}$. For instance, if the magnet is one inch long, while the coil has a diameter of 10 inches, the intensity of the field will be different from that assumed, by terms of the order of $(\frac{1}{10})^3$, so that the reading will be subject to an error of about one part in a thousand.

By replacing the single coil of the tangent galvanometer by two or more parallel coils, it is possible to make the field in the region in which the magnet moves, as uniform as we please. It is therefore possible, although at the expense of great complication, to make a tangent galvanometer which shall read to any required degree of accuracy

The Sine Galvanometer.

494. The sine galvanometer differs from the tangent galvanometer in having its coil adjusted so that it can be turned about a vertical axis. Before the current is sent through the coil, the instrument is turned until the needle is at rest in the plane of the coil. The coil is then in the direction of the earth's field at the point.

As soon as a current is sent through the coil, the needle is deflected, as in the tangent galvanometer. The coil is now slowly turned in the direction in which the needle has moved, until it overtakes the needle, and as soon as the needle is again at rest in the plane of the coil, a reading is taken, giving the angle through which the coil has been turned. Let θ be this angle, then the earth's field may be resolved into components, $H \cos \theta$ in the plane of the coil and $H \sin \theta$ at right angles to this plane. Since the needle rests in the plane of the coil, the latter component must be just neutralized by the field set up by the current, this being, as we have seen, entirely at right angles to the plane of the coil. We accordingly have

$$H \sin \theta = \frac{2\pi i n}{a},$$

so that we must have

$$i = \frac{H}{G} \sin \theta \quad \dots\dots\dots\dots\dots\dots\dots\dots\dots(418),$$

where G, the galvanometer constant, has the same meaning as before.

This instrument has the disadvantage that it cannot be used to measure currents greater than $\frac{H}{G}$. It is, however, sensitive over the whole range through which it can be used: if $d\theta$ is the increase in θ caused by a change di in i, we have

$$d\theta = \frac{G}{H} \sec \theta \, di,$$

so that the greater the current the more sensitive the instrument.

The great advantage of this form of galvanometer, however, is that when the reading is taken the magnet is always in the same position relative to the field set up by the current in the coil. Thus the deviations from uniformity of intensity at the centre of the field do not produce any error in the readings obtained: they result only in the galvanometer constant having a value different from that which it has so far been supposed to have. But when once the right value has been assigned to the constant G, equation (418) will be true absolutely, no matter how large the movable needle may be in comparison with the coil.

Other galvanometers.

495. There are various other types of galvanometers in use to serve various purposes other than the exact measurement of a current. For full descriptions of these the reader may be referred to books treating the theory of electricity and magnetism from the more experimental side. The following may be briefly mentioned here.

I. *The D'Arsonval Galvanometer.* This instrument is typical of a class of galvanometer in which there is no moving needle, the moving part being the coil itself, which is free to turn in a strong magnetic field. The coil is suspended by a torsion fibre between the poles of a powerful horseshoe magnet. When a current is sent through the coil, the coil itself produces the same field as a magnetic shell, and so tends to set itself across the lines of force of the permanent magnet, this motion being resisted by no forces except the torsion of the fibre.

II. *The Mirror Galvanometer.* This is a galvanometer originally designed by Lord Kelvin for the measurement of the small currents used in the transmission of signals by submarine cables. The design is, in its main outlines, identical with that of the tangent galvanometer, but, to make the instrument as sensitive as possible, the coil is made of a great number of turns of fine wire, wound as closely as possible round the space in which the needle moves, and the needle is suspended as delicately as possible by a fine torsion-thread. To make the instrument still more sensitive, permanent magnets can be arranged so as to neutralize part of the intensity of the earth's field. The instrument is read by observing the motion of a ray of light reflected from a small mirror which moves with the needle: it is from this that the instrument takes its name. In the most sensitive form of this instrument a visible motion of the spot of light can be produced by a current of 10^{-10} ampères.

III. *The Ballistic Galvanometer.* This instrument does not measure the current passing at a given instant, but the total flow of electricity which passes during an infinitesimal interval. If the needle is at rest in the plane of the coil, a current sent through the coil will establish a magnetic field tending to turn the needle out of this plane. So long as the needle is approximately in the plane of the coil, the couple acting on the needle will be proportional to the current in the coil: let it be denoted by ci, where i is the current.

Then if ω is the angular velocity of the needle at any instant, we shall have an equation of the form

$$mk^2 \frac{d\omega}{dt} = ci,$$

where mk^2 is the moment of inertia of the needle. Integrating through the small interval of time during which the current may be supposed to flow, we obtain

$$mk^2 \Omega = c \int i\, dt.$$

Here Ω is the angular velocity with which the needle starts into motion, and $\int i\, dt$ is the total current which passes through the coil. Thus the total flow $\int i\, dt$ can be obtained by measuring Ω, and this again can be obtained by observing the angle through which the needle swings before coming to rest at the end of its oscillation.

Vector-potential of a Field due to Currents.

496. From the formulae obtained in § 446 for the vector-potential of a uniform magnetic shell, we can at once write down expressions for the vector-potential of a field due to currents.

For, by § 483, the field due to any system of currents may be regarded as the field due to a number of shells of uniform strength, so that the vector-potential at any point will be the sum of the vector-potentials due to these different shells. Hence if ϕ, ϕ', ... are the strengths of the various shells, the vector-potential at any point P has components (cf. § 446)

$$F = \Sigma \int \frac{\phi}{r} \frac{dx'}{ds'} \, ds', \text{ etc.,}$$

where the summation is over all the shells, and dx', ds' refer to an element of the edge of a shell of strength ϕ, this element being at a distance r from the point P.

The equations just found may clearly be replaced by

$$\left. \begin{aligned} F &= \int \frac{i}{r} \frac{dx}{ds} \, ds \\[2mm] G &= \int \frac{i}{r} \frac{dy}{ds} \, ds \\[2mm] H &= \int \frac{i}{r} \frac{dz}{ds} \, ds \end{aligned} \right\} \quad \dots\dots\dots\dots\dots\dots(419),$$

where ds is now an element of any wire or linear conductor in which a current of strength i is flowing, and the integration is now along all the conductors in the field.

By the use of equations (376), we may at once obtain the components of magnetic force or induction at any point x', y', z' in the forms

$$a = \frac{\partial H}{\partial y'} - \frac{\partial G}{\partial z'}$$

$$= \int i \left\{ \frac{\partial}{\partial y'} \left(\frac{1}{r} \right) \frac{dz}{ds} - \frac{\partial}{\partial z'} \left(\frac{1}{r} \right) \frac{dy}{ds} \right\} ds, \text{ etc. } \dots\dots\dots(420).$$

Mechanical Action in the Field.

Ampère's rule for the force from a circuit.

497. Let $O\,(x, y, z)$ be the position of any element ds of a circuit, and let P be any point (x', y', z') in free space.

From equations (420) it follows that the magnetic force at P may be regarded as made up of contributions from each element of the circuit such that the contribution from the element ds at O has components

$$i \left\{ \frac{\partial}{\partial y'} \left(\frac{1}{r} \right) \frac{dz}{ds} - \frac{\partial}{\partial z'} \left(\frac{1}{r} \right) \frac{dy}{ds} \right\} ds, \text{ etc., etc.}$$

On putting $r^2 = (x - x')^2 + (y - y')^2 + (z - z')^2$, and differentiating, these components become

$$\frac{ids}{r^2}\left\{\frac{y - y'}{r}\frac{dz}{ds} - \frac{z - z'}{r}\frac{dy}{ds}\right\}, \quad \frac{ids}{r^2}\left\{\frac{z - z'}{r}\frac{dx}{ds} - \frac{x - x'}{r}\frac{dz}{ds}\right\}, \text{ etc. } \dots(421).$$

Let us denote $\dfrac{x - x'}{r}, \dfrac{y - y'}{r}, \dfrac{z - z'}{r}$ by l_1, m_1, n_1, these being the direction-cosines of the line OP, and let $\dfrac{dx}{ds}, \dfrac{dy}{ds}, \dfrac{dz}{ds}$ be denoted by l_2, m_2, n_2, these being the direction-cosines of ds. Then the components of force (421) become

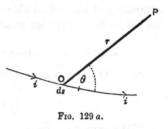

$$\frac{ids}{r^2}(m_1 n_2 - m_2 n_1), \quad \frac{ids}{r^2}(n_1 l_2 - n_2 l_1),$$

$$\frac{ids}{r^2}(l_1 m_2 - l_2 m_1) \dots(422).$$

Fig. 129 a.

Clearly the resultant is a force at right angles both to OP and to ds, and of amount

$$\frac{ids \sin\theta}{r^2} \quad\dots\dots\dots\dots\dots\dots\dots\dots\dots(423),$$

where θ is the angle between OP and ds.

Thus the total force at P may be regarded as made up of contributions such as (423) from each element of the circuit. This is known as Ampère's law.

Mechanical action on a circuit.

498. We are at present assuming the currents to be steady, so that action and reaction may be supposed to be equal and opposite. It follows that the force exerted at a unit pole at P upon the circuit of which the element ds is part, may be regarded as made up of forces of amount

$$\frac{i\sin\theta}{r^2}$$

per unit length, acting at right angles to OP and to ds. If we have poles of strength m at P, m' at P', etc., the resultant force on the circuit may be regarded as made up of contributions

$$\frac{im\sin\theta}{r^2}, \quad \frac{im'\sin\theta'}{r'^2}, \dots$$

per unit length. The resultant of these forces may be put in the form

$$iH\sin\chi$$

where H is the resultant magnetic intensity at O of all the poles m, m', etc., and χ is the angle between the direction of this intensity and ds. This resultant force acts at right angles to the directions of H and of ds.

499. We have found that the force from a whole circuit is the same as if each element $i\,ds$ contributed a force $i\,ds \sin\theta/r^2$, and the force on a whole circuit is the same as if each element were acted on by a force $iH \sin\chi$. But so long as we are dealing only with complete closed currents, it is impossible to discover what the actual force from or on a single element of the current will be. In a later chapter we shall regard a current as a stream of electrons in motion. The element $i\,ds$ will then be treated as a small number of moving electrons, and we shall be able to shew that the actual forces associated with the single element $i\,ds$ are exactly identical with those just found.

Energy of a System of Circuits carrying Currents.

500. The energy of a magnetic field, as we have seen (§ 470), is

$$\frac{1}{8\pi} \iiint \mu\,(\alpha^2 + \beta^2 + \gamma^2)\,dx\,dy\,dz \quad \dots\dots\dots\dots(424).$$

If the energy resides in the medium, this expression may be regarded as the energy of the field, no matter how this field is produced. If the field is produced wholly by currents, expression (424) may be regarded as the energy of the system of currents. As we shall now see, it can be transformed in a simple way, so as to express the energy of the field in terms of the currents by which the field is produced.

The integral through all space, as given by expression (424), may be regarded as the sum of the integrals taken over all the tubes of induction by which space is filled. The lines of induction, as we have seen, will be closed curves, so that the tubes are closed tubular spaces.

If ds is an element of length, and dS the cross-section at any point, of a tube of unit strength, we may replace $dx\,dy\,dz$ by $dS\,ds$, and instead of integrating with respect to dS we may sum over all tubes. Thus expression (424) becomes

$$\frac{1}{8\pi}\, \Sigma \int \{\mu\,(\alpha^2 + \beta^2 + \gamma^2)\,dS\}\,ds,$$

where the summation is over all unit tubes of induction. If $H^2 = \alpha^2 + \beta^2 + \gamma^2$, we have, by the definition of a unit tube, $\mu H dS = 1$, so that

$$\mu\,(\alpha^2 + \beta^2 + \gamma^2)\,dS = \mu H^2 dS = H,$$

and the integral becomes

$$\frac{1}{8\pi}\, \Sigma \int H ds.$$

Now $\int H ds$ is the work performed on a unit pole in taking it once round the tube of induction, and this we know is equal to $4\pi \Sigma'i$, where $\Sigma'i$ is the sum of all the currents threaded by the tube, taken each with its proper sign. Thus the energy becomes $\frac{1}{2}\Sigma\,(\Sigma'i)$.

This indicates that for every time that a unit tube threads a current i, a contribution $\frac{1}{2}i$ is added to the energy. Thus the whole energy is

$$\tfrac{1}{2}\Sigma i N \dots\dots\dots\dots\dots\dots\dots(425),$$

where the summation is over all the currents in the field, and N is the number of unit tubes which thread the current i.

501. We have seen that a shell of strength ϕ is equivalent, as regards the field produced at all external points, to a current i, if $\phi = i$. The energy of a system of currents has however been found to be $\frac{1}{2}\Sigma i N$, whereas the energy of a system of shells was found (§ 450) to be

$$-\tfrac{1}{2}\Sigma \phi N \dots\dots\dots\dots\dots\dots(426).$$

The difference of sign can readily be accounted for. Let us consider a single shell of strength ϕ, and let dS be an element of area, and dn an element of length inside the shell measured normally to the shell. At any point just outside the shell, let the three components of magnetic force be α, β, γ, the first being a component normal to the shell, and the others being components in directions which lie in the shell. On passing to the inside of the shell, the normal induction is discontinuous owing to the permanent magnetism which must be supposed to reside on the surface of the shell. Thus inside the shell, we may suppose the components of force to be $S + \dfrac{\alpha}{\mu}$, β, γ, where μ is the permeability of the matter of which the shell is composed, and S is the force originating from the permanent magnetism of the shell.

The contribution to the energy of the field which is made by the space inside the shell is

$$\frac{1}{8\pi}\iiint \mu \left\{ \left(S + \frac{\alpha}{\mu}\right)^2 + \beta^2 + \gamma^2 \right\} dx\,dy\,dz,$$

where the integral is taken throughout the interior of the shell; or

$$\frac{1}{8\pi}\iiint \mu \left\{ \left(S + \frac{\alpha}{\mu}\right)^2 + \beta^2 + \gamma^2 \right\} dn\,dS.$$

This can be regarded as the sum of three integrals,

$$\text{(i)} \quad \frac{1}{8\pi}\iiint \mu S^2 dn\,dS$$

$$\text{(ii)} \quad \frac{1}{8\pi}\iiint \left(\frac{\alpha^2}{\mu} + \mu\beta^2 + \mu\gamma^2\right) dn\,dS \Bigg\} \quad \dots\dots\dots\dots(427).$$

$$\text{(iii)} \quad \frac{1}{4\pi}\iiint S\alpha\,dn\,dS$$

On reducing the thickness of the shell indefinitely, S becomes infinite, for at any point of the shell,

$$\int S dn = - \text{(difference of potential between the two forces of shell)}$$

$$= - 4\pi\phi,$$

so that S becomes infinite when the thickness vanishes.

Thus on passing to the limit, the first integral

$$\frac{1}{8\pi} \iiint \mu S^2 dn \, dS$$

becomes infinite. This quantity is, however, a constant, for it represents the energy required to separate the shell into infinitesimal poles scattered at infinity.

The second integral vanishes on passing to the limit, and so need not be further considered.

The third integral can be simplified. We have

$$\frac{1}{4\pi} \iiint S\alpha \, dn \, dS = \frac{1}{4\pi} \iint \alpha \left(\int S dn \right) dS.$$

Now $\int S dn = - 4\pi\phi$, while $\iint \alpha \, dS$ is the integral of normal induction over the shell, and may therefore be replaced by N, the number of unit tubes of induction from the external field, which pass through the shell. Thus the third integral is seen to be equal to

$$- \phi N.$$

In calculating expression (424) when the energy is that of a system of currents, the contribution from the space occupied by the equivalent magnetic shells is infinitesimal. Thus all the terms which we have discussed represent differences between the energies of shells and of circuits.

Terms such as the first integrals of scheme (427) represent merely that the energies are measured from different standard positions. In the case of the shells, we suppose the shells to have a permanent existence, and merely to be brought into position. The currents, on the other hand, have to be created, as well as placed in position. Beyond this difference, there is an outstanding difference of amount ϕN for each circuit, and this exactly accounts for the difference between expressions (425) and (426).

502. Let us suppose that we have a system of circuits, which we shall denote by the numbers 1, 2, ... Let us suppose that when a unit current flows through 1, all the other circuits being devoid of currents, a magnetic field is produced such that the numbers of tubes of induction which cross circuits 1, 2, 3, ... are

$$L_{11}, \ L_{12}, \ L_{13}, \ ....$$

Similarly, when a unit current flows through 2, let the numbers of tubes of induction be

$$L_{21}, \ L_{22}, \ L_{23}, \ \dots$$

The theorem of § 446 shews at once that

$$L_{12} = L_{21} = \iint \frac{\cos \epsilon}{r} \, ds \, ds', \text{ etc.} \dots\dots\dots\dots\dots(428).$$

If currents $i_1, i_2, \dots$ flow through the circuits simultaneously, and if the numbers of tubes of induction which cut the circuits are $N_1, N_2, N_3, \dots$, we have

$$\left.\begin{aligned} N_1 &= L_{11} i_1 + L_{12} i_2 + L_{13} i_3 + \dots \\ N_2 &= L_{21} i_1 + L_{22} i_2 + L_{23} i_3 + \dots, \text{ etc.} \end{aligned}\right\} \ \dots\dots\dots(429).$$

The energy of the system of currents is

$$\begin{aligned} E &= \tfrac{1}{2} \Sigma i N, \\ &= \tfrac{1}{2} \Sigma i_1 (L_{11} i_1 + L_{12} i_2 + \dots), \\ &= \tfrac{1}{2} L_{11} i_1^2 + L_{12} i_1 i_2 + \tfrac{1}{2} L_{22} i_2^2 + \dots \ \dots\dots\dots(430). \end{aligned}$$

COEFFICIENTS OF INDUCTION.

503. The coefficient L_{11} is commonly called the coefficient of self-induction (or, more briefly, the self-inductance) of circuit 1, while L_{12} is called the coefficient of mutual induction of the two circuits 1 and 2. The value of L_{12} for any pair of circuits can be calculated from formula (428).

As an example, consider the important case of two circular wires, radii a, a' in parallel planes, the line joining their centres being perpendicular to the planes and of length d, b. Formula (428) gives

$$\begin{aligned} L_{12} &= \int_0^{2\pi} \int_0^{2\pi} \frac{aa' \cos(\theta - \theta') \, d\theta \, d\theta'}{[a^2 + a'^2 + b^2 - 2aa' \cos(\theta - \theta')]^{\frac{1}{2}}} \\ &= 2\pi \int_0^{2\pi} \frac{aa' \cos \psi \, d\psi}{[a^2 + a'^2 + b^2 - 2aa' \cos \psi]^{\frac{1}{2}}}. \end{aligned}$$

Put

$$c^2 = \frac{4aa'}{(a + a')^2 + b^2}, \qquad \phi = \tfrac{1}{2}(\pi - \psi)$$

and we readily find

$$\begin{aligned} L_{12} &= 4\pi (aa')^{\frac{1}{2}} c \int_0^{\frac{1}{2}\pi} \frac{2 \sin^2 \phi - 1}{(1 - c^2 \sin^2 \phi)^{\frac{1}{2}}} \, d\phi \\ &= 4\pi (aa')^{\frac{1}{2}} \left[\left(\frac{2}{c} - c\right) K(c) - \frac{2}{c} E(c) \right], \end{aligned}$$

where $K(c)$, $E(c)$ are the complete elliptic functions to modulus c.

When the circles nearly coincide, b is small and a and a' are nearly equal. Thus c is nearly equal to unity, and $E(c)$ approximates to unity. Put

$$c' = (1 - c^2)^{\frac{1}{2}},$$

so that c' is small, then

$$K(c) = \int_0^{\frac{1}{2}\pi} \frac{d\phi}{(1 - c^2 \sin^2 \phi)^{\frac{1}{2}}} = \int_0^{\frac{1}{2}\pi} \frac{d\phi}{(\cos^2 \phi + c'^2 \sin^2 \phi)^{\frac{1}{2}}},$$

of which the approximate value is found to be $\log (4/c')$.

If r is the nearest distance apart of the two circles, we have, when r is small, $c' = r/2a$, so that

$$K(c) = \log (8a/r),$$

and

$$L_{12} = 4\pi a \left(\log \frac{8a}{r} - 2 \right) \quad \dots\dots\dots\dots\dots(430a).$$

504. It might be expected that we could obtain the value of L_{11} in any problem by making the two circuits 1 and 2 coincide, but this proves not to be the case; the value of the integral in equation (428), where the integral is taken twice round the same circuit, is always infinite. As an instance, we may notice that on putting $r = 0$ in the formula just obtained, we find $L_{12} = \infty$.

We can readily see why this must be. When there is only one current flowing, we have

$$\tfrac{1}{2} L_{11} i_1^2 = \frac{\mu}{8\pi} \iiint (\alpha^2 + \beta^2 + \gamma^2)\, dx\, dy\, dz,$$

each side of this equation representing the energy of the current. Near to the wire, at a small distance r from it, the magnetic force is $2i/r$ so that $\alpha^2 + \beta^2 + \gamma^2 = 4i^2/r^2$. Thus the energy contained within a thin ring formed of coaxal cylinders of radii r_1, r_2, bent so as to follow the wire conveying the current, will be

$$\frac{\mu}{8\pi} \iiint \frac{4i^2}{r^2} r\, dr\, d\theta\, ds,$$

where the integration with respect to r is from r_1 to r_2, that with respect to θ is from 0 to 2π, and that with respect to s is along the wire. Integrating, we find energy

$$\mu i^2 \log (r_2/r_1)$$

per unit length, and on taking $r_1 = 0$, the energy is seen to be infinite.

Suppose that the wire has a circular cross-section of radius a, and that the current is uniformly distributed over this cross-section. A circle of radius r inside the wire will enclose a current ir^2/a^2, so that the magnetic force at distance r from the centre will be $2ir/a^2$, and

$$\alpha^2 + \beta^2 + \gamma^2 = \frac{4i^2 r^2}{a^4}.$$

On integrating this from $r = 0$ to $r = a$ we find that there is magnetic energy inside the wire of amount $\tfrac{1}{4}\mu' i^2$ per unit length, where μ' is the magnetic permeability of the material of the wire. Hence the total energy per unit length inside a cylinder of radius r_2 enclosing the wire is

$$\tfrac{1}{4}\mu' i^2 + \mu i^2 \log (r_2/a) \quad \dots\dots\dots\dots\dots(430b).$$

Even when a is finite this still becomes infinite when r_2 is made infinite—
i.e. when the magnetic field extends to infinity. Thus the self-induction per
unit length of a straight wire in free space is infinite except when the
magnetic field is limited by the presence of other conductors.

Suppose that the return current is carried by a concentric cylinder of
radius b surrounding the wire. The total flow of current through a circle of
radius greater than b is zero, so that there will be no magnetic force outside
the cylindrical conductor, and the magnetic field will be limited by the cylinder
$r = b$. The energy per unit length is now given by formula $(430b)$ with r_2 put
equal to b, so that the coefficient of self-induction per unit length is

$$L = \tfrac{1}{2}\mu' + 2\mu \log (b/a) \quad \dots\dots\dots\dots\dots(430c),$$

and this is finite for all finite values of b and a.

505. The energy of the magnetic field produced by a current i in a wire
will always be the sum of the energies of the magnetic field in the wire and
of the magnetic field outside the wire. If the current is uniformly distributed
in the wire, the former energy will always be $\tfrac{1}{4}\mu'i^2$ as in § 504. Thus L, the
self-induction of a wire of length l, will always be of the form

$$L = \tfrac{1}{2}\mu'l + L' \quad \dots\dots\dots\dots\dots\dots(430d),$$

where the term $\tfrac{1}{2}\mu'l$ arises from the field inside the wire, and L' arises from
the field outside the wire.

When the circuit lies entirely in one plane and the radius of cross-section
of the wire is small a simple value can be obtained for L'. Let S denote the
curve formed by the centres of the cross-sections of the wire, and let S' denote
the curve formed by the inner edge of the wire in the plane in which
the circuit lies. Then it will be easily verified that the magnetic force at any
point inside S' is the same as if the whole current i flowed along the curve S.
Hence the number of tubes of induction which flow through S' when the
current flows in the wire is the same as if a current i flowed in S, and so is
equal to i times L'_{12} where L'_{12} is the coefficient of mutual induction between
S and S'. Thus in formula $(430d)$, L' will be the coefficient of mutual induction between the circuits S and S'.

As an example, let us find the coefficient of self-induction in a wire of
length $2\pi a$ whose cross-section is a circle of radius r, bent into a circle of
radius a. The curve S is a circle of radius a, the curve S' is a concentric
circle of radius $a - r$. By formula $(430a)$,

$$L' = 4\pi a \left(\log \frac{8a}{r} - 2 \right),$$

so that

$$L = \pi a\mu' + 4\pi a \left(\log \frac{8a}{r} - 2 \right).$$

As a second example, let us find the coefficient of self-induction of a rectangular circuit of sides a, b made of wire of circular cross-section of radius r. In this case the circuit S will be a rectangle of sides a, b, while the circuit S' is a coplanar concentric rectangle of sides $a - r$, $b - r$. We evaluate L' the coefficient of mutual induction of S and S' from formula (428). There is no contribution from pairs of elements on sides perpendicular to one another, since for these $\cos \epsilon = 0$; the whole value of L' is contributed by parallel pairs of elements.

For two parallel lines of lengths l, l' at distance h apart, we find

$$\int \frac{ds\,ds'}{r} = \int_{-\frac{1}{2}l}^{\frac{1}{2}l} \int_{-\frac{1}{2}l'}^{\frac{1}{2}l'} \frac{dx\,dx'}{[(x'-x)^2 + h^2]^{\frac{1}{2}}}$$

$$= \int_{-\frac{1}{2}l}^{\frac{1}{2}l} \left| \sinh^{-1} \frac{x'-x}{h} \right|_{x'=-\frac{1}{2}l'}^{x'=\frac{1}{2}l'} dx$$

$$= (l + l') \sinh^{-1} \frac{l + l'}{2h} - (l - l') \sinh^{-1} \frac{l - l'}{2h} - [4h^2 + (l + l')^2]^{\frac{1}{2}}$$
$$+ [4h^2 + (l - l')^2]^{\frac{1}{2}}.$$

On making $l - l'$ small, and replacing $\sinh^{-1}$ by its logarithmic value, this becomes

$$2l \log \frac{l + (l^2 + h^2)^{\frac{1}{2}}}{h} - 2 (l^2 + h^2)^{\frac{1}{2}} + 2h.$$

By repeated use of this formula we find

$$L' = - 8 (a + b) + 8 (a^2 + b^2)^{\frac{1}{2}} - 4a \log [a + (a^2 + b^2)^{\frac{1}{2}}]$$
$$- 4b \log [b + (a^2 + b^2)^{\frac{1}{2}}] + 4 (a + b) \log \frac{2ab}{r},$$

and the coefficient of self-induction is now given by

$$L = (a + b) \mu' + L'.$$

505 A. Formula (430c), expressing the self-induction per unit length of a circular wire with a concentric return, can be put in the form $L = \frac{1}{2}\mu' + L'$, where

$$L' = 2\mu \log (b/a).$$

If K is the electrostatic capacity per unit length of the condenser formed by the wire and its surrounding cylinder, we have, from § 82,

$$K = \frac{\kappa}{2 \log (b/a)},$$

where κ is the inductive capacity of the insulating material surrounding the wire. Thus

$$L' = \frac{\kappa \mu}{K} \qquad \dots\dots\dots\dots\dots\dots\dots\dots(430\,e).$$

It is not a mere accident that this simple relation holds. Suppose we solve the electrostatic problem by the method of conjugate functions (§ 312).

The appropriate transformation is readily found to be (cf. § 318)

$$U + iV = \text{Cons.} + 2 \log r + 2i\theta,$$

where $x = r \cos \theta$, $y = r \sin \theta$. In this transformation U may be taken to be the electrostatic potential due to unit charge per unit length, and V will clearly be the magnetic potential due to unit current. It follows at once that the value of $X^2 + Y^2$ at any point when there is unit charge per unit length is the same as the value of $\alpha^2 + \beta^2$ at the same point when there is unit current flowing, and relation (430 e) is at once seen to be true.

The argument can be applied equally well to any conjugate-function transformation whatever. Thus relation (430 a) is seen to be universally true for any straight conductor accompanied by a parallel return.

EXAMPLES.

1. A current i flows in a very long straight wire. Find the forces and couples it exerts upon a small magnet.

Shew that if the centre of the small magnet is fixed at a distance c from the wire, it has two free small oscillations about its position of equilibrium, of equal period

$$2\pi \Big/ \sqrt{\frac{2\mu i}{Mk^2 c}},$$

where Mk^2 is the moment of inertia, and μ the magnetic moment, of the magnet.

2. Two parallel straight infinite wires convey equal currents of strength i in opposite directions, their distance apart being $2a$. A magnetic particle of strength μ and moment of inertia mk^2 is free to turn about a pivot at its centre, distant c from each of the wires. Shew that the time of a small oscillation is that of a pendulum of length l given by

$$4ial\mu = mgk^2 c^2.$$

3. Two equal magnetic poles are observed to repel each other with a force of 40 dynes when at a decimetre apart. A current is then sent through 100 metres of thin wire wound into a circular ring eight decimetres in diameter and the force on one of the poles placed at the centre is 25 dynes. Find the strength of the current in ampères.

4. Regarding the earth as a uniformly and rigidly magnetised sphere of radius a, and denoting the intensity of the magnetic field on the equator by H, shew that a wire surrounding the earth along the parallel of south latitude λ, and carrying a current i from west to east, would experience a resultant force towards the south pole of the heavens of amount

$$6\pi a i H \sin \lambda \cos^2 \lambda.$$

5. Shew that at any point along a line of force, the vector potential due to a current in a circle is inversely proportional to the distance between the centre of the circle and the foot of the perpendicular from the point on to the plane of the circle. Hence trace the lines of constant vector potential.

6. A current i flows in a circuit in the shape of an ellipse of area A and length l. Shew that the force at the centre is $\pi i l / A$.

7. A current i flows round a circle of radius a, and a current i' flows in a very long straight wire in the same plane. Shew that the mutual attraction is $4\pi i i' (\sec a - 1)$, where $2a$ is the angle subtended by the circle at the nearest point of the straight wire.

8. If, in the last question, the circle is placed perpendicular to the straight wire with its centre at distance c from it, shew that there is a couple tending to set the two wires in the same plane, of moment $2\pi i i' a^2/c$ or $2\pi i i' c$, according as $c >$ or $< a$.

9. A long straight current intersects at right angles a diameter of a circular current, and the plane of the circle makes an acute angle a with the plane through this diameter and the straight current. Shew that the coefficient of mutual induction is

$$4\pi \left\{ c \sec a - (c^2 \sec^2 a - a^2)^{\frac{1}{2}} \right\} \text{ or } 4\pi c \tan\left(\frac{\pi}{4} - \frac{a}{2} \right),$$

according as the straight current passes within or without the circle, a being the radius of the circle, and c the distance of the straight current from its centre.

10. Prove that the coefficient of mutual induction between a pair of infinitely long straight wires and a circular one of radius a in the same plane and with its centre at a distance $b (> a)$ from each of the straight wires, is

$$8\pi \left(b - \sqrt{b^2 - a^2} \right).$$

11. A circuit contains a straight wire of length $2a$ conveying a current. A second straight wire, infinite in both directions, makes an angle a with the first, and their common perpendicular is of length c and meets the first wire in its middle point. Prove that the additional electromagnetic forces on the first straight wire, due to the presence of a current in the second wire, constitute a wrench of pitch

$$2 \left(a \sin a - c \tan^{-1} \frac{a \sin a}{c} \right) \bigg/ \sin 2a \tan^{-1} \frac{a \sin a}{c}.$$

12. Two circular wires of radii a, b have a common centre, and are free to turn on an insulating axis which is a diameter of both. Shew that when the wires carry currents i, i', a couple of magnitude

$$\frac{2\pi^2 b^2}{a} \left(1 - \tfrac{9}{16} \frac{b^2}{a^2} \right) i i'$$

is required to hold them with their planes at right angles, it being assumed that b/a is so small that its fifth power may be neglected.

13. Two circular circuits are in planes at right angles to the line joining their centres. Shew that the coefficient of induction

$$= 2\pi (a^2 - c^2) \int_0^{\frac{\pi}{2}} \frac{\cos 2\theta \, d\theta}{\sqrt{a^2 \sin^2 \theta + c^2 \cos^2 \theta}},$$

where a, c are the longest and shortest lines which can be drawn from one circuit to the other. Find the force between the circuits.

14. Two currents i, i' flow round two squares each of side a, placed with their edges parallel to one another and at right angles to the distance c between their centres. Shew that they attract with a force

$$8 i i' \left\{ \frac{c \sqrt{2a^2 + c^2}}{a^2 + c^2} + 1 - \frac{a^2 + 2c^2}{c \sqrt{a^2 + c^2}} \right\}.$$

15. A current i flows in a rectangular circuit whose sides are of lengths $2a$, $2b$, and the circuit is free to rotate about an axis through its centre parallel to the sides of length $2a$. Another current i' flows in a long straight wire parallel to the axis and at a distance *d* from it. Prove that the couple required to keep the plane of the rectangle inclined at an angle ϕ to the plane through its centre and the straight current is

$$\frac{8ii'abd(b^2+d^2)\sin\phi}{b^4+d^4-2b^2d^2\cos 2\phi}.$$

16. Two circular wires lie with their planes parallel on the same sphere, and carry opposite currents inversely proportional to the areas of the circuits. A small magnet has its centre fixed at the centre of the sphere, and moves freely about it. Shew that it will be in equilibrium when its axis either is at right angles to the planes of the circuits, or makes an angle $\tan^{-1}\frac{1}{2}$ with them.

17. An infinitely long straight wire conveys a current and lies in front of and parallel to an infinite block of soft iron bounded by a plane face. Find the magnetic potential at all points, and the force which tends to displace the wire.

18. A small sphere of radius b is placed in the neighbourhood of a circuit, which when carrying a current of unit strength would produce magnetic force H at the point where the centre of the sphere is placed. Shew that, if κ is the coefficient of induced magnetization for the sphere, the presence of the sphere increases the coefficient of self-induction of the wire by an amount approximately equal to

$$\frac{8\pi b^3\kappa(3+2\pi\kappa)H^2}{(3+4\pi\kappa)^2}.$$

19. A circular wire of radius a is concentric with a spherical shell of soft iron of radii b and c. If a steady unit current flow round the wire, shew that the presence of the iron increases the number of lines of induction through the wire by

$$\frac{2\pi^2a^4(c^3-b^3)(\mu-1)(\mu+2)}{b^3\{(2\mu+1)(\mu+2)c^3-2(\mu-1)^2b^2\}}$$

approximately, where a is small compared with b and c.

20. A right circular cylindrical cavity is made in an infinite mass of iron of permeability μ. In this cavity a wire runs parallel to the axis of the cylinder carrying a steady current of strength I. Prove that the wire is attracted towards the nearest part of the surface of the cavity with a force per unit length equal to

$$\frac{2(\mu-1)I^2}{(\mu+1)d},$$

where a is the distance of the wire from its electrostatic image in the cylinder.

21. A steady current C flows along one wire and back along another one, inside a long cylindrical tube of soft iron of permeability μ, whose internal and external radii are a_1 and a_2, the wires being parallel to the axis of the cylinder and at equal distance a on opposite sides of it. Shew that the magnetic potential outside the tube will be

$$V=\frac{B_1}{r}\sin\theta+\frac{B_3}{r^3}\sin 3\theta+\frac{B_5}{r^5}\sin 5\theta+\ldots,$$

where $\qquad B_{2n+1}=\frac{16\mu Ca^{2n+1}}{2n+1}\Big/\left\{(\mu+1)^2-\left(\frac{a_1}{a_2}\right)^{2n}(\mu-1)^2\right\}.$

Hence shew that a tube of soft iron, of 150 cm. radius and 5 cm. thickness, for which the effective value of μ is 1200 c.g.s., will reduce the magnetic field at a distance, due to the current, to less than one-twentieth of its natural strength.

22.   A wire is wound in a spiral of angle a on the surface of an insulating cylinder of radius a, so that it makes n complete turns on the cylinder. A current i flows through the wire. Prove that the resultant magnetic force at the centre of the cylinder is

$$\frac{2\pi i n}{a\,(1+\pi^2 n^2 \tan^2 a)^{\frac{1}{2}}}$$

along the axis.

23.   A current of strength i flows along an infinitely long straight wire, and returns in a parallel wire. These wires are insulated and touch along generators the surface of an infinite uniform circular cylinder of material whose coefficient of induction is k. Prove that the cylinder becomes magnetized as a lamellar magnet whose strength is $2\pi k i/(1+2\pi k)$.

24.   A fine wire covered with insulating material is wound in the form of a circular disc, the ends being at the centre and the circumference. A current is sent through the wire such that I is the quantity of electricity that flows per unit time across unit length of any radius of the disc. Shew that the magnetic force at any point on the axis of the disc is

$$2\pi I \{\cosh^{-1}(\sec a) - \sin a\},$$

where a is the angle subtended at the point by any radius of the disc.

25.   Coils of wire in the form of circles of latitude are wound upon a sphere and produce a magnetic potential $A r^n P_n$ at internal points when a current is sent through them. Find the mode of winding and the potential at external points.

---

26.   A tangent galvanometer is to have five turns of copper wire, and is to be made so that the tangent of the angle of deflection is to be equal to the number of ampères flowing in the coil. If the earth's horizontal force is ·18 dynes, shew that the radius of the coil must be about 17·45 cms.

27.   A given current sent through a tangent galvanometer deflects the magnet through an angle θ. The plane of the coil is slowly rotated round the vertical axis through the centre of the magnet. Prove that if $\theta > \frac{1}{4}\pi$, the magnet will describe complete revolutions, but if $\theta < \frac{1}{4}\pi$, the magnet will oscillate through an angle $\sin^{-1}(\tan\theta)$ on each side of the meridian.

28.   Prove that, if a slight error is made in reading the angle of deflection of a tangent galvanometer, the percentage error in the deduced value of the current is a minimum if the angle of deflection is $\frac{1}{4}\pi$.

29.   The circumference of a sine galvanometer is 1 metre: the earth's horizontal magnetic force is ·18 c.g.s. units. Shew that the greatest current which can be measured by the galvanometer is 4·56 ampères approximately.

30.   The poles of a battery (of electromotive force 2·9 volts and internal resistance 4 ohms) are joined to those of a tangent galvanometer whose coil has 20 turns of wire and is of mean radius 10 cms. : shew that the deflection of the galvanometer is approximately 45°. The horizontal intensity of the earth's magnetic force is 1·8 and the resistance of the galvanometer is 16 ohms.

31.   A tangent galvanometer is incorrectly fixed, so that equal and opposite currents give angular readings a and β measured in the same sense. Shew that the plane of the coil, supposed vertical, makes an angle ϵ with its proper position such that

$$2\tan\epsilon = \tan a + \tan\beta.$$

32.   If there be an error a in the determination of the magnetic meridian, find the true strength of a current which is i as ascertained by means of a sine galvanometer.

33. In a tangent galvanometer, the sensibility is measured by the ratio of the increment of deflection to the increment of current, estimated per unit current. Shew that the galvanometer will be most sensitive when the deflection is $\frac{\pi}{4}$, and that in measuring the current given by a generator whose electromotive force is E, and internal resistance R, the galvanometer will be most sensitive if there be placed across the terminals a shunt of resistance

$$\frac{HRr}{E-H(R+r)},$$

where r is the resistance of the galvanometer, and H is the constant of the instrument.

What is the meaning of the result if the denominator vanishes or is negative?

34. A tangent galvanometer consists of two equal circles of radius 3 cms. placed on a common axis 8 cms. apart. A steady current sent in opposite directions through the two circles deflects a small needle placed on the axis midway between the two circles through an angle a. Shew that if the earth's horizontal magnetic force be H in c.g.s. units, then the strength of the current in c.g.s. units will be $125\,H \tan a/36\pi$.

35. A galvanometer coil of n turns is in the form of an anchor-ring described by the revolution of a circle of radius b about an axis in its plane distant a from its centre. Shew that the constant of the galvanometer

$$=\frac{8n}{a} \int_0^K \mathrm{cn}^2 u\, \mathrm{dn}^2 u\, du \qquad\qquad (k=b/a)$$

$$=(8n/3k^2a)[(1+k^2)\,E-(1-k^2)\,K].$$

CHAPTER XIV

INDUCTION OF CURRENTS IN LINEAR CIRCUITS

PHYSICAL PRINCIPLES.

506. IT has been seen that, on moving a magnetic pole about in the presence of electric currents, there is a certain amount of work done on the pole by the forces of the field. If the conservation of energy is to be true of a field of this kind, the work done on the magnetic pole must be represented by the disappearance of an equal amount of energy in some other part of the field. If all the currents in the field remain steady, there is only one store of energy from which this amount of work can be drawn, namely the energy of the batteries which maintain the currents, so that these batteries must, during the motion of the magnetic poles, give up more than sufficient energy to maintain the currents, the excess amount of energy representing work performed on the poles. Or again, if the batteries supply energy at a uniform rate, part of this energy must be used in performing work on the moving poles, so that the currents maintained in the circuits will be less than they would be if the moving poles were at rest.

Let us suppose that we have an imaginary arrangement by which additional electromotive forces can be inserted into, or removed from, each circuit as required, and let us suppose that this arrangement is manipulated so as to keep each current constant.

Consider first the case of a single movable pole of strength m and a single circuit in which the current is maintained at a uniform strength i. If ω is the solid angle subtended by the circuit at the position of the pole at any instant, the potential energy of the pole in the field of the current is $mi\omega$, so that in an infinitesimal interval dt of the motion of the pole, the work performed on the pole by the forces of the field is $mi\dfrac{d\omega}{dt}\,dt$. The total charge which has flowed in this time is $i\,dt$, so that the extra work done by the additional batteries is the same as that of an additional electromotive force $m\dfrac{d\omega}{dt}$.

Thus the motion of the pole must have set up an additional electromotive force in the circuit of amount $-m\dfrac{d\omega}{dt}$, to counteract which the additional electromotive forces are needed. The electromotive force $-m\dfrac{d\omega}{dt}$ which appears to be set up by the motion of the magnets is called the electromotive force due to induction.

The number of tubes of induction which start from the pole of strength m is $4\pi m$, and of these a number $m\omega$ pass through the circuit. Thus if n is the number of tubes of induction which pass through the circuit at any instant, the electromotive force may be expressed in the form $-\dfrac{dn}{dt}$.

So also if we have any number of magnetic poles, or any magnetic system of any kind, we find, by addition of effects such as that just considered, that there will be an electromotive force $-\dfrac{dN}{dt}$ arising from the motion of the whole system, where N is the total number of tubes of induction which cut the circuit.

It will be noticed that the argument we have given supplies no reason for taking N to be the number of tubes of induction rather than tubes of force. But if the number of tubes crossing the circuit is to depend only on the boundary of the circuit we must take tubes of induction and not tubes of force, for the induction is a solenoidal vector while the force, in general, is not.

507. The electromotive force of induction $-\dfrac{dN}{dt}$ has been supposed to be measured in the same direction as the current, and on comparing this with the law of signs previously given in § 483, we obtain the relation between the directions of the electromotive force round the circuit, and of the lines of induction across the circuit. The magnitude and direction of the electromotive force are given in the two following laws:

NEUMANN'S LAW. *Whenever the number of tubes of magnetic induction which are enclosed by a circuit is changing, there is an electromotive force acting round the circuit, in addition to the electromotive force of any batteries which may be in the circuit, the amount of this additional electromotive force being equal to the rate of diminution of the number of tubes of induction enclosed by the circuit.*

LENZ'S LAW. *The positive direction of the electromotive force $\left(-\dfrac{dN}{dt}\right)$ and the direction in which a tube of force must pass through the circuit in order to be counted as positive, are related in the same way as the forward motion and rotation of a right-handed screw.*

If there is no battery in the circuit, the total electromotive force will be $-\frac{dN}{dt}$, and the current originated by this electromotive force is spoken of as an "induced" current.

508. In order that the phenomena of induced currents may be consistent with the conservation of energy, it must obviously be a matter of indifference whether we cause the magnetic lines of induction to move across the circuit, or cause the circuit to move across the lines of induction. Thus Neumann's Law must apply equally to a circuit at rest and a circuit in motion. So also if the circuit is flexible, and is twisted about so as to change the number of lines of induction which pass through it, there will be an induced current of which the amount will be given by Neumann's Law.

509. For instance if a metal ring is spun about a diameter, the number of lines of induction from the earth's field which pass through it will change continuously, so that currents will flow in it. Furthermore, energy will be consumed by these currents so that work must be expended to keep the ring in rotation. Again the wheels and axles of two cars in motion on the same line of rails, together with the rails themselves, may be regarded as forming a closed circuit of continually changing dimensions in the earth's magnetic field. Thus there will be currents flowing in the circuit, and there will be electromagnetic forces tending to retard or accelerate the motions of the cars.

510. If, as we have been led to believe, electromagnetic phenomena are the effect of the action of the medium itself, and not of action at a distance, it is clear that the induced current must depend on the motion of the lines of force, and cannot depend on the manner in which these lines of force are produced. Thus induction must occur just the same whether the magnetic field originates in actual magnets or in electric currents in other parts of the field. This consequence of the hypothesis that the action is propagated through the medium is confirmed by experiment—indeed in Faraday's original investigations on induction, the field was produced by a second current.

511. Let us suppose that we have two circuits 1, 2, of which 1 contains a battery and a key by which the circuit can be closed and broken, while circuit 2 remains permanently closed, and contains a galvanometer but no battery. On closing the circuit 1, a current flows through circuit 1, setting up a magnetic field. Some of the tubes of induction of this field pass through circuit 2, so that the number of these tubes changes as the current establishes itself in circuit 1, and the galvanometer in 2 will accordingly shew a current. When the current in 1 has reached its steady

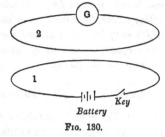

Fig. 130.

value, as given by Ohm's Law, the number of tubes through circuit 2 will no longer vary with the time, so that there will be no electromotive force in circuit 2, and the galvanometer will shew no current. If we break the circuit 1, there is again a change in the number of tubes of induction passing through the second circuit, so that the galvanometer will again shew a momentary current.

GENERAL EQUATIONS OF INDUCTION IN LINEAR CIRCUITS.

512. Let us suppose that we have any number of circuits 1, 2, ... Let their resistances be R_1, R_2, ..., let them contain batteries of electromotive forces E_1, E_2, ..., and let the currents flowing in them at any instant be i_1, i_2, ...

The numbers of tubes of induction N_1, N_2, ... which cross these circuits are given by (cf. equations (429))

$$N_1 = L_{11} i_1 + L_{12} i_2 + L_{13} i_3 + ..., \text{ etc.}$$

In circuit 1 there is an electromotive force E_1 due to the batteries, and an electromotive force $-\dfrac{dN_1}{dt}$ due to induction. Thus the total electromotive force at any instant is $E_1 - \dfrac{dN_1}{dt}$, and this, by Ohm's Law, must be equal to $R_1 i_1$. Thus we have the equation

$$E_1 - \frac{d}{dt}(L_{11} i_1 + L_{12} i_2 + L_{13} i_3 + ...) = R_1 i_1 \quad ....(431).$$

Similarly for the second circuit,

$$E_2 - \frac{d}{dt}(L_{21} i_1 + L_{22} i_2 + L_{23} i_3 + ...) = R_2 i_2 \quad ....(432),$$

and so on for the other circuits.

Equations (431), (432), ... may be regarded as differential equations from which we can derive the currents i_1, i_2, ... in terms of the time and the initial conditions. We shall consider various special cases of this problem.

INDUCTION IN A SINGLE CIRCUIT.

513. If there is only a single circuit, of resistance R and self-induction L, equation (431) becomes

$$E - \frac{d}{dt}(L i_1) = R i_1 ....(433).$$

Let us use this equation first to find the effect of closing a circuit previously broken. Suppose that before the time $t = 0$ the circuit has been open, but that at this instant it is suddenly closed with a key, so that the current is free to flow under the action of the electromotive force E.

The first step will be to determine the conditions immediately after the circuit is closed. Since $\frac{d}{dt}(Li_1)$ is, by equation (433), a finite quantity, it follows that Li_1 must increase or decrease continuously, so that immediately after closing the circuit the value of Li_1 must be zero.

To find the way in which i_1 increases, we have now to solve equation (433), in which E, L and R are all constants, subject to the initial condition that $i_1 = 0$ when $t = 0$. Writing the equation in the form

$$E - Ri_1 = -\frac{L}{R}\frac{d}{dt}(E - Ri_1),$$

we see that the general solution is

$$E - Ri_1 = Ce^{-\frac{R}{L}t},$$

where C is a constant, and in order that i_1 may vanish when $t = 0$, we must have $C = E$, so that the solution is

$$i_1 = \frac{E}{R}\left(1 - e^{-\frac{R}{L}t}\right) \quad \dots\dots\dots\dots\dots\dots(434).$$

The graph of i_1 as a function of t is shewn in fig. 131. It will be seen that the current rises gradually to its final value E/R given by Ohm's Law, this rise being rapid if L is small, but slow if L is great. Thus we may say that the increase in the current is retarded by its self-induction. We can see why this should be. The energy of the current i_1 is $\frac{1}{2}Li_1^2$, and this is large when L is large. This energy represents work performed by the electric forces: when the current

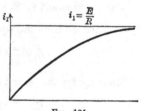

Fig. 131.

is i_1, the rate at which these forces perform work is Ei_1, a quantity which does not depend on L. Thus when L is large, a great time is required for the electric forces to establish the great amount of energy Li_1^2.

A simple analogy may make the effect of this self-induction clearer. Let the flow of the current be represented by the turning of a mill-wheel, the action of the electric forces being represented by the falling of the water by which the mill-wheel is turned. A large value of L means large energy for a finite current, and must therefore be represented by supposing the mill-wheel to have a large moment of inertia. Clearly a wheel with a small moment of inertia will increase its speed up to its maximum speed with great rapidity, while for a wheel with a large moment of inertia the speed will only increase slowly.

Alternating Current.

514. Let us next suppose that the electromotive force in the circuit is not produced by batteries, but by moving the circuit, or part of the circuit, in a magnetic field. If N is the number of tubes of induction of the external magnetic field which are enclosed by the circuit at any instant, the equation is

$$-\frac{d}{dt}(Li_1 + N) = Ri_1 \quad \dots\dots\dots\dots\dots\dots(435).$$

The simplest case arises when N is a simple-harmonic function of the time, proportional let us say to $\cos pt$. We can simplify the problem by supposing that N is of the form $C(\cos pt + i\sin pt)$. The real part of N will give rise to a real value of i_1, and the imaginary part of N to an imaginary value of i_1. Thus if we take $N = Ce^{ipt}$ we shall obtain a value for i_1 of which the real part will be the true value required for i_1.

Assuming $N = C(\cos pt + i\sin pt) = Ce^{ipt}$, the equation becomes

$$-\frac{d}{dt}(Li_1 + Ce^{ipt}) = Ri_1,$$

and clearly the solution will be proportional to e^{ipt}. Thus the differential operator $\frac{d}{dt}$ will act only on a factor e^{ipt}, and will accordingly be equivalent to multiplication by ip. We may accordingly write the equation as

$$-ip(Li_1 + Ce^{ipt}) = Ri_1,$$

a simple algebraic equation of which the solution is

$$i_1 = \frac{-pi\, Ce^{ipt}}{R + Lip}.$$

Let the modulus and argument of this expression be denoted by ρ and χ, so that the value of the whole expression is $\rho(\cos\chi + i\sin\chi)$. The value of ρ, the modulus, is equal (§ 311) to the product of the moduli of the factors, so that

$$\rho = \frac{pC}{\sqrt{R^2 + L^2 p^2}},$$

while the argument χ, being equal (§ 311) to the sum of the arguments of the factors, is given by

$$\chi = pt - \frac{\pi}{2} - \tan^{-1}\left(\frac{Lp}{R}\right).$$

The solution required for i_1 is the real term $\rho\cos\chi$, so that

$$i_1 = \rho\cos\chi$$
$$= \frac{pC}{\sqrt{R^2 + L^2 p^2}}\sin\left\{pt - \tan^{-1}\left(\frac{Lp}{R}\right)\right\} \quad \dots\dots\dots\dots(436).$$

The electromotive force produced by the change in the number of tubes of the external field is

$$-\frac{dN}{dt} = -\frac{d}{dt}(C\cos pt) = pC\sin pt.$$

Thus, if self-induction were neglected, the current, as given by Ohm's Law, would be

$$\frac{pC}{R} \sin pt,$$

and this of course would agree with that which would be given by equation (436) if L were zero.

The modifications produced by the existence of self-induction are represented by the presence of L in expression (436), and are two in number. In the first place the phase of the current lags behind that of the impressed electromotive force by $\tan^{-1} \dfrac{Lp}{R}$, and in the second place the apparent resistance is increased from R to $\sqrt{R^2 + L^2 p^2}$.

515. The conditions assumed in this problem are sufficiently close to those which occur in the working of a dynamo to illustrate this working. A coil which forms part of a complete circuit is caused to rotate rapidly in a magnetic field in such a way as to cut a varying number of lines of induction.

The quantity $\dfrac{p}{2\pi}$ may be supposed to represent the number of alternations per second. In the simple case of a two-pole alternator this will be equal to the number of revolutions of the engine by which the dynamo is driven, so that the current sent through the circuit will be an "alternating" current of frequency equal to that of the engine. In the example given, the rate at which heat is generated is $(\rho \cos \chi)^2 R$, and the average rate, averaged over a large number of alternations, is $\tfrac{1}{2}\rho^2 R$ or

$$\tfrac{1}{2} \frac{p^2 C^2 R}{R^2 + L^2 p^2}.$$

This, then, would be the rate at which the engine driving the dynamo would have to perform the work.

Discharge of a Condenser.

516. A further example of the effect of induction in a single circuit which is of extreme interest is supplied by the phenomenon of the discharge of a condenser.

Let us suppose that the charges on the two plates at any instant are Q and $-Q$, the plates being connected by a wire of resistance R and of self-induction L. If C is the capacity of the condenser, the difference of potential of the two plates will be $\dfrac{Q}{C}$, and this will now play the same part as the electromotive force of a battery. The equation is accordingly

$$\frac{Q}{C} - \frac{d}{dt}(Li) = Ri \quad \dotfill (437).$$

The quantities Q and i are not independent, for i measures the rate of flow of electricity to or from either plate, and therefore the rate of diminution of Q. We accordingly have $i = -\dfrac{dQ}{dt}$, and on substituting this expression for i, equation (437) becomes

$$L\frac{d^2Q}{dt^2} + R\frac{dQ}{dt} + \frac{Q}{C} = 0.$$

The solution is known to be

$$Q = Ae^{-\lambda_1 t} + Be^{-\lambda_2 t} \quad\quad\quad\quad\quad\quad(438),$$

where A, B are arbitrary constants, and λ_1, λ_2 are the roots of

$$Lx^2 - Rx + \frac{1}{C} = 0 \quad\quad\quad\quad\quad\quad(439).$$

If the circuit is completed at time $t = 0$, the charge on each plate being initially Q_0, we must have, at time $t = 0$,

$$Q = Q_0, \quad\quad i \equiv -\frac{dQ}{dt} = 0,$$

and these conditions determine the constants A and B. The equations giving these quantities are

$$A + B = Q_0, \quad\quad A\lambda_1 + B\lambda_2 = 0.$$

If the roots of equation (439) are real, it is clear, since both their sum and their product are positive, that they must themselves be positive quantities. Thus the value of Q given by equation (438) will gradually sink from Q_0 to zero. The current at any instant is

$$i = -\frac{dQ}{dt} = A\lambda_1 e^{-\lambda_1 t} + B\lambda_2 e^{-\lambda_2 t}$$
$$= A\lambda_1 e^{-\lambda_1 t}(1 - e^{-(\lambda_2 - \lambda_1) t}),$$

and this starts by being zero, rises to a maximum and then falls again to zero. The current is always in the same direction, so that Q is always of the same sign.

It is, however, possible for equation (439) to have imaginary roots. This will be the case if

$$R^2 - \frac{4L}{C}$$

is negative. Denoting $R^2 - \dfrac{4L}{C}$, when negative, by $-\kappa^2$, the roots will be

$$\lambda_1, \lambda_2 = \frac{R \pm i\kappa}{2L},$$

so that the solution (438) becomes

$$Q = e^{-\frac{Rt}{2L}} \left(A e^{\frac{i\kappa t}{2L}} + B e^{-\frac{i\kappa t}{2L}} \right)$$

$$= e^{-\frac{Rt}{2L}} D \cos\left(\frac{\kappa t}{2L} - \epsilon \right),$$

where D, ϵ are new constants. In this case the discharge is oscillatory. The charge Q changes sign at intervals $\dfrac{2\pi L}{\kappa}$, so that the charges surge backwards and forwards from one plate to the other. The presence of the exponential $e^{-\frac{Rt}{2L}}$ shews that each charge is less than the preceding one, so that the charges ultimately die away. The graphs for Q and i in the two cases of

    (i) $R^2 > \dfrac{4L}{C}$ (discharge continuous),

    (ii) $R^2 < \dfrac{4L}{C}$ (discharge oscillatory),

are given in figs. 132 and 133.

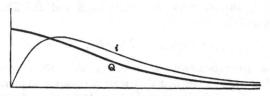

Fɪɢ. 132.

(i)   discharge continuous.

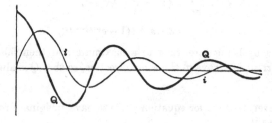

Fɪɢ. 133.

(ii)   discharge oscillatory.

    The existence of the oscillatory discharge is of interest, as the possibility of a discharge of this type was predicted on purely theoretical grounds by Lord Kelvin in 1853. Four years later the actual oscillations were observed by Feddersen.

517. It is of value to compare the physical processes in the two kinds of discharge.

Let us consider first the continuous discharge of which the graphs are shewn in fig. 132. The first part of the discharge is similar to the flow already considered in § 513. At first we can imagine that the condenser is exactly equivalent to a battery of electromotive force $E = \dfrac{Q}{C}$, and the act of discharging is equivalent to completing a circuit containing this battery. After a time the difference between the two cases comes into effect. The battery would maintain a constant electromotive force, so that the current would reach a constant final value $\dfrac{E}{R}$, whereas the condenser does not supply a constant electromotive force. As the discharge occurs, the potential difference between the plates of the condenser diminishes, and so the electromotive force, and consequently the current, also diminish. Thus the graph for i in fig. 132, can be regarded as shewing a gradual increase towards the value $\dfrac{E}{R}\left(\text{where } E = \dfrac{Q}{C}\right)$ in the earlier stages, combined with a gradual falling off of the current, consequent on the diminution of E, in the later stages.

For the oscillatory discharge to occur, the value of L must be greater than for the continuous discharge. The energy of a current of given amount is accordingly greater, while the rate at which this is dissipated by the generation of heat, namely Ri^2, remains unaltered by the greater value of L. Thus for sufficiently great values of L the current may persist even after the condenser is fully discharged, a continuation of the current meaning that the condenser again becomes charged, but with electricity of different signs from the original charges. In this way we get the oscillatory discharge.

INDUCTION IN A PAIR OF CIRCUITS.

518. If L, M, N are the coefficients of induction (L_{11}, L_{12}, L_{22}) of a pair of circuits of resistances R, S, in which batteries of electromotive forces E_1, E_2 are placed, the general equations become

$$E_1 - \frac{d}{dt}(Li_1 + Mi_2) = Ri_1 \quad \ldots\ldots\ldots\ldots\ldots(440),$$

$$E_2 - \frac{d}{dt}(Mi_1 + Ni_2) = Si_2 \quad \ldots\ldots\ldots\ldots\ldots(441).$$

Sudden Completing of Circuit.

519. Let us consider the conditions which must hold when one of the circuits is suddenly completed, the process occupying the infinitesimal interval from $t = 0$ to $t = \tau$. Let the changes which occur in i_1 and i_2 during this interval be denoted by Δi_1 and Δi_2. Equations (440) and (441) shew that during the interval from $t = 0$ to $t = \tau$ the values of $\dfrac{d}{dt}(Li_1 + Mi_2)$ and of $\dfrac{d}{dt}(Mi_1 + Ni_2)$ are finite, so that when τ is infinitesimal, the changes in $Li_1 + Mi_2$ and $Mi_1 + Ni_2$ must vanish. Thus we must have

$$L\Delta i_1 + M\Delta i_2 = 0,$$

$$M\Delta i_1 + N\Delta i_2 = 0.$$

Except in the special case in which $LN - M^2 = 0$ (a case of importance, which will be considered later), these equations can be satisfied only by $\Delta i_1 = \Delta i_2 = 0$. Thus the currents remain unaltered by suddenly making a circuit, and the change in the currents is gradual and not instantaneous.

520. Suppose, for instance, that before the instant $t = 0$ circuit 2 is closed but contains no battery, while circuit 1, containing a battery, is broken. Let circuit 1 be closed at the instant $t = 0$, then the initial conditions are that at time $t = 0$, $i_1 = i_2 = 0$. The equations to be solved are

$$\left(R + L\,\frac{d}{dt}\right)i_1 + M\,\frac{d}{dt}\,i_2 = E_1 \quad\dotfill(442),$$

$$M\,\frac{d}{dt}\,i_1 + \left(S + N\,\frac{d}{dt}\right)i_2 = 0 \quad\dotfill(443).$$

The solution is known to be

$$i_1 = Ae^{-\lambda t} + A'e^{-\lambda' t} + \frac{E_1}{R},$$

$$i_2 = Be^{-\lambda t} + B'e^{-\lambda' t},$$

where A, A', B, B' are constants, and λ, λ' are the roots of

$$(R - L\lambda)(S - N\lambda) - M^2\lambda^2 = 0,$$

or of $$RS - (RN + SL)\lambda + (LN - M^2)\lambda^2 = 0 \quad\dotfill(444).$$

The energy of the currents, namely

$$\tfrac{1}{2}(Li_1^2 + 2Mi_1 i_2 + Ni_2^2),$$

being positive for all values of i_1 and i_2, it follows that $LN - M^2$ is necessarily positive. Since RS and $RN + SL$ are also necessarily positive, we see that all the coefficients in equation (444) are positive, so that the roots λ, λ' are both positive.

When $t = 0$, we must have

$$(i_1)_{t=0} = A + A' + \frac{E_1}{R} = 0 \quad\dotfill(445),$$

$$(i_2)_{t=0} = B + B' = 0 \dotfill(446),$$

and in order that equation (443) may be satisfied at every instant, we must have

$$- MA\lambda e^{-\lambda t} - MA'\lambda' e^{-\lambda' t} + (S - N\lambda) B e^{-\lambda t} + (S - N\lambda') B' e^{-\lambda' t} = 0,$$

for all values of t, and for this to be satisfied the coefficients of $e^{-\lambda t}$ and $e^{-\lambda' t}$ must vanish separately. Thus we must have

$$(S - N\lambda) B = MA\lambda \quad ....(447),$$

$$(S - N\lambda') B' = MA'\lambda' \quad ....(448),$$

and if these relations are satisfied, and λ, λ' are the roots of equation (444), then equation (442) will be satisfied identically. From equations (445), (446), (447) and (448), we obtain

$$\frac{B}{M} = \frac{-B'}{M} = \frac{A\lambda}{S - N\lambda} = \frac{-A'\lambda'}{S - N\lambda'} = \frac{-E_1}{RS(\lambda^{-1} - \lambda'^{-1})},$$

and the solution is found to be

$$i_1 = \frac{(S - N\lambda) E_1}{RS\lambda(\lambda^{-1} - \lambda'^{-1})} (1 - e^{-\lambda t}) + \frac{(S - N\lambda') E_1}{RS\lambda'(\lambda'^{-1} - \lambda^{-1})} (1 - e^{-\lambda' t}),$$

$$i_2 = - \frac{ME_1}{RS(\lambda^{-1} - \lambda'^{-1})} e^{-\lambda t} - \frac{ME_1}{RS(\lambda'^{-1} - \lambda^{-1})} e^{-\lambda' t}.$$

We notice that the current in 1 rises to its steady value $\dfrac{E}{R}$, the rise being similar in nature to that when only a single circuit is concerned (§ 513). The rise is quick if λ and λ' are large—*i.e.* if the coefficients of induction are small, and conversely. The current in 2 is initially zero, rises to a maximum and then sinks again to zero. The changes in this current are quick or slow according as those of current 1 are quick or slow.

Sudden Breaking of Circuit.

521. The breaking of a circuit may be represented mathematically by supposing the resistance to become infinite. Thus if circuit 1 is broken, the process occurring in the interval from $t = 0$ to $t = \tau$, the value of R will become infinite during this interval, while the value of i_1 becomes zero. The changes in i_1 and i_2 are still determined by equations (440) and (441), but we can no longer treat R as a constant, and we cannot assert that in the interval from 0 to τ the value of Ri_1 is always finite.

It follows, however, from equation (441) that $\dfrac{d}{dt}(Mi_1 + Ni_2)$ remains finite throughout the short interval, so that we have, with the same notation as before,

$$M\Delta i_1 + N\Delta i_2 = 0.$$

Suppose for instance that before the circuit 1 was broken we had a steady current $\frac{E_1}{R}$ in circuit 1, and no current in circuit 2. We shall then have

$$\Delta i_1 = -\frac{E_1}{R},$$

so that

$$\Delta i_2 = \frac{M E_1}{N R},$$

and therefore immediately after the break, the initial current in circuit 2 is

$$i_2 = \frac{M E_1}{N R}.$$

This current simply decays under the influence of the resistance of the circuit. Putting $E_2 = 0$ and $i_1 = 0$ in equation (441) we obtain

$$\frac{di_2}{dt} = -\frac{S}{N} i_2,$$

and the solution which gives $i_2 = \frac{M E_1}{N R}$ initially is

$$i_2 = \frac{M E_1}{N R} e^{-\frac{S}{N} t}.$$

The changes in the current i_1 during the infinitesimal interval τ are of interest. These are governed by equation (440), the value of R not being constant.

The value of E_1 is finite, and may accordingly be neglected in comparison with the other terms of equation (440), which are very great during the interval of transition. Thus the equation becomes, approximately,

$$\frac{d}{dt}(L i_1 + M i_2) = -R i_1 \quad \ldots\ldots\ldots\ldots\ldots\ldots\ldots(449).$$

The value of $\frac{d}{dt}(M i_1 + N i_2)$ is, as we have already seen, finite, so that we may subtract $\frac{M}{N}$ times this quantity from the left-hand member of equation (449) and the equation remains true. By doing this we eliminate i_2, and obtain

$$\left(L - \frac{M^2}{N}\right) \frac{di_1}{dt} = -R i_1.$$

The solution which gives to i_1 the initial value $(i_1)_0$ is

$$i_1 = (i_1)_0 \, e^{-\frac{N}{LN - M^2} \int_0^t R \, dt},$$

giving the way in which the current falls to zero. We notice that if $LN - M^2$ is very small, the current falls off at once, while if $LN - M^2$ is large, the current will persist for a longer time. In the former case the breaking of the circuit is accompanied only by a very slight spark, in the latter case by a stronger spark.

One Circuit containing a Periodic Electromotive Force.

522. Let us suppose next that the circuits contain no batteries, but that circuit 1 is acted upon by a periodic electromotive force, say $E \cos pt$, such as might arise if this circuit contained a dynamo.

As in § 514, it is simplest to assume an electromotive force $E e^{ipt}$: the solution actually required will be obtained by ultimately rejecting the imaginary terms in the solution obtained.

The equations to be solved are now

$$E e^{ipt} - \frac{d}{dt}(L i_1 + M i_2) = R i_1 \quad \dots\dots\dots\dots(450),$$

$$-\frac{d}{dt}(M i_1 + N i_2) = S i_2 \quad \dots\dots\dots\dots(451).$$

As before both i_1 and i_2, as given by these equations, will involve the time only through a factor e^{ipt}, so that we may replace $\frac{d}{dt}$ by ip, and the equations become

$$R i_1 + L i p i_1 + M i p i_2 = E e^{ipt},$$

$$S i_2 + M i p i_1 + N i p i_2 = 0,$$

from which we obtain

$$\frac{i_1}{S + Nip} = \frac{i_2}{-Mip} = \frac{E e^{ipt}}{(R + Lip)(S + Nip) + M^2 p^2}.$$

The current i_1 in the primary is given, from these equations, by

$$i_1 = \frac{E e^{ipt}}{R + Lip + \dfrac{M^2 p^2}{S + Nip}}$$

$$= \frac{E e^{ipt}}{R + Lip + \dfrac{M^2 p^2 (S - Nip)}{S^2 + N^2 p^2}}$$

$$= \frac{E e^{ipt}}{R' + L'ip},$$

where

$$R' = R + \frac{S M^2 p^2}{S^2 + N^2 p^2}, \qquad L' = L - \frac{N M^2 p^2}{S^2 + N^2 p^2}.$$

The case of no secondary circuit being present is obtained at once by putting $S = \infty$, and the solution for i_1 is seen to be the same as if no secondary circuit were present, except that R', L' are replaced by R and L. Thus the current in the primary circuit is affected by the presence of the secondary in just the same way as if its resistance were increased from R to R', and its coefficient of self-induction decreased from L' to L.

The amplitudes of the two currents are $|i_1|$ and $|i_2|$, so that the ratio of the amplitude of the current in the secondary to that in the primary is

$$\frac{|i_2|}{|i_1|} = \left|\frac{-Mip}{S + Nip}\right|$$

$$= \frac{Mp}{\sqrt{S^2 + N^2 p^2}} \quad\dots\dots\dots\dots\dots\dots(452).$$

The difference of phase of the two currents

$$= \arg i_2 - \arg i_1$$

$$= \arg (i_2/i_1)$$

$$= \arg \left(\frac{-Mip}{S + Nip}\right)$$

$$= \pi - \tan^{-1}\left(\frac{S}{Np}\right) \quad\dots\dots\dots\dots\dots\dots(453).$$

523. The analysis is of practical importance in connection with the theory of transformers. In such applications, the current usually is of very high frequency, so that p is large, and we find that approximately the ratio of the amplitudes (cf. expression (452)) is $\frac{M}{N}$, while the difference of phase (cf. expression (453)) is π. These limiting results, for the case of p infinite, can be obtained at a glance from equation (451). The right-hand member, Si_2, is finite, so that $\frac{\partial}{\partial t}(Mi_1 + Ni_2)$ is finite in spite of the infinitely rapid variations in i_1 and i_2 separately. In other words, we must have approximately $Mi_1 + Ni_2$ constant, and clearly the value of this constant must be zero, giving at once the two results just obtained.

524. Whatever the value of p, the result expressed in equation (452) can be deduced at once from the principle of energy. The current in the primary is the same as it would be if the secondary circuit were removed and R, L changed to R', L'. Thus the rate at which the generator performs work is $R'i_1^2$, or averaged over a great number of periods (since i_1 is a simple-harmonic function of the time) is $\frac{1}{2}R'|i_1|^2$. Of this an amount $\frac{1}{2}R|i_1|^2$ is consumed in the primary, so that the rate at which work is performed in the secondary is $\frac{1}{2}(R' - R)|i_1|^2$, or

$$\frac{1}{2}\frac{SM^2p^2}{S^2 + N^2p^2}|i_1|^2.$$

This rate of performing work is also known to be $\frac{1}{2}S|i_2|^2$, and on equating these two expressions we obtain at once the result expressed by equation (452).

Case in which $LN - M^2$ is small.

525. The energy of currents i_1, i_2 in the two circuits is

$$\tfrac{1}{2}(Li_1^2 + 2Mi_1i_2 + Ni_2^2) \quad\dots\dots\dots\dots\dots\dots(454),$$

and since this must always be positive, it follows that $LN - M^2$ must necessarily be positive. The results obtained in the special case in which $LN - M^2$ is so small as to be negligible in comparison with the other quantities involved are of special interest, so that we shall now examine what special features are introduced into the problems when $LN - M^2$ is very small.

Expression (454) can be transformed into

$$\tfrac{1}{2}(Li_1 + Mi_2)^2 + \frac{LN - M^2}{2L}i_2^2,$$

so that when $LN - M^2$ is neglected the energy becomes

$$\tfrac{1}{2}(Li_1 + Mi_2)^2,$$

and this vanishes for the special case in which the currents are in the ratio $i_1/i_2 = -M/L$. This enables us to find the geometrical meaning of the relation $LN - M^2 = 0$. For since the energy of the currents, as in § 501, is

$$\frac{1}{8\pi}\iiint \mu(\alpha^2 + \beta^2 + \gamma^2),$$

we see that this energy can only vanish if the magnetic force vanishes at every point. This requires that the equivalent magnetic shells must coincide and be of strengths which are equal and opposite. Thus the two circuits must coincide geometrically. The number of turns of wire in the circuits may of course be different: if we have r turns in the primary and s in the secondary, we must have

$$\frac{L}{M} = \frac{M}{N} = \frac{r}{s},$$

and when the currents are such as to give a field of zero energy, each fraction is equal to $-i_2/i_1$.

526. Let us next examine the modifications introduced into the analysis by the neglect of $LN - M^2$ in problems in which the value of this quantity is small. We have the general equations (§ 518),

$$E_1 - \frac{d}{dt}(Li_1 + Mi_2) = Ri_1 \quad\dots\dots\dots\dots\dots(455),$$

$$E_2 - \frac{d}{dt}(Mi_1 + Ni_2) = Si_2 \quad\dots\dots\dots\dots\dots(456).$$

If we multiply equation (455) by M and equation (456) by L and subtract, we obtain

$$ME_1 - LE_2 = RMi_1 - SLi_2 \quad\dots\dots\dots\dots\dots(457),$$

an equation which contains no differentials.

527.   To illustrate, let us consider the sudden making of one circuit, discussed in the general case in § 519.   The general equations there obtained, namely

$$L\Delta i_1 + M\Delta i_2 = 0,$$

$$M\Delta i_1 + N\Delta i_2 = 0,$$

now become identical.   We no longer can deduce the relations $\Delta i_1 = \Delta i_2 = 0$, but have only the single initial conditions

$$\frac{\Delta i_1}{\Delta i_2} = -\frac{M}{L} \quad \dots\dots\dots\dots\dots\dots(458).$$

But by supposing equations (455) and (456) replaced by equations (455) and (457) we have only one differential coefficient and therefore only one constant of integration in the solution, and this can be determined from the one initial condition expressed by equation (458).

Let us, for instance, consider the definite problem discussed (for the general case) in § 520.   Circuit 2 contains no battery so that $E_2 = 0$, and at time $t = 0$ circuit 1 is suddenly closed, so that the electromotive force E_1 comes into play in the first circuit.   The initial currents are given by

(from equation (458)), $Li_1 + Mi_2 = 0 \dots\dots\dots\dots\dots\dots(459),$

(from equation (457)), $ME_1 = RMi_1 - SLi_2 \quad \dots\dots\dots\dots\dots(460),$

so that $\dfrac{i_1}{M} = \dfrac{i_2}{-L} = \dfrac{ME_1}{RM^2 + SL^2} = \dfrac{ME_1}{L(RN + SL)}.$

Thus finite currents come into existence at once, but the system of currents is one of zero energy, since equation (459) is satisfied.   To find the subsequent changes, we multiply equation (455) by $\dfrac{L}{R}$ and equation (456) by $\dfrac{M}{S}$ (putting $E_2 = 0$), and find on addition

$$\frac{LE_1}{R} - \left(\frac{L}{R} + \frac{N}{S}\right)\frac{d}{dt}(Li_1 + Mi_2) = Li_1 + Mi_2,$$

of which the solution, subject to the initial condition $Li_1 + Mi_2 = 0$, is

$$Li_1 + Mi_2 = \frac{LE_1}{R}\left(1 - e^{-\frac{RS}{RN + LS}t}\right).$$

From this and equation (460) we obtain

$$\frac{i_1}{N + \dfrac{LS}{R}\left(1 - e^{-\frac{RS}{RN + LS}t}\right)} = \frac{-i_2}{Me^{-\frac{RS}{RN + LS}t}} = \frac{E_1}{RN + LS},$$

and these equations give the currents at any time.

These results can of course be deduced also by examining the limiting form assumed by the solution of § 520, when $LN - M^2$ vanishes.

The problem of the breaking of a circuit, discussed in § 521, can be examined in a similar way in the special case in which $LN - M^2 = 0$. The current i_1 in the broken circuit is found to disappear instantaneously, its energy immediately reappearing as that of a current Li_1/M in circuit (2); this latter current then decays under the resistance of the circuit.

EXAMPLES.

1. A coil is rotated with constant angular velocity ω about an axis in its plane in a uniform field of force perpendicular to the axis of rotation. Find the current in the coil at any time, and shew that it is greatest when the plane of the coil makes an angle $\tan^{-1}\left(\dfrac{L\omega}{R}\right)$ with the lines of magnetic force.

2. The resistance and self-induction of a coil are R and L, and its ends A and B are connected with the electrodes of a condenser of capacity C by wires of negligible resistance. There is a current $I \cos pt$ in a circuit connecting A and B, and the charge of the condenser is in the same phase as this current. Shew that the charge at any time is $\dfrac{LI}{R} \cos pt$, and that $C(R^2 + p^2 L^2) = L$. Obtain also the current in the coil.

3. The ends B, D of a wire (R, L) are connected with the plates of a condenser of capacity C. The wire rotates about BD which is vertical with angular velocity ω, the area between the wire and BD being A. If H is the horizontal component of the earth's magnetism, shew that the average rate at which work must be done to maintain the rotation is

$$\tfrac{1}{2} H^2 A^2 C^2 R \omega^4 / [R^2 C^2 \omega^2 + (1 - CL\omega^2)^2].$$

4. A closed solenoid consists of a large number N of circular coils of wire, each of radius a, wound uniformly upon a circular cylinder of height $2h$. At the centre of the cylinder is a small magnet whose axis coincides with that of the cylinder, and whose moment is a periodic quantity $\mu \sin pt$. Shew that a current flows in the solenoid whose intensity is approximately

$$\frac{2\pi\mu Np}{\{(a^2 + h^2)(R^2 + L^2 p^2)\}^{\frac{1}{2}}} \sin (pt + a),$$

where R, L are the resistance and self-induction of the solenoid, and $\tan a = R/Lp$.

5. A circular coil of n turns, of radius a and resistance R, spins with angular velocity ω round a vertical diameter in the earth's horizontal magnetic field H: shew that the average electromagnetic damping couple which resists its motion is $\dfrac{H^2 n^2 \pi^2 a^4 \omega R}{2(R^2 + \omega^2 L^2)}$. Given $H = 0.17$, $n = 50$, $R = 1$ ohm, $a = 10$ cm., and that the coil makes 20 turns per second, express the couple in dyne-centimetres, and the mean square of the current in ampères.

6. A condenser, capacity C, is discharged through a circuit, resistance R, induction L, containing a periodic electromotive force $E \sin nt$. Shew that the "forced" current in the circuit is

$$E \sin (nt - \theta) \left[R^2 + \left(nL - \frac{1}{Cn}\right)^2 \right]^{-\frac{1}{2}},$$

where $\tan \theta = (n^2 CL - 1)/nCR$.

7. Two circuits, resistances R_1 and R_2, coefficients of induction L, M, N, lie near each other, and an electromotive force E is switched into one of them. Shew that the total quantity of electricity that traverses the other is EM/R_1R_2.

8. A current is induced in a coil B by a current $I \sin pt$ in a coil A. Shew that the mean force tending to increase any coordinate of position θ is

$$-\tfrac{1}{2}\frac{I^2p^2LM}{R^2+L^2p^2}\frac{\partial M}{\partial \theta},$$

where L, M, N are the coefficients of induction of the coils, and R is the resistance of B.

9. A plane circuit, area S, rotates with uniform velocity ω about the axis of z, which lies in its plane at a distance h from the centre of gravity of the area. A magnetic molecule of strength μ is fixed in the axis of x at a great distance a from the origin, pointing in the direction Ox. Prove that the current at time t is approximately

$$\frac{2S\omega\mu}{a^3(R^2+L^2\omega^2)^{\frac{1}{2}}}\cos(\omega t - \epsilon) + \frac{9S\omega\mu h}{a^4(R^2+4L^2\omega^2)^{\frac{1}{2}}}\cos(2\omega t - \eta),$$

where η, ϵ are determinate constants.

10. Two points A, B are joined by a wire of resistance R without self-induction; B is joined to a third point C by two wires each of resistance R, of which one is without self-induction, and the other has a coefficient of induction L. If the ends A, C are kept at a potential difference $E \cos pt$, prove that the difference of potentials at B and C will be $E' \cos(pt - \gamma)$, where

$$E' = E\left\{\frac{R^2+p^2L^2}{9R^2+4p^2L^2}\right\}^{\frac{1}{2}}, \quad \tan\gamma = \frac{pLR}{2p^2L^2+3R^2}.$$

11. A condenser, capacity C, charge Q, is discharged through a circuit of resistance R, there being another circuit of resistance S in the field. If $LN = M^2$, shew that there will be initial currents $-NQ/C(RN+SL)$ and $MQ/C(RN+SL)$, and find the currents at any time.

12. Two insulated wires A, B of the same resistance have the same coefficient of self-induction L, while that of mutual induction is slightly less than L. The ends of B are connected by a wire of small resistance, and those of A by a battery of small resistance, and at the end of a time t a current i is passing through A. Prove that except when t is very small,

$$i = \tfrac{1}{2}(i_0 + i'),$$

approximately, where i_0 is the permanent current in A, and i' is the current in each after a time t, when the ends of both are connected in multiple arc by the battery.

13. The ends of a coil forming a long straight uniform solenoid of m turns per unit length are connected with a short solenoidal coil of n turns and cross-section A, situated inside the solenoid, so that the whole forms a single complete circuit. The latter coil can rotate freely about an axis at right angles to the length of the solenoid. Shew that in free motion without any external field, the current i and the angle θ between the cross-sections of the coils are determined by the equations

$$Ri = -\frac{d}{dt}(L_1 i + L_2 i + 8\pi mnAi\cos\theta),$$

$$I\frac{d^2\theta}{dt^2} + 4\pi mnAi^2\sin\theta = 0,$$

where L_1, L_2 are the coefficients of self-induction of the two coils, I is the moment of inertia of the rotating coil, R is the resistance of the whole circuit, and the effect of the ends of the long solenoid is neglected.

14. Two electrified conductors whose coefficients of electrostatic capacity are γ_1, γ_2, Γ are connected through a coil of resistance R and large inductance L. Verify that the frequency of the electric oscillations thus established is

$$\frac{1}{2\pi}\left(\frac{2\Gamma+\gamma_1+\gamma_2}{\gamma_1\gamma_2-\Gamma^2}\frac{1}{L}-\frac{R^2}{4L^2}\right)^{\frac{1}{2}}.$$

15. An electric circuit contains an impressed electromotive force which alternates in an arbitrary manner and also an inductance. Is it possible, by connecting the extremities of the inductance to the poles of a condenser, to arrange so that the current in the circuit shall always be in step with the electromotive force and proportional to it?

16. Two coils (resistances R, S; coefficients of induction L, M, N) are arranged in parallel in such positions that when a steady current is divided between the two, the resultant magnetic force vanishes at a certain suspended galvanometer needle. Prove that if the currents are suddenly started by completing a circuit including the coils, then the initial magnetic force on the needle will not in general vanish, but that there will be a "throw" of the needle, equal to that which would be produced by the steady (final) current in the first wire flowing through that wire for a time interval

$$\frac{M-L}{R}-\frac{M-N}{S}.$$

17. A condenser of capacity C is discharged through two circuits, one of resistance R and self-induction L, and the other of resistance R' and containing a condenser of capacity C'. Prove that if Q is the charge on the condenser at any time,

$$LR'\frac{d^3Q}{dt^3}+\left(\frac{L}{C}+\frac{L}{C'}+RR'\right)\frac{d^2Q}{dt^2}+\left(\frac{R}{C}+\frac{R}{C'}+\frac{R'}{C}\right)\frac{dQ}{dt}+\frac{Q}{CC'}=0.$$

18. A condenser of capacity C is connected by leads of resistance r, so as to be in parallel with a coil of self-induction L, the resistance of the coil and its leads being R. If this arrangement forms part of a circuit in which there is an electromotive force of period $\frac{2\pi}{p}$, shew that it can be replaced by a wire without self-induction if

$$(R^2-L/C)=p^2LC(r^2-L/C),$$

and that the resistance of this equivalent wire must be $(Rr+L/C)/(R+r)$.

19. Two coils, of which the coefficients of self- and mutual-induction are L_1, L_2, M, and the resistances R_1, R_2, carry steady currents C_1, C_2 produced by constant electromotive forces inserted in them. Shew how to calculate the total extra currents produced in the coils by inserting a given resistance in one of them, and thus also increasing its coefficients of induction by given amounts.

In the primary coil, supposed open, there is an electromotive force which would produce a steady current C, and in the secondary coil there is no electromotive force. Prove that the current induced in the secondary by closing the primary is the same, as regards its effects on a galvanometer and an electrodynamometer, and also with regard to the heat produced by it, as a steady current of magnitude

$$-\tfrac{1}{2}\frac{CMR_1}{R_1L_2+R_2L_1},$$

lasting for a time

$$\frac{R_1L_2+R_2L_1}{\tfrac{1}{2}R_1R_2},$$

while the current induced in the secondary by suddenly breaking the primary circuit may be represented in the same respects by a steady current of magnitude $CM/2L_2$ lasting for a time $2L_2/R_2$.

20.   Two conductors ABD, ACD are arranged in multiple arc.   Their resistances are R, S and their coefficients of self- and mutual-induction are L, N, and M.   Prove that when placed in series with leads conveying a current of frequency p, the two circuits produce the same effect as a single circuit whose coefficient of self-induction is

$$\frac{NR^2 + LS^2 + 2MRS + p^2(LN - M^2)(L + N - 2M)}{(L + N - 2M)^2 p^2 + (R + S)^2},$$

and whose resistance is

$$\frac{RS(S + R) + p^2\{R(N - M)^2 + S(L - M)^2\}}{(L + N - 2M)^2 p^2 + (R + S)^2}.$$

21.   A condenser of capacity C containing a charge Q is discharged round a circuit in the neighbourhood of a second circuit.   The resistances of the circuits are R, S, and their coefficients of induction are L, M, N.

Obtain equations to determine the currents at any moment.

If $\dot{x}$ is the current in the primary, and the disturbance be over in a time less than τ, shew that

$$\left\{\frac{N^2 R}{C} + S\left(NR^2 + \frac{M^2}{C}\right) + S^2 LR\right\} \int_0^\tau \ddot{x}^2 dt = \frac{1}{2}\frac{Q^2}{C^2}\left\{S^2 + \frac{N^2 RS}{LN - M^2} + \frac{N^3}{C(LN - M^2)}\right\}$$

and that

$$\left\{\frac{N^2 R}{C} + S\left(NR^2 + \frac{M^2}{C}\right) + S^2 LR\right\} \int_0^\tau \dot{x}^2 dt = \frac{1}{2}\frac{Q^2}{C^2}\{CS^2 L + CSNR + N^2\}.$$

Examine how $\int_0^\tau \ddot{x}^2 dt$ varies with S.

CHAPTER XV

GENERAL EQUATIONS.

528. WE have seen that when the number N, of tubes of induction, which cross any circuit, is changing, there is an electromotive force $-\dfrac{dN}{dt}$ acting round the circuit. Thus a change in the magnetic field brings into play certain electric forces which would otherwise be absent.

We have now abandoned the conception of action at a distance, so that we must suppose that the electric force at any point depends solely on the changes in the magnetic field at that point. Thus at a point at which the magnetic field is changing, we see that there must be electric forces set up by the changes in the magnetic field, and the amount of these forces must be the same whether the point happens to coincide with an element of a closed conducting circuit or not.

Let ds be an element of any closed circuit drawn in the field, either in a conducting medium or not, and let X, Y, Z denote the components of electric intensity at this point. Then the work done by the electric forces on a unit electric charge in taking it round this circuit is

$$\int\left(X\frac{dx}{ds} + Y\frac{dy}{ds} + Z\frac{dz}{ds}\right)ds \quad\ldots\ldots\ldots\ldots\ldots\ldots(461),$$

and this, by the principle just explained, must be equal to $-\dfrac{dN}{dt}$ where N is the number of tubes of induction which cross this circuit.

529. We have (cf. § 437)

$$N = \iint (la + mb + nc)\,dS \quad\ldots\ldots\ldots\ldots\ldots\ldots(462),$$

so that on equating expression (461) to $-\dfrac{dN}{dt}$, we have

$$\int\left(X\frac{dx}{ds} + Y\frac{dy}{ds} + Z\frac{dz}{ds}\right)ds = -\iint\left(l\frac{da}{dt} + m\frac{db}{dt} + n\frac{dc}{dt}\right)dS \quad\ldots(463).$$

The left-hand member is equal, by Stokes' Theorem (§ 438), to

$$\iint \left\{ l \left(\frac{\partial Z}{\partial y} - \frac{\partial Y}{\partial z} \right) + m \left(\frac{\partial X}{\partial z} - \frac{\partial Z}{\partial x} \right) + n \left(\frac{\partial Y}{\partial x} - \frac{\partial X}{\partial y} \right) \right\} dS,$$

the integration being over the same area as that on the right hand of equation (463). Hence we have

$$\iint \left\{ l \left(\frac{\partial Z}{\partial y} - \frac{\partial Y}{\partial z} + \frac{da}{dt} \right) + m \left(\frac{\partial X}{\partial z} - \frac{\partial Z}{\partial x} + \frac{db}{dt} \right) + n \left(\frac{\partial Y}{\partial x} - \frac{\partial X}{\partial y} + \frac{dc}{dt} \right) \right\} dS = 0.$$

This equation is true for every surface, so that not only must each integrand vanish, but it must vanish for all possible values of l, m, n. Hence each coefficient of l, m, n must vanish separately. We must accordingly have

$$-\frac{da}{dt} = \frac{\partial Z}{\partial y} - \frac{\partial Y}{\partial z} \quad \dots\dots\dots\dots\dots(464),$$

$$-\frac{db}{dt} = \frac{\partial X}{\partial z} - \frac{\partial Z}{\partial x} \quad \dots\dots\dots\dots\dots(465),$$

$$-\frac{dc}{dt} = \frac{\partial Y}{\partial x} - \frac{\partial X}{\partial y} \quad \dots\dots\dots\dots\dots(466).$$

530. The components F, G, H of the magnetic vector-potential are given, as in equations (376), by

$$a = \frac{\partial H}{\partial y} - \frac{\partial G}{\partial z}, \text{ etc. } \quad \dots\dots\dots\dots(467).$$

On comparing these equations with equations (464)—(466), it is clear that the simplest solution for the vector-potential is given by the relations

$$\frac{\partial F}{\partial t} = -X, \quad \frac{\partial G}{\partial t} = -Y, \quad \frac{\partial H}{\partial t} = -Z \quad \dots\dots\dots(468).$$

If F, G, H is the most general vector-potential, we must have relations of the form (cf. equations (375))

$$\frac{\partial F}{\partial t} = -X - \frac{\partial \Psi}{\partial x}, \text{ etc. } \quad \dots\dots\dots\dots(469),$$

where Ψ is an arbitrary function replacing the $-\chi$ of equations (375).

531. Writing these relations in the form

$$X = -\frac{dF}{dt} - \frac{\partial \Psi}{\partial x} \quad \dots\dots\dots\dots\dots(470),$$

$$Y = -\frac{dG}{dt} - \frac{\partial \Psi}{\partial y} \quad \dots\dots\dots\dots\dots(471),$$

$$Z = -\frac{dH}{dt} - \frac{\partial \Psi}{\partial z} \quad \dots\dots\dots\dots\dots(472),$$

we have equations giving the electric forces explicitly.

The function Ψ has, so far, had no physical meaning assigned to it. Equations (470), (471), (472) shew that the electric force $(X,\ Y,\ Z)$ can be regarded as compounded of two forces:

(i) a force $\left(-\dfrac{dF}{dt},\ -\dfrac{dG}{dt},\ -\dfrac{dH}{dt}\right)$ arising from the changes in the magnetic field;

(ii) a force of components $\left(-\dfrac{\partial \Psi}{\partial x},\ -\dfrac{\partial \Psi}{\partial y},\ -\dfrac{\partial \Psi}{\partial z}\right)$ which is present when there are no magnetic changes occurring.

We now see that the second force is the force arising from the ordinary electrostatic field, so that we may identify Ψ with the electrostatic potential when no changes are occurring. The meaning to be assigned to Ψ when changes are in progress is discussed below (Chapter xx).

532. If the medium is a conducting medium, the presence of the electric forces sets up currents, and the components $u,\ v,\ w$ of the current at any point are, as in § 374, connected with the currents by the equations

$$X = \tau u, \qquad Y = \tau v, \qquad Z = \tau w,$$

these equations being the expression of Ohm's Law, where τ is the specific resistance of the conductor at the point.

On substituting these values for $X,\ Y,\ Z$ in equations (464)—(466) or (470)—(472), we obtain a system of equations connecting the currents in the conductor with the changes in the magnetic field.

533. There is, however, a further system of equations expressing relations between the currents and the magnetic field. We have seen (§ 480) that a current sets up a magnetic field of known intensity, and since the whole magnetic field must arise either from currents or from permanent magnets, this fact gives rise to a second system of equations.

In a field arising solely from permanent magnetism, we can take a unit pole round any closed path in the field, and the total work done will be *nil*. Hence on taking a unit pole round a closed circuit in the most general magnetic field, the work done will be the same as if there were no permanent magnetism, and the whole field were due to the currents present. The amount of this work, as we have seen, is $4\pi\Sigma i$, where Σi is the sum of all the currents which flow through the circuit round which the pole is taken. If $u,\ v,\ w$ are the components of current at any point, we have

$$\Sigma i = \iint (lu + mv + nw)\, dS,$$

the integration being over any area which has the closed path as boundary. Hence our experimental fact leads to the equation

$$\int \left(\alpha \frac{dx}{ds} + \beta \frac{dy}{ds} + \gamma \frac{dz}{ds} \right) ds = 4\pi \iint (lu + mv + nw)\, dS.$$

Transforming the line integral into a surface integral by Stokes' Theorem (§ 438), we obtain the equation in the form

$$\iint \left\{ l \left(\frac{\partial \gamma}{\partial y} - \frac{\partial \beta}{\partial z} - 4\pi u \right) + m \left(\frac{\partial \alpha}{\partial z} - \frac{\partial \gamma}{\partial x} - 4\pi v \right) + n \left(\frac{\partial \beta}{\partial x} - \frac{\partial \alpha}{\partial y} - 4\pi w \right) \right\} dS = 0.$$

As with the integral of § 529, each integrand must vanish for all values of l, m, n, so that we must have

$$4\pi u = \frac{\partial \gamma}{\partial y} - \frac{\partial \beta}{\partial z} \quad\dots\dots\dots\dots\dots(473),$$

$$4\pi v = \frac{\partial \alpha}{\partial z} - \frac{\partial \gamma}{\partial x} \quad\dots\dots\dots\dots\dots(474),$$

$$4\pi w = \frac{\partial \beta}{\partial x} - \frac{\partial \alpha}{\partial y} \quad\dots\dots\dots\dots\dots(475).$$

534. If we differentiate these three equations with respect to x, y, z respectively and add, we obtain

$$\frac{\partial u}{\partial x} + \frac{\partial v}{\partial y} + \frac{\partial w}{\partial z} = 0 \quad\dots\dots\dots\dots\dots(476),$$

of which the meaning (cf. § 375, equation (311)) is that no electricity is destroyed or created or allowed to accumulate in the conductor.

The interpretation of this result is not that it is a physical impossibility for electricity to accumulate in a conductor, but that the assumptions upon which we are working are not sufficiently general to cover cases in which there is such an accumulation of electricity. It is easy to see directly how this has come about. The supposition underlying our equations is that the work done in taking a unit pole round a circuit is equal to 4π times the total current flow through the circuit. It is only when equation (476) is satisfied by the current components that the expression "total flow through a circuit" has a definite significance: the current flow across every area bounded by the circuit must be the same. We shall see later (Chapter XVII) how the equations must be modified to cover the case of an electric flow in which the condition is not satisfied. For the present we proceed upon the supposition that the condition is satisfied.

Currents in homogeneous media.

535. Let us now suppose that we are considering the currents in a homogeneous non-magnetised medium. We write

$$a = \mu \alpha, \text{ etc.}, \qquad X = \tau u, \text{ etc.},$$

in which μ and τ are constant. The systems of equations of §§ 529 and 533 now become

$$-\mu \frac{da}{dt} = \tau \left(\frac{\partial w}{\partial y} - \frac{\partial v}{\partial z} \right), \text{ etc.} \quad\dots\dots\dots\dots(477),$$

$$4\pi u = \frac{\partial \gamma}{\partial y} - \frac{\partial \beta}{\partial z}, \text{ etc.}\dots\dots\dots\dots\dots(478).$$

Differentiating equation (478) with respect to the time, we obtain

$$4\pi\mu \frac{du}{dt} = \frac{\partial}{\partial y}\left(\mu \frac{d\gamma}{dt}\right) - \frac{\partial}{\partial z}\left(\mu \frac{d\beta}{dt}\right)$$

$$= -\tau\left\{\frac{\partial}{\partial y}\left(\frac{\partial v}{\partial x} - \frac{\partial u}{\partial y}\right) - \frac{\partial}{\partial z}\left(\frac{\partial u}{\partial z} - \frac{\partial w}{\partial x}\right)\right\}$$

$$= \tau\left\{\left(\frac{\partial^2 u}{\partial x^2} + \frac{\partial^2 u}{\partial y^2} + \frac{\partial^2 u}{\partial z^2}\right) - \frac{\partial}{\partial x}\left(\frac{\partial u}{\partial x} + \frac{\partial v}{\partial y} + \frac{\partial w}{\partial z}\right)\right\}$$

$$= \tau \nabla^2 u,$$

in virtue of equation (476).

Similar equations are satisfied by the other current-components, so that we have the system of differential equations

$$\left.\begin{aligned} \frac{4\pi\mu}{\tau} \frac{du}{dt} &= \nabla^2 u \\[4pt] \frac{4\pi\mu}{\tau} \frac{dv}{dt} &= \nabla^2 v \\[4pt] \frac{4\pi\mu}{\tau} \frac{dw}{dt} &= \nabla^2 w \end{aligned}\right\} \quad\dots\dots\dots\dots\dots\dots(479).$$

If we eliminate the current-components from the system of equations (477) and (478), we obtain

$$\frac{4\pi\mu}{\tau} \frac{da}{dt} = \nabla^2 a \quad\dots\dots\dots\dots\dots\dots(480),$$

and similar equations are satisfied by b and c.

536. The equation which has been found to be satisfied by u, v, w, a, β and γ is the well-known equation of conduction of heat. Thus we see that the currents induced in a mass of metal, as well as the components of the magnetic field associated with these currents, will diffuse through the metal in the same way as heat diffuses through a uniform conductor.

Rapidly alternating currents.

537. The equations assume a form of special interest when the currents are alternating currents of high frequency. We may assume each component of current to be proportional to e^{ipt} (cf. § 514), and may then replace the operator $\frac{d}{dt}$ by the multiplier ip. The equations now assume the form

$$\frac{4\pi\mu ip}{\tau} u = \nabla^2 u \quad\dots\dots\dots\dots\dots\dots(481),$$

$$\frac{4\pi\mu ip}{\tau} a = \nabla^2 a, \text{ etc.,}$$

and if p is so large that it may be treated as infinite, these equations assume the simple form

$$u = v = w = 0,$$

$$a = b = c = 0.$$

Thus for currents of infinite frequency, there is neither current nor magnetic field in the interior. The currents are confined to the surface, and the only part of the conductor which comes into play at all is a thin skin on the surface.

Equations (481) enable us to form an estimate of the thickness of this skin when the frequency of the currents is very great without being actually infinite.

At a point O on the surface of the conductor, let us take rectangular axes so that the direction of the current is that of Ox while the normal to the surface is Oz. If the thickness of the skin is very small, we need not consider any region except that in the immediate neighbourhood of the origin, so that the problem is practically identical with that of current flowing parallel to Ox in an infinite slab of metal having the plane Oxy for a boundary.

Equation (481) reduces in this case to

$$\frac{4\pi\mu i p}{\tau} u = \frac{d^2 u}{dz^2},$$

and if we put $\dfrac{4\pi\mu i p}{\tau} = \kappa^2$, the solution is

$$u = A e^{-\kappa z} + B e^{\kappa z}.$$

The value of κ is found to be

$$\kappa = \sqrt{\frac{2\pi\mu p}{\tau}}\,(1 + i),$$

so that
$$u = A e^{-\sqrt{\frac{2\pi\mu p}{\tau}} z} e^{-\sqrt{\frac{2\pi\mu p}{\tau}} iz} + B e^{\sqrt{\frac{2\pi\mu p}{\tau}} z} e^{\sqrt{\frac{2\pi\mu p}{\tau}} iz},$$

and the condition that the current is to be confined to a thin skin may now be expressed by the condition that $u = 0$ when $z = \infty$, and is accordingly $B = 0$. The multiplier A is independent of z, but will of course involve the time through the factor e^{ipt}; let us put $A = u_0 e^{ipt}$, and we then have the solution

$$u = u_0 e^{-\sqrt{\frac{2\pi\mu p}{\tau}} z} e^{i\left(pt - \sqrt{\frac{2\pi\mu p}{\tau}} z\right)}.$$

Rejecting the imaginary part, we are left with the real solution

$$u = u_0 e^{-\sqrt{\frac{2\pi\mu p}{\tau}}z}\cos\left(pt - \sqrt{\frac{2\pi\mu p}{\tau}}z\right),$$

from which we see that as we pass inwards from the surface of the conductor, the phase of the current changes at a uniform rate, while its amplitude decreases exponentially.

We can best form an idea of the rate of decrease of the amplitude by considering a concrete case. For copper we may take (in c.g.s. electromagnetic units) $\mu = 1$, $\tau = 1600$. Thus for a current which alternates 1000 times per second, we have

$$p = 2\pi \times 1000, \qquad \sqrt{\frac{2\pi\mu p}{\tau}} = 5 \text{ approximately.}$$

It follows that at a depth of 1 cm. the current will be only e^{-5} or ·0067 times its value at the surface. Thus the current is practically confined to a skin of thickness 1 cm.

The total current per unit width of the surface at a time t is $\int_{z=0}^{z=\infty} u\, dz$, of which the value is found to be

$$\frac{u_0 \cos\left(pt - \dfrac{\pi}{4}\right)}{\sqrt{\dfrac{4\pi\mu p}{\tau}}}.$$

Thus, if we denote the amplitude of the aggregate current by U, the value of u_0 will be $U\sqrt{\dfrac{4\pi\mu p}{\tau}}$.

The heat generated per unit time in a strip of unit width and unit length is

$$\tau \int_{t=0}^{t=1}\int_{z=0}^{z=\infty} u^2\, dt\, dz$$

$$= \tfrac{1}{2}\tau u_0^2 \int_{z=0}^{z=\infty} e^{-2\sqrt{\frac{2\pi\mu p}{\tau}}z}\, dz$$

$$= \tfrac{1}{2}\tau U^2 \sqrt{\frac{2\pi\mu p}{\tau}}.$$

Thus the resistance of the conductor is the same as would be the resistance for steady currents of a skin of depth $2\Big/\sqrt{\dfrac{2\pi\mu p}{\tau}}$.

The results we have obtained will suffice to explain why it is that the conductors used to convey rapidly alternating currents are made hollow, as also why it is that lightning conductors are made of strips, rather than cylinders, of metal.

PLANE CURRENT-SHEETS.

538. We next examine the phenomenon of the induction of currents in a plane sheet of metal.

Let the plane of the current-sheet be taken to be $z = 0$. Let us introduce a current-function Φ, which is to be defined for every point in the sheet by the statement that the total strength of all the currents which flow between the point and the boundary is Φ. Then the currents in the sheet are known when the value of Φ is known at every point of the sheet. If we assume that no electricity is introduced into, or removed from, the current-sheet, or allowed to accumulate at any point of it, then clearly Φ will be a single-valued function of position on the sheet.

The equation of the current-lines will be $\Phi = \text{constant}$, and the line $\Phi = 0$ will be the boundary of the current-sheet. Between the lines Φ and $\Phi + d\Phi$ we have a current of strength $d\Phi$ flowing in a closed circuit. The magnetic field produced by this current is the same as that produced by a magnetic shell of strength $d\Phi$ coinciding with that part of the current-sheet which is enclosed by this circuit, so that the magnetic effect of the whole system of currents in the sheet is that of a shell coinciding with the sheet and of variable strength Φ. This again may be replaced by a distribution of magnetic poles of surface density Φ/ϵ on the positive side of the sheet, together with a distribution of surface density $-\Phi/\epsilon$ on the negative side of the sheet, where ϵ is the thickness of the sheet.

Let P denote the potential at any point of a distribution of poles of strength Φ, so that

$$P = \iint \frac{\Phi}{r}\, dx'\, dy' \quad\dots\dots\dots\dots\dots\dots(482),$$

where $dx'\, dy'$ is any element of the sheet. The magnetic potential at any point outside the current-sheet of the field produced by the currents is then

$$\Omega = -\frac{\partial P}{\partial z} \quad\dots\dots\dots\dots\dots\dots\dots(483).$$

If σ is the resistance of a unit square of the sheet at any point, and u, v the components of current, we have, by Ohm's Law,

$$X = \sigma u, \qquad Y = \sigma v.$$

The components u, v are readily found to be given by

$$u = \frac{\partial \Phi}{\partial y}, \qquad v = -\frac{\partial \Phi}{\partial x},$$

so that we have the equations

$$X = \sigma \frac{\partial \Phi}{\partial y}, \qquad Y = -\sigma \frac{\partial \Phi}{\partial x} \quad \dots\dots\dots\dots\dots(484)$$

true at every point of the sheet.

Hence, by equation (466),

$$-\frac{dc}{dt} = \frac{\partial Y}{\partial x} - \frac{\partial X}{\partial y} = -\sigma \left(\frac{\partial^2 \Phi}{\partial x^2} + \frac{\partial^2 \Phi}{\partial y^2} \right) \quad \dots\dots\dots\dots(485).$$

The total magnetic field consists of the part of potential Ω due to the currents and a part of potential (say) Ω', due to the magnetic system by which the currents are induced. Thus the total magnetic potential is $\Omega + \Omega'$, and at a point just outside the current-sheet (taking $\mu = 1$)

$$-\frac{dc}{dt} = \frac{d}{dt} \frac{\partial}{\partial z} (\Omega + \Omega'),$$

the equation (485) becomes

$$\frac{d}{dt} \frac{\partial}{\partial z} (\Omega + \Omega') = -\sigma \left(\frac{\partial^2 \Phi}{\partial x^2} + \frac{\partial^2 \Phi}{\partial y^2} \right) \dots\dots\dots\dots\dots(486).$$

The function P (equation (482)) is the potential of a distribution of poles of surface density Φ on the sheet. Hence P satisfies Laplace's equation at all points outside the sheet, and at a point just outside the sheet and on its positive face $-\dfrac{\partial P}{\partial z} = 2\pi\Phi.$

Hence, at a point just outside the positive face of the sheet,

$$\begin{aligned}
\frac{\partial^2 \Phi}{\partial x^2} + \frac{\partial^2 \Phi}{\partial y^2} &= -\frac{1}{2\pi} \left(\frac{\partial^3 P}{\partial x^2 \partial z} + \frac{\partial^3 P}{\partial y^2 \partial z} \right) \\
&= \frac{1}{2\pi} \frac{\partial^3 P}{\partial z^3} \\
&= -\frac{1}{2\pi} \frac{\partial^2 \Omega}{\partial z^2},
\end{aligned}$$

by equation (483), so that equation (486) becomes

$$\frac{d}{dt} \frac{\partial}{\partial z} (\Omega + \Omega') = \frac{\sigma}{2\pi} \frac{\partial^2 \Omega}{\partial z^2} \dots\dots\dots\dots\dots\dots(487),$$

and similarly, at the negative face of the sheet, we have the equation

$$\frac{d}{dt} \frac{\partial}{\partial z} (\Omega + \Omega') = -\frac{\sigma}{2\pi} \frac{\partial^2 \Omega}{\partial z^2} \quad \dots\dots\dots\dots\dots(488).$$

Finite Current-sheets.

539. Suppose that in an infinitesimal interval any pole of strength m moves from P to Q. This movement may be represented by the creation of a pole of strength $-m$ at P and of one of strength $+m$ at Q. Thus the most general motion of the inducing field may be replaced by the creation of a series of poles. The simplest problem arises when the inducing field is produced by the sudden creation of a single pole, and the solution of the most general problem can be obtained from the solution of this simple problem by addition.

From equations (487) and (488) it is clear that $\dfrac{d}{dt}\dfrac{\partial}{\partial z}(\Omega + \Omega')$ remains finite on both surfaces of the sheet during the sudden creation of a new pole, so that $\dfrac{\partial}{\partial z}(\Omega + \Omega')$ remains unaltered in value over the whole surface of the sheet. Let the increment in $\dfrac{\partial}{\partial z}(\Omega + \Omega')$ at any point in space be denoted by Δ, then Δ is a potential of which the poles are known in the space outside the sheet, and of which the value is known to be zero over the surface of the sheet. The methods of Chapter VIII are accordingly available for the determination of Δ: the required value of Δ is the electrostatic potential when the current-sheet is put to earth in the presence of the point charges which would give a potential $\dfrac{\partial \Omega'}{\partial z}$ if the sheet were absent.

Physically, the fact that $\dfrac{\partial}{\partial z}(\Omega + \Omega')$ remains unaltered over the whole surface of the sheet means that the field of force just outside the sheet remains unaltered, and hence that currents are instantaneously induced in the sheet such that the lines of force at the surfaces of the sheet remain unaltered.

The induced currents can be found for any shape of current-sheet for which the corresponding electrostatic problem can be solved*, but in general the results are too complicated to be of physical interest.

Infinite Plane Current-sheet.

540. Let the current-sheet be of infinite extent, and occupy the whole of the plane of xy, and let the moving magnetic system be in the region in which z is negative. Then throughout the region for which z is positive the potential $\Omega + \Omega'$ has no poles, and hence the potential

$$\frac{d}{dt}\frac{\partial}{\partial z}(\Omega + \Omega') - \frac{\sigma}{2\pi}\frac{\partial^2 \Omega}{\partial z^2}$$

---

* See a paper by the author, "Finite Current-sheets," *Proc. Lond. Math. Soc.* Vol. XXXI. p. 151.

has no poles. Moreover this potential is a solution of Laplace's equation, and vanishes over the boundary of the region, namely at infinity and over the plane $z = 0$ (cf. equation (487)). Hence it vanishes throughout the whole region (cf. § 186), so that equation (487) must be true at every point in the region for which z is positive. We may accordingly integrate with respect to z and obtain the equation in the form

$$\frac{d}{dt}(\Omega + \Omega') = \frac{\sigma}{2\pi} \frac{\partial \Omega}{\partial z} \quad\quad\quad\quad\quad\text{......................(489)},$$

no arbitrary function of x, y being added because the equation must be satisfied at infinity.

The motion of the system of magnets on the negative side of the sheet may be replaced, as in § 539, by the instantaneous creation of a number of poles. At the creation of a single pole currents are set up in the sheet such that $\Omega + \Omega'$ remains unaltered (cf. equation (489)) on the positive side of the sheet. Thus these currents form a magnetic screen and shield the space on the positive side of the sheet from the effects of the magnetic changes on the negative side.

To examine the way in which these currents decay under the influence of resistance and self-induction, we put $\Omega' = 0$ in equation (489), and find that Ω must be a solution of the equation

$$\frac{d\Omega}{dt} = \frac{\sigma}{2\pi} \frac{\partial \Omega}{\partial z}.$$

The general solution of this equation is

$$\Omega = f\left(x,\ y,\ z + \frac{\sigma}{2\pi} t\right),$$

and this corresponds to the initial value

$$\Omega = f(x, y, z).$$

Thus the decay of the currents can be traced by taking the field of potential Ω at time $t = 0$ and moving it parallel to the axis of z with a velocity $\frac{\sigma}{2\pi}$.

EXAMPLES.

1.   Prove that the currents induced in a solid with an infinite plane face, owing to magnetic changes near the face, circulate parallel to it, and may be regarded as due to the diffusion into the solid of current-sheets induced at each instant on the surface so as to screen off the magnetic changes from the interior.

Shew that for periodic changes, the current penetrates to a depth proportional to the square root of the period.   Give a solution for the case in which the strength of a fixed inducing magnet varies as $\cos pt$.

2.   A magnetic system is moving towards an infinite plane conducting sheet with velocity w.   Shew that the magnetic potential on the other side of the sheet is the same as it would be if the sheet were away, and the strengths of all the elements of the magnetic system were changed in the ratio $R/(R+w)$, where $2\pi R$ is the specific resistance of the sheet per unit area.   Shew that the result is unaltered if the system is moving away from the sheet, and examine the case of $w=-R$.

If the system is a magnetic particle of mass M and moment m, with its axis perpendicular to the sheet, prove that if the particle has been projected at right angles to the sheet, then when it is at a distance z from the sheet, its velocity $\dot{z}$ is given by

$$\tfrac{1}{2}M(\dot{z}-R)^2=C-m^2/8z^3.$$

3.   A small magnet horizontally magnetised is moving with a velocity u parallel to a thin horizontal plate of metal.   Shew that the retarding force on the magnet due to the currents induced in the plate is

$$\frac{m^2}{(2c)^4}\frac{uR}{Q(Q+R)},$$

where m is the moment of the magnet, c its distance above the plate, $2\pi R$ the resistance of a sq. cm. of the plate, and $Q^2=u^2+R^2$.

4.   A slowly alternating current $I\cos pt$ is traversing a small circular coil whose magnetic moment for a unit current is M.   A thin spherical shell, of radius a and specific resistance σ, has its centre on the axis of the coil at a distance f from the centre of the coil.   Shew that the currents in the shell form circles round the axis of the coil, and that the strength of the current in any circle whose radius subtends an angle $\cos^{-1}\mu$ at the centre is

$$\frac{M}{4\pi f^2}\frac{I(1-\mu^2)^{\frac{1}{2}}}{a}\Sigma(2n+1)\left(\frac{a}{f}\right)^n\frac{\partial P_n}{\partial\mu}\cos\epsilon_n\cos(pt-\epsilon_n),$$

where

$$\tan\epsilon_n=\frac{(2n+1)\sigma}{4\pi pa}.$$

5.   An infinite iron plate is bounded by the parallel planes $x=h$, $x=-h$; wire is wound uniformly round the plate, the layers of wire being parallel to the axis of y.   If an alternating current is sent through the wire producing outside the plate a magnetic force $H_0\cos pt$ parallel to z, prove that H, the magnetic force in the plate at a distance x from the centre, will be given by

$$H=H_0\left(\frac{\cosh 2mx+\cos 2mx}{\cosh 2mh+\cos 2mh}\right)^{\frac{1}{2}}\cos(pt+\beta),$$

$$\tan\beta=\frac{\sinh m(h+x)\sin m(h-x)-\sinh m(h-x)\sin m(h+x)}{\cosh m(h+x)\cos m(h-x)+\cosh m(h-x)\cos m(h+x)},$$

where

$$m^2=2\pi\mu p/\sigma.$$

Discuss the special cases of (i) mh small, (ii) mh large.

CHAPTER XVI

DYNAMICAL THEORY OF CURRENTS

General Theory of Dynamical Systems.

541. We have so far developed the theory of electromagnetism by starting from a number of simple data which are furnished or confirmed by experiment, and examining the mathematical and physical consequences which can be deduced from these data.

There are always two directions in which it is possible for a theoretical science to proceed. It is possible to start from the simple experimental data and from these to deduce the theory of more complex phenomena. And it may also be possible to start from the experimental data and to analyse these into something still more simple and fundamental. We may, in fact, either advance from simple phenomena to complex, or we may pass backwards from simple phenomena to phenomena which are still simpler, in the sense of being more fundamental.

As an example of a theoretical science of which the development is almost entirely of the second kind may be mentioned the Dynamical Theory of Gases. The theory starts with certain simple experimental data, such as the existence of pressure in a gas, and the relation of this pressure to the temperature and density of a gas. And the theory is developed by shewing that these phenomena may be regarded as consequences of still more fundamental phenomena, namely the motion of the molecules of the gas.

In our development of electromagnetic theory there has so far been but little progress in this second direction. It is true that we have seen that the phenomena from which we started—such as the attractions and repulsions of electric charges, or the induction of electric currents—may be interpreted as the consequences of other and more fundamental phenomena taking place in the ether by which the material systems are surrounded. We have even obtained formulae for the stresses and the energy in the ether. But it has not been possible to proceed any further and to explain the existence of these stresses and energy in terms of the ultimate mechanism of the ether.

The reason why we have been brought to a halt in the development of electromagnetic theory will become clear as soon as we contrast this theory with the theory of gases. The ultimate mechanism with which the theory of gases is concerned is that of molecules in motion, and we know (or at least can provisionally assume that we know) the ultimate laws by which this motion is governed. On the other hand the ultimate mechanism with which electromagnetic theory is concerned is that of action in the ether, and we are in utter ignorance of the ultimate laws which govern action in the ether. We do not know how the ether behaves, and so can make no progress towards explaining electromagnetic phenomena in terms of the behaviour of the ether.

542. There is a branch of dynamics which attempts to explain the relation between the motions of certain known parts of a mechanism, even when the nature of the remaining parts is completely unknown. We turn to this branch of dynamics for assistance in the present problem. The whole mechanism before us consists of a system of charged conductors, magnets, currents, etc., and of the ether by which all these are connected. Of this mechanism one part (the motion of the material bodies) is known to us, while the remainder (the flow of electric currents, the transmission of action by the ether, etc.) is unknown to us, except indirectly by its effect on the first part of the mechanism.

543. An analogy, first suggested by Professor Clerk Maxwell, will explain the way in which we are now attacking the problem.

Imagine that we have a complicated machine in a closed room, the only connection between this machine and the exterior of the room being by means of a number of ropes which hang through holes in the floor into the room beneath. A man who cannot get into the room which contains the machine will have no opportunity of actually inspecting the mechanism, but he can manipulate it to a certain extent by pulling the different ropes. If, on pulling one rope, he finds that others are set into motion, he will understand that the ropes must be connected by some kind of mechanism above, although he may be unable to discover the exact nature of this mechanism.

In this analogy, the concealed mechanism is supposed to represent those parts of the universe which do not directly affect our senses—*e.g.* the ether—while the ropes represent those parts of which we can observe the motion—*e.g.* material bodies. In nature, there are certain acts which we can perform (analogous to the pulling of certain ropes), and these are invariably followed by certain consequences (analogous to the motion of other ropes), but the ultimate mechanism by which the cause produces the effect is unknown. For instance we can close an electric circuit by pressing a key, and the needle of a distant galvanometer may be set into motion. We infer that there must be some mechanism connecting the two, but the nature of this mechanism is almost completely unknown.

Suppose now that an observer may handle the ropes, but may not penetrate into the room above to examine the mechanism to which they are attached. He will know that whatever this mechanism may be, certain laws must govern the manipulation of the ropes, provided that the mechanism is itself subject to the ordinary laws of mechanics.

To take the simplest illustration, suppose that there are two ropes only, A and B, and that when rope A is pulled down a distance of one inch, it is found that rope B rises through two inches. The mechanism connecting A and B may be a lever or an arrangement of pulleys or of clockwork, or something different from any of these. But whatever it is, provided that it is subject to the laws of dynamics, the experimenter will know, from the mechanical principle of "virtual work," that the downward motion of rope A can be restrained on applying to B a force equal to half of that applied to A.

544. The branch of dynamics of which we are now going to make use enables us to predict what relation there ought to be between the motions of the accessible parts of the mechanism. If these predictions are borne out by experiment, then there will be a presumption that the concealed mechanism is subject to the laws of dynamics. If the predictions are not confirmed by experiment, we shall know that the concealed mechanism is not governed by the laws of dynamics.

Hamilton's Principle.

545. Suppose, first, that we have a dynamical system composed of discrete particles, each of which moves in accordance with Newton's Laws of Motion. Let any typical particle of mass m_1 have at any instant t coordinates x_1, y_1, z_1 and components of velocity u_1, v_1, w_1, and let it be acted on by forces of which the resultant has components X_1, Y_1, Z_1. Then, since the motion of the particle is assumed to be governed by Newton's Laws, we have

$$m_1 \frac{du_1}{dt} = X_1 \quad\dots\dots\dots\dots\dots\dots\dots(490),$$

$$m_1 \frac{dv_1}{dt} = Y_1 \quad\dots\dots\dots\dots\dots\dots\dots(491),$$

$$m_1 \frac{dw_1}{dt} = Z_1 \quad\dots\dots\dots\dots\dots\dots\dots(492).$$

Let us compare this motion with a slightly different motion, in which Newton's Laws are not obeyed. At the instant t let the coordinates of this same particle be $x_1 + \delta x_1$, $y_1 + \delta y_1$, $z_1 + \delta z_1$ and let its components of velocity be $u_1 + \delta u_1$, $v_1 + \delta v_1$, $w_1 + \delta w_1$. Let us multiply equations (490), (491) and (492) by δx_1, δy_1, δz_1 respectively, and add. We obtain

$$m_1 \left(\frac{du_1}{dt} \delta x_1 + \frac{dv_1}{dt} \delta y_1 + \frac{dw_1}{dt} \delta z_1 \right) = X_1 \delta x_1 + Y_1 \delta y_1 + Z_1 \delta z_1 \ \dots(493).$$

Now $\qquad \dfrac{du_1}{dt} \delta x_1 = \dfrac{d}{dt}(u_1 \delta x_1) - u_1 \dfrac{d}{dt}(\delta x_1)$

$$= \frac{d}{dt}(u_1 \delta x_1) - u_1 \delta u_1.$$

If we sum equation (493) for all the particles of the system, replacing the terms on the left by their values as just obtained, we arrive at the equation

$$\frac{d}{dt} \Sigma m_1 (u_1 \delta x_1 + v_1 \delta y_1 + w_1 \delta z_1) - \Sigma m_1 (u_1 \delta u_1 + v_1 \delta v_1 + w_1 \delta w_1)$$
$$= \Sigma (X_1 \delta x_1 + Y_1 \delta y_1 + Z_1 \delta z_1) \ \dots\dots(494).$$

Let T denote the kinetic energy of the actual motion, and $T + \delta T$ that of the slightly varied motion, then

$$T = \tfrac{1}{2} \Sigma m_1 (u_1^2 + v_1^2 + w_1^2),$$

so that $\qquad \delta T = \Sigma m_1 (u_1 \delta u_1 + v_1 \delta v_1 + w_1 \delta w_1),$

and this is the value of the second term in equation (494).

If W and $W + \delta W$ are the potential energies of the two configurations (assuming the forces to form a conservative system), we have

$$W = - \Sigma \int^{x_1, y_1, z_1} (X_1 dx_1 + Y_1 dy_1 + Z_1 dz_1),$$

and $\qquad \delta W = - \Sigma (X_1 \delta x_1 + Y_1 \delta y_1 + Z_1 \delta z_1),$

and so the value of the right-hand member of equation (494) is $- \delta W$.

We may now rewrite equation (494) in the form

$$\delta (T - W) = \frac{d}{dt} \Sigma m_1 (u_1 \delta x_1 + v_1 \delta y_1 + w_1 \delta z_1).$$

This equation is true at every instant of the motion. Let us integrate it throughout the whole of the motion, say from $t = 0$ to $t = \tau$. We obtain

$$\delta \int_0^\tau (T - W) \, dt = \left[\Sigma m_1 (u_1 \delta x_1 + v_1 \delta y_1 + w_1 \delta z_1) \right]_{t=0}^{t=\tau} \ \dots\dots(495).$$

The displaced motion has been supposed to be any motion which differs only slightly from the actual motion. Let us now limit it by the restriction that the configurations at the beginning and end of the motion are to coincide with those of the actual motion, so that the displaced motion is now to be one in which the system starts from the same configuration as in the actual motion at time $t = 0$, and, after passing through a series of configurations slightly different from those of the actual motion, finally ends in the same configuration at time $t = \tau$ as that of the actual motion. Mathematically this new restriction is expressed by saying that at times $t = 0$ and $t = \tau$ we must have $\delta x = \delta y = \delta z = 0$ for each particle. Equation (495) now becomes

$$\delta \int_0^\tau (T - W) \, dt = 0 \ \dots\dots\dots\dots\dots(496).$$

546. Speaking of the two parts of the mechanism under discussion as the "accessible" and "concealed" parts, let us suppose that the kinetic and potential energies T and W depend only on the configuration of the accessible parts of the mechanism. Then throughout any imaginary motion of the accessible parts of the system we shall have a knowledge of T and W at every instant, and hence shall be able to calculate the value of

$$\int_0^\tau (T - W)\, dt \dots\dots\dots\dots\dots\dots\dots(497).$$

We can imagine an infinite number of motions which bring the system from one configuration A at time $t = 0$ to a second configuration B at time $t = \tau$, and we can calculate the value of the integral for each. Equation (496) shews that those motions for which the value of the integral is stationary would be the motions actually possible for the system. Having found which these motions were, we should have a knowledge of the changes in the accessible parts of the system, although the concealed parts remained unknown to us, both as regards their nature and their motion.

547. Equation (496) has been proved to be true only for a system consisting of discrete material particles. At the same time the equation itself contains, in its form, no reference to the existence of discrete particles. It is at least possible that the equation may be the expression of a general dynamical principle which is true for all systems whether they consist of discrete particles or not. We cannot of course know whether or not this is so. What we have to do in the present chapter is to examine whether the phenomena of electric currents are in accordance with this equation. We shall find that they are, but we shall of course have no right to deduce from this fact that the ultimate mechanism of electric currents is to be found in the motion of discrete particles. Before setting to work on this problem, however, we shall express equation (496) in a different form.

Lagrange's Equations for Conservative Systems of Forces.

548. Let $\theta_1, \theta_2, \dots \theta_n$ be a set of quantities associated with a mechanical system such that when their value is known, the configuration of the system is fully determined. Then $\theta_1, \theta_2, \dots \theta_n$ are known as the *generalised coordinates* of the system.

The velocity of any moving particle of the system will depend on the values of $\dfrac{d\theta_1}{dt}, \dfrac{d\theta_2}{dt}$, etc. Let us denote these quantities by $\dot{\theta}_1, \dot{\theta}_2$, etc. Let x be a Cartesian coordinate of any moving particle. Then by hypothesis x is a function of $\theta_1, \theta_2, \dots$, say

$$x = f(\theta_1, \theta_2, \dots),$$

so that by differentiation,

$$\frac{dx}{dt} = \frac{\partial f}{\partial \theta_1} \dot{\theta}_1 + \frac{\partial f}{\partial \theta_2} \dot{\theta}_2 + \dots.$$

Thus each component of velocity of each moving particle will be a linear function of $\dot{\theta}_1$, $\dot{\theta}_2$, ..., from which it follows that the kinetic energy of motion of the system must be a quadratic function of $\dot{\theta}_1$, $\dot{\theta}_2$, ..., the coefficients in this function being of course functions of θ_1, θ_2, ...

Let us denote $T - W$ by L, so that L is a function of θ_1, θ_2, ... θ_n, and of $\dot{\theta}_1$, $\dot{\theta}_2$, ... $\dot{\theta}_n$, say

$$L = \phi\,(\theta_1,\ \theta_2,\ ...\ \theta_n,\ \dot{\theta}_1,\ \dot{\theta}_2,\ ...\ \dot{\theta}_n).$$

If $L + \delta L$ is the value of L in the displaced configuration $\theta_1 + \delta\theta_1$, $\theta_2 + \delta\theta_2$, ... $\theta_n + \delta\theta_n$, we have

$$\delta L = \frac{\partial \phi}{\partial \theta_1}\,\delta\theta_1 + ... + \frac{\partial \phi}{\partial \theta_n}\,\delta\theta_n + \frac{\partial \phi}{\partial \dot{\theta}_1}\,\delta\dot{\theta}_1 + ...,$$

so that equation (496), which may be put in the form

$$\int_0^\tau \delta L\,dt = 0,$$

now assumes the form

$$\int_0^\tau \left(\sum_1^n \frac{\partial L}{\partial \theta_1}\,\delta\theta_1 + \sum_1^n \frac{\partial L}{\partial \dot{\theta}_1}\,\delta\dot{\theta}_1 \right) dt = 0 \quad ....(498).$$

We have
$$\delta\dot{\theta}_1 = (\dot{\theta}_1 + \delta\dot{\theta}_1) - \dot{\theta}_1$$
$$= \frac{d}{dt}\,(\theta_1 + \delta\theta_1) - \frac{d\theta_1}{dt}$$
$$= \frac{d}{dt}\,(\delta\theta_1),$$

so that
$$\int_0^\tau \frac{\partial L}{\partial \dot{\theta}_1}\,\delta\dot{\theta}_1\,dt = \int_0^\tau \frac{\partial L}{\partial \dot{\theta}_1}\,\frac{d}{dt}\,(\delta\theta_1)\,dt$$
$$= -\int_0^\tau \frac{d}{dt}\left(\frac{\partial L}{\partial \dot{\theta}_1}\right)\delta\theta_1\,dt + \left[\frac{\partial L}{\partial \dot{\theta}_1}\,\delta\theta_1\right]_0^\tau.$$

The last term vanishes since, by hypothesis, $\delta\theta_1$ vanishes at the beginning and end of the motion, and equation (498) now assumes the form

$$\int_0^\tau \sum_1^n \left\{ \frac{\partial L}{\partial \theta_1} - \frac{d}{dt}\left(\frac{\partial L}{\partial \dot{\theta}_1}\right) \right\}\,\delta\theta_1\,dt = 0.$$

Let us denote the integrand, namely

$$\sum_1^n \left\{ \frac{\partial L}{\partial \theta_1} - \frac{d}{dt}\left(\frac{\partial L}{\partial \dot{\theta}_1}\right) \right\}\,\delta\theta_1$$

by I, so that the equation becomes

$$\int_0^\tau I\,dt = 0.$$

The varied motion is entirely at our disposal, except that it must be continuous and must be such that the configurations in the varied motion coincide with those in the actual motion at the instants $t = 0$ and $t = \tau$. Thus the values of $\delta\theta_1$, $\delta\theta_2$, ... at every instant may be any we please which are permitted by the mechanism of the system, except that they must be continuous functions of t and must vanish when $t = 0$ and when $t = \tau$. Whatever series of values we assign to $\delta\theta_1$, $\delta\theta_2$, ..., we have seen that the equation

$$\int_0^\tau I \, dt = 0$$

is true. Hence the value of I must vanish at every instant, and we must have

$$\sum_1^n \left\{ \frac{\partial L}{\partial \theta_1} - \frac{d}{dt} \left(\frac{\partial L}{\partial \dot{\theta}_1} \right) \right\} \delta\theta_1 = 0 \dots\dots\dots\dots\dots(499).$$

549. At this stage there are two alternatives to be considered. It may be that whatever values are assigned to $\delta\theta_1$, $\delta\theta_2$, ... $\delta\theta_n$, the new configuration $\theta_1 + \delta\theta_1$, $\theta_2 + \delta\theta_2$, ... $\theta_n + \delta\theta_n$ will be a possible configuration—that is to say, will be one in which the system can be placed without violating the constraints imposed by the mechanism of the system. In this case equation (499) must be true for all values of $\delta\theta_1$, $\delta\theta_2$, ... $\delta\theta_n$, so that each term must vanish separately, and we have the system of equations

$$\frac{\partial L}{\partial \theta_s} - \frac{d}{dt} \left(\frac{\partial L}{\partial \dot{\theta}_s} \right) = 0, \qquad (s = 1, 2, \dots n) \dots\dots\dots\dots(500).$$

There are n equations between the n variables θ_1, θ_2, ... θ_n and the time. Hence these equations enable us to trace the changes in θ_1, θ_2, ... θ_n and to express their values as functions of the time and of the initial values of θ_1, θ_2, ... θ_n, $\dot{\theta}_1$, $\dot{\theta}_2$, ... $\dot{\theta}_n$.

550. Next, suppose that certain constraints are imposed on the values of θ_1, θ_2, ... θ_n by the mechanism of the system. Let these be m in number, and let them be such that the small increments $\delta\theta_1$, $\delta\theta_2$, ... $\delta\theta_n$ are connected by equations of the form

$$a_1 \delta\theta_1 + a_2 \delta\theta_2 + \dots + a_n \delta\theta_n = 0 \quad \dots\dots\dots\dots(501),$$
$$b_1 \delta\theta_1 + b_2 \delta\theta_2 + \dots + b_n \delta\theta_n = 0 \quad \dots\dots\dots\dots(502),$$
etc.

Then equation (499) must be true for all values of $\delta\theta_1$, $\delta\theta_2$, ... which are such as also to satisfy equations (501), (502), etc. Let us multiply equations (501), (502), ... by λ, μ, ... and add to equation (499).

We obtain an equation of the form

$$\sum_1^n \left\{ \frac{\partial L}{\partial \theta_1} - \frac{d}{dt} \left(\frac{\partial L}{\partial \dot{\theta}_1} \right) + \lambda a_1 + \mu b_1 + \dots \right\} \delta\theta_1 = 0 \dots\dots\dots\dots(503).$$

Let us assign arbitrary values to $\delta\theta_{m+1}$, $\delta\theta_{m+2}$, ... $\delta\theta_n$, and then assign to the m quantities $\delta\theta_1$, $\delta\theta_2$, ... $\delta\theta_m$ the values given by the m equations (501), (502), etc. In this way we obtain a system of values for $\delta\theta_1$, $\delta\theta_2$, ... $\delta\theta_n$ which is permitted by the constraints of the system.

The m multipliers λ, μ, ... are at our disposal: let these be supposed to be chosen so that the m equations

$$\frac{\partial L}{\partial\theta_s} - \frac{d}{dt}\left(\frac{\partial L}{\partial\dot{\theta}_s}\right) + \lambda a_s + \mu b_s + ... = 0, \qquad (s = 1, 2, ...\ m)......(504)$$

are satisfied. Then equation (503) reduces to

$$\sum_{m+1}^{n}\left\{\frac{\partial L}{\partial\theta_s} - \frac{d}{dt}\left(\frac{\partial L}{\partial\dot{\theta}_s}\right) + \lambda a_s + \mu b_s + ...\right\}\delta\theta_s = 0 \quad ....(505),$$

and since arbitrary values have been assigned to $\delta\theta_{m+1}$, ... $\delta\theta_n$, it follows that each coefficient in this equation must vanish separately. Combining the system of equations so obtained with equations (504), we obtain the complete system of equations

$$\frac{\partial L}{\partial\theta_s} - \frac{d}{dt}\left(\frac{\partial L}{\partial\dot{\theta}_s}\right) + \lambda a_s + \mu b_s + ... = 0, \qquad (s = 1, 2, ...\ n)......(506).$$

Lagrange's Equations for General (including Non-conservative) Forces.

551. If the system of forces is not a conservative system, we cannot replace the expression

$$\Sigma\ (X_1\delta x_1 + Y_1\delta y_1 + Z_1\delta z_1)$$

in § 545 by $-\delta W$ where W is the potential energy. We may, however, still denote this expression for brevity by $-\{\delta W\}$, no interpretation being assigned to this symbol, and equation (496) will assume the form

$$\int_0^\tau (\delta T - \{\delta W\})\,dt = 0 \quad ....(507).$$

By the transformation used in § 548, we may replace $\displaystyle\int_0^\tau \delta T\,dt$ by

$$\int_0^\tau \sum_1^n \left\{\frac{\partial T}{\partial\theta_1} - \frac{d}{dt}\left(\frac{\partial T}{\partial\dot{\theta}_1}\right)\right\}\delta\theta_1\,dt.$$

Now $-\{\delta W\}$ is, by definition, the work done in moving the system from the configuration θ_1, θ_2, ... θ_n to the configuration $\theta_1 + \delta\theta_1$, $\theta_2 + \delta\theta_2$, ... $\theta_n + \delta\theta_n$. It is therefore a linear function of $\delta\theta_1$, $\delta\theta_2$, ... $\delta\theta_n$, and we may write

$$-\{\delta W\} = \Theta_1\delta\theta_1 + \Theta_2\delta\theta_2 + ... + \Theta_n\delta\theta_n,$$

where Θ_1, Θ_2, ... Θ_n are functions of θ_1, θ_2, ... θ_n.

We now have equation (507) in the form

$$\int_0^\tau \sum_1^n \left\{ \frac{\partial T}{\partial \theta_1} - \frac{d}{dt}\left(\frac{\partial T}{\partial \dot\theta_1}\right) + \Theta_1 \right\} \delta\theta_1 \, dt = 0.$$

As before each integrand must vanish. We have therefore at every instant

$$\sum_1^n \left\{ \frac{\partial T}{\partial \theta_1} - \frac{d}{dt}\left(\frac{\partial T}{\partial \dot\theta_1}\right) + \Theta_1 \right\} \delta\theta_1 = 0.$$

If the coordinates θ_1, θ_2, ... θ_n are all capable of independent variation, this leads at once to the system of equations

$$\frac{d}{dt}\left(\frac{\partial T}{\partial \dot\theta_s}\right) - \frac{\partial T}{\partial \theta_s} = \Theta_s, \qquad (s = 1, 2, \ldots n)\ldots\ldots\ldots\ldots(508),$$

while if the variations in θ_1, θ_2, ... are connected by the constraints implied in equations (501), (502), ... we obtain, as before, the system of equations

$$\frac{d}{dt}\left(\frac{\partial T}{\partial \dot\theta_s}\right) - \frac{\partial T}{\partial \theta_s} = \Theta_s + \lambda a_s + \mu b_s + \ldots, \qquad (s = 1, 2, \ldots n) \ldots(509).$$

The quantities Θ_1, Θ_2, ... are called the "generalised forces" corresponding to the coordinates θ_1, θ_2, ...

Lagrange's Equations for Impulsive Forces.

552. Let us now suppose that the system is acted on by a series of impulsive forces, these lasting through the infinitesimal interval from $t = 0$ to $t = \tau$. If we multiply equations (508) by dt and integrate throughout this interval we obtain

$$\left[\frac{\partial T}{\partial \dot\theta_s}\right]_{t=0}^{t=\tau} - \int_0^\tau \frac{\partial T}{\partial \theta_s}\, dt = \int_0^\tau \Theta_s \, dt.$$

The interval τ is to be considered as infinitesimal, and $\frac{\partial T}{\partial \theta_s}$ is finite. Thus the second term may be neglected and the equation becomes

$$\text{change in } \frac{\partial T}{\partial \dot\theta_s} = \int_0^\tau \Theta_s \, dt \ldots\ldots\ldots\ldots\ldots\ldots(510).$$

We call $\int_0^\tau \Theta_s \, dt$ the generalised impulse corresponding to the generalised force Θ_s, and then, from the analogy between equation (510) and the equation

$$\text{change in momentum} = \text{impulse},$$

we call $\frac{\partial T}{\partial \dot\theta_s}$ the generalised momentum corresponding to the generalised coordinate θ_s.

APPLICATION TO ELECTROMAGNETIC PHENOMENA.

553. We have already obtained expressions for the energy of an electrostatic system, a system of magnets, of currents, etc., and in every case this energy can be expressed in terms of coordinates associated with "accessible" parts of the mechanism. We can also find the work done in any small change in the system, so that we can obtain the values of the quantities denoted in the last section by Θ_1, Θ_2, ... All that remains to be done before we can apply Lagrange's equations provisionally (cf. § 547) to the interpretation of electromagnetic phenomena is to determine whether the different kinds of energy are to be regarded as kinetic energy or potential energy.

Kinetic and Potential Energy.

554. At first sight it might be thought obvious that the energy of electric charges at rest and of magnets at rest ought to be treated as potential energy, while that of electric charges or magnets in motion ought to be treated as kinetic. On this view the energy of a steady electric current, being the energy of a series of charges in motion, ought to be regarded as kinetic energy. We have also seen that this energy is to be regarded as being spread throughout the medium surrounding the circuit in which the current flows, and not as concentrated in the circuit itself. Thus we must regard the medium as possessing kinetic energy at every point, the amount of this energy being, as we have seen, $\dfrac{\mu H^2}{8\pi}$ per unit volume.

But we have also been led to suppose that the medium is in just the same condition whether the magnetic force is produced by steady currents or by magnetic shells at rest. Thus, on the simple view which we are now considering, we are driven to treat the energy of magnets at rest as kinetic—a result which is inconsistent with the simple conceptions from which we started. Having arrived at this contradictory result, there is no justification left for treating electrostatic energy, any more than magnetostatic energy, as potential rather than kinetic.

555. Abandoning this simple but unsatisfactory hypothesis, let us turn our attention in the first place to the definite discussion of the nature of the energy of a steady electric current.

Let us suppose that we have two currents i, i' flowing in small circuits at a distance r apart. As a matter of experiment we know that these circuits exert mechanical forces upon one another as if they were magnetic shells of strengths i, i'. Let us suppose that a force R is required to keep them apart, so that initially the circuits attracted one another with a force R, but are now in equilibrium under the action of their mutual attraction and this force R acting in the direction of r increasing.

If M is the quantity $\iint \dfrac{\cos \epsilon}{r} \, ds \, ds'$, we know that the value of R is

$$R = -ii' \frac{\partial M}{\partial r} \quad \dotfill (511),$$

this value being found directly from the experimental fact that the circuits attract like their equivalent magnetic shell (cf. § 499).

The energy of the two currents is known to be

$$E = \tfrac{1}{2}(Li^2 + 2Mii' + Ni'^2) \quad \dotfill (512).$$

Let us suppose, for the sake of generality, that this consists of kinetic energy T and potential energy W. Then, assuming for the moment that the mechanism of these currents is dynamical, in the sense that Lagrange's equations may be applied, we shall have a dynamical system of energy $T + W$, and one of the coordinates may be taken to be r, the distance apart of the circuits.

The Lagrangian equation corresponding to the coordinate r is found to be (cf. equation (508)),

$$\frac{d}{dt}\left(\frac{\partial T}{\partial \dot{r}}\right) - \frac{\partial(T - W)}{\partial r} = R \quad \dotfill (513),$$

and since we know that, in the equilibrium configuration,

$$\frac{d}{dt}\left(\frac{\partial T}{\partial \dot{r}}\right) = 0, \qquad R = -ii' \frac{\partial M}{\partial r},$$

we obtain on substitution in equation (513),

$$\frac{\partial(T - W)}{\partial r} = ii' \frac{\partial M}{\partial r}.$$

From equation (512) we see that the right-hand member is the value of $\dfrac{\partial E}{\partial r}$, or of $\dfrac{\partial(T + W)}{\partial r}$. Hence our equation shews that $\dfrac{\partial W}{\partial r} = 0$, from which we deduce that $W = 0$. In other words, assuming that a system of steady currents forms a dynamical system, the energy of this system must be wholly kinetic.

This result compels us also to accept that the energy of a system of magnets at rest must also be wholly kinetic. We shall discuss this result later. For the present we confine our attention to the case of electric phenomena only. We have found that if the mechanism of these phenomena is dynamical (the hypothesis upon which we are going to work), then the energy of electric currents must be kinetic.

Induction of Currents.

556. Let us consider a number of currents flowing in closed circuits. Let the strengths of the currents be i_1, i_2, ... and let the number of tubes of induction which cross these circuits at any instant be N_1, N_2, ..., so that if the magnetic field arises entirely from the currents, we have (cf. § 502)

$$\left. \begin{array}{l} N_1 = L_{11} i_1 + L_{12} i_2 + \ldots \\ N_2 = L_{21} i_1 + L_{22} i_2 + \ldots, \text{ etc.} \end{array} \right\} \ldots\ldots\ldots\ldots\ldots(514).$$

The energy of the currents is wholly kinetic so that we may take

$$T = \tfrac{1}{2} (N_1 i_1 + N_2 i_2 + \ldots)$$
$$= \tfrac{1}{2} (L_{11} i_1{}^2 + 2 L_{12} i_1 i_2 + \ldots)$$

as before (§ 503).

In the general dynamical problem, it will be remembered that T was a quadratic function of the velocities. Thus i_1, i_2, ... must now be treated as velocities and we must take as coordinates quantities x_1, x_2, ..., defined by

$$i_1 = \frac{dx_1}{dt}, \qquad i_2 = \frac{dx_2}{dt}, \text{ etc.}$$

Clearly x_1 measures the quantity of electricity which has flowed past any point in circuit 1 since a given instant, and so on. Thus in terms of the coordinates x_1, x_2, ... we have

$$T = \tfrac{1}{2} (L_{11} \dot{x}_1{}^2 + 2 L_{12} \dot{x}_1 \dot{x}_2 + \ldots)\ldots\ldots\ldots\ldots\ldots(515).$$

There is no potential energy in the present system, but the system is acted on by external forces, namely the electromotive forces in the batteries and the reaction between the currents and the material of the circuits which shews itself in the resistance of the circuits. We have therefore to evaluate the generalised forces Θ_1, Θ_2, ...

Consider a small change in the system in which x_1 is increased by δx_1, so that the current i_1 flows for a time dt given by $i_1 dt = \delta x_1$. The work performed by the battery is $E_1 \delta x_1$, the work performed by the reaction with the matter of the circuit, being equal and opposite to the heat generated in the circuit, is $- R_1 i_1{}^2 dt$. Thus if X_1 is the generalised force corresponding to the coordinate x_1, we have

$$X_1 \delta x_1 = E_1 \delta x_1 - R_1 i_1{}^2 dt,$$

so that

$$X_1 = E_1 - R_1 i_1.$$

The Lagrangian equation corresponding to the coordinate x_1 is

$$\frac{\partial}{\partial t} \left(\frac{\partial T}{\partial i_1} \right) - \frac{\partial T}{\partial x_1} = X_1,$$

or

$$\frac{\partial}{\partial t} (L_{11} i_1 + L_{12} i_2 + \ldots) = E_1 - R_1 i_1 \ldots\ldots\ldots\ldots\ldots(516),$$

or again

$$E_1 - \frac{\partial N_1}{\partial t} = \dot{R}_1 i_1.$$

The equations corresponding to the coordinates x_2, x_3, ... are

$$E_2 - \frac{\partial N_2}{\partial t} = R_2 i_2, \text{ etc.}$$

Thus the Lagrangian equations are found to be exactly identical with the equations of current-induction already obtained, shewing not only that the phenomenon of induction is consistent with the hypothesis that the whole mechanism is a dynamical system, but also that this phenomenon follows as a direct consequence of this hypothesis. In this system the accessible parts of the mechanism are the currents flowing in the wires; the inaccessible parts consist of the ether which transmits the action from one circuit to another.

556 a. On the electron theory, the kinetic energy must be supposed made up partly of magnetic energy, as before, and partly of the kinetic energy of the motion of the electrons by which the current is produced.

Let the average forward velocity of the electrons at any point be v_0 (cf. § 345 a), and let $v + v_0$ be the actual velocity of any single electron, so that the average value of v is *nil*. The kinetic energy of motion of the electrons, say T_e, is then

$$T_e = \Sigma \tfrac{1}{2} m (v + v_0)^2$$
$$= \Sigma \tfrac{1}{2} m v^2 + \tfrac{1}{2} N m v_0^2.$$

The first term represents part of the heat-energy of the matter, and this does not depend on the values of the currents $\dot{x}_1$, $\dot{x}_2$, ... To evaluate the second term we use equation (b) of § 345 a,

$$N e v_0 = i = \dot{x},$$

and obtain the kinetic energy of the electrons in the complete system of currents in the form

$$T_e = \tfrac{1}{2} \left(\dot{x}_1^2 \int \frac{m}{N e^2} ds + \dot{x}_2^2 \int \frac{m}{N e^2} ds + ... \right).$$

Thus the total kinetic energy may still be expressed in the form (515) if we take

$$L_{11} = L'_{11} + \int \frac{m}{N e^2} ds, \text{ etc.} \dots\dots\dots\dots\dots\dots(517),$$

and in this the first term is the contribution from the magnetic energy (cf. § 503), and the second term is the contribution from the kinetic energy of the electrons.

Equation (516) assumes the form

$$\frac{\partial}{\partial t} (L'_{11} i_1 + L_{12} i_2 + ...) = E_1 - R_1 i_1 - \left(\int \frac{m}{N e^2} ds \right) \frac{d i_1}{dt} \dots\dots(517\, a).$$

If the induction terms on the left are omitted, we have as the equation of a circuit in which induction is negligible

$$E_1 - R_1 i_1 - \left(\int \frac{m}{Ne^2} \, ds \right) \frac{di_1}{dt} = 0.$$

This, with the help of the formulae of § 345 a, may be expressed in the form

$$\int X \, ds - i_1 \int \frac{\gamma}{Ne^2} \, ds - \frac{di_1}{dt} \int \frac{m}{Ne^2} \, ds = 0,$$

which in turn is seen to be exactly identical with equation (c) of § 345 a, integrated round the circuit.

Thus we see that the analysis of § 556 applies perfectly to the electron theory of matter, provided L_{11}, L_{22}, ... are supposed to have the values given by equation (517), and equation (517 a) is then the general equation of induction of currents, when the inertia of the electrons is taken into account.

Electrokinetic Momentum.

557. The generalised momentum corresponding to the coordinate x_1 is $\frac{\partial T}{\partial \dot{x}_1}$ or N_1. Thus the generalised momenta corresponding to the currents in the different circuits are N_1, N_2, ..., the numbers of tubes of induction which cross the circuits. The quantity N_1 is accordingly sometimes called the electrokinetic momentum of circuit 1, and so on.

If we give to L_{11} the value obtained in equation (517) of § 556 a, the value of the electrokinetic momentum is (cf. equations (514))

$$(L'_{11} i_1 + L_{12} i_2 + \dots) + i_1 \int \frac{m}{Ne^2} \, ds,$$

in which clearly the last term comes from the momentum of the electrons, and the remaining terms from the momentum of the magnetic field.

EXAMPLES.

I. *Discharge of a Condenser.*

558. As a further illustration of the dynamical theory, let us consider the discharge of a condenser. Let Q be the charge on the positive plate at any instant, and let this be taken as a Lagrangian coordinate. The current i is given by $i = -\frac{\partial Q}{\partial t} = -\dot{Q}$. In the notation already employed (§ 516) we have

$$T = \tfrac{1}{2} L i^2 = \tfrac{1}{2} L \dot{Q}^2, \quad W = \tfrac{1}{2} \frac{Q^2}{C},$$

and Lagrange's equation is

$$\frac{d}{dt}\left(\frac{\partial T}{\partial \dot{Q}}\right) - \frac{\partial T}{\partial Q} + \frac{\partial W}{\partial Q} = Ri,$$

or

$$L\frac{\partial^2 Q}{\partial t^2} + R\frac{\partial Q}{\partial t} + \frac{Q}{C} = 0,$$

which is the equation already obtained in § 516, and leads to the solution already found.

II. *Oscillations in a network of conductors*

559. The equations governing the currents flowing in any network of conductors when induction is taken into account can be obtained from the general dynamical theory.

Let us suppose that the currents in the different conductors are $i_1, i_2, \ldots i_n$, and let the corresponding coordinates be $x_1, x_2, \ldots x_n$, these being given by $i_1 = \frac{dx_1}{dt}$, etc. If any conductor, say 1, terminates on a condenser plate, let x_1 denote the actual charge on the plate, and let the current be measured towards the plate, so that the relations $i_1 = \frac{dx_1}{dt}$, etc. will still hold. Let conductor 1 contain an electromotive force E_1 and be of resistance R_1.

The quantities $x_1, x_2, \ldots$ may be taken as Lagrangian coordinates, but they are not, in general, independent coordinates. If any number of the conductors, say 2, 3, $\ldots$ s meet in a point, the condition for no accumulation of electricity at the point is, by Kirchhoff's first law,

$$i_2 \pm i_3 \pm \ldots \pm i_s = 0,$$

from which we find that variations in $x_2, x_3, \ldots$ are connected by the relations

$$\delta x_2 \pm \delta x_3 \pm \ldots \pm \delta x_s = 0.$$

Let us suppose that there are m junctions. The corresponding constraints on the values of $\delta x_1, \delta x_2, \ldots$ can be expressed by m equations of the form

$$\left. \begin{array}{l} a_1\delta x_1 + a_2\delta x_2 + \ldots + a_n\delta x_n = 0 \\ b_1\delta x_1 + b_2\delta x_2 + \ldots + b_n\delta x_n = 0 \end{array} \right\} \quad \ldots\ldots\ldots\ldots\ldots(518),$$

etc., in which each of the coefficients $a_1, a_2, \ldots a_n, b_1, \ldots$ has for its value either 0, $+1$ or -1.

The kinetic energy T will be a quadratic function of $\dot{x}_1, \dot{x}_2$, etc., while the potential energy W (arising from the charges, if any, on the condensers) will be a quadratic function of x_1, x_2, ... The dynamical equations are now n in number, these being of the form (cf. equations (509))

$$\frac{d}{dt}\left(\frac{\partial T}{\partial \dot{x}_s}\right) - \frac{\partial T}{\partial x_s} + \frac{\partial W}{\partial x_s} = E_s - R_s \dot{i}_s + \lambda a_s + \mu b_s + \ldots \quad (s = 1, 2, \ldots n) \ldots (519).$$

These equations, together with the m equations obtained by applying Kirchhoff's first law to the different junctions, form a system of $m + n$ equations, from which we can eliminate the m multipliers λ, μ, ..., and then determine the n variables x_1, x_2, ... x_n.

560. As an example of the use of these equations, let us imagine that a current I arrives at A and divides into two parts i_1, i_2, which flow along arms

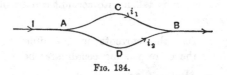

FIG. 134.

ACB, ADB and reunite at B. Neglecting induction between these arms and the leads to A and B, we may suppose that the part of the kinetic energy which involves i_1 and i_2 is

$$\tfrac{1}{2}L i_1^2 + M i_1 i_2 + \tfrac{1}{2} N i_2^2.$$

There are no batteries and no condenser in the arms in which the currents i_1 and i_2 flow. The currents are, however, connected by the relation

$$i_1 + i_2 = I$$

so that the corresponding coordinates x_1 and x_2 are connected by

$$\delta x_1 + \delta x_2 = 0.$$

The dynamical equations are now found to be (cf. equations (519))

$$\frac{d}{dt}\left(L i_1 + M i_2\right) = -R i_1 + \lambda,$$

$$\frac{d}{dt}\left(M i_1 + N i_2\right) = -S i_2 + \lambda.$$

If we subtract and replace i_2 by $I - i_1$, we eliminate λ and obtain

$$(L + N - 2M)\frac{d i_1}{dt} + (M - N)\frac{dI}{dt} = SI - (R + S)i_1.$$

If I is given as a function of the time, this equation enables us to determine i_1, and thence i_2.

For instance, suppose that the current I is an alternating current of frequency $p/2\pi$. If we put $I = i_0 e^{ipt}$, the solution of the equation is

$$i_1 = \frac{S - (M - N)ip}{(L + N - 2M)ip + (R + S)} I,$$

while similarly

$$i_2 = \frac{R - (M - L)ip}{(L + N - 2M)ip + (R + S)} I.$$

When $p = 0$, the solution of course reduces to that for steady currents. As p increases, we notice that the three currents i_1, i_2 and I become, in general, in different phases, and that their amplitudes assume values which depend upon the coefficients of induction as well as on the resistances. Finally, for very great values of p, the values of i_1 and i_2 are given by

$$\frac{i_1}{N - M} = \frac{i_2}{L - M} = \frac{I}{L + N - 2M},$$

shewing that the currents are now in the same phase and are divided in a ratio which depends only on their coefficients of induction. For instance, if the arms ACB, ADB are arranged so as to have very little mutual induction (M very small), the current will distribute itself between the two arms in the inverse ratio of the coefficients of self-induction.

It is possible to arrange for values for L, M and N such that the two currents i_1 and i_2 shall be of opposite sign. In such a case the current in one at least of the branches is greater than that in the main circuit. Let us, for instance, suppose that the branches consist of two coils having r and s turns respectively, arranged so as to have very little magnetic leakage. Then $LN - M^2$ is negligible (cf. § 525) and we have approximately

$$\frac{L}{r^2} = \frac{M}{rs} = \frac{N}{s^2}.$$

The equations become

$$\frac{i_1}{s} = \frac{i_2}{-r} = \frac{I}{s - r},$$

so that the currents will flow in opposite directions, and either may be greater than the current in the main circuit. By making s nearly equal to r and keeping the magnetic leakage as small as possible, we can make both currents large compared with the original current.

III. *Rapidly alternating currents.*

561. This last problem illustrates an important point in the general theory of rapidly alternating currents. In the general equations (519),

$$\frac{d}{dt}\left(\frac{\partial T}{\partial \dot{x}_s}\right) - \frac{\partial T}{\partial x_s} + \frac{\partial W}{\partial x_s} = E_s - R_s i_s + \lambda a_s + \mu b_s + \dots,$$

let us suppose that the whole system is oscillating with frequency $p/2\pi$, which is so great that it may be treated as infinite. We may assume that every variable is proportional to e^{ipt}, and may accordingly replace $\frac{d}{dt}$ by the multiplier ip. The equations now become

$$ip\left(\frac{\partial T}{\partial \dot{x}_s}\right) - \frac{\partial T}{\partial x_s} + \frac{\partial W}{\partial x_s} - E_s + R_s i_s = \lambda a_s + \mu b_s + \ldots,$$

and all the terms on the left hand may be neglected in comparison with the first, which contains the factor ip. The terms on the right cannot legitimately be neglected because λ, μ, ... are entirely undetermined, and may be of the same large order of magnitude as the terms retained. If we replace λ, μ, ... by $ip\lambda'$, $ip\mu'$, ..., the equations become

$$\frac{\partial T}{\partial \dot{x}_s} + \lambda' a_s + \mu' b_s + \ldots = 0, \text{ etc.}$$

in which λ', μ', ... are now undetermined multipliers. These, however, are exactly the equations which express that T is a maximum or a minimum for values of $\dot{x}_1$, $\dot{x}_2$, ... which are consistent with the relations (cf. § 559) necessary to satisfy Kirchhoff's first law. Since T can be made as large as we please, the solution must clearly make T a minimum. Thus we see that

As the frequency of a system of alternating currents becomes very great, the currents tend to distribute themselves in such a way as to make the kinetic energy of the currents a minimum subject only to the relations imposed by Kirchhoff's first law.

This result may be compared with that previously obtained (§ 357) for steady currents. We see that while the distribution of steady currents is determined entirely by the resistance of the conductors, that of rapidly alternating currents is, in the limit in which the frequency is infinite, determined entirely by the coefficients of induction.

It follows that, in a continuous medium of any kind, the distribution of rapidly alternating currents will depend only on the geometrical relations of the medium, and not on its conducting properties. In point of fact, we have already seen that the current tends to flow entirely in the surface of the conductor (§ 537). We now obtain the further result that it will, in the limit, distribute itself in the same way over the surface of this conductor, no matter in what way the specific resistance varies from point to point of the surface.

IV. *Transmission of Signals along a wire.*

562. Imagine a signal being sent along a wire, initially free from all electrical disturbance. At any instant let i denote the current at a point distant x from the end of the wire, and let q denote the total quantity of electricity which has flowed past this point. Then i and q are functions of x and t.

Let q be measured in electrostatic units, but let i be measured in electro-magnetic units. Then the rate of flow past any point will be iC electrostatic units per second, where C denotes the number of electrostatic units in one electromagnetic unit (cf. §484). Thus

$$Ci = \frac{\partial q}{\partial t}.$$

If L is the self-induction of the wire per unit length, the total kinetic energy of the currents is

$$T = \tfrac{1}{2}L \int i^2\,dx = \frac{L}{2C^2}\int \dot{q}^2 dx,$$

where the integral is taken along the wire. In any element dx of the wire the charge is $-\frac{\partial q}{\partial x}\,dx$, so that if K is the electrostatic capacity of the wire per unit length, the potential energy W is given by

$$W = \frac{1}{2K}\int \left(\frac{\partial q}{\partial x}\right)^2 dx.$$

Let R be the resistance of the wire per unit length in electromagnetic units, then the rate of generation of heat is

$$R\int i^2 dx.$$

The values of q at different points of the wire may be taken as Lagrangian coordinates, for they suffice to specify the position of each element of current. The Lagrangian equation corresponding to the coordinate q at any distance x will be (cf. §556)

$$\frac{d}{dt}\left(\frac{\partial T}{\partial \dot{q}}\right) - \frac{\partial T}{\partial q} + \frac{\partial W}{\partial q} = -Ri,$$

in which we have

$$\frac{\partial T}{\partial \dot{q}} = \frac{L}{C^2}\dot{q} \quad \text{and} \quad \frac{\partial T}{\partial q} = 0.$$

To evaluate $\frac{\partial W}{\partial q}$, let us imagine q changed to $q + \delta q$ at every point of the wire, subject to δq vanishing at the two ends. The increment in W, say δW, is given by

$$\delta W = \frac{1}{2K}\int\left[\left(\frac{\partial(q+\delta q)}{\partial x}\right)^2 - \left(\frac{\partial q}{\partial x}\right)^2\right]dx = \frac{1}{K}\int\frac{\partial q}{\partial x}\frac{\partial}{\partial x}(\delta q)\,dx,$$

and, on integrating by parts, this becomes

$$\delta W = -\frac{1}{K}\int\frac{\partial^2 q}{\partial x^2}\,\delta q\,dx.$$

Thus at any point x,

$$\frac{\partial W}{\partial q} = -\frac{1}{K}\frac{\partial^2 q}{\partial x^2}.$$

The Lagrangian equation accordingly becomes

$$\frac{L}{C^2}\frac{\partial^2 q}{\partial t^2} - \frac{1}{K}\frac{\partial^2 q}{\partial x^2} = -R\frac{\partial q}{\partial t}.$$

Since $Ci = \frac{\partial q}{\partial t}$, it is at once seen that the current i at any point satisfies the same differential equation, and this is also true of the potential V, since $\frac{\partial q}{\partial x} = -KV$. Thus q, i and V all satisfy the same differential equation, namely

$$\frac{KL}{C^2}\frac{\partial^2 \phi}{\partial t^2} + KR\frac{\partial \phi}{\partial t} = \frac{\partial^2 \phi}{\partial x^2} \qquad \ldots\ldots\ldots\ldots\ldots\ldots(520).$$

This equation is the general equation for the transmission of electric signals along a wire. It is called the "Telegraphic equation" by Poincaré and others.

We have seen in § 505 a, how to calculate the self-induction per unit length of any wire. If the wire is sufficiently thin in comparison with its distance from other conductors, the self-induction L per unit length becomes identical with the quantity denoted by L' in § 505, and we accordingly have the relation (cf. equation (430 e)),

$$KL = \kappa\mu,$$

where κ is the dielectric constant, and μ the magnetic permeability of the insulator surrounding the wire. Let us put

$$a^2 = \frac{C^2}{\kappa\mu}$$

so that a depends only on the properties of the insulating material, and the telegraphic equation becomes

$$\frac{1}{a^2}\frac{\partial^2 \phi}{\partial t^2} + KR\frac{\partial \phi}{\partial t} = \frac{\partial^2 \phi}{\partial x^2}.$$

For slow signals, the first term in this equation, which arises from the inertia of the electric current, may be neglected. The equation then reduces to equation (303) of Chapter IX which was obtained as the equation of transmission of signals along a submarine cable. Under practical conditions signals along a submarine cable are so retarded by the high electrostatic capacity of the cable that this inertia term may legitimately be neglected, but the term has to be retained when the equation is applied to telegraph and telephone problems.

When the wire is far removed from other conductors, the electrostatic capacity K will be small. If K is neglected entirely, the equation becomes

$$\frac{\partial^2 \phi}{\partial t^2} = a^2\frac{\partial^2 \phi}{\partial x^2}.$$

The solution of this equation is

$$\phi = f(x - at) + \Phi(x + at),$$

where f, Φ are arbitrary functions, and the solution is seen to represent the transmission of a signal without change of type or loss of intensity, the velocity of transmission being a.

In practical telephony and telegraphy it is not usually possible to neglect entirely the value of K in the second term of the equation. Solutions of the general three-term equation have been obtained by Heaviside[*], Poincaré[†], Picard[‡], Boussinesq[§], and Riemann[||].

It is found that the signal is still transmitted with the same velocity a, but that there is a change of type and loss of intensity; there is also an electric field and current left trailing behind each signal; these would of course tend to confuse the succeeding signal if the signals are sent without sufficient interval.

Thus for rapid transmission or clear speaking it is necessary to reduce the value of KR (cf. § 369); the smaller this term is made, the smaller the amount of blurring or indistinctness will be. We see at once why telephone wires are kept as far as possible from other conductors, and can understand the difficulty of clear speaking or rapid signalling through a submarine cable.

Mechanical Force acting on a Circuit.

563. Let θ be any geometrical coordinate, and let Θ be the generalised force tending to increase the coordinate θ, so that to keep the system of circuits at rest we must suppose it acted on by an external force $-\Theta$. Then Lagrange's equation for the coordinate θ is

$$\frac{d}{dt}\left(\frac{\partial T}{\partial \dot{\theta}}\right) - \frac{\partial T}{\partial \theta} = -\Theta,$$

and therefore, when the system is in equilibrium, we must have

$$\Theta = \frac{\partial T}{\partial \theta} \quad\dots\dots\dots\dots\dots\dots\dots\dots\dots\dots\dots(521).$$

If the energy of the system were wholly potential and of amount W, the force Θ would be given by

$$\Theta = -\frac{\partial W}{\partial \theta}.$$

Thus the mechanical forces acting are just the same as they would be if the system had potential energy of amount $-T$.

---

[*] *Phil. Mag.* 1888 and *Coll. Papers.*   [†] *C. R.* 117 (1893), p. 1027.
[‡] *C. R.* 118 (1894), p. 16.   [§] *C. R.* 118 (1894), p. 162.
[||] Riemann-Weber, *Die partielle Differentialgleichungen der Math. Physik*, 4th edn. (1901), II. p. 322.

564. Let us suppose that any geometrical displacement takes place, this resulting in increases $\delta\theta_1$, $\delta\theta_2$, ... in the geometrical coordinates θ_1, θ_2, ..., and let the currents in the circuits remain unaltered, additional energy being supplied by the batteries when needed.

The increase in the kinetic energy of the system of currents is

$$\Sigma \frac{\partial T}{\partial \theta}\, d\theta,$$

while the work done by the electrical forces during displacement is $\Sigma\Theta d\theta$ which, by equation (521), is also equal to

$$\Sigma \frac{\partial T}{\partial \theta}\, d\theta.$$

These two quantities would be equal and opposite if the system were a conservative dynamical system acted on by no external forces. In point of fact they are seen to be equal and of the same sign. The inference is that the batteries supply during the motion an amount of energy equal to *twice* the increase in the energy of the system. Of this supply of energy half appears as an increase in the energy of the system, while the other half is used in the performance of mechanical work.

This result should be compared with that obtained in § 120.

565. As an example of the use of formula (521), let us examine the force acting on an element of a circuit. Let the components of the mechanical force acting on any element ds of a circuit carrying a current i be denoted by X, Y, Z.

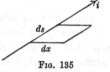

Fig. 135

To find the value of X, we have to consider a displacement in which the element ds is displaced a distance dx parallel to itself, the remainder of the circuit being left unmoved. Let the component of magnetic induction perpendicular to the plane containing ds and dx be denoted by N, then if T denotes the kinetic energy of the whole system, the increase in T caused by displacement will be equal to i times the increase in the number of tubes of induction enclosed by the circuit, and therefore

$$dT = iN\,ds\,dx.$$

Thus, using equation (521),

$$X = \frac{\partial T}{\partial x} = iN\,ds,$$

and there are similar equations giving the values of the components Y and Z.

If B is the total induction and if $B \cos \epsilon$ is the component at right angles to ds, then the resultant force acting on ds is seen to be a force of amount $iB \cos \epsilon\, ds$, acting at right angles to the plane containing B and ds, and in such a direction as to *increase* the kinetic energy of the system. This is a generalisation of the result already obtained in § 498.

MAGNETIC ENERGY.

566. We have seen that the energy of the field of force set up by a system of electric currents must be supposed to be kinetic energy. We know also that this field is identical with that set up by a certain system of magnets at rest. These two facts can be reconciled only by supposing that the energy of a system of magnets at rest is kinetic energy—a suggestion originally due to Ampère.

Weber's theory of magnetism (§ 476) has already led us to regard any magnetic body as a collection of permanently magnetised particles. Ampère imagined the magnetism of each particle to arise from an electric current which flowed permanently round a non-resisting circuit in the interior of the particle. The phenomena of magnetism, on this hypothesis, become in all respects identical with those of electric currents, and in particular the energy of a magnetic body must be interpreted as the kinetic energy of systems of electric currents circulating in the individual molecules. For instance two magnetic poles of opposite sign attract because two systems of currents flowing in opposite directions attract.

We have seen that the mechanical forces in a system of energy E are $-\dfrac{\partial E}{\partial \theta}$, etc., if the energy is potential, but are $+\dfrac{\partial E}{\partial \theta}$, etc., if the energy is kinetic. It might therefore be thought that the acceptance of the hypothesis that all magnetic energy is kinetic would compel us to suppose all mechanical forces in the magnetic system to be the exact opposites of what we have previously supposed them to be. This, however, is not so, because accepting this hypothesis compels us also to suppose the energy to be exactly opposite in amount to what we previously supposed it to be. Instead of supposing that we have potential energy E and forces $-\dfrac{\partial E}{\partial x}$, etc., we now suppose that we have kinetic energy $-E$ and forces $+\dfrac{\partial (-E)}{\partial x}$, etc., so that the amounts of the forces are unaltered.

To understand how it is that the amount of the magnetic energy must be supposed to change sign as soon as we suppose it to originate from a series of molecular currents, we need only refer back to § 501.

567. The molecular currents by which we are now supposing magnetism to be originated must be supposed to be acted on by no resistance and by no batteries, but if the assemblage of currents is to constitute a true dynamical system we must suppose them capable of being acted upon by induction whenever the number of tubes of force or induction which crosses them is changed. In the general dynamical equation

$$\frac{d}{dt}\left(\frac{\partial T}{\partial \dot{x}}\right) - \frac{\partial T}{\partial x} = E - R\dot{x},$$

we may put E and R each equal to zero, and $\dfrac{\partial T}{\partial x}$ is already known to vanish.

Thus the equation expresses that $\dfrac{\partial T}{\partial \dot{x}}$ remains unaltered.

We now see that the strengths of the molecular currents will be changed by induction in such a way that the electrokinetic momentum of each remains unaltered. If the molecule is placed in a magnetic field whose lines of force run in the same direction as those from the molecule, then induction will decrease the strength of the molecular current until the aggregate number of tubes of force which cross it is equal to the number originally crossing it. This effect of induction is of the opposite kind from that required to explain the phenomenon of induced magnetism in iron and other paramagnetic substances. It has, however, been suggested by Weber that it may account for the phenomenon of diamagnetism.

568. Modern views as to the structure of matter compel us to abandon Ampère's conception of molecular currents, but this conception can be replaced by another which is equally capable of accounting for magnetic phenomena. On the modern view all electric currents are explained as the motion of streams of electrons. The flow of Ampère's molecular current may accordingly be replaced by the motion of rings of electrons. The rotation of one or more rings of electrons would give rise to a magnetic field exactly similar to that which would be produced by the flow of a current of electricity in a circuit of no resistance.

It is on these lines that it appears probable that an explanation of magnetic phenomena will be found in the future. No complete explanation has so far been obtained, for the simple and sufficient reason that the arrangement and behaviour of the electrons in the molecule or atom is still unknown.

EXAMPLES.

1. Two wires are arranged in parallel, their resistances being R and S, and their coefficients of induction being L, M, N. Shew that for an alternating current of frequency p the pair of wires act like a single conductor of resistance **R** and self-induction **L**, given by

$$\frac{\mathbf{R}}{RS(R+S)+p^2\{R(N-M)^2+S(L-M)^2\}}$$
$$=\frac{\mathbf{L}}{NR^2+LS^2+2MRS+p^2(LN-M^2)(L+N-2M)}=\frac{1}{(R+S)^2+p^2(L+N-2M)^2}.$$

2. A conductor of considerable capacity S is discharged through a wire of self-induction L. At a series of points along the wire dividing it into n equal parts, $(n-1)$ equal conductors each of capacity S' are attached. Find an equation to determine the periods of oscillations in the wire, and shew that if the resistance of the wire may be neglected, the equation may be written

$$2\tan\tfrac{1}{2}\phi\,(S-\tfrac{1}{2}S')=S'\cot n\phi,$$

where the current varies as $e^{-i\lambda t}$, and $\sin^2\phi=S'\lambda^2L/4n$.

3. A Wheatstone bridge arrangement is used to compare the coefficient of mutual induction M of two coils with the coefficient of self-induction L of a third coil. One of the coils of the pair is placed in the battery circuit AC, the other is connected to B, D as a shunt to the galvanometer, and the third coil is placed in AD. The bridge is first balanced for steady currents, the resistances of AB, BC, CD, DA being then R_1, R_2, R_3, R_4: the resistance of the shunt is altered till there is no deflection of the galvanometer needle at make and break of the battery circuit, and the total resistance of the shunt is then R. Prove that

$$LRR_1 = M(R_1 + R_4)^2.$$

4. Two circuits each containing a condenser, having the same natural frequency when at a distance, are brought close together. Shew that, unless the mutual induction between the circuits is small, there will be in each circuit two fundamental periods of oscillation given by

$$p^2 = \frac{1}{\sqrt{C_1 C_2}} \frac{1}{(\sqrt{L_1 L_2} \pm M)},$$

where C_1, C_2 are the capacities, L_1, L_2 the coefficients of self-induction, and M the coefficient of mutual induction, of the circuits.

5. Let a network be formed of conductors A, B, ... arranged in any order. Prove that when a periodic electromotive force $F \cos pt$ is placed in A the current in B is the same in amplitude and phase as the current is in A when an electromotive force $F \cos pt$ is placed in B.

CHAPTER XVII

DISPLACEMENT CURRENTS AND ELECTROMAGNETIC WAVES

MAXWELL'S EQUATIONS.

569. OUR development of the theory of electromagnetism has been based upon the experimental fact that the work done in taking a unit magnetic pole round any closed path in the field is equal to 4π times the aggregate current enclosed by this path. But it has already been seen (§ 534) that this development of the theory is not sufficiently general to take account of phenomena in which the flow of current is not steady: "the aggregate current enclosed by a path" is an expression which has a definite meaning only when the flow of current is steady. Before proceeding to a more general theory, which is to cover all possible cases of current flow, it is necessary to determine in what way the experimental basis is to be generalised, in order to provide material for the construction of a more complete theory.

The answer to this question has been provided by Maxwell. According to Maxwell's displacement theory (§ 171), the motion of electric charges is accompanied by a "displacement" of the surrounding medium. The motion produced by this displacement will be spoken of as a "displacement-current," and we have seen that the total flow which is obtained by compounding the displacement-current with the current produced by the motion of electric charges (which will be called the *conduction-current*), will be such that the total flow into any closed surface is, under all circumstances, zero. Thus if S_1, S_2 are any two surfaces bounded by the same closed path s, the total flow of current across S_1 is the same as the total flow, in the same direction, across S_2, so that either may be taken to be the flow through the circuit s. Maxwell's theory proceeds on the supposition that in

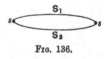

FIG. 136.

any flow of current, the work done in taking a unit magnetic pole round s is equal to 4π times the total flow of current, including the displacement-current, through s. The justification for this supposition is obtained as soon as it is seen how it brings about a complete agreement between electromagnetic theory and innumerable facts of observation.

570. Let us first put the hypothesis of the existence of displacement-currents into mathematical language. Let u, v, w be the components of the ordinary current at any point which is produced by the motion of electric charges, and let this be measured, as before, in electromagnetic units (cf. § 484). Let the components of the displacement, which has been shewn to be identical with Faraday's polarisation (§ 172), be denoted as before by f, g, h. On Maxwell's theory of displacement, f, g, h are the quantities of electricity of the second kind which have crossed unit areas perpendicular to the coordinate axes at any point. The corresponding rates of current-flow, or quantities which cross unit area per unit time, are of course

$$\frac{df}{dt}, \frac{dg}{dt}, \frac{dh}{dt}.$$

These are accordingly the components of Maxwell's "displacement-current." They are, however, measured in electrostatic units. If we suppose there to be C electrostatic units of charge in one electromagnetic unit, the displacement current, measured in electromagnetic units, will have components

$$\frac{1}{C}\frac{df}{dt}, \frac{1}{C}\frac{dg}{dt}, \frac{1}{C}\frac{dh}{dt} \quad\dots\dots\dots\dots\dots\dots(522),$$

and Maxwell's total current, measured in electromagnetic units, will have components

$$u+\frac{1}{C}\frac{df}{dt}, \quad v+\frac{1}{C}\frac{dg}{dt}, \quad w+\frac{1}{C}\frac{dh}{dt}.$$

Maxwell's hypothesis is that the work done in taking a unit magnetic pole round a closed circuit is equal to 4π times the total current flowing through that circuit. This hypothesis is, as we have seen, self-consistent, because the total current behaves like an incompressible fluid, and consequently the total flow through a circuit has a definite meaning which is independent of the particular surface we select, closing up the circuit, over which to measure the current.

The hypothesis may be transformed into mathematical language by following the procedure of § 533. It is found to be represented by the equations

$$\left.\begin{aligned} 4\pi\left(u+\frac{1}{C}\frac{df}{dt}\right) &= \frac{\partial\gamma}{\partial y}-\frac{\partial\beta}{\partial z} \\ 4\pi\left(v+\frac{1}{C}\frac{dg}{dt}\right) &= \frac{\partial\alpha}{\partial z}-\frac{\partial\gamma}{\partial x} \\ 4\pi\left(w+\frac{1}{C}\frac{dh}{dt}\right) &= \frac{\partial\beta}{\partial x}-\frac{\partial\alpha}{\partial y} \end{aligned}\right\} \quad\dots\dots\dots\dots(523).$$

These are the equations which must replace equations (473)—(475) in the most general motion of electricity. If we differentiate the three equations with respect to x, y, z and add, we obtain

$$\frac{\partial u}{\partial x}+\frac{\partial v}{\partial y}+\frac{\partial w}{\partial z} = -\frac{1}{C}\frac{d}{dt}\left(\frac{\partial f}{\partial x}+\frac{\partial g}{\partial y}+\frac{\partial h}{\partial z}\right).$$

Since, by equation (63),

$$\frac{\partial f}{\partial x} + \frac{\partial g}{\partial y} + \frac{\partial h}{\partial z} = \rho$$

this may be written in the form

$$C\left(\frac{\partial u}{\partial x} + \frac{\partial v}{\partial y} + \frac{\partial w}{\partial z}\right) = -\frac{d\rho}{dt} \quad\dots\dots\dots\dots\dots(524).$$

Now $C\left(\dfrac{\partial u}{\partial x} + \dfrac{\partial v}{\partial y} + \dfrac{\partial w}{\partial z}\right) dx\,dy\,dz$ simply expresses the rate at which currents of ordinary electricity, measured in electrostatic units, flow out of a small element of volume $dx\,dy\,dz$, and so is necessarily equal to $-\dfrac{d}{dt}(\rho\,dx\,dy\,dz)$.

We accordingly see that equation (524) is true, quite independently of the truth of Maxwell's displacement-theory. It follows that equations (523) form a consistent scheme, independently of the truth of the hypothesis from which they have been derived. The displacement-theory may be regarded merely as scaffolding, and Maxwell's theory may be regarded as being simply the theory expressed by equations (523), independently of any physical interpretation that may be assigned to the various terms in these equations. Although we may, if we please, discard Maxwell's interpretation, it will be convenient to continue to use the name "displacement-current" to designate the vector whose components are given by formula (522).

We proceed to examine the consequences implied in Maxwell's equations (523). Since the truth of the equations must ultimately rest on something more substantial than the displacement-theory by the help of which they were derived, it is important to seize every opportunity of comparing the results of the theory with observation.

MAXWELL'S EQUATIONS FOR A NON-CONDUCTING MEDIUM.

571. In a non-conducting medium there can be no ordinary currents of electricity, so that we put $u = v = w = 0$, and Maxwell's equations assume the form

$$\left.\begin{array}{l} \dfrac{4\pi}{C}\dfrac{df}{dt} = \dfrac{\partial\gamma}{\partial y} - \dfrac{\partial\beta}{\partial z} \\[2mm] \dfrac{4\pi}{C}\dfrac{dg}{dt} = \dfrac{\partial\alpha}{\partial z} - \dfrac{\partial\gamma}{\partial x} \\[2mm] \dfrac{4\pi}{C}\dfrac{dh}{dt} = \dfrac{\partial\beta}{\partial x} - \dfrac{\partial\alpha}{\partial y} \end{array}\right\} \quad\dots\dots\dots\dots\dots\dots(525).$$

We notice that the whole of the left-hand members arise entirely from the "displacement-current." If the displacement-current were omitted, we should have

$$\frac{\partial\gamma}{\partial y} = \frac{\partial\beta}{\partial z}, \text{ etc.}$$

so that the magnetic forces (α, β, γ) would be derivable from a potential, and the only magnetic field in a dielectric in which no currents flowed would be one arising from permanent magnetism.

Maxwell's hypothesis, as expressed in equations (525), implies that there will be a magnetic field in a dielectric whenever the electric field changes, and enables us to calculate the forces in this field.

Magnetic Field of a Moving Charge.

572. As a simple but important example of the use of Maxwell's equations (525) let us calculate the magnetic field produced by a single point charge e moving with a velocity v.

Let the direction of motion of the charge at any instant t be taken for axis of x, the position of the charge being taken for origin.

Let O (fig. 137) be the position of the charge at time t, and O' its position at time $t - dt$; then $O'O = vdt$.

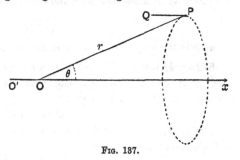

Fig. 137.

Let P be the point at which we wish to evaluate the magnetic force. Draw PQ parallel and equal to OO'. Then the electric field at P at time t will be the same as the electric field at Q at time $t - dt$, so that the increase in the electric field at P in time dt will be the same as the increase produced by moving a distance $- v\,dt$ parallel to the axis of x. Thus we have

$$\frac{\partial f}{\partial t} = - v \frac{\partial f}{\partial x} \text{ etc.}$$

and equations (525) may be put in the form

$$- \frac{4\pi v}{C} \frac{\partial f}{\partial x} = \frac{\partial \gamma}{\partial y} - \frac{\partial \beta}{\partial z}$$

$$- \frac{4\pi v}{C} \frac{\partial g}{\partial x} = \frac{\partial \alpha}{\partial z} - \frac{\partial \gamma}{\partial x}$$

$$- \frac{4\pi v}{C} \frac{\partial h}{\partial x} = \frac{\partial \beta}{\partial x} - \frac{\partial \alpha}{\partial y}.$$

We have here three equations from which to determine the three components of magnetic force, α, β and γ.

A solution which obviously satisfies the last two equations is

$$\alpha = 0, \quad \beta = - \frac{4\pi v}{C} h, \quad \gamma = \frac{4\pi v}{C} g.$$

This solution is also seen to satisfy the first equation in virtue of the relation (cf. Equation (64)),

$$\frac{\partial f}{\partial x} = -\left(\frac{\partial g}{\partial y} + \frac{\partial h}{\partial z}\right);$$

it is therefore the required solution of the problem.

For the electric field of a single point charge, we have[*]

$$4\pi f = \frac{ex}{r^3}, \quad 4\pi g = \frac{ey}{r^3}, \quad 4\pi h = \frac{ez}{r^3},$$

and on substituting these values for f, g, h, the solution becomes

$$\alpha = 0, \quad \beta = -\frac{U}{C}\frac{ez}{r^3}, \quad \gamma = \frac{U}{C}\frac{ey}{r^3} \ldots\ldots\ldots\ldots\ldots(526).$$

These equations give the components of magnetic force at any point. The lines of magnetic force are circles about the path of the electron, and the intensity at distance r from the electron is

$$\frac{eU}{C}\frac{\sin\theta}{r^2} \ldots\ldots\ldots\ldots\ldots\ldots\ldots\ldots\ldots\ldots\ldots(527),$$

where θ is the angle between the distance r and the direction of motion.

572 a. If a small element ds of a circuit in which a current i (measured in electromagnetic units) is flowing contains $N\,ds$ electrons moving with an average forward velocity U_0, we have (cf. equation (b) of § 345)

$$NeU_0 = Ci.$$

The magnetic force at distance r produced by the motion of the electrons in the element ds of the circuit is (cf. expression 527))

$$N ds \frac{eU_0}{C}\frac{\sin\theta}{r^2} \quad \text{or} \quad i\,ds\frac{\sin\theta}{r^2}.$$

This is exactly identical with the force given by Ampère's Law (§ 497). But Ampère's formula was only proved to be true when integrated round a closed circuit, whereas it is now seen that Maxwell's theory implies that the formula is true for every element of a circuit.

Experimental Confirmation.

573. The possibility that a moving electric charge might produce a magnetic field occurred to Faraday and was noted by him in his Experimental Researches (1837); the effect was observed by Rowland in 1876 and again by Röntgen in 1885. Maxwell's equations, as we have just seen, predict the actual amount of this effect. The only quantity other than the measurable electric charge which appears in Maxwell's formulae is C, the ratio of the electric units, and this can be determined in other ways (cf. § 582 below), its value being found to be almost exactly 3×10^{10}.

---

[*] This is not quite accurate, for the motion of the magnetic field (α, β, γ) induces an electric field which ought to be taken into account in evaluating (f, g, h). Equations (526) are, however, very nearly accurate except for very rapidly moving charges. The exact solution will be given later (cf. §§ 624, 647, 656).

The first attempt to measure the effect quantitatively was made by Rowland and Hutchinson in 1889. They used discs charged to a potential of 5000 volts, which were made to rotate at 125.revolutions a second. The motion of the charged discs may be regarded as the motion of a succession of electric charges, and the magnetic force predicted by Maxwell's theory can be calculated from formula (527). On comparing the observed effect with that predicted by theory, values for C were found which varied from $2 \cdot 26 \times 10^{10}$ to $3 \cdot 74 \times 10^{10}$, the mean being $3 \cdot 19 \times 10^{10}$. More exact experiments of a similar type performed by H. Pender in 1901 gave for C an average value of $3 \cdot 05 \times 10^{10}$; a second set, with slightly modified apparatus, gave $C = 2 \cdot 96 \times 10^{10}$. These values will be seen to agree very closely with the known value for C, $3 \cdot 00 \times 10^{10}$, so that the experiments not only prove the existence of the magnetic field produced by moving charges, but also confirm Maxwell's theory quantitatively.

It may be objected that the foregoing experiments only test the magnetic field produced by a continuous chain of electric charges moving in a closed circuit, but this objection cannot be urged against experiments performed by E. P. Adams in 1901. In these experiments charged brass spheres were made to pass a suspended magnetic needle at the rate of about 800 per second and the apparatus was arranged so that the effect of one sphere had almost disappeared before the needle came under the influence of the next. From a series of such experiments Adams determined values for C ranging from $2 \cdot 6 \times 10^{10}$ to $3 \cdot 1 \times 10^{10}$, the mean being $2 \cdot 8 \times 10^{10}$.

Further confirmation of the existence of the displacement-current is provided in a great number of indirect ways, particularly through the electromagnetic theory of light and the electromagnetic mass of the electron. For the present we shall assume the truth of Maxwell's hypothesis and proceed to examine its consequences.

The General Equations of the Electromagnetic Field.

574. In § 529, we obtained the system of equations

$$-\frac{da}{dt} = \frac{\partial Z}{\partial y} - \frac{\partial Y}{\partial z} \text{ etc.}$$

in which all the quantities were expressed in electromagnetic units. If the electric forces are expressed in electrostatic units, X, Y, Z must be replaced in these equations by CX, CY, CZ, and the system of equations becomes

$$\left. \begin{array}{l} -\dfrac{1}{C}\dfrac{da}{dt} = \dfrac{\partial Z}{\partial y} - \dfrac{\partial Y}{\partial z} \\[2mm] -\dfrac{1}{C}\dfrac{db}{dt} = \dfrac{\partial X}{\partial z} - \dfrac{\partial Z}{\partial x} \\[2mm] -\dfrac{1}{C}\dfrac{dc}{dt} = \dfrac{\partial Y}{\partial x} - \dfrac{\partial X}{\partial y} \end{array} \right\} \quad \ldots\ldots\ldots\ldots\ldots\ldots(528).$$

These three equations together with equations (523), namely

$$4\pi \left(u + \frac{1}{C} \frac{df}{dt} \right) = \frac{\partial \gamma}{\partial y} - \frac{\partial \beta}{\partial z}$$

$$4\pi \left(v + \frac{1}{C} \frac{dg}{dt} \right) = \frac{\partial \alpha}{\partial z} - \frac{\partial \gamma}{\partial x} \Biggr\} \quad \dots\dots\dots\dots(529),$$

$$4\pi \left(w + \frac{1}{C} \frac{dh}{dt} \right) = \frac{\partial \beta}{\partial x} - \frac{\partial \alpha}{\partial y}$$

constitute a system of six equations giving the rate of changes in the electric and magnetic fields in terms of the field at any instant. With them may be associated the two equations (63) and (362), namely

$$\frac{\partial f}{\partial x} + \frac{\partial g}{\partial y} + \frac{\partial h}{\partial z} = \rho \quad \dots\dots\dots\dots\dots\dots(530),$$

$$\frac{\partial a}{\partial x} + \frac{\partial b}{\partial y} + \frac{\partial c}{\partial z} = 0 \quad \dots\dots\dots\dots\dots\dots(531).$$

The eight equations (528)—(531) form the most general system of equations of the electromagnetic field. In these equations u, v, w, a, b, c, α, β, γ are expressed in electromagnetic units, while f, g, h, X, Y, Z are expressed in electrostatic units.

Localisation and Flow of Energy.

575. We have already considered the hypothesis that electromagnetic energy may not be confined to the regions occupied by electric charges, magnets and currents, but may be spread through the whole of space. On this hypothesis the kinetic (magnetic) energy T and the potential (electric) energy W of an isotropic medium are given by

$$T = \frac{1}{8\pi} \iiint \mu \left(\alpha^2 + \beta^2 + \gamma^2 \right) dx\,dy\,dz,$$

$$W = \frac{1}{8\pi} \iiint K \left(X^2 + Y^2 + Z^2 \right) dx\,dy\,dz,$$

and the energy is supposed to be localised in space in the way indicated by these integrals. Knowing the kinetic and potential energies of the system, it ought to be possible to determine its equations of motion by the general dynamical methods explained in Chapter XVI.

The quantities α, β, γ which enter in the kinetic energy must be fundamentally of the nature of velocities. Let us denote them by $\dot{\xi}$, $\dot{\eta}$, $\dot{\zeta}$, so that ξ, η, ζ may be treated as positional coordinates.

Similarly u, v, w which express the rates of flow of electricity at any point are of the nature of velocities. If q_x, q_y, q_z denote the total quantity of electricity, measured in electrostatic units, which have crossed unit areas perpendicular to Ox, Oy, Oz at any point since a specified instant, then

$$Cu = \dot{q}_x, \quad Cv = \dot{q}_y, \quad Cw = \dot{q}_z.$$

Maxwell's equations (529) now assume the form

$$\frac{4\pi}{C}\left(\dot{q}_x + \frac{df}{dt}\right) = \frac{\partial\zeta}{\partial y} - \frac{\partial\dot{\eta}}{\partial z} \text{ etc.}$$

giving on integrating, and replacing $4\pi f$ by KX,

$$\frac{1}{C}(4\pi q_x + KX) = \frac{\partial\zeta}{\partial y} - \frac{\partial\eta}{\partial z} \quad\dots\dots\dots\dots\dots(532).$$

This relation connects the various positional coordinates q_x, X (regarded as a "displacement"), ξ etc.

The principle of least action can be expressed, as in equation (507), in the form

$$\int_0^\tau (\delta T - \{\delta W\})\, dt = 0,$$

where the value of $\{\delta W\}$ in the present problem is

$$\{\delta W\} = \delta W + \iiint (X\delta q_x + Y\delta q_y + Z\delta q_z)\, dx\, dy\, dz,$$

which again, on substituting for W, can be put in the form

$$\{\delta W\} = \frac{1}{4\pi}\iiint [X(K\delta X + 4\pi\delta q_x) + Y(K\delta Y + 4\pi\delta q_y)$$
$$+ Z(K\delta Z + 4\pi\delta q_z)]\, dx\, dy\, dz$$
$$= \frac{C}{4\pi}\iiint\left[X\left(\frac{\partial\delta\zeta}{\partial y} - \frac{\partial\delta\eta}{\partial z}\right) + Y\left(\frac{\partial\delta\xi}{\partial z} - \frac{\partial\delta\zeta}{\partial x}\right) + Z\left(\frac{\partial\delta\eta}{\partial x} - \frac{\partial\delta\xi}{\partial y}\right)\right] dx\, dy\, dz$$

on using relations (532). On further transforming by Green's Theorem, this becomes

$$\{\delta W\} = \frac{C}{4\pi}\iint [X(m\delta\zeta - n\delta\eta) + \dots]\, dS$$
$$- \frac{C}{4\pi}\iiint\left[\left(\frac{\partial X}{\partial y}\delta\zeta - \frac{\partial X}{\partial z}\delta\eta\right) + \dots\right] dx\, dy\, dz.$$

Similarly on varying T, we find

$$\delta T = \frac{1}{4\pi}\iiint [\mu\alpha\delta\alpha + \mu\beta\delta\beta + \mu\gamma\delta\gamma]\, dx\, dy\, dz$$
$$= \frac{1}{4\pi}\iiint [a\delta\xi + b\delta\dot{\eta} + c\delta\zeta]\, dx\, dy\, dz,$$

giving

$$\int_0^\tau \delta T\, dt = \frac{1}{4\pi}\left|\iiint (a\delta\xi + b\delta\eta + c\delta\zeta)\, dx\, dy\, dz\right|_0^\tau$$
$$- \frac{1}{4\pi}\int_0^\tau dt \iiint (\dot{a}\delta\xi + \dot{b}\delta\eta + \dot{c}\delta\zeta)\, dx\, dy\, dz.$$

As in § 545, we suppose the values of $\delta\xi$, $\delta\eta$, $\delta\zeta$ all to vanish at the instants $t = 0$ and $t = \tau$, so that the top line on the right hand vanishes.

Collecting terms, we now obtain

$$\int_0^\tau (\delta T - \{\delta W\})\,dt = -\frac{C}{4\pi}\int_0^\tau dt \iiint \left\{ \left(\frac{\dot{a}}{C} + \frac{\partial Z}{\partial y} - \frac{\partial Y}{\partial z}\right)\delta\xi \right.$$

$$\left. + \left(\frac{\dot{b}}{C} + \frac{\partial X}{\partial z} - \frac{\partial Z}{\partial x}\right)\delta\eta + \left(\frac{\dot{c}}{C} + \frac{\partial Y}{\partial x} - \frac{\partial X}{\partial y}\right)\delta\zeta \right\}dx\,dy\,dz$$

$$-\frac{C}{4\pi}\int_0^\tau dt\, \Sigma \iint \{(nY - mZ)\,\delta\xi + \ldots\}\,dS.$$

If our suppositions as to the localisation of the kinetic and potential energies are correct, then ξ, η, ζ may be regarded as independent coordinates at every point of the field. Thus the variations $\delta\xi$, $\delta\eta$, $\delta\zeta$ may have all possible values at all points of the field. It follows that their coefficients must vanish separately; hence at every point of the field, we must have

$$\frac{\dot{a}}{C} + \frac{\partial Z}{\partial y} - \frac{\partial Y}{\partial z} = 0, \text{ etc.}$$

These are the equations which the principle of least action gives as the equations of motion when we assume Maxwell's equations (529). We see at once that they are identical with equations (528), so that the two sets of equations (528) and (529) are related through the principle of least action.

Poynting's Theorem.

576. If we still assume the energy to be localised in the medium in the way imagined by Maxwell, the total energy in any closed region will be given by

$$T + W = \iiint \left\{ \frac{K}{8\pi}(X^2 + Y^2 + Z^2) + \frac{\mu}{8\pi}(\alpha^2 + \beta^2 + \gamma^2) \right\}dx\,dy\,dz,$$

whence, on differentiating, and replacing $\mu\alpha$ by a, KX by $4\pi f$, etc.,

$$\frac{d(T+W)}{dt} = \iiint \left\{ \left(X\frac{df}{dt} + Y\frac{dg}{dt} + Z\frac{dh}{dt} \right) + \frac{1}{4\pi}\left(\alpha\frac{da}{dt} + \beta\frac{db}{dt} + \gamma\frac{dc}{dt} \right) \right\}dx\,dy\,dz.$$

On substituting from equations (528) and (529), this becomes

$$\frac{d(T+W)}{dt} = \frac{C}{4\pi}\iiint \left\{ X\left(\frac{\partial\gamma}{\partial y} - \frac{\partial\beta}{\partial z}\right) + \ldots - \alpha\left(\frac{\partial Z}{\partial y} - \frac{\partial Y}{\partial z}\right) - \ldots \right\}dx\,dy\,dz$$

$$- C\iiint (uX + vY + wZ)\,dx\,dy\,dz.$$

In this equation, the last line represents exactly the rate at which work is performed or energy dissipated by the flow of currents, so that the first line must represent the rate at which energy flows into the region from outside.

By Green's Theorem (§179), the first line

$$= \frac{C}{4\pi} \iiint \left\{ \frac{\partial}{\partial x}(Z\beta - Y\gamma) + \frac{\partial}{\partial y}(X\gamma - Z\alpha) + \frac{\partial}{\partial z}(Y\alpha - X\beta) \right\} dx\,dy\,dz$$

$$= -\frac{C}{4\pi} \iint \{ l(Z\beta - Y\gamma) + m(Y\gamma - Z\alpha) + n(Y\alpha - X\beta) \}\, dS,$$

l, m, n being the direction-cosines of the normal inwards into the region.

Thus if we put

$$\Pi_x = \frac{C}{4\pi}(Y\gamma - Z\beta), \text{ etc.} \quad \ldots\ldots\ldots\ldots\ldots\ldots(533),$$

it appears that the value of $\frac{d}{dt}(T + W)$ is the same as if there were a flow of energy in the direction l, m, n of amount $l\Pi_x + m\Pi_y + n\Pi_z$. The vector Π of which Π_x, Π_y, Π_z are components is of amount

$$\Pi = \sqrt{(\Pi_x^2 + \Pi_y^2 + \Pi_z^2)} = \frac{C}{4\pi} RH \sin\theta,$$

where R, H are the electric and magnetic intensities and θ is the angle between them. The direction of the vector Π is at right angles to both R and H, and the flow of energy into or out of the surfaces is the same as if there were a flow equal to Π in magnitude and direction at every point of space. This vector Π is called the "Poynting flux of energy."

The integral of the Poynting Flux over a closed surface gives the total flow of energy into or out of a surface, but it has not been proved, and we are not entitled to assume, that there is an actual flow of energy at every point equal to the Poynting Flux. For instance if an electrified sphere is placed near to a bar magnet, this latter assumption would require a perpetual flow of energy at every point in the field except the special points at which the electric and magnetic lines of force are tangential to one another. It is difficult to believe that this predicted circulation of energy can have any physical reality. On the other hand it is to be noticed that such a circulation of energy is almost meaningless. The circulation of a fluid is a definite conception because it is possible to identify the different particles of a fluid; we can say for instance whether or not the particles entering a small element of volume are identical or not with an equal number of particles coming out, but the same is not true of energy.

Equations for a Uniform Isotropic Dielectric.

577. We return now to the general equations of § 574, and proceed to examine the form they assume in a uniform isotropic dielectric. Since there can be no electric current we put $u = v = w = 0$. We also put

$$4\pi f = KX \text{ etc.}, \quad a = \mu\alpha \text{ etc.},$$

and the equations assume the form

$$\frac{K}{C}\frac{dX}{dt} = \frac{\partial \gamma}{\partial y} - \frac{\partial \beta}{\partial z} \left.\begin{array}{c} \\ \\ \end{array}\right\} \quad\dots\dots(A), \qquad -\frac{\mu}{C}\frac{d\alpha}{dt} = \frac{\partial Z}{\partial y} - \frac{\partial Y}{\partial z} \left.\begin{array}{c} \\ \\ \end{array}\right\}$$

$$\frac{K}{C}\frac{dY}{dt} = \frac{\partial \alpha}{\partial z} - \frac{\partial \gamma}{\partial x} \qquad\qquad -\frac{\mu}{C}\frac{d\beta}{dt} = \frac{\partial X}{\partial z} - \frac{\partial Z}{\partial x} \quad\dots\dots(B).$$

$$\frac{K}{C}\frac{dZ}{dt} = \frac{\partial \beta}{\partial x} - \frac{\partial \alpha}{\partial y} \qquad\qquad -\frac{\mu}{C}\frac{d\gamma}{dt} = \frac{\partial Y}{\partial x} - \frac{\partial X}{\partial y}$$

From the first equation of system (A), we have

$$\frac{K\mu}{C^2}\frac{d^2X}{dt^2} = \frac{\partial}{\partial y}\left(\frac{\mu}{C}\frac{d\gamma}{dt}\right) - \frac{\partial}{\partial z}\left(\frac{\mu}{C}\frac{d\beta}{dt}\right),$$

and on substituting the values of $\dfrac{\mu}{C}\dfrac{d\gamma}{dt}$ and $\dfrac{\mu}{C}\dfrac{d\beta}{dt}$ from the last two equations of system (B), this equation becomes

$$\frac{K\mu}{C^2}\frac{d^2X}{dt^2} = -\frac{\partial}{\partial y}\left(\frac{\partial Y}{\partial x} - \frac{\partial X}{\partial y}\right) + \frac{\partial}{\partial z}\left(\frac{\partial X}{\partial z} - \frac{\partial Z}{\partial x}\right)$$

$$= \frac{\partial^2 X}{\partial y^2} + \frac{\partial^2 X}{\partial z^2} - \frac{\partial}{\partial x}\left(\frac{\partial Y}{\partial y} + \frac{\partial Z}{\partial z}\right).$$

Since the medium is supposed to be uncharged, we have

$$\frac{\partial X}{\partial x} + \frac{\partial Y}{\partial y} + \frac{\partial Z}{\partial z} = 0,$$

so that the last term may be replaced by $+\dfrac{\partial^2 X}{\partial x^2}$, and the equation becomes

$$\frac{K\mu}{C^2}\frac{d^2X}{dt^2} = \nabla^2 X.$$

By exactly similar analysis we can obtain the differential equation satisfied by Y, Z, α, β and γ, and in each case this differential equation is found to be identical with that satisfied by X. Thus the three components of electric force and the three components of magnetic force all satisfy exactly the same differential equation, namely

$$\frac{d^2\chi}{dt^2} = a^2 \nabla^2 \chi \quad\dots\dots\dots\dots\dots\dots\dots\dots\dots(534),$$

where a stands for $C/\sqrt{K\mu}$. This equation, for reasons which will be seen from its solution, is known as the "equation of wave-propagation."

SOLUTIONS OF $\dfrac{d^2\chi}{dt^2} = a^2 \nabla^2 \chi$.

Solution for spherical waves.

578. The general solution of the equation of wave-propagation is best approached by considering the special form assumed when the solution χ is spherically symmetrical. If χ is a function of r only, where r is the distance from any point, we have

$$\frac{d^2\chi}{dt^2} = a^2\nabla^2\chi = \frac{a^2}{r^2}\frac{d}{dr}\left(r^2\frac{d\chi}{dr}\right),$$

which may be transformed into

$$\frac{d^2(r\chi)}{d(at)^2} = \frac{d^2(r\chi)}{dr^2} \quad \dots\dots\dots\dots\dots\dots(535),$$

and the solution is

$$r\chi = f(r - at) + \Phi(r + at) \quad \dots\dots\dots\dots\dots(536),$$

where f and Φ are arbitrary functions.

The form of solution shews that the value of χ at any instant over a sphere of any radius r depends upon its values at a time t previous over two spheres of radii $r - at$ and $r + at$. In other words, the influence of any value of χ is propagated backwards and forwards with velocity a. For instance, if at time $t = 0$ the value of χ is zero except over the surface of a sphere of radius r, then at time t the value of χ is zero everywhere except over the surfaces of the two spheres of radii $r \pm at$; we have therefore two spherical waves, converging and diverging with the same velocity a.

General solution (Liouville).

579. The general solution of the equation can be obtained in the following manner, originally due to Liouville.

Expressed in spherical polars, r, θ and ϕ, the equation to be solved is

$$\frac{1}{a^2}\frac{d^2\chi}{dt^2} = \frac{1}{r^2}\frac{\partial}{\partial r}\left(r^2\frac{\partial\chi}{\partial r}\right) + \frac{1}{r^2\sin\theta}\frac{\partial}{\partial\theta}\left(\sin\theta\frac{\partial\chi}{\partial\theta}\right) + \frac{1}{r^2}\frac{\partial^2\chi}{\partial\phi^2} = 0.$$

Let us multiply by $\sin\theta\, d\theta\, d\phi$ and integrate this equation over the surface of a sphere of radius r surrounding the origin. If we put

$$\lambda = \iint\chi\sin\theta\, d\theta\, d\phi\dots\dots\dots\dots\dots\dots(537),$$

the equation becomes

$$\frac{1}{a^2}\frac{d^2\lambda}{dt^2} = \frac{1}{r^2}\frac{\partial}{\partial r}\left(r^2\frac{\partial\lambda}{\partial r}\right),$$

the remaining terms vanishing on integration. The solution of this equation (cf. equation (536)) is

$$\lambda = \frac{1}{r}\{f(at - r) + \Phi(at + r)\} \quad \dots\dots\dots\dots (538).$$

For small values of r this assumes the form

$$\lambda = \frac{1}{r}\left[\{f(at) + \Phi(at)\} - r\{f'(at) - \Phi'(at)\} + \frac{r^2}{2}\{f''(at) + \Phi''(at)\} + \dots\right]$$
$$\dots\dots\dots(539).$$

In order that λ may be finite at the origin through all time, we must have

$$f(at) + \Phi(at) = 0$$

at every instant, so that the function Φ must be identical with $-f$. On putting $r = 0$, equation (539) becomes

$$(\lambda)_{r=0} = -2f'(at),$$

and from equation (537), putting $r = 0$, we have

$$(\lambda)_{r=0} = 4\pi (\chi)_{r=0},$$

so that

$$4\pi (\chi)_{r=0} = -2f'(at) \quad \dotfill (540).$$

Equation (538) may now be written as

$$r\lambda = f(at - r) - f(at + r).$$

On differentiating this equation with respect to r and t respectively,

$$\frac{\partial}{\partial r}(r\lambda) = -f'(at - r) - f'(at + r),$$

$$\frac{1}{a}\frac{\partial}{\partial t}(r\lambda) = f'(at - r) - f'(at + r),$$

and on addition we have

$$-2f'(at + r) = \frac{\partial}{\partial r}(r\lambda) + \frac{1}{a}\frac{\partial}{\partial t}(r\lambda).$$

This equation is true for all values of r and t: putting $t = 0$, we have

$$-2f'(r) = \frac{\partial}{\partial r}(r\lambda_{t=0}) + \frac{r}{a}\dot\lambda_{t=0}$$

as an equation which is true for all values of r. Giving to r the special value $r = at$, the equation becomes

$$-2f'(at) = \frac{\partial}{\partial t}(t\lambda_{t=0}) + t\dot\lambda_{t=0}.$$

The left hand is, by equation (520), equal to $4\pi (\chi)_{r=0}$. If we use $\overline{\chi}, \overline{\dot\chi}$ to denote the mean values of χ and $\dot\chi$ averaged over a sphere of radius at at any instant, the equation becomes

$$(\chi)_{r=0} = \frac{\partial}{\partial t}(t\overline{\chi}_{t=0}) + t\overline{\dot\chi}_{t=0} \quad \dotfill (541).$$

Thus the value of χ at any point (which we select to be the origin) at any instant t depends only on the values of χ and $\dot\chi$ at time $t = 0$ over a sphere of radius at surrounding this point. The solution is of the same nature as that obtained in § 578, but is no longer limited to spherical waves.

General solution (Kirchhoff).

580. A still more general form of solution has been given by Kirchhoff. Let Φ and Ψ be any two independent solutions of the original equation, so that

$$\frac{d^2\Phi}{dt^2} = a^2\nabla^2\Phi, \quad \frac{d^2\Psi}{dt^2} = a^2\nabla^2\Psi \quad \dotfill (542).$$

By Green's Theorem (equation (101))

$$-\Sigma \iint \left(\Phi\frac{\partial\Psi}{\partial n} - \Psi\frac{\partial\Phi}{\partial n}\right)dS = \iiint (\Phi\nabla^2\Psi - \Psi\nabla^2\Phi)\,dx\,dy\,dz$$

$$= \frac{1}{a^2}\iiint \left(\Phi\frac{d^2\Psi}{dt^2} - \Psi\frac{d^2\Phi}{dt^2}\right)dx\,dy\,dz$$

by equations (542). The volume integrations extend through the interior of any space bounded by the closed surfaces $S_1, S_2, \ldots$, and the normals to $S_1, S_2, \ldots$ are drawn, as usual, into the space. If we integrate the equation just obtained throughout the interval of time from $t = -t'$ to $t = +t''$, we obtain

$$-\Sigma \int_{-t'}^{t''} dt \iint \left(\Phi \frac{\partial \Psi}{\partial n} - \Psi \frac{\partial \Phi}{\partial n} \right) dS$$

$$= \frac{1}{a^2} \left[\iiint \left(\Phi \frac{d\Psi}{dt} - \Psi \frac{d\Phi}{dt} \right) dx \, dy \, dz \right]_{-t'}^{t''} \quad \ldots \ldots \ldots (543).$$

So far Ψ has denoted any solution of the differential equation. Let us now take it to be $\dfrac{1}{r} F(r + at)$, this being a solution (cf. equation (536)) whatever function is denoted by F, and let $F(x)$ be a function of x such that it and all its differential coefficients vanish for all values of x except $x = 0$, while

$$\int_{-\infty}^{+\infty} F(x) \, dx = 1.$$

Such a function, for instance, is $F(x) = \underset{c=0}{\mathrm{Lt}} \dfrac{c}{\pi (x^2 + c^2)}$.

We can choose t' so that, for all values of r considered, the value of $r - at'$ is negative. The value of $r + at''$ is positive if t'' is positive. Thus $F(r + at)$ and all its differential coefficients vanish at the instants $t = t''$ and $t = -t'$, so that the right-hand member of equation (543) vanishes, and the equation becomes

$$-\Sigma \int_{-t'}^{t''} dt \iint \left(\Phi \frac{\partial \Psi}{\partial n} - \Psi \frac{\partial \Phi}{\partial n} \right) dS = 0 \quad \ldots \ldots \ldots \ldots (544).$$

Let us now suppose the surfaces over which this integral is taken to be two in number. First, a sphere of infinitesimal radius r_0, surrounding the origin, which will be denoted by S_1, and second, a surface, as yet unspecified, which will be denoted by S. Let us first calculate the value of the contribution to equation (544) from the first surface. We have, on this first surface,

$$\Psi = \frac{1}{r_0} F(r_0 + at),$$

$$\frac{\partial \Psi}{\partial n} = \frac{\partial \Psi}{\partial r} = -\frac{1}{r_0^2} F(r_0 + at) + \frac{1}{r_0} F'(r_0 + at),$$

so that when r_0 is made to vanish in the limit, we have

$$\iint \left(\Phi \frac{\partial \Psi}{\partial n} - \Psi \frac{\partial \Phi}{\partial n} \right) dS_1 = -4\pi \Phi_{r=0} F(at),$$

and therefore

$$\int_{-t'}^{t''} dt \iint \left(\Phi \frac{\partial \Psi}{\partial n} - \Psi \frac{\partial \Phi}{\partial n} \right) dS_1 = -4\pi \int_{-t'}^{t''} \Phi_{r=0} F(at) \, dt$$

$$= -\frac{4\pi}{a} \Phi_{\substack{r=0, \\ t=0}},$$

since the integrand vanishes except when $t = 0$.

Thus equation (544) becomes

$$\Phi_{\substack{r=0 \\ t=0}} = -\frac{a}{4\pi} \int_{-t'}^{t''} dt \iint \left(\Phi \frac{\partial \Psi}{\partial n} - \Psi \frac{\partial \Phi}{\partial n} \right) d\mathbf{S}$$

$$= -\frac{a}{4\pi} \iint d\mathbf{S} \int_{-t'}^{t''} \left\{ \frac{\Phi}{r} \frac{\partial r}{\partial n} F'(r+at) \right.$$

$$\left. + \Phi \frac{\partial}{\partial n} \left(\frac{1}{r} \right) F(r+at) - \frac{1}{r} F(r+at) \frac{\partial \Phi}{\partial n} \right\} dt \dots\dots(545).$$

Integrating by parts, we have, as the value of the first term under the time integral,

$$\int_{-t'}^{t''} \frac{\Phi}{r} \frac{\partial r}{\partial n} F'(r+at) \, dt$$

$$= \frac{1}{a} \frac{\Phi}{r} \frac{\partial r}{\partial n} \bigg| F(r+at) \bigg|_{t=-t'}^{t=t''} - \int_{-t'}^{t''} \frac{1}{ar} \frac{\partial r}{\partial n} \frac{d\Phi}{dt} F(r+at) \, dt.$$

The first term vanishes at both limits, and equation (545) now becomes

$$\Phi_{\substack{r=0 \\ t=0}} = \frac{a}{4\pi} \iint d\mathbf{S} \int_{-t'}^{t''} F(r+at) \left\{ \frac{1}{ar} \frac{\partial r}{\partial n} \frac{d\Phi}{dt} - \Phi \frac{\partial}{\partial n} \left(\frac{1}{r} \right) + \frac{1}{r} \frac{\partial \Phi}{\partial n} \right\} dt.$$

We can now integrate with respect to the time, for $F(r+at)$ exists only at the instant $t=-r/a$. Thus the equation becomes

$$\Phi_{\substack{r=0 \\ t=0}} = \frac{1}{4\pi} \iint \left[\frac{1}{ar} \frac{\partial r}{\partial n} \frac{d\Phi}{dt} - \Phi \frac{\partial}{\partial n} \left(\frac{1}{r} \right) + \frac{1}{r} \frac{\partial \Phi}{\partial n} \right]_{t=-\frac{r}{a}} d\mathbf{S},$$

giving the value of Φ at the time $t=0$ in terms of the values of Φ and $\dot{\Phi}$ taken at previous instants over any surface surrounding the point. The solution reduces to that of Liouville on taking the surface $\mathbf{S}$ to be a sphere, so that $\dfrac{\partial}{\partial n} = -\dfrac{\partial}{\partial r}$.

As with the former solutions, the result obtained clearly indicates propagation in all directions with uniform velocity a.

PROPAGATION OF ELECTROMAGNETIC WAVES.

581. It is now clear that the system of equations

$$\frac{K\mu}{C^2} \frac{d^2 X}{dt^2} = \nabla^2 X,$$

etc., obtained in § 577 indicate that, in a homogeneous isotropic dielectric, all electromagnetic effects ought to be propagated with the uniform velocity $\dfrac{C}{\sqrt{K\mu}}$. This may be compared with the result obtained in § 562. It was there shewn that electric signals propagated along a wire would advance with a velocity $\dfrac{C}{\sqrt{K\mu}}$ where K, μ were the inductive capacity and magnetic permeability of the medium surrounding the wire. It now appears that the velocity of signals along a wire is identical with the velocity of waves in the medium outside the wire.

Maxwell's displacement theory gives a simple explanation of this. A current flowing in a wire is accompanied by a displacement current in the ether. This sets up a magnetic field which is propagated with velocity $C/\sqrt{K\mu}$ in the dielectric and this in turn induces a further current in the wire. On this view the actual process of propagation takes place in the medium, the wire directs the path of the electromagnetic disturbance and absorbs some of the energy.

It is to be noticed that the velocity of propagation along wires was obtained in § 562 before we had introduced the conception of "displacement-currents" at all. That the result is not inconsistent with the velocity obtained on the hypothesis of displacement-currents will be understood from the result of § 575.

Numerical Values.

582. We notice that in free air, in which $K = \mu = 1$, the velocity of propagation of electric waves, whether along wires or in the air, ought to be the same as C, the ratio of the electric units. This enables us to apply a severe test to the truth of the theory which has so far been developed, for both the value of C and the velocity of propagation of electric waves admit of direct experimental determination.

The best determinations of C, the ratio of the two units, are the following:

Rosa and Dorsey (1907)	...	...	$2 \cdot 9971 \times 10^{10}$
Pcrot and Fabry (1898)	...	...	$2 \cdot 9973 \times 10^{10}$
Hurmuzescu (1896) ...	...	...	$3 \cdot 0010 \times 10^{10}$
Abraham (1890)	...	...	$2 \cdot 9913 \times 10^{10}$

The true value is probably very close to the value obtained by Rosa and Dorsey, namely $C = \mathbf{2 \cdot 9971} \times \mathbf{10^{10}}$.

Recent determinations of the velocity of propagation of electromagnetic waves in air are as follows:

Maclean (1899)	...	...	...	$2 \cdot 991 \times 10^{10}$
Saunders (1897)	...	...	...	$2 \cdot 997 \times 10^{10}$
Trowbridge and Duane (1895)	...	...	$3 \cdot 003 \times 10^{10}$

The mean of these values is $\mathbf{2 \cdot 997} \times \mathbf{10^{10}}$.

In the determinations of Saunders and of Trowbridge and Duane the waves were guided by copper wires, while the experiments of Maclean dealt with waves propagated through air without wires. The equality of velocities is of course a consequence, and also a confirmation, of the results obtained in § 562.

The ratio of the units, C, is also equal, or at least very nearly equal, to the velocity of light in air, and this confirmed Maxwell in his suggestion that light propagation is a special case of the propagation of electromagnetic waves. Out of this suggestion, amply borne out by the results of further experiments, has grown the electromagnetic theory of light of which a short account is given in the next chapter. The best determination of the velocity of light in air at present available is based on the experiments of Michelson (1927), and is

$$2\cdot9977 \times 10^{10} \text{ cms. a second.}$$

Except for small differences, which are well within the errors of the various experiments, the quantities previously mentioned are seen to agree with this in value.

Thus we may say, that the ratio of units C is identical with the velocity of propagation of electromagnetic waves, and this again is identical with the velocity of light.

EQUATIONS FOR A UNIFORM ISOTROPIC CONDUCTOR.

583. In an isotropic conductor the current (u, v, w) is proportional at every point to the electric force (X, Y, Z). We are supposing u, v, w to be measured in electromagnetic units. The values of the components of electric force, measured in electromagnetic units, are CX, CY, CZ, these being of course the forces acting on an electromagnetic unit of electrical charge. Thus by Ohm's Law,

$$u = \frac{CX}{\tau}, \text{ etc.}$$

where τ is the specific resistance measured in electromagnetic units. If we further put $4\pi f = KX$, etc., equations (529) become

$$\left(\frac{4\pi C}{\tau} + \frac{K}{C}\frac{d}{dt}\right) X = \frac{\partial \gamma}{\partial y} - \frac{\partial \beta}{\partial z} \quad \dots\dots\dots\dots\dots(546)$$

and two similar equations.

On replacing equations (529) by these, the equations of § 574 become the general equations of an isotropic conducting medium.

If we differentiate the three equations of the system (546) with respect to x, y, z and add, we obtain

$$\left(\frac{4\pi C}{\tau} + \frac{K}{C}\frac{d}{dt}\right)\left(\frac{\partial X}{\partial x} + \frac{\partial Y}{\partial y} + \frac{\partial Z}{\partial z}\right) = 0.$$

From equation (530) we have

$$\frac{\partial X}{\partial x} + \frac{\partial Y}{\partial y} + \frac{\partial Z}{\partial z} = \frac{4\pi\rho}{K}$$

so that our equation becomes

$$\frac{d\rho}{dt} = -\frac{4\pi C^2}{K\tau}\rho.$$

If ρ_0 is the value of ρ at time $t = 0$, the solution of this equation is

$$\rho = \rho_0 e^{-\frac{4\pi C^2}{K\tau}t},$$

shewing that ρ falls away exponentially, no matter what electric or magnetic fields may be acting. This equation is identical with that already obtained in § 396, the factor C^2 simply corresponding to a change of units. Thus inside a conducting medium any initial charge will rapidly disappear, and we may suppose that

$$\frac{\partial X}{\partial x} + \frac{\partial Y}{\partial y} + \frac{\partial Z}{\partial z} = 0 \; ; \; \rho = 0.$$

583 a. Multiply both sides of equation (546) by μ and differentiate with respect to the time. We find

$$\frac{K\mu}{C}\frac{d^2X}{dt^2} + \frac{4\pi\mu C}{\tau}\frac{dX}{dt} = \frac{\partial}{\partial y}\left(\frac{dc}{dt}\right) - \frac{\partial}{\partial z}\left(\frac{db}{dt}\right).$$

The right-hand member of this equation may by equations (528) be replaced by

$$-C\frac{\partial}{\partial y}\left(\frac{\partial Y}{\partial x} - \frac{\partial X}{\partial y}\right) + C\frac{\partial}{\partial z}\left(\frac{\partial X}{\partial z} - \frac{\partial Z}{\partial x}\right)$$

or

$$C\left[\nabla^2 X - \frac{\partial}{\partial x}\left(\frac{\partial X}{\partial x} + \frac{\partial Y}{\partial y} + \frac{\partial Z}{\partial z}\right)\right]$$

and this is equal to $C\nabla^2 X$, in virtue of the relation

$$\frac{\partial X}{\partial x} + \frac{\partial Y}{\partial y} + \frac{\partial Z}{\partial z} = 0.$$

Thus the equation becomes, on dividing through by C,

$$\frac{K\mu}{C^2}\frac{d^2X}{dt^2} + \frac{4\pi\mu}{\tau}\frac{dX}{dt} = \nabla^2 X.$$

This equation involves X only, and so is the differential equation satisfied by X when electromagnetic waves are propagated in a conductor. Naturally Y, Z satisfy similar equations, and equations (528) shew that a, b, c or α, β, γ again satisfy similar equations. Thus X, Y, Z, α, β, γ all satisfy the same differential equation, namely

$$\frac{d^2X}{dt^2} + \frac{4\pi C^2}{K\tau}\frac{dX}{dt} = a^2\nabla^2 X,$$

where a stands for $C/\sqrt{K\mu}$. The complete solution of this equation has been given by Riemann[*].

---

[*] *Die partielle Differentialgleichungen der Math. Physik,* 4th edn. (1901), II. p. 399.

We may notice that in a dielectric, $\tau = \infty$, so that the second term disappears. The equation then reduces, as it ought, to equation (534) already obtained in § 577. In many problems, the second term is more important than the first. When the first term is omitted, the equation reduces to the well-known equation of conduction of heat, already obtained in § 535 (equation (480)).

To form an estimate of the relative importance of the two terms on the left, let us examine the case of an alternating current in which the time-factor is e^{ipt}. We may as usual replace $\dfrac{d}{dt}$ by ip, and the equation becomes

$$\left(-p^2 + \frac{4\pi C^2}{K\tau} ip\right)\chi = a^2 \nabla^2 \chi.$$

The neglect of the first term, which is of course the same thing as neglecting the displacement-current, is clearly permissible if $4\pi C^2/K\tau p$ is numerically large. When this ratio is not large, the error produced by the neglect of the first term will be greatest in problems in which τ is large (conductors of high resistance) and in which p is large (rapidly changing fields). On substituting numerical values it will be found that in problems of conduction through metals, the neglect of the factor $K\tau p/4\pi C^2$ produces a quite inappreciable error unless p is comparable with 10^{15}—*i.e.* unless we are dealing with oscillating fields of which the frequency is comparable with that of light-waves. Thus the effect of the displacement-current in metals has been inappreciable in the problems so far discussed, so that the neglect of this effect may be regarded as justifiable. The matter stands differently as regards the problems to be discussed in the next chapter, in which the oscillations of the field are identical with those of light-waves.

UNITS.

584. We may at this stage sum up all that has been said about the different systems of electrical units.

There are three different systems of units to be considered, of which two are theoretical systems, the electrostatic and the electromagnetic, while the third is the practical system. We shall begin by discussing the two theoretical systems and their relation to one another.

585. In the Electrostatic System the fundamental unit is the unit of electric charge, this being defined as a charge such that two such charges at unit distance apart in air exert unit force upon one another. There will, of course, be different systems of electrostatic units corresponding to different units of length, mass and time, but the only system which need be considered is that in which these units are taken to be the centimetre, gramme and second respectively.

In the Electromagnetic System the fundamental unit is the unit magnetic pole, this being defined to be such that two such poles at unit distance apart in air exert unit force upon one another. Again the only system which need be considered is that in which the units of length, mass and time are the centimetre, gramme and second.

From the unit of electric charge can be derived other units—*e.g.* of electric force, of electric potential, of electric current, etc.—in which to measure quantities which occur in electric phenomena. These units will of course also be electrostatic units, being derived from the fundamental electrostatic unit.

So also from the unit magnetic pole can be derived other units—*e.g.* of magnetic force, of magnetic potential, of strength of a magnetic shell, etc.— in which to measure quantities which occur in magnetic phenomena. These units will belong to the electromagnetic system.

If electric phenomena were entirely dissociated from magnetic phenomena, the two entirely different sets of units would be necessary, and there could be no connection between them. But the discovery of the connection between electric currents and magnetic forces enables us at once to form a connection between the two sets of units. It enables us to measure electric quantities— *e.g.* the strength of a current—in electromagnetic units, and conversely we can measure magnetic quantities in electrostatic units.

We find, for instance, that a magnetic shell of unit strength (in electromagnetic measure) produces the same field as a current of certain strength. We accordingly take the strength of this current to be unity in electromagnetic measure, and so obtain an electromagnetic unit of electric current. We find, as a matter of experiment, that this unit is not the same as the electrostatic unit of current, and therefore denote its measure in electrostatic units of current by C. This is the same as taking the electromagnetic unit of charge to be C times the electrostatic unit, for current is measured in either system of units as a charge of electricity per unit time.

In the same way we can proceed to connect the other units in the two systems. For instance, the electromagnetic unit of electric intensity will be the intensity in a field in which an electromagnetic unit of charge experiences a force of one dyne. An electrostatic unit of charge in the same field would of course experience a force of $1/C$ dynes, so that the electrostatic measure of the intensity in this field would be $1/C$. Thus the electromagnetic unit of intensity is $1/C$ times the electrostatic. The following table of the ratios of the units can be constructed in this way:

Ratios of Units.

Charge of Electricity.   One electromag. unit $= C$ electrostat. units.

Electromotive Force.	„	„	„	$= 1/C$	„	„
Electric Intensity.	„	„	„	$= 1/C$	„	„
Potential.	„	„	„	$= 1/C$	„	„
Electric Polarisation.	„	„	„	$= C$	„	„
Capacity.	„	„	„	$= C^2$	„	„
Current.	„	„	„	$= C$	„	„
Resistance of a conductor.	„	„	„	$= 1/C^2$	„	„
Strength of magnetic pole.	„	„	„	$= 1/C$	„	„
Magnetic Intensity.	„	„	„	$= C$	„	„
„ Induction.	„	„	„	$= 1/C$	„	„
Inductive Capacity.	„	„	„	$= C^2$	„	„
Magnetic Permeability.	„	„	„	$= 1/C^2$	„	„

586. The value of C, as we have said, is equal to about 3×10^{10} in c.g.s. units. If units other than the centimetre, gramme and second are taken, the value of C will be different. Since we have seen that C represents a velocity, it is easy to obtain its value in any system of units.

For instance a velocity 3×10^{10} in c.g.s. units $= 6\cdot71 \times 10^8$ miles per hour, so that if miles and hours are taken as units the value of C will be $6\cdot71 \times 10^8$.

Practical Units.

587. The practical system of units is derived from the electromagnetic system, each practical unit differing only from the corresponding electro-magnetic unit by a certain power of ten, the power being selected so as to make the unit of convenient size. The actual measures of the practical units are as follows:

Quantity	Name of Unit	Measure in electromag. units	Measure in electrostatic units (Taking $C = 3 \times 10^{10}$)
Charge of Electricity	Coulomb	10^{-1}	3×10^9
Electromotive Force⎫ Electric Intensity ⎬ Potential ⎭	Volt	10^8	$\frac{1}{800}$
Capacity	Farad	10^{-9}	9×10^{11}
„	Microfarad	10^{-15}	9×10^5
Current	Ampère	10^{-1}	3×10^9
Resistance	Ohm	10^9	$\dfrac{1}{9 \times 10^{11}}$

For legal and commercial purposes, the units are defined in terms of material standards. Thus according to the resolutions of the International Conference of 1908 the legal (International) ohm is defined to be the resistance offered to a steady current by a uniform column of mercury of length 106·300 cms., the temperature being 0° C., and the mass being 14·4521 grammes, this resistance being equal, as nearly as can be determined by experiment, to 10^9 electromagnetic units. Similarly the legal (International) ampère is defined to be the current which, when passed through a solution of silver nitrate in water, deposits silver at the rate of ·00111800 grammes per second.

Physical Dimensions of Units.

588. As explained in § 18, all the electric and magnetic units will have apparent dimensions in mass, length and time. These are shewn in the following table:

		Electrostatic	Electromagnetic
Charge of Electricity	e	$M^{\frac{1}{2}} L^{\frac{3}{2}} T^{-1}$	$M^{\frac{1}{2}} L^{\frac{1}{2}}$
Density ,, ,,	ρ	$M^{\frac{1}{2}} L^{-\frac{3}{2}} T^{-1}$	$M^{\frac{1}{2}} L^{-\frac{5}{2}}$
Electromotive Force	E	$M^{\frac{1}{2}} L^{\frac{1}{2}} T^{-1}$	$M^{\frac{1}{2}} L^{\frac{3}{2}} T^{-2}$
Electric Intensity	$R\,(X,\,Y,\,Z)$	$M^{\frac{1}{2}} L^{-\frac{1}{2}} T^{-1}$	$M^{\frac{1}{2}} L^{\frac{1}{2}} T^{-2}$
Potential	V	$M^{\frac{1}{2}} L^{-\frac{1}{2}} T^{-1}$	$M^{\frac{1}{2}} L^{\frac{3}{2}} T^{-2}$
Electric Polarisation	$P\,(f,\,g,\,h)$	$M^{\frac{1}{2}} L^{-\frac{1}{2}} T^{-1}$	$M^{\frac{1}{2}} L^{-\frac{3}{2}}$
Capacity	C	L	$L^{-1} T^2$
Current	i	$M^{\frac{1}{2}} L^{\frac{3}{2}} T^{-2}$	$M^{\frac{1}{2}} L^{\frac{1}{2}} T^{-1}$
Current per unit area	$(u,\,v,\,w)$	$M^{\frac{1}{2}} L^{-\frac{1}{2}} T^{-2}$	$M^{\frac{1}{2}} L^{-\frac{3}{2}} T^{-1}$
Resistance	R	$L^{-1} T$	$L T^{-1}$
Specific resistance	τ	T	$L^2 T^{-1}$
Strength of magnetic pole	m	$M^{\frac{1}{2}} L^{\frac{1}{2}}$	$M^{\frac{1}{2}} L^{\frac{3}{2}} T^{-1}$
Magnetic Force	$H\,(\alpha,\,\beta,\,\gamma)$	$M^{\frac{1}{2}} L^{\frac{1}{2}} T^{-2}$	$M^{\frac{1}{2}} L^{-\frac{1}{2}} T^{-1}$
,, Induction	$B\,(a,\,b,\,c)$	$M^{\frac{1}{2}} L^{-\frac{3}{2}}$	$M^{\frac{1}{2}} L^{-\frac{1}{2}} T^{-1}$
Inductive Capacity	K	1	$L^{-2} T^2$
Magnetic Permeability	μ	$L^{-2} T^2$	1

CHAPTER XVIII

THE ELECTROMAGNETIC THEORY OF LIGHT

VELOCITY OF LIGHT IN DIFFERENT MEDIA.

589. IT has been seen that, on the electromagnetic theory of light, the propagation of waves of light *in vacuo* ought to take place with a velocity equal, within limits of experimental error, to the actual observed velocity of light. A further test can be applied to the theory by examining whether the observed and calculated velocities are in agreement in other media.

According to the electromagnetic theory, if V is the velocity in any medium, and V_0 the velocity *in vacuo*, we ought to have the relation

$$\frac{V}{V_0} = \frac{1}{\sqrt{K\mu}} \bigg/ \frac{1}{\sqrt{K_0\mu_0}},$$

where K_0, μ_0 refer to free space.

For free space and all media which will be considered, we may take $\mu = 1$. Also if ν is the refractive index for a plane wave of light passing from free space to any medium, we have from optical theory the relation

$$\frac{V_0}{V} = \nu,$$

so that, according to the electromagnetic theory, the refractive index of any medium ought to be connected with its inductive capacity by the relation

$$\nu = \sqrt{\frac{K}{K_0}}.$$

One difficulty appears at once. According to this equation there ought to be a single definite refractive index for each medium, whereas the phenomenon of dispersion shews that the refractive index of any medium varies with the wave-length of the light. It is easy to trace this difficulty to its source. The phenomenon of dispersion is supposed to arise from the periodic motion of charged electrons associated with the molecules of the medium (cf. § 610, below), whereas the theoretical value which has been obtained for the velocity of light has been deduced on the supposition that there are no moving charges at any point of the dielectric (cf. § 577). A correction to the value just obtained for ν will be needed to represent the effect of the motion of charged electrons in the medium. When this motion is infinitely slow, the correction disappears, so that our equation ought to give the true value of ν in the limiting case of light, or other electromagnetic waves, of infinite wave-length. It is impossible to deal experimentally with waves of infinite wave-length, but the following tables* shew that as the wave-length increases, the refractive index ν approximates to $\sqrt{K/K_0}$.

<table>
<tr><td colspan="2" style="text-align:center">WATER.

$\sqrt{\dfrac{K}{K_0}} = \sqrt{80} = 8{\cdot}94.$</td><td colspan="2" style="text-align:center">ETHYL ALCOHOL.

$\sqrt{\dfrac{K}{K_0}} = 5{\cdot}1.$</td></tr>
<tr><td>Wave-length (cms.)</td><td>ν (observed)</td><td>Wave-length (cms.)</td><td>ν (observed)</td></tr>
<tr><td>65</td><td>8·88</td><td>65</td><td>4·89</td></tr>
<tr><td>8·8</td><td>8·89</td><td>5·7</td><td>3·4</td></tr>
<tr><td>5·7</td><td>8·79</td><td>0·8</td><td>2·57</td></tr>
<tr><td>3·7</td><td>8·10</td><td>0·4</td><td>2·24</td></tr>
<tr><td>1·75</td><td>7·82</td><td>·0000589‡</td><td>1·36</td></tr>
<tr><td>0·8</td><td>8·97</td><td></td><td></td></tr>
<tr><td>0·4</td><td>9·50</td><td></td><td></td></tr>
<tr><td>·000126†</td><td>1·32</td><td></td><td></td></tr>
<tr><td>·0000589‡</td><td>1·33</td><td></td><td></td></tr>
</table>

590. For gases there is quite good agreement between theory and experiment, in spite of the failure of the theory to take all the facts into account.

In the following table, the values of $\sqrt{\dfrac{K}{K_0}}$ are mean values taken from the table already given on p. 132 of the inductive capacities of gases. The values of ν refer to sodium light.

Gas	Mean $\sqrt{\dfrac{K}{K_0}}$ §	ν (observed) Sodium D-line ‖
Hydrogen ...	1·000132	1·000138
Air	1·000293	1·000292
Carbon Monoxide	1·000347	1·000334
Carbon Dioxide	1·000492	1·000450
Nitrous Oxide ...	1·000495	1·000515
Ethylene ...	1·00073	1·000719

---

* From material collected by Pidduck, *A Treatise on Electricity* (Camb. Univ. Press, 1916), p. 451.

† Infra-red radiation.   ‡ Sodium light.

§ From table on p. 132.

‖ Kaye and Laby, *Physical and Chemical Constants* (eighth edition), p. 75.

WAVES OF LIGHT IN NON-CONDUCTING MEDIA.

Solution of Differential Equation for Plane Waves.

591. The equation of wave-propagation

$$\frac{d^2\chi}{dt^2} = a^2 \nabla^2 \chi$$

has, as a particular solution,

$$\chi = A e^{i\kappa \, (lx+my+nz-at)} \quad\dots\dots\dots\dots\dots(547),$$

provided $l^2 + m^2 + n^2 = 1$. This value of χ is a complex quantity of which the real and imaginary parts separately must be solutions of the original equation. Thus we have the two solutions

$$\chi = A \cos \kappa \, (lx + my + nz - at)\dots\dots\dots\dots\dots(548),$$
$$\chi = A \sin \kappa \, (lx + my + nz - at).$$

Either of these solutions represents the propagation of a plane wave. The direction-cosines of the direction of propagation are l, m, n, and the velocity of propagation is a. Usually it will be found simplest to take the value of χ given by equation (547) as the solution of the equation and reject imaginary terms after the analysis is completed. This procedure will be followed throughout the present chapter; it will of course give the same result as would be obtained by taking equation (548) as the solution of the differential equation.

Propagation of a Plane Wave.

592. Let us now consider in detail the propagation of a plane wave of light, the direction of propagation being taken, for simplicity, to be the axis of x. The values of $X, Y, Z, \alpha, \beta, \gamma$ must all be solutions of the differential equation, each being of the form

$$\chi = A e^{i\kappa \, (x-at)} \quad\dots\dots\dots\dots\dots\dots\dots(549).$$

The six values of $X, Y, Z, \alpha, \beta, \gamma$ are not independent, being connected by the six equations of § 577, namely

$$\left. \frac{K}{C} \frac{dX}{dt} = \frac{\partial \gamma}{\partial y} - \frac{\partial \beta}{\partial z} \atop \frac{K}{C} \frac{dY}{dt} = \frac{\partial \alpha}{\partial z} - \frac{\partial \gamma}{\partial x} \atop \frac{K}{C} \frac{dZ}{dt} = \frac{\partial \beta}{\partial x} - \frac{\partial \alpha}{\partial y} \right\} \dots\dots(A),$$

$$\left. -\frac{\mu}{C} \frac{d\alpha}{dt} = \frac{\partial Z}{\partial y} - \frac{\partial Y}{\partial z} \atop -\frac{\mu}{C} \frac{d\beta}{dt} = \frac{\partial X}{\partial z} - \frac{\partial Z}{\partial x} \atop -\frac{\mu}{C} \frac{d\gamma}{dt} = \frac{\partial Y}{\partial x} - \frac{\partial X}{\partial y} \right\} \dots\dots(B).$$

From the form of solution (equation (549)), it is clear that all the differential operators may be replaced by multipliers. We may put

$$\frac{d}{dt} = -i\kappa a, \qquad \frac{\partial}{\partial x} = i\kappa, \qquad \frac{\partial}{\partial y} = \frac{\partial}{\partial z} = 0.$$

The equations now become

$$X = 0 \atop \left. \begin{array}{r} \dfrac{Ka}{C} Y = -\gamma \\[2mm] \dfrac{Ka}{C} Z = \quad \beta \end{array} \right\} \dots (A'), \qquad \begin{array}{r} \alpha = 0 \\[2mm] \dfrac{\mu a}{C} \beta = \quad Z \\[2mm] \dfrac{\mu a}{C} \gamma = -Y \end{array} \right\} \dots (B').$$

Since $K\mu a^2 = C^2$, it is clear that the second and third equations in (A') are identical with the third and second equations respectively in (B').

Since $X = 0$, $\alpha = 0$, it appears that both the electric and magnetic forces are, at every instant, at right angles to the axis of x, *i.e.* to the direction of propagation. From the last two equations of system (A') we obtain

$$\beta Y + \gamma Z = 0,$$

shewing that the electric force and the magnetic force are also at right angles to one another.

On comparing the results obtained from the electromagnetic theory of light, with those obtained from physical optics, it is found that the wave of light which we have been examining is a plane-polarised ray whose plane of polarisation is the plane containing the magnetic force and the direction of propagation. Thus the magnetic force is in the plane of polarisation, while the electric force is at right angles to this plane.

CRYSTALLINE DIELECTRIC MEDIA.

592 a. Let us consider the propagation of light, on the electromagnetic theory, in a crystalline medium in which the ratio of the polarisation to the electric force is different in different directions.

By equation (92), the electric energy W per unit volume in such a medium is given by

$$W = \frac{1}{8\pi} (K_{11}X^2 + 2K_{12}XY + \dots).$$

If we transform axes, taking as new axes of reference the principal axes of the quadric

$$K_{11}x^2 + 2K_{12}xy + \dots = 1,$$

then the energy per unit volume assumes the form

$$W = \frac{1}{8\pi} (K_1 X^2 + K_2 Y^2 + K_3 Z^2).$$

The components of polarisation are now given by (cf. equations (89))

$$4\pi f = K_1 X, \qquad 4\pi g = K_2 Y, \qquad 4\pi h = K_3 Z,$$

so that the general equations (529) and (528) of § 574 assume the forms

$$
\left.
\begin{aligned}
\frac{K_1}{C}\frac{dX}{dt} &= \frac{\partial \gamma}{\partial y} - \frac{\partial \beta}{\partial z} \\[2mm]
\frac{K_2}{C}\frac{dY}{dt} &= \frac{\partial \alpha}{\partial z} - \frac{\partial \gamma}{\partial x} \\[2mm]
\frac{K_3}{C}\frac{dZ}{dt} &= \frac{\partial \beta}{\partial x} - \frac{\partial \alpha}{\partial y}
\end{aligned}
\right\} \dots\dots \text{(A'')},
\qquad
\left.
\begin{aligned}
-\frac{\mu}{C}\frac{d\alpha}{dt} &= \frac{\partial Z}{\partial y} - \frac{\partial Y}{\partial z} \\[2mm]
-\frac{\mu}{C}\frac{d\beta}{dt} &= \frac{\partial X}{\partial z} - \frac{\partial Z}{\partial x} \\[2mm]
-\frac{\mu}{C}\frac{d\gamma}{dt} &= \frac{\partial Y}{\partial x} - \frac{\partial X}{\partial y}
\end{aligned}
\right\} \dots\dots \text{(B'')}.
$$

These may be compared with equations (A) and (B) of § 592.

If we differentiate the system of equations (A'') with respect to the time, and substitute the values of $\dfrac{d\alpha}{dt}$, $\dfrac{d\beta}{dt}$, $\dfrac{d\gamma}{dt}$ from system (B''), we obtain

$$\frac{K_1\mu}{C^2}\frac{d^2 X}{dt^2} = \nabla^2 X - \frac{\partial}{\partial x}\left(\frac{\partial X}{\partial x} + \frac{\partial Y}{\partial y} + \frac{\partial Z}{\partial z}\right), \text{ etc.}$$

On assuming a solution in which X, Y, Z are each proportional to

$$e^{i\kappa\,(lx + my + nz - Vt)}$$

these equations become

$$\frac{K_1\mu}{C^2} V^2 X = X - l\,(lX + mY + nZ) = 0, \text{ etc.}$$

On eliminating X, Y and Z from these three equations, we obtain

$$\frac{l^2}{V^2\dfrac{K_1\mu}{C^2} - 1} + \frac{m^2}{V^2\dfrac{K_2\mu}{C^2} - 1} + \frac{n^2}{V^2\dfrac{K_3\mu}{C^2} - 1} + 1 = 0.$$

If we put $\dfrac{C^2}{K_1\mu} = v_1^2$, etc., and simplify, this becomes

$$\frac{l^2}{V^2 - v_1^2} + \frac{m^2}{V^2 - v_2^2} + \frac{n^2}{V^2 - v_3^2} = 0.$$

This equation gives the velocity of propagation V in terms of the direction-cosines l, m, n of the normal to the wave-front. The equation is identical with that found by Fresnel to represent the results of experiment. It can be shewn that the corresponding wave-surface is the well-known Fresnel wave-surface, and all the geometrical phenomena of the propagation of light in a crystalline medium follow directly. For the development of this part of the theory, the reader is referred to books on physical optics.

Assuming that α, β, γ as well as X, Y, Z are proportional to the exponential $e^{i\kappa(lx + my + nz - Vt)}$, the equations of system (A'') become

$$-\frac{K_1\mu V}{C} X = m\gamma - n\beta,$$

and two similar equations.

If we multiply these three equations by l, m, n respectively and add, we obtain

$$lK_1X + mK_2Y + nK_3Z = 0,$$

shewing that the electric polarisation is in the wave-front.

The system (B'') of equations reduce to

$$\mu \frac{V}{C} \alpha = mZ - nY,$$

and two similar equations, and on again multiplying by l, m, n and adding, we obtain

$$l\alpha + m\beta + n\gamma = 0,$$

which shews that the magnetic force also is in the wave-front.

We shall not discuss crystalline media in detail in the present book since their special peculiarities are the same on the electromagnetic as on any other theory of light. The discussion of these peculiarities is a branch of the science of optics rather than of electromagnetism.

MECHANICAL ACTION.

Energy in Light-waves.

592 b. For a wave of light propagated along the axis of Ox, and having the electric force parallel to Oy, we have (cf. § 592) the solution

$$X = Z = 0; \quad Y = Y_0 \cos \kappa (x - at),$$

$$\alpha = \beta = 0; \quad \gamma = \gamma_0 \cos \kappa (x - at),$$

and this satisfies all the electromagnetic equations, provided the ratio of γ_0 to Y_0 is given by

$$-\frac{\gamma_0}{Y_0} = \frac{Ka}{C} = \frac{C}{\mu a} = \sqrt{\frac{K}{\mu}}.$$

The energy per unit volume at the point x is

$$\frac{1}{8\pi}(KY^2 + \mu\gamma^2) = \frac{1}{8\pi}(KY_0^2 + \mu\gamma_0^2)\cos^2 \kappa (x - at).$$

Since $\mu\gamma_0^2 = KY_0^2$, it appears that the electric energy is equal to the magnetic at every point of the wave. The average value of $\cos^2 \kappa (x - at)$, averaged with respect either to x or to t, is $\frac{1}{2}$, so that the average energy per unit volume

$$= \frac{KY_0^2}{8\pi} = \frac{\mu\gamma_0^2}{8\pi}.$$

As Maxwell has pointed out*, these formulae enable us to determine the magnitude of the electric and magnetic forces involved in the propagation of

---

* Maxwell, *Electricity and Magnetism* (Third Edition), § 793.

light. According to the determination of Langley, the mean energy of sun-light, after allowing for partial absorption by the earth's atmosphere, is 4.3×10^{-5} ergs per unit volume. This gives, as the maximum value of the electric intensity,

$$Y_0 = .33 \text{ c.g.s. electrostatic units} = 9.9 \text{ volts per centimetre,}$$

and, as the maximum value of the magnetic force,

$$\gamma_0 = .033 \text{ c.g.s. electromagnetic units,}$$

which is about one-sixth of the horizontal component of the earth's field in England.

The Pressure of Radiation.

592 c. In virtue of the existence of the electric intensity Y, there is in any medium (§ 165) a pressure $\dfrac{KY^2}{8\pi}$ per unit area at right angles to the lines of electric force. There is therefore a pressure of this amount per unit area over each wave-front. Similarly the magnetic field results (§471) in a pressure of amount $\dfrac{\mu\gamma^2}{8\pi}$ per unit area.

Thus the total pressure per unit area

$$= \frac{KY^2 + \mu\gamma^2}{8\pi} = \frac{KY_0^2 + \mu\gamma_0^2}{8\pi} \cos^2 \kappa (x - at).$$

This is exactly the expression just found for the energy per unit volume. Thus we see that over every wave-front there ought, on the electromagnetic theory, to be a pressure of amount per unit area equal to the energy of the wave per unit volume at that point. The existence of this pressure has been demonstrated experimentally by Lebedew [*] and by Nichols and Hull [†], and their results agree quantitatively with those predicted by Maxwell's Theory.

REFRACTION AND REFLECTION.

Conditions at a Boundary between two different media.

593. Let us next consider what happens when a wave meets a boundary between two different dielectric media 1, 2. Let the suffix 1 refer to quantities evaluated in the first medium, and the suffix 2 to quantities evaluated in the second medium. For simplicity let us suppose the boundary to coincide with the plane of yz.

---

[*] *Annalen der Physik*, 6, p. 433.　　　　[†] *Physical Review*, 13, p. 307.

At the boundary, the conditions to be satisfied are (§§ 137, 467):

(1) the tangential components of electric force must be continuous,

(2) the normal components of electric polarisation must be continuous,

(3) the tangential components of magnetic force must be continuous,

(4) the normal components of magnetic induction must be continuous.

Analytically, these conditions are expressed by the equations

$$K_1 X_1 = K_2 X_2, \qquad Y_1 = Y_2, \qquad Z_1 = Z_2 \ \dots\dots\dots(550),$$

$$\mu_1 \alpha_1 = \mu_2 \alpha_2, \qquad \beta_1 = \beta_2, \qquad \gamma_1 = \gamma_2 \ \dots\dots\dots(551).$$

It will be at once seen that these six equations are not independent: if the last two of equations (550) are satisfied, then the first of equations (551) is necessarily satisfied also, as a consequence of the relation

$$-\frac{\mu}{C}\frac{d\alpha}{dt} = \frac{\partial Z}{\partial y} - \frac{\partial Y}{\partial z}$$

being satisfied in each medium, while similarly, if the last two of equations (551) are satisfied, then the first of equations (550) is necessarily satisfied. Thus there are only four independent conditions to be satisfied at the boundary, and each of these must be satisfied for all values of y, z and t. It is most convenient to suppose the four boundary conditions to be the continuity of Y, Z, β, γ.

Refraction of a Wave polarised in plane of incidence.

594. Let us now imagine a wave of light to be propagated through medium (1), and to meet the boundary, this wave being supposed polarised in the plane of incidence. Let the boundary, as before, be the plane of yz, and let the plane of incidence be supposed to be the plane of xy. Since the wave is supposed to be polarised in the plane of incidence, the magnetic force must be in the plane of xy, and the electric force must be parallel to the axis of z. Hence for this wave, we may take

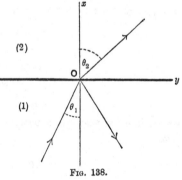

Fig. 138.

$$X = Y = 0,$$

$$Z = Z' \ e^{i\kappa_1 \,(x\cos\theta_1 + y\sin\theta_1 - V_1 t)},$$

$$\alpha = \alpha' \ e^{i\kappa_1 \,(x\cos\theta_1 + y\sin\theta_1 - V_1 t)},$$

$$\beta = \beta' \ e^{i\kappa_1 \,(x\cos\theta_1 + y\sin\theta_1 - V_1 t)},$$

$$\gamma = 0,$$

and it is found that the six equations (A), (B) of p. 534 are satisfied if

$$\frac{\alpha'}{\sin \theta_1} = \frac{\beta'}{-\cos \theta_1} = \frac{Z'}{\sqrt{\dfrac{\mu_1}{K_1}}} \quad \dots\dots\dots\dots(552).$$

The angle θ_1 is seen to be the "angle of incidence" of the wave, namely, the angle between its direction of propagation and the normal (Ox) to the boundary.

Let us suppose that in the second medium there is a refracted wave, given by

$$X = Y = 0,$$
$$Z = Z'' \, e^{i\kappa_2 (x\cos\theta_2 + y\sin\theta_2 - V_2 t)},$$
$$\alpha = \alpha'' \, e^{i\kappa_2 (x\cos\theta_2 + y\sin\theta_2 - V_2 t)},$$
$$\beta = \beta'' \, e^{i\kappa_2 (x\cos\theta_2 + y\sin\theta_2 - V_2 t)},$$
$$\gamma = 0,$$

where, in order that the equations of propagation may be satisfied, we must have

$$\frac{\alpha''}{\sin \theta_2} = \frac{\beta''}{-\cos \theta_2} = \frac{Z''}{\sqrt{\dfrac{\mu_2}{K_2}}} \quad \dots\dots\dots\dots(553).$$

It will be found on substitution in the boundary equations (550) and (551) that the presence of an incident and refracted wave is not sufficient to enable these equations to be satisfied. The equations can, however, all be satisfied if we suppose that in the first medium, in addition to the incident wave, there is a reflected wave given by

$$X = Y = 0,$$
$$Z = Z''' \, e^{i\kappa_3 (x\cos\theta_3 + y\sin\theta_3 - V_1 t)},$$
$$\alpha = \alpha''' \, e^{i\kappa_3 (x\cos\theta_3 + y\sin\theta_3 - V_1 t)},$$
$$\beta = \beta''' \, e^{i\kappa_3 (x\cos\theta_3 + y\sin\theta_3 - V_1 t)},$$
$$\gamma = 0,$$

where, in order that the equations of propagation may be satisfied, we must have

$$\frac{\alpha'''}{\sin \theta_3} = \frac{\beta'''}{-\cos \theta_3} = \frac{Z'''}{\sqrt{\dfrac{\mu_1}{K_1}}} \quad \dots\dots\dots\dots(554).$$

The boundary conditions must be satisfied for all values of y and t. Since y and t enter only through exponentials in the different waves, this requires that we have

$$\kappa_1 \sin \theta_1 = \kappa_2 \sin \theta_2 = \kappa_3 \sin \theta_3 \quad \dots\dots\dots\dots(555),$$
$$\kappa_1 V_1 = \kappa_2 V_2 = \kappa_3 V_1 \dots\dots\dots\dots(556).$$

From (556) we must have $\kappa_1 = \kappa_3$, and hence from (555), $\sin \theta_1 = \sin \theta_3$. Since θ_1 and θ_3 must not be identical, we must have $\theta_1 = \pi - \theta_3$. Thus

The angle of incidence is equal to the angle of reflection.

We further have, from equations (555) and (556),

$$\frac{\sin \theta_1}{\sin \theta_2} = \frac{V_1}{V_2} = \nu \quad \dots\dots\dots\dots\dots\dots(557),$$

where ν is the index of refraction on passing from medium 1 to medium 2, so that *the sine of the angle of incidence is equal to ν times the sine of the angle of refraction.*

Thus the geometrical laws of reflection and refraction can be deduced at once from the electromagnetic theory. These laws can, however, be deduced from practically any *undulatory* theory of light. A more severe test of a theory is its ability to predict rightly the relative intensities of the incident, reflected and refracted waves, and this we now proceed to examine.

595. The only boundary conditions to be satisfied are the continuity, at the boundary, of Z and β (cf. § 593). Thus we must have

$$Z' + Z''' = Z'' \quad \dots\dots\dots\dots\dots\dots\dots(558),$$

$$\beta' + \beta''' = \beta'' \quad \dots\dots\dots\dots\dots\dots\dots(559).$$

On substituting from equations (552), (553) and (554), the last relation becomes

$$\sqrt{\frac{K_1}{\mu_1}} \cos \theta_1 (Z' - Z''') = \sqrt{\frac{K_2}{\mu_2}} \cos \theta_2 Z'' \quad \dots\dots\dots(560),$$

so that all the boundary conditions are satisfied if

$$\frac{Z'}{1+u} = \frac{Z''}{2} = \frac{Z'''}{1-u} \quad \dots\dots\dots\dots\dots\dots(561),$$

where
$$u^2 = \frac{K_2}{\mu_2} \frac{\mu_1}{K_1} \frac{\cos^2 \theta_2}{\cos^2 \theta_1} \quad \dots\dots\dots\dots\dots\dots(562).$$

For all media in which light can be propagated, we may take $\mu = 1$, so that

$$u = \sqrt{\frac{K_2}{K_1}} \frac{\cos \theta_2}{\cos \theta_1} = \frac{\sin \theta_1 \cos \theta_2}{\sin \theta_2 \cos \theta_1} = \frac{\tan \theta_1}{\tan \theta_2} \quad \dots\dots\dots(563).$$

Thus the ratio of the amplitude of the reflected to the incident ray is

$$\frac{Z'''}{Z'} = \frac{1-u}{1+u} = \frac{\tan \theta_2 - \tan \theta_1}{\tan \theta_2 + \tan \theta_1} = \frac{\sin(\theta_2 - \theta_1)}{\sin(\theta_2 + \theta_1)} \quad \dots\dots\dots(564).$$

This prediction of the theory is in good agreement with experiment. This being so, the predicted ratio of $\dfrac{Z''}{Z'}$ is necessarily in agreement with experiment, since both in theory and experiment the energy of the incident wave must be equal to the sum of the energies of the reflected and refracted waves.

Total reflection.

596. We have seen (equation (557)) that the angle θ_2 is given by

$$\sin \theta_2 = \frac{1}{\nu} \sin \theta_1,$$

where ν is the index of refraction for light passing from medium 1 to medium 2. If ν is less than unity, the value of $\frac{1}{\nu} \sin \theta_1$ may be either greater or less than unity according as $\theta_1 >$ or $< \sin^{-1} \nu$. In the former case $\sin \theta_2$ is greater than unity, so that the value of θ_2 is imaginary.

This circumstance does not affect the value of the foregoing analysis in a case in which $\theta_1 > \sin^{-1} \nu$, but the geometrical interpretation no longer holds.

Let us denote $\frac{1}{\nu} \sin \theta_1$ by p, and $\sqrt{p^2 - 1}$ by q. Then in the analysis we may replace $\sin \theta_2$ by p, and $\cos \theta_2$ by iq, both p and q being real quantities. The exponential which occurs in the refracted wave is now

$$e^{i\kappa_2 (x \cos \theta_2 + y \sin \theta_2 - V_2 t)}$$

$$= e^{i\kappa_2 (iqx + py - V_2 t)}$$

$$= e^{-\kappa_2 q x} e^{i\kappa_2 (py - V_2 t)}.$$

Thus the refracted wave is propagated parallel to the axis of y, *i.e.* parallel to the boundary, and its magnitude decreases proportionally to the factor $e^{-\kappa_2 q x}$. At a small distance from the boundary the refracted wave becomes imperceptible.

Algebraically, the values of Z', Z'' and Z''' are still given by equations (561), but we now have

$$u = \sqrt{\frac{K_2 \mu_1}{\mu_2 K_1}} \frac{\cos \theta_2}{\cos \theta_1} = i \sqrt{\frac{K_2 \mu_1}{\mu_2 K_1}} \frac{q}{\cos \theta_1},$$

so that u is an imaginary quantity, say $u = iv$, and, from equations (561),

$$\frac{Z'''}{Z'} = \frac{1 - u}{1 + u} = \frac{1 - iv}{1 + iv}.$$

Since v is real, we have $\left| \dfrac{1 - iv}{1 + iv} \right| = 1$, so that we may take

$$Z''' = Z' e^{i\chi},$$

where

$$\chi = \arg \left(\frac{1 - iv}{1 + iv} \right) = -2 \tan^{-1} v.$$

In the reflected wave, we now have

$$Z = Z''' e^{i\kappa_1 (-x \cos \theta_1 + y \sin \theta_1 - V_1 t)}$$

$$= Z' e^{i\kappa_1 (-x \cos \theta_1 + y \sin \theta_1 - V_1 t - 2 \tan^{-1} v)}.$$

Comparing with the incident wave, in which

$$Z = Z' \, e^{i\kappa_1 \, (x \cos \theta_1 + y \sin \theta_1 - V_1 t)},$$

we see that reflection is now accompanied by a change of phase $- 2\kappa \tan^{-1} v$, but the amplitude of the wave remains unaltered, as obviously it must from the principle of energy.

Refraction of a Wave polarised perpendicular to plane of incidence.

597.   The analysis which has been already given can easily be modified so as to apply to the case in which the polarisation of the incident wave is perpendicular to the plane of incidence. All that is necessary is to interchange corresponding electric and magnetic quantities: we then have an incident wave in which the *magnetic* force is perpendicular to the plane of incidence, and this is what is required.

Clearly all the geometrical laws which have already been obtained will remain true without modification, and the analysis of § 596 (total reflection) will also hold without modification.

Formula (563), giving the amplitude of the reflected ray, will, however, require alteration. We have, as in equation (564), for the ratio of the amplitudes of the incident and reflected rays,

$$\frac{\gamma'''}{\gamma'} = \frac{1 - u}{1 + u} \quad \dots\dots\dots\dots\dots\dots\dots\dots\dots(565),$$

but the value of u, instead of being given by equation (563), must now be supposed to be given by

$$u^2 = \frac{\mu_2}{K_2} \frac{K_1}{\mu_1} \frac{\cos^2 \theta_2}{\cos^2 \theta_1},$$

this equation being obtained by the interchange of electric and magnetic terms in equation (562). Taking $\mu_2 = \mu_1 = 1$, we obtain

$$u = \sqrt{\frac{K_1}{K_2}} \frac{\cos \theta_2}{\cos \theta_1} = \frac{\sin \theta_2}{\sin \theta_1} \frac{\cos \theta_2}{\cos \theta_1} = \frac{\sin 2\theta_2}{\sin 2\theta_1},$$

whence, from equation (565),

$$\frac{\gamma'''}{\gamma'} = \frac{\tan (\theta_2 - \theta_1)}{\tan (\theta_2 + \theta_1)} \quad \dots\dots\dots\dots\dots\dots\dots\dots(566),$$

giving the ratio of the amplitudes of the incident and reflected waves. This result also agrees well with experiment.

598.   We notice that if $\theta_1 + \theta_2 = 90°$, then $\gamma''' = 0$. Thus there is a certain angle of incidence such that no light is reflected. Beyond this angle γ''' is negative, so that the reflected light will shew an abrupt change of phase of 180°. This angle of incidence is known as the polarising angle, because if a beam of non-polarised light is incident at this angle, the reflected beam will consist entirely of light polarised in the plane of incidence, and will accordingly be plane-polarised light.

It has been found by Jamin that formula (566) is not quite accurate at and near to the polarising angle. It appears from experiment that a certain small amount of light is reflected at all angles, and that instead of a sudden change of phase of 180° occurring at this angle there is a gradual change, beginning at a certain distance on one side of the polarising angle and not reaching 180° until a certain distance on the other side. Lord Rayleigh shewed that this discrepancy between theory and experiment can often be attributed largely to the presence of thin films of grease and other impurities on the reflecting surface. Drude found that the outstanding discrepancy could be accounted for by supposing the phenomena of reflection and refraction to occur, not actually at the surface between the two media, but throughout a small transition layer of which the thickness must be supposed finite, although small compared with the wave-length of the light.

WAVES IN METALLIC AND CONDUCTING MEDIA.

599. In a metallic medium of specific resistance τ, equations (A) of § 592, namely

$$\frac{K}{C}\frac{dX}{dt} = \frac{\partial \gamma}{\partial y} - \frac{\partial \beta}{\partial z} \quad \dotfill (567),$$

etc., must be replaced (cf. equation (546)) by

$$\left(\frac{4\pi C}{\tau} + \frac{K}{C}\frac{d}{dt} \right) X = \frac{\partial \gamma}{\partial y} - \frac{\partial \beta}{\partial z} \quad \dotfill (568),$$

etc.

For a plane wave of light, the time may be supposed to enter through the complex imaginary e^{ipt} and we may replace $\frac{d}{dt}$ by ip. Thus the left-hand of equation (567) becomes $\frac{Kip}{C} X$, while the left-hand of equation (568) becomes $\left(\frac{4\pi C}{\tau} + \frac{Kip}{C} \right) X$. It accordingly appears that the conducting power of the medium can be allowed for by replacing K by $K + \frac{4\pi C^2}{ip\tau}$.

600. In a non-conducting medium, the equation $\frac{K\mu}{C^2}\frac{d^2\chi}{dt^2} = \nabla^2\chi$, satisfied by each of the quantities $X, Y, Z, \alpha, \beta, \gamma$ (cf. § 577), reduces to

$$\frac{-p^2 K\mu}{C^2}\chi = \nabla^2\chi$$

when the wave is of frequency $p/2\pi$. The corresponding equation for a conducting medium must, by what has just been said, be

$$-p^2\left(\frac{K\mu}{C^2} + \frac{4\pi\mu}{ip\tau}\right)\chi = \nabla^2\chi \quad\dots\dots\dots\dots(569),$$

an equation which has already been obtained in § 583 a.

For a plane wave propagated in a direction which, for simplicity, we shall suppose to be the axis of x, the solution of this equation will be

$$\chi = Ae^{ipt}e^{\pm(q+ir)x} \quad\dots\dots\dots\dots\dots(570),$$

where

$$(q+ir)^2 = -\frac{K\mu p^2}{C^2} + \frac{4\pi i\mu p}{\tau} \quad\dots\dots\dots(571).$$

Clearly the solution (570) represents the propagation of waves with a velocity V equal to p/r, the amplitude of these waves falling off with a modulus of decay q per unit length.

On equating imaginary parts of equation (571) we obtain

$$qr = \frac{2\pi\mu p}{\tau} \quad\dots\dots\dots\dots\dots(572),$$

so that q is given by

$$q = \frac{2\pi\mu}{\tau}\frac{p}{r} = \frac{2\pi V\mu}{\tau} \quad\dots\dots\dots\dots(573).$$

601. For a good conductor τ is small, so that q is large, shewing that good conductors are necessarily bad transmitters of light. For a wave of light in silver or copper we may take as approximate values in C.G.S. units (remembering that τ as given on p. 342 is measured in practical units)

$$\tau = 1.6 \times 10^{-6}\ \text{ohms} = 1.6 \times 10^3\ (\text{electromag.}),\quad \mu = 1,\quad V = 3 \times 10^{10},$$

from which we obtain $q = 1.2 \times 10^6$. It appears that, according to this theory, a ray of light in a good conductor ought to be almost extinguished before traversing more than a small portion of a wave-length. This prediction of the theory is not borne out by experiment.

We shall see below (§ 600) that the difficulty is to some extent removed on taking account of the presence of electrons in the metal. Before passing to the more general theory in which the electrons are taken into account we shall examine the phenomenon of metallic reflection according to our present simple theory, and shall again find that the simple theory fails to agree with the facts.

Metallic Reflection.

602. Let us suppose, as in fig. 138, that we have a wave of light incident at an angle θ_1 upon the boundary between two media, and let us suppose medium 2 to be a conducting medium of inductive capacity K_2'. Then (cf. § 599) all the analysis which has been given in §§ 593—597 will still hold if we take K_2 to be a complex quantity given by

$$K_2 = K_2' + \frac{4\pi C^2}{ip\tau} \quad\quad\quad\dots\dots\dots\dots\dots\dots(574).$$

Since K_2 is complex, it follows at once that V_2 is complex, being given by

$$V_2^2 = \frac{C^2}{K_2 \mu_2},$$

and hence that the angle θ_2 is complex, being given (cf. equation (557)) by

$$\sin^2 \theta_2 = \frac{\sin^2 \theta_1}{V_1^2} V_2^2 = \frac{\sin^2 \theta_1}{V_1^2} \frac{C^2}{K_2 \mu_2} = \sin^2 \theta_1 \frac{K_1 \mu_1}{K_2 \mu_2} \quad\dots\dots\dots(575).$$

The value of u is now given, from equation (562), by

$$u^2 = \frac{K_2}{\mu_2} \frac{\mu_1}{K_1} \frac{\cos^2 \theta_2}{\cos^2 \theta_1}$$

$$= \frac{K_2 \mu_1}{\mu_2 K_1} \sec^2 \theta_1 - \frac{\mu_1^2}{\mu_2^2} \tan^2 \theta_1 \dots\dots\dots\dots\dots\dots(576)$$

(cf. equation (575)) for light polarised in the plane of incidence. For light polarised perpendicular to the plane of incidence, the value of u is found, as before, by interchanging electric and magnetic symbols.

On putting $u = \alpha + i\beta$, we have, as before (equation (564)),

$$\frac{Z'''}{Z'} = \frac{1-u}{1+u} = \frac{1-\alpha-i\beta}{1+\alpha+i\beta}.$$

If we put this fraction in the form $\rho e^{i\chi}$, then the reflected wave is given by

$$Z = Z''' e^{i\kappa_1(-x\cos\theta_1 + y\sin\theta_1 - V_1 t)} = Z' \rho e^{i\kappa_1(-x\cos\theta_1 + y\sin\theta_1 - V_1 t + x/K_1)}.$$

Comparing this with the incident wave, for which

$$Z = Z' e^{i\kappa_1(x\cos\theta_1 + y\sin\theta_1 - V_1 t)},$$

we see that there is a change of phase χ at reflection, and the amplitude is changed in the ratio $1 : \rho$. The electric force in the refracted wave is accompanied by a system of currents, and these dissipate energy, so that the amplitude of the reflected wave must be less than that of the incident wave.

We have

$$\rho e^{i\chi} = \frac{1-\alpha-i\beta}{1+\alpha+i\beta},$$

so that
$$\rho^2 = \frac{(1-\alpha)^2 + \beta^2}{(1+\alpha)^2 + \beta^2} = 1 - \frac{4\alpha}{(1+\alpha)^2 + \beta^2} \quad \ldots\ldots\ldots\ldots\ldots(577)$$

shewing that $\rho < 1$, as it ought to be. Also

$$\chi = -\tan^{-1}\frac{\beta}{1-\alpha} - \tan^{-1}\frac{\beta}{1+\alpha} = -\tan^{-1}\frac{2\beta}{1-\alpha^2-\beta^2} \quad \ldots\ldots(578).$$

603. Experimental determinations of the values of ρ and χ have been obtained, but only for light incident normally, the first medium being air. For this reason we shall only carry on the analysis for the case of $\theta = 0$. It is now a matter of indifference whether the light is polarised in or at right angles to the plane of incidence; indeed it is easily verified that the values given for ρ and χ by equations (577) and (578) are the same in either case.

Taking for simplicity the analysis appropriate to light polarised in the plane of incidence, and putting $\theta = 0$, $\mu_1 = 1$, $K_1 = 1$, we have from equation (576)

$$u^2 = \frac{K_2}{\mu_2} = \frac{K_2'}{\mu_2} + \frac{4\pi C^2}{ip\tau\mu_2},$$

and, since $u = \alpha + i\beta$, this gives

$$\alpha^2 - \beta^2 = \frac{K_2'}{\mu_2} \quad \ldots\ldots\ldots\ldots\ldots\ldots\ldots\ldots(579)$$

$$\alpha\beta = -\frac{2\pi C^2}{p\tau\mu_2} \quad \ldots\ldots\ldots\ldots\ldots\ldots\ldots\ldots(580).$$

604. Let us consider the results as applied to light of great wave-length, for which p is very small. For such values of p, $\alpha\beta$ is clearly very large compared with $\alpha^2 - \beta^2$, so that α and β are nearly equal numerically, and we may suppose as an approximation that (cf. equation (580))

$$\alpha = -\beta = \sqrt{\frac{2\pi C^2}{p\tau\mu_2}} \quad \ldots\ldots\ldots\ldots\ldots\ldots(581).$$

When α and β are equal and large, equation (577) becomes

$$\rho^2 = 1 - \frac{2}{\alpha} = 1 - 2\sqrt{\frac{p\tau\mu_2}{2\pi C^2}} \quad \ldots\ldots\ldots\ldots\ldots(582).$$

Let us suppose that an incident beam has intensity denoted by 100, and that of this a beam of intensity R is reflected from the surface of the metal, while a beam of intensity $100 - R$ enters the metal. Then R may be called the reflecting power of the metal.

The intensity of the absorbed beam is

$$100 - R = 100\,(1 - \rho^2)$$

$$= 200\sqrt{\frac{p\tau\mu_2}{2\pi C^2}} \quad \ldots\ldots\ldots\ldots\ldots(583).$$

We notice that for waves of very great wave-length (p very small) R approximates to 100, so that for waves of very great wave-length all metals become perfect reflectors. This is as it should be, for these waves of very long period may ultimately be treated as slowly-changing electrostatic fields, and the electrons at the surface of the metal screen its interior from the effects of the electric disturbances falling upon it (cf. § 114).

Equation (583) predicts the way in which $100 - R$ ought to increase as p increases, and an extremely important series of experiments have been conducted by Hagen and Rubens* to test the truth of the formula for light of great wave-length. The following table will illustrate the results obtained†:

Metal				$100 - R$ for $\lambda = 12\mu$	
				observed	calculated
Silver	...	...	...	1·15	1·3
Copper	...	...	...	1·6	1·4
Gold	...	...	...	2·1	1·6
Platinum	...	...	...	3·5	3·5
Nickel	...	...	...	4·1	3·6
Steel	...	...	...	4·9	4·7
Bismuth	...	...	...	17·8	11·5
Patent Nickel P	...	...	...	5·7	5·4
,, ,, M	...	...	...	7·0	6·2
Constantin	...	...	...	6·0	7·4
Rosse's alloy	...	...	...	7·1	7·3
Brande's and Schünemann's alloy				9·1	8·6

In the calculated values, the value of K is assumed to be unity, and an error is of course introduced from the fact that the wave-length dealt with, $\lambda = 12\mu$, although large is still finite.

It will be seen that the agreement between the calculated and the observed values is surprisingly good, when allowance is made for the extreme difficulty of the experiments and for the roughness of some of the approximations which have to be made.

605. Hagen and Rubens also conducted experiments for light of wave-lengths $\lambda = 25 \cdot 5\,\mu$, 8μ, and 4μ. On comparing the whole series it is found that the differences between observed and calculated values become progressively greater on passing to light of shorter wave-length. Drude has conducted a series of experiments on visible light, from which it appears that the simple theory so far given fails entirely to agree with observation for wave-lengths as short as those of visible light.

* *Annalen der Physik*, 11, p. 873; *Phil. Mag.* 7, p. 157.
† *Phil. Mag.* 7, p. 168.

Electron Theory.

606. We have now reached a stage in the development of electromagnetic theory in which it is clear that the simple conceptions which have so far been employed are no longer adequate to give a complete explanation of the phenomena. The conceptions on which the preceding analysis has been based have been the original conceptions of Maxwell's theory: it is natural now to examine in what way the theory can be modified or improved by the introduction of the more modern conceptions of the electron theory. Instead of regarding a current as a continuous flow of electricity, we shall take definite account of the presence of electrons. We shall have to consider two sets of electrons, the "free" and "bound" electrons of § 345 a, these being the mechanisms respectively of conduction and of inductive capacity.

The application of an electric force X will result in a motion of free electrons similar to that investigated in § 345 a, and in a motion of the bound electrons similar to that discussed in § 151. But if X is variable with the time, the inertia of the electrons will come into play and the resulting motions will be different from those given by Ohm's law and Faraday's law. We shall suppose that at any instant the current produced by the motion of the free electrons is u_f, and that that produced by the motion of the bound electrons is u_b.

607. We may consider first the evaluation of u_f. Taking N to be the number of free electrons per unit volume, and allowing for change of notation, equation (c) of § 345 a may be re-written in the form

$$CX = \tau' u_f + \frac{m}{Ne^2} \frac{du_f}{dt} \quad \dots\dots\dots\dots\dots\dots(584),$$

in which, as throughout this chapter, X is expressed in electrostatic units, while u_f is in electromagnetic units, and τ' stands for γ/Ne^2, so that τ' becomes identical with the specific resistance τ when the currents are steady.

This equation is applicable to our present investigation if we suppose X to be periodic in the time of frequency $p/2\pi$. Taking $X = X_0 e^{ipt}$, the solution of equation (584) is

$$u_f = \frac{CX_0 e^{ipt}}{\tau' + \frac{m}{Ne^2} ip} \quad \dots\dots\dots\dots\dots\dots\dots(585).$$

The quantity τ' here may depend on p, and without a full knowledge of the structure of matter it is impossible to decide how important the dependence of τ' on p may be. We are therefore compelled to retain it as an unknown quantity in our equations, remembering that it becomes identical with τ when $p = 0$, and is probably numerically comparable with τ for all values of p.

We may note that the real part of the current, corresponding to the force $X = X_0 \cos pt$, is

$$\frac{CX_0}{\tau'} \cos (pt - \epsilon) \cos \epsilon,$$

in which $\tan \epsilon = \dfrac{mp}{Ne^2\tau'}$, shewing that the inertia of the electrons, as represented in the last term of equation (584), results in a lag ϵ in the phase of the current, accompanied by a change in amplitude. The rate of generation of heat by the current u_f, being equal to the average value of $u_f X_0 \cos pt$, is found to be $\frac{1}{2} \dfrac{C^2 X_0^2}{\tau'} \cos^2 \epsilon$ or $\frac{1}{2} \dfrac{C^2 X_0^2}{\tau_p}$, where

$$\tau_p = \tau' \sec^2 \epsilon = \tau' + \frac{m^2 p^2}{N^2 e^4 \tau'} \quad \dots\dots\dots\dots\dots(586).$$

It is worth noticing that for light of short wave-length the last term in τ_p may be more important than the first term τ'. Thus τ_p may be largest for good conductors, and smallest for bad conductors.

608. We turn to the evaluation of u_b, the current produced by the small excursions of the bound electrons, as they oscillate under the periodic electric forces.

We shall regard a molecule (or atom), as in § 151, as a cluster of electrons, and these electrons will be supposed capable of performing small excursions about their positions of equilibrium. As has already been said (§ 192) it is probable that this conception of the structure of the molecule represents only a half-way house towards the truth, but it provides a picture or model of the structure with the help of which many properties may be explained.

Let θ_1, θ_2, ... be generalised coordinates (cf. § 548) determining the positions of the electrons in the molecule, these being chosen so as to be measured from the position of equilibrium. So long as we consider only small vibrations, the kinetic energy T and the potential energy W of the molecule can be expressed in the forms

$$2W = a_{11}\theta_1^2 + 2a_{12}\theta_1\theta_2 + a_{22}\theta_2^2 + \dots \quad \dots\dots\dots\dots(587),$$
$$2T = b_{11}\dot\theta_1^2 + 2b_{12}\dot\theta_1\dot\theta_2 + b_{22}\dot\theta_2^2 + \dots \quad \dots\dots\dots\dots(588),$$

in which the coefficients a_{11}, a_{12}, a_{22}, ..., b_{11}, ... may be treated as constants. By a known algebraic process, new variables ϕ_1, ϕ_2, ... can be found, such that equations (587), (588) when expressed in terms of these variables assume the forms

$$2W = \alpha_1\phi_1^2 + \alpha_2\phi_2^2 + \dots \quad \dots\dots\dots\dots\dots(589),$$
$$2T = \beta_1\dot\phi_1^2 + \beta_2\dot\phi_2^2 + \dots \quad \dots\dots\dots\dots\dots(590),$$

these equations involving only squares of the new coordinates ϕ_1, ϕ_2, ... The coordinates found in this way for any dynamical system are spoken of as the "principal coordinates" of the system.

The equation of motion of the molecule, when acted on by no external forces, is readily found to be (cf. equations (500))

$$\beta_s\ddot{\phi}_s = -\alpha_s\phi_s, \quad (s = 1, 2, \ldots) \quad\ldots\ldots\ldots\ldots(591).$$

These equations are known to represent simply periodic changes in $\phi_1, \phi_2, \ldots$ of frequencies $n_1/2\pi, n_2/2\pi, \ldots$ given by

$$n_s^2 = \frac{\alpha_s}{\beta_s} \quad\ldots\ldots\ldots\ldots\ldots\ldots\ldots\ldots(592).$$

It is possible that we have evidence of the frequencies of molecular vibration in certain of the lines of the spectrum emitted by the substance under consideration; if so equations (592) connect the frequencies of these spectral lines with the coefficients of the principal coordinates of the molecule.

609. If the molecule is now supposed to vibrate under the influence of externally applied forces (such, for instance, as would occur during the passage of a wave of light through the medium), equation (591) must be replaced (cf. equation (508)) by

$$\beta_s\ddot{\phi}_s = -\alpha_s\phi_s + \Phi_s \ldots\ldots\ldots\ldots\ldots\ldots(593),$$

where Φ_s is that part of the "generalised force" corresponding to the coordinate ϕ_s, which originates in the externally applied forces.

If X is the electromotive force in the wave of light at any instant, each electron will experience a force Xe, and there will be a contribution of the form $\zeta_s Xe$ to Φ_s.

Again the electrostatic field created by the displacements of the electrons in the various neighbouring molecules will contribute a further term to Φ_s. The displacement of any electron through a distance ξ will produce the same field as the creation of a doublet of strength $e\xi$. Thus if there are M molecules per unit volume, the total strength of the doublets per unit volume, say Γ, may be supposed to be of the form

$$\Gamma = Me(\gamma_1\phi_1 + \gamma_2\phi_2 + \ldots) \quad\ldots\ldots\ldots\ldots\ldots(594),$$

and these will produce an electric intensity of which the average value may be taken to be (cf. § 145) $\kappa\Gamma$, which must be added to the original intensity X of the wave.

The total value of Φ_s is therefore $\zeta_s e(X + \kappa\Gamma)$, so that on replacing α_s by its value from equation (592), equation (593) becomes

$$\beta_s(\ddot{\phi}_s + n_s^2\phi_s) = \zeta_s e(X + \kappa\Gamma) \quad\ldots\ldots\ldots\ldots(595).$$

If we suppose X to depend on the time through the factor e^{ipt}, then ϕ will clearly depend on the time through the same factor, and we may replace $\ddot{\phi}_s$ by $-p^2\phi_s$. Equation (595) now becomes

$$\phi_s = \frac{\zeta_s e(X + \kappa\Gamma)}{\beta_s(n_s^2 - p^2)} \quad\ldots\ldots\ldots\ldots\ldots\ldots(596),$$

whence, by equation (594),

$$\Gamma = Me^2 \Sigma \frac{\gamma_s \zeta_s}{\beta_s (n_s^2 - p^2)} (X + \kappa \Gamma) \quad \dots\dots\dots\dots(597),$$

and if we write

$$\theta = Me^2 \Sigma \frac{\gamma_s \zeta_s}{\beta_s} \frac{1}{(n_s^2 - p^2)} \quad \dots\dots\dots\dots\dots(598),$$

this gives, as the value of Γ,

$$\Gamma = \frac{\theta}{1 - \kappa\theta} X \dots\dots\dots\dots\dots\dots(599).$$

The current produced by the motion of the bound electrons is u_b in electromagnetic, and therefore Cu_b in electrostatic units. Its value in electrostatic units is also (cf. §345 a) Neu or $\Sigma e \frac{\partial \xi}{\partial t}$, where the summation is taken through a unit volume, and this in turn is equal to $\dot{\Gamma}$. Thus

$$u_b = \frac{\dot{\Gamma}}{C} = \frac{ip\theta}{1 - \kappa\theta} \frac{X}{C}.$$

The total current, expressed in electromagnetic units, is

$$\frac{1}{C} \frac{df}{dt} + u_b + u_f.$$

In calculating f we must remember that the polarisation produced by the motion of the bound electrons is already allowed for in the presence of the term u_b. We accordingly take f equal simply to $X/4\pi$, and on further replacing u_b and u_f by the values found for them, the total current becomes

$$\frac{ipX}{4\pi C} \left(1 + \frac{4\pi\theta}{1 - \kappa\theta}\right) + \frac{CX}{\tau' + \frac{m}{Ne^2} ip} \quad \dots\dots\dots\dots(600).$$

In place of equation (569), the equation of propagation is

$$-p^2 \left\{ \frac{\mu}{C^2} \left(1 + \frac{4\pi\theta}{1 - \kappa\theta}\right) + \frac{4\pi\mu}{ip \left(\tau' + \frac{m}{Ne^2} ip\right)} \right\} \chi = \nabla^2 \chi.$$

As in §600, the solution is

$$\chi = A e^{ipt} e^{\pm (q+ir)x} \quad \dots\dots\dots\dots\dots\dots(601),$$

where

$$(q + ir)^2 = -\frac{\mu p^2}{C^2} \left(1 + \frac{4\pi\theta}{1 - \kappa\theta}\right) + \frac{4\pi\mu ip}{\tau' + \frac{m}{Ne^2} ip} \quad \dots\dots\dots(602).$$

Non-conducting media.

610. For a non-conducting medium $\tau' = \infty$, so that the last term in equation (602) vanishes, and the right-hand member becomes wholly real. For certain values of θ, this right-hand member is negative, so that $q = 0$, shewing that light is transmitted without diminution; the medium is perfectly transparent.

For transparent media we may take $\mu = 1$, and the velocity of propagation V is given by

$$\frac{1}{V^2} = \frac{r^2}{p^2} = \frac{1}{C^2}\left(1 + \frac{4\pi\theta}{1 - \kappa\theta}\right).$$

If ν is the refractive index of the medium, as compared with that of a vacuum, $V = C/\nu$, so that

$$\nu^2 = 1 + \frac{4\pi\theta}{1 - \kappa\theta} \quad \dots\dots\dots\dots\dots\dots(603),$$

whence

$$\frac{\nu^2 - 1}{\nu^2 + a} = \kappa\theta = M\Sigma \frac{c_s}{n_s^2 - p^2}, \quad \dots\dots\dots\dots\dots(604),$$

in which $a = \dfrac{4\pi}{\kappa} - 1$, $c_s = \dfrac{e^2\kappa\gamma_s\zeta_s}{\beta_s}$, so that a and c_s are constants.

Clearly (cf. § 609) the value of a can be calculated if we make assumptions as to the arrangement of the molecules in the medium. On assuming that the molecules are regularly arranged in cubical piling, κ is found to have the value $\frac{4}{3}\pi$, so that a becomes equal to 2.

Formula (604) in which a is neglected altogether becomes exactly identical with the well-known Sellmeyer or Ketteler-Helmholtz formula for the dispersion of light, of which the accuracy is known to be very considerable. If a is put equal to 2, the formula becomes identical with dispersion formulae which have been suggested by Larmor and Lorentz.

It has been shewn by Maclaurin* that formula (604) will give results in almost perfect agreement with experiment, at least for certain solids, if a is treated as an adjustable constant. The agreement of the formula is so very good that little doubt can be felt that it is founded on a true basis. Maclaurin finds for a values widely different from 2 (for rocksalt $a = 5\cdot51$, for fluorite $a = 1\cdot04$), the differences between these numbers and 2 pointing perhaps to the crystalline arrangement of the molecules. For liquids and gases we should expect to find a equal to 2.

Since M is proportional to ρ, the density of the substance, formula (604) indicates that $\dfrac{\nu^2 - 1}{\nu^2 + a}$ ought to vary directly as ρ when ρ varies. This law, with a taken equal to 2, was announced by H. A. Lorentz† of Leyden and L. Lorenz‡ of Copenhagen in 1880. Its truth has been verified by various observers, and, in particular, by Magri§ for a large range of densities of air.

From equation (604) it also follows that the values of $\dfrac{\nu^2 - 1}{\nu^2 + a}$ for a mixture of liquids or gases ought to be equal to the sum of the values of $\dfrac{\nu^2 - 1}{\nu^2 + a}$ for its

* *Proc. Roy. Soc.* A, 81, p. 367 (1908).   † *Wied. Ann.* 9, p. 641 (1880).
‡ *Wied. Ann.* 11, p. 70 (1880).   § *Phys. Zeitschrift*, 6, p. 629 (1905).

ingredients, a law which is also found to agree closely with observation on taking $a = 2$.

611. For certain other values of θ, the right hand of equation (602) (in which τ' is taken infinite) is found to be real and positive. We now have $r = 0$ and the solution (601) becomes

$$\chi = A e^{ipt} e^{\pm qx} \dots\dots\dots\dots\dots\dots\dots(605),$$

shewing that there is no wave-motion proper, but simply extinction of the light. Thus there are certain ranges of values of p (namely those which make $(q + ir)^2$ positive in equation (601)) for which light cannot be transmitted at all; these must represent absorption bands in the spectrum of the substance.

Clearly $(q + ir)^2$ becomes positive when θ is large and negative. It will be noticed that θ, as given by equation (598), becomes infinite when p has any of the values $n_1, n_2, \dots$, changing from $-\infty$ to $+\infty$ as p passes through these values. Thus the absorption bands will occur close to the frequencies of the natural vibrations of the molecule. But just in these regions we have to consider certain new physical agencies which cannot legitimately be neglected when p has values near to $n_1, n_2, \dots$, although probably negligible in other regions of the spectrum.

612. Equation (593) is not strictly true with the value we have assigned to Φ_s. For, in the first place the vibrations represented by the changes in ϕ_s are subject to dissipation on account of the radiation of light, and of this no account has been taken. In the second place there must be sudden forces acting in liquids and gases occasioned by molecular impacts and requiring the addition of terms to Φ_s throughout the short periods of these impacts. There must be analogous changes to be considered in the case of a solid, although our ignorance of the processes of molecular motion in a solid makes it impossible to specify them with any precision.

The effect of these agencies must be to throw the ϕ_s's of the different molecules out of phase with one another and also out of phase with X and Γ. The analysis of § 609 has made the ratios of $X : \Gamma : \phi_s$ wholly real (cf. equations (596) and (597)), indicating that X, Γ and ϕ_s are exactly in the same phase. The considerations just brought forward shew that these ratios ought also to contain small imaginary parts.

The process of separating real and imaginary parts in equation (602) now becomes much more complicated, but it will be obvious that for all values of p, both q and r will have some value different from zero. Thus there is always some extinction of light and some transmission, for all values of p, and there is no longer the sudden change from total extinction to perfect transmission. The edges of the absorption band become gradual and not sharp.

But the molecular model now in use probably does not represent the details of molecular action with sufficient truthfulness to make it worth trying to represent the conditions now under discussion in exact analysis.

Conducting media.

613. For a conducting medium we retain τ in equation (602), and on equating imaginary parts we obtain, in place of equation (572) of § 600,

$$qr = \frac{2\pi p \mu \tau'}{\tau'^2 + \dfrac{m^2 p^2}{N^2 e^4}} = \frac{2\pi \mu p}{\tau_p} \quad \ldots\ldots\ldots\ldots\ldots(606),$$

where τ_p is given by equation (586). Thus equation (573) of § 600 becomes replaced by

$$q = \frac{2\pi V \mu}{\tau_p} \quad \ldots\ldots\ldots\ldots\ldots\ldots\ldots\ldots(607).$$

For visible light this gives a very much smaller value of q than that discussed in § 600, and the value of q will obviously be still further modified by the considerations mentioned in § 612.

614. On comparing the total current, as given by formula (600), with the value $\dfrac{ipKX}{4\pi C}$ assigned to it in the analysis of §§ 594—598, we see that all this earlier analysis will apply to the present problem if we suppose K to be a complex quantity given by

$$K = \nu^2 + \frac{4\pi C^2}{ip\left(\tau' + \dfrac{m}{Ne^2}ip\right)} \quad \ldots\ldots\ldots\ldots\ldots(608),$$

where ν is given by formula (603).

If, as in § 603, we put

$$u^2 = \frac{K_2}{\mu_2} = (\alpha + i\beta)^2,$$

we find

$$\left.\begin{aligned}
\alpha^2 - \beta^2 &= \frac{1}{\mu_2}\left[\nu^2 - \frac{m}{Ne^2}\frac{4\pi C^2}{\tau'\tau_p}\right] \\
\alpha\beta &= -\frac{2\pi C^2}{p\tau_p\mu_2}
\end{aligned}\right\} \quad \ldots\ldots\ldots\ldots(609),$$

so that the reflecting power R of a metal may be calculated from equation (577) in terms of τ_p.

615. On comparing formulae (609) with experiment, the general result appears to emerge, that, in order to account for the optical properties of conductors in this way, the number of free electrons in conductors must be comparable with the number of atoms. According to a paper by Schuster, published in 1904*, the ratio of the number of free electrons to atoms ought to range from 1 to 3 in various substances; Nicholson†, as the result of a more elaborate investigation, obtains values for this ratio ranging from 2 to 7.

This result discloses a difficulty from which the electron theory, in the form in which we have so far considered it, has shewn little power of extricating itself.

Specific Heats and Electrical Conductivity.

616. According to the well-known law of Dulong and Petit the atomic heats of a large number of elements have values which are approximately all equal. Nernst and Lindemann have recently determined the specific heats of a large number of elements, and have found that, for all the elements they have examined, the atomic heats measured for constant volume (*i.e.* after correction for expansion arising out of change of temperature) have all the same value 5·95. Now the atomic heat represents the increase per unit rise of temperature in the energy of the solid measured per atom of its structure. This energy can be regarded as the sum of two contributions, namely the energy of the atoms and the energy of the free electrons. The energy of the atoms can be calculated by the well-known methods of the Kinetic Theory of matter, and it is found that this energy will provide a contribution to the atomic heat equal exactly to the total amount of the atomic heat, namely 5·95; in other words the contribution from the energy of the free electrons is as small as the experimental error. But the contribution from a given number of free electrons also admits of theoretical calculation if we make the assumption that their motion conforms to the ordinary dynamical laws. If there were as many free electrons as one-tenth of the number of atoms, the contribution to the atomic heat would be ·30, so that the total atomic heat would be 6·25, a number much too large to be reconciled with the experiments of Nernst and Lindemann.

617. The foregoing figures refer only to matter at comparatively high temperatures. The specific heats of the elements have however been determined by Nernst and Lindemann through a very wide range of temperatures, namely from normal temperatures down to the lowest temperatures now available in the laboratory. And it has recently been shewn by Debye that the atomic heats found by these experiments are, at all temperatures, almost exactly equal to those to be expected on theoretical grounds on the supposition that the free electrons contribute nothing to the specific heat. The observed atomic heats agree so well with those calculated from theory, for all substances examined and at all temperatures available, that the conclusion seems to be inevitable that the number of free electrons is very small compared with the number of atoms.

* *Phil. Mag.* February 1904.   † *Phil. Mag.* Aug. 1911.

618. Thus we are led to the conclusion that although the present electron theory may shew a certain power of explaining the optical properties of metals, qualitatively at least, yet this explanation demands the presence of a far greater number of free electrons than can be reconciled with the values of the specific heats.

If the present electron theory were in other respects satisfactory, the difficulty just revealed might be thought to constitute a serious defect in the electromagnetic theory of light. But the present electron theory is far from satisfactory in other respects; indeed a difficulty very similar to that just disclosed has been found to arise in connection with a much simpler phenomenon, namely the conductivity of metals. We have seen (§ 345 a) that the electron theory requires that in a good conductor the number of free electrons should be large; approximately how large it must be is a matter which can also be determined by further analysis. The requisite analysis has been given by Drude.

619. We may suppose, as a rough approximation to the truth, that in a conductor each free electron moves freely for a certain length of time t between two consecutive collisions with molecules. In the notation already used in § 345 a, the momentum gained in this time will be Xet. If we suppose this momentum to be entirely checked at each collision (cf. §§ 355, 373), the average forward momentum of all the electrons at any instant will be $\frac{1}{2}Xet$, and since this is equal to mu in the notation of § 345 a, we have

$$u = \frac{1}{2}\frac{Xe}{m}t \quad \dots\dots\dots\dots\dots\dots\dots(610),$$

and hence (by equation (b), § 345 a)

$$i = Neu = \frac{1}{2}\frac{Ne^2}{m}tX \quad \dots\dots\dots\dots\dots(611).$$

Thus the quantity γ of § 345 a is, as regards order of magnitude at least, equal to $\frac{2m}{t}$, and the specific resistance τ of a substance will be given by

$$\frac{1}{\tau} = \frac{1}{2}\frac{Ne^2}{m}t \quad \dots\dots\dots\dots\dots\dots(612),$$

where N is the number of free electrons per cubic centimetre. Now for silver or copper $\tau = 1\cdot6 \times 10^{-6}$ ohms $= 1\cdot8 \times 10^{-18}$ in electrostatic units. The value of $\frac{1}{2}\frac{e^2}{m}$ in electrostatic units is $1\cdot26 \times 10^8$, and hence to give to τ the value appropriate for silver or copper we must have $Nt = 5 \times 10^9$ approximately. In silver or copper the number of atoms per cubic centimetre is of the order of 10^{23}, so that if the observed values of the specific heats do not allow of N being more than one-hundredth part of this we must at most suppose that N is of the order of 10^{21}, and this requires t to be comparable with 5×10^{-12} at least.

Since the average velocity of the free electrons is believed to be about 10^7 cms. per second (§ 345 a), this would require each electron to travel an average distance of 5×10^{-5} cms. between consecutive violent collisions. This appears to be too large to be reconciled with present beliefs as to the structure of matter.

The difficulty becomes much worse when we consider the phenomenon at low temperatures. Kamerlingh Onnes has found for silver at a temperature of $13 \cdot 88°$ abs. a resistance only equal to $0 \cdot 7$ per cent. of that at $0°$ C. Thus in silver at this low temperature we must have Nt of the order of 10^{12}, so that if we take $N = 10^{21}$ as above, $t = 10^{-9}$. This velocity of free electrons at this low temperature is of the order of 2×10^6, so that the average distance travelled would be about $\frac{1}{500}$ cm.

620.   We have now found that contradictions exist in connection with the Electromagnetic Theory of Light, the theory of Specific Heats of metals, and the theory of Electric Conductivity, so long as we treat these questions in terms of ordinary dynamical laws and Maxwell's electromagnetic equations. A large accumulation of evidence, of which our discussion has touched only on a small fringe, suggests that a new system of dynamics and a new electron theory is needed. So far as can be seen the special feature of this new theory must be that the interaction between electrons and radiation is of an entirely different nature from that imagined by the classical laws. The new theory is in existence and is generally known as the Quantum-theory. A brief introduction to it will be found in the last chapter of the present book.

CHAPTER XIX

THE MOTION OF ELECTRONS

GENERAL EQUATIONS.

621. THE motion of an electron or other electric charge gives rise to a system of displacement currents, which in turn produce a magnetic field. The changes in this magnetic field give rise to new electric forces, and so on. Thus the motion of electrons or other charges is accompanied by magnetic and electric fields, mutually interacting. To examine the nature and effects of these fields is the object of the present chapter.

The necessary equations have already been obtained in § 574, but the current u, v, w will now be regarded as produced by the motion of charged bodies. If at any point x, y, z there is a volume density ρ of electricity moving with a velocity of components U, V, W, then the current at x, y, z has components $\rho U, \rho V, \rho W$ in electrostatic units. Since u, v, w in equations (529) are measured in electromagnetic units, they must be replaced by $\rho U/C, \rho V/C, \rho W/C$, and the equations become

$$\frac{4\pi}{C}\left(\rho U + \frac{df}{dt}\right) = \frac{\partial \gamma}{\partial y} - \frac{\partial \beta}{\partial z}, \text{ etc.} \quad \ldots\ldots\ldots\ldots\ldots(613).$$

Equations (528), namely

$$-\frac{1}{C}\frac{da}{dt} = \frac{\partial Z}{\partial y} - \frac{\partial Y}{\partial z}, \text{ etc.} \quad \ldots\ldots\ldots\ldots\ldots(614),$$

remain unaltered, and the two sets of equations (613) and (614) provide the material for our present discussion.

When we had these same equations under review in § 574, C was regarded merely as the ratio of the units. We may now regard C as being the velocity of light, this being also the velocity of any other electromagnetic disturbance in free space.

622. If we differentiate equations (613) with respect to x, y, z and add, we obtain

$$\frac{\partial}{\partial x}(\rho U) + \frac{\partial}{\partial y}(\rho V) + \frac{\partial}{\partial z}(\rho W) = -\frac{d}{dt}\left(\frac{\partial f}{\partial x} + \frac{\partial g}{\partial y} + \frac{\partial h}{\partial z}\right).$$

We have also, as an equation of continuity, expressing that the increase in ρ in any small element is accounted for by the flow of electricity across the faces by which the element is bounded,

$$\frac{\partial}{\partial x}(\rho U) + \frac{\partial}{\partial y}(\rho V) + \frac{\partial}{\partial z}(\rho W) + \frac{d\rho}{dt} = 0.$$

By comparison with the equation just obtained, we have

$$\frac{d}{dt}\left(\frac{\partial f}{\partial x}+\frac{\partial g}{\partial y}+\frac{\partial h}{\partial z}\right)=\frac{d\rho}{dt},$$

of which the integral is our former equation (63), namely

$$\frac{\partial f}{\partial x}+\frac{\partial g}{\partial y}+\frac{\partial h}{\partial z}=\rho \quad\ldots\ldots\ldots\ldots\ldots\ldots\ldots(615).$$

Similarly, on differentiating equations (614) with respect to x, y, z and adding, we obtain

$$\frac{d}{dt}\left(\frac{\partial a}{\partial x}+\frac{\partial b}{\partial y}+\frac{\partial c}{\partial z}\right)=0,$$

of which the integral is our former equation (362), namely

$$\frac{\partial a}{\partial x}+\frac{\partial b}{\partial y}+\frac{\partial c}{\partial z}=0 \quad\ldots\ldots\ldots\ldots\ldots\ldots\ldots(616).$$

623. At a point at which there is no electric charge ($\rho = 0$), equations (613) and (614) become identical with the systems of equations (A) and (B) of § 577, and the quantities X, Y, Z, α, β, γ must all satisfy the differential equation (534), namely

$$\frac{d^2\chi}{dt^2}=a^2\nabla^2\chi \quad\ldots\ldots\ldots\ldots\ldots\ldots\ldots(617).$$

FORCE OF A MOVING ELECTRON.

624. Consider afresh the problem of which a preliminary discussion has already been given in § 572, of a single electron moving with a velocity v parallel to Ox. Since the field necessarily moves with the electron, the rate of change of any quantity χ as we follow it in its motion must be *nil*. Thus we must have

$$\left(\frac{d}{dt}+v\frac{\partial}{\partial x}\right)\chi=0$$

so that $\dfrac{d}{dt}$ may be replaced by $-v\dfrac{\partial}{\partial x}$ throughout our equations.

Equation (617) becomes

$$v^2\frac{\partial^2\chi}{\partial x^2}=a^2\left(\frac{\partial^2\chi}{\partial x^2}+\frac{\partial^2\chi}{\partial y^2}+\frac{\partial^2\chi}{\partial z^2}\right),$$

or, since $a^2 = C^2/K\mu$,

$$\left(1-\frac{K\mu v^2}{C^2}\right)\frac{\partial^2\chi}{\partial x^2}+\frac{\partial^2\chi}{\partial y^2}+\frac{\partial^2\chi}{\partial z^2}=0 \quad\ldots\ldots\ldots\ldots\ldots(618).$$

Also equations (613), (614) assume the forms

$$\frac{4\pi v}{C}\left(\rho-\frac{\partial f}{\partial x}\right)=\frac{\partial\gamma}{\partial y}-\frac{\partial\beta}{\partial z} \quad\ldots\ldots\ldots\ldots\ldots(619),$$

$$\frac{v}{C}\frac{\partial a}{\partial x}=\frac{\partial Z}{\partial y}-\frac{\partial Y}{\partial z} \quad\ldots\ldots\ldots\ldots\ldots(620).$$

625. In most problems, the velocity of motion v is small compared with the velocity of light, so that v/C may be treated as a small quantity. Equation (619) shews that the magnetic field set up by a moving charge may be regarded as small if v/C is small. The same is of course true of the field set up by any number of moving charges provided all their velocities are small compared with that of light.

When v/C is small, equation (620) shews that

$$\frac{\partial Z}{\partial y} - \frac{\partial Y}{\partial z}$$

will be a small quantity of the second order. Let us suppose, until the contrary is stated, that v/C is so small for each moving charge that v^2/C^2 may legitimately be neglected. Then

$$\frac{\partial Z}{\partial y} - \frac{\partial Y}{\partial z} = 0, \text{ etc.,}$$

so that the forces X, Y, Z are derivable from a potential Ω. When v^2/C^2 is neglected equation (618) reduces to $\nabla^2 \chi = 0$. This equation is satisfied by X, Y, Z separately, and therefore also by Ω. Since X, Y, Z also satisfy equation (615), or

$$\frac{\partial X}{\partial x} + \frac{\partial Y}{\partial y} + \frac{\partial Z}{\partial z} = 4\pi\rho,$$

it is clear that the values of X, Y, Z are exactly the same as if the moving charge were instantaneously at rest.

626. This is exactly the assumption we made in § 572 in calculating the magnetic force from a moving charge. The forces there calculated, namely

$$\alpha = 0, \quad \beta = -\frac{v}{C}\frac{ez}{r^3}, \quad \gamma = \frac{v}{C}\frac{ey}{r^3} \quad \ldots\ldots\ldots\ldots\ldots(621),$$

are now seen to be accurate provided v^2/C^2 may be neglected, but not otherwise.

THE FORCE ACTING ON A MOVING ELECTRON.

627. The assumption we have made that v/C is small is the same as assuming to a first approximation that C is so great that the medium may be supposed to adjust itself instantaneously to changes occurring in it, just as an incompressible fluid would do. The time taken for action to pass from one point to another may be neglected. We may accordingly assume that at any instant the mechanical actions of any two parts of the field upon one another are such that action and reaction are equal and opposite.

From equations (621), it appears that an electron moving with velocity $v, 0, 0$ at the origin will exert a force of components

$$0, \quad -\frac{Ve}{C}\frac{mz}{r^3}, \quad \frac{Ve}{C}\frac{my}{r^3}$$

upon a magnetic pole of strength m at x, y, z. It follows that a magnetic pole of strength m at x, y, z will exert a force of components

$$0, \quad \frac{Ve}{C}\frac{mz}{r^3}, \quad -\frac{Ve}{C}\frac{my}{r^3} \quad \dots\dots\dots\dots\dots\dots(622)$$

upon the moving electron at the origin.

628. If we have a number of magnetic poles, the resultant force upon the moving electron has components

$$0, \quad \frac{Ve}{C}\Sigma\frac{mz}{r^3}, \quad -\frac{Ve}{C}\Sigma\frac{my}{r^3} \quad \dots\dots\dots\dots\dots(623)$$

and the components of magnetic force at the origin are given by (cf. § 408)

$$\alpha = -\Sigma\frac{mx}{r^3}, \text{ etc.}$$

Thus the force on the moving electron may be put in the form

$$0, \quad -\frac{Ve}{C}\gamma, \quad \frac{Ve}{C}\beta \dots\dots\dots\dots\dots\dots(624).$$

Plainly the force on the electron will be given by formulae (624), whether the magnetic field arises from poles of permanent magnetism or not. It is clearly a force at right angles both to the direction of motion of the electron, and to the magnetic force α, β, γ at the point. If H is the resultant magnetic force, and θ the angle between the directions of H and the axis of x, then the resultant of the mechanical force is $veH \sin\theta/C$.

629. If the electron has components of velocity u, v, w, the component of the mechanical force on it will be

$$\frac{e}{C}(\gamma v - \beta w), \quad \frac{e}{C}(\alpha w - \gamma u), \quad \frac{e}{C}(\beta u - \alpha v)\dots\dots\dots(625).$$

Since the mechanical force is always perpendicular to the direction of motion, it does no work on the moving particle; and, in particular, if a charged particle moves freely in a magnetic field, its velocity remains constant.

The existence of this force explains the mechanism by which an induced current is set up in a wire moved across magnetic lines of force. The force (625) has its direction along the wire and so sets each electron into motion, producing a current proportional jointly to the velocity and strength of the field—*i.e.* to dN/dt.

The "Hall Effect."

630. Very direct evidence of the existence of this force is provided by the "Hall Effect." Hall[*] found that when a metallic conductor conveying a current is placed in a magnetic field, the lines of flow rearrange themselves as they would under a superposed electromotive force at right angles both to the direction of the current and of the magnetic field. The same effect has also been detected in electrolytes and in gases.

The Hall Effect is of interest as exhibiting a definite point of divergence between Maxwell's original theory and the modern electron theory. According to Maxwell's theory, a magnetic field could act only on the material conductor conveying a current, and not on the current itself, so that if the conductor was held at rest the lines of flow ought to remain unaltered[†]. The electron theory, confirmed by the experimental evidence of the Hall Effect, shews that this is not so, and that the lines of flow must be altered in the presence of a transverse magnetic field.

Motion of a charged particle in a uniform magnetic field.

631. Let a particle of charge e move freely in a uniform magnetic field of intensity H. Let its velocity be resolved into a component A parallel to the lines of force, and a component B in the plane perpendicular to them. By what has just been said (§ 629) both A and B must remain constant throughout the motion, and there will be a force eHB/C acting on the particle in a direction perpendicular to that of B, and in the plane perpendicular to the lines of force. Thus if m is the mass of the particle, its acceleration must be eHB/mC in this same direction.

Considering only the motion in a plane perpendicular to the lines of force, we have a velocity B and an acceleration eHB/mC perpendicular to it. This latter must be equal to B^2/ρ, where ρ is the curvature of the path. Thus $\rho = \dfrac{BmC}{eH}$, a constant, shewing that the motion in question is circular.

Combining this circular motion with the motion parallel to the lines of force we find that the complete orbit is a circular helix, of radius BmC/eH, described about one of the lines of magnetic force as axis.

632. By measuring the curvature of an orbit described in this manner, it is found possible to determine e/m experimentally for electrons and other charged particles (cf. § 665 below). Incidentally the fact that curvature is observed at all provides experimental confirmation of the existence of the force acting on a moving electron.

* *Phil. Mag.* 9 (1880), p. 225.
† Maxwell, *Electricity and Magnetism*, § 501.

The Zeeman Effect.

633. When a source of light emitting a line-spectrum is placed in a strong magnetic field, the lines of the spectrum are observed to undergo certain striking modifications. The simplest form assumed by the phenomenon is as follows.

If the light is examined in a direction parallel to the lines of magnetic force, each of the spectral lines appears split into two lines, on opposite sides of, and equidistant from, the position of the original line, and the light of these two lines is found to be circularly polarised, the direction of polarisation being different for the two.

If the light is examined across the lines of force, these same two lines appear, accompanied now by a line at the original position of the line, so that the original line now appears split into three. The side lines are observed to be plane polarised in a plane through the line of sight and the lines of force, while the middle line is plane polarised in a plane perpendicular to the lines of force.

634. These various phenomena were observed by Zeeman in 1896, and an explanation in terms of the electron theory was at once suggested by Lorentz.

Let us first examine a simple artificial case in which the spectrum contains one line only, assumed to be produced by the oscillations of a single electron about a position of equilibrium.

If the frequency of this oscillation is $p/2\pi$, the equations of motion of the electron must be of the form

$$m \frac{d^2x}{dt^2} = - mp^2x, \text{ etc.} \dots\dots\dots\dots\dots\dots\dots(626),$$

in which x, y, z are the coordinates of the electron referred to its position of equilibrium.

Next suppose the electron to move in a field of force of intensity H parallel to the axis of x. In addition to the force of restitution of components $- mp^2x$, $- mp^2y$, $- mp^2z$, the electron will be acted on by a force (cf. formulae (625)) of components

$$0, \quad \frac{eH}{C} \frac{dz}{dt}, \quad -\frac{eH}{C} \frac{dy}{dt}.$$

In place of the former equations, the equations of motion are now

$$\left.\begin{aligned}
m \frac{d^2x}{dt^2} &= - mp^2x \\
m \frac{d^2y}{dt^2} &= - mp^2y + \frac{eH}{C} \frac{dz}{dt} \\
m \frac{d^2z}{dt^2} &= - mp^2z - \frac{eH}{C} \frac{dy}{dt}
\end{aligned}\right\} \dots\dots\dots\dots\dots\dots(627).$$

The solutions of these equations are

$$x = A \cos (pt - \epsilon),$$

$$y = A_1 \cos (q_1 t - \epsilon_1) + A_2 \cos (q_2 t - \epsilon_2),$$

$$z = A_1 \sin (q_1 t - \epsilon_1) + A_2 \sin (q_2 t - \epsilon_2),$$

in which A, A_1, A_2, ϵ, ϵ_1, ϵ_2 are constants, and q_1, q_2 are the roots of

$$- mq^2 = - mp^2 + \frac{eH}{C} q \quad \dots\dots\dots\dots\dots\dots(628).$$

For even the strongest fields which are available in the laboratory, the value of the last term in this equation is small compared with that of the other terms, so that the solution of equation (628) may be taken to be

$$q = \pm p - \frac{eH}{2mC}.$$

The original vibrations of the electron, all of frequency p, may now be replaced by the three following vibrations:

I. $x = A \cos (pt - \epsilon)$, $y = 0$, $z = 0$.

II. $x = 0$, $y = A_1 \cos \left[\left(p + \dfrac{eH}{2mC} \right) t - \epsilon_1 \right]$, $z = A_1 \sin \left[\left(p + \dfrac{eH}{2mC} \right) t - \epsilon_1 \right]$.

III. $x = 0$, $y = A_2 \cos \left[\left(p - \dfrac{eH}{2mC} \right) t - \epsilon_2 \right]$, $z = -A_2 \sin \left[\left(p - \dfrac{eH}{2mC} \right) t - \epsilon_2 \right]$.

Vibration I of frequency p is a linear motion of the electron parallel to Ox, the direction of the lines of magnetic force. The magnetic force in the emitted radiation is accordingly always parallel to the plane of yz and vanishes immediately behind and in front of the electron (cf. § 618). Thus there is no radiation emitted in the direction of the axis of x, and the radiation emitted in the plane of yz will be polarised (§ 592) in this plane.

Vibrations II and III represent circular motions in the plane of yz of frequencies $p \pm \dfrac{eH}{2mC}$. Clearly the radiation emitted along the axis of x will be circularly polarised, while that emitted in the plane of yz will be plane polarised in a plane through the line Ox and the line of sight (the motion along the line of sight sending no radiation in this direction). Thus the observed appearances are accounted for.

635.   The analysis just given explains the observed facts of the normal Zeeman Effect, but only in terms of a model which is known not to be in accordance with the actual facts of atomic structure. As was pointed out by Larmor, the explanation just given can be easily generalised so that the atomic model shall at least accord better with the facts of nature than that we have just had under discussion.

If an electron is moving in a field of magnetic force of intensity H parallel to the axis of x, its equations of motion will be

$$\left.\begin{array}{l} m\dfrac{d^2x}{dt^2} = F_x \\[2mm] m\dfrac{d^2y}{dt^2} = F_y + \dfrac{eH}{C}\dfrac{dz}{dt} \\[2mm] m\dfrac{d^2z}{dt^2} = F_z - \dfrac{eH}{C}\dfrac{dy}{dt} \end{array}\right\} \quad \ldots\ldots\ldots\ldots\ldots\ldots(629),$$

where F_x, F_y, F_z are the components of the force which acts on the electron apart from the superimposed magnetic field H. These equations of course contain equations (627) as a special case.

If x, y, z were coordinates measured with reference to a system of axes rotating with uniform angular velocity ω about the axis of x in the direction from Oy to Oz, the component of the velocity of the point x, y, z in space would be given by

$$\left.\begin{array}{l} u = \dfrac{dx}{dt} \\[2mm] v = \dfrac{dy}{dt} - \omega z \\[2mm] w = \dfrac{dz}{dt} + \omega y \end{array}\right\} \quad \ldots\ldots\ldots\ldots\ldots\ldots(630),$$

and the accelerations in space by

$$\frac{Dv}{Dt} = \frac{dv}{dt} + u\frac{\partial v}{\partial x} + v\frac{\partial v}{\partial y} + w\frac{\partial v}{\partial z} = \frac{d^2y}{dt^2} - 2\omega\frac{dz}{dt} - \omega^2 y \ldots\ldots\ldots(631),$$

and similar equations.

When the angular velocity is so small that ω^2 may be neglected, the system of accelerations, as given by equations such as (631), reduce to

$$\left.\begin{array}{l} \dfrac{Du}{Dt} = \dfrac{d^2x}{dt^2} \\[2mm] \dfrac{Dv}{Dt} = \dfrac{d^2y}{dt^2} - 2\omega\dfrac{dz}{dt} \\[2mm] \dfrac{Dw}{Dt} = \dfrac{d^2z}{dt^2} + 2\omega\dfrac{dy}{dt} \end{array}\right\} \quad \ldots\ldots\ldots\ldots\ldots\ldots(632).$$

Thus if ω is defined by the equation

$$\omega = \frac{eH}{2mC} \quad \ldots\ldots\ldots\ldots\ldots\ldots\ldots(633),$$

the equations of motion (629) of the electron in the superimposed magnetic field become

$$m\frac{Du}{Dt} = F_x,$$

$$m\frac{Dv}{Dt} = F_y,$$

$$m\frac{Dw}{Dt} = F_z,$$

which would be precisely the equations of motion of the electron referred to axes at rest with the magnetic field non-existent. Thus the superposition of the magnetic field parallel to the axis of x is seen to have had precisely the same effect on electronic motion as the setting of the axes in rotation with an angular velocity ω defined by equation (633).

Before the magnetic field is superposed, let the electron describe a path such that when its coordinates are resolved into simple-harmonic terms by Fourier's theorem, one of the constituent simple-harmonic vibrations is of the form

$$x = 0, \quad y = A_1 \cos(pt - \epsilon_1), \quad z = A_1 \sin(pt - \epsilon_1).$$

Assume that one of the lines in the spectrum of the atom when in its natural state corresponds to a frequency $p/2\pi$. The superposition of a magnetic field has the same effect on the coordinates x, y, z as the setting of the axes in rotation with an angular velocity ω, so that when this field is superposed the coordinates of the electron may be taken to be

$$x = 0, \quad y = A_1 \cos[(p+\omega)t - \epsilon_1], \quad z = A_1 \sin[(p+\omega)t - \epsilon_1].$$

It is at once seen that the vibration is identical in general type with the vibration II that we found in § 634, so that the discussion of the polarisation and change of frequency there given will apply to the present case also.

636. The discussion of the last section is applicable to any electron describing an orbit such that its motion can be resolved into oscillations of definite frequencies. It shews that each spectral line ought in general to be resolved into a triplet of three equidistant lines, a line initially at p giving place to lines at $p \pm \delta p$ where

$$\delta p = \frac{e}{2mC} H \quad\quad\quad\quad\quad\quad\quad\quad\quad\quad\quad\text{.....(634).}$$

This represents what is normally observed, and a formation of triplets of this type is commonly spoken of as the normal Zeeman Effect. Certain lines separate out in a more complex way in the presence of a magnetic field, these lines having generally appeared as multiple lines (doublets, triplets, etc.) even before the magnetic field was turned on. This is known as the complex or abnormal Zeeman Effect, and is not covered by our simple theory.

In the normal Zeeman Effect, the frequency difference, predicted by equation (634), is constant for all the lines of the spectrum. Observationally this is found to be the case, and equation (634) makes it possible to determine a value of e/m from the observed separation of spectral lines in a magnetic field of known strength. The value so obtained proves to be in good agreement with values for e/m measured by other and more direct methods.

The General Equations of Moving Electrons.

637. We now return to the general equations of § 621, namely

$$\frac{4\pi}{C}\left(\rho U + \frac{df}{dt}\right) = \frac{\partial \gamma}{\partial y} - \frac{\partial \beta}{\partial z}, \text{ etc. } \quad \ldots\ldots\ldots\ldots(635),$$

$$-\frac{1}{C}\frac{da}{dt} = \frac{\partial Z}{\partial y} - \frac{\partial Y}{\partial z}, \text{ etc. } \quad \ldots\ldots\ldots\ldots(636),$$

and discuss the field set up by the motion of electric charges when there is no restriction as to the smallness of their velocities.

On multiplying both sides of equation (635) by μ and differentiating with respect to the time, we obtain

$$\frac{4\pi\mu}{C}\frac{d}{dt}\left(\rho U + \frac{df}{dt}\right) = \frac{d}{dt}\left(\frac{\partial c}{\partial y} - \frac{\partial b}{\partial z}\right).$$

Using relations (636) we readily find that the right-hand member

$$= -C\left[\frac{\partial}{\partial y}\left(\frac{\partial Y}{\partial x} - \frac{\partial X}{\partial y}\right) - \frac{\partial}{\partial z}\left(\frac{\partial X}{\partial z} - \frac{\partial Z}{\partial x}\right)\right]$$

$$= C\left[\nabla^2 X - \frac{\partial}{\partial x}\left(\frac{\partial X}{\partial x} + \frac{\partial Y}{\partial y} + \frac{\partial Z}{\partial z}\right)\right].$$

Putting $4\pi f = KX$ and $\dfrac{\partial X}{\partial x} + \dfrac{\partial Y}{\partial y} + \dfrac{\partial Z}{\partial z} = 4\pi\dfrac{\rho}{K}$, this becomes

$$\nabla^2 X - \frac{K\mu}{C^2}\frac{d^2 X}{dt^2} = \frac{4\pi\mu}{C^2}\frac{d}{dt}(\rho U) + \frac{4\pi}{K}\frac{\partial \rho}{\partial x}.$$

638. This is the differential equation satisfied by X. Similar equations are of course satisfied by Y and Z. If we divide both sides of equation (636) by μ and differentiate with respect to the time, we readily find that α satisfies the differential equation

$$\nabla^2 \alpha - \frac{K\mu}{C^2}\frac{d^2 \alpha}{dt^2} = -\frac{4\pi}{C}\left[\frac{\partial}{\partial y}(\rho W) - \frac{\partial}{\partial z}(\rho V)\right].$$

We shall shortly obtain these differential equations in another way.

Introduction of the Potentials.

639. With equations (636) we may combine the relation

$$\frac{\partial a}{\partial x} + \frac{\partial b}{\partial y} + \frac{\partial c}{\partial z} = 0 \quad \dots\dots\dots\dots\dots(637)$$

(equation (616)), and it follows, as in § 443, that we can find a vector-potential of components F, G, H connected with a, b, c by the relations

$$a = \frac{\partial H}{\partial y} - \frac{\partial G}{\partial z}, \text{ etc.} \dots\dots\dots\dots\dots(638),$$

and with X, Y, Z by the relations (cf. § 530)

$$X + \frac{1}{C}\frac{dF}{dt} = -\frac{\partial \Psi}{\partial x}, \text{ etc.} \dots\dots\dots\dots\dots(639),$$

in which Ψ is a function, at present undetermined in the general case, which becomes identical with the electrostatic potential when there is no motion.

640. We have seen (§ 442) that equations (638) are not adequate to determine F, G, H completely, and hence Ψ also (cf. equation (639)) is not fully determined.

Let F_0, G_0, H_0, Ψ_0 be any special set of values satisfying equations (638) and (639). Then the most general values of F, G, H are given by (cf. § 442)

$$F = F_0 + \frac{\partial \chi}{\partial x}, \text{ etc.} \dots\dots\dots\dots\dots(640),$$

where χ is any arbitrary single-valued function.

To find the most general value of Ψ, we have from equation (639)

$$-\frac{\partial \Psi}{\partial x} = X + \frac{1}{C}\left(\frac{dF_0}{dt} + \frac{\partial^2 \chi}{\partial x \partial t}\right) = -\frac{\partial \Psi_0}{\partial x} + \frac{1}{C}\frac{\partial^2 \chi}{\partial x \partial t},$$

and similar equations in y and z, so that, on integration,

$$\Psi = \Psi_0 - \frac{1}{C}\frac{\partial \chi}{\partial t} + f(t), \text{ a function of } t \text{ only} \dots\dots\dots(641).$$

From (640) and (641) we obtain

$$\frac{\partial F}{\partial x} + \frac{\partial G}{\partial y} + \frac{\partial H}{\partial z} + \frac{K\mu}{C}\frac{\partial \Psi}{\partial t} = \frac{\partial F_0}{\partial x} + \frac{\partial G_0}{\partial y} + \frac{\partial H_0}{\partial z} + \frac{K\mu}{C}\frac{\partial \Psi_0}{\partial t} + \nabla^2 \chi - \frac{K\mu}{C^2}\frac{\partial^2 \chi}{\partial t^2}$$
$$+ \frac{K\mu}{C}f'(t)\dots\dots\dots(642).$$

The function χ is entirely at our disposal, so that

$$\nabla^2 \chi - \frac{K\mu}{C^2}\frac{\partial^2 \chi}{\partial t^2}$$

may have any value we please to assign to it. Let us agree to give to χ such a value, for every instant of time and all values of x, y, z, as shall make the right-hand member of equation (642) vanish.

The value of χ is now fixed, except for a set of values of χ such that

$$\nabla^2\chi - \frac{K\mu}{C^2}\frac{\partial^2\chi}{\partial t^2} = 0$$

at every instant and point, these values of χ representing of course contributions that might arise from a set of disturbances propagated through the medium from outside.

Except for such additional values of χ, the values of F, G, H, Ψ are now uniquely determined by equations (640) and (641). The vector-potential will in future mean the special vector of which these values of F, G, H are the components, while the corresponding special value of Ψ will be called the " Electric Potential."

From equation (642) it follows that the vector-potential and the electric potential are connected by the relation

$$\frac{\partial F}{\partial x} + \frac{\partial G}{\partial y} + \frac{\partial H}{\partial z} = -\frac{K\mu}{C}\frac{\partial\Psi}{\partial t} \quad\ldots\ldots\ldots\ldots\ldots(643).$$

Differential Equations satisfied by the Potentials.

641. If we differentiate equations (639) with respect to x, y, z and add, we obtain

$$\left(\frac{\partial X}{\partial x} + \frac{\partial Y}{\partial y} + \frac{\partial Z}{\partial z}\right) + \frac{1}{C}\frac{d}{dt}\left(\frac{\partial F}{\partial x} + \frac{\partial G}{\partial y} + \frac{\partial H}{\partial z}\right) = -\nabla^2\Psi,$$

which, on substituting from equations (643) and (639), becomes

$$\nabla^2\Psi - \frac{K\mu}{C^2}\frac{d^2\Psi}{dt^2} = -\frac{4\pi\rho}{K} \quad\ldots\ldots\ldots\ldots\ldots(644),$$

the differential equation satisfied by Ψ. We notice that for a steady field it becomes identical with Poisson's equation, while in regions in which there are no charges it becomes identical with the equation of wave-propagation.

642. To obtain the differential equation satisfied by F, we transform equation (635) by the use of equations (638). We have

$$\frac{4\pi\mu}{C}\left(\rho U + \frac{df}{dt}\right) = \frac{\partial c}{\partial y} - \frac{\partial b}{\partial z}$$

$$= \frac{\partial}{\partial y}\left(\frac{\partial G}{\partial x} - \frac{\partial F}{\partial y}\right) - \frac{\partial}{\partial z}\left(\frac{\partial F}{\partial z} - \frac{\partial H}{\partial x}\right)$$

$$= \frac{\partial}{\partial x}\left(\frac{\partial F}{\partial x} + \frac{\partial G}{\partial y} + \frac{\partial H}{\partial z}\right) - \nabla^2 F,$$

whence, from equations (643) and (639),

$$\nabla^2 F - \frac{K\mu}{C^2}\frac{d^2 F}{dt^2} = -\frac{4\pi\mu}{C}\rho U \quad\ldots\ldots\ldots\ldots\ldots(645),$$

the differential equation satisfied by F. Similar equations are of course satisfied by G and H.

Differential Equations satisfied by the Forces.

643. Operating on equation (639) with the operator $\nabla^2 - \dfrac{K\mu}{C^2}\dfrac{d^2}{dt^2}$, we have

$$\nabla^2 X - \frac{K\mu}{C^2}\frac{d^2 X}{dt^2} = -\frac{1}{C}\frac{d}{dt}\left(\nabla^2 F - \frac{K\mu}{C^2}\frac{d^2 F}{dt^2}\right) - \frac{\partial}{\partial x}\left(\nabla^2 \Psi - \frac{K\mu}{C^2}\frac{d^2\Psi}{dt^2}\right)$$

$$= \frac{4\pi\mu}{C^2}\frac{d}{dt}(\rho U) + \frac{4\pi}{K}\frac{\partial\rho}{\partial x} \quad\ldots\ldots\ldots\ldots\ldots\ldots\ldots(646).$$

This is the differential equation satisfied by X, and similar equations are satisfied by Y and Z. These same equations were obtained by a more direct method in § 637.

644. For the differential equation satisfied by α, β, γ we have, from equations (638) and (645),

$$\nabla^2\alpha - \frac{K\mu}{C^2}\frac{d^2\alpha}{dt^2} = \frac{1}{\mu}\left(\nabla^2 - \frac{K\mu}{C^2}\frac{d^2}{dt^2}\right)\left(\frac{\partial H}{\partial y} - \frac{\partial G}{\partial z}\right)$$

$$= -\frac{4\pi}{C}\left\{\frac{\partial(\rho W)}{\partial y} - \frac{\partial(\rho V)}{\partial z}\right\} \quad\ldots\ldots\ldots\ldots(647),$$

and similar equations for β and γ. These equations agree with those already obtained in § 638.

Solution of the Differential Equations.

645. It will be seen that all the differential equations are of the same general form, namely

$$\nabla^2\chi - \frac{1}{a^2}\frac{d^2\chi}{dt^2} = -4\pi\sigma \quad\ldots\ldots\ldots\ldots\ldots(648),$$

where σ arises from electric charges, at rest or in motion.

Clearly the value of χ may be regarded as the sum of contributions from the values of σ in the different small elements of volume. The simplest solution for χ is that arising from a distribution of σ at and close to the origin, σ being zero everywhere else.

For this special solution χ is a function of r only, which must satisfy

$$\nabla^2\chi = \frac{1}{a^2}\frac{d^2\chi}{dt^2}$$

everywhere except at the origin. Proceeding as in § 578, and rejecting the term which represents convergent waves, as having no physical importance, we obtain the solution (cf. equation (536))

$$\chi = \frac{1}{r}f(r - at) \quad\ldots\ldots\ldots\ldots\ldots\ldots(649),$$

where f is so far a perfectly arbitrary function.

Close to the origin, this reduces to

$$\chi = \frac{1}{r} f(-at) \dots\dots\dots\dots\dots\dots(650),$$

and it now appears that in equation (648) the middle term becomes insignificant near the origin in comparison with the first term $\nabla^2\chi$. Thus close to the origin the equation becomes identical with Poisson's equation, and the integral is

$$\chi = \frac{\iiint \sigma\, dx\, dy\, dz}{r} = \frac{\tau}{r} \dots\dots\dots\dots\dots(651),$$

where the integral is taken only through the element of volume at the origin in which σ exists, and τ represents the integral of σ taken through this element of volume.

On comparing solutions (650) and (651), both of which are true near the origin, we find that

$$f(-at) = \tau \dots\dots\dots\dots\dots\dots(652),$$

and this determines the function f completely. The general solution (649) is now fully known, and by summation of such solutions the general solution of equation (648) is obtained.

Let P, Q be any points distant r apart; let t be any instant of time, and let t_0 denote the instant of time r/a previous to it, so that $t_0 = t - r/a$. Clearly t_0 is the instant of departure from P of a disturbance reaching Q at t. We may speak of t_0 as the "retarded time" at P corresponding to the time t at Q.

With this meaning assigned to t_0, we have

$$f(r - at) = f\left\{ -a\left(t - \frac{r}{a} \right) \right\} = f(-at_0) = \tau,$$

where τ is evaluated at time t_0 (cf. equation (652)). If we agree to denote by $[\phi]$ the value of ϕ estimated at the retarded time at the point at which ϕ occurs, then this value of τ will be expressed by $[\tau]$, and solution (649) becomes

$$\chi = \frac{[\tau]}{r} \dots\dots\dots\dots\dots\dots(653).$$

The most general solution of equation (648), obtained by the summation of solutions such as (653), is

$$\chi = \iiint \frac{[\sigma]\, dx\, dy\, dz}{r} = \Sigma \frac{[\tau]}{r} \dots\dots\dots\dots(654),$$

the last form applying when the distribution of σ occurs only at points or in small regions so small that the variations of the retardation of time through each region are negligible.

The analogy of Poisson's equation and its solution in electrostatics (cf. §§ 49, 40, 41) is obvious.

646. From equations (644) and (645) it follows that the potentials are given by

$$\Psi = \frac{1}{K}\iiint \frac{[\rho]\,dx\,dy\,dz}{r} \quad \ldots\ldots\ldots\ldots\ldots\ldots\ldots(655),$$

$$F = \frac{\mu}{C}\iiint \frac{[\rho U]\,dx\,dy\,dz}{r}\,,\ \text{etc.}\ \ldots\ldots\ldots\ldots\ldots(656).$$

These potentials are commonly spoken of as "Retarded Potentials." They differ from the ordinary potentials, in which the finite velocity of propagation is not taken into account, only in that the quantities in the numerators must be evaluated at the retarded times appropriate to the point.

The solution of equations (646) and (647) may be similarly written down, but it is usually easier to evaluate the forces by differentiation of the potentials.

If the moving electrons in formula (656) are conveying currents in linear circuits, the formula becomes (on taking $\mu = 1$)

$$F = \Sigma \int \frac{[i_x]\,ds}{r}\,,$$

where the summation is over the different circuits and i_x denotes the x-component of the current, which may also be expressed as $i\dfrac{dx}{ds}$. This formula may be compared with (419), from which it differs only in that it takes account of the finite time required for the propagation of electromagnetic action.

The Field set up by Moving Electrons.

647. An electron is a charge of total amount e spread through a very small volume. When we attempt to apply the equations just obtained to the motion of electrons, a complication arises. We must not integrate ρ or ρU through the space occupied by the electron because the retarded time varies from one part of the electron to another. And this complication does not disappear even when we pass to the limit and suppose the electron to be of infinitesimal size.

Let the electron be moving with a velocity (not necessarily uniform) of which the components at any instant are U, V, W. Suppose we wish to evaluate the potentials at x', y', z' at time t.

Let x, y, z be the position of any element of the electron at the retarded time t_0, defined by

$$t_0 = t - \frac{r}{a} \text{ where } r^2 = (x' - x)^2 + (y' - y)^2 + (z' - z)^2.$$

We may speak of x, y, z as the effective position of the element of the electron under consideration, since the element contributes to the potentials we are in search of, only when it is at x, y, z.

The retarded time t_0 will be different for different parts of the electron. Let its value at the centre of the electron be θ_0. Let the position of the element under consideration at time θ_0 be x_0, y_0, z_0. Then the element which is at x_0, y_0, z_0 at time θ_0 has moved to x, y, z by time t_0, so that

$$x = x_0 + U(t_0 - \theta_0) + \tfrac{1}{2}\dot{U}(t_0 - \theta_0)^2 + \dots \text{ etc.,}$$

where U, V, W refer to the velocity of the electron at time θ_0.

Remembering that t_0 is a function of x, y, z, we obtain on differentiation with respect to x,

$$\frac{\partial x_0}{\partial x} = 1 - \frac{\partial t_0}{\partial x}\{U + \dot{U}(t_0 - \theta_0) + \tfrac{1}{2}\ddot{U}(t_0 - \theta_0)^2 + \dots\},$$

and similarly,

$$\frac{\partial x_0}{\partial y} = \qquad - \frac{\partial t_0}{\partial y}\{U + \dot{U}(t_0 - \theta_0) + \tfrac{1}{2}\ddot{U}(t_0 - \theta_0)^2 + \dots\}.$$

Those elements of the electron which have their effective positions inside a small element of volume $dxdydz$ occupy at the fixed time θ_0 an element of volume $dx_0dy_0dz_0$. The ratio of these elements of volume is given by the usual Jacobian determinant

$$\frac{dx_0dy_0dz_0}{dxdydz} = \begin{vmatrix} \dfrac{\partial x_0}{\partial x}, & \dfrac{\partial x_0}{\partial y}, & \dfrac{\partial x_0}{\partial z} \\[2mm] \dfrac{\partial y_0}{\partial x}, & \dfrac{\partial y_0}{\partial y}, & \dfrac{\partial y_0}{\partial z} \\[2mm] \dfrac{\partial z_0}{\partial x}, & \dfrac{\partial z_0}{\partial y}, & \dfrac{\partial z_0}{\partial z} \end{vmatrix}.$$

On inserting the values of the differential coefficients as just calculated, and expanding the determinant, we readily obtain

$$\frac{dx_0dy_0dz_0}{dxdydz} = 1 - \sum_{x,y,z} \frac{\partial t_0}{\partial x}\{U + \dot{U}(t_0 - \theta_0) + \tfrac{1}{2}\ddot{U}(t_0 - \theta_0)^2 + \dots\},$$

all terms in U, V, W of degree higher than the first being found to disappear.

If the electron is small in comparison with its distance from the point x', y', z', local variations of x, y, z throughout the electron may be neglected, so that $t_0 - \theta_0$ may be neglected, and in the above expression x, y, z, r may be supposed to refer to the centre of the electron. In this case we have

$$\frac{dx_0dy_0dz_0}{dxdydz} = 1 - \sum_{x,y,z} \frac{\partial t_0}{\partial x}U = 1 + \frac{1}{a}\left(U\frac{\partial r}{\partial x} + V\frac{\partial r}{\partial y} + W\frac{\partial r}{\partial z}\right),$$

or, if v_r denote the radial velocity of the electron towards the point x', y', z' at the instant $t = \theta_0$,

$$\frac{dx_0dy_0dz_0}{dxdydz} = 1 - \frac{v_r}{a}.$$

Equation (655) may now be written in the form

$$\Psi = \frac{1}{K}\iiint \frac{\rho\, dx_0dy_0dz_0}{r\left(1 - \dfrac{v_r}{a}\right)},$$

where all quantities are evaluated at the time $t = \theta_0$, or since

$$\iiint \rho \, dx_0 \, dy_0 \, dz_0 = e,$$

$$\Psi = \frac{e}{K \left[r \left(1 - \dfrac{V_r}{a} \right) \right]},$$

where square brackets signify that the quantity inside is to be evaluated at the retarded time as estimated at the electron. Similarly equation (656) becomes*

$$F = \frac{\mu e}{C} \left[\frac{U}{r \left(1 - \dfrac{V_r}{a} \right)} \right].$$

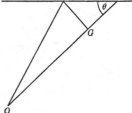

Suppose it is required to calculate the field at a point O at time t. Let E be the position of one of the electrons in the field at a time t_0 such that

$$t_0 = t - \frac{r}{a},$$

where $r = EO$. Then the quantities in square brackets must be calculated for this electron in the position E at the time t_0.

Let the velocity of the electron at the time t_0 be V in a direction EF making an angle θ with

Fig. 139.

EO, and let EF be the distance $V(t - t_0)$ which the electron would describe by the time t if its velocity remained constant.

If FG is the perpendicular from F on to EO, the intercept EG is given by

$$EG = EF \cos \theta = V \cos \theta \, (t - t_0).$$

Now $V \cos \theta$ is simply the component V_r of velocity along EO, while $t - t_0 = r/a$. Thus $EG = r V_r / a$ and

$$OG = r - EG = r \left(1 - \frac{V_r}{a} \right).$$

The formulae for the potentials now become

$$\Psi = \frac{1}{K} \frac{e}{OG}; \quad F = \frac{\mu}{C} \frac{eV}{OG}.$$

If squares of V/C are neglected, the angle FOE in figure 139 is a small angle and OG is approximately equal to OF. Thus as far as terms of the first

---

* These formulae for Ψ and F were first given by Lienard (*L'Eclairage Electrique*, 1898) and E. Wiechert (*Arch. Néerland.* 5, 1900, p. 549). Our proof has followed closely the method given by Lorentz (*Theory of Electrons*, p. 254); an alternative proof is given by Schott (*Electromagnetic Radiation*, 1912, p. 22).

order in V/C the potentials are

$$\Psi = \frac{1}{K}\frac{e}{OF}; \quad F = \frac{\mu}{C}\frac{eV}{OF},$$

where F is the position of the electron at the instant at which the potentials are evaluated, except for a correction arising from accelerations or sudden changes in the motion of the electron.

648. In the case in which u, v, w are treated as small we can also write down the potentials directly from equations (655) and (656). For in this case $dx_0\, dy_0\, dz_0$ becomes equal to $dx\, dy\, dz$ and the equations assume the forms

$$\Psi = \frac{[e]}{Kr}, \qquad F = \frac{\mu}{C}\frac{[ev]}{r},$$

where r is the distance from the point x', y', z' at which the forces are measured to the effective position of the electron. Thus the magnetic forces are given by

$$\alpha = \frac{1}{\mu}\left(\frac{\partial H}{\partial y'} - \frac{\partial G}{\partial z'}\right) = \frac{1}{C}\left(\frac{\partial}{\partial y'}\frac{[ew]}{r} - \frac{\partial}{\partial z'}\frac{[ev]}{r}\right), \text{ etc.} \quad\ldots\ldots(657).$$

Since $[ew]$ is a function of $t - r/a$, we have

$$\frac{\partial}{\partial r}[ew] = -\frac{1}{a}\left[\frac{\partial}{\partial t}(ew)\right] = -\frac{1}{a}[e\dot w],$$

so that

$$\frac{\partial}{\partial y'}\frac{[ew]}{r} = \frac{y'-y}{r}\frac{\partial}{\partial r}\frac{[ew]}{r} = -\frac{y'-y}{r}\left\{\frac{1}{a}\frac{[e\dot w]}{r} - \frac{[ew]}{r^2}\right\},$$

and on substitution in equations (657) we obtain formulae for α, β, γ.

These formulae are seen to contain terms both in r^{-1} and r^{-2}. At a great distance from the electron the former alone are of importance, and the components of force become

$$\alpha = -\frac{1}{aC}\left\{\frac{y'-y}{r^2}[e\dot w] - \frac{z'-z}{r^2}[e\dot v]\right\}, \text{ etc.} \quad\ldots\ldots\ldots(658).$$

Similarly we find for the electric forces at a great distance

$$X = -\frac{\mu}{C^2}\frac{[e\dot u]}{r}, \text{ etc.} \quad\ldots\ldots\ldots\ldots\ldots(659).$$

For a single electron moving along the axis of x with an acceleration $\dot u$, in free space for which $\mu = K = 1$, the components of force assume the simple forms

$$\left.\begin{array}{l} \alpha = 0, \quad \beta = -\dfrac{z'-z}{C^2 r^2}[e\dot u], \quad \gamma = \dfrac{y'-y}{C^2 r^2}[e\dot u] \\[2mm] X = -\dfrac{1}{C^2 r}[e\dot u], \quad Y = 0, \quad Z = 0 \end{array}\right\} \quad\ldots\ldots(660),$$

these being accurate only at great distances from the electron.

RADIATION OF ENERGY.

649. We saw in § 576 that the flow of energy across any closed surface is given by

$$\iint (l\Pi_x + m\Pi_y + n\Pi_z)\, dS \quad\ldots\ldots\ldots\ldots\ldots(661),$$

where

$$\Pi_x = \frac{C}{4\pi}(Y\gamma - Z\beta),\ \text{etc.}$$

In proving this the energy was assumed to be localised in the medium in the way imagined by Maxwell, but if we identify our closed surface with a sphere at infinity this assumption is no longer necessary. For independently of this assumption, the total energy in the whole of space is given by

$$T + W = \iiint \left\{ \frac{K}{8\pi}(X^2 + Y^2 + Z^2) + \frac{\mu}{8\pi}(\alpha^2 + \beta^2 + \gamma^2) \right\} dx\,dy\,dz$$

and from this we can deduce formula (661) directly. On assigning to α, β, γ, X, Y, Z the value obtained in equations (660) for the forces from a single electron, we find

$$\Pi_x = 0,\quad \Pi_y = -\frac{C}{4\pi} X\gamma,\quad \Pi_z = \frac{C}{4\pi} X\beta,$$

$$l\Pi_x + m\Pi_y + n\Pi_z = \frac{(y'-y)^2 + (z'-z)^2}{4\pi C^3 r^4}(e\dot{v})^2,$$

whence the flow of energy across a sphere of infinite radius is readily found to be

$$\frac{2}{3}\frac{e^2\dot{v}^2}{C^3} \quad\ldots\ldots\ldots\ldots\ldots\ldots\ldots\ldots(662).$$

This is Larmor's formula for the rate at which a single moving electron radiates energy. We notice that a steady velocity v contributes nothing to the radiation; energy is radiated away from an electron which is undergoing acceleration but not from one in steady motion.

It must be added that the new dynamics referred to in § 620 seems to throw doubt on this formula for emission of radiation. Many physicists now question whether any emission of radiation is produced by the acceleration of an electron, except under certain special conditions. Bearing this caution in mind, we may proceed to examine some of the consequences of the formulae just obtained.

650. If each of a cluster of electrons is so near to the point x, y, z that differences of retardation of time may be neglected throughout the cluster, the radiation from the cluster is easily seen to be the same as that from a single electron of charge E moving with components of acceleration $\dot{U}$, $\dot{V}$, $\dot{W}$, such that

$$E\dot{U} = \Sigma e\dot{v},\ \text{etc.}$$

The condition that there shall be no radiation from such a cluster is

$$\Sigma e \dot{u} = \Sigma e \dot{v} = \Sigma e \dot{w} = 0.$$

If this condition is not satisfied, the rate of emission of radiation is (cf. formula (662))

$$\frac{2}{3C^3} \{ (\Sigma e \dot{u})^2 + (\Sigma e \dot{v})^2 + (\Sigma e \dot{w})^2 \} \quad \dotfill (663).$$

651. Consider next the field produced by a particle of charge E oscillating along the axis of x with simple harmonic motion, its coordinate at any instant being $x_0 \cos pt$. We have

$$E \dot{u} = - E p^2 x_0 \cos pt; \quad [E \dot{u}] = - E p^2 x_0 \cos p \left(t - \frac{r}{a} \right),$$

and the field can be written down by substitution in formulae (660).

From formula (662) the average rate of emission of radiation is found to be

$$\frac{1}{3} \frac{p^4 E^2 x_0^2}{C^3} = \frac{16 \pi^4 E^2 x_0^2 C}{3 \lambda^4},$$

where λ is the wave-length of the emitted light.

A particle moving in this way is spoken of as a simple Hertzian vibrator. Its motion was taken by Hertz to represent the oscillating flow of current in an oscillatory discharge of a condenser. Such an oscillation formed the source of the waves in Hertz's original experiments (1888), and forms the source of the waves used in modern wireless telegraphy.

652. A case of great interest is that in which the velocity of a moving electron undergoes a very sudden change, such as would occur during a collision with matter of any kind. Let us represent such a sudden change by supposing that $e\dot{u}$, $e\dot{v}$, $e\dot{w}$ vanish except through a very small interval surrounding the time $t = 0$, during which they are very great. At a point at distance r, $[e\dot{u}]$, $[e\dot{v}]$ and $[e\dot{w}]$ will vanish except through a small interval of time surrounding the instant $t = r/a$. During this short interval, the electric and magnetic forces will be very great; before and after this interval they will have the smaller values arising from the steady motion of the electron. Thus the sudden check on the motion of the electron results in the outward spread of a thin sheet of electric and magnetic force, the forces being very intense but only of brief duration.

The radiation which is emitted when rapidly moving electrons impinge on matter is generally called X-radiation or Röntgen-radiation. It was suggested by Stokes that this consists of thin sheets or "pulses" of electric and magnetic force of the type we have just investigated. Although there is no doubt that this is true in a general way, yet the growth of the new dynamics already referred to has made it clear that there is far more in the problem of X-radiation than can be explained by the theories of Maxwell and Stokes.

MECHANICAL FORCES ON MOVING CHARGES.

653. Whether we assume Maxwell's localisation of energy in the medium or not, the total energy of an electromagnetic field, as we noticed in § 649, will be $T + W$, where

$$W = \iiint \frac{K}{8\pi} (X^2 + Y^2 + Z^2)\, dx\, dy\, dz \quad\ldots\ldots\ldots\ldots(664),$$

$$T = \iiint \frac{\mu}{8\pi} (\alpha^2 + \beta^2 + \gamma^2)\, dx\, dy\, dz \quad\ldots\ldots\ldots\ldots(665),$$

and the integrals extend through the whole of space.

Let us suppose that, on account of the electromagnetic forces at work, each element of charge experiences a mechanical force of components Ξ, H, Z per unit charge. We can find the forces Ξ, H, Z by the methods of § 196 and the general principle of least action.

Let us imagine a small displaced motion in which the coordinates of any point x, y, z are displaced to $x + \delta x$, $y + \delta y$, $z + \delta z$, while the components of electric polarisation are changed from f, g, h to $f + \delta f$, $g + \delta g$, $h + \delta h$, these new components of polarisation as well as the old satisfying relation (615). Thus if ρ is the density of electricity at any point in the original motion and $\rho + \delta\rho$ the corresponding density in the displaced motion, we must have

$$\frac{\partial f}{\partial x} + \frac{\partial g}{\partial y} + \frac{\partial h}{\partial z} = \rho,$$

$$\frac{\partial \delta f}{\partial x} + \frac{\partial \delta g}{\partial y} + \frac{\partial \delta h}{\partial z} = \delta\rho.$$

Let us denote the total work performed by the mechanical forces in this small displacement by $-\{\delta U\}$ (cf. § 551), so that

$$\{\delta U\} = \iiint \rho\, (\Xi\, \delta x + \mathrm{H}\, \delta y + \mathrm{Z}\, \delta z)\, dx\, dy\, dz \quad\ldots\ldots\ldots\ldots(666).$$

Then the equations of motion are contained in (cf. equation (507))

$$\int_0^\tau (\delta T - \delta W - \{\delta U\})\, dt = 0 \quad\ldots\ldots\ldots\ldots\ldots(667).$$

We have
$$\delta T = \frac{1}{4\pi} \iiint (a\,\delta\alpha + b\,\delta\beta + c\,\delta\gamma)$$

$$= \frac{1}{4\pi} \iiint \left\{ \delta\alpha \left(\frac{\partial H}{\partial y} - \frac{\partial G}{\partial z} \right) + \ldots \right\} dx\, dy\, dz$$

$$= \frac{1}{4\pi} \iiint \left\{ F \left(\frac{\partial \delta\gamma}{\partial y} - \frac{\partial \delta\beta}{\partial z} \right) + \ldots \right\} dx\, dy\, dz$$

on applying Green's Theorem; and on further using equation (635), this becomes

$$\delta T = \frac{1}{C} \iiint \left\{ F\delta\left(\rho U + \frac{df}{dt} \right) + G\delta\left(\rho V + \frac{dg}{dt} \right) + H\delta\left(\rho W + \frac{dh}{dt} \right) \right\} dx\, dy\, dz.$$

Let δ, $\dfrac{d}{dt}$ refer to a point fixed in space, and let Δ, $\dfrac{D}{Dt}$ refer to a point moving with the moving material. Then we have the two formulae for Δv,

$$\Delta v = \frac{D}{Dt}\,\delta x = \frac{d}{dt}\,\delta x + v\frac{\partial}{\partial x}\,\delta x + v\frac{\partial}{\partial y}\,\delta x + w\frac{\partial}{\partial z}\,\delta x,$$

$$\Delta v = \delta v + \frac{\partial v}{\partial x}\,\delta x + \frac{\partial v}{\partial y}\,\delta y + \frac{\partial v}{\partial z}\,\delta z,$$

so that on comparison

$$\delta v = \frac{d}{dt}\,\delta x + v\frac{\partial}{\partial x}\,\delta x + v\frac{\partial}{\partial y}\,\delta x + w\frac{\partial}{\partial z}\,\delta x - \left(\frac{\partial v}{\partial x}\,\delta x + \frac{\partial v}{\partial y}\,\delta y + \frac{\partial v}{\partial z}\,\delta z\right).$$

We now have

$$\delta\left(\rho v + f\right) = v\delta\rho + \rho\delta v + \frac{d}{dt}\,\delta f$$

$$= v\delta\rho + \frac{d}{dt}\left(\rho\,\delta x + \delta f\right) - \delta x\frac{d\rho}{dt}$$

$$+ \rho\left(v\frac{\partial}{\partial x} + v\frac{\partial}{\partial y} + w\frac{\partial}{\partial z}\right)\delta x - \rho\left(\frac{\partial v}{\partial x}\,\delta x + \frac{\partial v}{\partial y}\,\delta y + \frac{\partial v}{\partial z}\,\delta z\right).$$

On substituting for $d\rho/dt$ and $\delta\rho$ their values (cf. § 622)

$$\frac{d\rho}{dt} = -\left\{\frac{\partial}{\partial x}\left(\rho v\right) + \frac{\partial}{\partial y}\left(\rho v\right) + \frac{\partial}{\partial z}\left(\rho w\right)\right\},$$

$$\delta\rho = -\left\{\frac{\partial}{\partial x}\left(\rho\,\delta x\right) + \frac{\partial}{\partial y}\left(\rho\,\delta y\right) + \frac{\partial}{\partial z}\left(\rho\,\delta z\right)\right\},$$

and simplifying, we obtain

$$\delta\left(\rho v + f\right) = \frac{d}{dt}\left(\rho\,\delta x + \delta f\right) + \frac{\partial}{\partial y}\left(\rho v\delta x - \rho v\delta y\right) - \frac{\partial}{\partial z}\left(\rho v\delta z - \rho w\delta x\right),$$

whence

$$\delta T = \frac{1}{C}\iiint F\frac{d}{dt}\left(\rho\,\delta x + \delta f\right)dx\,dy\,dz + \text{terms in } G,\ H$$

$$+ \frac{1}{C}\iiint F\left\{\frac{\partial}{\partial y}\left(\rho v\delta x - \rho v\delta y\right) - \frac{\partial}{\partial z}\left(\rho v\delta z - \rho w\delta x\right)\right\}dx\,dy\,dz + \dots.$$

Transforming by Green's Theorem, the second line in δT becomes

$$\frac{1}{C}\iiint\left\{\left(\frac{\partial H}{\partial y} - \frac{\partial G}{\partial z}\right)\left(\rho w\delta y - \rho v\delta z\right) + \dots\right\}dx\,dy\,dz$$

$$= \frac{1}{C}\iiint\left\{\rho\,\delta x\left(cv - bw\right) + \rho\,\delta y\left(aw - cv\right) + \rho\,\delta z\left(bv - av\right)\right\}dx\,dy\,dz.$$

On integrating with respect to the time, and transforming the first term on integration by parts, we have

$$\int_0^\tau \delta T\,dt = \int_0^\tau dt\left[-\frac{1}{C}\iiint\frac{dF}{dt}\left(\rho\,\delta x + \delta f\right) + \rho\,\delta x\left(cv - bw\right) + \dots\right]dx\,dy\,dz.$$

We have from variation of equation (664),

$$\delta W = \iiint (X\,\delta f + Y\,\delta g + Z\,\delta h)\,dx\,dy\,dz.$$

Hence, freed from the integration with respect to the time, equation (667) becomes

$$\frac{1}{C}\iiint \left[-\frac{dF}{dt}(\rho\,\delta x + \delta f) + \rho\,\delta x\,(cv - bw) + \dots \right] dx\,dy\,dz$$

$$-\iiint (X\,\delta f + Y\,\delta g + Z\,\delta h)\,dx\,dy\,dz$$

$$-\iiint \rho\,(\Xi\,\delta x + \mathrm{H}\,\delta y + \mathrm{Z}\,\delta z)\,dx\,dy\,dz = 0 \quad\dots\dots\dots\dots\dots\dots(668).$$

We may not equate coefficients of the differentials, for δf, δg, δh are not independent, being connected by

$$\frac{\partial \delta f}{\partial x} + \frac{\partial \delta g}{\partial y} + \frac{\partial \delta h}{\partial z} = \delta\rho = -\frac{\partial}{\partial x}(\rho\,\delta x) - \frac{\partial}{\partial y}(\rho\,\delta y) - \frac{\partial}{\partial z}(\rho\,\delta z).$$

We multiply this by an undetermined multiplier Ψ, a function of x, y, z, and integrate through all space. We obtain

$$\iiint \Psi \left(\frac{\partial \delta f}{\partial x} + \frac{\partial \delta g}{\partial y} + \frac{\partial \delta h}{\partial z} + \frac{\partial}{\partial x}(\rho\,\delta x) + \frac{\partial}{\partial y}(\rho\,\delta y) + \frac{\partial}{\partial z}(\rho\,\delta z) \right) dx\,dy\,dz = 0,$$

or, after integration by parts,

$$-\iiint \left(\frac{\partial \Psi}{\partial x}\,\delta f + \frac{\partial \Psi}{\partial y}\,\delta g + \frac{\partial \Psi}{\partial z}\,\delta h + \rho\,\delta x\,\frac{\partial \Psi}{\partial x} + \rho\,\delta y\,\frac{\partial \Psi}{\partial y} + \rho\,\delta z\,\frac{\partial \Psi}{\partial z} \right) dx\,dy\,dz = 0.$$

Adding this integral to the left hand of equation (668), we may equate coefficients, and obtain

$$X = -\frac{1}{C}\frac{dF}{dt} - \frac{\partial \Psi}{\partial x},\ \text{etc.} \quad\dots\dots\dots\dots\dots\dots(669),$$

$$\Xi = -\frac{1}{C}\frac{dF}{dt} - \frac{\partial \Psi}{\partial x} + \frac{1}{C}(cv - bw)$$

$$= X + \frac{1}{C}(cv - bw),\ \text{etc.}\dots\dots\dots\dots\dots\dots(670).$$

The first equation is simply equation (639), of which we have now obtained a proof direct from the principle of least action (cf. § 575); the second gives us the mechanical forces acting on moving charges. It will be seen that the forces given by formula (670) are identical with those obtained in § 629, but they have now been obtained without any limitation as to the smallness or steadiness of the velocities.

Stresses in the Medium.

654. We can next evaluate the stresses in the medium, following the method of §193 and assuming the medium to be free ether.

Let X be the total x-component of force acting on any finite region of the medium, so that

$$\mathsf{X} = \iiint \Xi \rho \, dx \, dy \, dz = \iiint \rho X \, dx \, dy \, dz + \frac{1}{C} \iiint (\gamma \rho v - \beta \rho w) \, dx \, dy \, dz.$$

On substituting for ρv, ρw from equations (635), the last term becomes

$$-\frac{1}{4\pi C} \iiint \left(\gamma \frac{dY}{dt} - \beta \frac{dZ}{dt} \right) dx \, dy \, dz$$

$$+ \frac{1}{4\pi} \iiint \left\{ \gamma \left(\frac{\partial \alpha}{\partial z} - \frac{\partial \gamma}{\partial x} \right) - \beta \left(\frac{\partial \beta}{\partial x} - \frac{\partial \alpha}{\partial y} \right) \right\} dx \, dy \, dz$$

$$= \frac{1}{4\pi C} \frac{d}{dt} \iiint (\beta Z - \gamma Y) \, dx \, dy \, dz - \frac{1}{4\pi C} \iiint \left(Z \frac{\partial \beta}{\partial t} - Y \frac{\partial \gamma}{\partial t} \right) dx \, dy \, dz + \dots.$$

On substituting for ρ from equation (615), and for $d\beta/dt$, $d\gamma/dt$ from equations (636), and collecting terms, this becomes

$$\mathsf{X} = \frac{1}{4\pi} \iiint \left[\left(\frac{\partial X}{\partial x} + \frac{\partial Y}{\partial y} + \frac{\partial Z}{\partial z} \right) X - Y \left(\frac{\partial Y}{\partial x} - \frac{\partial X}{\partial y} \right) + Z \left(\frac{\partial X}{\partial z} - \frac{\partial Z}{\partial x} \right) \right] dx \, dy \, dz$$

$$+ \frac{1}{4\pi} \iiint \left[\qquad\qquad - \beta \left(\frac{\partial \beta}{\partial x} - \frac{\partial \alpha}{\partial y} \right) + \gamma \left(\frac{\partial \alpha}{\partial z} - \frac{\partial \gamma}{\partial x} \right) \right] dx \, dy \, dz$$

$$- \frac{1}{C^2} \frac{d}{dt} \iiint \Pi_x \, dx \, dy \, dz \dots \dots \dots \dots \dots (671),$$

in which Π_x as in §576 denotes the x-component of the Poynting Flux.

On transforming the volume integrals in the first two lines into surface integrals, this becomes

$$\mathsf{X} = -\frac{1}{4\pi} \iint \{ \tfrac{1}{2} l (X^2 - Y^2 - Z^2) + mXY + nXZ \} \, dS$$

$$- \frac{1}{4\pi} \iint \{ \tfrac{1}{2} l (\alpha^2 - \beta^2 - \gamma^2) + m\alpha\beta + n\alpha\gamma \} \, dS$$

$$- \frac{1}{C^2} \frac{d}{dt} \iiint \Pi_x \, dx \, dy \, dz \dots \dots \dots \dots \dots \dots (672).$$

Since the last volume integral cannot be transformed into a surface integral, it is clear that the mechanical action is not such as can be transmitted by a system of stresses in a medium at rest.

655. On the other hand it is clear that if we suppose the medium to possess momentum of components

$$\frac{\Pi_x}{C^2}, \quad \frac{\Pi_y}{C^2}, \quad \frac{\Pi_z}{C^2} \quad \dots \dots \dots \dots \dots (673)$$

per unit volume, then equation (672) would become exactly the equation of motion of this medium, if it is supposed to be acted on by a system of stresses defined by

$$P_{xx} = \frac{1}{8\pi}(X^2 - Y^2 - Z^2 + \alpha^2 - \beta^2 - \gamma^2) \text{ etc.} \atop P_{xy} = \frac{1}{4\pi}(XY + \alpha\beta) \text{ etc.} \Bigg\} \quad \dots \dots \dots (674).$$

Thus the mechanical action is such as can be transmitted by a medium in motion, the momentum per unit volume being given by formula (673). The vector whose components are given by formula (673) is commonly called the "electromagnetic momentum." We see that it is of amount equal to $1/C^2$ times the Poynting Flux, and in the same direction.

For an electrostatic or magnetostatic field existing alone, the electromagnetic momentum vanishes, and the stresses reduce to those previously found in §§ 193 and 471.

Motion with Uniform Velocity.

656. Let us again return to the general equations, and examine the special form they assume for a system moving with uniform velocity. This may for convenience be supposed to be a velocity v parallel to the axis of x.

As in § 624 we may replace $\dfrac{d}{dt}$ by $-v\dfrac{d}{dx}$ and the general equation (648) becomes

$$\left(1 - \frac{v^2}{a^2}\right)\frac{\partial^2 \chi}{\partial x^2} + \frac{\partial^2 \chi}{\partial y^2} + \frac{\partial^2 \chi}{\partial z^2} = -4\pi\sigma.$$

Let us now write κ for $\left(1 - \dfrac{v^2}{a^2}\right)^{-\frac{1}{2}}$, and the equation becomes

$$\frac{1}{\kappa^2}\frac{\partial^2 \chi}{\partial x^2} + \frac{\partial^2 \chi}{\partial y^2} + \frac{\partial^2 \chi}{\partial z^2} = -4\pi\sigma,$$

or, if we write x' for κx,

$$\frac{\partial^2 \chi}{\partial x'^2} + \frac{\partial^2 \chi}{\partial y^2} + \frac{\partial^2 \chi}{\partial z^2} = -4\pi\sigma \quad \dots \dots \dots \dots \dots (675).$$

We may conveniently speak of x', y, z as the "contracted" coordinates corresponding to the original coordinates x, y, z, since if two surfaces have the same equation, one in x', y, z and the other in x, y, z coordinates, the former will be identical with the latter contracted in the ratio $1/\kappa$ parallel to the axis of x.

Equation (675) is Poisson's equation in contracted coordinates. Its solution is

$$\chi = \iiint \frac{\sigma \, dx' \, dy \, dz}{r'} = \kappa \iiint \frac{\sigma \, dx \, dy \, dz}{r'} = \kappa \Sigma \frac{\tau}{r'},$$

where r' denotes distance measured in the contracted space.

Hence (cf. equations (644), (645)) the values of Ψ and F, G, H are given by

$$\Psi = \frac{\kappa}{K} \Sigma \frac{e}{r'}$$

$$F = \frac{\kappa \mu U}{C} \Sigma \frac{e}{r'}, \quad G = H = 0 \qquad \Bigg\} \quad \ldots\ldots\ldots\ldots(676),$$

so that the potentials are the same in contracted coordinates as they would be in ordinary coordinates if the system were at rest, multiplied by the factor κ.

Motion of a uniformly electrified sphere.

657. To illustrate the method just explained, we shall examine the field produced by a uniformly electrified sphere of radius a, moving with velocity v.

The surface in the contracted space is a sphere of radius a, so that that in the uncontracted space is a prolate spheroid of semi-axes κa, a, a, and therefore of eccentricity v/C. To find the distribution of electricity, we imagine the charge on the sphere to be uniformly spread between the spheres $r = a$ and $r = a + \epsilon$, where ϵ is infinitesimal. The charge on the spheroid is now seen to be uniformly spread between the spheroid itself and another similar spheroid of semi-axes $\kappa (a + \epsilon)$, $a + \epsilon, a + \epsilon$. Thus the distribution of electricity in the spheroid in the uncontracted space is just what it would be if the spheroid were a freely charged conductor, and is given by the analysis of §§ 283, 284.

658. The field has been discussed in detail by Searle[*] and Abraham[†]. The electric and magnetic energies W and T are found to be given by

$$W = \frac{e^2}{8a} \left\{ \frac{3C^2 - U^2}{CU} \log \frac{C + U}{C - U} - 2 \right\},$$

$$T = \frac{e^2}{8a} \left\{ \frac{C^2 + U^2}{CU} \log \frac{C + U}{C - U} - 2 \right\},$$

$$T + W = \frac{e^2}{2a} \left\{ \frac{C}{U} \log \frac{C + U}{C - U} - 1 \right\},$$

while the total electromagnetic momentum G in the whole of space is given by

$$G = \frac{e^2}{4a} \left\{ \frac{C^2 + U^2}{CU^2} \log \frac{C + U}{C - U} - \frac{2}{U} \right\},$$

this direction of G being of course that of v.

Motion of any system in equilibrium.

659. When a material system moves with any velocity v, the electric field produced by its charges is different from the field when at rest. The difference between these fields must shew itself in a system of forces which must act on the moving system and in some way modify its configuration.

---

[*] *Phil. Trans.* A, 187 (1896), p. 165.

[†] *Phys. Zeitschrift*, 5 (1904), p. 576, or *Theorie der Elektrizität* (2nd ed.), p. 165.

Let us consider first a simple system which we shall call S in which all the forces are electrostatic, and all the charges are supposed concentrated in points (*e.g.* electrons). Let us suppose that when the system is at rest there is equilibrium when a charge e_1 is at $x = x_1$, $y = y_1$, $z = z_1$; e_2 at $x = x_2$, $y = y_2$, $z = z_2$, and so on.

Let us compare this with a second system S' consisting of the same electrons but moving with a uniform velocity U, and having the charges e_1 at $x' = x_1$, $y = y_1$, $z = z_1$; e_2 at $x' = x_2$, $y = y_2$, $z = z_2$, etc., so that each electron has the position in the contracted space which corresponds to its original position in the original space. Then if V denotes the electrostatic potential in the original system, the potentials in the moving system are (cf. equations (676))

$$\Psi = \kappa V, \quad F = \frac{K \mu U}{C} \Psi, \quad G = 0, \quad H = 0,$$

and the forces in the moving system are

$$X = -\frac{\partial \Psi}{\partial x} - \frac{1}{C} \frac{dF}{dt}$$

$$= -\frac{\partial \Psi}{\partial x} + \frac{U}{C} \frac{\partial F}{\partial x}$$

$$= -\frac{\partial \Psi}{\partial x} \left(1 - \frac{K \mu U^2}{C^2}\right) = -\frac{1}{\kappa} \frac{\partial V}{\partial x},$$

$$Y = -\frac{\partial \Psi}{\partial y} = -\kappa \frac{\partial V}{\partial y}, \text{ etc.}$$

We notice that the electrostatic forces in S' are $1/\kappa$ times those in S as regards their x-components, but κ times those in S as regards their y-components. As a special case we notice that if the system S was in electrical equilibrium, then S' will also be in electrical equilibrium, so that a system which is in equilibrium when at rest can regain equilibrium after being set in motion with velocity U by contracting in a ratio $1/\kappa$.

ELECTROMAGNETIC MASS.

660. Consider a charged body, which will ultimately be identified with an electron, moving with a uniform velocity U parallel to the axis of x. Let us first consider the simple case in which U is so small that U^2/C^2 may be neglected.

The moving charge creates a magnetic field. If the charged body is supposed to be a sphere of radius a, whose surface is uniformly electrified to a total charge e, then there is no field inside the sphere, and the components of magnetic force outside the sphere are given by

$$\alpha = 0, \quad \beta = -\frac{U}{C} \frac{ez}{r^3}, \quad \gamma = \frac{U}{C} \frac{ey}{r^3}.$$

If we assume localisation of energy in the medium, then at a distance r greater than a from the centre of the sphere there will be magnetic energy per unit volume of amount

$$\frac{1}{8\pi}(\alpha^2 + \beta^2 + \gamma^2) = \frac{e^2\,U^2}{8\pi\,C^2}\frac{\sin^2\theta}{r^4},$$

where θ denotes the angle between the radius r and the axis of x. On integration, the total energy of this magnetic field is found to be

$$\frac{e^2\,U^2}{8\pi\,C^2}\iiint \frac{\sin^2\theta}{r^4}\,r^2\sin\theta\,d\theta\,d\phi\,dr = \frac{e^2}{3a\,C^2}\,U^2 \quad\ldots\ldots\ldots\ldots(677).$$

This result is of course only true provided we suppose the energy to reside in the medium as imagined by Maxwell. In this case the energy, being magnetic, must be supposed to be kinetic energy.

Thus if the charged body is supposed to be of mass m_0, the total kinetic energy of its forward movement will be

$$\tfrac{1}{2}\left(m_0 + \frac{2}{3}\frac{e^2}{aC^2}\right)U^2 \quad\ldots\ldots\ldots\ldots\ldots\ldots\ldots\ldots(678),$$

in which the first term arises from the ordinary mass of the body and the second from the kinetic energy of the medium.

An analogy from hydrodynamics will illustrate the result at which we have arrived. Suppose we have a balloon of mass m moving in air with a velocity v and displacing a mass m' of air. If the velocity v is small compared with the velocity of propagation of waves in air, the motion of the balloon will set up currents in the air surrounding it, such that the velocity of these currents will be proportional to v at every point. The whole kinetic energy of the motion will accordingly be

$$\tfrac{1}{2}(m+M)\,v^2,$$

the term $\tfrac{1}{2}mv^2$ being contributed by the motion of the matter of the balloon itself, and the term $\tfrac{1}{2}Mv^2$ by the air currents outside the balloon. The value of M is comparable with m', the mass of air displaced—for instance if the balloon is spherical, and if the motion of the air is irrotational, the value of M is known to be $\tfrac{1}{2}m'$ (cf. Lamb, *Hydrodynamics*, § 91).

661. Strictly speaking, formula (678) is true only when U remains steady through the motion. Any change in the value of U will be accompanied by magnetic disturbances in the ether which spread out with velocity C from the sphere. An examination of integral (677) will, however, shew that the energy is concentrated round the sphere—the energy outside a sphere of radius R is only a fraction a/R of the whole, and if R is taken to be a large multiple of a this may be disregarded. The time required for the energy to readjust itself after a change of velocity is now comparable with R/C.

Thus if we exclude sudden changes in U, and limit our attention to gradual changes extending over periods great compared with R/C, we may take expression (678) to represent the kinetic energy, both for steady and variable motion.

The problem gains all its importance from its application to the electron. For this a is of the order of 2×10^{-13} cms. (see below, § 666), so that all except one per cent. of the magnetic energy is contained within a sphere of radius $R = 2 \times 10^{-11}$ cms. Since $C = 3 \times 10^{10}$, the time of readjustment of this energy is $\cdot 66 \times 10^{-21}$ seconds, an interval small enough to be disregarded in almost all physical problems.

662. We shall now consider the same problem in a different manner, and shall remove the restriction that v/C is to be a small quantity. The electron will still be supposed to move with a uniform velocity u, v, w which may be of any amount. The field arising from its motion may be calculated as explained in § 647. So long as the electron has no acceleration, the forces X, Y, Z, α, β, γ fall off at infinity as $1/r^2$, so that the stresses defined by equations (674) fall off as $1/r^4$.

If we now apply equation (672) to the field of the single electron, allowing the closed surface S to recede to infinity, the equation becomes

$$X = -\frac{1}{C^2}\frac{d}{dt}\iiint \Pi_x \, dx\, dy\, dz \dots\dots\dots\dots\dots(679),$$

where the integral is taken through the whole of space. Here X will now represent the x-component of the ponderomotive force on the electron from the field set up by its own motion through the ether.

When the electron moves with uniform velocity, the integral on the right retains a constant value. In this case $X = Y = Z = 0$; there is no resultant force acting on the electron from the ether.

Now suppose that the electron has not only a velocity u, v, w but also an acceleration $\dot{u}$, $\dot{v}$, $\dot{w}$. The forces X, Y, Z, α, β, γ now contain terms in $1/r$, but these depend only on the accelerations. When the surface S recedes to infinity in equation (672), the surface integrals will no longer vanish, but will contain terms dependent on the squares and products of the accelerations. If we suppose the accelerations to be so small that their squares and products may be neglected, then equation (679) remains true even for an accelerated electron.

We have seen that Π_x will depend on the values of u, v, w, $\dot{u}$, etc., both at the instant t under consideration and also at preceding instants. Thus we may in general suppose that

$$\frac{1}{C^2}\iiint \Pi_x dx\, dy\, dz = f_x(u, v, w, \dot{u}, \dots \ddot{u}, \dots \text{etc.}).$$

Each side of this equation represents the x-component of electromagnetic momentum, and equation (679) assumes the form

$$X = -\frac{df_x}{dt} = -\left[\dot{u}\frac{\partial f_x}{\partial u} + \dot{v}\frac{\partial f_x}{\partial v} + \dot{w}\frac{\partial f_x}{\partial w} + \ddot{u}\frac{\partial f_x}{\partial \dot{u}} + \dots\right] \dots\dots(680).$$

It is clear that the force X will depend on all the accelerations and their differential coefficients with respect to the time.

Consider first the case in which all the accelerations are steady and so small that their squares may be neglected. Then $\dot{U}$, $\ddot{v}$ etc. all vanish and equation (680) reduces to

$$X = - \left[\dot{U}\frac{\partial f_x}{\partial U} + \dot{v}\frac{\partial f_x}{\partial V} + \dot{w}\frac{\partial f_x}{\partial W} \right] \quad \ldots\ldots\ldots\ldots\ldots(681).$$

In general $\partial f_x/\partial U$ etc. may depend on $\dot{U}$, $\dot{v}$, $\dot{w}$, but if we agree that squares of $\dot{U}$, $\dot{v}$, $\dot{w}$ may be neglected in calculating X, then we may calculate $\partial f_x/\partial U$ etc. on the supposition that $\dot{U}$, $\dot{v}$, $\dot{w}$ all vanish. In other words f_x etc. may be calculated as if the motion were steady.

When the motion is steady the whole electromagnetic momentum G is clearly in the direction of the motion and its amount will depend only on c, where $c^2 = U^2 + V^2 + W^2$. Thus we may put

$$f_x = \frac{U}{c} G,$$

where G is the whole electromagnetic momentum in the whole of space, a function of c only. On differentiation, we obtain

$$\frac{\partial f_x}{\partial U} = \frac{G}{c} + \frac{U^2}{c}\frac{\partial}{\partial c}\left(\frac{G}{c}\right); \quad \frac{\partial f_x}{\partial V} = \frac{UV}{c}\frac{\partial}{\partial c}\left(\frac{G}{c}\right), \text{ etc.}$$

Now suppose the whole motion to be in the direction of Ox, so that $c = U$, $V = W = 0$. The three equations such as (680) now assume the forms

$$X = -\dot{U}\frac{\partial G}{\partial c}, \quad Y = -\dot{v}\frac{G}{c}, \quad Z = -\dot{w}\frac{G}{c}.$$

When $\dot{U}$ exists alone, $\dot{v} = \dot{w} = 0$, so that $Y = Z = 0$. Thus the electromagnetic field exerts a force on the electron in the direction opposite to $\dot{U}$. This force is the same as would be exerted if the electron possessed an additional mass equal to dG/dc. This is called the longitudinal electromagnetic mass of the electron. An electron of mass m_0 will respond to a force in the direction of its motion in the same way as an electron, unencumbered by an electromagnetic field, of mass

$$m_0 + \frac{dG}{dc} \quad \ldots\ldots\ldots\ldots\ldots\ldots\ldots\ldots\ldots(682).$$

Similarly if $\dot{v}$ exists, along the opposing force of the electromagnetic field is $-\dot{v}(G/c)$. By a similar interpretation, G/c is called the transverse electromagnetic mass. The electron will respond to a force transverse to its motion in the same way as an electron, unencumbered by a magnetic field, of mass

$$m_0 + \frac{G}{c} \quad \ldots\ldots\ldots\ldots\ldots\ldots\ldots\ldots\ldots(683).$$

663. Abraham suggested in 1904 that the electron might be treated as a rigid sphere of radius a, uniformly electrified over its surface. If so, the longitudinal and transverse masses m_l and m_t would be given, from the formulae of § 658, by

$$m_l = \frac{e^2}{2a} \frac{C}{v^3} \left(\frac{2vC}{C^2 - v^2} - \log \frac{C+v}{C-v} \right),$$

$$m_t = \frac{e^2}{4a} \frac{C}{v^3} \left(\frac{C^2 + v^2}{C^2} \log \frac{C+v}{C-v} - 2\frac{v}{C} \right).$$

664. Lorentz brought forward an alternative conception of the electron according to which it is spherical in shape only when at rest. The electricity is not supposed to be rigidly fixed in a spherical configuration, so that when the electron is set in motion with a velocity v, it contracts, in accordance with the theorem of § 659, in the ratio $1 : \kappa$ in its direction of motion and so assumes the form of an oblate spheroid. Against this conception of the electron Abraham has brought the objection that the original electron cannot be simply a distribution of electric charges acted on by their own mutual repulsions; there must be other forces at work to keep the charges from flying apart. When these other forces are taken into account, there is no reason for supposing that the contracted electron would be in equilibrium, or if it were in equilibrium, that the equilibrium would be stable. We shall return to this point later.

The electromagnetic field of Lorentz's electron is readily calculated by the method of § 656, for the configuration, when expressed in terms of contracted coordinates, is spherically symmetrical.

If W is the electrostatic energy of the system of charges which constitute the electron when at rest, it is readily found that the electromagnetic momentum G of the contracted electron moving with velocity v is

$$G = \frac{4}{3} \frac{\kappa v}{C^2} W,$$

so that the longitudinal and transverse masses are

$$\left. \begin{aligned} m_l &= \frac{4}{3} \frac{W}{C^2} \kappa^3 \\ m_t &= \frac{4}{3} \frac{W}{C^2} \kappa \end{aligned} \right\} \quad \dots\dots\dots\dots\dots\dots\dots\dots\dots\dots(684).$$

665. The formulae for the transverse mass can be tested experimentally. It was shewn in § 631 that an electron in a uniform magnetic field H would describe a path of constant curvature mvC/eH, where v is the velocity perpendicular to the magnetic lines of force. When electromagnetic mass is taken into account, m in this formula must be replaced by $m_0 + m_t$, where m_0 is the mass of the electron apart from its electromagnetic mass. Experiments to determine the variation of $m_0 + m_t$ with the velocity were first undertaken by Kaufmann in 1906. More recent experiments by Bucherer, Bestelmeyer and others shew that $m_0 + m_t$ varies precisely as $(1 - U^2/C^2)^{-\frac{1}{2}}$ or κ. This is in exact agreement with the transverse mass of the Lorentz contractile electron if m_0 is taken to be zero—*i.e.* if the mass of the electron is supposed to be wholly electromagnetic.

666. All experiments agree in giving a value for e/m at zero velocity very nearly equal to $1·757 \times 10^7$ in Electromagnetic Units. Combining this with the value for e, namely $4·803 \times 10^{-10}$ in Electrostatic Units, we find for the mass of the electron at rest

$$m = 9·12 \times 10^{-28} \text{ grammes.}$$

The mass of the electron at rest is, from formulae (684),

$$m = \frac{4}{3} \frac{W}{C^2}.$$

If the charge e of the electron is supposed spread uniformly over the surface of a sphere of radius a, the value of W, the electrostatic energy, is $e^2/2a$, so that

$$m = \frac{2}{3} \frac{e^2}{aC^2} \quad \dots\dots\dots\dots\dots\dots\dots\dots(685)$$

in agreement with formula (678). In this equation we know the values of m, e and C, so can deduce

$$a = 1·80 \times 10^{-13} \text{ cms.}$$

This must be the radius of the electron if its charge is spread uniformly over the surface of a sphere. If the charge is spread uniformly through the volume of a sphere, $W = 3e^2/5a$, giving

$$m = \frac{4}{5} \frac{e^2}{aC^2}; \qquad a = 2·16 \times 10^{-13} \text{ cms.}$$

Other distributions of charge would give other values for a but always of the same order. We conclude that the value of a is of the order of 2×10^{-13} cms.

The Internal Mechanics of the Electron.

667. Let us regard the electron as a contractile sphere of radius a whose surface is uniformly charged with electricity. Then m is given by formula (685), and the electromagnetic energy of the electron, when moving with a velocity v, is found to be

$$T = mC^2 \left(\kappa - \frac{1}{4\kappa} \right) + \text{a constant} \quad \dots\dots\dots\dots(686).$$

Suppose an acceleration $\dot{v}$ to operate for an instant dt. Since the longitudinal mass is $m\kappa^3$, the work done by the force producing the acceleration is

$$m\kappa^3 \dot{v}\, dt$$

which may be written as $\dfrac{d}{dt}(mC^2\kappa)\, dt$. The increment in the electromagnetic energy (686) is, however,

$$\frac{d}{dt}(mC^2\kappa)\, dt - \frac{1}{4}\frac{d}{dt}\left(\frac{mC^2}{\kappa}\right) dt,$$

and this is not equal to the work done on the electron.

To satisfy the conservation of energy it appears that in addition to its electromagnetic energy T the electron must have energy U of some type unknown but of amount

$$U = \frac{1}{4}\frac{mC^2}{\kappa} + \text{a constant} \dots\dots\dots\dots\dots\dots\dots(687).$$

Then $T + U = m\kappa +$ a constant, and the work done by external forces is equal to the increment of $T + U$.

668. If a charge e is spread over a conducting sphere of radius a, the force per unit area on its conducting surface is

$$R = 2\pi\sigma^2 = \frac{e^2}{8\pi a^4}.$$

The electron is not a charged conductor, but the above formula makes it clear that the electron at rest could be held in equilibrium by the action of a normal tension R of amount $e^2/8\pi a^4$ per unit area. Poincaré[*] has shewn that the electron in its contracted state would still be in equilibrium if tensions of this amount continued to act while the electron was in motion. Now if v is the volume of the electron at any instant, the work done on these tensions as the electron changes shape will be $R\,dv$. When the electron is moving with velocity v, its volume is $\frac{4}{3}\pi a^3/\kappa$, so that

$$Rv = \frac{e^2}{6a\kappa} = \frac{1}{4}\frac{mC^2}{\kappa}.$$

Thus if U is taken to be Rv in formula (687), the conservation of energy will be exactly satisfied.

There is no evidence as to whether these tensions do or do not exist; the possibility of their existence suggests a mechanism by which the electron can be held in equilibrium at all velocities, while its motion conforms to the conservation of energy.

---

[*] *Rendiconti del Circolo Matem. di Palermo*, 21 (1906), p. 129.

The Reaction on an Accelerated Electron.

669. The whole force acting on a moving electron is given by equation (680), in which we have so far neglected all terms beyond those in $\dot{u}$, $\dot{v}$, $\dot{w}$. Lorentz [*] has calculated the effect of the terms in $\ddot{u}$, $\ddot{v}$, $\ddot{w}$ etc., and finds that they give rise to a force acting on the electron of components F_x, F_y, F_z given by

$$F_x = -\frac{2}{3}\frac{e^2}{C^3}\ddot{u}, \text{ etc.} \dots\dots\dots\dots\dots\dots\dots(688).$$

Lorentz also gives formulae from which the remaining terms in equation (680) can be calculated, but these terms are of little physical interest.

The force given by formula (688) may be regarded as a frictional resistance opposing the motion of the electron through the ether. The rate at which the electron does work to overcome this force is

$$uF_x + vF_y + wF_z,$$

so that the work done by the electron in an interval from $t=0$ to $t=\tau$ will be

$$-\frac{2}{3}\frac{e^2}{C^3}\int_0^\tau (u\ddot{u} + v\ddot{v} + w\ddot{w})\,dt.$$

On integrating by parts, this becomes

$$-\frac{2}{3}\frac{e^2}{C^3}\left| u\dot{u} + v\dot{v} + w\dot{w} \right|_0^\tau + \frac{2}{3}\frac{e^2}{C^3}\int_0^\tau (\dot{u}^2 + \dot{v}^2 + \dot{w}^2)\,dt.$$

The last term represents the radiation emitted by the electron as calculated by Larmor's formula (662); the first term must represent changes in the energy stored in the ether.

---

[*] *The Theory of Electrons*, p. 251.

CHAPTER XX

THE THEORY OF RELATIVITY

MOTION THROUGH THE ETHER.

The Michelson-Morley Experiment.

670. WHEN we have spoken of a system at rest we have so far meant, for all practical purposes, a system at rest in our laboratories. But if we have been right in conjecturing that all electromagnetic phenomena have their seat in the ether, then a system at rest would most naturally be taken to mean a system at rest in the ether. We have so far made no clear distinction between the conceptions of rest in the ether and rest relative to the walls of a laboratory.

The view was at one time held that a moving body drags the ether along with it. If this were a true view the distinction just referred to would not arise; a body at rest relative to the walls of a laboratory would also be at rest in the ether. But in time it was found that this was not a true view; it could not be reconciled simultaneously with results of laboratory experiments such as Fizeau's water-tube experiment (cf. § 687 below), and with the astronomical theory of the aberration of light* (cf. § 689 below). Finally it became established that the ether, if one existed at all, could not share in the motion of moving bodies; it must be stagnant, and moving bodies must simply move through it without setting up mass-motions in it.

The earth's velocity in its orbit is about 30 kms. a second, so that the velocity of the earth relative to the supposed ether must at some season of the year be at least 30 kms. a second. If an ether exists, there must be a stream of ether flowing through every laboratory which must attain velocities at least as great as 30 kms. a second.

Starting in 1887, Michelson and Morley conducted experiments with a view to measuring the actual velocity of this supposed stream of ether relative to their laboratory, or, what is the same thing, the velocity of the earth through the ether. The principle of the experiment is easily explained. Let the laboratory be moving with velocity u through the ether, then a ray of light travelling against the stream of ether will move with an actual velocity C in the ether, and so will have an apparent velocity $C - u$ if measured relatively

---

* For a fuller account the reader is referred to special treatises—Larmor's *Ether and Matter* (Camb. Univ. Press, 1900) or Cunningham's *Relativity* (Camb. Univ. Press, 1914).

to the moving laboratory. Similarly a ray of light made to travel in the reverse direction will have an apparent velocity $C + u$. If a ray travel over a path l and is then reflected back to its starting-point, the time t_1 taken will be given by

$$t_1 = \frac{l}{C - u} + \frac{l}{C + u} = \frac{2l}{C}\left(1 - \frac{u^2}{C^2}\right)^{-1} \quad \dots\dots\dots\dots\dots(689).$$

Suppose next that a ray is made to travel a distance L across the direction of motion and back to its starting-point, the system moving with velocity u as before. Let the whole time be t_2, then the distance travelled by the system is ut_2. The actual path of the ray through the ether consists of two equal parts, one before reflection and one after; each part is the hypotenuse of a right-angled triangle of sides L and $\frac{1}{2}ut_2$, and the time of describing each part is $\frac{1}{2}t_2$. Hence

$$\tfrac{1}{2}t_2 C = (L^2 + \tfrac{1}{4}u^2 t_2^2)^{\frac{1}{2}},$$

whence
$$t_2 = \frac{2L}{C}\left(1 - \frac{u^2}{C^2}\right)^{-\frac{1}{2}} \quad \dots\dots\dots\dots\dots\dots(690).$$

From formulae (689) and (690) it appears that the times taken by a ray of light to travel a distance l and be reflected back, while the laboratory is in motion through the ether, will be different according as the path of the rays is along or across the direction of motion of the system. This time difference admits of measurement by optical means, and from such measurements it ought to be possible to determine u.

When the experiment was performed no time difference could be observed. The obvious explanation would be that, at the moment of performing the experiment, the laboratory was at rest in the ether, but this explanation was not found to be tenable, since no time difference could be discovered at any season of the year.

The Fitzgerald-Lorentz Contraction Hypothesis.

671. Fitzgerald in 1893 and Lorentz in 1895 suggested independently that the reason why no time difference was observed might be because the arm l of the apparatus which moved with velocity u longitudinally through the ether was contracted in a ratio $(1 - u^2/C^2)^{\frac{1}{2}}$ as a result of its motion. In such a case the arm l would have shrunk from an initial length l_0 given by

$$l_0 = l\left(1 - \frac{u^2}{C^2}\right)^{-\frac{1}{2}},$$

measured in the system when at rest. Equation (689), expressed in terms of l_0, now becomes

$$t_1 = \frac{2l_0}{C}\left(1 - \frac{u^2}{C^2}\right)^{-\frac{1}{2}},$$

and so agrees with formula (690).

Thus the Fitzgerald-Lorentz contraction hypothesis would account completely for the null result of the Michelson-Morley experiment. The hypothesis in itself is not unreasonable, for we have already seen (§ 659) that an electrostatic system set in motion with a velocity u would only regain its equilibrium after contracting longitudinally in exactly the ratio $(1 - u^2/C^2)^{\frac{1}{2}}$ assumed by the hypothesis. It is true that the arms of sandstone and pine used by Michelson and Morley were not purely electrostatic systems. But neither is the electron (cf. § 664), and yet Lorentz's hypothesis that this contracts longitudinally in exactly the same ratio is found to lead to a value for the electromagnetic mass which is entirely confirmed by experiment (§ 665).

672. According to the contraction hypothesis, the Michelson-Morley experiment failed to detect the velocity of motion through the ether because this motion was exactly concealed by the shrinkage of the apparatus. If this were so, the velocity ought of course to become measurable if we could in any way measure the amount of this shrinkage.

It is at once obvious that the shrinkage could not be measured, or even detected, by any process of direct measurement, for any material measuring-rod would shrink in exactly the same ratio as the apparatus to be measured. Indirect means might, however, be expected to reveal the amount of shrinkage.

673. Lord Rayleigh* pointed out that an isotropic medium ought to become anisotropic when shrunk, so that ordinary transparent matter ought to be doubly refracting for a ray of light crossing it in a direction oblique to its motion through the ether. But no trace of double refraction was found either by Lord Rayleigh or by Brace† who repeated the experiment with apparatus so sensitive that a fiftieth part of the expected effect would have been detected.

Following a similar train of thought, Trouton and Rankine‡ tried to detect changes in the resistance of a bar of metal as it was turned in various directions, but found no measurable change.

These experiments do not prove either that there is no motion through the ether, or that the Fitzgerald-Lorentz contraction does not occur. They prove that if there is motion through an ether, and if the contraction does occur, then the effect of this contraction is somehow veiled or compensated by some other effect. Thus Lorentz shewed§ that the null result of the experiments of Rayleigh and Brace would be exactly accounted for on his own theory of the constitution of the electron, on which the electrons would be contracted in just the same ratio as the transparent matter. And Trouton and Rankine shewed, in their original paper, that the null result of their experiment is an inevitable consequence of the electron theory of conduction through matter

---

* *Phil. Mag.* 4 (1902), p. 678.   † *Ibid.* 7 (1904), p. 317.

‡ *Proc. R. S.* 80 (1908), p. 420.   § *Theory of Electrons*, p. 217.

(cf. § 345 a), provided the electron has the transverse and longitudinal masses assigned to it by Lorentz (§ 664). Thus these experiments, undertaken originally in order to find velocity through the ether, resulted finally in providing confirmation of Lorentz's theory of the constitution of the electron.

In other experiments, the compensatory effect is still more easily discovered. A charged body moving through the ether ought to set up a magnetic field, so that every charged body in a laboratory ought to be surrounded by a magnetic field proportional to u/C. Every other charged body in the laboratory is moving across the lines of force of this magnetic field with velocity u and so ought to be acted on by a mechanical force proportional to u^2/C^2. Trouton and Noble* suspended a parallel plate condenser by a torsion thread and looked for a couple, proportional to u^2/C^2, tending to turn the plates parallel to the direction of motion through the ether. No such couple was observed.

The null result of this experiment is readily explained as a consequence of the Fitzgerald-Lorentz contraction. A shrinkage of the distance between the plates decreases the energy of the condenser. There is therefore a mechanical couple tending to turn the system into its position of minimum potential energy—*i.e.* into a position in which the plates are at right angles to the direction of motion through the ether. It is readily verified that this couple exactly neutralises the couple of magnetic origin, which the original experiment tried to detect. Indeed, granted the Fitzgerald-Lorentz shrinkage, the theorem proved in § 659 shews at once that the system would be in equilibrium in all orientations.

The Relativity-Condition.

674. These and similar experiments have one and all failed to detect motion through an ether. They have not proved that there is no motion through an ether, but shew that if this motion exists, its effects are in every case veiled by some other effect, and, in every case, it has proved possible to discover this veiling effect as an effect predicted by general electromagnetic theory.

The question arises whether there must always and of necessity be a veiling effect in every experiment. In other words, are the electromagnetic equations of such a nature that it is inherently impossible to detect motion through an ether by electromagnetic means?

It is well known that the ordinary Newtonian equations of dynamics are of this nature. For the equations

$$m \frac{d^2x}{dt^2} = F$$

* *Phil. Trans.* A, 202 (1903), p. 165 and *Proc. R. S.* 72 (1903), p. 132.

do not change their form when referred to axes moving with a uniform velocity u—i.e. when x is replaced by $x - ut$. Thus all phenomena governed by these equations are the same on an earth moving with a uniform velocity u as they would be on an earth at rest, so that it is necessarily futile to attempt to determine the earth's velocity in space by means of such phenomena.

Systems of equations or natural laws which are such as to make it impossible to determine absolute motion may be said to satisfy the "Relativity-condition." The characteristic of such equations will be that they do not change their form when referred to axes moving with a uniform velocity relative to the axes to which they were originally referred. We have seen that the Newtonian equations satisfy the relativity-condition, and the continual failure of experiment to determine the earth's velocity through the ether leads us to consider whether the electromagnetic laws may not also satisfy the relativity-condition.

If we simply change the electromagnetic laws by replacing x by $x - ut$, it is at once seen that a change of form results. But the hypothesis of the Fitzgerald-Lorentz contraction has already given grounds for suspecting that the required change may not be so simple as this. For instance, it may be that on changing to moving axes, all lengths parallel to the x-axis ought to be contracted in the ratio $(1 - u^2/C^2)^{\frac{1}{2}}$. In this case the transformation would be from x to a new coordinate x' defined by $x' = \kappa(x - ut)$, where κ denotes $(1 - u^2/C^2)^{-\frac{1}{2}}$. The analysis of § 659 has already shewn that all electrostatic phenomena conform to the relativity-condition when this transformation is made.

This change really amounts to a change in the measurement of the unit of length, as regards lengths parallel to the axis of x, when we change the velocity of motion parallel to the axis of x. Following a method originated by Einstein[*] we proceed to examine whether similar changes in all the units can result in the electromagnetic laws conforming to the relativity-condition.

675. Consider first the condition that the simple phenomenon of the transmission of a light-signal shall satisfy the relativity-condition. Imagine an experimenter S moving with an unknown but uniform velocity, and using coordinates x, y, z, t to record the result of his observations. If the phenomenon of light-transmission satisfies the relativity-condition, a signal started from the origin at any instant $t = 0$ will after time t have reached points lying on a sphere

$$x^2 + y^2 + z^2 - C^2 t^2 = 0 \quad \dots\dots\dots\dots\dots\dots(691),$$

where C is the velocity of light determined by the observer S.

Let a second observer S' move with a different velocity, and let him use coordinates x', y', z', t' to record the result of his observations. The sphere

[*] *Ann. d. Phys.* 17 (1905), p. 891.

whose equation is (691) for S will have an equation expressed in terms of x', y', z', t' for S', and if the relativity-condition is satisfied, this equation must be

$$x'^2 + y'^2 + z'^2 - C'^2 t'^2 = 0 \quad \dots\dots\dots\dots\dots(692),$$

where C' is the velocity of light determined by S'.

If S' changes his units of length or time he will change his value of C', which is the distance light appears to him to travel in unit time. We may without any loss of generality suppose S' to use units which make C' equal to C.

We may also suppose that light will appear, both to S and to S', to travel in straight lines with uniform velocity*. Thus for S the equation connecting the position x, y, z of a light-signal with the time t must be linear in x, y, z and t. The similar equation for S' will be linear in x', y', z' and t'. Thus x', y', z' and t' will necessarily be linear functions of x, y, z and t. And we have already supposed that the equations of transformation from x', y', z', t' to x, y, z, t must be such that equation (691) transforms into equation (692), C' being equal to C.

Let us introduce new variables τ, τ' in place of t, t', these being given by $\tau = iCt$, $\tau' = iCt'$ where $i = \sqrt{(-1)}$. Then equations (691) and (692) become

$$x^2 + y^2 + z^2 + \tau^2 = 0,$$
$$x'^2 + y'^2 + z'^2 + \tau'^2 = 0.$$

The relativity-condition is satisfied if a linear transformation transforms the one equation into the other. Since the equations of transformation are linear this requires that

$$x^2 + y^2 + z^2 + \tau^2 = k(x'^2 + y'^2 + z'^2 + \tau'^2)\dots\dots\dots\dots(693),$$

where k is a constant.

Imagine a four-dimensional space constructed in which x, y, z, τ are orthogonal rectilinear coordinates. On account of the linearity of the equations of transformation, x', y', z', τ' may also be regarded as rectilinear coordinates in this same space, but these have not yet been required to be orthogonal. Now $x^2 + y^2 + z^2 + \tau^2$ is the square of the distance of the point x, y, z, τ from the origin when expressed in x, y, z, τ coordinates, so that, by equation (693), $k(x'^2 + y'^2 + z'^2 + \tau'^2)$ must be the square of the distance of x', y', z', τ' from the origin. It follows at once that x', y', z', τ' must be orthogonal coordinates; if

---

* According to Einstein's theory of generalised relativity, to which we shall return below (§ 702), light does not travel in straight lines in the presence of a gravitational field. The assumption we have just made marks the parting of the ways between the old physical theory of relativity and the new generalised theory. On the new theory the assumption just made is strictly true only at an infinite distance from all matter; it may nevertheless be regarded as a very accurate first approximation to the truth except in gravitational fields enormously more intense than any of which we have experience.

they were not orthogonal, cross products $x'y'$, $x'\tau'$ etc. would enter into the expression for the square of the distance from x', y', z', τ' to the origin. Thus the axes of x', y', z', τ' can be obtained from those of x, y, z, τ by a pure rotation in the four-dimensional space.

We have already fixed the ratio of S''s units of length and time by making $C' = C$. If we further change the absolute values of these units, we can alter the value of k, and we may agree to fix these absolute values so that $k = 1$. The change from coordinates x, y, z, τ to x', y', z', τ', or conversely, is now effected by a pure rigid body rotation of the axes.

We may notice in passing that if the relativity-condition is satisfied as regards the transmission of light-signals, no set of axes x, y, z, τ in the four-dimensional space is geometrically more fundamental than any other. A change of velocity of translation is merely effected by turning the axes about, and no observer can claim on purely geometrical grounds that his system of axes provides a standard set from which all other positions of the axes ought to be measured.

676. The simplest case of rotation of the axes occurs when every point moves parallel to one of the coordinate planes, say x, τ. The formulae of transformation then assume the simple forms

$$\left.\begin{aligned} x' &= x\cos\theta + \tau\sin\theta \\ \tau' &= \tau\cos\theta - x\sin\theta \\ y' &= y; \quad z' = z \end{aligned}\right\} \quad \dots\dots\dots\dots\dots\dots(694).$$

To determine what physical meaning is to be assigned to θ, we notice that $x' = 0$ when

$$x = -\tau\tan\theta = -iCt\tan\theta \quad \dots\dots\dots\dots\dots(695).$$

Thus a point which the experimenter S regards as moving along the axis of x with a velocity $-iC\tan\theta$ will appear to S' to be at rest. In other words the axes of S' move relative to those of S with a velocity $-iC\tan\theta$ along the axis of x. Let us put

$$u = -iC\tan\theta \quad \dots\dots\dots\dots\dots\dots(696),$$

then the transformation (694) is that appropriate to the case in which the axes of S' have a velocity $(u, 0, 0)$ relative to those of S.

Put
$$\kappa = \left(1 - \frac{u^2}{C^2}\right)^{-\frac{1}{2}},$$

then $\kappa = \cos\theta$, and the formulae of transformation (694) become

$$x' = \kappa(x - ut), \quad y' = y, \quad z' = z, \quad t' = \kappa\left(t - \frac{xu}{C^2}\right) \dots\dots\dots(697).$$

677. Following Einstein we have found that the transformation relations (697) express the necessary and sufficient condition that the propagation of light-signals shall conform to the relativity-condition. We have already noticed (§ 674) that the first relation $x' = \kappa (x - ut)$ is simply an expression of the Fitzgerald-Lorentz contraction which is necessary if the Michelson-Morley experiment is to conform to the relativity-condition. We now have the further information that if *all* experiments of light transmission are to satisfy the relativity-condition, we must have the further relation

$$t' = \kappa \left(t - \frac{xu}{C^2} \right).$$

The transformation (697), although we have obtained it by a method due mainly to Einstein, is commonly known as Lorentz's transformation. For Lorentz had shewn*, before the appearance of Einstein's paper, that precisely the same transformation expresses the condition that the ordinary electrodynamical equations shall conform to the relativity-condition.

678. Before proving this, let us examine some of the purely kinematical properties of the Lorentz transformation expressed by equations (697).

Transforming to axes moving with a relative velocity u is equivalent, as we have seen, to turning the axes through an angle θ in the x, τ plane, where θ is given by equation (696). Transforming to axes moving with a velocity u' relative to these new axes is equivalent to turning through a further angle θ' given by

$$u' = - iC \tan \theta'.$$

But these last axes can be obtained from the original axes on turning through an angle $\theta + \theta'$, and we have

$$- iC \tan (\theta + \theta') = \frac{- iC (\tan \theta + \tan \theta')}{1 - \tan \theta \tan \theta'} = \frac{u + u'}{1 + \frac{uu'}{C^2}}.$$

Thus the velocity of the last set of moving axes relative to the first is not $u + u'$; it is U, given by

$$U = \frac{u + u'}{1 + \frac{uu'}{C^2}}, \quad\dots\dots\dots\dots\dots\dots\dots(698),$$

and we notice that U is necessarily less than $u + u'$ when both u and u' are positive. We should only have a right to expect that U would be equal to $u + u'$ if both S and S' measured their lengths, times and velocities in similar ways, and this, under the Lorentz transformation, they do not do.

As a direct consequence of equation (698),

$$C - U = \frac{(C - u)(C - u')}{C \left(1 + \frac{uu'}{C^2} \right)},$$

so that if u and u' are each less than C, then U is necessarily less than C.

* *Amsterdam Proc.* (1904), p. 809.

Thus no possible compounding of velocities less than C can ever give a resultant velocity v greater than C. As a special case if $u' = C$ then $v = C$, regardless of the value of u; the resultant of the velocity of light and any other velocity is the velocity of light.

Similar analysis will give the result of superposing two velocities not in the same direction, but the required formulae can be obtained rather more directly from the formulae of transformation (697), as we shall now see.

679. Let a point move with velocity u, v, w relative to the axes used by S, so that

$$x = x_0 + ut, \quad y = y_0 + vt, \quad z = z_0 + wt \dots\dots\dots\dots(699),$$

and let the velocity of the same point relative to the axes used by S' be u', v', w', so that

$$x' = x_0' + u't', \quad y' = y_0' + v't', \quad z' = z_0' + w't' \dots\dots\dots(700).$$

In these last equations, let us substitute Lorentz's values for x', y', z', t', as given by equations (697). We obtain

$$\kappa\,(x - ut) = x_0' + u'\kappa\left(t - \frac{xu}{C^2}\right),$$

$$y = y_0' + v'\kappa\left(t - \frac{xu}{C^2}\right),$$

and a similar equation for z. Differentiate these three equations with respect to t, putting $\dfrac{dx}{dt} = u$, etc., in accordance with equations (699), and we find

$$\left.\begin{aligned} u - u &= u'\left(1 - \frac{vu}{C^2}\right) \\ v &= v'\kappa\left(1 - \frac{vu}{C^2}\right) \\ w &= w'\kappa\left(1 - \frac{vu}{C^2}\right) \end{aligned}\right\} \quad \dots\dots\dots\dots(701),$$

from which follows directly

$$\left.\begin{aligned} u &= \frac{u' + u}{1 + \dfrac{u'u}{C^2}} \\ &= \frac{v'}{\kappa\left(1 + \dfrac{u'u}{C^2}\right)} \\ w &= \frac{w'}{\kappa\left(1 + \dfrac{u'u}{C^2}\right)} \end{aligned}\right\} \quad \dots\dots\dots\dots(702).$$

In these equations u, v, w may be regarded as the resultant velocity obtained by compounding velocities u, 0, 0 and u', v', w'.

From equations (701) we obtain directly

$$U' = \frac{U - u}{1 - \dfrac{Uu}{C^2}}$$

$$V' = \frac{V}{\kappa\left(1 - \dfrac{Uu}{C^2}\right)}$$

$$W' = \frac{W}{\kappa\left(1 - \dfrac{Uu}{C^2}\right)}$$

............(703).

These are also a necessary consequence of equations (702), for U', V', W' is the velocity obtained by compounding velocities U, V, W and $-u$, 0, 0.

ELECTROMAGNETIC EQUATIONS.

680. Following Lorentz[*] and Einstein[†], let us now proceed to transform the general electrodynamical equations of Chap. XIX (§§ 621, 622), namely

$$\frac{4\pi}{C}\left(\rho U + \frac{df}{dt}\right) = \frac{\partial \gamma}{\partial y} - \frac{\partial \beta}{\partial z}, \text{ etc. } ....(704),$$

$$-\frac{1}{C}\frac{da}{dt} = \frac{\partial Z}{\partial y} - \frac{\partial Y}{\partial z}, \text{ etc. } ....(705),$$

$$\frac{\partial f}{\partial x} + \frac{\partial g}{\partial y} + \frac{\partial h}{\partial z} = \rho ....(706),$$

$$\frac{\partial a}{\partial x} + \frac{\partial b}{\partial y} + \frac{\partial c}{\partial z} = 0 ....(707),$$

to the new variables x', y', z', t' connected with x, y, z, t by relations (697).

If χ is any function whatever of x, y, z, t we have

$$\frac{\partial \chi}{\partial x} = \frac{\partial \chi}{\partial x'}\frac{\partial x'}{\partial x} + \frac{\partial \chi}{\partial t'}\frac{\partial t'}{\partial x} = \kappa\left(\frac{\partial \chi}{\partial x'} - \frac{u}{C^2}\frac{\partial \chi}{\partial t'}\right)$$

$$\frac{\partial \chi}{\partial t} = \frac{\partial \chi}{\partial x'}\frac{\partial x'}{\partial t} + \frac{\partial \chi}{\partial t'}\frac{\partial t'}{\partial t} = \kappa\left(\frac{\partial \chi}{\partial t'} - u\frac{\partial \chi}{\partial x'}\right)$$

$$\frac{\partial \chi}{\partial y} = \frac{\partial \chi}{\partial y'}; \ \frac{\partial \chi}{\partial z} = \frac{\partial \chi}{\partial z'}$$

............(708).

The three equations (704) and equation (706) accordingly assume the form

$$\frac{4\pi}{C}\left[\rho U + \kappa\left(\frac{df}{dt'} - u\frac{\partial f}{\partial x'}\right)\right] = \frac{\partial \gamma}{\partial y'} - \frac{\partial \beta}{\partial z'}, ...(709),$$

$$\frac{4\pi}{C}\left[\rho V + \kappa\left(\frac{dg}{dt'} - u\frac{\partial g}{\partial x'}\right)\right] = \frac{\partial \alpha}{\partial z'} - \kappa\left(\frac{\partial \gamma}{\partial x'} - \frac{u}{C^2}\frac{\partial \gamma}{\partial t'}\right) ......(710),$$

* *l.c. ante.* † *Ann. d. Physik*, 17 (1905), p. 916.

$$\frac{4\pi}{C}\left[\rho W + \kappa\left(\frac{dh}{dt'} - u\frac{\partial h}{\partial x'}\right)\right] = \kappa\left(\frac{\partial \beta}{\partial x'} - \frac{u}{C^2}\frac{\partial \beta}{\partial t'}\right) - \frac{\partial \alpha}{\partial y'} \quad\ldots\ldots(711),$$

and
$$\kappa\frac{\partial f}{\partial x'} + \frac{\partial g}{\partial y'} + \frac{\partial h}{\partial z'} - \frac{\kappa u}{C^2}\frac{df}{dt'} = \rho \quad\ldots\ldots\ldots\ldots(712).$$

If we introduce $\alpha', \beta', \gamma', f', g', h'$, defined by

$$\left.\begin{aligned}
\alpha' &= \alpha, & f' &= f \\
\beta' &= \kappa\left(\beta + \frac{4\pi u}{C}h\right), & g' &= \kappa\left(g - \frac{u}{4\pi C}\gamma\right) \\
\gamma' &= \kappa\left(\gamma - \frac{4\pi u}{C}g\right), & h' &= \kappa\left(h + \frac{u}{4\pi C}\beta\right)
\end{aligned}\right\} \quad\ldots\ldots(713);$$

then equations (710) and (711) may be written in the form

$$\frac{4\pi}{C}\left(\rho V + \frac{dg'}{dt'}\right) = \frac{\partial \alpha'}{\partial z'} - \frac{\partial \gamma'}{\partial x'} \quad\ldots\ldots\ldots\ldots(714),$$

$$\frac{4\pi}{C}\left(\rho W + \frac{dh'}{dt'}\right) = \frac{\partial \beta'}{\partial x'} - \frac{\partial \alpha'}{\partial y'} \quad\ldots\ldots\ldots\ldots(715).$$

681. We still require to transform $\rho U, \rho V, \rho W$ to the new coordinates. The density ρ' in the new coordinates must be such that

$$\rho'\, dx'\, dy'\, dz' = \rho\, dx\, dy\, dz.$$

Since the coordinates x', y', z', τ' are derived from x, y, z, τ by a pure rotation in four-dimensional space, we have

$$\frac{\partial(x', y', z', \tau')}{\partial(x, y, z, \tau)} = 1, \text{ or } dx'\, dy'\, dz'\, d\tau' = dx\, dy\, dz\, d\tau.$$

Thus
$$\frac{dx\, dy\, dz}{dx'\, dy'\, dz'} = \frac{d\tau'}{d\tau} = \frac{dt'}{dt} = \kappa\left(1 - \frac{Uu}{C^2}\right) \quad\ldots\ldots\ldots(716),$$

so that
$$\rho' = \rho\kappa\left(1 - \frac{Uu}{C^2}\right) \quad\ldots\ldots\ldots\ldots\ldots(717).$$

Combining this with equations (703), we find at once that $\rho(U - u) = \rho'U'$, $\rho V = \rho'V'$ and $\rho W = \rho'W'$. Thus equations (714) and (715) become

$$\frac{4\pi}{C}\left(\rho'V' + \frac{dg'}{dt'}\right) = \frac{\partial \alpha'}{\partial z'} - \frac{\partial \gamma'}{\partial x'} \quad\ldots\ldots\ldots\ldots(718),$$

$$\frac{4\pi}{C}\left(\rho'W' + \frac{dh'}{dt'}\right) = \frac{\partial \beta'}{\partial x'} - \frac{\partial \alpha'}{\partial y'} \quad\ldots\ldots\ldots\ldots(719).$$

On multiplying throughout by κ and using relations (713), equation (709) becomes

$$\frac{4\pi}{C}\left[\kappa\rho U + \kappa^2\frac{df}{dt'} - u\kappa\left(\kappa\frac{\partial f}{\partial x'} + \frac{\partial g}{\partial y'} + \frac{\partial h}{\partial z'}\right)\right] = \frac{\partial \gamma'}{\partial y'} - \frac{\partial \beta'}{\partial z'},$$

which, by the use of equation (712), reduces further to

$$\frac{4\pi}{C}\left[\kappa\rho(U - u) + \frac{df}{dt'}\right] = \frac{\partial \gamma'}{\partial y'} - \frac{\partial \beta'}{\partial z'}.$$

Using the relation $\rho\,(U-u)=\rho'U'$ just obtained, and also the relation $f=f'$, this becomes

$$\frac{4\pi}{C}\left(\rho'\,U'+\frac{df'}{dt'}\right)=\frac{\partial\gamma'}{\partial y'}-\frac{\partial\beta'}{\partial z'} \quad\dots\dots\dots\dots\dots(720).$$

Finally, again using relations (713), equation (712) transforms into

$$\kappa\frac{\partial f'}{\partial x'}+\frac{1}{\kappa}\left(\frac{\partial g'}{\partial y'}+\frac{\partial h'}{\partial z'}\right)+\frac{u}{4\pi C}\left(\frac{\partial\gamma}{\partial y'}-\frac{\partial\beta}{\partial z'}\right)-\frac{\kappa u}{C^2}\frac{df}{dt'}=\rho.$$

Using the relation (720), which has just been obtained, this becomes

$$\kappa\left[\frac{\partial f'}{\partial x'}\left(1-\frac{u^2}{C^2}\right)\right]+\frac{1}{\kappa}\left(\frac{\partial g'}{\partial y'}+\frac{\partial h'}{\partial z'}\right)=\rho\left(1-\frac{Uu}{C^2}\right),$$

or, multiplying throughout by κ and using equation (717),

$$\frac{\partial f'}{\partial x'}+\frac{\partial g'}{\partial y'}+\frac{\partial h'}{\partial z'}=\rho' \quad\dots\dots\dots\dots\dots(721).$$

682. We have now seen that if the new quantities α', β', γ', f', g', h' are defined by equations (713), then the electric equations (704) and (706), when transformed to coordinates x', y', z', t', resume their original form exactly. By precisely similar analysis we find that if

$$\left.\begin{aligned}
a'&=a, & X'&=X\\
b'&=\kappa\left(b+\frac{u}{C}Z\right), & Y'&=\kappa\left(Y-\frac{u}{C}c\right)\\
c'&=\kappa\left(c-\frac{u}{C}Y\right), & Z'&=\kappa\left(Z+\frac{u}{C}b\right)
\end{aligned}\right\}\dots\dots\dots\dots(722),$$

then the magnetic equations (705) and (707), when transformed to the new coordinates x', y', z', t', will also resume their original form exactly.

Thus it appears that the relativity-condition will be satisfied by all electromagnetic phenomena, if the relation between the forces as estimated by S and those estimated by S' moving with a velocity $(u, 0, 0)$ relative to S can be supposed to be those given by relations (713) and (722). If these relations are found, in actual fact, to be satisfied, it will be impossible to determine absolute motion by any electromagnetic means whatever.

683. Consider first the form assumed by the relations in free space, for which we may take $K=\mu=1$. Here a, b, c become identical with α, β, γ, and a', b', c' with α', β', γ'. Also f, g, h become the same as $X/4\pi$, $Y/4\pi$, $Z/4\pi$ and similarly for f', g', h'. The two sets of equations (713) and (722) are now seen to become identical, each reducing to

$$\left.\begin{aligned}
\alpha'&=\alpha, & X'&=X\\
\beta'&=\kappa\left(\beta+\frac{u}{C}Z\right). & Y'&=\kappa\left(Y-\frac{u}{C}\gamma\right)\\
\gamma'&=\kappa\left(\gamma-\frac{u}{C}Y\right), & Z'&=\kappa\left(Z+\frac{u}{C}\beta\right)
\end{aligned}\right\}\dots\dots\dots\dots(723).$$

If u^2/C^2 is neglected, κ may be put equal to unity, and the forces X', Y', Z' are exactly those which we found in § 628 for the forces on a unit charge moving with velocity $(u, 0, 0)$. Similarly the forces α', β', γ' are easily shewn by the method of § 572 to be precisely those which would be acting on a unit magnetic pole moving with a velocity $(u, 0, 0)$. Thus there is direct experimental verification of these equations when u^2/C^2 is neglected.

When u^2/C^2 is not neglected, it is naturally impossible to obtain direct experimental evidence of the accuracy of the equations. A complication arises from the fact that S and S' are using different units of length in the direction of Ox, and on allowing for this and treating the problem in the manner of § 656, it is at once found that the presence of the factors κ in equations (723) exactly represents the complication introduced by the finiteness of u^2/C^2.

Thus it appears, by what is not far short of absolute proof, that the relativity-condition is satisfied by all electromagnetic phenomena.

684. The problem presented by phenomena in dielectric and magnetic media is naturally môre complex. Various hypotheses have been put forward as to the relation between a', b', c' and α', β', γ' in moving magnetic media, as also regarding the relation between f', g', h' and X', Y', Z' in moving dielectrics. Some of these hypotheses are in agreement with relations (713) and (722), while some are not. Experiments have been conducted by various physicists, and in particular by H. A. Wilson and A. Eichenwald, with a view to discriminating between these rival hypotheses. In each case the victorious hypothesis is found to be in conformity with equations (713) and (722) above.

Wilson[*] moved a dielectric body through a magnetic field and found that there was an electric polarisation (f', g', h') set up of which the amount agreed very closely with that demanded by equations (713). And Eichenwald[†] set a polarised dielectric in motion and found that it produced a magnetic field similar to that demanded by equations (713). Thus there seems to be experimental confirmation for every term in equations (713). Experiments on moving magnetic media have not been performed, but there seems to be little room for doubt that they would similarly confirm equations (722).

If, on the strength of this evidence, we assume equations (713) and (722) to be fully confirmed, then we have shewn that the electromagnetic equations conform to the relativity-condition. In other words, all experiments to determine velocity through the ether are necessarily futile. If for the moment we assume that we are moving through the ether with a velocity u in a direction which we call Ox, then we may consider that we are playing the rôle of our observer S', while an imaginary observer at rest in the ether may be supposed to be playing the rôle of our observer S. But we have seen that

[*] *Phil. Trans.* A, 204 (1904), p. 121.   [†] *Ann. d. Phys.* 11 (1904), p. 421.

all electromagnetic phenomena would be exactly the same for us as for S. If we could deduce a velocity u through the ether for our motion, S would necessarily deduce a velocity u for his own motion, which would be contrary to the facts. By this argument, here put in the form of a *reductio ad absurdum*, we see the impossibility of determining our velocity through the ether.

If at any time equations (713) and (722) are proved to be untrue—and, as we have seen, the remaining opportunities for proving these equations untrue are very few—then it will become possible, in theory at least, to determine our velocity through the ether. But for the present we shall assume, as a working hypothesis, that it is in no way possible to determine velocity through the ether. This is commonly called the Hypothesis of Relativity. We proceed to examine some of the consequences of this hypothesis.

The Relativity Hypothesis.

685. This hypothesis commits us, generally speaking, to all the equations of the present chapter. It does not commit us to any special physical interpretations of them. For instance, the first equation of the Lorentz transformation, namely $x' = \kappa (x - ut)$, may if we please be interpreted in terms of the Fitzgerald-Lorentz contraction-hypothesis; we may postulate a fixed ether and the equation is then taken to shew that any length moving with a velocity $(u, 0, 0)$ through the ether will be contracted in the direction of the x-axis in the ratio $1/\kappa$. Alternatively we may interpret the same equation in such a way as not to assume a fixed ether at all. Any observer S measures out a sphere which remains at rest relative to him; to a second observer S' moving relative to S with a velocity $(u, 0, 0)$, this sphere will appear to be contracted in the ratio $1/\kappa$ along Ox.

In a similar way all the other equations can be interpreted so as to have no reference to a fixed ether: they may be taken merely as expressing relations between quantities as measured by one observer S and another observer S' moving with a velocity u relative to S.

The kinematical relations of Einstein, namely equations (702), may, on this interpretation, be regarded merely as laws for the composition of velocities. It appears that the simple laws of composition of velocities and of vector-addition—the so-called "parallelogram of velocities"—are no longer true if the hypothesis of relativity is true. The simple laws are true if u^2/C^2 and u'^2/C^2 are small, but not otherwise.

Startling though this result may appear, there is almost direct experimental confirmation of it, as we shall soon see.

OPTICAL CONSEQUENCES OF THE RELATIVITY HYPOTHESIS.

686. The relativity hypothesis makes no claim to explain the nature of phenomena, it merely proposes, tentatively, a general law of a restrictive nature, which so far has appeared to dominate all known phenomena. All explanations of phenomena which conform to the limits of this restriction are equally permitted by this hypothesis, but the hypothesis serves to rule out, tentatively, all explanations which do not conform to the condition. Consequently it is only in rare cases that the principle of relativity by itself enables us to obtain a full solution of a problem. As an instance of such a case, we have seen that it enables us to determine the electric and magnetic forces in ponderable media. Other instances occur in optical phenomena, and to these we now turn.

Fizeau's Water Tube Experiment.

687. In Fizeau's water tube experiment, a stream of water was made to flow through a tube, its velocity of flow being u relative to the earth, and a ray of light was passed through the water in the direction of its motion. To an observer moving with the stream, the water would appear to be at rest, so that the light would be propagated relative to this observer with a velocity u' connected with the refractive-index ν of the water by the relation $u' = C/\nu$.

According to the classical laws of kinematics, the light ought to travel, relative to an observer at rest on the earth, with a velocity $u + u'$ or

$$\frac{C}{\nu} + u \quad\dots\dots\dots\dots\dots\dots\dots\dots\dots\dots\dots(724).$$

Fizeau found it possible to measure the actual velocity by an interference method and formula (724) was not confirmed. The formula

$$\frac{C}{\nu} + u\left(1 - \frac{1}{\nu^2}\right) \quad\dots\dots\dots\dots\dots\dots\dots\dots(725)$$

was found to represent the velocity accurately both for water and other transparent media.

As we shall now see, formula (725) is not only consistent with the theory of relativity, but could also have been fully predicted by this theory. For the velocity in question is simply that which results from compounding the velocities u and C/ν, and the resultant velocity obtained by the relativity formula (702) is

$$U = \frac{\dfrac{C}{\nu} + u}{1 + \dfrac{u}{C^2}\left(\dfrac{C}{\nu}\right)} = \frac{C}{\nu} + \frac{u\left(1 - \dfrac{1}{\nu^2}\right)}{1 + \left(\dfrac{u}{C^2}\dfrac{C}{\nu}\right)}.$$

If u^2/C^2 is neglected this reduces the formula (725). In this experiment we have very direct experimental confirmation of Einstein's formula for the composition of velocities.

Reflection of Light from a Moving Mirror and Emission from a Moving Source.

688. According to the relativity hypothesis, the velocity of light in free space is always equal to C. If the light be observed by an observer moving with a velocity u relative to the source, the velocity is still equal to C, for we have seen in § 678 that the velocity obtained by compounding a velocity C with any other velocity u is itself equal to C.

This consequence of the relativity hypothesis has been tested by Majorana. He first examined the light reflected by a moving mirror and found its velocity to be exactly equal to C independently of the velocity of the mirror[*]. In a later investigation[†] he tested the velocity of light emitted by a rapidly-moving source and found this to be equal to C independently of the velocity of the source.

These experimental results are of very great importance, for it will be seen that the Michelson-Morley experiment and the experiments of Majorana taken in combination establish the Lorentz transformation equations (697) as a fact of observation. The Michelson-Morley experiment shewed that the average to-and-fro velocity of light reflected from a mirror back to the source was the same for all directions in space. The Majorana experiments now shew that the result is true for the separate paths before and after reflection, so that the velocity of light, as measured by any observer, is the same for all directions in space. We now have as an experimental fact that, independently of the velocities of the source and observer, the wave-surface is a sphere *having the observer as centre*. This is precisely the supposition from which we started in § 675; it was found to lead directly to the Lorentz transformation (697).

Aberration and the Doppler effect.

689. As in § 591, the equation of wave-propagation in free space, namely

$$\frac{d^2\chi}{dt^2} = C^2 \nabla^2 \chi,$$

has a solution

$$\chi = A \cos \frac{2\pi}{\nu C} (lx + my + nz - Ct) \dots\dots\dots\dots(726),$$

where $l^2 + m^2 + n^2 = 1$, and this corresponds to the propagation of a plane wave of light of frequency ν in a direction l, m, n. Suppose that the same ray of light appears to the observer S' to be of frequency ν' and to be propagated in a direction l', m', n', so that the solution of the wave-equation for S' will be

$$\chi' = A' \cos \frac{2\pi}{\nu' C} (l'x' + m'y' + n'z' - Ct') \dots\dots\dots\dots(727).$$

[*] *Phil. Mag.* 35 (1918), p. 163.   [†] *Phil. Mag.* 37 (1919), p. 145.

On substituting for x', y', z', t' in terms of x, y, z, t from equations (697), this becomes

$$\chi' = A' \cos \frac{2\pi}{\nu' C} \left[l'\kappa (x - ut) + m'y + n'z - C\kappa \left(t - \frac{xu}{C^2} \right) \right].$$

This expression must be identical with (726), so that by comparison we obtain

$$\frac{\kappa \left(l' + \dfrac{u}{C} \right)}{l} = \frac{m'}{m} = \frac{n'}{n} = \frac{\kappa (C + l'u)}{C} = \frac{\nu'}{\nu} \quad \dots\dots\dots(728).$$

Aberration. Equating the first and fourth fractions in equations (728) we find

$$l = \frac{l' + \dfrac{u}{C}}{1 + l' \dfrac{u}{C}} \quad \dots\dots\dots\dots\dots\dots\dots(729).$$

This must, according to the hypothesis of relativity, be the exact formula for astronomical aberration. Let the observer S be at rest relative to any system of axes in uniform motion, while S' moves relative to these axes with a velocity u along Ox. Then light which appears to S to arrive in a direction l, m, n will appear to S' to arrive in a direction l', m', n', where l, l' are related by equation (729). Put $l = \cos\phi$, $l' = \cos\phi'$; then

$$\cos\phi' - \cos\phi = l' - l = - \sin^2 \phi' \frac{u}{C \left(1 + \dfrac{u}{C} \cos\phi' \right)}.$$

If u/C is small, this reduces to the ordinary formula of practical astronomy,

$$\phi' - \phi = \sin\phi' \left(\frac{u}{C} \right).$$

Doppler Effect. Equating the last two fractions in equations (728), we find

$$\frac{\nu'}{\nu} = \left(1 + \frac{u \cos\phi'}{C} \right) \left(1 - \frac{u^2}{C^2} \right)^{-\frac{1}{2}} \quad \dots\dots\dots(730).$$

This is the full expression for the Doppler effect. If u^2/C^2 is neglected, the right-hand member reduces to

$$1 + \frac{u}{C} \cos\phi',$$

which is the Doppler factor usually assumed. If the observer is moving directly towards the source of light with velocity u, we have $\cos\phi' = 1$, and equation (730) becomes

$$\frac{\nu'}{\nu} = \left[\frac{1 + \dfrac{u}{C}}{1 - \dfrac{u}{C}} \right]^{\frac{1}{2}} \quad \dots\dots\dots\dots\dots(731).$$

ACCELERATION, MASS AND FORCE.

690. In formulae (702) we obtained equations for the velocity U, V, W obtained by compounding a velocity u, 0, 0 with a velocity U', V', W'. When the velocity U', V', W' is so small that its square may be neglected in comparison with C^2, these formulae reduce to

$$U = u + U' - \frac{u^2}{C^2} U' = u + \frac{U'}{\kappa^2} \left. \vphantom{\begin{array}{c} a \\ b \end{array}} \right\} \quad\quad\quad (732).$$
$$V = \frac{V'}{\kappa}; \quad W = \frac{W'}{\kappa}$$

Suppose that u, 0, 0 is the velocity relative to S of a moving particle at an instant $t = 0$. Let it appear to an observer S', moving with a uniform velocity u, 0, 0, to have accelerations

$$\frac{dU'}{dt'}, \quad \frac{dV'}{dt'}, \quad \frac{dW'}{dt'},$$

these being measured in the coordinates used by S'. In formulae (732) let us put

$$U' = \frac{dU'}{dt'} dt', \quad V' = \frac{dV'}{dt'} dt', \quad W' = \frac{dW'}{dt'} dt' \quad\quad\quad (733),$$

then U, V, W will be the velocities, as measured by S at the end of a small interval dt' as measured by S'.

The times t, t' used by S and S' are connected by equation (697), namely,

$$t' = \kappa \left(t - \frac{xu}{C^2} \right),$$

so that on differentiation with respect to t following the particle in its motion,

$$\frac{dt'}{dt} = \kappa \left(1 - \frac{u}{C^2} \frac{dx}{dt} \right) = \kappa \left(1 - \frac{u^2}{C^2} \right) = \frac{1}{\kappa} \quad\quad\quad (734).$$

Thus relations (733) may be replaced by

$$U' = \frac{1}{\kappa} \frac{dU'}{dt'} dt, \text{ etc.,}$$

and equations (732) become

$$U = u + \frac{1}{\kappa^3} \frac{dU'}{dt'} dt, \quad V = \frac{1}{\kappa^2} \frac{dV'}{dt'} dt, \quad W = \frac{1}{\kappa^2} \frac{dW'}{dt'} dt \quad\quad (735).$$

If $\dfrac{dU}{dt}$, $\dfrac{dV}{dt}$, $\dfrac{dW}{dt}$ are the accelerations as measured by S, we must have

$$U = u + \frac{dU}{dt} dt, \text{ etc.,}$$

whence by comparison with equations (735),

$$\frac{dU}{dt} = \frac{1}{\kappa^3} \frac{dU'}{dt'}; \quad \frac{dV}{dt} = \frac{1}{\kappa^2} \frac{dV'}{dt'}; \quad \frac{dW}{dt} = \frac{1}{\kappa^2} \frac{dW'}{dt'} \quad\quad\quad (736).$$

These formulae give the accelerations as measured by S in terms of the accelerations as measured by an observer S' moving with the particle.

691. Let the accelerations be supposed to originate from the action of a force. Since the particle is supposed to be at rest relative to S' at the instant $t' = 0$, its equations of motion, in terms of the coordinates used by S', will be

$$m\frac{du'}{dt'} = P', \quad m\frac{dv'}{dt'} = Q', \quad m\frac{dw'}{dt'} = R',$$

where m is the mass, as estimated by S', and P', Q', R' are the components of the force, also as estimated by S'.

From these equations and equations (736), we obtain

$$m\kappa^3\frac{du}{dt} = P'; \quad m\kappa^2\frac{dv}{dt} = Q'; \quad m\kappa^2\frac{dw}{dt} = R' \quad \dots\dots(737),$$

and these will be the equations of motion as observed by S.

If the particle is an electron of charge e, the values of P', Q', R' will be eX', eY', eZ', where X', Y', Z' are given by equations (723). Substituting these, we find for the equations of motion of the electron as observed by S,

$$\left. \begin{aligned} m\kappa^3\frac{du}{dt} &= eX \\ m\kappa^2\frac{dv}{dt} &= e\kappa\left(Y - \frac{u}{C}\gamma\right) \\ m\kappa^2\frac{dw}{dt} &= e\kappa\left(Z + \frac{u}{C}\beta\right) \end{aligned} \right\} \quad \dots\dots\dots\dots(738).$$

The observer S will suppose the electron moving with velocity u to have longitudinal and transverse masses m_l and m_t; and his equations of motion for the electron will be

$$\left. \begin{aligned} m_l\frac{du}{dt} &= eX \\ m_t\frac{dv}{dt} &= e\left(Y - \frac{u}{C}\gamma\right) \\ m_t\frac{dw}{dt} &= e\left(Z + \frac{u}{C}\beta\right) \end{aligned} \right\} \quad \dots\dots\dots\dots(739).$$

By comparison with equations (738)

$$m_l = m\kappa^3; \quad m_t = m\kappa \quad \dots\dots\dots\dots(740).$$

692. These are precisely Lorentz's expressions for the longitudinal and transverse mass of a moving electron (§ 664), which we have seen to be fully confirmed by experiment (§ 665). In deducing these expressions in § 664 we supposed the inertia to be produced by a magnetic field in the ether, so that u denoted the velocity relative to the ether, but the theory of relativity shews that u may legitimately be supposed to mean merely the velocity relative to the observer by whom the accelerations are measured. A further difference between the two calculations may also be noticed. When the mass was regarded as arising from an ethereal magnetic field, it was possible to estimate the radius of the electron from a knowledge of the value of m; the relativity calculation does not make any such estimate possible.

We may notice that

$$m\kappa^3 \frac{du}{dt} = \frac{d}{dt}(m\kappa u); \quad m\kappa \frac{dv}{dt} = \frac{d}{dt}(m\kappa v), \text{ etc.,}$$

whence it follows that the equation of motion of an electron moving with any velocity u, v, w relative to any observer must be

$$\left.\begin{aligned}
\frac{d}{dt}(m\kappa u) &= e\left(X + \frac{v}{C}\gamma - \frac{w}{C}\beta\right) \\
\frac{d}{dt}(m\kappa v) &= e\left(Y + \frac{w}{C}\alpha - \frac{u}{C}\gamma\right) \\
\frac{d}{dt}(m\kappa w) &= e\left(Z + \frac{u}{C}\beta - \frac{v}{C}\alpha\right)
\end{aligned}\right\} \quad\dots\dots\dots\dots(741),$$

where κ is now given by

$$\kappa = \left(1 - \frac{u^2 + v^2 + w^2}{C^2}\right)^{-\frac{1}{2}} \quad\dots\dots\dots\dots\dots(742).$$

To shew that these are the true equations, it is sufficient to notice that they are invariant as regards transformations of the x, y, z axes, and reduce to equations (738) for one special direction of these axes.

MOMENTUM AND ENERGY.

693. The expressions on the right of equations (741) are the components of force on the moving electron as they would be measured by S (cf. §§ 629, 653). If we denote these by P, Q, R, and if we regard

$$m\kappa u, \quad m\kappa v, \quad m\kappa w$$

as the components of momentum of the moving electron, then the equations of motion (741) assume the form

$$(Force) = (rate\ of\ change\ of\ momentum) \quad\dots\dots\dots(743).$$

The rate at which work is done on the electron will be

$$Pu + Qv + Rw = mu\frac{d}{dt}(\kappa u) + mv\frac{d}{dt}(\kappa v) + mw\frac{d}{dt}(\kappa w),$$

and after simple algebraic transformation this becomes

$$Pu + Qv + Rw = \frac{d}{dt}(m\kappa C^2) \dots\dots\dots\dots\dots(744).$$

Thus if we suppose the energy of the electron to be $m\kappa C^2 + a\ constant$ we have the equation

$$(Rate\ of\ doing\ work) = (rate\ of\ increase\ of\ energy).$$

In the equation,
$$\text{Energy} = m\kappa C^2 + \text{cons.} \dots\dots\dots\dots\dots\dots(745),$$
the additive constant is entirely at our disposal. If we take it equal to $-mC^2$, the energy is given by
$$\text{Energy} = mC^2(\kappa - 1) \dots\dots\dots\dots\dots\dots(746),$$
and this reduces to the Newtonian kinetic energy $\frac{1}{2}m(U^2 + V^2 + W^2)$ when the velocity is small compared with that of light. But it is generally more convenient to put the additive constant equal to zero, so that the energy is simply $m\kappa C^2$. This may be regarded as representing kinetic energy $mC^2(\kappa - 1)$ and intrinsic electronic energy mC^2.

694. Let us put
$$T' = mC^2\left(1 - \frac{1}{\kappa}\right) = mC^2\left[1 - \left(1 - \frac{U^2 + V^2 + W^2}{C^2}\right)^{\frac{1}{2}}\right],$$
then $\dfrac{\partial T'}{\partial U} = m\kappa U$, etc. and equations (741) become
$$\frac{d}{dt}\left(\frac{\partial T'}{\partial U}\right) = P; \quad \frac{d}{dt}\left(\frac{\partial T'}{\partial V_{\prime}}\right) = Q, \text{ etc.}$$

These are analogous to the classical Lagrangian equations of motion of a particle. We must note, however, that T' is not the kinetic energy, but is equal to the kinetic energy divided by κ.

Conservation of Mass, Momentum and Energy.

695. In defining the energy of an electron to be $m\kappa C^2$, we have already arranged, by definition, that conservation of energy shall hold. The total energy of a system of electrons is $\Sigma m\kappa C^2$, and from equation (744) it is at once apparent that if no work is done from outside the total energy remains constant.

As a system of electrons change their velocities under their mutual interactions, the values of κ for the different electrons will be continually changing. If the total energy remains constant, it is clear from equation (745) that $\Sigma m\kappa$ must remain constant. Thus if in future we agree to define the mass of a moving electron as $m\kappa$—the "transverse" mass of § 664—then it appears that conservation of energy will imply conservation of mass.

The full principle of conservation of energy states that as energy is interchanged between different modes of energy the sum total of energy always remains constant. Hence in order that the sum total of masses shall remain constant—*i.e.* to secure complete conservation of mass—it is necessary that all forms of energy should possess mass, and energy E of any kind whatever must possess mass of amount E/C^2.

For instance suppose that a system of electrons of energy E, and therefore of total mass $\Sigma m\kappa$ equal to E/C^2, radiates away energy of amount R. The final energy of the electrons is $E - R$, so that their final mass is $E/C^2 - R/C^2$. But the radiation possessing energy R must also possess mass R/C^2, so that the total mass of electrons and radiation remains equal to E/C^2, and the total mass is entirely conserved.

696. Our typical electron has been supposed to move with a velocity of components U, V, W relative to an observer S. Let us suppose its velocity relative to a second observer S' to be U', V' W', when S' moves relative to S with a velocity $u, 0, 0$. Then the two sets of velocities are related by the kinematical equations (702) and (703) of § 679.

Let us write

$$\left.\begin{aligned}\kappa &= \left(1 - \frac{U^2 + V^2 + W^2}{C^2}\right)^{-\frac{1}{2}} \\ \kappa' &= \left(1 - \frac{U'^2 + V'^2 + W'^2}{C^2}\right)^{-\frac{1}{2}} \\ \kappa_0 &= \left(1 - \frac{u^2}{C^2}\right)^{-\frac{1}{2}}\end{aligned}\right\} \quad\quad\quad\text{......................(747),}$$

so that κ_0 is identical with the κ of § 679. On using the values of U', V', W' given by equations (703), we find

$$\begin{aligned}1 - \frac{U'^2 + V'^2 + W'^2}{C^2} &= 1 - \frac{(U-u)^2 + \left(1 - \frac{u^2}{C^2}\right)(V^2 + W^2)}{C^2\left(1 - \frac{Uu}{C^2}\right)^2} \\ &= \frac{\left(1 - \frac{u^2}{C^2}\right)\left(1 - \frac{U^2 + V^2 + W^2}{C^2}\right)}{\left(1 - \frac{Uu}{C^2}\right)^2}\end{aligned}$$

so that, on raising each side to the power $-\frac{1}{2}$,

$$\kappa' = \kappa\kappa_0\left(1 - \frac{Uu}{C^2}\right) \quad\quad\quad\quad\text{.........................(748),}$$

or, again using the first of equations (703),

$$\kappa'U' = \kappa\kappa_0(U - u).$$

Multiplying both sides by m, and summing over all the particles in the field,

$$\Sigma m\kappa'U' = \kappa_0\,\Sigma m\kappa U - u\kappa_0\,\Sigma m\kappa.$$

Let M, μ_x denote the total mass and total x-momentum as observed by S, and let M', μ_x' denote the same quantities as observed by S'. Then our equation may be written

$$\mu_x' = \kappa_0\,\mu_x - u\kappa_0\,M \quad\quad\quad\quad\text{........................(749).}$$

Similarly, since S is moving relative to S' with a velocity $-u$,

$$\mu_x = \kappa_0\,\mu_x' + u\kappa_0\,M' \quad\quad\quad\quad\text{..........................(750).}$$

If the total energy of the system remains constant throughout any motion, M and M' must remain constant, so that μ_x and μ_x' necessarily remain constant. Thus conservation of energy implies conservation of momentum.

697. As a particular case, let us suppose the velocity of the axes of S' chosen so that $\mu_x' = 0$. Thus S' moves with the centre of gravity of the system, and this centre of gravity moves relative to S with a velocity u along Ox. Putting $\mu_x' = 0$ in equations (749) and (750), we find

$$\mu_x = uM,$$

$$M = \kappa_0 M'.$$

Thus the x-momentum observed by S is u times the total mass observed by S. The total mass observed by S is κ_0 times the total mass observed by S'. And, if we take the energy equal to MC^2, the total energy observed by S is κ times the total energy observed by S'.

698. Returning to the analysis of §696, let us suppose that the system emits a beam of radiation along Ox. Let the energy of this beam as estimated by S be E, so that its mass is E/C^2, and let its momentum as estimated by S be R_x along Ox. Let accented letters denote the same quantities as estimated by S'. Then in order that equation (749) may be true both before and after the emission of the radiation, both mass and momentum being assumed to be conserved, we must have

$$R_x' = \kappa_0 R_x - u\kappa_0 E/C^2.$$

This equation is true for all values of u. Take u equal to C, so that S' moves with the beam of light. Then $R_x' = 0$, and the equation becomes

$$R_x = \frac{E}{C} \quad\dots\dots\dots\dots\dots\dots\dots\dots\dots\dots(751).$$

Thus the momentum of a beam of light, as measured by any observer whatever, is equal to $(1/C)$ times its energy. Or, again, the momentum is equal to C times the mass. If W is the energy of the beam per unit volume, the momentum per unit volume will be W/C, and the flow of momentum per unit area of cross-section will be C times this, and so equal to W. Thus the pressure of radiation is equal to the energy per unit volume.

This result was obtained in §592c as a consequence of the hypothesis that electric action was transmitted by an ether. It now appears that the result is in accordance with the hypothesis of relativity, and can be deduced as a direct consequence of this hypothesis.

THE ENERGY AND MOMENTUM OF RADIATION.

699. In free space the fundamental equations (613) and (614) of p. 559 assume the form

$$\frac{4\pi\rho U}{C} + \frac{1}{C}\frac{dX}{dt} = \frac{\partial\gamma}{\partial y} - \frac{\partial\beta}{\partial z}, \text{ etc. } \quad\dots\dots\dots(752),$$

$$-\frac{1}{C}\frac{d\alpha}{dt} = \frac{\partial Z}{\partial y} - \frac{\partial Y}{\partial z}, \text{ etc. } \quad\dots\dots\dots(753).$$

Multiply these six equations by $X, Y, Z, -\alpha, -\beta, -\gamma$ respectively and add corresponding sides. We obtain

$$\frac{4\pi}{C}\rho\,(XU + YV + ZW) + \frac{1}{2C}\frac{d}{dt}(X^2 + Y^2 + Z^2 + \alpha^2 + \beta^2 + \gamma^2)$$

$$= -\frac{\partial}{\partial x}(Y\gamma - Z\beta) - \frac{\partial}{\partial y}(Z\alpha - X\gamma) - \frac{\partial}{\partial z}(X\beta - Y\alpha).$$

Multiply both sides by $C/4\pi$ and integrate throughout any closed space. Assuming that the distribution of electric density ρ arises entirely from electrons, we obtain

$$\Sigma e\,(XU + YV + ZW) + \frac{d}{dt}\left[\frac{1}{8\pi}\iiint(X^2 + Y^2 + Z^2 + \alpha^2 + \beta^2 + \gamma^2)\,dx\,dy\,dz\right]$$

$$= \frac{C}{4\pi}\iint\left[l\,(Y\gamma - Z\beta) + m\,(Z\alpha - X\gamma) + n(X\beta - Y\alpha)\right]dS \quad\dots\dots(754).$$

In §693, we put

$$e\left(X + \frac{V}{C}\gamma - \frac{W}{C}\beta\right) = P, \text{ etc.}$$

Multiplying these relations by U, V, W and adding we find, after a further use of equation (744),

$$e\,(XU + YV + ZW) = PU + QV + RW = \frac{d}{dt}\,(m\kappa C^2).$$

Thus equation (754) becomes

$$\frac{d}{dt}\left[\Sigma m\kappa C^2 + \frac{1}{8\pi}\iiint(X^2 + Y^2 + Z^2 + \alpha^2 + \beta^2 + \gamma^2)\,dx\,dy\,dz\right]$$

$$= \frac{C}{4\pi}\iint\left[l\,(Y\gamma - Z\beta) + \dots\right]dS \dots(755).$$

Now let the closed space be allowed to extend to infinity, so that the integration is through the whole of space. The surface integral on the right of equation (755) now vanishes, so that the left-hand member must also vanish. In other words, throughout the motion of the system of electrons,

$$\Sigma m\kappa C^2 + \frac{1}{8\pi}\iiint(X^2 + Y^2 + Z^2 + \alpha^2 + \beta^2 + \gamma^2)\,dx\,dy\,dz = \text{cons.} \dots(756).$$

If R is the energy radiated away from a system of electrons, we have already had the relation

$$\Sigma m\kappa C^2 + R = \text{cons.},$$

whence it appears that the volume-integral in (756) must represent radiated energy.

If we assume the radiated energy to be localised in space according to the distribution of the integral, then the flow of energy into any closed surface must be represented by the surface-integral on the right hand of equation (755), and this flow is precisely that given by the Poynting Flux of § 576.

700. Again, let us multiply the six equations (752) and (753) by 0, γ, $-\beta$, 0, Z, $-Y$ and add. We obtain

$$4\pi\rho\left(\frac{v}{C}\gamma - \frac{w}{C}\beta\right) + \frac{1}{C}\frac{d}{dt}(Y\gamma - Z\beta)$$

$$= \frac{1}{2}\frac{\partial}{\partial x}(X^2 - Y^2 - Z^2 + \alpha^2 - \beta^2 - \gamma^2) + \frac{\partial}{\partial y}(XY + \alpha\beta) + \frac{\partial}{\partial z}(XZ + \alpha\gamma)$$

$$- \alpha\left(\frac{\partial\alpha}{\partial x} + \frac{\partial\beta}{\partial y} + \frac{\partial\gamma}{\partial z}\right) - X\left(\frac{\partial X}{\partial x} + \frac{\partial Y}{\partial y} + \frac{\partial Z}{\partial z}\right).$$

Dividing throughout by 4π, and using equations (615) and (616), this becomes

$$\rho\left(X + \frac{v}{C}\gamma - \frac{w}{C}\beta\right) + \frac{1}{4\pi C}\frac{d}{dt}(Y\gamma - Z\beta)$$

$$= \frac{\partial}{\partial x}\left[\frac{1}{8\pi}(X^2 - Y^2 - Z^2 + \alpha^2 - \beta^2 - \gamma^2)\right] + \frac{\partial}{\partial y}\left[\frac{1}{4\pi}(XY + \alpha\beta)\right]$$

$$+ \frac{\partial}{\partial z}\left[\frac{1}{4\pi}(XZ + \alpha\gamma)\right] \quad \dots(757).$$

Integrating through a closed surface as before, this becomes

$$\Sigma e\left(X + \frac{v}{C}\gamma - \frac{w}{C}\beta\right) + \frac{1}{4\pi C}\frac{d}{dt}\iiint(Y\gamma - Z\beta)\,dx\,dy\,dz$$

$$= -\iint\left[\frac{l}{8\pi}(X^2 - Y^2 - Z^2 + \alpha^2 - \beta^2 - \gamma^2) + \frac{m}{4\pi}(XY + \alpha\beta)\right.$$

$$\left. + \frac{n}{4\pi}(XZ + \alpha\gamma)\right]dS.$$

Using the first of equations (741), the left-hand member becomes

$$\frac{d}{dt}\left[\Sigma m\kappa U + \frac{1}{4\pi C}\iiint(Y\gamma - Z\beta)\,dx\,dy\,dz\right] \quad \dots\dots\dots(758).$$

If the integral is taken through the whole of space, this expression must vanish, so that the quantity inside square brackets must remain constant. The first term $\Sigma m\kappa U$ is the x-momentum of the electrons, so that the second term must represent the x-momentum of the emitted radiation. This is identical with the formula found for the momentum in the ether in § 655; it has now been obtained, independently of the assumption of an ether, as a general expression for the momentum of radiation.

If we assume the momentum to be localised in space according to the distribution of the integral, then equation (757) can be interpreted as shewing that there is a flow of x-momentum per unit area at any point, whose components will be P_{xx}, P_{xy}, P_{xz}, where

$$P_{xx} = \frac{1}{8\pi} (X^2 - Y^2 - Z^2 + a^2 - \beta^2 - \gamma^2),$$

$$P_{xy} = \frac{1}{4\pi} (XY + a\beta), \text{ etc.}$$

These are precisely the quantities we obtained in § 655 to represent the components of stress in an assumed ether. On the relativity-theory, they appear in a much more general way as representing the flow of momentum in space.

The Existence of an Ether.

701. Throughout the earlier chapters of this book, we treated the existence of an ether as a working hypothesis. Maxwell and Faraday appear to have had no doubt that the ether had a real objective existence, but no proof that it exists outside our own minds has ever been obtained, and it seems best to regard it merely as a working hypothesis, to be discarded if it is found to lead to contradictory or impossible results, and to be retained if it proves to be useful as well as self-consistent.

The considerations which have seemed to favour the hypothesis of an objective ether are mainly the following:

(i) That light and other forms of electromagnetic action are propagated with a uniform velocity C, which is most naturally interpreted as a velocity of propagation in a medium of some sort.

(ii) That the hypothesis of an objective ether explains electrical forces with comparative simplicity as arising from systems of stresses transmitted by the ether.

(iii) That the hypothesis of an objective ether gives a simple account of electromagnetic energy as being the energy of a medium in a state of strain and stress.

The force of the first consideration is very much weakened by the discovery, resulting from the Michelson-Morley experiment, that the velocity of propagation is the same for an observer moving through the supposed ether as for one at rest. We have seen in the present chapter that this fact, whether we assume an ether to exist or not, requires us to adopt systems of kinematics and dynamics which are different from the old classical systems. The consequences of these new systems of kinematical and dynamical laws are found to be confirmed by experiment. If they had not been confirmed, we should have reached an *impasse*; the circumstance that they are confirmed provides no definite information on the question of the existence of an ether. But the hypothesis of an ether is weakened to this extent, that the theory of relativity has shewn that the results in question, although possibly admitting of an explanation in terms of an ether, are necessary consequences of the simpler supposition that phenomena are the same for all observers, no matter with what velocity they are moving. And the simplest way of all of arranging that phenomena shall be the same for all moving observers, is to suppose that there is no ether at all; all moving observers then necessarily stand on the same footing, for there is no fixed framework by which their motion can be estimated. The hypothesis that there is an ether may give a possible explanation of the phenomena, but the hypothesis that there is no ether provides an equally possible and very much simpler explanation.

If we still wish to retain the hypothesis of an ether through which light and electromagnetic phenomena are propagated, we must adjust the properties of this ether to agree with experiment. Now we have seen (§ 688) that, no matter how an observer and a source of light move, the wave-surface formed by the light emitted at any instant will be a sphere having the observer as its centre. If the observed constant velocity of light is simply the constant velocity of propagation through an ethereal medium, it would seem to follow that each observer must carry a complete ether about with him. This at least robs the ether of the greater part of its reality. We cannot quite go so far as to assert that the ether is reduced to a subjective imagination, as a simple analogy will shew. A number of travellers may all see what they would describe in ordinary language as being the same rainbow. The angle of the rainbow would be the same for each traveller, and no amount of travelling towards the rainbow would cause it to subtend a greater angle. If the travellers compared observations they would have to conclude that each traveller carried his own rainbow about with him. This would not, however, prove the rainbow to be merely a subjective illusion; when the rainbow disappeared for one traveller it would disappear for all. Considerations such as we have mentioned do not prove in strictness that light cannot be propagated through an ether; what they prove is that if an ether exists, it must be something very different from the absolutely objective ether imagined by Maxwell and Faraday.

The hypothesis of an ether shewed great aptitude for explaining either electric or magnetic forces in systems at rest; a simple system of pressures and tensions was found to account perfectly for the observed forces. On the other hand the same explanation cannot account for both electric and magnetic forces simultaneously. If, as is usually assumed, the electric forces are accounted for by simple pressures and tensions, then some other explanation must be found for magnetic forces, and the hypothesis of ether-stresses loses its principal advantage. Further it has been found that the hypothesis of an ether at rest fails entirely to account for the forces in systems in motion (§ 655). To account for these forces, it appears that the ether must be supposed endowed with momentum. On the relativity-theory also we have seen (§ 700) that the forces can be explained in terms of a flow of momentum. The relativity-theory has thus shewn that what is essential to the ethereal explanation is not the ether but the momentum with which it was supposed to be endowed. It is quite easy to imagine a flow of momentum without there being an ether to carry it, and the conception of forces and pressures arising from a flow of momentum is one with which we have become familiar in other branches of physics, as for example the Kinetic Theory of Gases.

Almost similar remarks apply to the interpretation of electromagnetic energy. Stress in the ether would account quite simply for either electric or magnetic energy, but not for both. Usually the energy of ethereal stress is regarded as electrostatic energy, so that kinetic energy of the ether must be invoked to account for magnetic energy. Again the ether has to be supposed endowed with motion, but the motion requisite to account for the magnetic energy is something quite different from that corresponding to the momentum required to account for electromagnetic forces. So far from the ether providing a simple explanation of all phenomena, it is found that highly complex properties must be ascribed to it in order to account for electrical and magnetic properties simultaneously.

If an ether existed, it would provide a fixed set of axes relative to which all positions and velocities could be measured. To account for the result of the Michelson-Morley experiment, it would be necessary to postulate a real shrinkage of all bodies moving through the ether. This shrinkage could not be detected by mechanical means, for a measuring rod would shrink in precisely the same ratio as the body to be measured, but it could be detected by gravitational means unless every gravitational field of force shrunk in just such a way as to conceal the shrinkage of matter. For instance, if the gravitational field did not shrink, the geoid, or surface of mean sea-level on the earth, might be a gravitational equipotential for some one velocity through the ether, but could not remain an equipotential as the earth's velocity through the ether changed from point to point of its orbit. Thus we might anticipate seasonal and daily tidal surgings as a result of the earth's motion through the ether. No such events are observed. It is true that even if these occurred the earth's motion through the ether might not be sufficiently rapid for them to be capable of observation, but the generalised theory of relativity, explained in the next section, makes it clear that such events could not be observed whatever the earth's motion might be. There is no longer any room for reasonable doubt that gravitational phenomena conform to the relativity condition.

If, then, we continue to believe in the existence of an ether we are compelled to believe not only that all electromagnetic phenomena are in a conspiracy to conceal from us the speed of our motion through the ether, but also that gravitational phenomena, which so far as is known have nothing to do with the ether, are parties to the same conspiracy. The simpler view seems to be that there is no ether. If we accept this view, there is no conspiracy of concealment for the simple reason that there is no longer anything to conceal.

GENERALISED RELATIVITY.

702. In discussing the transmission of light-signals in § 675, we made the assumption that light travelled in straight lines with uniform velocity. If space and time were known to be uniform throughout their whole extent, no such assumption would be needed; the uniformity of velocity and the straightness of path would be direct consequences of the uniformity of time and space.

The theory of relativity has however developed in such a direction that space and time can no longer be supposed to be everywhere uniform. In the early days of the theory it was noticed that Newton's inverse-square law of gravitation did not conform to the relativity-condition, and in 1915 Einstein put forward a theory of generalised relativity according to which all gravitational phenomena are the consequences merely of departures from uniformity of time and space.

According to Einstein's theory, the properties of both time and space in the neighbourhood of a gravitating mass differ from those in regions far removed from all matter. The properties of space in the latter regions can be adequately described by the geometry of Euclid, but those in the neighbourhood of gravitating matter need a new geometry for their description.

In ordinary space, as described by Euclid's geometry, parallel lines never meet. Other geometries and other spaces can, however, be imagined. For instance, two lines drawn upon the earth's surface, through two points on the earth's equator, both running due north, and so running exactly parallel to one another, will ultimately meet in a point—the North Pole. We see that the geometry of a spherical surface is different from that of a plane, so much so that almost all the ordinary theorems of Euclidean geometry fail when applied to a spherical surface.

The geometry required by Einstein's gravitational theory is much less simple than the spherical geometry we have just used as an illustration. It is not concerned with two-dimensional surfaces, or even with a three-dimensional space. It is concerned with a four-dimensional continuum, of the type considered in § 675, in which three space-coordinates and one time-coordinate are plotted parallel to four rectangular axes. In the neighbourhood of gravitating matter this four-dimensional continuum is supposed to be curved somewhat in the way in which the earth's two-dimensional surface is curved. A circle of diameter 1000 miles drawn round a point on the earth's surface will not have a circumference of 1000π miles, as it would be if the circle were drawn on a plane, but of only about 997π miles. In the same way, according to Einstein's geometry, the circumference of a circle of radius r drawn about a gravitating mass is not precisely $2\pi r$, but is less by a fraction which is proportional to the mass and falls off as we recede from it.

As a consequence of the special geometry of a spherical surface, it is not possible to draw a map on a plane surface so as to shew all the parts of the earth's surface simultaneously in their proper shapes and relative sizes. This results from its not being possible to select coordinates x, y on the earth's surface such that the element of length ds is given by

$$ds^2 = dx^2 + dy^2 \quad\ldots\ldots\ldots\ldots\ldots\ldots\ldots(759)$$

at all points of the surface. The simplest coordinates it is possible to select are the ordinary θ, ϕ of spherical polar coordinates, in terms of which the element of length on the surface of a sphere of unit radius is given by

$$ds^2 = d\theta^2 + \sin^2\theta\, d\phi^2 \quad\ldots\ldots\ldots\ldots\ldots\ldots(760).$$

According to Einstein's generalised relativity, the element of length in the four-dimensional continuum can be expressed in the form*

$$ds^2 = dx^2 + dy^2 + dz^2 + d\tau^2 \quad\ldots\ldots\ldots\ldots\ldots(761)$$

only in regions which are far removed from all gravitating matter. This form for ds^2 is of course analogous to expression (759) for the value of ds^2 on a plane surface. As in § 675, τ stands for iCt where $i = \sqrt{(-1)}$.

If we replace τ by its value iCt and transform from the space coordinates x, y, z to the usual spherical polar coordinates r, θ, ϕ, equation (761) assumes the form

$$ds^2 = dr^2 + r^2 d\theta^2 + r^2\sin^2\theta\, d\phi^2 - C^2 dt^2 \quad\ldots\ldots\ldots\ldots(762).$$

Einstein's theory requires that in the neighbourhood of a gravitating particle of mass m, equation (762) shall be replaced by

$$ds^2 = \frac{dr^2}{1 - \dfrac{2\gamma m}{rC^2}} + r^2 d\theta^2 + r^2\sin^2\theta\, d\phi^2 - C^2\left(1 - \frac{2\gamma m}{rC^2}\right) dt^2 \quad\ldots(763).$$

A particle describing a "geodesic" or most direct path—defined by $\delta\int ds = 0$—in this space, can be shewn to change its coordinates x, y, z, t in very approximately the same way as a particle describing an ordinary ellipse or hyperbola in ordinary space about a mass m under a law of attractive force $\gamma m/r^2$. The agreement, however, is not quite exact, and Einstein's theory is found to require three phenomena which were not predicted by, and are in

---

* In technical investigations on Relativity $- ds^2$ is usually written for our ds^2.

fact inconsistent with, the classical Newtonian theory; first, the perihelia of all the planets ought to advance at a rate which should be easily detected in the case of Mercury; second, light passing near to the sun ought to shew an appreciable deflection; and third, the spectral lines emitted in a strong gravitational field such as that of the sun ought to be seen shifted slightly to the red when compared with the corresponding lines emitted in the weak gravitational field of the earth. Of these three phenomena, the first had been observed by Leverrier long before any explanation was forthcoming, the second was observed as soon as it was looked for, first in the solar eclipse of 1919 and subsequently in that of 1923, while the third is still in doubt, on account of the extreme difficulty of the observation and the smallness of the quantity to be measured. But the quantitative agreement in the case of the first two phenomena is so good that no doubt is felt as to the substantial truth and accuracy of Einstein's theory.

In brief Einstein supposes a particle in a gravitational field to describe a straight path through a curved space* whereas Newton had imagined it to describe a curved path through a straight space. Newton imagined the curvature of path to result from the action of "forces" which emanated from the gravitating mass, and tried, although without success, to interpret these forces as stresses transmitted by a gravitational ether. Einstein's theory abolishes the conception of gravitational "force" and so escapes altogether the dilemma of having to suppose these forces either to be transmitted through a medium or by direct action at a distance.

Weyl's Electromagnetic Theory.

703. This dilemma of action through a medium or action at a distance is precisely that which has led to most confusion in electromagnetic theory (cf. § 154). If it can be avoided in gravitational theory, it would seem reasonable to hope that an electromagnetic theory could be constructed which should also avoid it.

This has in actual fact been attempted. Weyl in 1918, followed by Eddington in 1921, shewed that Einstein's geometry is far from being the most general geometry in which the relativity-condition is satisfied. In the expression (763) which specifies the element of length ds on Einstein's theory, the coefficients of the differentials $dr, r\,d\theta, r\sin\theta\,d\phi$ and dt are functions solely of the position of ds in space. With certain conventions as to the meaning and

---

* It must always be remembered that the space in question is four-dimensional; otherwise the statement appears nonsensical. It would be absurd to say that the approximate semicircle described by the earth between perihelion and aphelion is the most direct path between these two points; it is only when the six months interval in time is taken into account that the statement begins to appear reasonable. We can get rid of the time-interval by supposing the particle to move with infinite velocity in which case the path becomes a straight line even in ordinary three-dimensional space.

method of the measurement of length*, we may deduce from this expression that the length of a measuring rod would change as it was moved about from place to place in a gravitational field, but that its length at any instant would depend solely on its position in space. Indeed we may be even more precise, for the coefficients of the differentials in expression (763) all depend on the single quantity $1 - \dfrac{2\gamma m}{rC^2}$ which in turn depends only on the gravitational potential $\gamma m/r$, so that the length in question depends only on the gravitational potential at the place.

In the most general geometry possible in a space of coordinates x, y, z, τ, a measuring rod of length l moved parallel to itself through a displacement $dx, dy, dz, d\tau$ may be expected to experience a change of length dl defined by

$$dl = l\,(Fdx + Gdy + Hdz + Kd\tau) \dots\dots\dots\dots(764),$$

where F, G, H, K may be the most general functions of the position of the point. If the rod is moved from one point P to any other point Q its whole change of length will be given by

$$\log \frac{l_Q}{l_P} = \int_P^Q (Fdx + Gdy + Hdz + Kd\tau) \dots\dots\dots\dots(765).$$

In Einstein's geometry, l_Q and l_P depend only on the positions of P and Q, so that the integrand on the right is necessarily a perfect differential. The condition that this integrand shall be a perfect differential is expressed by the six equations

$$\left.\begin{array}{ll} \dfrac{\partial H}{\partial y} - \dfrac{\partial G}{\partial z} = 0, & \dfrac{\partial K}{\partial x} - \dfrac{\partial F}{\partial \tau} = 0 \\[2mm] \dfrac{\partial F}{\partial z} - \dfrac{\partial H}{\partial x} = 0, & \dfrac{\partial K}{\partial y} - \dfrac{\partial G}{\partial \tau} = 0 \\[2mm] \dfrac{\partial G}{\partial x} - \dfrac{\partial F}{\partial y} = 0, & \dfrac{\partial K}{\partial z} - \dfrac{\partial H}{\partial \tau} = 0 \end{array}\right\} \dots\dots\dots\dots(766).$$

In Weyl's geometry, on the other hand, the integrand on the right hand is not in general a perfect differential, and the six quantities which constitute

---

* Unfortunately it is round these conventions that the difficulties of the subject mainly centre. It is meaningless to speak of a measuring rod changing its length unless there is something more absolute against which it can be measured, and the complexities of the theory, especially of Weyl's theory, turn on the properties of the imaginary gauges or meshes against which material objects may be measured. It is impossible to give a full discussion in the present book. The student who wishes to pursue the subject further may be referred to

     Eddington, *Space, Time and Gravitation.*
     Eddington, *The Mathematical Theory of Relativity.*
     Weyl, *Raum, Zeit, Materie* (French Translation, *Temps, Espace, Matière*).

It ought to be added that the brief sketch in the present book follows the exposition of Eddington rather than that of Weyl.

the left-hand members of equations (766), instead of vanishing, have values a, b, c, d, e, f so that

$$\left.\begin{array}{ll}
\dfrac{\partial H}{\partial y} - \dfrac{\partial G}{\partial z} = a, & \dfrac{\partial K}{\partial x} - \dfrac{\partial F}{\partial \tau} = d \\[2mm]
\dfrac{\partial F}{\partial z} - \dfrac{\partial H}{\partial x} = b, & \dfrac{\partial K}{\partial y} - \dfrac{\partial G}{\partial \tau} = e \\[2mm]
\dfrac{\partial G}{\partial x} - \dfrac{\partial F}{\partial y} = c, & \dfrac{\partial K}{\partial z} - \dfrac{\partial H}{\partial \tau} = f
\end{array}\right\} \quad \dots\dots\dots\dots\dots(767).$$

704. If we restore its value iCt to τ, and replace K by $i\Psi$ and id, ie, if by X, Y, Z, the system of equations assumes the form

$$\left.\begin{array}{ll}
\dfrac{\partial H}{\partial y} - \dfrac{\partial G}{\partial z} = a, & -\dfrac{\partial \Psi}{\partial x} - \dfrac{1}{C}\dfrac{\partial F}{\partial t} = X \\[2mm]
\dfrac{\partial F}{\partial z} - \dfrac{\partial H}{\partial x} = b, & -\dfrac{\partial \Psi}{\partial y} - \dfrac{1}{C}\dfrac{\partial G}{\partial t} = Y \\[2mm]
\dfrac{\partial G}{\partial x} - \dfrac{\partial F}{\partial y} = c, & -\dfrac{\partial \Psi}{\partial z} - \dfrac{1}{C}\dfrac{\partial H}{\partial t} = Z
\end{array}\right\} \quad \dots\dots\dots\dots(768).$$

On differentiating the three equations on the left with respect to x, y, z and adding, we obtain

$$\dfrac{\partial a}{\partial x} + \dfrac{\partial b}{\partial y} + \dfrac{\partial c}{\partial z} = 0 \quad \dots\dots\dots\dots\dots\dots(769).$$

The seven equations (768) and (769) are formally identical with the seven equations from which we proceeded to develop the general equations of the electromagnetic field in § 639. Weyl's electromagnetic theory supposes that the identity is one not only of form but also of reality; he supposes an equation of the form

$$dl = \kappa l \left(F dx + G dy + H dz - C \Psi dt \right) \quad \dots\dots\dots\dots(770),$$

to connect the change undergone by a rod on displacement with the three components of magnetic vector potential F, G, H and the electric potential Ψ. This equation is identical with our former equation (764) except for the occurrence of the constant κ which is necessary to secure that F, G, H and Ψ shall be of the appropriate physical dimensions. The only result of introducing this factor κ is that the left-hand member of equation (765) must be replaced by

$$\dfrac{1}{\kappa} \log \dfrac{l_Q}{l_P};$$

after this κ disappears and we reach equations (768) and (769) as before. We see that on Weyl's theory the values of F, G, H, Ψ at any point are determined by the rate at which a unit measuring rod changes its length as it passes through the point.

On Weyl's theory a material body has no permanent size associated with it. By many this has been regarded as an objection to the theory. The various attempts which have been made to avoid it, while retaining the obvious advantages of the theory, can hardly be discussed here.

705. From equations (768) and (769) it is easy to develop the whole of the classical electromagnetic theory.

We notice first that equations (768) are identical with the six equations (767) which form a system symmetrical with respect to the four coordinates x, y, z, τ. This of course ensures that the equations satisfy the relativity condition. The first three of this system of six equations contain only the three coordinates x, y, z; τ is entirely absent from this set of three. From the set of three equations from which τ was absent we obtained equation (769). From the corresponding three sets of equations from which x, y and z in turn are absent, it is of course possible to obtain three other equations of similar type. These are found to be

$$-\frac{1}{C}\frac{\partial a}{\partial t} = \frac{\partial Z}{\partial y} - \frac{\partial Y}{\partial z}$$
$$-\frac{1}{C}\frac{\partial b}{\partial t} = \frac{\partial X}{\partial z} - \frac{\partial Z}{\partial x} \qquad \dots\dots\dots\dots(771),$$
$$-\frac{1}{C}\frac{\partial c}{\partial t} = \frac{\partial Y}{\partial x} - \frac{\partial X}{\partial y}$$

which are simply Maxwell's equations of current induction (cf. § 574).

On differentiating the three equations on the right of the system (768) with respect to x, y, z and adding, we find

$$\frac{\partial X}{\partial x} + \frac{\partial Y}{\partial y} + \frac{\partial Z}{\partial z} = -\left(\frac{\partial^2 \Psi}{\partial x^2} + \frac{\partial^2 \Psi}{\partial y^2} + \frac{\partial^2 \Psi}{\partial z^2}\right) - \frac{1}{C}\frac{\partial}{\partial t}\left(\frac{\partial F}{\partial x} + \frac{\partial G}{\partial y} + \frac{\partial H}{\partial z}\right) \dots(772).$$

Precisely as in § 640, it can be shewn that F, G, H and Ψ are not uniquely determined when the electric and magnetic forces are given. We again suppose them to conform to the additional equation (643), namely

$$\frac{\partial F}{\partial x} + \frac{\partial G}{\partial y} + \frac{\partial H}{\partial z} = -\frac{1}{C}\frac{\partial \Psi}{\partial t} \qquad \dots\dots\dots\dots(773),$$

which was imposed on them before. This equation satisfies the relativity condition, being symmetrical in the four coordinates x, y, z, τ; in terms of the notation used in equation (764) it has the form

$$\frac{\partial F}{\partial x} + \frac{\partial G}{\partial y} + \frac{\partial H}{\partial z} + \frac{\partial K}{\partial \tau} = 0,$$

and so expresses that the vector (F, G, H, K) is of zero divergence.

If we now introduce ρ, defined by

$$\frac{\partial X}{\partial x} + \frac{\partial Y}{\partial y} + \frac{\partial Z}{\partial z} = 4\pi\rho \qquad \dots\dots\dots\dots(774),$$

equation (772) assumes the form

$$\frac{\partial^2 \Psi}{\partial x^2} + \frac{\partial^2 \Psi}{\partial y^2} + \frac{\partial^2 \Psi}{\partial z^2} - \frac{1}{C^2}\frac{\partial^2 \Psi}{\partial t^2} = -4\pi\rho \qquad \dots\dots\dots(775),$$

of which the solution, obtained as in § 645, is

$$\Psi = \iiint \frac{[\rho]\,dx\,dy\,dz}{r} \qquad \text{......................(776).}$$

Since our equations have all satisfied the relativity condition, these formulae must remain true for any rotation of the axes in the x, y, z, τ space. Writing

$$\nabla_4^2 = \frac{\partial^2}{\partial x^2} + \frac{\partial^2}{\partial y^2} + \frac{\partial^2}{\partial z^2} - \frac{1}{C^2}\frac{\partial^2}{\partial t^2},$$

we notice that ∇_4^2 in terms of the coordinates x, y, z, τ assumes the form

$$\nabla_4^2 = \frac{\partial^2}{\partial x^2} + \frac{\partial^2}{\partial y^2} + \frac{\partial^2}{\partial z^2} + \frac{\partial^2}{\partial \tau^2},$$

which is obviously invariant for any rotation of the axes.

Let us now take equation (775) which, in x, y, z, τ coordinates, has the form

$$\nabla_4^2 \Psi = -4\pi\rho \qquad \text{..........................(777),}$$

and transform to a new set of orthogonal axes, obtained by rotating the old axes in the x, y, z, τ space. As has just been seen the operator ∇_4^2 will be the same in the new axes as in the old. Ψ is the component along the axis of τ of the vector

$$-iF,\ -iG,\ -iH,\ -iK,$$

while ρ is the corresponding component of the vector

$$\rho\frac{dx}{d\tau},\ \ \rho\frac{dy}{d\tau},\ \ \rho\frac{dz}{d\tau},\rho$$

or of the vector

$$-i\rho\frac{U}{C},\ \ -i\rho\frac{V}{C},\ \ -i\rho\frac{W}{C},\rho,$$

where U, V, W are the components of velocity of ρ. Thus after transformation, equation (777) becomes

$$\nabla_4^2(-il_1F - il_2G - il_3H + l_4\Psi) = -4\pi\left(-i\rho\frac{U}{C}l_1 - i\rho\frac{V}{C}l_2 - i\rho\frac{W}{C}l_3 + \rho l_4\right),$$

where l_1, l_2, l_3, l_4 are the direction cosines of the old axis of τ.

Since this equation must be true for all values of l_1, l_2, l_3 and l_4, we deduce at once

$$\nabla_4^2 F = -4\pi\rho\frac{U}{C} \qquad \text{..........................(778),}$$

and two similar equations. Returning to the coordinates x, y, z, t, these assume the form

$$\frac{\partial^2 F}{\partial x^2} + \frac{\partial^2 F}{\partial y^2} + \frac{\partial^2 F}{\partial z^2} - \frac{1}{C^2}\frac{\partial^2 F}{\partial t^2} = -4\pi\rho\frac{U}{C} \qquad \text{..............(779),}$$

and similar equations in G and H. These together with equation (775) constitute the equations of propagation of the four potentials F, G, H, Ψ.

By direct solution of equation (779) we obtain the value of F in the form

$$F = \iiint \frac{[\rho U] \, dx \, dy \, dz}{rC} \quad \dots\dots\dots\dots\dots\dots(780),$$

with similar equations for G and H.

Using equations (779) and (768) we find

$$4\pi\rho \frac{U}{C} + \frac{1}{C}\frac{\partial X}{\partial t} = \frac{1}{C^2}\frac{\partial^2 F}{\partial t^2} - \left(\frac{\partial^2 F}{\partial x^2} + \frac{\partial^2 F}{\partial y^2} + \frac{\partial^2 F}{\partial z^2}\right) + \frac{1}{C}\frac{\partial X}{\partial t}$$

$$= -\frac{1}{C}\frac{\partial^2 \Psi}{\partial x \partial t} - \left(\frac{\partial^2 F}{\partial x^2} + \frac{\partial^2 F}{\partial y^2} + \frac{\partial^2 F}{\partial z^2}\right),$$

and, by use of equation (773), this

$$= \frac{\partial}{\partial x}\left(\frac{\partial F}{\partial x} + \frac{\partial G}{\partial y} + \frac{\partial H}{\partial z}\right) - \left(\frac{\partial^2 F}{\partial x^2} + \frac{\partial^2 F}{\partial y^2} + \frac{\partial^2 F}{\partial z^2}\right)$$

$$= \frac{\partial}{\partial y}\left(\frac{\partial G}{\partial x} - \frac{\partial F}{\partial y}\right) - \frac{\partial}{\partial z}\left(\frac{\partial F}{\partial z} - \frac{\partial H}{\partial x}\right)$$

$$= \frac{\partial c}{\partial y} - \frac{\partial b}{\partial z} \quad \dots\dots\dots\dots\dots\dots\dots\dots\dots\dots\dots(781),$$

giving Maxwell's equations of magnetic force.

Differentiating this and the two similar equations with respect to x, y, z and adding, we obtain, by the use of equation (774),

$$\frac{d}{dx}(\rho U) + \frac{d}{dy}(\rho V) + \frac{d}{dz}(\rho W) + \frac{d\rho}{dt} = 0,$$

which is the equation of continuity (cf. § 622), shewing that electric charges have a permanent existence.

706. Einstein's gravitational theory was accepted as soon as the phenomena it predicted in opposition to the classical Newtonian theory were actually observed. It is impossible for Weyl's electromagnetic theory to establish itself in a similar manner since the phenomena it predicts are precisely identical with those of the classical theory of Maxwell. As a direct observational test is impossible, Weyl's theory can only be judged by its inherent plausibility. It may be said to be the only theory at present in the field, Maxwell's mechanism of stresses and strains in an ether having, for all practical purposes, received its deathblow by the establishment of the restricted relativity theory. In its favour may be said that it gives a consistent account of electromagnetic phenomena on lines which, in view of the convincing experimental confirmation obtained for the parallel theory of gravitation, must be admitted to be in accordance with the general workings of nature. The principal objection which can be brought against it has already been mentioned (§ 704).

CHAPTER XXI

THE ELECTRICAL STRUCTURE OF MATTER

707. BY the end of the nineteenth century, it was generally believed that all physical phenomena, with the possible exception of gravitation, were of electric origin. Associated with this was the belief that matter was a purely electrical structure. Positive and negative charges, arranged in combination in different ways, were supposed to give rise to all the various kinds of matter in the universe, changes in the positions and arrangements of these charges being regarded as the origin of all the phenomena of physics and chemistry.

Various conjectures were made as to the actual arrangement of the positive and negative charges in matter, but positive knowledge was only obtained when the new experimental methods made available by the discovery of radioactive substances were brought into action.

708. The special properties of radioactive substances originate from their spontaneously and continuously emitting rays of various kinds. If a beam of the emitted rays is allowed to traverse a strong magnetic field, it is found to be split up into three distinct beams, two of which are deflected in opposite directions, while the third passes straight on. The three types of rays in these three beams are known as α rays, β rays and γ rays respectively. We have seen (§ 631) that a charged electric particle traversing a magnetic field will describe a circle of radius vmC/eH. The curvature of the paths of the α rays is found to be the same as if the rays were positively charged particles, that of the paths of the β rays is curved as though they were negatively charged particles, while the absence of curvature of the paths of the γ rays suggests that they are not charged particles at all. The charges can be measured by shooting the rays into an electrometer. It is found that the α rays are rapidly moving particles each with a positive charge equal to twice the charge on an electron, and with a mass almost exactly equal to that of the helium atom. The β rays prove simply to be negative electrons moving with velocities comparable with that of light. The γ rays are found to be radiation of the same general nature as light or X-rays, but of exceedingly short wave-length.

If a thin piece of metal foil is placed in the path of a beam of α particles, the majority of the particles pass through without their paths shewing any appreciable deflection, but a small fraction of the total number are substantially deflected. From experiments under varied conditions, the deflections are found to be such as would be expected if only isolated small areas of the foil had the power of appreciably deflecting the particles, and the number of such areas is found to be equal to the number of atoms in the foil. Moreover the deflections observed are precisely those which would be expected if each of these areas had at its centre a fixed particle which repelled the α particle according to the law of the inverse square of the distance.

Investigations of this type led Sir E. Rutherford to put forward in 1911 a theory of the structure of the atom, generally known as the nuclear theory, which has stood the test of time and has now won universal acceptance. According to this theory, an atom consists of a positively charged central nucleus surrounded by a number of negative electrons, the charge on the central nucleus being such that the total charge of the atom is zero.

The simplest atom is the hydrogen atom, consisting of only one electron and the positive nucleus. If $-e$ is the charge of an electron, that of the positive nucleus of the hydrogen atom is of course $+e$. Next in order comes the helium atom consisting of two electrons and a positive nucleus of charge $+2e$. This positive nucleus is found to be exactly identical with the α particles of radium radiations.

With insignificant exceptions chemical elements are known having respectively 1, 2, 3, ... electrons, and consequently nuclei of charges $+e$, $+2e$, $+3e$, ..., up to 60 electrons and a nuclear charge $+60e$ (Neodymium). After this gaps appear in the sequence, which appears to end altogether at Uranium with 92 electrons and a charge $+92e$. The number which fixes the position of an element in this sequence is called its "atomic number"; for the elements of low atomic number, the atomic number is approximately equal to half the atomic weight. The first few elements with their atomic numbers are as follows:

Atomic Number	Element	Atomic Weight	Isotopes
1	Hydrogen	1·008	1, 2, 3
2	Helium	4·00	3, 4, 5 (?)
3	Lithium	6·94	6, 7
4	Beryllium	9·02	8 (?), 9, 10
5	Boron	10·82	10, 11
6	Carbon	12·00	12, 13
7	Nitrogen	14·01	14, 15
8	Oxygen	16	16, 17, 18
9	Fluorine	19·00	19
10	Neon	20·18	20, 21, 22
11	Sodium	23·00	23
12	Magnesium	24·32	24, 25, 26
13	Aluminium	26·97	27
14	Silicon	28·06	28, 29, 30
15	Phosphorus	31·02	31
16	Sulphur	32·06	32, 33, 34
17	Chlorine	35·46	35, 37
18	Argon	39·94	36, 38, 40
19	Potassium	39·10	39, 41
20	Calcium	40·08	40, 42, 43, 44

709. As we have seen the mass of the negative electron is $9·00 \times 10^{-28}$ grammes. The mass of the hydrogen atom is about 1845 times this, so that the nucleus has about 1844 times the mass of the electron. In the helium atom the mass of the nucleus is about 7320 times the mass of a single electron and so outweighs the two attendant electrons in the ratio of about 3660 to one. As the ratio of atomic weight to number of electrons is about the same in all elements except hydrogen, it follows that in all elements other than hydrogen only about one part in 3660 of the total mass resides outside the central nucleus. For most purposes we may think of the centre of gravity of an atom as coinciding with its nucleus.

Assuming the mass of the negative electron to be wholly electromagnetic, we have seen that its radius must be of the order of 2×10^{-13} cms. If the mass of the nucleus also is wholly electromagnetic, its radius must be much smaller than that of the negative electron; that of the nucleus of the hydrogen atom, for instance, would be about 10^{-16} cms. There is direct evidence that the nucleus is exceedingly small, experiments on the scattering of α rays having shewn that α particles can pass within 2×10^{-13} cms. of the centre of an atomic nucleus, and yet be deflected in accordance with the ordinary law of the inverse square.

These figures shew that both nuclei and electrons are very small in comparison with atoms. The hydrogen atom whose radius is approximately 0.53×10^{-8} cms. is made up of only two constituent parts, each of radius 2×10^{-13} cms. or less. Since a positive and a negative charge cannot stand in statical equilibrium at a distance apart equal to several thousands of times the radius of either, we must suppose that the two charges maintain their distance as a consequence of orbital motion. The negative electron does not fall onto the positive electron for the same reason for which the earth does not fall onto the sun. In many respects an atom may be compared to a solar system, the heavy positive nucleus at the centre of the atom representing the sun and the electrons representing planets, the law of force between the nucleus and the electrons being the same as that between sun and planets, namely a force of attraction varying as the inverse square of the distance.

710. According to the analysis of § 650, a negative electron describing an orbit about a nucleus must radiate energy. If E, e are the charges on the nucleus and electron respectively, and m the mass of the electron, the acceleration of the electron towards the nucleus is Ee/mr^2, while the acceleration of the nucleus may be neglected in comparison, on account of its much greater mass. From formula (663), the rate of emission of radiation per unit time is

$$\frac{2}{3C^3}\{(\Sigma e\dot{u})^2 + (\Sigma e\dot{v})^2 + (\Sigma e\dot{w})^2\} = \frac{2E^2e^4}{3C^3\,m^2r^4} \quad\ldots\ldots\ldots\ldots(782).$$

For the hydrogen atom, consisting only of one electron and a nucleus of equal charge, we may put

$$E = -e = 4.803 \times 10^{-10} \text{ el. stat. units,}$$
$$r = 0.53 \times 10^{-8} \text{ cms.,}$$

from which the rate of radiation is found to be 0·46 ergs per second. As a result of this loss of energy, the radius of the orbit ought to decrease. When the radius is r, the energy of the orbit is readily found to be $-e^2/2r$, so that the rate of decrease of energy is

$$-\frac{e^2}{2r^2}\frac{dr}{dt}.$$

Putting this equal to 0·46 ergs a second, we find that $-dr/dt$ must be equal to about 112 cms. a second. Since the radius of the orbit initially is only $0·53 \times 10^{-8}$ cms., the distance between the two constituents of the hydrogen atom ought to vanish altogether in a fraction of a millionth of a second.

Even in the light of common sense such a conclusion is preposterous; it is more so in the light of exact knowledge. So far as we know all hydrogen atoms, no matter how or where selected, are identical structures, all giving the same spectrum and all having precisely the same radius except for a reservation which will shortly be explained. There is not the slightest indication of any secular change in their properties, and a change of the rapidity of that just calculated is utterly out of the question.

The conclusion to which we are driven is not merely that a normal hydrogen atom of the type we have been considering does not radiate as rapidly as is predicted by equation (782), but that it does not radiate at all. In some way the whole theory which has led to the conclusion that an accelerated electron must radiate energy is in need of amendment.

711. This discovery, if it stood alone, would be extremely disconcerting. In actual fact it does not stand alone; it is only one of a long series of discoveries, each of which has indicated, with very little room for doubt, that the classical mechanics of Newton and the classical electrodynamics of Maxwell both fail when applied to atomic phenomena. Since the beginning of the present century the need has been recognised for a wholly new system of dynamics, such as shall be applicable to physical phenomena on atomic and sub-atomic scales, and shall merge into the Newtonian and Maxwellian dynamics in the case of larger scale phenomena. In spite of much labour, this system of dynamics has not yet been found in its entirety. Fragments are known with fair certainty, although it is of course impossible to feel absolute confidence in any part of the system until the whole has been pieced together and seen to form a consistent structure. Fortunately the parts which are known with the nearest approximation to certainty are precisely those which are necessary in the discussion of the subject of the present chapter, the electrical structure of matter.

712. In discussing the motion of a dynamical system as predicted by the classical mechanics, the usual procedure is to start from the general equations of motion, which are differential equations of the second degree, and attempt in the first place to discover one or more first integrals of these equations. In many problems, for instance, the equations of energy and of linear and angular momentum figure as first integrals of the equations of motion. Each time a first integral is derived from the equations of motion a constant of integration is introduced, different values of this constant representing different states of the dynamical system. Under the classical mechanics these constants of integration could usually have any value we chose to assign to them, or, if this was not possible, there was at least a finite continuous range of values open to each constant of integration.

The distinguishing feature of the new dynamics is that there are no longer continuous ranges of values open to the constants of integration, but only certain definite discrete values. Generally speaking, the values available for each constant of integration are an infinite set associated with the natural integers 1, 2, 3, ... as, for instance, the set of values obtained by taking integral multiples of a given constant. Thus the constants of integration shew a sort of atomicity.

A parallel can be found in the atomicity of electricity. In the earlier chapters of this book, we treated electric charges as being capable of continuous variation; for instance, in calculating the energy of a condenser in § 97 we treated the electric charge e as changing continuously and integrated with respect to de. In actual fact we know that in charging a condenser, the charge must move by whole electrons at a time. The procedure of treating the charge as capable of continuous variation was, nevertheless, legitimate so long as we were dealing with charges of billions of electrons; our step de could be quite insignificant in comparison with the total value of e, although representing perhaps a million electrons. The same procedure would, however, lead to disastrous errors if it were followed in problems of atomic physics. An atom normally is an electrically neutral structure, the total charge of the positive nucleus and the negative electrons being zero. It is possible to charge it positively by withdrawing one, two, three or more electrons. Thus the atom can have a positive charge E, but E is restricted to having the discrete values e, $2e$, $3e$, ... etc.; it may not be regarded as capable of continuous variation.

In precisely the same way, although for reasons not clearly understood, the constants of integration in the new dynamics are limited to definite discrete values which may perhaps, in a similar manner, be integral multiples of a fundamental constant. It may for instance be possible for a constant of integration to have any one of the values 0, c, $2c$, $3c$, ... but no more possible for it to have the values $\frac{1}{2}c$ or $\frac{2}{3}c$ than it is possible for an atom to carry the charges $\frac{1}{2}e$ or $\frac{2}{3}e$.

There is a further difference between the new dynamics and the old. Under the old dynamics a constant of integration was a true constant, and retained its value absolutely unchanged until the conditions of the problem altered. The new dynamics has no knowledge of such absolute constancy. If a constant of integration can have any one of the values c, $2c$, $3c$, ... and has one of these values, say $2c$, at a given instant, there is a possibility of its value taking a jump from the value $2c$ to some other value. It is usual to think of these jumps as occurring absolutely spontaneously, although this conception is probably only a cover for our ignorance of some underlying mechanism.

The new mechanics was originally developed by Planck and others from a study of the phenomena of black-body radiation. In 1913 Prof. N. Bohr applied the new system to the problem of atomic motions and was led to a theory of the nature of these motions which gained immediate acceptance and which has stood the test of time. This theory we shall now explain.

BOHR'S THEORY.

713. Let us consider the simplest case of a single electron of charge e describing an orbit about a nucleus of charge E, which, on account of its much greater mass, is treated as a fixed centre of force. In ordinary polar coordinates the equations of motion of the electron are

$$m\left(\ddot{r} - r\dot{\theta}^2\right) = -\frac{eE}{r^2} \quad\dots\dots\dots\dots\dots(783),$$

$$m\frac{d}{dt}\left(r^2\dot{\theta}\right) = 0 \quad\dots\dots\dots\dots\dots\dots(784).$$

Equation (784) at once yields the integral

$$mr^2\dot{\theta} = \text{cons.} \quad\dots\dots\dots\dots\dots\dots(785),$$

but in accordance with the principles just explained, we may not suppose all values to be permissible for the constant on the right. We shall suppose that it is restricted to being an integral multiple of a fundamental constant which, to agree with an established notation, we shall denote by $h/2\pi$. Thus equation (785) must be written in the form

$$mr^2\,\dot{\theta} = \tau h/2\pi \quad\dots\dots\dots\dots\dots(786),$$

where τ is an integer. Using this value for $\dot{\theta}$ to eliminate the angle θ from equation (783), we obtain

$$m\ddot{r} = \frac{1}{r^3 m}\left(\frac{\tau h}{2\pi}\right)^2 - \frac{eE}{r^2} \quad\dots\dots\dots\dots(787),$$

which, on integration with respect to r, yields the integral

$$\tfrac{1}{2}\left[m\dot{r}^2 + \frac{1}{r^2 m}\left(\frac{\tau h}{2\pi}\right)^2\right] - \frac{eE}{r} = \text{cons.} \quad\dots\dots\dots(788).$$

Utilising equation (786), this assumes the form

$$\tfrac{1}{2}m\left(\dot{r}^2 + r^2\dot{\theta}^2\right) - \frac{eE}{r} = \text{cons.} \quad\dots\dots\dots\dots(789),$$

which is at once seen to be the integral of energy, but again the constant on the right must be restricted to certain definite values, just as was the case with the right-hand member of equation (785).

714. Before proceeding to the comparatively complicated general case, let us consider the simple problem of circular orbits. Assuming that circular orbits are possible, these will be obtained by putting $\ddot{r} = 0$ in equation (787) giving

$$r = \frac{\tau^2 h^2}{4\pi^2 eEm} \qquad \qquad \text{.............................(790)}.$$

As has been seen, the hydrogen atom consists of a single electron describing an orbit about a nucleus of charge e, while the helium atom consists of two electrons describing orbits about a nucleus of charge $2e$. Thus on putting $E = e$ the foregoing analysis is applicable to a normal hydrogen atom, while on putting $E = 2e$ it becomes applicable to a helium atom from which one electron has been removed—*i.e.*, to a positively charged helium atom of charge e.

Thus the circular orbits in both these atoms are obtained by giving various integral values to τ in equation (790). The radii of the various circular orbits which are possible for either atom are seen to be proportional to the squares of the natural numbers, and so to 1, 4, 9, 16, 25, ..., while the radii possible for the positively charged helium atom are exactly half those which are possible for the hydrogen atom.

715. Mention has already been made of the possibility of what appear to be spontaneous jumps taking place in the values of the constants of integration, and therefore also in the value of τ in equation (790). We proceed to discuss these changes.

Corresponding to the values of r and $\dot{\theta}$ determined by equations (786) and (790), the negative energy W of a circular orbit is found to be given by

$$W = -\tfrac{1}{2}mr^2\dot{\theta}^2 + \frac{eE}{r} = \frac{2\pi^2 e^2 E^2 m}{\tau^2 h^2} \qquad \text{...................(791)}.$$

The values of W form a discrete series, proportional to the inverse squares of the natural numbers. If a spontaneous change occurs in τ it can only be from a higher value of τ to a lower value, since any change in the reverse direction would lessen the value of W and so increase the energy of the system. If τ_1 is greater than τ_2, both τ_1 and τ_2 being integral numbers, a spontaneous jump from $\tau = \tau_1$ to $\tau = \tau_2$ results in the system losing energy of amount

$$\frac{2\pi^2 e^2 E^2 m}{h^2}\left(\frac{1}{\tau_2^2} - \frac{1}{\tau_1^2}\right) \qquad \text{.......................(792)}.$$

According to Bohr's theory this lost energy leaves the system in the form of radiation; indeed the theory supposes that the atoms we have been considering do not emit radiation at all except on the occasion of jumps of the kind we have been considering.

A change in the value of τ, then, results in a change in the amount of radiant energy in the space surrounding the atom. Now Planck, from his study of black-body radiation to which we have already referred, had concluded that the radiant energy of an enclosure, if it changed at all, must change by jumps. The radiation in any space or enclosure can be analysed by Fourier's theorem into trains of waves of different frequencies, and the energy of the radiation can be regarded as the sum of the energies of these trains of waves. The total energy of radiation may accordingly be thought of as the sum of the contributions from disturbances of different frequencies. Planck's conclusion was that the energy of radiation of each frequency must be an integral multiple of a certain unit, this unit being equal to a fundamental constant h multiplied by the frequency ν of the radiation in question. Planck called this unit a "quantum." Thus the quantum of energy of frequency ν was equal to $h\nu$, and the total energy of frequency ν, being an integral number of quanta, was restricted to being of amount $\tau h\nu$ where τ was an integral number. It followed that any change in the field of radiation must consist of a jump in the energy of the radiation of a definite frequency and must be equal in amount to an integral number of quanta of this radiation.

In view of these results of Planck, it was natural for Bohr to suppose that when energy of the amount specified by formula (792) was set free into space, it formed one quantum of energy. This supposition by itself suffices to determine the frequency of the radiated energy. For $h\nu$, the quantum of energy, must be equal to expression (792) in order to satisfy the principle of the conservation of energy, and this gives the relation

$$\nu = N\left(\frac{1}{\tau_2^2} - \frac{1}{\tau_1^2}\right) \dots\dots\dots\dots\dots\dots(793),$$

where

$$N = \frac{2\pi^2 e^2 E^2 m}{h^3} \dots\dots\dots\dots\dots\dots(794).$$

Thus Bohr's theory restricts the radiation emitted from a hydrogen atom, or from a positively charged helium atom, to one of the frequencies specified by formula (793) where τ_2 and τ_1 are positive integers. This gives a series of detached frequencies, or what the spectroscopist calls a "line-spectrum," whereas it is easily seen that the classical system of electrodynamics would have predicted a continuous range of frequencies or a "continuous spectrum."

716. The spectrum of hydrogen, as observed in the light from an ordinary vacuum tube, consists of a series of lines, the strongest of which (H_a) lies at the red end of the spectrum while the remainder ($H_\beta, H_\gamma, H_\delta, \dots$) spread out, at ever diminishing distances, towards the violet. As far back as 1885 Balmer had found that the frequencies of the different members of this series could be represented, with very great precision (about one part in 200,000), by giving to n the values 3, 4, 5, ... in the formula

$$\nu = N \left(\frac{1}{2^2} - \frac{1}{n^2}\right) \quad \ldots\ldots\ldots\ldots\ldots\ldots\ldots\ldots(795),$$

with $N = 3.2902 \times 10^{15}$. It is clear that if we are free to assign this value to the N which is defined by equation (794), then Bohr's theoretical formula (793) will contain Balmer's observational formula as the special series of lines obtained on taking $\tau_2 = 2$. We are not free to assign any arbitrary value to the N of equation (794) since the value of every quantity which enters into N is known. The values of e and m have already been given (§ 28); for hydrogen E is equal to e, and the value of h can be obtained from a study of the spectrum of black body radiation and in a variety of other ways. The best determinations of h give

$$h = 6.62 \times 10^{-27},$$

and on substituting these values for h, e and m, the value of N given by equation (794) is found to agree, to within the errors in the determination of e and h, with the observed value $N = 3.2902 \times 10^{15}$.

It is, then, clear that N has precisely the required value in equation (793), and that Bohr's theoretical formula (793) includes Balmer's observational formula (795) as a special case. It was this success of Bohr's theory that brought about its immediate acceptance by the majority of physicists. The theory, however, requires that Balmer's series should be only one of an infinite number. Other series are obtained by putting τ_2 equal to 1, 3, 4, 5, ... ∞ in equation (793), and, if Bohr's theory is correct, these series ought equally to appear in the hydrogen spectrum. The majority of the lines of these series were unknown when Bohr's theory was first published, but all the predicted lines have been found which lie in the region of the spectrum which is accessible to observation.

717. According to this theory the spectrum of positively charged helium is the same as that of hydrogen except that E in equation (794) must be replaced by $2e$ instead of by e, as for hydrogen. Or, if we denote the value of N for hydrogen by N_H, the value of N for helium will be $4N_H$ and the spectrum will be given by

$$\nu = 4N_H \left(\frac{1}{\tau_2^2} - \frac{1}{\tau_1^2}\right).$$

In this formula even values for both τ_2 and τ_1 give values of ν which are identical with the whole system of values of ν given by equation (793) for the hydrogen spectrum. The spectrum of ionised helium ought accordingly to shew all the lines of the normal hydrogen spectrum and, in addition, the various lines which are obtained by giving odd values to τ_1 or τ_2 or both in the above formula. This is in actual fact found to be the case, except for a reservation which must now be explained.

In deducing formula (793) we assumed the nucleus to be so massive that it could be treated as a fixed centre of force. In actual fact the nucleus of the hydrogen atom has about 1844 times the mass of the negative electron, so that this assumption leads to an error of the order of one in 1844 in the value of N for hydrogen. For helium the ratio of the two masses is about 7300 to one, and the predicted value of N for helium is in error by about one part in 7300. When allowance is made for these errors, the value of N for helium is not exactly four times the value for hydrogen, and the difference is shewn spectroscopically by the hydrogen spectrum not coinciding exactly with the corresponding lines of the helium spectrum. By measuring the distance between corresponding lines Fowler deduced the value 1836 for the mass-ratio hydrogen nucleus and the negative electron. Allowing for the relativity correction and other refinements Paschen subsequently amended this to 1843·7, a value which is in excellent agreement with the values of this ratio determined by other methods.

718. So far we have considered only circular orbits. The orbit of an electron about a nucleus is, however, in no way restricted to being circular; as the law of force is that of the inverse square the orbit may be elliptic, parabolic or hyperbolic. But, just as the radius of a circular orbit is restricted to having certain definite values, so the eccentricity of an elliptic orbit, as well as its major axis, is restricted to having certain values.

An adequate discussion of the manner of calculating these restrictions would carry us too far outside the scope of the present book. The following will, however, suffice for the immediate purpose in hand.

Let $q_1, q_2, q_3, \ldots$ be the generalised coordinates of any dynamical system, defined as in § 548, and let $p_1, p_2, p_3, \ldots$ be the corresponding momenta defined by

$$p_1 = \frac{\partial E}{\partial \dot{q}_1} \text{ etc. } \ldots\ldots\ldots\ldots\ldots\ldots\ldots(796),$$

where E is the energy expressed as a function of $q_1, q_2, q_3, \ldots$ and $\dot{q}_1, \dot{q}_2, \dot{q}_3, \ldots$. For certain dynamical systems it is possible to deduce, by ordinary mechanics, a number of equations of motion such that only one coordinate and the corresponding momentum, e.g., q_1 and p_1, enter in each. As we shall at once see, equation (784) is an equation of this type involving only the momentum corresponding to the coordinate θ, and equation (787) is another, for it involves only the coordinate r and the corresponding momentum $m\dot{r}$. The solution of such an equation will be the same as if the whole system had only the one degree of freedom corresponding to this one coordinate, so that either the coordinate will increase or decrease beyond limit, or will oscillate repeatedly between two constant extreme values. In the latter case it is found that the proper restriction to apply to the motion of any coordinate q_1 is that given by

$$\int p_1 dq_1 = \tau h \quad \dots\dots\dots\dots\dots\dots(797),$$

where the integration extends throughout a whole oscillation in the value of q, h is the constant already defined, and τ is any integral number.

For instance if q_1 is the coordinate θ in the orbit of an electron about a nucleus, the momentum, as defined by equation (796), is $p_1 = mr^2\dot{\theta}$. Equation (785) may accordingly be written in the form

$$p_1 = \text{constant},$$

which is of the required form, since it contains no coordinates or momenta other than p_1 and q_1. A complete oscillation extends from $q_1 = 0$ to $q_1 = 2\pi$, so that equation (797) assumes the form

$$2\pi p_1 = \tau h,$$

or

$$mr^2\dot{\theta} = \tau h/2\pi,$$

which is identical with our equation (786).

Similarly, if q_2 is the coordinate r of the same orbit, $p_2 = m\dot{r}$ so that equation (787) is of the required form. The first integral of this equation is equation (788), and this may be written in the form

$$\tfrac{1}{2}\left[\frac{p_2^2}{m} + \frac{1}{r^2 m}\left(\frac{\tau h}{2\pi}\right)^2\right] - \frac{eE}{r} = -W,$$

where W is constant, the negative energy of the orbit. This equation gives p_2^2 as a quadratic function of $1/r$; it may be written in the form

$$p_2^2 = \left(\frac{\tau h}{2\pi}\right)^2\left(\frac{1}{r_1} - \frac{1}{r}\right)\left(\frac{1}{r} - \frac{1}{r_2}\right),$$

where r_1, r_2 are determined by

$$\left(\frac{\tau h}{2\pi}\right)^2 \frac{1}{r_1 r_2} = 2Wm, \quad \left(\frac{\tau h}{2\pi}\right)^2\left(\frac{1}{r_1} + \frac{1}{r_2}\right) = 2eEm \quad \dots\dots\dots(798).$$

In the course of a complete oscillation r varies from r_1 to r_2 and then back to r_1. Thus the path of integration in equation (797) may be taken to be twice the range from r_1 to r_2, and the equation assumes the form

$$2\left(\frac{\tau h}{2\pi}\right)\int_{r_1}^{r_2}\left[\left(\frac{1}{r_1} - \frac{1}{r}\right)\left(\frac{1}{r} - \frac{1}{r_2}\right)\right]^{\frac{1}{2}} dr = \tau' h \quad \dots\dots\dots(799),$$

where τ' is a new integer. Evaluating the integral by the transformation

$$\frac{1}{r} = \frac{1}{r_1}\cos^2\theta + \frac{1}{r_2}\sin^2\theta,$$

we readily obtain

$$2\int_{r_1}^{r_2}\left[\left(\frac{1}{r_1} - \frac{1}{r}\right)\left(\frac{1}{r} - \frac{1}{r_2}\right)\right]^{\frac{1}{2}} dr = \frac{\pi}{\sqrt{(r_1 r_2)}}[\sqrt{r_2} - \sqrt{r_1}]^2.$$

Equation (799) now becomes

$$\frac{\tau}{2\sqrt{(r_1 r_2)}} = \frac{\tau'}{(\sqrt{r_2} - \sqrt{r_1})^2},$$

and each side is at once seen to be equal to

$$\frac{\tau + \tau'}{r_1 + r_2}.$$

The elimination of r_1 and r_2 from this equation and the two equations (798) gives

$$W = \frac{2\pi^2 m e^2 E^2}{(\tau + \tau')^2 h^2} \quad ....(800).$$

We obtain all possible values for W by giving integral values to τ and τ'. It is at once seen that no new values are introduced beyond those already discovered in equation (791).

By a well-known formula the eccentricity ϵ of the orbit is given by

$$1 - \epsilon^2 = \frac{2 W p_1^2}{m e^2 E^2} = \frac{\tau^2}{(\tau + \tau')^2} \quad ....(801),$$

and, since τ and τ' are necessarily integrals, it is clear that only definite values are permissible for the eccentricity. The semi-axes a, b of the orbit are related by

$$\frac{b^2}{a^2} = 1 - \epsilon^2,$$

so that equation (801) shews that b/a will be commensurable for all orbits which can be described.

By a well-known theorem, the energy of an elliptic orbit is equal to that of a circular orbit whose radius is equal to the semi-major-axis of the ellipse. It follows from equation (790) that the semi-major-axis a of an elliptic orbit is given by

$$a = \frac{n^2 h^2}{4\pi^2 e E m} \quad ....(802),$$

where n is written for $\tau + \tau'$, while from equation (801) the eccentricity is given by

$$\epsilon^2 = 1 - \frac{\tau^2}{n^2} \quad ....(803).$$

The orbits $n = 1$ $(\tau = 1)$, $n = 2$ $(\tau = 1$ or $2)$, $n = 3$ $(\tau = 1, 2$ or $3)$ and $n = 4$ $(\tau = 1, 2, 3$ or $4)$ are shewn in fig. 140*. The distance of closest approach to the nucleus is $a(1 - \epsilon)$ which is given by

$$a(1 - \epsilon) = \frac{h^2}{4\pi^2 e E m} n (n - \sqrt{n^2 - \tau^2}).$$

---

* Reproduced by permission from a paper by Bohr (*Nature*, July 7, 1923).

The expression on the right has its minimum value when $\tau = 1$ and $n = \infty$, namely

$$a(1 - \epsilon) = \frac{h^2}{8\pi^2 eEm}.$$

Thus the electron never approaches the nucleus to within a distance less than half of the radius of the smallest circular orbit.

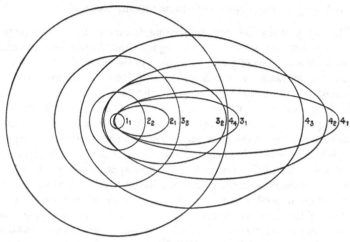

Fig. 140.

719. Since the extension to elliptical orbits has introduced no new values of W, it follows that the spectrum will consist of those lines which were predicted by the simple theory of circular orbits and no others. Nevertheless the possibility of elliptical orbits has introduced an essential difference into the spectrum. Since an orbit of any permissible energy W_1 or W_2 can now be described in more than one way, it follows that a fall from energy W_2 to energy W_1 can occur in more than one way, so that the spectral line which is produced by a fall from energy W_2 to energy W_1 may appropriately be thought of as the superposition of a number of lines, all of which, although having precisely the same frequency, are produced by different events.

720. It is possible to separate out these coincident lines in a variety of ways. Perhaps the simplest is by placing the radiating atoms in a magnetic field. Each electronic orbit is affected by the field, and different orbits, even though of the same energy before the field was put on, will be affected in different ways. It follows that the spectral lines which were originally coincident will be displaced to different extents, as is observed in the Zeeman effect. It can be shewn that the explanation of the normal Zeeman effect which has already been given in § 635 holds valid even when the quantum-restrictions are applied to the electronic orbits. The anomalous Zeeman effect presents a more complicated problem which can hardly yet be said to have been satisfactorily solved.

The placing of the radiating matter in a powerful electrostatic field also results in a separation of the originally coincident lines, this being known as the Stark effect. The dynamical theory just explained provides a calculation of the separations to be expected in this case, and the predicted separations are found to agree very closely with those actually observed.

721. But perhaps the most interesting feature of all is that there is a slight separation even when external magnetic and electric fields are entirely absent. In the analysis of § 713, we treated m the mass of the electron as an absolute constant, although we already knew (§ 660) that the mass varies with the velocity of motion of the electron. For circular orbits this does not matter much; the mass of course remains strictly constant throughout the description of any single circular orbit, although varying slightly from one orbit to another. But in an elliptic orbit the mass varies from one part to another of the same orbit. When allowance is made for this, the orbit is no longer strictly elliptical, and formula (800) only provides a first approximation to its energy. When the necessary additional terms are included, it is found that the value of W no longer depends solely on $\tau + \tau'$, but on τ and τ' separately. It follows that each of the lines which our simple theory treated as a superposition of coincident lines must in actual fact shew a "fine-structure" of adjacent slightly separated lines. Such "fine-structures" are easily observed in a powerful spectroscope. The theoretical separations to be expected have been calculated by Sommerfeld and others, and although the observed separations are so small as to make exact measurement exceedingly difficult, there seems no room for doubt that they agree with those predicted by theory.

The dynamical theory of these phenomena is not given in the present book. The reader who wishes to study it is referred to the original papers of Bohr, Sommerfeld and others, or to the author's "Dynamical Theory of Gases."

722. The new dynamics, as has now been seen, allots a definite size to the atom and so provides a mechanism by which atoms have a permanent existence, instead of radiating away all their energy and collapsing. For the hydrogen atom the minimum energy is found by taking $\tau + \tau' = 1$ in equation (800), and since τ cannot be zero, this requires that $\tau = 1$ and $\tau' = 0$. Thus the orbit of minimum energy is the circular orbit of radius (cf. equation (802))

$$a = \frac{h^2}{4\pi^2 e^2 m} \quad \dots\dots\dots\dots\dots\dots\dots\dots\dots(804).$$

The electrons in hydrogen atoms can describe circular orbits of radii 4, 9, 16, ... times this and a variety of non-circular orbits as well, but this equation defines the hydrogen atom in its normal state of minimum energy.

On inserting the numerical values already given, we find $a = 0.53 \times 10^{-8}$ cms., which is in good agreement with the radius of the hydrogen atom as found in other ways.

Dewar found the density of solid hydrogen at $13.2°$ absolute to be 0.0763. Thus a cubic centimetre of hydrogen at this temperature has mass 0.0763 grammes and consists of atoms of hydrogen each of which is known to have a mass of 1.662×10^{-24} grammes. It follows that the number of hydrogen atoms in a cubic centimetre of solid hydrogen is 4.59×10^{22}, so that the space occupied by each is 2.18×10^{-23} cubic centimetres. This is the space that would be occupied if the atoms were spheres arranged in cubical packing, each being of radius 1.40×10^{-8} cms. In § 150 we found that the dielectric constant of hydrogen is the same as if the molecules were spheres of radius 0.916×10^{-8} cms. A similar calculation would have suggested that the hydrogen atom might be regarded as having a radius equal to $1/\sqrt[3]{2}$ times this or 0.723×10^{-8} cms. Neither of these calculations can lay claim to great accuracy; a far more accurate determination of atomic dimensions is obtained from the Kinetic Theory of Gases. If the molecules of hydrogen are regarded as spheres, the three phenomena of viscosity, conduction of heat and diffusion agree in assigning to these spheres a radius of 0.68×10^{-8} cms., while observations on the deviations from Boyle's law suggest the slightly lower value of 0.64×10^{-8} cms. Again the hydrogen atom may be supposed to have a radius equal to $1/\sqrt[3]{2}$ times that of the molecule, so that the two values of the atomic radius are respectively 0.54×10^{-8} and 0.51×10^{-8} cms., in close agreement with the value 0.53×10^{-8} required by the electrical structure of the atom.

It must, however, be noticed that the Kinetic Theory requires the atom to occupy a three-dimensional volume, whereas on Bohr's theory the hydrogen atom is at most a disc. If we imagine this disc, the orbit of the negative electron, to be continually changing its orientation in space we pass naturally to the conception of the hydrogen atom reserving for itself, or perhaps clearing for itself, a spherical space equal in radius to the orbit of the electron. This conception is in accordance with the known facts of crystal structure.

723. The theory of structures of more than two constituent parts is far less advanced, no satisfactory mechanism having yet been devised for either the helium atom or the hydrogen molecule. A large amount of consistent evidence suggests that the electrons of complex atoms are arranged in shells or rings corresponding to different quantum numbers, but the method of arrangement has not yet been brought within the scope of mathematical treatment.

Questions of electrical conductivity and of the optical and dispersive properties of substances are clearly subjects for treatment by the new dynamics, but only meagre progress has so far been made. The same applies to the problem of the nature of radiation. There is at present a divergence of opinion as to whether radiation is propagated in accordance with Maxwell's equations or in the form of "atomic" packets of energy which travel through space without spreading out in the manner demanded by the classical electromagnetic theory.

The Correspondence Principle of Bohr.

724. In conclusion we may refer to a procedure which holds out some hope of bridging the gulf between the classical electrodynamics and the new electrodynamics of quanta.

When an electron describes a circular orbit about a nucleus, the number of revolutions per second in this orbit, n, is equal to $\dot{\theta}/2\pi$, whence, from equations (786) and (790),

$$n = \frac{4\pi^2 e^2 E^2 m}{\tau^3 h^3} \qquad \dots\dots\dots\dots\dots\dots(805).$$

The period of an elliptic orbit is known to be the same as that of a circular orbit of the same energy, so that the same equation will give the frequency of revolution in an elliptic orbit of total quantum number τ.

The frequencies of the radiation which can be emitted on the electron dropping from this orbit to one of lower quantum number τ' are, from equation (792),

$$\nu = \frac{2\pi^2 e^2 E^2 m}{h^3} \left(\frac{1}{\tau'^2} - \frac{1}{\tau^2} \right) \dots\dots\dots\dots\dots(806).$$

If the integers τ and τ' are both large, and differ only by a small number s, which must of course also be integral, the approximate value of $\left(\dfrac{1}{\tau'^2} - \dfrac{1}{\tau^2} \right)$ will be

$$\frac{1}{\tau'^2} - \frac{1}{\tau^2} = -(\tau - \tau') \frac{\partial}{\partial \tau} \left(\frac{1}{\tau^2} \right) = \frac{2s}{\tau^3},$$

and equations (805) and (806) now shew that the possible frequencies of radiation are given by

$$\nu = sn \qquad \dots\dots\dots\dots\dots\dots\dots(807).$$

If the motion of the electron, describing its orbit with frequency n, had been analysed by Fourier's theorem, and the resulting radiation calculated by the classical electrodynamics, we should have found radiations of frequencies

$$n, \ 2n, \ 3n, \ \dots,$$

so that according to the classical electrodynamics also, the frequencies of the emitted radiation would be given by equation (807).

It accordingly appears that in the limiting case in which τ and τ' are both large, the old classical electrodynamics and the new quantum dynamics agree in predicting the same frequencies for the spectrum of emitted radiation. In this limiting case, the radius of the electron orbit is infinite, successive radii only differ by an infinitesimal fraction of each, and orbits of all eccentricities are possible. Thus the electron is just on the verge of becoming a free electron. This limiting case provides a bridge between the old mechanics and the new; on one side of the bridge the classical electrodynamics holds undisputed sway, but as we cross the bridge and advance into the territory on the other side, the additional restrictions imposed by the quantum dynamics become ever more important until finally they may be considered to govern the whole situation. The exploration of the territory on the far side of the bridge will provide work for a new generation of mathematical physicists; the present work attempts only to bring the reader as far as the bridge, and to make clear to him that if he crosses it he must expect to find different conditions prevailing on the other side.

INDEX

The numbers refer to the pages.

[*pp.* 1—299, *Electrostatic Problems. pp.* 300—*end, Current and Magnetic.*]

Printed in the United States
By Bookmasters